Macrophage Targeted Delivery Systems

Swati Gupta • Yashwant V. Pathak
Editors

Macrophage Targeted Delivery Systems

Basic Concepts and Therapeutic Applications

Springer

Editors
Swati Gupta
Department of Pharmaceutics
Amity Institute of Pharmacy
Amity University Uttar Pradesh
Noida, India

Yashwant V. Pathak
University of South Florida
Taneja College of Pharmacy
Tampa, FL, USA

Adjunct Professor
University of Airlangga
Surabaya, Indonesia

ISBN 978-3-030-84163-8 ISBN 978-3-030-84164-5 (eBook)
https://doi.org/10.1007/978-3-030-84164-5

This Springer imprint is published by the registered company Springer Nature Switzerland AG
The registered company address is: Gewerbestrasse 11, 6330 Cham, Switzerland

To my beloved parents, my husband Manish who was always there to support me throughout my academic and research career, and my children Krishna and Advay who are my lifeline and filled my life with happiness.
Swati Gupta

To the loving memories of my parents, and Dr. Keshav Baliram Hedgewar, who showed the right direction, my wife Seema who gave my life positive meaning, and my son Sarvadaman who gave a golden lining to my life.
Yashwant V. Pathak

Foreword

During the past decade, the rapid growth in understanding of macrophages (Mϕ), as a specialized host defense and homeostatic system, has begun to offer Mϕ as attractive targets for therapeutic intervention. Mϕ play central role in a wide range of disease processes, from genetically determined lysosomal storage diseases to acute sepsis, chronic inflammation and repair, tissue injury, and cell death. Under reactivity or over reactivity of Mϕ clearance, immune effector functions and responses to metabolic abnormalities contribute to common disorders such as autoimmunity, atherosclerosis, allergy, Alzheimer's disease, and major infections including AIDS and tuberculosis.

This book, entitled *Macrophage Targeted Delivery Systems: Basic Concepts & Therapeutic Applications*, emphasizes on the applications of targeted delivery systems in Mϕ-associated disorders. It reflects thoroughness and high work ethic which can only be accomplished through perseverance, diligence, and collaborative efforts of the editors as well as individual chapter contributors. The edition is judicious because there is unparalleled pursuit in understanding the molecular basis of the Mϕ-targeted delivery. The editors of this volume, Drs. Swati Gupta and Yashwant V. Pathak, have carefully chosen an incongruent and archetypal collection of 25 contributions written by leading authors and scientists in the field of drug delivery and targeting. It comprises empathizing of the surface modification of nanoparticles and their applications as Mϕ-targeted delivery systems, as well as functionalization of lipid nanoparticles for site-specific Mϕ delivery.

A full part of the book has been devoted specifically to targeting Mϕ-associated disorders ranging from tumor-associated Mϕ, autoimmune diseases, cardiovascular diseases, inflammatory diseases, respiratory, and ocular disorders. This book imparts far-reaching knowledge of nanomedicine with fastidious emphasis on drug/nucleic acid targeting, theranostics, chronic disorders, and clinical studies related to Mϕ.

As a primary pharmaceutical and medicinal tool of the twenty-first century, nanotechnology offers a new area of drug discovery, delivery, and research. This concept opens an advanced therapeutic option: nanomedicine that will be more targeted, cost-effective, and less toxic to the patients. A new hope for a better outcome in disease management. This book lays down a solid foundation regarding nanotechnology and nanomedicine. The volume offers a diverse compilation of outlooks and notions expressed by leading scientists, medical researchers, and educators in the field. The chapters assimilate various prospects and accentuate the prominence of nano-biomedical coalition. I encourage the readers to reconnoiter the interesting points of view, methods, and trends congregated in this well-documented collection of research works. I am confident that the book will be beneficial to pharmaceutical industry and academia, medical and pharmaceutical professionals, undergraduate and postgraduate students, research scholars, and Ph.D. students and postdocs working in the field of medical and pharmaceutical sciences. I would like to congratulate both Prof. Dr. Swati Gupta and Prof. Dr. Yashwant V. Pathak for successfully bringing out this meaningful and timely volume, and I am positive it will be very well received by the scientific community worldwide.

Enjoy the journey of discovery!

Founding Dean, School of Pharmacy
Professor of Pharmacology, School of Medicine
University of Texas Rio Grande Valley
Edinburg, TX, USA

Hieu T. Tran

Preface

Macrophages (MFs) are chief cells exhibiting non-specific immune response, persistently located in various tissues throughout the body, and play an essential role in host defense, immunology, physiology, and homeostatis and consequently perform imperative functions in management and advancement of various ailments. Apart from their beneficial role, MFs are considered unfavorable to host because of their central role in pathogenesis of almost each and every disease, including respiratory diseases, cancer, ocular diseases, metabolic syndrome, cardiovascular diseases, sepsis, allergy, immunodeficiency, and autoimmune disease, to name a few. In view of these decrees, MFs and their associated cells of immune system have been ascertained as promising targets for therapeutic relevance. During the last two decades, a prodigious level of research has allowed us to upgrade our knowledge of MFs' origin; their anatomy, physiology, and functions; polarization states; and their study in vivo in various disease settings and succession.

Development of colloidal carriers with variable physicochemical properties for targeting them specifically towards MFs has posed diverse confronts considering phagocytosis a central element of host immune response against pathogenic invaders and injury. Various pattern recognition receptors including scavenger receptors, Fc and complement receptors, and dectins and mannose receptors involved in phagocytic clearance of MFs offer a viable approach for the development and functionalization of nanocarrier systems such as polymeric nano- and microparticles, liposomes, nanocapsules, and polymer micelles to contend MF-associated disorders. The surface of these nanocarriers can be easily manipulated with macrophage/monocyte receptor specific ligands apart from undergoing opsonization events in the biological milieu. Undeniably, various nanosized delivery systems have already been researched for codelivery of drugs and diagnostics to diverse tissue/cell specific MFs to combat obstinate infections, synchronize macrophage auxiliary functions, treat macrophage-associated genetic disorders, persuade macrophage killing, and reveal local pathogenesis (e.g., cancer). Incorporation of drug/nucleic acid into the aforementioned nanocarriers protects the cargo against deterioration en route to the MFs with simultaneous reduction of therapeutic dose required to acquire a pharmacological response, and may effectually diminish adverse effects associated with encapsulated drug therein. Nevertheless, possible toxicity/undesirable effects of nanocarriers must also be considered while utilizing them as site-directed delivery vehicle towards MFs. Aforementioned aspects have been critically discussed along with application of various delivery systems in the treatment of macrophage-associated disorders.

The first part of the book contemplates on anatomy, physiology, and pharmacology of MFs from drug delivery perspective with special reference to pathophysiology of macrophage-associated disorders, their treatment challenges, and various approaches and pathways of macrophage delivery using novel delivery systems. The second part of the book focuses on the surface modifications and basic and therapeutic role of polymeric nanoparticles in targeted drug and gene delivery to MFs for therapeutic applications. The third part of the book spotlights lipid nanoparticles, specifically liposomes and solid lipid nanoparticles-based drug and gene delivery to MFs with special reference to in vivo fate of nanoparticles and nanoparticle-based theranostics undergoing macrophage targeting. The fourth part of the book is specifically devoted to role of MFs in pathogenesis of variety of diseases and targeting MFs for therapy of various diseases like cancer, autoimmune diseases, cardiovascular diseases, inflammatory diseases, tuberculosis and respiratory diseases other than tuberculosis, intraocular diseases, and HIV. Finally, the last part of the book outlines specific therapeutics like nano- and microparticles-based dry powder inhalers, delivery of siRNA to MFs, MFs and immunotherapy in wound healing, effects of mycotoxins on MFs, and clinical trials on macrophage-targeted delivery.

In totality, we anticipate that this book will be a very good reference source for academicians, industry experts, and, more importantly, students of nanomedicine. This book can also be used as a resource for teaching the graduate classes where nanotechnology applications in medicine are part of day-to-day discussion.

This book is an endeavor of many scientists and chapter authors whose responses were overwhelming to this specific book title and they submitted the quality work in record time so we could get the book out in very short time. Our sincere thanks to all the authors who have contributed to this book.

Nothing is possible without the support of our family, as this work is in a way an encroachment on their (specifically kids') time during COVID lockdown, but they have been always very considerate toward our research and academic activities.

We would like to acknowledge the excellent assistance of Carolyn Spence, Kate Lazaro, Jeffrey Newton, Radha Lakshmanan and all the editorial staff at Springer-Nature.

Department of Pharmaceutics
Amity Institute of Pharmacy
Amity University Uttar Pradesh
Noida, India
Swati Gupta

University of South Florida
Taneja College of Pharmacy
Tampa, FL, USA
Adjunct Professor University of Airlangga
Surabaya, Indonesia
Yashwant V. Pathak

Contents

Part III Lipid Nanoparticles for Macrophage Targeting

Part IV Targeting Macrophages for Treatment of Various Diseases

Contributors

Bahareh Asadi Aghbolagh California Northstate University College of Pharmacy, Elk Grove, CA, USA

Mansoor M. Amiji Department of Pharmaceutical Sciences, School of Pharmacy, Northeastern University, Boston, MA, USA

Nagendra Bhuwane University Institute of Pharmacy, Pt. Ravishankar Shukla University, Raipur, Chhattisgarh, India

Papiya Bigoniya DSKM College of Pharmacy, RKDF University, Bhopal, MP, India

Largee Biswas Nanobiotech Lab, Department of Zoology, Kirori Mal College, University of Delhi, Delhi, India

Amisha Chauhan Department of Pharmaceutics, Amity Institute of Pharmacy, Amity University Uttar Pradesh, Noida, India

Ishwari Choudhary University Institute of Pharmacy, Pt. Ravishankar Shukla University, Raipur, Chhattisgarh, India

David Cleary Sullivan University, College of Pharmacy and Health Sciences, Louisville, Kentucky, USA

Thanh Ba Duong Department of Pharmaceutical & Biomedical Sciences, College of Pharmacy, California Northstate University, Elk Grove, CA, USA

Laxmikant Gautam Drug Delivery Research Laboratory, Department of Pharmaceutical Sciences, Dr. Harisingh Gour Vishwavidyalaya, Sagar, MP, India

Mahima Gupta Department of Pharmaceutics, Amity Institute of Pharmacy, Amity University Uttar Pradesh, Noida, India

Swati Gupta Department of Pharmaceutics, Amity Institute of Pharmacy, Amity University Uttar Pradesh, Noida, India

Narayan Hemnani University Institute of Pharmacy, Pt. Ravishankar Shukla University, Raipur, Chhattisgarh, India

Linh Ho Department of Pharmaceutical & Biomedical Sciences, College of Pharmacy, California Northstate University, Elk Grove, CA, USA

Zhuqiu Jin Department of Pharmaceutical & Biomedical Sciences, College of Pharmacy, California Northstate University, Elk Grove, CA, USA

Srinivas Reddy Jitta Department of Pharmaceutics, Manipal College of Pharmaceutical Sciences, Manipal Academy of Higher Education, Udupi, Karnataka, India

Divya Kaushal The University of South Florida, Judy Genshaft Honors College, Tampa, FL, USA

Seema Kohli Department of Pharmaceutical Sciences, Kalaniketan Polytechnic College, Jabalpur, MP, India

Lalit Kumar Department of Pharmaceutics, Manipal College of Pharmaceutical Sciences, Manipal Academy of Higher Education, Udupi, Karnataka, India

Uyen Le Department of Pharmaceutical & Biomedical Sciences, California Northstate University College of Pharmacy, Elk Grove, CA, USA

Yubin Li Department of Dermatology, Perelman School of Medicine, University of Pennsylvania, Philadelphia, PA, USA

Corporal Michael J. Crescenz VA Medical Center, Philadelphia, PA, USA

Department of Neurology, Xinqiao Hospital, Third Military Medical University, Chongqing, People's Republic of China

Asiya Mahtab Department of Pharmaceutics, School of Pharmaceutical Education and Research, Jamia Hamdard, New Delhi, India

Ashim Malhotra California Northstate University College of Pharmacy, Elk Grove, CA, USA

Priyanka Maurya Department of Pharmaceutical Sciences, Babasaheb Bhimrao Ambedkar University, Vidyavihar, Lucknow, Uttar Pradesh, India

Nidhi Mishra Department of Pharmaceutical Sciences, Babasaheb Bhimrao Ambedkar University, Vidyavihar, Lucknow, Uttar Pradesh, India

Sakshi Nainwani Department of Pharmaceutics, Amity Institute of Pharmacy, Amity University Uttar Pradesh, Noida, India

Ashley Oake Morsani College of Medicine, Tampa, FL, USA

Dhaval Oza Department of Pharmaceutical Sciences, School of Pharmacy, Northeastern University, Boston, MA, USA

Alnylam Pharmaceuticals, Inc., Cambridge, MA, USA

Ravi Raj Pal Department of Pharmaceutical Sciences, Babasaheb Bhimrao Ambedkar University, Vidyavihar, Lucknow, Uttar Pradesh, India

Poonam Parashar Department of Pharmaceutical Sciences, Babasaheb Bhimrao Ambedkar University, Vidyavihar, Lucknow, Uttar Pradesh, India

Atul Pathak BioMedica Diagnostics Inc., London, Ontario, Canada

Yashwant V. Pathak University of South Florida, Taneja College of Pharmacy, Tampa, FL, USA

Adjunct Professor University of Airlangga, Surabaya, Indonesia

Madhulika Pradhan Rungta College of Pharmaceutical Sciences and Research, Bhilai, Chhattisgarh, India

Syed Arman Rabbani Department of Clinical Pharmacy and Pharmacology, RAK College of Pharmaceutical Sciences, RAK Medical and Health Sciences University, Ras Al Khaimah, UAE

Shweta Ramkar University Institute of Pharmacy, Pt. Ravishankar Shukla University, Raipur, Chhattisgarh, India

Abhishek K. Sah Department of Pharmacy, Shri Govindram Seksariya Institute of Technology & Science (SGSITS), Indore, MP, India

Fitsum Feleke Sahle Department of Pharmaceutical and Biomedical Sciences, College of Pharmacy, California Northstate University, Elk Grove, CA, USA

GlaxoSmithKline, Richmond, VA, USA

Shubhini A. Saraf Department of Pharmaceutical Sciences, Babasaheb Bhimrao Ambedkar University, Vidyavihar, Lucknow, Uttar Pradesh, India

Priya Shrivastava Drug Delivery Research Laboratory, Department of Pharmaceutical Sciences, Dr. Harisingh Gour Vishwavidyalaya, Sagar, MP, India

Lubna Siddiqui Department of Pharmaceutics, School of Pharmaceutical Education and Research, Jamia Hamdard, New Delhi, India

Deependra Singh University Institute of Pharmacy, Pt. Ravishankar Shukla University, Raipur, Chhattisgarh, India

Manju Rawat Singh University Institute of Pharmacy, Pt. Ravishankar Shukla University, Raipur, Chhattisgarh, India

Neelu Singh Department of Pharmaceutical Sciences, Babasaheb Bhimrao Ambedkar University, Vidyavihar, Lucknow, Uttar Pradesh, India

Priya Singh Department of Pharmaceutical Sciences, Babasaheb Bhimrao Ambedkar University, Vidyavihar, Lucknow, Uttar Pradesh, India

Priyanka Singh Nanobiotech Lab, Department of Zoology, Kirori Mal College, University of Delhi, Delhi, India

Samipta Singh Department of Pharmaceutical Sciences, Babasaheb Bhimrao Ambedkar University, Vidyavihar, Lucknow, Uttar Pradesh, India

Alka Sonkar Department of Pharmaceutical Sciences, Babasaheb Bhimrao Ambedkar University, Vidyavihar, Lucknow, Uttar Pradesh, India

Preeti K. Suresh University Institute of Pharmacy, Pt. Ravishankar Shukla University, Raipur, Chhattisgarh, India

Sushma Talegaonkar Department of Pharmaceutics, Delhi Pharmaceutical Sciences and Research University, New Delhi, India

Anushka Tyagi Department of Pharmaceutics, Amity Institute of Pharmacy, Amity University Uttar Pradesh, Noida, India

Anita Kamra Verma Nanobiotech Lab, Department of Zoology, Kirori Mal College, University of Delhi, Delhi, India

Sonal Vyas Director, Sampoorna Path Care Labs, Sagar, MP, India

Suresh P. Vyas Drug Delivery Research Laboratory, Department of Pharmaceutical Sciences, Dr. Harisingh Gour Vishwavidyalaya, Sagar, MP, India

Hongbin Wang Department of Pharmaceutical and Biomedical Sciences College of Pharmacy; Master of Pharmaceutical Sciences College of Graduate Studies; Department of Basic Science College of Medicine, California Northstate University, Elk Grove, CA, USA

Shaofei Wang Department of Cellular and Genetic Medicine, School of Basic Medical Sciences, Fudan University, Shanghai, People's Republic of China

Krishna Yadav University Institute of Pharmacy, Pt. Ravishankar Shukla University, Raipur, Chhattisgarh, India

Monika Yadav Nanobiotech Lab, Department of Zoology, Kirori Mal College, University of Delhi, Delhi, India

Xian Zeng Department of Biological Medicine & Shanghai Engineering Research Center of Immunotherapeutics, School of Pharmacy, Fudan University, Shanghai, People's Republic of China

Xuyao Zhang Department of Biological Medicine & Shanghai Engineering Research Center of Immunotherapeutics, School of Pharmacy, Fudan University, Shanghai, People's Republic of China

Part I

Anatomy, Physiology and Pharmacology of Macrophages

Introduction to Pharmacology of Macrophages with Drug Delivery Perspective

Mahima Gupta, Atul Pathak, Yashwant V. Pathak, and Swati Gupta

Abstract

Ubiquitously present macrophages are an integral member of the innate immune system and are crucially involved in host defense and homeostasis. They aid host defense by inducing inflammation, exhibiting microbicidal and tumoricidal properties, controlling the activation of adaptive immunity, alleviating inflammation, and boosting tissue repair while also performing important trophic functions in organs/tissues (such as brain, mammary glands, bones) that lead to developmental processes. Based on the function that needs to be performed, macrophages can polarize to evident functional phenotypes and respond efficiently to tissue micro-environmental signs. Hence, they are perceived to be immensely multi-faceted-functional plasticity and diversity being major attributes of these cells.

M. Gupta · S. Gupta (✉)
Department of Pharmaceutics, Amity Institute of Pharmacy, Amity University Uttar Pradesh, Noida, India

A. Pathak
BioMedica Diagnostics Inc., London, Ontario, Canada

Y. V. Pathak
University of South Florida, Taneja College of Pharmacy, Tampa, FL, USA

Adjunct Professor University of Airlangga, Surabaya, Indonesia

A critical function in pathogenesis of almost every major disease including cancer, immunodeficiency, allergy, sepsis, metabolic disorders, atherosclerosis, inflammatory diseases, etc. suggests that macrophages can play a detrimental role as well. Being key players in homeostasis, host defense, and disease, they can also be possible targets for therapeutic applications. Developing novel strategies to target monocyte–macrophage lineage cells can pave the way for innovative treatments for a spectrum of ailments.

In the last 15 years, a phenomenal amount of research has culminated in a massive improvement in our understanding of macrophages, including fresh notions regarding their origin; their trophic functions; their contribution to homeostasis; relationship with the newly emerging myelomonocytic subsets; in vivo study of monocytes and macrophages including fate mapping and novel transgenic models; definition and concept and regulation of polarization in macrophages including characterization of transcriptional and post-transcription networks; and in vivo profiling of macrophages under different conditions of disease and their contribution to disease progression.

These discoveries call for a reconsideration of our perception of macrophages. In order to enhance the therapeutic potential of the enclosed medication, selective drug distribution to the macrophages seems to be an appeal-

S. Gupta, Y. V. Pathak (eds.), *Macrophage Targeted Delivery Systems*,
https://doi.org/10.1007/978-3-030-84164-5_1

ing idea. Thus, for selective drug distribution, macrophages may be manipulated. Nanocarriers can transit from across various membrane barriers at infection sites and release their drug cargo. The pharmacological modulation of macrophages and their function in immunomodulation in various diseases are discussed in this chapter.

Keywords

Macrophage · Activation · Active targeting · Passive targeting

1 Introduction

Macrophages were first identified in the larvae of starfish after insertion of tangerine tree thorns by Elia Metchnikoff in 1882 and then as cells responsible for the process of phagocytosis of foreign particles in *Daphea magna* or common water flea infected with fungal spores.

Elia Metchnikoff received the Noble Prize (Physiology and Medicine) for his discovery and the explanation of the phagocytosis mechanism in 1908. More than 130 years have passed, and researchers have identified different subtypes of macrophages and their functions as innate immune cells (Kumar 2019).

Immunity traditionally means defense against diseases, more precisely, infectious diseases. The immune system is made up of cells and molecules that are responsible for immunity, and their collective and organized response to the foreign substances is referred to as the immune response [Fig. 1]. Protection against infectious microbes is the physiological feature of the immune framework. However, immune responses can be elicited even by noninfectious foreign substances. In addition, in some cases, processes that usually keep people safe from infection and remove foreign substances are often capable of causing tissue injury and disease. Therefore, reactions to components of microbes as well as macromolecules such as proteins and polysaccharides, and small chemicals that are recognized as alien, irrespective of the physiological or pathological consequences, are a more inclusive description of the immune response.

Immunology is the study of the immune responses in a wider context, as well as the cellular and molecular occurrences after microbes and other foreign macromolecules are encountered by an organism. Immunology is an experimental science in which theories of immunological phenomena are based on the experimental findings and their conclusions. Our capacity to regulate the immune system's role under controlled circumstances has been the basis of evolution of immunology as something of an experimental discipline (Abbas et al. 2007).

2 What Is a Macrophage?

Tissue macrophages, phenotypically and functionally, are an exceptionally heterogeneous collection of cells originating from circulating monocytes. They differ in appearance from the microglial cells of the dendritic-like type to the less abortive Kupffer cell. Fortunately, there is an intracellular membrane marker in humans where most macrophages can be classified as CD688 (macrosialin in mouse). Macrophages have long been considered to be a vital piece of the inherent immune response, but they are increasingly involved in tissue homeostasis, hematopoiesis control, systemic inflammation, atherosclerosis, wound healing, and tissue remodeling, as well as killing invading microorganisms.

While macrophage function depends, at least in part, on location, developmental state, and conditions of in vitro culture, among almost all macrophage populations observed to date, there are some properties that are retained (Table 1). The ability to digest particles through phagocytosis is among the most distinctive characteristics of macrophages. Using a number of germline-encoded pattern recognition receptors including lectins, toll-like receptors, and receptors for N-formyl methionine-containing peptides, macrophages are able to identify both pathogens and noninfectious agents. Macrophages are active in the healthy expulsion of apoptotic cells and, by phagocytosis, eliminate small numbers of potentially harmful microorganisms without causing a strong pro-inflammatory reaction. If perceived

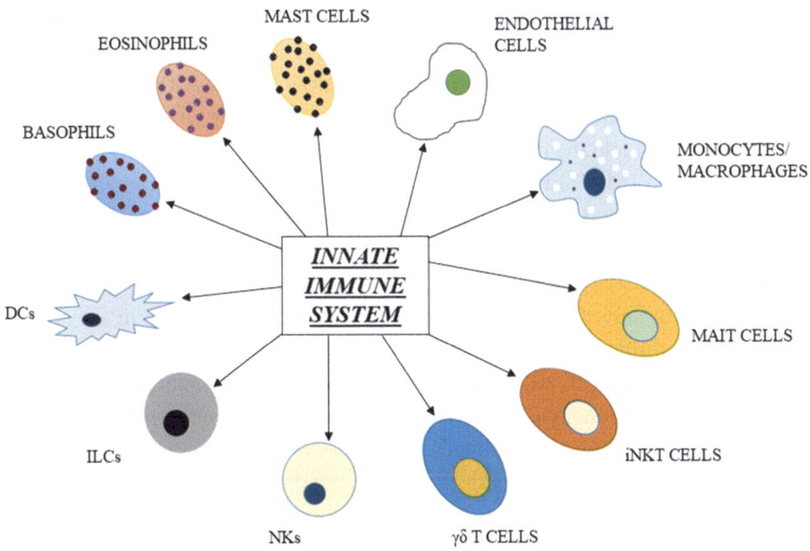

Fig. 1 Schematic representation of cellular components of innate immune system

Table 1 Origin of macrophages

Anatomical location	Cell name
Adipose tissue	Adipose tissue MΦ
Connective tissue	Histiocytes
Bone marrow/blood	Monocytes
Bone	Osteoclasts
Liver	Kupffer cells
Lymph nodes	Sinus histiocytes
Pulmonary alveoli	Alveolar MΦ
Central nervous system	Microglia
Placenta	Hofbauer cells
Kidney	Intraglomerular mesangial cells
Granulomas	Epithelioid cells
Red pulp of spleen	Red pulp MΦ
Peritoneal cavity	Peritoneal MΦ

risks are not apparent, an acute inflammatory response is mounted, which results in secretion of cytokines, chemokines, and antimicrobial agents. By binding cytokines to cytokine receptors or by recruiting cells involved in the adaptive immune response through secretion of chemokines, secretion of these mediators may lead to autocrine activation of macrophages. Using an arsenal of antimicrobial effector pathways encompassing enzymatic degradation, fermentation, nutrient restriction, and antimicrobial peptides, the macrophage kills invading microorganisms. The macrophage introduces foreign antigens to primed T lymphocytes upon internalization and absorption of the pathogen, thereby amplifying the adaptive immune response. Prolonged or chronic inflammation can occur when macrophage-based clearance is ineffective. A range of chronic infections and inflammatory diseases, including the development of atherosclerotic plaques and conditions such as rheumatoid arthritis, are intimately connected with macrophages (Taylor et al. 2005).

The origins of macrophages, variations in macrophage phenotypes, mechanisms of killing dependent on macrophages, and the subversion of this killing by intracellular parasites are described here.

2.1 Macrophage Origin

While it is evident that myeloid cells are formed from precursor cells located in the bone marrow, there is no full resolution of their developmental pathway. Confounding this is the fact that leukocyte growth studies are much more convenient in the mouse, and there is also some doubt as to the variations between the development of human and mouse leukocytes. $CD34^+$ precursors are widely recognized as giving birth to monocytes, granulocytes, erythrocytes, and thrombocytes. Myelomonocytic cells give birth exclusively to cells that express both CD34 and the macrophage colony-stimulating factor (M-CSFR) receptor (Blaschke et al. 2003). The blood contains circulating $CD34^+$ monocytes, and it is these circulating monocytes that produce tissue macrophages. There are no less than four subsets of monocytes in human peripheral blood that are distinguished by their CD64 expression levels (FCγRII), CD14, and CD16 (FCγRIII). These monocytes give rise to terminally differentiated cells, macrophages, and dendritic cells in response to stimuli (Fig. 2).

In healthy people, monocytes expressing elevated levels of CD14, CD64, and little to no CD16 ($CD14^+/CD64^+/CD16^+$) make up more than 80% of the circulating monocyte population. Such monocytes, when activated by bacterial components, develop elevated levels of pro-inflammatory cytokines. The in vitro reaction to a number of chemokines is believed to be an important factor in peripheral tissue migration during infection and inflammation; they can then be differentiated into macrophages with exceptional antimicrobial activity and the potential to associate with both B and T lymphocytes (Ancuta et al. 2003).

Less than 10% of circulating monocytes in humans are monocytes expressing CD16 as well as CD64 and CD14 ($CD14^+/CD64^+/CD16^+$), which produce high levels of pro-inflammatory cytokines, low IL-10 levels, have very high phagocytic ability, and are involved in antibody-dependent cell cytotoxicity (ADCC). It is suspected that they are precursors to "resident" macrophages (Smythies et al. 2005). $CD14^+/CD64^+/CD16^+$ expressing cells differentiate into macrophages or dendritic cells and have a distinctive DC1 phenotype in in vitro culture with cytokines and are increased in patients with Kawasaki disease and influenza and reduced in patients with rheumatoid arthritis (Gordon and Taylor 2005; Calder et al. 2005).

Less than 10% of circulating monocytes are monocytes that do not express CD64 but have medium to moderate CD14 levels and high CD16 levels ($CD14^{dim}/CD64^-/CD16^+$ or $CD14^{low}/CD64^{low}/CD16^+$). These monocytes have increased CD45 expression and costimulatory function but contain very little interferon type 1 or other pro-inflammatory cytokines. They have weaker phagocytic responses and ADCC but show an improved capacity to communicate with T or B lymphocytes and express greater chemotactic activity than their counterparts in $CD14^+/CD16^+$, especially in response to fractalkine, the endothelial cell tethered chemokine. The tendency of these cells to transmigrate in response to fractalkine (due to CX3CL1 expression), which is not expressed on CD16 cells, suggests that tis-

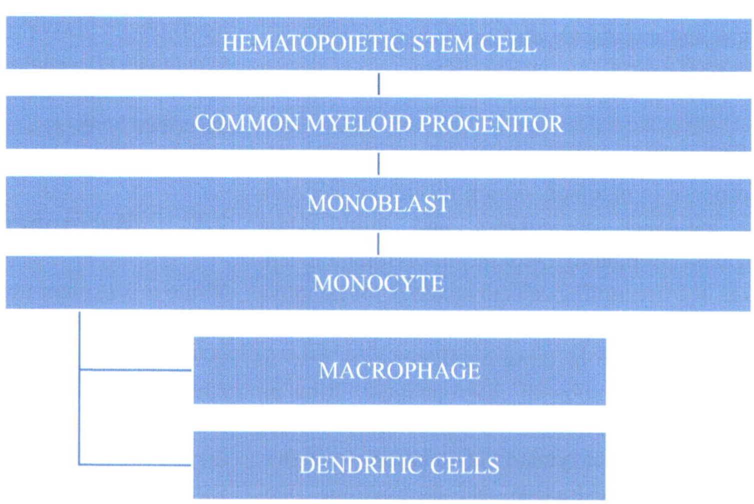

Fig. 2 Differentiation from monocytes to macrophages

sue macrophages may be precursors. Such cells differentiate into cells presenting myeloid antigen and are thought to play a role in the in vivo response to Th1 (Plata et al. 1987). In HIV-infected patients, and those suffering from pararheumatic systemic vasculitis and sepsis, elevated numbers of these cells are identified (Baszler et al. 1999). While these traits are useful to generalize widely, there are a variety of subsets of monocytes/macrophages that are not readily categorized. For instance, intestinal mucosa macrophages have a different receptor expression profile that does not include CD14, complement receptors, or Fc receptors (Schofield et al. 1987). This highlights the conclusion that no specific instructions for the detection and classification of subsets of macrophages are valid.

3 Activation States of Macrophages

Confusion surrounding the terms of macrophage activation states is plaguing macrophage biology. Macrophages have a variety of homeostatic functions, including the swallowing of apoptotic cells, erythrocyte clearance, and constitutive tissue regeneration in the absence of pro-inflammatory or contagious stimuli. The reaction of the macrophage to infection must be adapted to the microbial threat, and it has been revealed that the form of microbial or inflammatory stimulation leads to the development of macrophages with differing functions, mainly from in vitro studies. Four main classes, classical, innate, alternate, and deactivated (Table 2), of immunologically acquired macrophage activation have been proposed to date (Classen et al. 2009).

3.1 Classical Activation

As a consequence of the discovery that macrophages treated with bacterial components and interferon-γ (IFN-γ) produced an improved capacity to kill a wide variety of ingested pathogens, the idea of macrophage activation emerged. IFN-γ is formed by lymphocytes of $CD4^+$ and $CD8^+$ T, NK cells, and probably the infected macrophages themselves. This capacity is not bestowed by IFN-γ alone; it primes the macrophages for activation instead. A bacterial part usually LPS (lipopolysaccharide) is the second signal. While it has not been adequately analyzed, some studies indicate that it is the bacterial stimulation of TNF-α that provides the secondary signal rather than the LPS itself. Due to enhanced expression of MHC (major histocompatibility complex) class II as well as CD80/CD86 (B7.1/B7.2) and improved iNOS production, classically active macrophages have an increased capacity to present antigen. Because of an enhanced respiratory blast, they have an increased capacity to kill intracellular pathogens and develop the ability to mediate multiple inflammatory effects in the host through the secretion of a number of cytokines.

The significance of IFN-γ in parasite infection was shown in vivo when it was discovered that IFN-γ neutralization mediated by antibodies in infected mice caused them to die faster and have increased parasite loads (Skurkovich and Skurkovich 2006). Subsequent studies with mice that were deficient in the expression of IFN-γ or its receptor found that they were more prone to a number of infections from intracellular bacteria or protozoa (Daugherty et al. 2005). Activation caused by IFN-γ is a contributor to rheumatoid arthritis disease, hypersensitivity of delayed form, and may lead to atherosclerosis.

3.2 Innate Activation

Two stages are needed for classical activation, exposure to IFN-γ, and a bacterial compound, resulting in a macrophage with altered phenotypic and functional properties. Exposure to bacterial components such as LPS or CpG (5'-C-phosphate-G-3') alone has recently been shown to result in macrophages with altered phenotypes and functional properties. For example, macrophages treated with LPS or CpG have been shown to have an improved capacity to produce IL-12 in response to a second LPS exposure due to the expression of the collagenous structure macrophage receptor (MARCO). MARCO's

Table 2 Activation states of macrophages

	M1 (Classical)	M2 (Alternative)		
Subtype	M1	M2a	M2b	M2c
Stimulus	IFN-γ + LPS/TNF-α	IL-4 or IL-13	TLRs + immune complexes	IL-10
Functions	*Pro-inflammatory*	*Wound healing*	*Immunoregulatory*	*Immunosuppressive*
	Boosts inflammation, removal of apoptotic cells and debris, sterilization	Anti-inflammatory, cell proliferation, cell migration, removal of apoptotic cell	Cell maturation, stabilization of tissue, angiogenesis, synthesis of extracellular matrix	Tissue repair, inflammatory resolution, synthesis of extracellular matrix
Pathology	Chronic inflammation, tissue damage, autoimmunity	Cancer, fibrosis, epithelial hyperplasia		

TLR (toll-like receptors) agonist-induced expression has also been related to the macrophage's increased capacity to bind and clear Neisseria (Mukhopadhyay et al. 2006).

3.3 Alternative Activation

Early on, antigen-presenting cells retrieved from mice having experimental nematode infections (in which there is a Th2 cytokine environment) were capable to process and present antigen without causing the proliferation of T cells. Subsequently, exposure to Th2-associated cytokines, IL-13 and IL-4, was observed to result in macrophages with enhanced mannose receptor expression and class II MHC, which were unable to induce T cell proliferation. Endocytosis and antigen presentation are consistent with increased expression of the mannose receptor, but possibly less effectively than classically activated cells. An increased influx of internalized particles and ligands to lysosomes is also present. Alternatively, activated macrophages are critical in clearance of parasitic and extracellular pathogens, but they do not show an increased oxidative burst unlike classically activated macrophages and are thus not as effective in killing intracellular pathogens (Gordon 2003).

In parasitic and protozoan infections, the significance of alternatively triggered macrophages is now well known. *Schistosoma mansoni*, *Trypanosoma*, and *Leishmania* infection in vivo models indicate that there is a complex interplay between the formation of cytokines Th1 and Th2 and the subsequent creation of subsets of macrophages. To regulate the initial stages of infection by *T. cruzi, T. brucei,* and *S. mansoni*, an initial Th1 response (characterized by elevated levels of IFN-γ and IL-12) appears to be necessary; however, the cytokine balance switches to a Th2 reaction during the course of illness. This change to a Th2 preference is widely assumed to be important for the clearance and resolution of the infection as animals deficient in the development of Th2 cytokines do not survive and thus alternately active macrophages do not survive. There is also evidence that the change to a Th2-mediated response may contribute to dissemination of the parasite within the host for certain protozoan pathogens (Gordon and Martinez 2010).

Alternatively, activated macrophages, owing to their induction of arginase, an enzyme that counteracts the destructive effects of NO, do not contain significant quantities of nitric oxide (NO). Arginase leads to the biosynthesis of polyamine with proline and facilitates cell proliferation, the production of collagen, and tissue remodeling. This subclass of macrophages has been suggested to play a primary role in wound healing, angiogenesis, fibro-genesis, extracellular matrix synthesis, and the development of granuloma. Alternatively, active macrophages also tend to have an anti-inflammatory role, and they have been demonstrated to decrease T cell proliferation and develop the anti-inflammatory cytokine IL-1 receptor antagonist and IL-10. Compared to classically active macrophages, these cells have a small decrease in LPS-induced respiratory burst and cytokine production, consistent with this study. In Th2-mediated diseases such as asthma, allergies, and in the resolution of infectious diseases and parasitic infections, alternatively activated macrophages are essential (Merad et al. 2002).

3.4 Deactivation

Activated macrophages have powerful biological functions that are important for the host's infection response. However, it is important to stop the pro-inflammatory regimen until the infection is healed. A "deactivated" phenotype may be caused by exposure to a variety of anti-inflammatory molecules such as cytokines (e.g., IL-10, TGF-α), receptor ligation (e.g., CD200–CD200R), steroids, or apoptotic cell uptake. These cells can be characterized by CD163 (Town et al. 2005) expression and have reduced MHC class II expression, decreased respiratory bursts and pro-inflammatory cytokine production, as well as increased production of anti-inflammatory cytokines.

4 Types of Macrophages

Subpopulations of macrophages can be divided into a variety of ways. Phenotypical and functional differences occur between macrophages present in the body at various sites and between resident and recruited macrophages (Fig. 3). Due to the difficulties of separating the two sets in vivo, the distinction between resident and recruited macrophages is especially murky. There is some controversy regarding the role of newly recruited monocytes in the production of resident cells, but it has long been understood that circulating monocytes migrate to tissues where they become macrophages. Tissue macrophages were initially thought to be derived and replenished solely from circulating monocytes, but transplantation experiments in both mice and humans suggest that the replenishment of donor macrophages with resident tissue macrophages is exceedingly sluggish. This might arise due to extremely low recruitment and replenishment levels by circulating monocytes or because the recipient's tissue macrophages are capable of self-renewal. For epidermal Langerhans cells, identical findings were found. It is then thought that the tissues are populated with cells originating from circulating monocytes early in fetal or embryonic development. Such cells develop into resident macrophages, and replenishment of circulating cells is poor under steady-state conditions. They are able to reach the draining lymph nodes with the necessary chemotactic stimuli as these cells are triggered by infection or inflammation and transfer to the regions of B and T lymphocytes to present pathogens. Under such conditions, monocytes enter the tissues to replenish the macrophages that are activated and these cells become macrophages that are recruited. The monocytes which are deployed to the infection or inflammatory

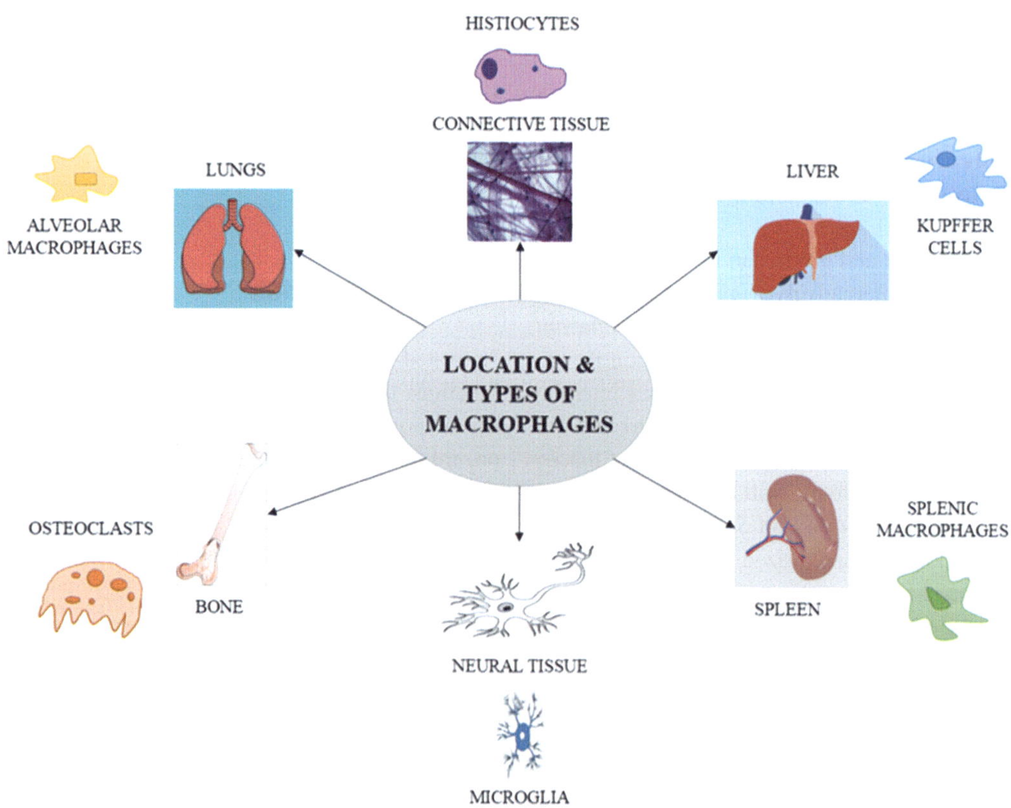

Fig. 3 Different types of macrophages and their locations

sites can be different from those that under steady-state conditions replenish resident cells.

Generally, tissue macrophages have stellate morphology and high endocytic capacity (including nonspecific uptake of particles and Fc receptor-mediated uptake). Although they proliferate, they have active RNA (ribonucleic acid) and protein synthesis while feeding very slowly. There are essential homeostatic roles of resident tissue macrophages and clear protein aggregates (e.g., protease–inhibitor complexes), biochemical molecules (e.g., lysosomal hydrolases), denatured molecules (e.g., modified lipoproteins), and intracellular cell apoptotic cells, either immunologically silent or tolerogenic. These cells are also major sentinels for the removal of microorganisms attacking them.

There are apparent functional distinctions between various subsets, considering the heterogeneity of macrophages, and it is helpful to characterize macrophage subpopulations on the basis of place. The attributes of macrophage subpopulations that are most often associated with parasitic infections are briefly summarized here. Although the discussion of macrophage subpopulations of the liver, brain, and bone has been omitted, macrophages have all been shown to host protozoan parasites at these locations and contribute in extreme cases to pathology (Gordon and Pluddemann 2017).

4.1 Kupffer Cells

Kupffer cells are considered resident macrophages of the liver. The liver is an integral and active part of the innate immune response. Local macrophages develop the cytokines IL-1, TNF-α, and IL-6 after infection with extrahepatic locations. IL-6 recognition induces hepatocytes to develop a variety of proteins in the acute process that are responsible for the systemic effects of inflammation and enhance the activation of opsonic phagocytosis and complement. Kupffer cells sense extrahepatic cytokines; the cells become stimulated and have improved antimicrobial properties, while resident Kupffer cells have a less intense respiratory burst and are thus less effective than other forms of macrophages in destroying such pathogens (Dixon et al. 2013; Bordon 2019).

High levels of phagocytic receptors are expressed by Kupffer cells. This includes Fc receptors by which soluble IgG complexes and antibody-coated particles or microorganisms are excluded, and complement receptors by which complement-coated bacteria and erythrocytes and scavenger and toll-like receptors are removed from the bloodstream by bacteria and endotoxins. Their zeal for clearing erythrocytes contributes to the distinctive deposition of iron in these cells.

Kupffer cells can be further subdivided into cells in the periportal, midzonal, and perivorous regions on the basis of their position in the liver. Macrophages have varying capacities for the secretion of TNF-α, prostaglandin E, nitric oxide, and IL-1 at various sites. The periportal region's Kupffer cells have the largest phagocytic activity and the highest activity of the lysosomal enzyme, which is thought to be because this is the entry point for blood and thus the first point of contact for any pathogens carried by blood. As they have been shown to host a variety of protozoan pathogens, Kupffer cells are involved in both clearance and transfer of pathogens.

In the systemic dissemination of *Plasmodium falciparum*, for instance, these cells may be particularly significant. Due to the recognition of at least two proteins, circumsporozoite protein and thrombospondin-related adhesive protein, sporozoites pass via the liver through the bloodstream and are phagocytized by Kupffer cells (TRAP). In the macrophage vacuoles, the sporozoites tend to be able to live and leave the macrophage at a later point in time. The parasites enter adjacent hepatocytes at this stage and cause their death, resulting in many of the disease's symptoms (Frevert 2004).

4.2 Splenic Macrophages

The spleen is a special lymphoid organ that requires the removal from the blood of infections and senescent erythrocytes as well as the presentation of antigens and the activation of an adaptive immune response. In general, spleen macrophages are subdivided on the basis of location; however, it is important to remember that the architecture of mouse and human spleen is very different and that murine studies derive the bulk of our knowledge of splenic macrophages. It seems as if the distribution of receptors and the activity of various macrophage species differs between humans and mice in the few comparative studies that have been carried out (Mebius and Kraal 2005). Nevertheless, the spleen functions both as a clearing site for blood-borne bacteria and as an interaction in both humans and mice between antigen-presenting cells and B and T lymphocytes. The description below is largely focused on mouse experiments.

A specialized region of lymphocyte aggregation which contains B and T lymphocytes is the white pulp. The white pulp is differentiated by the marginal zone from the red pulp, which is the main erythrocyte clearing region. The marginal zone includes the lymphocytes and dendritic cells of the marginal zone B and two groups of macrophages, the macrophages of the marginal zone adjacent to the red pulp, and the metallophilic macrophages of the marginal zone adjacent to the white pulp. In the clearance of apoptotic cells and microorganisms and the preservation of B lymphocytes, the macrophages of the marginal zone are active. Like Kupffer cells, these macrophages are involved in erythrocyte turnover and iron recycling. The role of metallophilic macrophages in the marginal zone is not completely clear, although it is known that they are involved in the response to viruses as they generate high IFN-alpha and IFN-alpha levels.

In the removal of blood-borne infections, including *Leishmania* spp. and *Plasmodium falciparum*, splenic macrophages have a number of pattern recognition receptors of essential significance. Any pathogens that are quickly cleared by the macrophages of the liver from circulation have virulence factors that prohibit easy clearance from the spleen. In laboratory models of visceral leishmaniasis, for example, the hepatic portion of the infection is self-limiting (probably due to the development of granuloma); however, growth of amastigote in the spleen cannot be contained and results in the loss of tissue. While it is understood that the marginal zone macrophages amastigotes avidly phagocytosis, it is not known if their failure to clear the parasite is attributable to discrepancies between the hepatic and splenic macrophages, such as differences in the pathogen entry process, differences in their capacity to inhibit the development of cytokines, or any other unexplained mechanism.

As splenectomized patients have a high risk of serious bacterial infections and may take prophylactic antibiotics, the significance of the spleen in the host's reaction to infection is obvious. In comparison, people who have had a splenectomy are more likely to die from malaria and to have elevated blood parasite levels (Henry et al. 2020; Liu and Uzonna 2012). In host protection against bacterial and parasitic infection, the macrophages of the spleen are thus essential.

4.3 Dendritic Cells

The proximate cousin of the macrophage is the dendritic cell (DC). Both macrophages and dendritic cells absorb and present non-antigens, while self-antigens are also present in the dendritic cell and are implicated in tolerance induction. The dendritic cell until it meets antigenic stimuli is referred to as an immature dendritic cell (iDC). These cells are located in non-lymphoid tissues, and dendritic cells are strongly phagocytic, like macrophages, a feature that is promoted by the involvement of antigen-sampling motile, long dendrite-like processes. Immature dendritic cells may pick up and process antigen in the absence of foreign or inflammatory factors but do not interact with T cells because they do not express large quantities of class II MHC or costimulatory molecules on their surface. If a "risk" signal (e.g., pathogen-associated molecules or pro-inflammatory cytokines) is obtained by the dendritic cell, it undergoes an activation mechanism that enhances the expression of MHC class II, costimulatory molecules (CD80 and CD86), and selected chemokine receptors that allow it to migrate to the lymph node where antigen is presented. Such dendritic cells are considered mature dendritic cells (mDCs). Unlike macrophages, dendritic cells are capable of supplying both naïve and activated T cells with antigen. By both introducing antigen to T cells, dendritic cells activate the adaptive immune response, but also by releasing a variety of cytokines and chemokines. Cytokine production by dendritic cells is also much higher than that of macrophages, resulting in higher T cell recruitment and activation.

While there can now seem to be as many as five classes of DCs, this discussion is restricted to the better described classes of dendritic cells most likely to be involved in the reaction to protozoan infections (Germic et al. 2019; Ghislat and Lawrence 2018).

The plasmacytoid DC (pDC), which is $CD64^-$/ $CD16^-$, is the most recent type of circulating precursor cell to be identified. These cells make up a relatively small percentage of the total circulation population, but considering those small numbers, they are important in the response of the host to viruses. It is also essential to receive plasmacytoid cells from the spleen. TLR7, TLR9, and TLR11 are expressed (in mice) and are not immune to agonists of TLR2 and TLR4, such as LPS and peptidoglycan. Plasmacytoid DCs create high IFN-alpha concentrations, but no or no IL-6 or TNF-α. Plasmacytoid DCs have minimal phagocytic ability relative to other DC subsets, do not participate in ADCC, and have very little contact with either B or T lymphocytes. They are not believed to play a role in the host defense against protozoan pathogens in genetic terms, although it has been shown that malaria blood stage schizonts can contribute to increased CD86 ex-pressure and stimulate in vitro pDC development of IFN-α.

In circulation, myeloid DCs can also be observed and are characterized by the expression of markers such as CD13, CD11c, and CD33. Upon stimulation via TLR1, TLR2, TLR5, and TLR8 with pathogen-associated microbial ligands, these cells do not develop IFN-α or IFN-β but rather the pro-inflammatory cytokines, IL-6 and TNF-α. In response to protozoan pathogens, myeloid DCs develop high levels of IL-12 in both toll-like receptor-dependent and receptor-independent forms. These cells generate primarily homeostatic chemokines as compared to pDCs and have a greater ability to migrate to chemokines such as MCP-1, RANTES, and IP-10 generated during protozoan infection (Penna et al. 2002). It is assumed that myeloid DCs that circulate under the right conditions will migrate to the tissues where they differentiate into tissue DCs.

The macrophages and dendritic cells of the skin and the gut are particularly important with regard to parasitic infection. Because of the interplay between macrophages and dendritic cells at these locations, their properties dependent on their position in response to infection are summarized.

4.4 Macrophages/Dendritic Cells of the Skin

In the removal of infections spread by insect bites or other skin breaks, the resident cells of the epidermis are essential. There are two dendritic cell groups in the skin, the Langerhans cells that are characterized by CD207 (Langerin) expression and dermal dendritic cells that are characterized by CD2088 expression (DC-SIGN). In binding *Leishmania amastigotes* and *Schistosoma mansoni* egg antigens, dermal DCs have been implicated. The capillaries and reticular dermis are contained in dermal dendritic cells, while the basal and supra-basal layers of the epidermis are located in Langerhans cells (Murray et al. 2005). These cells' long pathways are especially suited to the capture of anti-genes, which are regulated by C-type lectin and Fc receptor expression. In the sense of both MHC class I and class II, these DCs present antigens. Tissue macrophages and skin dendritic cells tend to have distinct capacities for particulate matter and pathogen phagocytosis. Phagocytose 0.5–1 μm beads of Langerhans cells when macrophages eat greater particles (>3.5 μm). Differences occur in the forms of pathogens that are preferentially phagocytized by diverse subsets of skin DCs and macrophages.

Over the process of parasite infection, a complex interplay occurs between tissue macrophages and dendritic cells. The skin is inoculated with swallowed promastigotes in the case of cutaneous *Leishmania* infection and is usually killed by resident macrophages by the processing of reactive oxygen and nitrogen species. In vitro experiments show that macrophages do not generally become activated by different surface markers or increase their surface expression. To present antigen and clear infection, sequential activation of skin dendritic cells through ingestion of amastigotes is necessary. In order to produce the Th1-promoting cytokines IL-12, IFN-γ, and TNF-α, $CD4^+$ T lymphocytes must be involved (Von Stebut and Tenzer 2018).

4.5 Macrophages and Dendritic Cells of the Gut/Intestine

Like Peyer's bands, mucosal lymphoid follicles, and lamina propria, the intestine comprises several distinct immunological niches. In all these regions, antigen-presenting cells can be identified. In order to deal with the heavy antigenic and bacterial load of the gut, macrophages of the mucosa and intestine are specially modified. While these cells are derived from circulating CD14 expressing monocytes, they do not express CD14 or other receptors for bacterial recognition and as such, are basically non-responsive to bacterial stimulation. The inability of these macrophages to respond to bacterial stimuli by developing cytokines such as IL-1, IL-6, IL-10, IL-12, RANTES, TGF-β, and TNF-α has resulted in the suggestion that they produce "inflammatory anergy" from circulating monocytes. It is suspected that when exposed to local cytokines such as TGF-β, recruited $CD14^+$ expressing monocytes evolve this phenotype and that this is an important adaptation to cope with the heavy load of the intestine's mainly commensal bacteria. It should be remembered that in their capacity to phagocytose or kill phagocytose bacteria, these macrophages are not deficient. A variety of other surface markers, including CR3 and LFA-1 and IgA, IgG receptors, are not expressed, but high levels of MHC class II and HLA-DR are expressed, suggesting that they have antigen-presenting potential. Macrophages can be present in the intestinal tract, but lamina propria tends to be the most widespread (Smith et al. 2005; Makala et al. 2004).

Dendritic cells in the gut and intestine are also important antigen-presenting cells. These cells are especially significant in inducing the differentiation of regulatory T cells under steady-state conditions in addition to their antigen-presenting functions. In Peyer's patches and in the lamina propria, iDCs are located. It is suspected that pathogens and parasites are transported through M cells to the dendritic cells of the Peyer patches, while lamina propria dendritic cells can sample pathogens using long dendrites that stretch between the close junctions of the epithelial cells into the lumen of the gut. The dendritic cells develop a mature phenotype following infection (e.g., they express MHC class II, CD40, CD80, etc.). It is suspected that they are involved in the spread of infections to remote locations in the body because of the ability of these cells to migrate (Kelsall and Leon 2005).

Through ingestion, several protozoan pathogens enter the host. It has been shown that Giardia spp., *Cryptosporidium parvum, Toxoplasma gondii,* and *Entamoeba histolytica* all reproduce within the gut. Pathogens bind to the intestinal epithelial layers and may cross epithelial boundaries in certain instances, at which stage they may be identified by intestinal macrophages and dendritic cells (Barragan and Sibley 2002; Lacroix-Lamande et al. 2002; Salata et al. 1985). Unless there is a violation of the integrity of the epithelial membrane or pro-inflammatory cytokines or chemokines are identified, a macrophage and dendritic cell-mediated immune response is not mounted. For both the mobilization and activation of macrophages and dendritic cells, chemokine and cytokine outputs from epithelial cells and resident leukocytes are important. Macrophages, neutrophils, and dendritic cells, once mobilized, create IL-12 and activate a Th1 response. In order to protect against protozoan pathogens in the gut, IL-12 development is important since it inhibits the production of IFN-γ and activates macrophages. IFN-γ induced macrophage activation is in effect, so important for host defense that the ability to decrease or remove its development is an integral determinant of virulence for protozoan pathogens (Gazzinelli et al. 1994).

5 Recognition and Destruction by Macrophages and Subversion by Pathogens

Despite the exquisite adaptation of macrophages to the destruction of intracellular bacterial and protozoan parasites, such infections occur at disturbing rates, especially in the developed world. Malaria (*Plasmodium spp.*), Chagas disease (*Trypanosoma cruzi*), leishmaniasis (*Leishmania spp.*), and toxoplasmosis (*Toxoplasma gondii*) are identified as significant health threats in developed nations by the World Health Organization (WHO). It is crucial to understand the processes of identification, ingestion, and degradation of pathogens by macrophages in order to understand what makes these pathogens so effective. The macrophage is capable of detecting, phagocytosing, and killing pathogens. With regard to the microorganism, the original binding and identification mechanism that causes phagocytosis varies. The result of opsonin-dependent phagocytosis and autonomous phagocytosis has variations. The amount of opsonin released by the host, both constitutively and in response to infection, illustrates the importance of opsonic phagocytosis. Opsonins such as C-reactive proteins are essential for a number of intracellular parasites to enhance phagocytosis (e.g., *Leishmania promastigotes*). However, complement activation-mediated phagocytosis does not result in a heavy oxidative burst from the macrophage, and therefore certain pathogens take advantage of this uptake process. For example, by expressing elongated lipidoglycans on its surface, *Leishmania* promotes complement-mediated absorption. These lipidoglycans do not inhibit the activation of the complement, but since the activated complement is far from the cell membrane, the parasite is not lysed. In addition, opsonization allows promastigotes via the complement pathway to join the macrophage, thereby preventing natural phagosome–lysosome fusion (Olivier et al. 2012).

In general, phagocytosis mediated by Fc receptors results in the phagosome maturing into an acidic, hydrolytically active compartment and the pathogen being killed. There are a variety of conserved techniques for subverting regular Fc receptor-mediated uptake by intracellular pathogens. The pathogens *Toxoplasma, Plasmodium, and Eimeria* have an intrusive motile process called zoetis, in which they can use a motile mechanism based on actinomyosin that mediates host cell invasion, subverting phagocytosis mediated by both complement and Fc receptor. This method is used by *Toxoplasma* to generate vacuoles which exist independently of the normal phagolysosomal pathway and are thus not exposed to the phagolysosome's destructive environment (Lodge and Descoteaux 2005).

The phagocytic vacuole undergoes various maturation steps that are followed by persistent remodeling of the composition of the phagosome membrane protein after Fc mediated phagocytosis. By measuring the aggregation of different surface markers, phagosomes sequentially fuse with the early endosomes, late endosomes, and lysosomes, and the maturation of the phagosome can be monitored. Upon fusion with the early endosomes, the pH decreases marginally (pH 6.2). This results in receptor/ligand pairs being uncoupled and receptor recycling regulated by the Rab proteins (Rab4 and Rab11) (Scianimanico et al. 1999). The membrane of the phagolysosome accumulates acid-resistant phospholipids and is distinguished by the expression of Lamp1 and Lamp2 after fusion with the late endosomes. The subsequent fusion with lysosomes results in a decrease in pH (4.7–5.2) resulting in the activation of the proteolytic enzymes contained within, such as cathepsins. These enzymes are important not only for microbial degradation, but also for the development of MHC molecules to produce antigens. Upon phagocyte activation, oxidative species such as O2- are rapidly generated. In a mechanism that is closely related to the cytoplasmic membrane and involves cytoskeletal elements and protein phosphorylation, the NADPH oxidase enzyme is essential for the catalysis of different oxidative compounds, including superoxide, hydrogen peroxide, and halogenated oxygen molecules. In antimicrobial death, nitric oxide species are also involved and are important for the destruction of a variety of intracellular parasites. By nitric oxide synthase from L-arginine and molecular oxygen, NO production is catalyzed. Hydrogen peroxide associations with myeloperoxidase, decreased iron, or NO contribute to the development of additional toxic intermediates such as hypochlorite anion, hydroxyl radicals, peroxynitrite, and nitrogen dioxide.

There are three choices after the pathogen has been phagocytized. In the intra-lysosomal environment, it may either exist and evolve pathways to cope with the acidic, hydrolytic environment in it, or it may exist in the vacuole, but it inhibits normal maturation from occurring and therefore remains shielded from the macrophage's microbicidal properties. Any pathogens escape from the vacuole entirely and survive in the cytosol's more permissive environment.

Phagolysosomal maturation is deliberately subverted by the majority of intracellular pathogens. The pathogen can avoid acidification (e.g., *Histoplasma capsulatum, Entamoeba histolytica*), turn the phagolysosome into a more permissive environment (e.g., *Salmonella*), or interrupt the production of the phagosome at an earlier or less destructive level (e.g., *L. donovani, M. tuberculosis*). Pathogens that have formed lysosome include *Leishmania* and *Coxiella* pathways for coping with life. By possessing a cell surface of resistant lipidoglycans, *Leishmania* prevents hydrolysis and can resist antigen presentation by controlling antigenic peptide expression and accessibility (Lodge and Descoteaux 2005).

Among intracellular parasites, including *T. cruzi, Listeria, Shigella, and Rickettsia*, escape from the phagocytic vacuole is a common theme. Pathogens have a variety of pathways through which they detach from the phagosomal membrane, such as pores (*Listeria spp.*), lysis (*Shigella flexneri*), and mechanisms that have not yet been established (*Rickettsia*). They are able to replicate in the more permissive cytosol setting until the pathogens have escaped.

The macrophage has elaborate mechanisms, such as iron and amino acids, to deprive the pathogen of vital ingredients for survival. The transferrin receptor by which they bind and internalize extracellular iron is expressed in an inactivated state by macrophages. When activated by IFN-γ, they downregulate the transferrin receptor, thereby reducing intracellular iron reserves. The enzyme indoleamine 2, 3-dioxygease, which catalyzes the degradation of L-tryptophan and thus restricts the supply of this amino acid to intracellular species, is also activated by IFN-γ-mediated activation. Survival in the phagosome is the most urgent concern of the intracellular parasite, but after the imminent danger of destruction is resolved, the parasite must gain insufficient nutrients to prevent the immune system's identification. A common theme used by *C. burnetti*, who has an active method for recruiting nutrients at an acidic but not neutral pH, is to remove nutrients from a hostile environment.

Cytokine release and involvement have a variety of secondary effects on the killing of macrophages. IFN-γ has a variety of indirect effects that increase antimicrobial function, in addition to being important for macrophage activation. IFN-γ penetration stimulates the development of a variety of chemokines, such as IP-10/CXCL10 and CXCL11, contributing to the mobilization of antimicrobial activity of additional leukocytes. The enhancement of antibacterial activity also leads to chemokines. RANTES, MIP-1α, and MIP-1β increase uptake and, by causing NO production, induce intracellular destruction by macrophages of trypomastigotes and *Rickettsia*.

By inhibiting or encouraging macrophage signaling, intracellular parasites often subvert host functions. This leads to disturbance of natural host processes such as apoptosis (e.g., *T. gondii*) and development of pro-inflammatory cytokines (e.g., *T. gondii*) (Butcher et al. 2001). In order to ensure safety, the abolition of pro-inflammatory cytokine production alone is not adequate, and certain parasites modify cell signaling in such a way that the equilibrium between production of Th1 and Th2 is altered, thereby inhibiting the natural anti-parasitic response of the host (Yun et al. 2001). This can be accomplished by modifying the signaling pathways by blocking or degrading main signaling components or by specifically degrading pivotal cytokines such as IL-12 more directly (Cameron et al. 2004).

6 Drug Delivery to Macrophages

Drug delivery to macrophages promises to be an appealing strategy for improving therapeutic efficacy of enclosed drug. As a result, macrophages may be utilized as Trojan horses for therapeutics. Nanocarriers can cross multiple membrane barriers and release their drug cargo at infection sites (Jain et al. 2013). To treat intracellular infections like tuberculosis, salmonellosis, and brucellosis, antibacterials are delivered to macrophages. Bacteria living in host cells cause these infections by replicating, surviving, and causing harm to the host. Macrophages serve as repositories for these intracellular pathogens and provide an immune-privileged niche. The lethal intracellular protozoan *Leishmania donovani* also infests and proliferates within macrophages, causing visceral leishmaniasis. Antibiotics are typically used to treat infections caused by these pathogens, but the clinical results are limited due to antibiotics' inadequate delivery to pathogens in macrophages. Antibiotics and intracellular pathogens are expected to co-localize better within drug delivery systems that target macrophages, enhancing therapeutic effects. Macrophages are dormant repositories for the human immunodeficiency virus type 1 (HIV-1) that can withstand the cytopathic effect of the virus a, rendering them an effective therapeutic target for HIV-1 infection treatment. Macrophages have also been studied as a potential therapeutic target for Gaucher disease, a genetic condition characterized by a lack of lysosomal enzyme activity. Since this enzyme deficiency primarily occurs in macrophages, researchers have focused their efforts on delivering replacement enzymes or enzyme activators to these cells. Furthermore, macrophages are crucial in rheumatoid arthritis (RA), as their overpopulation in inflamed synovial membranes causes acute and chronic joint destruction. As a result, in the field of RA therapy, drug delivery systems that can directly inactivate or destroy macrophages in the joints have piqued interest. Macrophages have traditionally been regarded as an unfavorable mechanism for the untimely clearing of drug delivery systems in cancer treatment, but with a growing understanding of TAMs' complex roles in tumor development, macrophages are being reintroduced as a possible target in cancer therapy. Because of their hypoxia sensitivity and capacity to migrate and penetrate tumors, macrophages and monocytes have also been investigated as potential carriers of anticancer drugs and imaging agents. Macrophages can also cross the blood–brain barrier to enter inflamed brain tissue in Alzheimer's and Parkinson's disease patients. Prior to administering, drugs are loaded into nanoparticles (NPs) that can potentiate their release, then bound to macrophages through non-covalent adsorption, ligand–receptor interactions, or covalent coupling, or internalization into macrophages (Pei and Yeo 2016).

By incorporating nanomedicine and cellular modulation, macrophage-mediated drug delivery approaches can more closely suit the word "drug delivery." Macrophages are highly mobile cells. Appropriate "cargo" (e.g., chemotherapeutic agents like doxorubicin or high-potency antimicrobials) can be loaded in host macrophages, thus potentially using their innate homing ability to enter weakened, sick, or malignant tissue and treat only the affected cellular areas. This approach will minimize the overall amount of drug available while further lowering or even eliminating the possibility of drug-related side effects (Fig. 4). (Gupta and Kumar 2012).

6.1 Active and Passive Targeting of Macrophages

Several methods to target pathological macrophages are currently being used. The manipulation of macrophage depletion and re-education poses a problem in the field of nanotechnology-

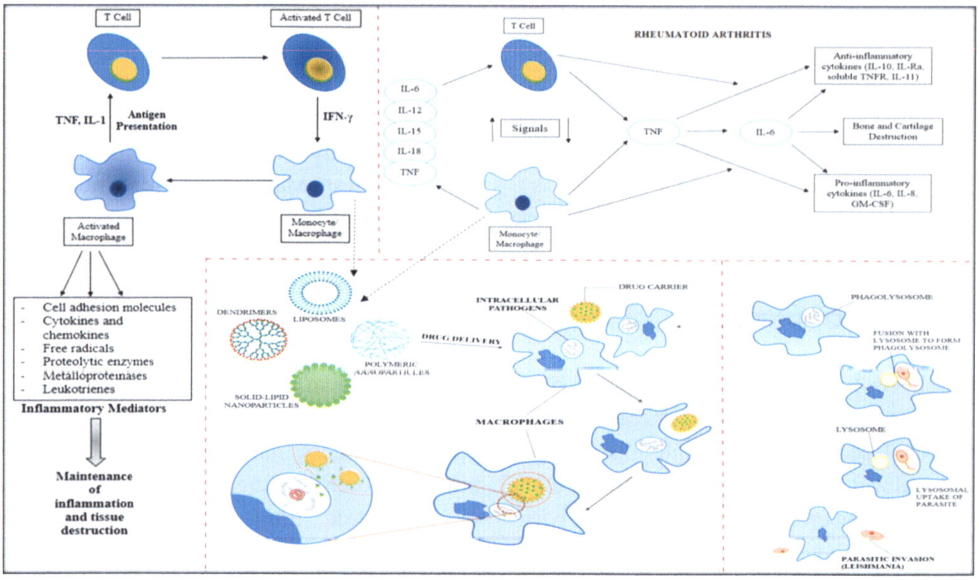

Fig. 4 Schematic representation of drug delivery to macrophages

based systems. Because of their multifunctional intracellular potential, macrophages are a target for a variety of nanoparticle-based therapies. Targeting can be done in two ways: passively or actively (Tables 3 and 4).

The leakiness of blood vessels in inflamed tissues is manipulated by passive targeting. This one-of-a-kind phenomenon is based on the passive enhanced permeability and retention (EPR effect), which is the basis for most passive nanoparticle-based therapies. Nanoparticles (size 50–100 nm) can passively infiltrate the vessels and accumulate at the target area due to leaky vasculature and impaired drainage from the tumor site. However, due to variations in tumor development, vascular distribution, and intratumoral blood flow, the EPR phenomenon appears to be highly unpredictable and heterogeneous among patients. This problem has an impact on nanoparticle distribution in tumors, especially when nanocarriers are used. Furthermore, tumor-associated M2-like macrophages were used as nanoparticle drug reservoirs because they demonstrated high nanoparticle absorption and the ability to release antitumor drugs into the surrounding tissue over time.

The active nano-targeting method involves attaching nanoparticles with ligands that bind directly to overexpressed surface cell receptors. There is a wide range of targeting ligands available, including monoclonal antibodies, peptides, oligomers, antibodies, and small molecules such as mannose and legumain. Therapeutic agents would be released into the particular region of interest as a result of this ligand–receptor interaction. As a result, active targeting takes advantage of the strong ligand–receptor interaction and decreases nonspecific recognition, preventing cargo delivery to non-target tissues (Gaspar et al. 2019).

6.2 Parasitic Infectious Diseases

Macrophage-specific delivery mechanisms are especially desirable because certain parasites and bacteria that give rise to the outbreak of too many deadly diseases serve as host cells for macrophages. In addition to the emergence of other classical drug carriers, polymeric nanoparticles on macrophage targeting have been researched for their effective use in experimental infectious diseases. For example, the symptoms of *Leishmania*, a parasite that causes multiple infectious diseases, have been documented in many studies; it can spread to the visceral organs such

Table 3 Engineered nanosystems to target macrophages actively

Type of NS (active)	Drug	Physicochemical features	Targeted macrophage/ cell line used	Modulation pathway	Ref.
1. Tuberculosis (mycobacterium tuberculosis)					
Solid lipid NP with low % of stearylamine/ mannose	Isoniazid	Size≈500 nm Z potential: +27 to +39 mV	Alveolar macrophage/ NCI-H441	Decrease in intrinsic toxicity toward epithelial cells and macrophages.	Costa et al. (2018)
2. *Leishmania* (*Leishmania*)					
Mannosylated thiolated chitosan-polyethyleneimine NPs	Glucantime	Size≈300 nm ζ potential: +23 mV	Human macrophage infected with *Leishmania*	Inhibition of trypanothione reductase and the P-gp efflux pump.	Sarwar et al. (2018)
Encapsulated PLGA-nanoparticles covered with mannose	Itraconazole	Size≈250 nm ζ potential: −1.1 mV	J774 cell	Encapsulating ITZ within PLGA NPs, with or without mannose, lowered cytotoxicity (15–20%) for J774 murine macrophages. The presence of mannose is protecting the THP-1 as well as J774 cells from toxicity.	Biswaro et al. (2019)
3. Atherosclerosis					
Mannosylated polyamidoamine dendrimers	Anti-inflammatory liver-x-receptor ligands	Size≈7 nm ζ potential: ≈ +7 mV	Macrophage of atherosclerotic plaques	Decrease in progression and necrosis as well as inflammation of plaques.	He et al. (2018)
NP	Hyaluronic acid	Size≈237–424 nm	L/Stab-2, RAW264.7 and Jurkat cells	By using an active targeting mechanism, HA-NPs are selectively accumulated into the atherosclerotic lesion, meaning that they could be used as nanocarriers for atherosclerosis diagnosis and therapy.	Lee et al. (2015)
4. Cancer					
Phospholipid-based and PEGylated NP	Anti-colony-stimulating factor-1 receptor (CSF-1R) small interfering (si)-RNA	Size≈30 nm	Tumor-associated macrophage (TAM)	Inhibition of production of immunosuppressive IL-10 and TGF-β, enhancement of production immune-stimulatory cytokines IL-12 and IFN-γ, and infiltration of antitumor $CD8^+$ T lymphocytes.	Qian et al. (2017)
TLR7/8-agonist-loaded nanoparticles	β-Cyclodextrin	Size≈30 nm ζ potential: −9.87 mV	RAW 264.7 macrophages/ C57BL/6 mice	Alteration in the functional orientation of the tumor immune microenvironment toward M1 phenotype, leading to controlled tumor growth and protecting the animals against tumor re-challenge.	Rodell et al. (2018)

(continued)

Table 3 (continued)

Type of NS (active)	Drug	Physicochemical features	Targeted macrophage/ cell line used	Modulation pathway	Ref.
5. HIV					
Gelatin nanoparticle encapsulated liposomes	Stavudine	Size≈200 nm	RAW 264.7 Macrophages	Dose-dependent toxicity having IC_{50} value of 59.21 μg/ml in comparison to the free drug (IC_{50} value Of 79.71 μg/ml) was exhibited. Sustained release.	Nayak et al. (2017)
Encapsulated glucan particles	Gallium	Size≈3–5 μm	Human primary Macrophages/293 T/17 cells	Receptor-mediated targeted delivery to macrophages and reduction in gallium dosage. A potential delivery system to block infection of macrophages in vivo.	Soto et al. (2016)
6. Parkinson's disease					
Nanozyme	Catalase		Bone marrow-derived macrophages of C57BL/6 mice	Improved bioavailability to the main organs compared to the cell-free catalase nano-formulation. Improved BBB permeability, increasing brain/ plasma ratio of cell-incorporated nanozyme compared to the free drug.	Zhao et al. (2011)

Table 4 Engineered nanosystems to target macrophages passively

Type of NS (passive)	Drug	Physicochemical features	Targeted macrophage/ cell line used	Modulation pathway	Ref.
1. *Mycobacterium avium* complex (MAC) infections					
Liposomes	Rifampicin	Size≈100 nm	Alveolar macrophage/ J774 cells	Liposomes inhibited the growth of MAC in infected macrophages and reached the lower airways in rats.	Zaru et al. (2009)
2. *Leishmania* (*Leishmania*)					
Chitosan microparticles	Doxorubicin	Size≈1.049 μm	J774.1	DOX-loaded chitosan microparticulate delivery system used for anti-VL using macrophage passive targeted drug delivery approach was efficacious.	Kunjachan et al. (2011), Kumar et al. (2010)
Albumin microspheres	Paromomycin	Size≈3 μm	RAW 264.7	The proposed drug delivery system was found suitable for targeting macrophages in vitro and may serve as an optimum carrier to target macrophages where *Leishmania* parasite resides.	Khan and Kumar (2011)
3. Cancer					
Long-circulating liposomes	Simvastatin	Size≈100 nm	TAM/B16.F10 murine melanoma cells	The antitumor activity of liposomal SIM in vivo is due to a strong inhibition of tumor oxidative stress mediated by TAM and strong inhibition of intratumor production of HIF-1α.	Qi et al. (2010)
PEGylated liposomes	Alendronate and doxorubicin	Size≈60–130 nm	Female Balb/C and Sabra	PLAD is a potent cytotoxic agent displaying long-circulating properties which allows passive tumor targeting.	Alupei et al. (2015)

as the liver and spleen, resulting in visceral leishmaniasis, or to the mucous membranes of the mouth and nose. These diseases cause high death rates if they are untreated (Gupta et al. 2016; Sharma and Gupta 2017a; Sharma and Gupta 2017b). The localization of such parasites inside the reticuloendothelial macrophage lysosomal vacuoles restricts the usability and bioavailability of antileishmanial drugs. This prompted many groups to design nanocarriers to treat macrophages with antileishmanial drugs in order to improve their bioavailability. Furthermore, to resolve their adverse side effects, the encapsulation of antileishmanial drugs is necessary (Gupta et al. 2013; Gupta et al. 2010). For example, Lowery and Greenberger have largely examined the implications of using amphotericin B (Lowery and Greenberger 2003). They demonstrate their significant role in the recruitment of inflammatory mediators in human and murine mononuclear cells, such as IL-1 and TNF-α. In combination with poly(e-caprolactone) nanospheres stabilized with poloxamer 188, the drug demonstrated not only an increase in effectiveness but also an inhibitory effect on the development of IL-1 and TNF-a cytokines in peritoneal macrophages of the mouse (Espuelas et al. 2002). Nano-encapsulated quercetin has also been shown to decrease the spleen parasite burden, as well as to decrease hepatotoxicity and renal toxicity (Sarkar et al. 2002).

In order to generate nanoparticles for intracellular transmission to mouse intraperitoneal macrophages, gentamycin, another well-known antibiotic with significant side effects, was bound to polybutylcyanoacrylate. The labeling of gentamycin tritium permitted intracellular radioactivity to be quantified, which was 6.34 times higher in the cells after 30 min of incubation (Zhang et al. 1996). Poly(D, L-lactide) nanoparticles were also used with primaquine against the behavior of *Leishmania donovani* after loading them. The drug's interaction with nanoparticles had an efficacy that was 3.3 times greater than the free form which resulted in a reduction in toxicity. For 600 mg/kg of PLA nanoparticles loaded with 30 mg of primaquine after intravenous injection in BALB/c mice, no systemic toxicity was found, but the same dose of the drug carrier reported a 15% loss of animal weight (Gaspar et al. 1992). For the in vitro targeting of primaquine to macrophages, PIHCA (polyisohexylcyanoacrylate) nanoparticles were also used. In this case, there was a 21-fold improvement in ED50 in the antileishmanial efficacy of a nanoparticle-loaded compound (Skidan et al. 2003).

Against *Staphylococcus aureus* and *Mycobacterium avium* infecting alveolar macrophages, rifampicin-loaded polybutylcyanoacrylate nanoparticles were used. In contrast to the free drug, the drug levels in the cells increased 2–3 times after incubation with rifampicin-loaded nanoparticles. Therapeutic effectiveness was also shown in vivo, where single administration of rifampicin nanoparticles resulted in 80% salmonellosis survival in mice, while double-free rifampicin concentrations were only able to provide 10% survival. The strong antibacterial potency of rifampicin bound to nanoparticles may be responsible for the successful delivery to macrophages.

For the transmission of ampicillin to mouse macrophages in vitro, PIHCA nanoparticles were used. Ampicillin-loaded PIHCA nanoparticles were found at intracellular level in Salmonella typhimurium-infected peritoneal and J-774 murine macrophages after labeling them with tritium. Using nanoparticles, the penetration and bactericidal activity of the medication were observed to improve, plus the phase of degradation of bacteria started within time intervals as short as 2–4 h. A close interaction within the phagosomes or phagolysosomes between ampicillin-loaded nanoparticles and bacteria at the intracellular level could clarify the effectiveness of bound ampicillin, which is otherwise absorbed by cells at very low levels (Balland et al. 1996).

The phagocytosis by J-744 macrophages of PACA (poly(alkylcyanoacrylate)) nanoparticles caused the initiation of a respiratory blast, resulting in enhanced antileishmanial action. As pentamidine was administered to a human macrophage cell line U 937 using methacrylate nanospheres, the compound was 25 times more potent than in the case of free form. Taken together the potency of nanotargeted drugs in macrophages during bacterial infection can be due to the maintenance within the infected monocyte/macrophagic system of higher drug levels.

As a new intravenous drug delivery mechanism for macrophage targeting, AmB was formulated in trilaurin-based nanosized lipid particles (emulsomes) stabilized by soya phosphatidylcholine. The proposed emulsomes-based systems showed great promise for targeting intracellular macrophages. The formulations had the potential to substantially alter the pharmacokinetics of AmB, offering sustained action at relatively low drug doses while reducing toxicity issues such as nephrotoxicity and cardiac arrhythmia (Kumar et al. 2016; Pal et al. 2012; Gupta and Vyas 2007).

6.3 Viral Infectious Disease: AIDS Therapy

Macrophages may be targeted in AIDS therapy because they constitute a RES (reticuloendothelial system) cell population that plays an important role in the disease's immunopathogenesis. The efficacy of nucleoside analog ((AZT (zidovudine) and ddC (zalcitabine))-loaded human serum albumin (HSA) and polyhexylcyanoacrylate (PHCA) nanoparticles were demonstrated on the basis of their ability to resist HIV infection in blood donor monocytes/macrophage cultures. In a rat model, colloidal ^{14}C-labeled AZT nanoparticles intravenously injected into the RES organs were found at concentrations as high as 18 times those of unbound AZT. Growing drug concentrations at particular locations where macrophages are abundant may therefore allow for a decrease in dose and as a consequence, a decrease in systemic toxicity. Similarly, AZT-loaded nanoparticles displayed a more effective distribution to macrophage-rich organs when delivered orally, in addition to an improved concentration and abundance of the medication in the blood.

Evaluating the ability of HIV-infected macrophages for phagocytosis of different forms of radiolabeled AZT-loaded nanoparticles (PACA, PHCA, polymethylmethacrylate, and HSA) has shown the effect of structure and scale on nanoparticle incorporation. It was observed that when PHCA and HSA nanoparticles with a diameter of around 200 nm were used the antiviral drug was administered more effectively (Lobenberg et al. 1998).

The protease–inhibitor saquinavir (Ro 31-8959) or nucleoside analog zalcitabine (20,30-dideoxycytidine)-loaded PHCA nanoparticles were attacked in vitro in another study in infected primary human monocytes/macrophages. Both formulations demonstrated a dose-dependent improvement in antiviral effectiveness, as determined by a decrease in the production of HIV type 1 antigen, while the saquinavir-loaded nanoparticles achieved substantially greater efficacy. Although an aqueous solution of saquinavir developed only a mild antiviral effect at 10 nM, drug-loaded nanoparticles showed a very potent 1 nM antiviral action. The level of antigen reduction remained important even at lower concentrations (0.1 nM): 0.39 nM of the nanoparticle-bound compound developed a 50% inhibitory effect as opposed to 4.23 nM of the free drug (Bender et al. 1996). There are promising signs of increased delivery to mononuclear phagocyte networks of antiviral agents that can ultimately resolve in vivo pharmacokinetic problems and increase the efficacy of medications to combat HIV infections and AIDS.

6.4 Inflammatory Diseases

The reason for the use of macrophage-targeted therapies is based on their direct or indirect involvement in a wide variety of diseases, particularly those based on inflammation. Indeed, in the initiation and maintenance of inflammation that contributes to tissue death, macrophages play a crucial role. Any inflammatory diseases that could be treated by macrophage targeting by nanocarriers are discussed in this section.

6.4.1 Atherosclerosis and Restenosis

Inflammatory modifications in the arterial wall play a central role in atherosclerosis growth. Atherosclerosis is characterized by a persistent inflammatory mechanism during which the vessel wall absorbs monocytes/macrophages. In fact, it has been shown that monocytes/macrophages are around 80% of leukocytes in animal and human atherosclerotic lesions. Atherosclerotic lesions and plaque susceptibility are promoted by their activation, expression of particular cell adhesion molecules (CAMs), and secretion of pro-inflammatory cytokine, chemokines, and matrix MMPs. These activities are governed by a multistep sequence of adhesive and signaling events.

Circulating leukocytes have to adhere to the vascular wall under shear forces with the cooperation of multiple CAMs in order to initiate these responses. The first reactions, which include leukocyte slowing and rolling on inflamed endothelium, primarily include the CAM selectin family (Gistera and Hansson 2017). These reactions induce, via the integrin family, activation and firm adhesion of leukocytes, followed by their transmigration to the sub-endothelial region. The macrophage phagocytosis of nanoparticles, which could cause their activation and regulation of the particular CAMs involved in their adhesion, could have a substantial effect on each of the various stages and as a result, could contribute to their aggregation in the tissue. This issue, however, is still unexplored, and further studies are needed to answer these questions adequately. Despite this, many scientists are involved in designing therapeutic methods for inflammatory process regulation. Of these methods, gene therapy has been used to suppress atherosclerosis by modulating inflammatory cytokines (Ito and Ikeda 2003). The use of gene therapy to inhibit the function of monocyte chemoattractant protein-1 (MCP-1) was tested in vivo by Egashira et al. (Egashira et al. 2000) In addition, monocyte recruitment has been reported to occur after free radicals oxidize low-density lipoprotein (LDL). LDL eventually accumulates and becomes atherogenic in macrophages. This oxidation facilitates vascular dysfunction by (i) exerting a direct cytotoxic effect on endothelial cells; (ii) enhancing the chemotactic properties of monocytes, thus increasing the proliferation and transformation of macrophages into foam cells; and (iii) increasing the proliferation of smooth muscle cells. A nanoparticle-based therapeutic strategy was suggested to shield LDL from oxidation and to target macrophages. The coenzyme Q10 (CoQ10), which exhibits an antiatherogenic effect, was used. This analysis shows a higher absorption by J-774 of CoQ10-loaded nanoparticles compared to unloaded ones. Another strategy has also been reported to stabilize the fragile plaque by reducing the lipid rate and the number of macrophages. This was accomplished by intravenous injection into mice of the apolipoprotein-A1 gene into the adenoviral vector, resulting in a temporary rise in high-density lipoprotein serum (HDL). As a result of the rise in HDL, the macrophages decline in number and activity levels, and potentially also the incidence of plaque rupture. Alternatively, genes encoding thrombolytic proteins or growth promoters may be used to restore endothelial cell antithrombic roles that are capable of inhibiting the development of arterial thrombus (Abdolmaleki et al. 2019).

Restenosis appears to be a concern in clinical cardiology after percutaneous coronary procedures (angioplasty or stent implantation). In this area, nanoparticle technology is seen as a promising solution to the delivery at the site of injury of antiproliferative and anti-remodeling drugs. Core–shell nanoparticles of polyethylene glycol (PEG)-based block copolymers for doxorubicin encapsulation have been developed for this reason and have been shown to accumulate with improved permeability in vascular lesions. Also used for loading paclitaxel, a potent antineoplastic drug, were albumin nanoparticles. A dose-dependent decrease of neointimal growth has been observed following systemic administration in rabbits (Uwatoku et al. 2003).

6.4.2 Rheumatoid Arthritis

TNF-α is one of the major cytokines which play a central role in the RA inflammatory process. RA is characterized by chronic joint synovium inflammation, where macrophages are stimulated and abnormally proliferate, contributing to the development of pro-inflammatory cytokines and MMPs. Ultimately, these conditions contribute to osteoclast formation and initiation, which has devastating effects on the bone and cartilage. Furthermore, since it applies to the mononuclear phagocyte system, the activation of monocytic lineage cells is not a condition confined to joints. Selective counteraction of macrophage activation seems to be an effective strategy in order to decrease inflammation and prevent permanent joint damage (Mishra and Gupta 2020).

Any pathophysiological effects can be clarified, in part, by insufficient macrophage apoptosis. Inducing their apoptosis may also be viewed as an interesting option (Siouti and Andreakos 2019). As an effective route to the treatment of RA, inhibition of pro-inflammatory cytokines such as TNF-α has also been proposed (Massey et al. 2018). This factor is released early in the cascade and induces a wide variety of cellular events, so its inhibition in many inflammatory models may avoid its deleterious effects. For various cell lines and for human peripheral blood mononuclear cells, antisense-mediated inhibition of TNF synthesis in vitro has been shown. To directly alter cellular gene expression, antisense oligodeoxynucleotide (ODN) could be used; once bound, ODN either disables or induces the degradation of the target RNA. With over 20 antisense drugs in clinical development, a wide variety of diseases seem to be important for clinical therapy; there is recent clinical evidence that antisense ODN can be used for the treatment of infections (viruses and bacteria), oncology, and inflammation. The presence of unique ODN binding cell surface receptors that are effectively transported through the plasma membrane has been shown (Crooke 2004).

Various derivatives such as phosphorothioate (PO) have been produced to prevent the degradation of unmodified phosphodiester ODN by nucleases, which are widespread in serum and intracellular settings. However, the use of other techniques in nuclease-rich media has been shown to be important to enhance their cellular uptake and stability. ODN substitution by lipophilic molecules and association with liposomes or polycations were some of the methods involved. Cationic lipids with unsaturated hydrocarbon chains, such as phosphatidylethanolamine, are efficient in promoting ODN transfection and may form a positively charged liposome. Due to electrostatic interactions, as ODNs are combined with lipids, complexes form randomly and form a compact and strong structure. The liposome triggers a disturbance to the endosomal membrane following endocytosis, resulting in the fusion and removal of ODN into the cytoplasm. Oligo–liposome complexes were discovered in the bloodstream in vivo for up to 24 h after injection.

ODN adsorption on nanoparticles has been proposed to be mediated by ion pair formation between the nucleic acid chain's negatively charged phosphate groups and the hydrophobic cations. It was observed that the adsorption efficiency of the nanoparticles was strongly dependent on the chain length of the ODN, the nature of the cyanoacrylic monomer, the hydrophobicity of the cations used as ion-pairing agents, and the medium's ionic concentration. Adsorption of ODN on PACA nanoparticles was performed, and their association with the cells appeared to improve, resulting in a substantial inhibition by multiple cell lines of their enzymatic digestion and cellular absorption. An investigation of the pharmacokinetic profile of the ODN–nanoparticle interaction also revealed an increase in ODN stability following intravenous administration of ODN-based nanoparticles in mice. Other formulations for ODN encapsulation using poly (isobutyl cyanoacrylate) and poly(lactic acid) nanoparticles, rather than simple electrostatic adsorption, have also been developed.

PEI derivatives were found to have good resistance against enzymatic degradation, among other polymeric vehicles for ODN. Poly(D, L-lactic acid) nanoparticles were synthesized and tested for the delivery of antisense ODN using the emulsification–diffusion process. The development of nanoparticles with a size close to 300 nm was enabled by loading with 2.5–10% ODN. However, purification steps aimed at removing residual solvents resulted in a small increase in the size of the ODN nanoparticles, and as ODN loading increased, the effect was more pronounced. In another analysis by Arnedo et al., the authors investigated the probability of holding PO-ODN adsorbed on the surface or stuck in the matrix using albumin nanoparticles. For both mechanisms, enzymatic stability in the presence of phosphodiesterase was assayed and revealed that they degraded more quickly than those caught in the matrix when PO-ODN was adsorbed on the surface of albumin nanoparticles (Morishita et al. 1994).

6.5 Neuroinflammatory Diseases

In macrophage-mediated neuroinflammatory disorders, nanoparticles may even find a use. Circulating macrophages could serve as carriers for therapeutic drug-loaded nanoparticles because of their potential to enter the central nervous system (CNS), representing a promising possibility for targeting therapies to treat neurological disorders.

The role of macrophages has been identified in the production of multiple sclerosis (MS), which is a CNS inflammatory disease. An elevated volume of nitric oxide (NO), which has neurotoxic effects and has been implicated in the permeability of the blood–brain barrier, is secreted during stimulation (BBB) (Hammer et al. 2017). Brosnan et al. found that macrophage depletion in a Lewis rat model would minimize the seriousness of the disease in a review of experimental allergic encephalomyelitis (EAE), which serves as an animal model for MS. (Brosnan et al. 1981) Later on, liposomes containing the drug dichloromethylene diphosphonate (Cl_2MDP) were used in the same animal model to remove macrophages. The consequence of this procedure was the complete avoidance of the clinical symptoms of EAE. Five years later, the same group found that the infiltrating macrophages but not the resident parenchymal microglia were successfully removed by Cl_2MDP liposomes. Therefore, while microglia have been documented to play a major role in the production of EAE, infiltrating macrophages, possibly via the release of inflammatory mediators, are required for their activation.

The movement of nanoparticulate structures could be encouraged because the permeability of the BBB is enhanced during MS. In order to assess their capacity to enter the CNS, PACA nanoparticles have been studied in the EAE. Their concentration in the CNS, particularly in white matter, has been found to increase. The pathological state under which the permeability of the BBB is improved was consistent with this finding. In higher concentrations than non-PEGylated PACA nanoparticles, long-term circulating PACA nanoparticles with PEG moieties were identified in the CNS. As for RA, there is no question about the significant role of TNF-α in the development and maintenance of inflammatory diseases. For example, during the height of EAE disease, TNF-a was shown to increase in the CNS and decrease upon remission (Chu et al. 2018).

7 Conclusion

For host protection against infectious diseases, the macrophage is of vital importance. There is a lot of studies to be done on understanding the subtleties of the infectious disease reaction of macrophages. First, macrophage variability and the intricacies of functional variations between subtypes and states of activation must be characterized and, second, slight differences in macrophage activity and sensitivity between persons must be examined. It is becoming clear that genetic variations, including subtle polymorphisms in macrophage receptor encoding genes, effector molecules, and signaling pathways, may contribute to the predisposition of the host to infectious disease. In order to convert in vitro findings to an in vivo understanding of pathogenesis, this information would be necessary. Recent advancements in the study of protozoa infection have provided insight into how these pathogens subvert host defenses and shown that the critical target for eradication of these pathogens is the macrophage. The creation of novel macrophage-based therapies requires a better understanding of these pathways.

References

Abbas AK, Lichtman AH, Pillai S. Cellular and molecular immunology. 6th ed. Saunders, an imprint of Elsevier Inc; 2007.

Abdolmaleki F, Hayat SMG, Bianconi V, Johnston TP, Sahebkar A. Atherosclerosis and immunity: a perspective. Trends Cardiovasc Med. 2019;29(6):363–71.

Alupei MC, Licarete E, Patras L, Banciu M. Liposomal simvastatin inhibits tumor growth via targeting tumor-associated macrophages-mediated oxidative stress. Cancer Lett. 2015;356(2 Pt B):946–52.

Ancuta P, Rao R, Moses A, Mehle A, Shaw SK, Luscinskas FW, Gabuzda D. Fractalkine preferentially mediates arrest and migration of CD16+ monocytes. J Exp Med. 2003;197(12):1701–7.

Balland O, Pinto-Alphandary H, Viron A, Puvion E, Andremont A, Couvreur P. Intracellular distribution of ampicillin in murine macrophages infected with Salmonella typhimurium and treated with (3H) ampicillin-loaded nanoparticles. J Antimicrob Chemother. 1996;37:105–15.

Barragan A, Sibley LD. Transepithelial migration of toxoplasma gondii is linked to parasite motility and virulence. J Exp Med. 2002;195(12):1625–33.

Baszler TV, Long MT, McElwain TF, Mathison BA. Interferon-gamma and interleukin-12 mediate protection to acute Neospora caninum infection in BALB/c mice. Int J Parasitol. 1999;29(10):1635–46.

Bender AR, von Briesen H, Kreuter J, Duncan IB, Rubsamen-Waigmann H. Efficiency of nanoparticles as a carrier system for antiviral agents in human immunodeficiency virus-infected human monocytes/macrophages in vitro. Antimicrob Agents Chemother. 1996;40:1467–71.

Biswaro LS, Garcia MP, da Silva JR, Neira Fuentes LF, Vera A, Escobar P, Azevedo RB. Itraconazole encapsulated PLGA-nanoparticles covered with mannose as potential candidates against leishmaniasis. J Biomed Mater Res B Appl Biomater. 2019;107(3):680–7.

Blaschke S, Koziolek M, Schwarz A, Benöhr P, Middel P, Schwarz G, Hummel KM, Müller

GA. Proinflammatory role of fractalkine (CX3CL1) in rheumatoid arthritis. J Rheumatol. 2003;30(9):1918–27.
Bordon Y. Kupffer cells: finding their niche. Nat Rev Immunol. 2019;19:660–1.
Brosnan CF, Bornstein MB, Bloom BR. The effects of macrophage depletion on the clinical and pathologic expression of experimental allergic encephalomyelitis. J Immunol. 1981;126:614–20.
Butcher BA, Kim L, Johnson PF, Denkers EY. Toxoplasma gondii tachyzoites inhibit proinflammatory cytokine induction in infected macrophages by preventing nuclear translocation of the transcription factor NF-kappa B. J Immunol. 2001;167(4):2193–201.
Calder CJ, Nicholson LB, Dick AD. A selective role for the TNF p55 receptor in autocrine signaling following IFN-gamma stimulation in experimental autoimmune uveoretinitis. J Immunol. 2005;175(10):6286–93.
Cameron P, McGachy A, Anderson M, Paul A, Coombs GH, Mottram JC, Alexander J, Plevin R. Inhibition of lipopolysaccharide-induced macrophage IL-12 production by Leishmania mexicana amastigotes: the role of cysteine peptidases and the NF-kappaB signaling pathway. J Immunol. 2004;173(5):3297–304.
Chu F, Shi M, Zheng C, Shen D, Zhu J, Zheng X, Cui L. The roles of macrophages and microglia in multiple sclerosis and experimental autoimmune encephalomyelitis. J Neuroimmunol. 2018;318:1–7.
Classen A, Lloberas J, Celada A. Macrophage activation: classical vs alternative. Methods Mol Biol. 2009;531:29–43.
Costa A, Sarmento B, Seabra V. Mannose-functionalized solid lipid nanoparticles are effective in targeting alveolar macrophages. Eur J Pharm Sci. 2018;114:103–13.
Crooke ST. Progress in antisense technology. Annu Rev Med. 2004;55:61–95.
Daugherty A, Webb NR, Rateri DL, King VL. Thematic review series: the immune system and atherogenesis. Cytokine regulation of macrophage functions in atherogenesis. J Lipid Res. 2005;46(9):1812–22.
Dixon LJ, Dixon LJ, Barnes M, Tang H, Pritchard MT, Nagy LE. Kupffer cells in the liver. Compr Physiol. 2013;3(2):785–97.
Egashira K, Koyanagi M, Kitamoto S, Ni W, Kataoka C, Morishita R, Kaneda Y, Akiyama C, Nishida KI, Sueishi K, Takeshita A. Anti-monocyte chemoattractant protein-1 gene therapy inhibits vascular remodeling in rats: blockade of MCP-1 activity after intramuscular transfer of a mutant gene inhibits vascular remodeling induced by chronic blockade of NO synthesis. FASEB J. 2000;14:1974–8.
Espuelas MS, Legrand P, Loiseau PM, Bories C, Barratt G, Irache JM. In vitro antileishmanial activity of amphotericin B loaded in poly(epsilon-caprolactone) nanospheres. J Drug Target. 2002;10:593–9.
Frevert U. Sneaking in through the back entrance: the biology of malaria liver stages. Trends Parasitol. 2004;20(9):417–24.
Gaspar R, Opperdoes FR, Preat V, Roland M. Drug targeting with polyalkylcyanoacrylate nanoparticles: in vitro activity of primaquine-loaded nanoparticles against intracellular Leishmania donovani. Ann Trop Med Parasitol. 1992;86:41–9.
Gaspar N, Zambito G, Löwik CMWG, Mezzanotte L. Active Nano-targeting of macrophages. Curr Pharm Des. 2019;25(17):1951–61.
Gazzinelli RT, Wysocka M, Hayashi S, Denkers EY, Hieny S, Caspar P, Trinchieri G, Sher A. Parasite-induced IL-12 stimulates early IFN-gamma synthesis and resistance during acute infection with toxoplasma gondii. J Immunol. 1994;153(6):2533–43.
Germic N, Frangez Z, Yousefi S, Simon HU. Regulation of the innate immune system by autophagy: monocytes, macrophages, dendritic cells and antigen presentation. Cell Death Differ. 2019;26:715–27.
Ghislat G, Lawrence T. Autophagy in dendritic cells. Cell Mol Immunol. 2018;15:944–52.
Gistera A, Hansson GK. The immunology of atherosclerosis. Nat Rev Neph. 2017;13(6):368.
Gordon S. Alternative activation of macrophages. Nat Rev Immunol. 2003;3(1):23–35.
Gordon S, Martinez FO. Alternative activation of macrophages: mechanism and functions. Immunity. 2010;32(5):593–604.
Gordon S, Pluddemann A. Tissue macrophages: heterogenecity and functions. BMC Biol. 2017;15:53.
Gordon S, Taylor PR. Monocyte and macrophage heterogeneity. Nat Rev Immunol. 2005;5(12):953–64.
Gupta S, Kumar P. Drug delivery using nanocarriers: Indian perspective. Proc Natl Acad Sci India Sect B Biol Sci. 2012;82(S1):167–206.
Gupta S, Vyas SP. Development and characterization of amphotericin B bearing emulsomes for passive and active macrophage targeting. J Drug Target. 2007;15:206–17.
Gupta S, Pal A, Vyas SP. Drug delivery strategies for therapy of visceral leishmaniasis. Expert Opin Drug Deliv. 2010;7(3):371–402.
Gupta S, Dube A, Vyas SP. Development and characterization of amphotericin B loaded solid lipid nanoparticles against experimental visceral leishmaniasis. Pharmac Nanotechnol. 2013;1(1):54–67.
Gupta S, Sharma P, Gupta MK. Leishmaniasis – drugs, nanotechnology based delivery systems and recent patents survey. Curr Nanomed. 2016;6:21–42.
Hammer A, Stegbauer J, Linker RA, Archiv P. Macrophages in neuroinflammation: role of the renin-angiotensin-system. Eur J Phys. 2017;469(3–4):431–44.
He H, Yuan Q, Bie J, Wallace RL, Yannie PJ, Wang J, Lancina MG 3rd, Zolotarskaya OY, Korzun W, Yang H, Ghosh S. Development of mannose functionalized dendrimeric nanoparticles for targeted delivery to macrophages: use of this platform to modulate atherosclerosis. Transl Res. 2018;193:13–30.
Henry B, Roussel C, Carucci M, Brousse V, Ndour PA, Buffet P. The human spleen in malaria: filter or shelter? Trends Parasitol. 2020;36(5):435–46.
Ito T, Ikeda U. Inflammatory cytokines and cardiovascular disease. Curr Drug Targets Inflamm Allergy. 2003;2:257–65.

Jain NK, Mishra V, Mehra NK. Targeted drug delivery to macrophages. Expert Opin Drug Deliv. 2013 Mar;10(3):353–67.

Kelsall BL, Leon F. Involvement of intestinal dendritic cells in oral tolerance, immunity to pathogens, and inflammatory bowel disease. Immunol Rev. 2005;206:132–48.

Khan W, Kumar N. Drug targeting to macrophages using paromomycin-loaded albumin microspheres for treatment of visceral leishmaniasis: an in vitro evaluation. J Drug Target. 2011;19(4):239–50.

Kumar V. Macrophages: the potent Immunoregulatory innate immune cells. In: Macrophage at the crossroads of innate and adaptive immunity; 2019.

Kumar N, Gupta S, Dube A, Vyas SP. Emerging role of vesicular carriers for therapy of visceral leishmaniasis: conventional vs novel. Crit Rev Ther Drug Carrier Syst. 2010;27(6):461–507.

Kumar N, Sharma P, Jaiswal A, Dube A, Gupta S. Development and evaluation of p-aminophenylmannopyranoside anchored emulsomes for treatment of experimental visceral leishmaniasis. Ann Clin Cytol Pathol. 2016;2(6):1042.

Kunjachan S, Gupta S, Dwivedi A, Dube A, Chourasia M. Chitosan-based macrophage-mediated drug targeting for the treatment of experimental visceral leishmaniasis. J Microencapsulation. 2011;28(4):301–10.

Lacroix-Lamande S, Mancassola R, Naciri M, Laurent F. Role of gamma interferon in chemokine expression in the ileum of mice and in a murine intestinal epithelial cell line after Cryptosporidium parvum infection. Infect Immun. 2002;70(4):2090–9.

Lee GY, Kim JH, Choi KY, Yoon HY, Kim K, Kwon IC, Choi K, Lee BH, Park JH, Kim IS. Hyaluronic acid nanoparticles for active targeting atherosclerosis. Biomaterials. 2015;53:341–8.

Liu D, Uzonna JE. The early interaction of Leishmania with macrophages and dendritic cells and its influence on the host immune response. Front Cell Infect Microbiol. 2012;2:83.

Lobenberg R, Araujo L, von Briesen H, Rodgers E, Kreuter J. Body distribution of azidothymidine bound to hexyl-cyanoacrylate nanoparticles after i.v. injection to rats. J Control Release. 1998;50:21–30.

Lodge R, Descoteaux A. Modulation of phagolysosome biogenesis by the lipophosphoglycan of Leishmania. Clin Immunol. 2005;114(3):256–65.

Lowery MM, Greenberger PA. Amphotericin-induced stridor: a review of stridor, amphotericin preparations, and their immunoregulatory effects. Ann Allergy Asthma Immunol. 2003;91(5):460–6.

Makala LH, Nishikawa Y, Suzuki N, Nagasawa H. Immunology. Antigen-presenting cells in the gut. J Biomed Sci. 2004;11(2):130–41.

Massey J, Plant D, Hyrich K, Morgan AW, Wilson AG, Spiliopoulou A, Colombo M, McKeigue P, Isaacs J, Cordell H, Pitzalis C, Barton A. Genome-wide association study of response to tumour necrosis factor inhibitor therapy in rheumatoid arthritis. Pharmacogenomics J. 2018;18(5):657–64.

Mebius RE, Kraal G. Structure and function of the spleen. Nat Rev Immunol. 2005;5(8):606–16.

Merad M, Manz MG, Karsunky H, Wagers A, Peters W, Charo I, Weissman IL, Cyster JG, Engleman EG. Langerhans cells renew in the skin throughout life under steady-state conditions. Nat Immunol. 2002;3(12):1135–41.

Mishra R, Gupta S. Novel nano carriers for the treatment of progressive auto immune disease rheumatoid arthritis. Curr Pharm Des. 2020;

Morishita R, Gibbons GH, Kaneda Y, Ogihara T, Dzau VJ. Pharmacokinetics of antisense oligodeoxyribonucleotides (cyclinB, and CDC2kinase) in the vessel wall in vivo: enhanced therapeutic utility for restenosis by HVJ-liposome delivery. Gene. 1994;149:13–9.

Mukhopadhyay S, Chen Y, Sankala M, Peiser L, Pikkarainen T, Kraal G, Tryggvason K, Gordon S. MARCO, an innate activation marker of macrophages, is a class a scavenger receptor for Neisseria meningitidis. Eur J Immunol. 2006;36(4):940–9.

Murray HW, Berman JD, Davies CR, Saravia NG. Advances in leishmaniasis. Lancet. 2005;366(9496):1561–77.

Nayak D, Boxi A, Ashe S, Thathapudi NC, Nayak B. Stavudine loaded gelatin liposomes for HIV therapy: preparation, characterization and in vitro cytotoxic evaluation. Mater Sci Eng C Mater Biol Appl. 2017;73:406–16.

Olivier M, Atayde VD, Isnard A, Hassani K. Leishmania virulence factors: focus on metalloprotease GP63. Microbes Infect. 2012;14(15):1377–89.

Pal A, Gupta S, Jaiswal A, Dube A, Vyas SP. Development and evaluation of tripalmitinemulsomes for the treatment of experimental visceral leishmaniasis. J Liposome Res. 2012;22(1):62–71.

Pei Y, Yeo Y. Drug delivery to macrophages: challenges and opportunities. J Control Release. 2016 Oct 28;240:202–11.

Penna G, Vulcano M, Roncari A, Facchetti F, Sozzani S, Adorini L. Cutting edge: differential chemokine production by myeloid and plasmacytoid dendritic cells. J Immunol. 2002;169(12):6673–6.

Plata F, Garcia-Pons F, Wietzerbin J. Immune resistance to Trypanosoma cruzi: synergy of specific antibodies and recombinant interferon gamma in vivo. Ann Inst Pasteur Immunol. 1987;138(3):397–415.

Qi XF, Kim DH, Yoon YS, Kim SK, Cai DQ, Teng YC, Shim KY, Lee KJ. Involvement of oxidative stress in simvastatin-induced apoptosis of murine CT26 colon carcinoma cells. Toxicol Lett. 2010;199(3):277–87.

Qian Y, Qiao S, Dai Y, Xu G, Dai B, Lu L, Yu X, Luo Q, Zhang Z. Molecular-targeted immunotherapeutic strategy for melanoma via dual-targeting nanoparticles delivering small interfering RNA to tumor-associated macrophages. ACS Nano. 2017;11(9):9536–49.

Rodell CB, Arlauckas SP, Cuccarese MF, Garris CS, Li R, Ahmed MS, Kohler RH, Pittet MJ, Weissleder R. TLR7/8-agonist-loaded nanoparticles promote the polarization of tumour-associated macrophages to enhance cancer immunotherapy. Nat Biomed Eng. 2018;2(8):578–88.

Salata RA, Pearson RD, Ravdin JI. Interaction of human leukocytes and Entamoeba histolytica. Killing of virulent amebae by the activated macrophage. J Clin Invest. 1985;76(2):491–9.

Sarkar S, Mandal S, Sinha J, Mukhopadhyay S, Das N, Basu MK. Quercetin: critical evaluation as an antileishmanial agent in vivo in hamsters using different vesicular delivery modes. J Drug Target. 2002;10:573–8.

Sarwar HS, Ashraf S, Akhtar S, Sohail MF, Hussain SZ, Rafay M, Yasinzai M, Hussain I, Shahnaz G. Mannosylated thiolated polyethylenimine nanoparticles for the enhanced efficacy of antimonial drug against Leishmaniasis. Nanomedicine (Lond). 2018;13(1):25–41.

Schofield L, Villaquiran J, Ferreira A, Schellekens H, Nussenzweig R, Nussenzweig V. Gamma interferon, CD8+ T cells and antibodies required for immunity to malaria sporozoites. Nature. 1987;330(6149):664–6.

Scianimanico S, Desrosiers M, Dermine JF. Impaired recruitment of the small GTPase rab7 correlates with the inhibition of phagosome maturation by Leishmania donovani promastigotes. Cell Microbiol. 1999;1(1):19–32.

Sharma P, Gupta S. Development and evaluation of macrophage targeted multidrug therapy against visceral leishmaniasis. Pharm Biomed Res. 2017a;3(1):9–18.

Sharma P, Gupta S. A convergent synthetic platform for nanoparticle with combination against visceral Leishmaniasis. Int J Drug Dev Res. 2017b;9(1):30–4.

Siouti E, Andreakos E. The many facets of macrophages in rheumatoid arthritis. Biochem Pharmacol. 2019;165:152–69.

Skidan IN, Gel'perina SE, Severin SE, Guliaev AE. Enhanced activity of rifampicin loaded with polybutyl cyanoacrylate nanoparticles in relation to intracellularly localized bacteria. Antibiot Khimioter. 2003;48:23–6.

Skurkovich B, Skurkovich S. Inhibition of IFN-gamma as a method of treatment of various autoimmune diseases, including skin diseases. Ernst Schering Res Found Workshop. 2006;56:1–27.

Smith PD, Ochsenbauer-Jambor C, Smythies LE. Intestinal macrophages: unique effector cells of the innate immune system. Immunol Rev. 2005;206:149–59.

Smythies LE, Sellers M, Clements RH, Mosteller-Barnum M, Meng G, Benjamin WH, Orenstein JM, Smith PD. Human intestinal macrophages display profound inflammatory anergy despite avid phagocytic and bacteriocidal activity. J Clin Invest. 2005;115(1):66–75.

Soto ER, O'Connell O, Dikengil F, Peters PJ, Clapham PR, Ostroff GR. Targeted delivery of glucan particle encapsulated gallium nanoparticles inhibits HIV growth in human macrophages. J Drug Deliv. 2016;2016:8520629.

Taylor PR, Martinez-Pomares L, Stacey M, Lin HH, Brown GD, Gordon S. Macrophage receptors and immune recognition. Annu Rev Immunol. 2005;23:901–44.

Town T, Nikolic V, Tan J. The microglial "activation" continuum: from innate to adaptive responses. J Neuroinflammation. 2005;2:24.

Uwatoku T, Shimokawa H, Abe K, Matsumoto Y, Hattori T, Oi K, Matsuda T, Kataoka K, Takeshita A. Application of nanoparticle technology for the prevention of restenosis after balloon injury in rats. Circ Res. 2003;92:62–9.

Von Stebut E, Tenzer S. Cutaneous leishmaniasis: distinct functions of dendritic cells and macrophages in the interaction of the host immune system with Leishmania major. Int J Med Microbiol. 2018;308(1):206–14.

Yun PL, Decarlo AA, Collyer C, Hunter N. Hydrolysis of interleukin-12 by Porphyromonas gingivalis major cysteine proteinases may affect local gamma interferon accumulation and the Th1 or Th2 T-cell phenotype in periodontitis. Infect Immun. 2001;69(9):5650–60.

Zaru M, Sinico C, De Logu A, Caddeo C, Lai F, Manca ML, Fadda AM. Rifampicin-loaded liposomes for the passive targeting to alveolar macrophages: in vitro and in vivo evaluation. J Liposome Res. 2009;19(1):68–76.

Zhang Q, Liao GT, Wei DP, Zhang CJ. Increase of gentamicin uptake in cultured mouse peritoneal macrophage and rat hepatocytes when used in the form of nanoparticles. Yao Xue Bao. 1996;31:375–80.

Zhao Y, Haney MJ, Mahajan V, Reiner BC, Dunaevsky A, Mosley RL, Kabanov AV, Gendelman HE, Batrakova EV. Active targeted macrophage-mediated delivery of catalase to affected brain regions in models of Parkinson's disease. J Nanomed Nanotechnol. 2011;S4:003.

Introduction to Anatomy and Physiology of Macrophages from Drug Delivery Perspective

Thanh Ba Duong and Linh Ho

Abstract

Macrophages are myeloid immune cells which present in most of the tissues of the body and have various functions in almost every aspect of mammalian biology. Understanding the diverse origins and functions of macrophages in disorders and disease states has gained increasing interest in modulating them for treatment. To develop drug delivery systems to macrophages, it is important to understand their anatomical and physiological roles and requirements for efficient targeting and delivering of therapeutic agents. This chapter provides an overview of biological roles and various strategies to target macrophages for treatment of diseases.

Keywords

Macrophages · Macrophage origin · Macrophage physiology · M1 macrophages · M2 macrophages · Erythropoiesis · Targeting macrophages · Drug delivery · Tumor-associated macrophages (TAMs)

T. B. Duong · L. Ho (✉)
Department of Pharmaceutical & Biomedical Sciences, College of Pharmacy, California Northstate University, Elk Grove, CA, USA
e-mail: linh.ho@cnsu.edu

1 Introduction

Macrophages are found widely in the body and have roles in development, tissue homeostasis, tissue repair, and immunity (Wynn et al. 2013). Macrophages can adopt different types of phenotypes responding to environmental changes to maintain homeostasis. However, the dysregulation of macrophages in altering their characteristics can lead to progression of many diseases such as intracellular infection, autoimmune diseases, and cancer (Martinez and Gordon 2014). Therefore, macrophages have become therapeutic target for numerous diseases. They are important hosts of intracellular pathogens in chronic infectious diseases, and thus they can be aimed at for intracellular delivery of antibiotics. Macrophages are latent reservoirs for human immunodeficiency virus type 1 (HIV-1); thus, long-term successful treatment of HIV-1 infection requires potent strategies to prevent HIV-1 from entering and persisting in these macrophages. Furthermore, targeting macrophages in cancer therapy has been revisited due to their complex roles and mounting evidence of their subpopulation identification in tumor microenvironment. Insights into the heterogeneity of macrophage lineages, identities, and regulations are essential for exploiting macrophages as important therapeutic targets in many human diseases.

The development of macrophages consists of different stages. Monocytes have the origin from

S. Gupta, Y. V. Pathak (eds.), *Macrophage Targeted Delivery Systems*,
https://doi.org/10.1007/978-3-030-84164-5_2

a common hematopoietic stem cell (HSC) in the bone marrow. The hematopoietic stem cells differentiate into the myeloid progenitor cells. In the presence of colony-stimulating factor (CSF), myeloid progenitor cells continue to divide and differentiate into monoblasts, pro-monocytes, and finally monocytes, which exit the bone marrow and go into the bloodstream (Mosser and Edwards 2008). Circulating monocytes are recruited to different tissues and differentiated into tissue-specific macrophages based on the environmental cues. For instance, macrophages of the bone, central nervous system, connective tissue, and liver are called osteoclasts, microglial cells, histiocytes, and Kupffer cells, respectively (Mosser and Edwards 2008). In short, tissue-resident macrophages possess different functions based on their anatomical locations.

Activated macrophages are known as M1, which is classical activation, or M2, which is alternative activation. M1 macrophages are pro-inflammatory and secrete pro-inflammatory cytokines such as IFN-1β, IL-12, IL-23, and TNF-α, along with reactive nitrogen intermediates and reactive oxygen species (ROS), which can destroy and kill pathogens (Arora et al. 2018). Conversely, M2 macrophages are anti-inflammatory and have functions in tissue remodeling and healing. M2 macrophages are differentiated into subtypes: M2a, M2b, M2c, and M2d depending on stimulators. M2a subtype is activated by IL-4 and IL-13, while M2b subtype is activated by immunocomplexes and LPS. IL-10, TGF-β, and glucocorticoids activate M2c subtype, while tumor-associated factors activate M2d subtype (Arora et al. 2018).

2 Anatomy and Physiology of Macrophages

2.1 Origin of Tissue Macrophages

Macrophages are found in mammals from mid-gestation, contributing to physiologic homeostasis throughout the life (Gordon and Martinez-Pomares 2017). Macrophages come from yolk sac and fetal liver progenitors during embryonic development and persist into adulthood in various organs as divergent, self-renewing tissue-resident macrophage populations (Gordon and Martinez-Pomares 2017; Ginhoux and Jung 2014). Circulating monocytes are recruited to various tissues after birth and differentiated into macrophages according to the environmental cues in response to inflammatory events, infection, and metabolic perturbations (Gordon and Martinez-Pomares 2017; Ginhoux and Jung 2014). According to the specific anatomical location where they are recruited, they have different names and diverse functions as alveolar macrophages in the lungs, microglia in the central nervous system (CNS), and Kupffer cells in the liver tissue (Gordon and Martinez-Pomares 2017) (Fig. 1). Kupffer cells located in the liver mainly regulate fatty acid oxidation in hepatocytes and waste disposal processes, such as clearance of microbes and cell remnants from the blood (Gordon and Martinez-Pomares 2017; Davies et al. 2013); alveolar macrophages in the lungs mediate surfactant lipid metabolism and serve as the first-line defender against inhaled pathogens (Davies et al. 2013; Maus et al. 2002); and red pulp macrophages in the spleen are responsible for iron metabolism and removal of senescent erythrocytes (Maus et al. 2002). In addition to the reticuloendothelial system, several other tissues have resident macrophages with distinct functions, such as osteoclasts (mineral disruption) and bone marrow macrophages (erythropoiesis support) in the bone, microglia cells (immune surveillance) in the brain, intestinal macrophages (intestinal homeostasis maintenance) in the gastrointestinal tract, intraocular macrophages protect the eye from infection and regulate healing processes of injuries, and Langerhans cells (interaction with T lymphocytes) in the skin (Ginhoux and Jung 2014) (Fig. 1).

2.2 M1 and M2 Phenotypes of Macrophages

Macrophages arise from various origins and local environments; thus, they respond differently to challenges and vary in their functions (Wynn

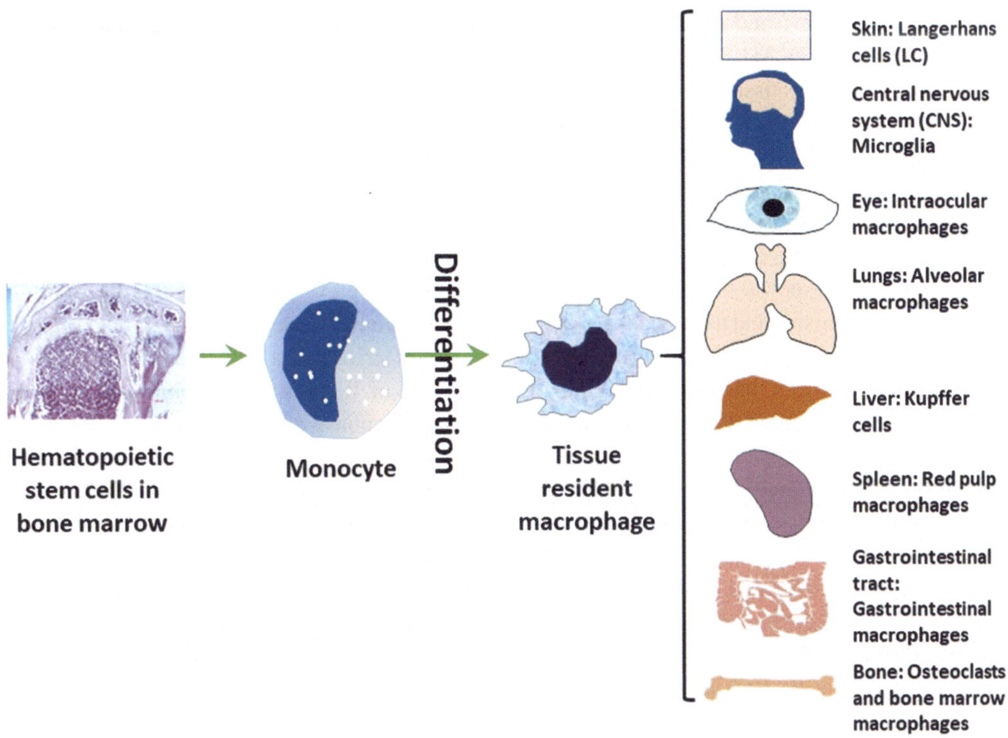

Fig. 1 Tissue-resident macrophages
Macrophages come from yolk sac and fetal liver progenitors during embryonic development and persist into adulthood in various organs as divergent, self-renewing tissue-resident macrophage populations. After birth, hematopoietic stem cells in bone marrow give rise to monocytes. Monocytes are mobilized to blood circulation and arrive at tissues via cellular signaling pathway where they become resident macrophages with their own features and names depending on each tissue.

et al. 2013). Based on function of macrophages, they might be categorized as classically activated macrophages and alternatively activated macrophages (Gordon 2003). Classically activated macrophages are dependent on the interferon-gamma (IFN-γ), tumor necrosis factor-alpha (TNF-α), and toll-like receptor 4 (TLR4) for their activation by T helper 1 (T_H1)-type responses as cell-mediated immunity to infection with intracellular pathogens, such as *Mycobacterium tuberculosis* and HIV (Gordon 2003). Alternatively activated macrophages are activated by the T_H2-type cytokines interleukin-4 (IL-4) and IL-13 as humoral immunity and repair (Gordon 2003). These two classes of macrophages were extended to M1 macrophages (classically activated macrophages) and M2 macrophages (alternatively activated macrophages) relating to their activation status (Mosser and Edwards 2008; Mills 2012). M1 macrophages typically arise from myeloid progenitors with murine macrophage-colony-stimulating factor (M-CSF, also known as CSF-1), while M2 macrophages are derived from mature M1 macrophages by treating with interleukin-4 (IL-4) or IL-13 (Mosser and Edwards 2008; Chavez-Galan et al. 2015). Activation into the M1 phenotype requires cellular or exogenous stimuli such as interferon-γ (IFN-γ), tumor necrosis factor (TNF), and lipopolysaccharide (LPS) (Mosser and Edwards 2008; Ho and Sly 2009). Basically, a TLR ligand acting in a MyD88-dependent manner or IFNβ (TRIF)-dependent pathways signal the transcription of TNF, which can then coordinate with IFNγ or IFNβ, respectively, in an autocrine manner to activate this macrophage population (Mosser and Edwards 2008; Yamamoto et al. 2003). M1 activity inhibits cell proliferation and causes tissue damage, while M2 activity stimulates cell proliferation and tissue repair (Mills 2012). M1 or "killer" macrophages produce pro-inflammatory cytokines like TNF-α, IL-15, IL-23, IL-1β, low level of IL-10, and high amount of IL-12, nitrogen radicals, and reactive oxygen species in response to stress or elimination of foreign organisms and tumor cells (Mosser and Edwards 2008; Ho and Sly 2009) (Fig. 2). In fact, classically activated macrophages are major regulators of the immunopathologic mechanism of several autoimmune diseases, including rheumatoid arthritis (Szekanecz and Koch 2007) and inflammatory bowel disease (Zhang and Mosser 2008). On the other hand, polarization into the M2 phenotype is associated with interleukin-4 (IL-4), IL-13, and transforming growth factor β (TGF-β), leading to the production of immunosuppressive cytokine IL-10 (Chavez-Galan et al. 2015). M2 macrophages possess high quantities of IL-10; however, low levels of IL-12 and IL-23 (Arora et al. 2018). There are four further subgroups of M2 macrophage M2a, M2b, M2c, and M2d based on their gene expression profiles (Martinez and Gordon 2014; Wang et al. 2010). IL-4 and IL-13 mainly produced in mast cells, T_{H2} cells, and basophils are in charge of M2a macrophage activation (Torocsik et al. 2005). M2b subset macrophages are stimulated by immune complexes together with LPS or IL-1β (Anderson and Mosser 2002). M2b macrophages are characterized by a reduction of IL-12 and an induction of IL-10 regulating inflammatory and immune response and T_{H2} activation (Anderson and Mosser 2002). The M2c subgroup is induced by glucocorticoids, TGF-β or IL-10. They enhance engulfing activity of cell debris by M1 macrophages, suppress the expression of pro-inflammatory intermediates, and function in remodeling and intercellular matrix production (Martinez and Gordon 2014). M2d subset of macrophages is activated by IL-6, adenosine, tumor-associated factors and characterized with decreased IL-12 production and increased IL-10

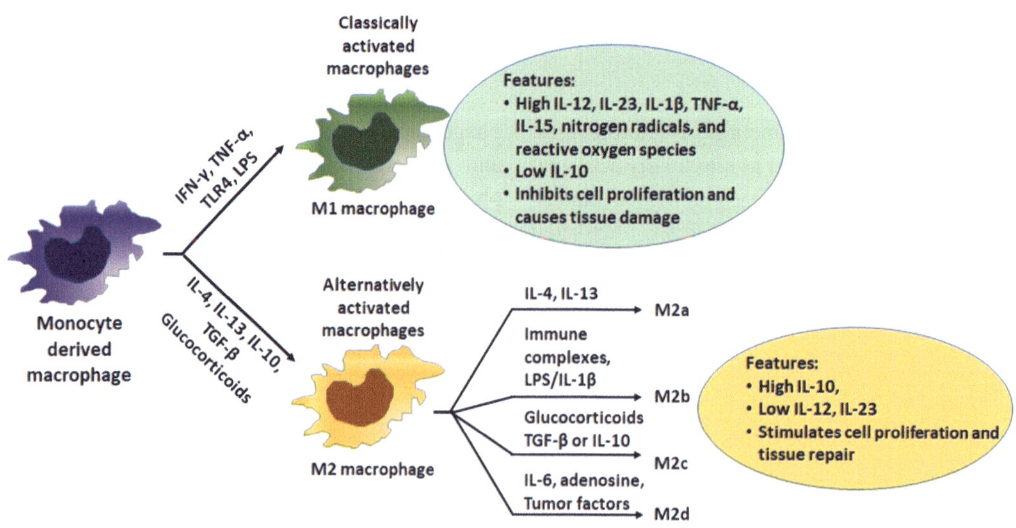

Fig. 2 Classically and alternatively activated macrophages
Macrophage polarization into M1 or M2 lineage in response to different challenges and signals from local environments. Classically activated macrophages are dependent on the interferon-gamma (IFN-γ), tumor necrosis factor-alpha (TNF-α), and toll-like receptor 4 (TLR4) for their activation by T helper 1 (T_{H1})-type responses as cell-mediated immunity to intracellular infection. Alternatively activated macrophages are activated by the T_{H2}-type cytokines interleukin-4 (IL-4) and IL-13 as humoral immunity and repair.

secretion, and with some features of tissue-associated macrophages that promote angiogenesis and progression of tumors (Wang et al. 2010; Qian and Pollard 2010; Zhukova et al. 2020) (Fig. 2). Alternatively activated M2 or "healer" macrophages are important in the resolution of inflammatory responses by participating in debris scavenging, wound healing, and angiogenesis (Mosser and Edwards 2008; Chavez-Galan et al. 2015; Ho and Sly 2009). M2 macrophages may also play critical roles in chronic infections, tumor formation, and tumor metastasis (Chavez-Galan et al. 2015; Ho and Sly 2009). Taken together, the characterization of these macrophage subpopulations at molecular level and their specific involvement in some human pathologies such as cancer, rheumatoid arthritis, and infectious diseases (tuberculosis, HIV) can be useful in designing new strategies including combinatory therapies for the treatment of disorders targeting macrophages.

2.3 Macrophages in Development

2.3.1 Macrophages in Erythropoiesis

Macrophages have an important role during red blood cells (RBC) development in the bone marrow and maturation as well as clearance of RBC (de Back et al. 2014; Chasis and Mohandas 2008) (Table 1). During steady-state hematopoiesis, approximately 10 RBC are produced per hour in erythroblastic islands in human beings (de Back et al. 2014). Mature RBC remain in circulation for about 120 days and are removed by macrophages residing in the spleen and the liver at the end of their life cycle (de Back et al. 2014). The interconnection between macrophage and red blood cells at various stages is critical for efficient erythropoiesis under various conditions including stress condition, to maintain RBC homeostasis or to ensure the correct removal of senescent or damaged RBC (de Back et al. 2014; Chasis and Mohandas 2008). During steady-state hematopoiesis, approximately 10 (10) RBC are produced per hour within erythroblastic islands in persons (de Back et al. 2014). Erythropoiesis is the process of forming red blood cells or erythrocytes in the bone marrow, especially in erythroblastic islands composed of erythroid precursors (as known as erythroblasts) surrounding a central macrophage (Chasis and Mohandas 2008). Central macrophages provide interactions that affect erythroid proliferation and/or differentiation as well as anchor erythroblasts within the island niches (Chasis and Mohandas 2008). Molecules partition in the interconnection between erythroid precursors, and macrophages include the erythroblast macrophage protein (Emp), the hemoglobin–haptoglobin receptor CD163, VLA-4 on erythroblasts and VCAM-1 on macrophages, and ICAM-4 on erythroblasts and αV integrin on macrophages, macrophage membrane protein Ephrin-2 binding erythroid receptor EphB4, and c-kit ligand interacting with c-kit on erythroblasts to promote erythroid proliferation (de Back et al. 2014; Chasis and Mohandas 2008).

It has been found that macrophages stimulate erythrocyte formation by directly transferring iron to erythroid progenitors (de Back et al. 2014; Bessis 1958). Macrophages release ferritin (an intracellular protein that stores iron) which is taken up by erythroblasts via endocytosis (de Back et al. 2014). Iron then is delivered from ferritin through acidification and proteolysis for heme synthesis in the erythroid precursor cells (de Back et al. 2014; Leimberg et al. 2008).

Macrophages also interact with erythroblasts to directly promoting their proliferation and survival which resulted from reduced transit time in the G0/G1 phase of cell cycle in response to erythropoietin (Rhodes et al. 2008). Retinoblastoma tumor suppressor (Rb) protein is a nuclear factor that prevents the progression from G1 to S phase of the cell division cycle and is critical for macrophage differentiation. Loss of Rb protein can lead to embryonic lethality due to anemia by failure of erythroblasts to enucleate in animal model (Iavarone et al. 2004).

Actin cytoskeletal-associated protein paladin has been found to stimulate actin cytoskeleton dynamics and interconnection of cell with extracellular matrix and has crucial function in definitive erythropoiesis and erythroblastic island formation and, especially, required for normal

Table 1 Physiological roles and diversity of macrophages

Tissue	Macrophages	Tissue-specific transcriptional factors	Functions
Bone	Bone marrow macrophages		Erythropoiesis in bone marrow, clearance of red blood cells
Brain	Microglia	Cx3cr1, CSF1R, CD33, HMGB1, α-synuclein	Brain development and protection, maintenance of neuronal networks, and injury repair, synaptic pruning
Lung	Alveolar macrophages as known and type I alveolar epithelial cells	PPARϒ	Removal of apoptotic cells and cellular debris as well as immune response in lung health and disease conditions, surfactant lipid metabolism
Liver	Kupffer cells	PPARδ	Metabolic adaptations of hepatocytes during increased caloric intake, regulating fatty acid oxidation in hepatocytes
Adipose tissue	Adipose-associated macrophages	PPARϒ	Insulin sensitivity and adaptive thermogenesis
Spleen	Red pulp macrophages	LXRα	Clearance of erythrocytes and iron metabolism

function of macrophages in fetal liver (Liu et al. 2007). Targeted disruption of the gene *Palld*, which is encoded for paladin (*Palld(−/−)*), shows a dramatic decrease in definitive erythrocytes arised from fetal liver but not primitive erythrocytes from yolk sac because of preventing effective erythroblast macrophage interactions (Liu et al. 2007). In addition, c-Maf transcription factor that involves in cellular proliferation and differentiation in both pathologic and physiologic situations has been identified as an essential component in definitive erythropoiesis in fetal liver (Kusakabe et al. 2011). Accordingly, deletion of *c-Maf* caused severe embryonic anemia, which is associated with a significant reduction in erythroblastic islands. Moreover, c-Maf deficiency macrophages decreased VCAM-1 expression leading to failure to maintain erythroblastic islands and subsequent embryonic anemia (Kusakabe et al. 2011).

There are various factors secreted by erythroblasts and macrophages that regulate erythropoiesis. Growth arrest-specific 6 (Gas-6) produced by erythroblasts has been suggested to stimulate erythropoiesis by positively regulating erythropoietin receptor via phosphoinositide 3 kinase (PI3K) and Akt activation (Angelillo-Scherrer et al. 2008). Erythropoietin has been known to enhance erythroid progenitor generation and maturation by human bone marrow or by murine spleen (Angelillo-Scherrer et al. 2008).

In contrast, interleukin 6 (IL-6), transforming growth factor-β (TGF-β), tumor necrosis factor-α (TNF-α), and interferon-γ (INF-γ), which are related to chronic inflammation and tumor progression, have been found to inhibit erythropoiesis (Gordon and Martinez-Pomares 2017). TNF-α secreted by macrophages suppresses erythropoiesis through caspase-mediated cleavage of GATA-1, a crucial transcriptional regulator of erythroblast development (de Back et al. 2014). TGF-β1 inhibits erythropoiesis by decreasing the entry of early progenitor cells in cell cycle and reducing the proliferation of intermediate and late erythroid progenitors and stimulating their differentiation toward enucleated erythrocytes by skipping cell divisions (Zermati et al. 2000).

The final stage of erythropoiesis involves phagocytosis by macrophages of nuclei extruded from erythroblasts for their maturation into reticulocytes. This process is enabled by the interaction between macrophages and the erythroblast/reticulocytes via adhesion molecules such as Emp and β1 integrin and exposure of phosphatidylserine at the surface of the membrane-enveloped nuclei (de Back et al. 2014).

Macrophages have a critical function in red blood cell generation, maintenance, and clearance. The interactions between macrophages and erythrocytes are important in these processes. Understanding the molecular mechanisms and regulation underlying these interactions may help to develop efficient therapeutic approaches for preventing unwanted destruction of erythrocytes in diseases or blood transfusion, drug-induced hemolytic anemia, or hemolytic anemia disorders.

2.3.2 Macrophages in Brain Development

Microglia are macrophage-like cells in the brain and spinal cord considered immune sentinels that regulate brain development and protection, maintenance of neuronal networks, and injury repair (Bachiller et al. 2018) (Table 1). Microglial cells derive from the yolk sac and move into the central nervous system (CNS) during embryogenesis (Bachiller et al. 2018). Microglia have been identified to actively phagocytose synaptic material and have major role in synaptic pruning during brain development (Paolicelli et al. 2011). Microglia have an emerging role in pathophysiology of the brain described as "microgliopathies," which includes mutations in the gene encoding CSF1R ascribed for HDLS (hereditary diffuse leukoencephalopathy with spheroids) and the deficiency in surface receptor TREM-2 associated with high risk of late-onset Alzheimer's disease (AD) (Amit et al. 2016; Yuan et al. 2016). Microglia constitute a protective barrier by wrapping around amyloid-β plaques preventing outward plaque expansion to reduced neuritic dystrophy (Condello et al. 2015). Another interesting link between microglia and AD is the deficiency in the microglial chemokine receptor (Cx3cr1) leading to activation of microglial cells and tau pathologic progression and memory impairment (Maphis et al. 2015). Cx3cr1 involves in microglial migration and in neuron/microglial activity modulation (Sheridan and Murphy 2013). In addition, an increase in cluster of differentiation 33 (CD33), a transmembrane receptor highly present in microglial cells in the brain that regulates innate immune response, causes elevated risk of AD due to reduced microglial phagocytosis and Aβ clearance, leading to an increase in plaque deposition (Griciuc et al. 2013).

Microglial activation has been identified to involve protein aggregation and neuroinflammation at the early stages of Frontotemporal Dementia (FTLD) or AD (Bachiller et al. 2018). Progranulin protein is constitutively expressed and secreted by microglia and participates in cell growth, injure repair, and inflammatory response in the brain. Loss of progranulin is associated with an age-dependent, progressive upregulation of lysosomal and innate immunity genes, increased complement production, and enhanced synaptic pruning in microglia in FTLD (Lui et al. 2016). Furthermore, in FTLD patients, microglial activation and high mobility group box 1 (HMGB1) have been shown co-localizing with tau oligomers initiating an inflammatory response by microglial cells leading to neuronal damage and inflammation (Nilson et al. 2017).

Furthermore, activation of microglia and their response to inflammation have major role in Parkinson's disease (PD) (Bachiller et al. 2018). The major histocompatibility complex class II (MHCII) is one of the most important factors linked to microglial activation in PD. Microglial MHCII expression has been frequently related to α-synuclein-positive Lewy neurites, TH-16-positive dopaminergic, and WH-3-positive serotonergic neurites (Imamura et al. 2003) as well as AAV2-synuclein-positive neurons (Harms et al. 2013).

Microglial macrophage-like cells are important in brain development, homeostasis, and pathology. Emerging evidence shed light on the role of microglial activation and response as well as innate immune system in the pathogenesis and progression of neurodegenerative diseases including AD, PD, and FTLD. Understanding the complex interaction of microglial activation, response, and regulation in brain diseases is promising for therapeutic targets.

2.3.3 Macrophages in Lung Homeostasis

Alveolar macrophages are present in the alveoli of the lung as known ad type I alveolar epithelial cells (Lopez-Rodriguez et al. 2017; Bhattacharya and Westphalen 2016) (Table 1). Type II alveolar epithelial cells are less abundant in the alveoli compared to type I cells and are in charge of production, secretion, and recycling of lung surfactant to prevent lung collapse by reducing surface tension during respiratory ventilation (Lopez-Rodriguez et al. 2017). Alveolar macrophages play key function in clearing apoptotic cells and cellular debris as well as immune response in lung health and disease conditions (Hussell and Bell 2014). The respiratory epithelium is also a major source of granulocyte–macrophage colony-stimulating factor (GM-CSF) which is required for the differentiation of fetal monocytes into alveolar macrophages (Hussell and Bell 2014). Alveolar macrophages regulate surfactant lipid metabolism which can explain the inefficient surfactant catabolism in the development of pulmonary alveolar proteinosis in GM-CSF-deficient mice and in mice lacking the ABC transporters (ABC-A1, ABCG1, or peroxisome proliferator-activating receptor-gamma (PPAR-ɣ)) (Lopez-Rodriguez et al. 2017). Signal-regulatory protein-α (SIRPα) and CD200 are protein interactors of alveolar macrophages as inhibitory receptors in the lung microenvironment. A deficiency of CD200 causes an elevation in alveolar macrophage numbers and activation (Hussell and Bell 2014). Alveolar macrophages with M2-like phenotype have been involved in rescuing lung inflammatory conditions but also contribute to fibrotic pathogenic mechanism by increased production of TGFβ, CCL18, resistin-like secreted protein, resistin-like-α, and chitinase-like secretory lectin YM1 (Hussell and Bell 2014). Asthma and progressive fibrotic lung disease have increased activity of alveolar macrophages which express significant levels of inflammatory cytokines including IL-13 and matrix metalloproteinases (MP1 and 12) leading to inflammation and damages to the lung tissue (Gibbons et al. 2011; Magnan et al. 1998). Alveolar macrophages have an important role in pulmonary immune homeostasis and pathologic mechanism of diseases. Insights into their interactions and regulation in the lung microenvironment, especially how alveolar macrophages switch their function between scavenging cellular debris and responding to pathogenic inflammatory events, provide therapeutic strategies improving efficacy of inhaled vaccines and airway conditions.

2.4 Macrophages in Metabolism

2.4.1 Macrophages in Metabolic Syndrome

Tissue macrophages in vertebrates are diverse in functions and the first line of defense mechanism against pathogens via phagocytosis of microorganism infection, particles, and cell debris (McNelis and Olefsky 2014; Verdeguer and Aouadi 2017). In addition to their immune protection, these macrophages also contribute to essential tissue metabolic homeostasis, development, and abnormalities (Verdeguer and Aouadi 2017) (Table 1). The association of macrophages in metabolic homeostasis has been identified by changes in macrophage numbers and the switch between anti-inflammatory macrophages as M2-like or alternatively activated macrophages (AAMs), and pro-inflammatory macrophages as M1-like or classically activated macrophages (CAMs) in metabolic tissues adipose, liver, and muscle tissue (McNelis and Olefsky 2014; Olefsky and Glass 2010). Inflammation caused by activation of pro-inflammatory macrophages has been shown as one of the connections between obesity and insulin resistance. In normal condition, insulin sensitivity in adipose tissue is modulated in part by the anti-inflammatory cytokine IL-10 secreted by resident macrophages acting as insulin sensitizer (Gordon and Martinez-Pomares 2017). Macrophage-specific peroxisome proliferator activator receptor-gamma (PPARγ), a key transcriptional regulator of mitochondrial function and fatty acid oxidation in macrophages, can potentiate IL4- and IL-13-driven alternative macrophage activation and ameliorate insulin resistance (Odegaard et al. 2007; Kratz et al. 2014). However, disruption of PPARγ in myeloid cells impairs alternative macrophage activation (Kratz et al. 2014). The increase of pro-inflammatory cytokines, tumor necrosis factor α (TNFα), and interleukin 1 β (IL-1β), by macrophages in obesity, have been associated with reduced insulin sensitivity and glucose uptake, glycogenesis (synthesis of glycogen from glucose), and lipogenesis (synthesis of lipid) by interfering with the insulin signaling pathway in adjacent insulin target cells (Verdeguer and Aouadi 2017). It has been shown that transcriptional factors involved in the inflammatory response regulate insulin sensitivity. As such, activation of nuclear factor kappa-B kinase subunit beta (IKK-β) and nuclear factor kappa-B (NF-κB) in the liver leads to increased production of inflammatory cytokines and insulin resistance (Cai et al. 2005); however, loss of IKK-β in hepatocytes and myeloid cells improves obesity-induced insulin resistance (Arkan et al. 2005). Particularly, macrophages can be self-renewed and have the capacity to proliferate within adipose tissue resulting in accumulation of alternatively activated macrophages in obesity (Wynn et al. 2013; Verdeguer and Aouadi 2017).

Eosinophils, as the source of IL-4 in adipocytes, have been shown to maintain adipose alternatively activated macrophages connected with glucose homeostasis (Wu et al. 2011). Deficiency of eosinophils has been associated with activation of adipose tissue macrophages and a predisposition of developing insulin resistance in response to obesity (Wu et al. 2011). Alterations in macrophage phenotype in adipocytes result in the secretion of chemokines CCL2, CCL5, and CCL8 that recruit Ly6Chi inflammatory monocytes from blood into adipose depot in response to chronic elevation in caloric intake (Wynn et al. 2013). These inflammatory monocytes differentiate into classically activated macrophages phagocytizing dead adipocytes and expressing inflammatory cytokines TNF-α and IL-6, which promote local and systemic insulin resistance (Lumeng et al. 2007).

Kupffer cells are the resident macrophages of the liver that regulate the metabolic adaptations of hepatocytes during increased caloric intake (Wynn et al. 2013; Samuel and Shulman 2012). Kupffer cells directly participate in regulating fatty acid oxidation in hepatocytes which is abnormal in obesity (Wynn et al. 2013; Samuel and Shulman 2012). It has been identified that loss of PPARδ impairs alternative activation of Kupffer cells, leading to the development of hepatic steatosis and insulin resistance (Huang et al. 2010). In addition, obesity-induced white adipose depot inflammation induces the expression of chemokines, such as CCL2 and CCCL1 resulting in the infiltration of inflammatory macrophages into the insulin-producing islets of the pancreas (Ehses et al. 2007; Eguchi et al. 2012). These infiltrated inflammatory macrophages secrete IL1β and TNFα causing an impairment in insulin secretion from β-cells (Ehses et al. 2007; Eguchi et al. 2012).

Macrophages go beyond their protective function against pathogens by adapting their tissue of residence responding to genetic and metabolic alterations. These changes trigger the switch between the pro-inflammatory and anti--inflammatory actions of macrophages. Identification of molecular interaction and regulation of these intermediate states could be valuable for the identification of molecular targets to treat immunometabolic abnormalities.

2.4.2 Macrophages and Iron Recycling

Resident spleen and liver macrophages are essential in phagocytizing and degrading senescent and damaged erythrocytes to recycle iron, majorly for the hemoglobin generation in new erythrocytes but also for other carriers and enzymes requiring iron (Ganz 2012) (Table 1). Upon uptake, heme is deteriorated via catalysis of heme oxygenase (HO; mostly HO-1) to free iron into the cytoplasm and then exported by plasma membrane ferroportin (Ganz 2012; Haldar and Murphy 2014). The release of iron from macrophages into plasma is regulated by the interaction of the hepatic hormone hepcidin with its receptor/iron exporter ferroportin (Ganz 2012). In humans, iron recycled from macrophages provides a major source for usage compared to iron obtained from diets and hepatic storage (Ganz 2012). Interestingly, splenic red pulp macrophages are specialized for erythrophagocytosis as F4/80hiCD68 + CD11blo/−, VCAMhi splenic cells through induction of the transcription factor Spi-C (Kohyama et al. 2009; Delaby et al. 2008). Red pulp macrophages also express Nramp1, HO, ferroportin, and the haptoglobin–hemoglobin receptor (CD163) which facilitate the uptake of hemoglobin, breakdown of heme, and the export of iron, and further confirm their critical role in systemic iron recycling

(Ganz 2012). Senescent erythrocytes can be recognized by macrophages for erythrophagocytosis as markers of erythrocyte aging including modifications of the most abundant erythrocyte membrane protein, Band 3, the presence of phosphatidylserine on the outer leaflet of the plasma membrane, increased membrane rigor, and the deprivation of sialic acid and the CD47 expression (de Back et al. 2014; Pantaleo et al. 2008; Lee et al. 2011; Bosman et al. 2005). The modified and clustered band 3 protein is recognized by opsonic natural antibodies and complement, triggering conventional antibody- and complement-mediated phagocytosis (Pantaleo et al. 2008). Loss of CD47 expression which means lack of an inhibitory signal by CD47-SIRPα interaction for phagocytosis and conformational changes in CD47 enables a communication between CD47 and thrombospondin-1 resulting in increased uptake by red pulp macrophages (de Back et al. 2014). The residential macrophages of the spleen also have a remarkable role in the removal of intracellular inclusions in erythrocytes, leaving the erythrocyte flawless. These inclusion bodies can be of denatured hemoglobin caused by oxidative damage or nuclear chromatin remnants or phagosomes eating excessive amounts of iron (de Back et al. 2014).

Iron uptake and recycling by macrophages are critical for maintaining regular functions and generation of red blood cells. Understanding the molecular mechanism and modulation of phagosomal iron deprivation to host defense mechanism against pathogens could be valuable for therapeutic strategies.

3 Macrophages as a Target for Drug Delivery

3.1 Targeting Macrophage for Intracellular Pathogen Therapy

Macrophages have been an important target for intracellular pathogen therapy. Even though macrophages play important roles in protecting body against foreign pathogens, many infectious diseases are caused by the dysregulation of macrophages. In some infectious diseases such as tuberculosis, salmonellosis, and brucellosis, bacteria find their ways to reside inside the macrophages where they adapt to the host intracellular environment to survive, replicate, and then damage to the host (Monack et al. 2004). A parasitic disease named Leishmaniasis is caused by the protozoan Leishmania parasites that interact and infect a variety of host cells, and infected macrophages are considered as the source that results in the outcome of infection. Leishmania parasites are able to survive in phagolysosomal environment and manipulate host cells to favor their replication and transmission (Liu and Uzonna 2012). Antibiotics have been utilized for the treatment of intracellular pathogens (Monack et al. 2004), but the outcomes of antibiotic therapy are not desirable in terms of efficacy and side effects because high doses of antibiotics must be employed to penetrate macrophages (Imbuluzqueta et al. 2010). Therefore, developing a drug delivery system (DDS) loaded with antibiotics that can target infected macrophages could be an effective solution for the treatment of intracellular infection.

Lipidic drug delivery systems have been studied as vehicles for the delivery of antimicrobial agents because of their sustained-drug release effect, toxicity minimization, and increased efficacy. Moreover, the incorporated drug is protected by the DDS against the immunological and enzymatic attacks (Prior et al. 2002). Liposomes are spherical vesicles that are usually made of phospholipids and cholesterol. Because the structure of liposomes is similar to the biological membranes, using liposomes as a drug carrier can potentially reduce toxicity and immunological reactions (Imbuluzqueta et al. 2010). Another advantage of using this lipidic DDS is that liposomes can deliver both hydrophilic and hydrophobic drugs to the target cells. Hydrophilic drugs are placed into the aqueous spaces, while hydrophobic drugs are incorporated into the lipid membrane (Imbuluzqueta et al. 2010). However, there are some limitations of employing liposome as DDS. After being administered intravenously, liposomes are opsonized and destabilized by lipoproteins, which removes phospholipid and disrupts the lipid membrane of liposomes (Kirby and Gregoriadis 1981). As a result, the incorporated drug is liberated before being engulfed by macrophages. Another limitation is that liposomes are sensitive to acidic pH and can be destabilized when encountering acidic environment. After entering cells via phagocytosis, liposomes are delivered to lysosome by endosome, but the mildly acidic environment in endosome can destabilize lysosomes, which causes them to release drug contents before reaching lysosome (Kirby and Gregoriadis 1981). It has been found that dioleoyl phosphatidylethanolamine and weakly acidic amphiphiles can help protect liposome from acidic environment when being incorporated into the lipid membrane (Chu et al. 1990). There have been number of drugs using liposomes to target macrophages. For instance, liposomes that are composed of phosphatidylcholine (PC), phosphatidylglycerol, cholesterol (Chol), and alpha-tocopherol were tested to deliver gentamicin to macrophages in the treatment of Salmonella Dublin infection (Fierer et al. 1990). Gentamycin-loaded liposomes showed more effectiveness and reduced the mortality significantly in comparison to the identical dose of free gentamycin (Fierer et al. 1990). Rifampicin-loaded liposome aerosol composed of PC, Chol, and phospholipid with methylated bovine serum albumin (MBSA), or O-stearoyl amylopectin were tested to target alveolar macrophages (Vyas et al. 2004). Rifampicin liposomes showed a greater lung distribution in comparison to free rifampicin in rats (Vyas et al. 2004).

3.2 Targeting Macrophage for Therapy in HIV-1 Infection

Macrophages are myeloid immune cells that play a critical role in clearing pathogens, dead cells, cellular debris, and foreign material as well as orchestrating inflammatory processes (Klimpel 1996). Macrophages protect against viruses through phagocytosis or inhibition of virus multiplication in neighboring cells by destroying virus-infected cells or by producing soluble factors (interferons) that act on these cells (Klimpel 1996). If a virus replicates in macrophages, the infected macrophages may aid in dissemination of the virus to other body cells. Macrophages act as the antigen presenting cells and present processed pathogen antigen peptides to the CD4+ T cells via MHC II pathway (Ackerman and Cresswell 2004; Koppensteiner et al. 2012). This exchange of information between macrophages and CD4+ T cells also has important role in the transmission of HIV-1 from macrophage to CD4+ T cells (Crowe et al. 1990, 1992; Groot et al. 2008). In addition, HIV-infected macrophages release soluble cytotoxic factors that can promote the apoptosis of bystander cells, for example, CD4+ and CD8+ T cells (Badley et al. 1997; Herbein et al. 1998).

HIV-1 infection leads to the lysis of T lymphocytes (CD4+ T and CD8+ T cells) resulting in their depletion, a hallmark of HIV-1 pathogenesis. In contrast, macrophages are relatively less sensitive to the cytopathic effect of the virus (Gendelman et al. 1988; Carter and Ehrlich 2008). Therefore, HIV-1-infected macrophages persist long in infected patients as a source of virus production for longer period of time (Kelly et al. 2008). In addition, macrophages are virtually present in every organ system (although with different names), thus can disseminate HIV-1 throughout the body of infected persons including brain (Gavegnano and Schinazi 2009).

Macrophages play a crucial role in different phases of HIV-1 infection and are latent reservoirs for HIV-1 (human immunodeficiency virus type 1) (Koppensteiner et al. 2012; Kumar and Herbein 2014) which is a significant hindrance in clearing virus from infected individuals (Gavegnano and Schinazi 2009). Macrophages allow the entry of HIV-1 via CD4, which interacts with the envelope glycoprotein gp120 of the virus (Crowe et al. 1990; Kumar and Herbein 2014)[26]. They resist the cytopathic effect of HIV-1 due to the presence of the host cell restriction factor SamHD1 and disseminate the virus throughout the body of infected individuals; thus, long-term successful treatment of HIV-1 infection requires potent strategies to prevent HIV-1 from entering and persisting in these macrophages (Koppensteiner et al. 2012; Kumar and Herbein 2014).

3.3 Targeting Macrophages for Rheumatoid Arthritis (RA)

Macrophages are central to the pathophysiology of inflammation process in RA due to complex causes and their response to successful antirheumatic treatment (McInnes and O'Dell 2010; Davignon et al. 2013). They have been discovered to be activated (Liote et al. 1996; Torsteinsdottir et al. 1999) and prominently present at inflammatory sites and synovial membranes and produce TNF-α in RA (Davignon et al. 2013; Kinne et al. 2007). In addition to their primary role in inflammation, monocytes/macrophages involve pathologic process of bone erosion due to their excessive differentiation into osteoclasts (OCs), which causes increased bone resorption (Takayanagi 2007). This differentiation is mediated by two major cytokines, M-CSF (monocyte/macrophage colony-stimulating factor) and RANKL (receptor activator of nuclear factor-kappa-B ligand) in which their expressions are induced by pro-inflammatory cytokines such as TNF-α, IL-1, IL-6, and IL-17 (Nakashima and Takayanagi 2008). In RA, bone destruction is caused by the enhanced osteoclast activity, which is mainly dependent on interleukin-17-producing helper T cells (T(H)17) (Nakashima and Takayanagi 2008). Therefore, pathological bone homeostasis as well as RA particularly is modulated by inflammatory and immunological events involved monocytes/macrophages which are ideal targets for treatment. Drug delivery systems that can specifically inactivate or deplete macrophages in the articulations (joints) have been advanced for RA treatment (Rollett et al. 2013; Bilthariya et al. 2015; Duan et al. 2014).

Rheumatoid arthritis is an autoimmune disease caused by the accumulation of macrophages in synovial tissues (Sack et al. 1994). Because the polarization of macrophages in RA favors M1 more than M2, it leads to more secretion of IL-1, IL-6, IL-12, and TNR-α, which causes cartilage destruction and bone erosion (Crowe et al. 1990). The standard care for rheumatoid arthritis treatment is methotrexate, steroids, and nonsteroidal anti-inflammatory drugs (NSAID). However, those drugs do not interfere with macrophages. Instead, they work by antagonizing the inflammatory effect of macrophages, which may assist in control and relieve the symptoms (Scott et al. 2010).

Because macrophages are thought to play a principal role in the pathogenesis of RA, elimination of inflammatory synovial macrophages and their mediators such as Il-1, IL-6, IL-12, and TNR-α could be a potential therapeutic approach to cure RA. Studies showed that administration of clodronate liposomes resulted in depletion of synovial macrophages along with the improvement of inflammatory symptoms (Van Lent et al. 1998). Clodronate, a bisphosphonate that lacks a nitrogen, is metabolized to beta–gamma-methylene (APPCp-type) analog of ATP, which involves competitive inhibition of ADP/ATP translocase, resulting in osteoclast apoptosis (Lehenkari et al. 2002). Another drug that shows to deplete macrophages is anti-miR-449a that inhibits SIRT1 expression (Opperman et al. 2019). SIRT1 has been known for its function in inducing macrophage self-renewal ability by regulating G1/S transition in cell cycle. Overexpression of SIRT1 during bone marrow-derived macrophage differentiation promotes the proliferation of macrophages. Conversely, when SIRT1 expression is reduced, macrophages' self-renewal ability is restricted (Imperatore et al. 2017). In addition to anti-miR-449a drug, immunotherapeutic agents targeting the Fc γ receptor (CD64), which is anti-CD64-Ricin A, also show a significant result in depleting macrophages. The studies showed that anti-CD64-Ricin A selectively induces apoptosis with high specificity in activated pro-inflammatory macrophages (Akinrinmade et al. 2017). Another strategy to treat RA is repolarizing M1 to M2 macrophages. Thapsigargin, which is a Notch inhibitor, can induce M1 to M2 macrophages. Notch inhibition reduces expression level of Notch-associated genes and M1 effector gene (CCL5) and increases M2 effector genes (IL-4, Arg1) (Sun et al. 2017).

Moreover, thapsigargin is also used as endoplasmic reticulum (ER) stress inducer, which plays a key role in macrophage differentiation (Sun et al. 2017).

Alternative therapeutic approach to treat rheumatoid arthritis is blocking cytokines and other pro-inflammatory factors. Anti-TNR-α such as infliximab, adalimumab, certolizumab, and golimumab have been used for RA treatment (Ma and Xu 2013). TNR-α is one of the key cytokines that induce inflammation, bone erosion, and cartilage destruction in RA. Therefore, TNR-α blockade by using those monoclonal antibodies can be an effective treatment for chronic RA (Ma and Xu 2013). Tofacitinib indirectly weakens the effect of pro-inflammatory factors by inhibiting the activity of Janus Kinases, which involves the expression of certain genes necessary for activation of inflammation. Tofacitinib was approved by FDA in 2012 for RA treatment (Ma and Xu 2013).

3.4 Targeting Macrophages for Cancer

3.4.1 Tumor-Associated Macrophages in Cancer Development

Even though macrophages help fight against foreign pathogens and diseases, tumor progression is highly correlated with macrophages' residence and polarization. Tumor-associated macrophages (TAMs) have been studied widely to investigate the roles of macrophages in cancer development and find appropriate therapies that can cure or prolong the lives of cancer patients. Based on many clinical and experimental evidence, macrophages can potentially promote cancer initiation and malignant progression. Studies showed that TAM can contribute up to 50% of the tumor mass and lead to poor patient prognosis (Qian and Pollard 2010). In addition, macrophages can lower the efficacy of certain immunotherapy in solid tumors (Pascual-Garcia et al. 2019). Therefore, targeting TAM could be a therapeutic approach in the fight against cancers.

Numerous cancers are associated with chronic inflammation although the detailed mechanisms behind it still remain unknown (Qian and Pollard 2010). However, we do know that chronic inflammation promotes and activates expression of key transcriptional factors such as nuclear factor-k-light-chain enhancer of activated B cell (NF-kB), STAT3, and HIF1α, which favors the development of M1 macrophage (Qian and Pollard 2010). Pro-inflammatory M1 macrophages secrete mediators that both provide mutagenic microenvironment for surrounding cells and recruit more myeloid cells such as monocytes that later differentiate into macrophages upon arrival (Qian and Pollard 2010). It is believed that tumor-associated macrophages are involved in all stages of cancer, such as promoting tumor initiation, tumor proliferation, and metastasis (Joyce and Pollard 2009). In mouse models of cancer, macrophages are found to stimulate angiogenesis, tumor cell migration, invasion, and extravasation. In addition, macrophages can also suppress antitumor immunity (Cassetta and Pollard 2018). At metastatic sites, macrophages prepare the target tissue for arrival of tumor cells, and then a different subpopulation of macrophages promotes tumor cell extravasation, survival, and subsequent growth (Qian and Pollard 2010). Therefore, specialized subpopulations of macrophages are targets for anticancer therapy.

Despite M1 and M2 forms of macrophage polarization being tempting, it is not the probable cause in the complexity of tumor microenvironment where multiple phenotypes of TAMs activity are involved in different biological functions in the tumor (Qian and Pollard 2010). Indeed macrophages have a mixed phenotype expressing both M1 and M2 markers in large-scale transcriptome analysis suggesting diverse phenotypes of macrophage subpopulations associated with different tumor types (Qian and Pollard 2010). Accordingly, depletion of specific subpopulations such as TIE2+ or metastasis-associated macrophages (MAMs) can impair angiogenesis or inhibit metastatic seeding (Qian and Pollard 2010). Thus, the challenges remain to identify and inhibit specific macrophage trophic phenotypes together with their immunosuppressive activities and enhance their activation that favors anticancerous activities (Qian and Pollard 2010; Ostuni et al. 2015).

3.4.2 Origins of Tumor-Associated Macrophages (TAMs)

The historical description of adult resident tissue macrophages as being solely arised from bone marrow (BM) is inadequate as macrophages arise from several origins during ontogenesis, and each of these different lineages displays great diversity in adulthood (Gautier et al. 2012). Macrophages involved in pathogen responses are proposed to come from circulating BM monocytes (Wynn et al. 2013). These different embryonic origins appear to respond differently to anti-macrophage therapies in which the recruited TAMs are able to survive in response to the inhibitor of colony-stimulating factor (CSF)-1 receptor as expected because macrophages depend upon CSF-1 for differentiation and survival (Pyonteck et al. 2013). Recent studies demonstrated that different TAM subsets derive from the Ly6C+ monocytes in implanted (grafted) tumors (Movahedi et al. 2010), primary mouse mammary tumors (Franklin et al. 2014), and lung metastases (Qian et al. 2011). BM and extramedullary hematopoiesis, particularly the spleen, can be reservoir for these monocytes (Cortez-Retamozo et al. 2012). However, splenic contribution is minor, and BM is the primary source of monocytes (Shand et al. 2014).

3.4.3 Therapeutic Approaches for Cancer

A therapeutical approach to treat cancers is inhibiting monocyte infiltration into solid tumor. Tumor-associated macrophages are not only arised from tissue-resident macrophages but also originated from bone marrow progenitor cells like circulating monocytes (Wynn et al. 2013). Monocytes are recruited into tumor cells by several CC chemokines, particularly CCL2 and CCL5, which are produced by tumor cells, fibroblast, endothelial cells, and TAMs themselves (Murdoch et al. 2004). In addition, CLL2 and CCL5 also stimulate monocytes to produce proteins that help with monocyte migration and also tumor progression. For instance, CCL5 stimulates monocytes to produce CCL2, CCL3, CCL4, CXCL8, and chemokine receptor CCR1, which in turn recruit more circulating monocytes to inflamed tissues or tumor cells (Locati et al. 2002). Moreover, both CCL2 and CCL5 induce monocytic cell lines, blood monocytes, and microglia to secrete matrix metalloproteinases-9 (MMP-9), MMP-19, and urokinase-type plasminogen activator receptor (uPA-R), which breakups the basement membrane and permit leukocytes to migrate into tissues, leading to the progress of cancer cells (Coussens and Werb 1996). Therefore, strategies to block the effects of CCL2/CCR2 and CCL5/CCR5 have been studied extensively in cancer therapy. There are a couple of drugs targeting CCL2/CCR2 which are currently undergoing clinical phase investigation, and one of them is carlumab. Carlumab is an antibody that binds and blocks CCL2 from binding to CCR2. Carlumab is expected to inhibit or reduce the infiltration of monocytes into the tumor microenvironment, which can potentially suppress cancer progression (He et al. 2020). In fact, some trials have shown that carlumab could induce regression of prostate cancer. In addition, a study showed that CCL2 inhibition in mice promoted the repression of metastatic breast cancer (Bonapace et al. 2014). Although anti-CCL2 agents have shown to have promising results, further research and trials need to be done to test the effectiveness and safety of carlumab.

Another therapeutic approach to battle against cancers is repolarizing tumor-associated macrophages to M1 phenotype macrophages. Restoration of pro-inflammatory phenotype of macrophages could help trigger antitumor response (Wynn et al. 2013). There are different strategies to repolarize TAM to M1 macrophages, and one of them is blocking CD47 immunoglobulin. CD47 immunoglobin is overexpressed on the surface of many types of cancer cells and related to poor patient prognosis (Zhang et al. 2020). CD47 binds to signal-regulatory protein α (SIRPα) on macrophages, triggering "not to eat" signal. In the other words, the binding of CD47 and SIRPα allows tumor cells to escape from macrophage-mediated phagocytosis (Zhang et al.

2020). Therefore, blocking CD47-SIRPα signaling pathway can stimulate the engulfing of tumor cells by macrophages. In addition to enhancing the phagocytic function of macrophages, some studies have shown that anti-CD47 antibodies can promote elimination of cancers via multiple mechanisms (Chao et al. 2012). One of the mechanisms is inducing the phagocytic uptake of tumor cells by dendritic cells, which then function as antigen presentation to CD4 and CD8 T cells (Chao et al. 2012). In another mechanism, anti-CD47 antibodies can activate natural killer cell-mediated antibody-dependent cytotoxicity and complement-dependent cytotoxicity to kill cancer cells (Chao et al. 2012). Lastly, anti-CD47 antibodies stimulate apoptosis of tumor cells via caspase-independent pathway (Chao et al. 2012).

Agonistic CD40 monoclonal antibodies are another therapeutic option to eliminate cancer cells. CD40 is a tumor necrosis factor receptor expressed broadly on antigen-presenting cells including dendritic cells, B cells, and monocytes (Vonderheide and Glennie 2013). When CD40 receptor binds to its ligand on T helper cells, it activates antigen-presenting cell (APC) and thus induces adaptive immunity (Vonderheide and Glennie 2013). In this context, agonistic CD40 mAb allows dendritic cells or macrophages to process and present tumor-associated antigens to local cytotoxic T lymphocyte (Vonderheide and Glennie 2013). Therefore, developing agonistic CD40 monoclonal antibodies is a promising therapy for cancers.

It has been discussed previously that tumor-associated macrophages (TAMs) have essential roles in tumor initiation and progression, including angiogenesis, immunosuppression, invasion, and metastasis. Increasing evidence demonstrate that TAMs may influence the antitumor efficacy of cytotoxic chemotherapy, cancer-cell targeting antibodies, and immunotherapeutic agents (De Palma and Lewis 2013). M1 macrophages need M-CSF/CSF-1 to be formed from myeloid progenitors. Consequently, blockade of M-CSF/CSF-1 or its receptor in combination with paclitaxel reduces TAMs presence and promotes T_H1 responses in late-stage mammary adenocarcinomas compared with paclitaxel treatment alone (DeNardo et al. 2011). CSF-1 responses signature genes (Beck et al. 2009) and the presence of M2 proliferating macrophages as poor clinical outcome to predict risk of recurrence and survival rate (Campbell et al. 2011), as well as response to chemotherapy in breast cancer (DeNardo et al. 2011). Mature M1 macrophages can become M2 macrophages by treatment with interleukin-4 (IL-4) or IL-13. Antagonist αIL-4 therapeutic antibodies reprogram M2 macrophages, monocytes, and other T_H2 cells toward T_H1 phenotypes in mammary cancer (DeNardo et al. 2009).

Early studies showed that doxorubicin (an anthracycline formerly known as adriamycin) enhanced the tumoricidal activity of macrophages ex vivo. In contrast, daunorubicin (formerly daunomycin) is less effective than doxorubicin as a cancer chemotherapeutic agent in vivo due to the greater elimination of macrophages by daunorubicin and a resultant reduction in the contribution of host immunity to the antitumor action (Mantovani 1977, 1979). These results suggested that doxorubicin is able to bolster the antitumor activities of TAMs.

Another approach in cancer immunotherapy would be to promote TAMs differentiation into a phenotype that no longer possesses immunosuppressive activity (Kodumudi et al. 2010). For example, docetaxel (a taxane) modulates the ablation of immunosuppressive (M2-like) TAMs and the concomitant promotion of antitumoral (M1-like) monocytes/MDSCs (Myeloid-Derived Suppressor Cells) in 4 T1-Neu tumor-bearing mice (Kodumudi et al. 2010). In fact, docetaxel enhances tumor-specific, cytotoxic T cell responses of treated monocytes/MDSCs in vitro (Kodumudi et al. 2010). In another example, depletion of TAM is a major component of the antitumor efficacy of some cytotoxic agents. Trabectedin, a DNA-damaging agent, caused selective depletion of circulating monocytes and TAMs together with reduction of angiogenesis in patients with soft-tissue sarcomas (Gnanadhas et al. 2013). Mechanistically, trabectedin activates caspase-8-dependent apoptosis selectively in monocytes/macrophages via TRAIL-R2, a death receptor not expressed by other leukocytes

(Gnanadhas et al. 2013). These findings provide strong evidence for the concept of TAMs reprogramming or TAMs ablating in macrophage-targeted therapy to enhance antitumor effects of chemotherapeutic agents (Kodumudi et al. 2010; Germano et al. 2013). TAMs can also limit the efficacy of chemotherapy (De Palma and Lewis 2013).

Insights into macrophage anatomy and physiology have been advancing with identification of their diverse origins and characteristics, transcriptional complexity and regulation, phenotypic polarization in response to homeostatic needs, and challenging environments. Sophisticated delineation of functions and regulations of macrophage subsets provides opportunities for novel mechanisms through which macrophages might be modulated for therapeutic applications of various diseases.

References

Ackerman AL, Cresswell P. Cellular mechanisms governing cross-presentation of exogenous antigens. Nat Immunol. 2004;5(7):678–84.

Akinrinmade OA, et al. CD64: an attractive immunotherapeutic target for M1-type macrophage mediated chronic inflammatory diseases. Biomedicine. 2017:5(3).

Amit I, Winter DR, Jung S. The role of the local environment and epigenetics in shaping macrophage identity and their effect on tissue homeostasis. Nat Immunol. 2016;17(1):18–25.

Anderson CF, Mosser DM. A novel phenotype for an activated macrophage: the type 2 activated macrophage. J Leukoc Biol. 2002;72(1):101 6.

Angelillo-Scherrer A, et al. Role of Gas6 in erythropoiesis and anemia in mice. J Clin Invest. 2008;118(2):583–96.

Arkan MC, et al. IKK-beta links inflammation to obesity-induced insulin resistance. Nat Med. 2005;11(2):191–8.

Arora S, et al. Macrophages: their role, activation and polarization in pulmonary diseases. Immunobiology. 2018;223(4–5):383–96.

Bachiller S, et al. Microglia in neurological diseases: a road map to brain-disease dependent-inflammatory response. Front Cell Neurosci. 2018;12:488.

Badley AD, et al. Macrophage-dependent apoptosis of CD4+ T lymphocytes from HIV-infected individuals is mediated by FasL and tumor necrosis factor. J Exp Med. 1997;185(1):55–64.

Beck AH, et al. The macrophage colony-stimulating factor 1 response signature in breast carcinoma. Clin Cancer Res. 2009;15(3):778–87.

Bessis M. Erythroblastic island, functional unity of bone marrow. Rev Hematol. 1958;13(1):8–11.

Bhattacharya J, Westphalen K. Macrophage-epithelial interactions in pulmonary alveoli. Semin Immunopathol. 2016;38(4):461–9.

Bilthariya U, et al. Folate-conjugated albumin nanoparticles for rheumatoid arthritis-targeted delivery of etoricoxib. Drug Dev Ind Pharm. 2015;41(1):95–104.

Bonapace L, et al. Cessation of CCL2 inhibition accelerates breast cancer metastasis by promoting angiogenesis. Nature. 2014;515(7525):130–3.

Bosman GJ, Willekens FL, Werre JM. Erythrocyte aging: a more than superficial resemblance to apoptosis? Cell Physiol Biochem. 2005;16(1–3):1–8.

Cai D, et al. Local and systemic insulin resistance resulting from hepatic activation of IKK-beta and NF-kappaB. Nat Med. 2005;11(2):183–90.

Campbell MJ, et al. Proliferating macrophages associated with high grade, hormone receptor negative breast cancer and poor clinical outcome. Breast Cancer Res Treat. 2011;128(3):703–11.

Carter CA, Ehrlich LS. Cell biology of HIV-1 infection of macrophages. Annu Rev Microbiol. 2008;62:425–43.

Cassetta L, Pollard JW. Targeting macrophages: therapeutic approaches in cancer. Nat Rev Drug Discov. 2018;17(12):887–904.

Chao MP, Weissman IL, Majeti R. The CD47-SIRPalpha pathway in cancer immune evasion and potential therapeutic implications. Curr Opin Immunol. 2012;24(2):225–32.

Chasis JA, Mohandas N. Erythroblastic islands: niches for erythropoiesis. Blood. 2008;112(3):470–8.

Chavez-Galan L, et al. Much more than M1 and M2 macrophages, there are also CD169(+) and TCR(+) macrophages. Front Immunol. 2015;6:263.

Chu CJ, et al. Efficiency of cytoplasmic delivery by pH-sensitive liposomes to cells in culture. Pharm Res. 1990;7(8):824–34.

Condello C, et al. Microglia constitute a barrier that prevents neurotoxic protofibrillar Abeta42 hotspots around plaques. Nat Commun. 2015;6:6176.

Cortez-Retamozo V, et al. Origins of tumor-associated macrophages and neutrophils. Proc Natl Acad Sci U S A. 2012;109(7):2491–6.

Coussens LM, Werb Z. Matrix metalloproteinases and the development of cancer. Chem Biol. 1996;3(11):895–904.

Crowe SM, et al. Full-length recombinant CD4 and recombinant gp120 inhibit fusion between HIV infected macrophages and uninfected CD4-expressing T-lymphoblastoid cells. AIDS Res Hum Retrovir. 1990;6(8):1031–7.

Crowe SM, et al. Human immunodeficiency virus-infected monocyte-derived macrophages express surface gp120 and fuse with CD4 lymphoid cells in vitro: a possible mechanism of T lymphocyte depletion in vivo. Clin Immunol Immunopathol. 1992;65(2):143–51.

Davies LC, et al. Tissue-resident macrophages. Nat Immunol. 2013;14(10):986–95.
Davignon JL, et al. Targeting monocytes/macrophages in the treatment of rheumatoid arthritis. Rheumatology (Oxford). 2013;52(4):590–8.
de Back DZ, et al. Of macrophages and red blood cells; a complex love story. Front Physiol. 2014;5:9.
De Palma M, Lewis CE. Macrophage regulation of tumor responses to anticancer therapies. Cancer Cell. 2013;23(3):277–86.
Delaby C, et al. Sequential regulation of ferroportin expression after erythrophagocytosis in murine macrophages: early mRNA induction by haem, followed by iron-dependent protein expression. Biochem J. 2008;411(1):123–31.
DeNardo DG, et al. CD4(+) T cells regulate pulmonary metastasis of mammary carcinomas by enhancing protumor properties of macrophages. Cancer Cell. 2009;16(2):91–102.
DeNardo DG, et al. Leukocyte complexity predicts breast cancer survival and functionally regulates response to chemotherapy. Cancer Discov. 2011;1(1):54–67.
Duan J, et al. Polyethyleneimine-functionalized iron oxide nanoparticles for systemic siRNA delivery in experimental arthritis. Nanomedicine (Lond). 2014;9(6):789–801.
Eguchi K, et al. Saturated fatty acid and TLR signaling link beta cell dysfunction and islet inflammation. Cell Metab. 2012;15(4):518–33.
Ehses JA, et al. Increased number of islet-associated macrophages in type 2 diabetes. Diabetes. 2007;56(9):2356–70.
Fierer J, et al. Successful treatment using gentamicin liposomes of salmonella Dublin infections in mice. Antimicrob Agents Chemother. 1990;34(2):343–8.
Franklin RA, et al. The cellular and molecular origin of tumor-associated macrophages. Science. 2014;344(6186):921–5.
Ganz T. Macrophages and systemic iron homeostasis. J Innate Immun. 2012;4(5–6):446–53.
Gautier EL, et al. Gene-expression profiles and transcriptional regulatory pathways that underlie the identity and diversity of mouse tissue macrophages. Nat Immunol. 2012;13(11):1118–28.
Gavegnano C, Schinazi RF. Antiretroviral therapy in macrophages: implication for HIV eradication. Antivir Chem Chemother. 2009;20(2):63–78.
Gendelman HE, et al. Efficient isolation and propagation of human immunodeficiency virus on recombinant colony-stimulating factor 1-treated monocytes. J Exp Med. 1988;167(4):1428–41.
Germano G, et al. Role of macrophage targeting in the antitumor activity of trabectedin. Cancer Cell. 2013;23(2):249–62.
Gibbons MA, et al. Ly6Chi monocytes direct alternatively activated profibrotic macrophage regulation of lung fibrosis. Am J Respir Crit Care Med. 2011;184(5):569–81.
Ginhoux F, Jung S. Monocytes and macrophages: developmental pathways and tissue homeostasis. Nat Rev Immunol. 2014;14(6):392–404.
Gnanadhas DP, et al. Chitosan-dextran sulphate nanocapsule drug delivery system as an effective therapeutic against intraphagosomal pathogen Salmonella. J Antimicrob Chemother. 2013;68(11):2576–86.
Gordon S. Alternative activation of macrophages. Nat Rev Immunol. 2003;3(1):23–35.
Gordon S, Martinez-Pomares L. Physiological roles of macrophages. Pflugers Arch. 2017;469(3–4):365–74.
Griciuc A, et al. Alzheimer's disease risk gene CD33 inhibits microglial uptake of amyloid beta. Neuron. 2013;78(4):631–43.
Groot F, Welsch S, Sattentau QJ. Efficient HIV-1 transmission from macrophages to T cells across transient virological synapses. Blood. 2008;111(9):4660–3.
Haldar M, Murphy KM. Origin, development, and homeostasis of tissue-resident macrophages. Immunol Rev. 2014;262(1):25–35.
Harms AS, et al. MHCII is required for alpha-synuclein-induced activation of microglia, CD4 T cell proliferation, and dopaminergic neurodegeneration. J Neurosci. 2013;33(23):9592–600.
He W, et al. Drug delivery to macrophages: a review of targeting drugs and drug carriers to macrophages for inflammatory diseases. Adv Drug Deliv Rev. 2020;165-166:15–40.
Herbein G, et al. Apoptosis of CD8+ T cells is mediated by macrophages through interaction of HIV gp120 with chemokine receptor CXCR4. Nature. 1998;395(6698):189–94.
Ho VW, Sly LM. Derivation and characterization of murine alternatively activated (M2) macrophages. Methods Mol Biol. 2009;531:173–85.
Huang W, et al. Depletion of liver Kupffer cells prevents the development of diet-induced hepatic steatosis and insulin resistance. Diabetes. 2010;59(2):347–57.
Hussell T, Bell TJ. Alveolar macrophages: plasticity in a tissue-specific context. Nat Rev Immunol. 2014;14(2):81–93.
Iavarone A, et al. Retinoblastoma promotes definitive erythropoiesis by repressing Id2 in fetal liver macrophages. Nature. 2004;432(7020):1040–5.
Imamura K, et al. Distribution of major histocompatibility complex class II-positive microglia and cytokine profile of Parkinson's disease brains. Acta Neuropathol. 2003;106(6):518–26.
Imbuluzqueta E, et al. Drug delivery systems for potential treatment of intracellular bacterial infections. Front Biosci (Landmark Ed). 2010;15:397–417.
Imperatore F, et al. SIRT1 regulates macrophage self-renewal. EMBO J. 2017;36(16):2353–72.
Joyce JA, Pollard JW. Microenvironmental regulation of metastasis. Nat Rev Cancer. 2009;9(4):239–52.
Kelly J, et al. Human macrophages support persistent transcription from unintegrated HIV-1 DNA. Virology. 2008;372(2):300–12.

Kinne RW, Stuhlmuller B, Burmester GR. Cells of the synovium in rheumatoid arthritis. Macrophages Arthritis Res Ther. 2007;9(6):224.
Kirby C, Gregoriadis G. Plasma-induced release of solutes from small unilamellar liposomes is associated with pore formation in the bilayers. Biochem J. 1981;199(1):251–4.
Klimpel GR. Medical Microbiology. 4th ed; 1996.
Kodumudi KN, et al. A novel chemoimmunomodulating property of docetaxel: suppression of myeloid-derived suppressor cells in tumor bearers. Clin Cancer Res. 2010;16(18):4583–94.
Kohyama M, et al. Role for Spi-C in the development of red pulp macrophages and splenic iron homeostasis. Nature. 2009;457(7227):318–21.
Koppensteiner H, Brack-Werner R, Schindler M. Macrophages and their relevance in human immunodeficiency virus type I infection. Retrovirology. 2012;9:82.
Kratz M, et al. Metabolic dysfunction drives a mechanistically distinct proinflammatory phenotype in adipose tissue macrophages. Cell Metab. 2014;20(4):614–25.
Kumar A, Herbein G. The macrophage: a therapeutic target in HIV-1 infection. Mol Cell Ther. 2014;2:10.
Kusakabe M, et al. c-Maf plays a crucial role for the definitive erythropoiesis that accompanies erythroblastic island formation in the fetal liver. Blood. 2011;118(5):1374–85.
Lee SJ, et al. Mechanism for phosphatidylserine-dependent erythrophagocytosis in mouse liver. Blood. 2011;117(19):5215–23.
Lehenkari PP, et al. Further insight into mechanism of action of clodronate: inhibition of mitochondrial ADP/ATP translocase by a nonhydrolyzable, adenine-containing metabolite. Mol Pharmacol. 2002;61(5):1255–62.
Leimberg MJ, et al. Macrophages function as a ferritin iron source for cultured human erythroid precursors. J Cell Biochem. 2008;103(4):1211–8.
Liote F, et al. Blood monocyte activation in rheumatoid arthritis: increased monocyte adhesiveness, integrin expression, and cytokine release. Clin Exp Immunol. 1996;106(1).13–9.
Liu D, Uzonna JE. The early interaction of Leishmania with macrophages and dendritic cells and its influence on the host immune response. Front Cell Infect Microbiol. 2012;2:83.
Liu XS, et al. Disruption of palladin leads to defects in definitive erythropoiesis by interfering with erythroblastic island formation in mouse fetal liver. Blood. 2007;110(3):870–6.
Locati M, et al. Analysis of the gene expression profile activated by the CC chemokine ligand 5/RANTES and by lipopolysaccharide in human monocytes. J Immunol. 2002;168(7):3557–62.
Lopez-Rodriguez E, et al. Lung surfactant metabolism: early in life, early in disease and target in cell therapy. Cell Tissue Res. 2017;367(3):721–35.
Lui H, et al. Progranulin deficiency promotes circuit-specific synaptic pruning by microglia via complement activation. Cell. 2016;165(4):921–35.
Lumeng CN, Bodzin JL, Saltiel AR. Obesity induces a phenotypic switch in adipose tissue macrophage polarization. J Clin Invest. 2007;117(1):175–84.
Ma X, Xu S. TNF inhibitor therapy for rheumatoid arthritis. Biomed Rep. 2013;1(2):177–84.
Magnan A, et al. Alveolar macrophage interleukin (IL)-10 and IL-12 production in atopic asthma. Allergy. 1998;53(11):1092–5.
Mantovani A. In vitro and in vivo cytotoxicity of adriamycin and daunomycin for murine macrophages. Cancer Res. 1977;37(3):815–20.
Mantovani A, et al. Role of host defense merchanisms in the antitumor activity of adriamycin and daunomycin in mice. J Natl Cancer Inst. 1979;63(1):61–6.
Maphis N, et al. Reactive microglia drive tau pathology and contribute to the spreading of pathological tau in the brain. Brain. 2015;138(Pt 6):1738–55.
Martinez FO, Gordon S. The M1 and M2 paradigm of macrophage activation: time for reassessment. F1000Prime Rep. 2014;6:13.
Maus UA, et al. Role of resident alveolar macrophages in leukocyte traffic into the alveolar air space of intact mice. Am J Physiol Lung Cell Mol Physiol. 2002;282(6):L1245–52.
McInnes IB, O'Dell JR. State-of-the-art: rheumatoid arthritis. Ann Rheum Dis. 2010;69(11):1898–906.
McNelis JC, Olefsky JM. Macrophages, immunity, and metabolic disease. Immunity. 2014;41(1):36–48.
Mills CD. M1 and M2 macrophages: oracles of health and disease. Crit Rev Immunol. 2012;32(6):463–88.
Monack DM, Mueller A, Falkow S. Persistent bacterial infections: the interface of the pathogen and the host immune system. Nat Rev Microbiol. 2004;2(9):747–65.
Mosser DM, Edwards JP. Exploring the full spectrum of macrophage activation. Nat Rev Immunol. 2008;8(12):958–69.
Movahedi K, et al. Different tumor microenvironments contain functionally distinct subsets of macrophages derived from Ly6C(high) monocytes. Cancer Res. 2010;70(14):5728–39.
Murdoch C, Giannoudis A, Lewis CE. Mechanisms regulating the recruitment of macrophages into hypoxic areas of tumors and other ischemic tissues. Blood. 2004;104(8):2224–34.
Nakashima T, Takayanagi H. The dynamic interplay between osteoclasts and the immune system. Arch Biochem Biophys. 2008;473(2):166–71.
Nilson AN, et al. Tau oligomers associate with inflammation in the brain and retina of Tauopathy mice and in neurodegenerative diseases. J Alzheimers Dis. 2017;55(3):1083–99.
Odegaard JI, et al. Macrophage-specific PPARgamma controls alternative activation and improves insulin resistance. Nature. 2007;447(7148):1116–20.
Olefsky JM, Glass CK. Macrophages, inflammation, and insulin resistance. Annu Rev Physiol. 2010;72:219–46.

Opperman KS, et al. Clodronate-liposome mediated macrophage depletion abrogates multiple myeloma tumor establishment in vivo. Neoplasia. 2019;21(8):777–87.
Ostuni R, et al. Macrophages and cancer: from mechanisms to therapeutic implications. Trends Immunol. 2015;36(4):229–39.
Pantaleo A, et al. Naturally occurring anti-band 3 antibodies and red blood cell removal under physiological and pathological conditions. Autoimmun Rev. 2008;7(6):457–62.
Paolicelli RC, et al. Synaptic pruning by microglia is necessary for normal brain development. Science. 2011;333(6048):1456–8.
Pascual-Garcia M, et al. LIF regulates CXCL9 in tumor-associated macrophages and prevents CD8(+) T cell tumor-infiltration impairing anti-PD1 therapy. Nat Commun. 2019;10(1):2416.
Prior S, et al. In vitro phagocytosis and monocyte-macrophage activation with poly(lactide) and poly(lactide-co-glycolide) microspheres. Eur J Pharm Sci. 2002;15(2):197–207.
Pyonteck SM, et al. CSF-1R inhibition alters macrophage polarization and blocks glioma progression. Nat Med. 2013;19(10):1264–72.
Qian BZ, Pollard JW. Macrophage diversity enhances tumor progression and metastasis. Cell. 2010;141(1):39–51.
Qian BZ, et al. CCL2 recruits inflammatory monocytes to facilitate breast-tumor metastasis. Nature. 2011;475(7355):222–5.
Rhodes MM, et al. Adherence to macrophages in erythroblastic islands enhances erythroblast proliferation and increases erythrocyte production by a different mechanism than erythropoietin. Blood. 2008;111(3):1700–8.
Rollett A, et al. HSA nanocapsules functionalized with monoclonal antibodies for targeted drug delivery. Int J Pharm. 2013;458(1):1–8.
Sack U, Stiehl P, Geiler G. Distribution of macrophages in rheumatoid synovial membrane and its association with basic activity. Rheumatol Int. 1994;13(5):181–6.
Samuel VT, Shulman GI. Mechanisms for insulin resistance: common threads and missing links. Cell. 2012;148(5):852–71.
Scott DL, Wolfe F, Huizinga TW. Rheumatoid arthritis. Lancet. 2010;376(9746):1094–108.
Shand FH, et al. Tracking of intertissue migration reveals the origins of tumor-infiltrating monocytes. Proc Natl Acad Sci U S A. 2014;111(21):7771–6.
Sheridan GK, Murphy KJ. Neuron-glia crosstalk in health and disease: fractalkine and CX3CR1 take Centre stage. Open Biol. 2013;3(12):130181.
Sun W, et al. Targeting notch-activated M1 macrophages attenuates joint tissue damage in a mouse model of inflammatory arthritis. J Bone Miner Res. 2017;32(7):1469–80.
Szekanecz Z, Koch AE. Macrophages and their products in rheumatoid arthritis. Curr Opin Rheumatol. 2007;19(3):289–95.
Takayanagi H. Osteoimmunology: shared mechanisms and crosstalk between the immune and bone systems. Nat Rev Immunol. 2007;7(4):292–304.
Torocsik D, et al. Identification of factor XIII-A as a marker of alternative macrophage activation. Cell Mol Life Sci. 2005;62(18):2132–9.
Torsteinsdottir I, et al. Monocyte activation in rheumatoid arthritis (RA): increased integrin, fc gamma and complement receptor expression and the effect of glucocorticoids. Clin Exp Immunol. 1999;115(3):554–60.
Van Lent PL, et al. Local removal of phagocytic synovial lining cells by clodronate-liposomes decreases cartilage destruction during collagen type II arthritis. Ann Rheum Dis. 1998;57(7):408–13.
Verdeguer F, Aouadi M. Macrophage heterogeneity and energy metabolism. Exp Cell Res. 2017;360(1):35–40.
Vonderheide RH, Glennie MJ. Agonistic CD40 antibodies and cancer therapy. Clin Cancer Res. 2013;19(5):1035–43.
Vyas SP, et al. Design of liposomal aerosols for improved delivery of rifampicin to alveolar macrophages. Int J Pharm. 2004;269(1):37–49.
Wang Q, et al. Fra-1 protooncogene regulates IL-6 expression in macrophages and promotes the generation of M2d macrophages. Cell Res. 2010;20(6):701–12.
Wu D, et al. Eosinophils sustain adipose alternatively activated macrophages associated with glucose homeostasis. Science. 2011;332(6026):243–7.
Wynn TA, Chawla A, Pollard JW. Macrophage biology in development, homeostasis and disease. Nature. 2013;496(7446):445–55.
Yamamoto M, et al. Role of adaptor TRIF in the MyD88-independent toll-like receptor signaling pathway. Science. 2003;301(5633):640–3.
Yuan P, et al. TREM2 Haplodeficiency in mice and humans impairs the microglia barrier function leading to decreased amyloid compaction and severe axonal dystrophy. Neuron. 2016;92(1):252–64.
Zermati Y, et al. Transforming growth factor inhibits erythropoiesis by blocking proliferation and accelerating differentiation of erythroid progenitors. Exp Hematol. 2000;28(8):885–94.
Zhang X, Mosser DM. Macrophage activation by endogenous danger signals. J Pathol. 2008;214(2):161–78.
Zhang W, et al. Advances in anti-tumor treatments targeting the CD47/SIRPalpha Axis. Front Immunol. 2020;11:18.
Zhukova OV, et al. Tumor-associated macrophages: role in the pathological process of tumorigenesis and prospective therapeutic use (review). Biomed Rep. 2020;13(5):47.

Mechanisms and Ways of Macrophage Delivery

Ashley Oake, Swati Gupta, and Yashwant V. Pathak

Abstract

The innate immune system consists of macrophages that maintain homeostasis by phagocytosis of foreign bacteria and viruses. Macrophages can be polarized toward either M1, which promotes inflammation, or M2, which inhibits inflammation through anti-inflammatory cytokines. Phagocytosis is the capability to destroy foreign pathogens by cellular degradation through lysosomes. A phagosome is formed around the target substance and releases components to initiate breakdown. The two types of receptors used are non-opsonic and opsonic that bind to the ligand initiating a signal cascade to initiate phagocytosis. A phagocytic cup is formed to engulf the foreign substance, and the actin/microtubule cytoskeleton is reorganized with myosin filaments to close and mature the phagosome. Macrophages may also act as antigen-presenting cells to signal memory T cells and other areas of the adaptive immunity for activation. Currently, the ability to use macrophages in antitumor and other drug delivery methods is being widely studied due to their mobility and ability to store organic material.

Keywords

Macrophages · Drug delivery · Phagocytosis · Immune system · Macrophage polarization

1 Introduction

Initially discovered by Elie Metchnikoff in 1893, macrophages are specialized white blood cells that regulate homeostatic maintenance through the phagocytosis and destruction of foreign bacteria, microbes, and debris invading the body (Ignacio Saldana n.d.; Haldar and Murphy 2014). The innate immune system, the initial defense against pathogens, includes neutrophils, macrophages, and dendritic cells originating from myeloid cells and can be followed up by the adaptive immune system for a more specific approach, which includes T cells and B cells (Hirayama et al. 2018). As monocytes migrate through the body, they differentiate into various macrophages as they enter specific tissues that determine their structure, pathogens they target,

A. Oake
Morsani College of Medicine, Tampa, FL, USA

S. Gupta
Department of Pharmaceutics, Amity Institute of Pharmacy, Amity University Uttar Pradesh, Noida, India

Y. V. Pathak (✉)
University of South Florida, Taneja College of Pharmacy, Tampa, FL, USA

Adjunct Professor University of Airlangga, Surabaya, Indonesia
e-mail: ypathak1@usf.edu

S. Gupta, Y. V. Pathak (eds.), *Macrophage Targeted Delivery Systems*,
https://doi.org/10.1007/978-3-030-84164-5_3

and the inflammatory cytokines released in order to initiate destruction (Ignacio Saldana n.d.). These monocytes are derived from either hematopoietic stem cells originating from the bone marrow or the embryonic yolk sacs (YS) and fetal liver (FL), which are maintained by self-renewal (Hirayama et al. 2018).

This self-renewal process includes proliferation and development of identical daughter cells in homeostatic conditions that are regulated by growth factors and cytokines including macrophage colony-stimulating factor (M-CSF) (Hirayama et al. 2018). The presence of Ly6C+ monocytes gives rise to tissue macrophages and assists in injury, infection, the gut, skin, and uterus (Italiani and Boraschi 2014). A reduction in CSF-1R leads to a rise in the Ly6C+ monocytes indicating that they help with monocyte maturation (Italiani and Boraschi 2014). The role that Ly6C- monocytes play throughout the body is currently unknown, but studies show that an increase in CSF-1 leads to an increase in Ly6C- (Italiani and Boraschi 2014). The rate of proliferation will increase when going to pro-inflammatory conditions from a homeostatic state or when there is a sudden reduction in macrophages (Italiani and Boraschi 2014).

Embryonic hematopoiesis can be considered primitive (myb-independent) while not involving monocytes progenitors or definitive (myb-dependent) which involves hematopoietic progenitors from the YS and hematopoietic stem cells (Haldar and Murphy 2014). Due to the discovery of the erythro-myeloid progenitor originating from the YS and contributing to the FL-derived macrophages, there is information promoting the theory of FL-derived macrophages originating from the YS as well (Haldar and Murphy 2014). Throughout development, the FL becomes the central origin for tissue macrophages but after birth the bone marrow dominates the production with blood monocyte precursors (Haldar and Murphy 2014). Microglia, macrophages in the central nervous system that eliminate dead cells and control the overall immunity, and macrophages throughout the intestinal tract are both originated from the embryonic yolk sac, but the cells in the central nervous system undergo continuous self-renewal as cells in the intestinal tract are replenished with those from the hematopoietic stem cells (Hirayama et al. 2018).

Determined by the microenvironment, macrophages have the ability to be polarized toward classically activated M1 that promote inflammation or alternatively activated M2 that inhibit inflammation (Martinez and Gordon 2014). Interferon-gamma (IFN-γ), a cytokine that is encoded by the IFNG gene and produced by Th1, regulates the activation of M1 macrophages and increases the antiviral of type 1 interferons (Martinez and Gordon 2014). M1 macrophages are associated with a positive immune response and release cytokines to initiate inflammation (Martinez and Gordon 2014). M2 macrophages are activated by interleukin-4 (IL-4) and are responsible for the anti-inflammatory response created throughout the body that allows the proliferation of cells and tissue repair (Martinez and Gordon 2014). Regarding tumor growth or inhibition, M1 macrophages produce tumor necrosis factor (TNF), IL-1, and interferons that inhibit the growth of cells and release lymphocytes to neutralize the cells, while M2 macrophages release tumor growth factor (TGF) that encourages the growth of tumor cells and kills the surrounding cells to increase proliferation (Martinez and Gordon 2014). Resident macrophages are normally polarized toward the M2 activation in their environment (Italiani and Boraschi 2015).

Ralph Van Furth introduced the theory of the Mononuclear Phagocyte System, which describes that all macrophages are derived from blood monocytes and are terminally differentiated, but further research almost 20 years later with Diesselhoff-den Dulk disproved that theory showing that macrophages were persistent in tissue (Italiani and Boraschi 2015). Macrophages are categorized according to tissue location, which include Kupffer cells, splenic microphages, Langerhans cells, osteoclasts, intraglomerular mesangial cells, and sinus histiocytes

(Italiani and Boraschi 2015). All macrophages display similar functions, maintaining homeostasis and developing an inflammatory response, but may acquire additional actions depending on their location. Alveolar macrophages, located in the lungs, assist with inspecting the inhaled air and eliminating pathogens but also are responsible for removing surfactants from the alveolar lining (Italiani and Boraschi 2015). Pulmonary surfactant prevents the alveoli from collapsing by decreasing the surface tension but when not enough is removed; it builds up and causes alveolar proteinosis (Italiani and Boraschi 2015). Microglia are located in the brain to prevent damage of the blood–brain barrier from harmful toxins but also are involved in the development of the brain and synaptic remodeling which can hinder neurodegeneration over time (Italiani and Boraschi 2014). Depending on the location, polarization, and action, macrophages will possess phenotypic markers (Haldar and Murphy 2014). Common M1 macrophage markers are CD80, CD86, CD64, and CD32, and M2 macrophage markers are CD163, CD206, and CD68 (Haldar and Murphy 2014).

When it comes to the bones, osteoclasts are the prominent macrophages due to the breakdown of excess bone and assistance with the production of blood (Italiani and Boraschi 2014). They are developed from a mononuclear phagocyte system, and their differentiation is determined by specific precursors and nuclear factors (Italiani and Boraschi 2014). These cells will dissolve and break down the minerals with the help of hydrogen ions to reabsorb and remodel bone consistently throughout an individual's lifetime (Italiani and Boraschi 2014). As excess bone is broken down, osteoblasts model new bone in its place (Italiani and Boraschi 2014). When osteoclasts are not functional, it can lead to a surplus dense bone, osteopetrosis, or can be over functional and create fragile, brittle bones referred to as osteoporosis (Italiani and Boraschi 2014). Bone marrow is also a location of stem cells that develop into erythrocytes or immune cells (Italiani and Boraschi 2014).

2 Differentiation into Macrophages

Signals specific to each tissue are sent to precursor cells in order to differentiate them into specific macrophages (Haldar and Murphy 2014). Splenic red pulp and bone marrow contain the metabolite heme in erythrocytes that are recycled into iron when damaged with F4/80^{hi}VCAM^{hi} macrophages (Haldar and Murphy 2014). Heme removes the *Bach1* suppressor through a proteasome-dependent mechanism which induces the transcription factor *Spic* differentiating monocytes into F4/80^{hi}VCAM^{hi} macrophages (Haldar and Murphy 2014). The excess increase of macrophages will suppress the transcription factor *Spic* to decrease the amount of heme and maintain equilibrium (Haldar and Murphy 2014). Bone marrow that is lacking *Spic* is unable to produce red pulp macrophages and bone marrow macrophages leading to the inability to process iron in order to maintain homeostasis (Haldar and Murphy 2014). The transcription factor NR4A1 regulates differentiation of Ly6C^{lo} macrophages that will cause the cells in the bone marrow to die if repressed (Haldar and Murphy 2014).

Located in the peritoneal cavity, which is a fluid-filled space protecting the organs located in the abdomen, large peritoneal macrophages express GATA6 that is important for localization and proliferation (Haldar and Murphy 2014). GATA6 is induced by retinoic acid produced by the omentum, which is positioned on the colon, and a shortage results in LPMs being absent in the peritoneal cavity but present in the momentum (Haldar and Murphy 2014). Small peritoneal macrophages take over during inflammation but have a shortage of GATA6 (Haldar and Murphy 2014). Many factors coordinate the activity of osteoclasts in the bone including PU.1, M-CSF, and C-fos which will cause a decrease in activity when disturbed (Haldar and Murphy 2014). In the lungs, the deletion of *Bach2* will disrupt the GM-CSF signaling pathway causing an excess of pulmonary surfactant (Haldar and Murphy 2014).

3 Signaling Pathways

Many factors coordinate the activation of these macrophages, determining their function and polarization state (Cheon et al. 2011). Th1 conducts the production of IFN-γ, a pro-inflammatory cytokine, that binds to receptors activating tyrosine kinases from the Janus kinase family (JAK1 and JAK2) (Cheon et al. 2011). The activation of kinases promotes the phosphorylation of signal transducers and activators of transcription 1 (STAT1) which control fundamental cellular processes (Cheon et al. 2011). To enter the nucleus, STAT1 will form dimers with interferon-stimulated gene factor 3 complex and IRF-9 which will affect viability (Cheon et al. 2011). Interferon regulatory factors (IRFs) regulate transcription of several genes controlling the development of Th cells (Cheon et al. 2011). IFN-γ also activates Major Histocompatibility Complex II, which presents antigens to CD4(+) T lymphocytes (Cheon et al. 2011). Interleukin (IL)-12, a heterodimer, will stimulate the production of IFN-γ as well as regulating the production of T cells, while suppressor of cytokine signaling (SOCS) negatively regulates IFN-γ (Cheon et al. 2011).

IFN-γ combined with lipopolysaccharide (LPS) is responsible for the initiation of classical activation by stimulating toll-like receptors (TLRs), which when activated will lead to two contrasting pathways regulating cytokines and other pro-inflammatory molecules (Cheon et al. 2011). The TIR-domain-containing adapter-inducing interferon-β can be induced causing a production of kinases and interferon-responsive factor 3 (IRF3), which controls the production of type 1 interferon (Cheon et al. 2011). STAT1 will be activated when type 1 interferons are then recognized and bound to interferon α/β receptor (Cheon et al. 2011). Myeloid differentiation primary response 88 (MyD88) pathway activates the nuclear factor kappa-B, involved in M1 polarization, and activator protein 1 (Cheon et al. 2011). M2 macrophage stimuli involve IL-4 and IL-13 instead of IFN-γ to activate STAT6 by binding to the interleukin-4 receptor (Martinez and Gordon 2014). The peroxisome proliferator-activated receptor γ is mobilized by the binding of free fatty acids and assists with regulating the M2 macrophage genes (Martinez and Gordon 2014). The complex between glucocorticoids and their receptor binds to DNA and produces anti-inflammatory IL-10 while communicating with other transcription factors (Martinez and Gordon 2014). To decrease inflammation, STAT3 is activated which promotes the expression of SOCS by the binding of IL-10 to its receptor (Martinez and Gordon 2014).

4 Pathways Affecting the Polarization of Macrophages

The Notch pathway is a highly conserved pathway that is important for development and homeostasis by controlling gene expression (Kopan 2012). There are four Notch receptors in humans designated 1–4, and they involve two types of ligands which are the "Jagged" ligand (JAG) family and DLL family (Kopan 2012). With the assistance of intercellular signaling reactions, the extracellular domain binds to an activated ligand (DLL) that has undergone post-translational modifications on another cell to release the Notch intracellular domain that binds to the CSL DNA-binding protein (Kopan 2012). This activating complex recruits Model-Agnostic Meta-Learning (MAML), which further recruits HAT p300, to target the M1-type genes resulting in Notch1 activation, thereby regulating the immune response (Kopan 2012). Interferon regulatory factor 5 can be associated with the M1 activation of macrophages as well, and nitrification of the protein can suppress the activity of the signaling gene (Kopan 2012).

The JAK-STAT signaling pathway is responsible for inducing the transcription of genes while also mediating signal transduction for cytokines that macrophages release (Rawlings et al. 2004). The receptors involved in this pathway are Gp130, Gp130, IFNR1, and IFNR2, while the different ligands are GM-CSF, EPo, prolactin, thrombopoietin, leptin, interleukins, and interferon (IFN) (Rawlings et al. 2004). The proteins comprising the JAK family are

JAK1, JAK2, JAK3, and TYK2 with two domains each that are the SH2 domain for binding to the phosphotyrosine and the FERM domain (F for 4.1 protein, E for ezrin, R for radixin and M for moesin) for binding to a receptor and other protein interactions (Seif et al. 2017).

As a receptor is bound to a ligand, the JAK tyrosine kinase bound to the receptor initiates an auto-phosphorylation cascade that leads to the stimulation of the Akt pathway and the Ras–Raf pathway (Rawlings et al. 2004). STAT is unphosphorylated when residing in the cytoplasm, but when JAK is phosphorylated it recruits STAT to phosphorylate and activate it enabling the binding of another STAT molecule (Rawlings et al. 2004). STAT can be activated by other kinases as well, and the molecule includes seven encoding genes (1, 2, 3, 4, 5a, 5b, 6) which are all transcription factors (Rawlings et al. 2004). This creates a homodimer or heterodimer where it can enter the nucleus with the help of nucleoprotein interactor through the Ran nuclear import pathway (Rawlings et al. 2004). It binds to gamma-activated sequence elements which coordinate the induction of the transcription of genes controlling viability, survivability, differentiation, and immunity (Rawlings et al. 2004). The ligand IFN-γ and IFN-α/IFN-β-mediated signaling pathways are also participating in the polarization of macrophages and affect inflammatory conditions (Rawlings et al. 2004).

The phosphatidylinositol 3′ –kinase (PI3K) pathway is responsible for regulating the cell cycle and plays a role in the overall polarization and durability of macrophages (Fruman et al. 2017). Like the JAK-STAT pathway, the PI3K pathway involves the activation of a protein tyrosine kinase resulting in an auto-phosphorylation that affects other intracellular pathways (Fruman et al. 2017). Phosphotyrosine residues on growth factor receptors recruit PI3K molecules to allosterically activate the catalytic subunit resulting in phosphatidylinositol-3, 4, 5-trisphosphate (PIP3) (Fruman et al. 2017). Signaling proteins PDK1 and Akt are then recruited with the pleckstrin homology domain of PIP3 (Fruman et al. 2017). PI3K activates Akt1 and Akt2, which mediate cell growth and proliferation, to polarize either M1-type macrophages or M2-type macrophages (Fruman et al. 2017).

Mutations or alterations of this pathway can lead to serious complications including cancer or diabetes (Fruman et al. 2017). This pathway has been targeted in many drug trials for oncology treatments as it is shown to contribute to tumorigenesis (Fruman et al. 2017). As the levels of Akt rise, there is an increase in growth factors that allow the entry of Mdm2 into the nucleus suppressing the effects of anti-proliferation signals and an inhibition of Bad and pro-caspase-9 which are pro-apoptotic factors (Fruman et al. 2017). This will also signal the cytoplasmic localization of p21Cip/Waf1 and p27Kip and the activation of IκB kinase (IKK) increasing proliferation and antiapoptotic factors (Fruman et al. 2017). In the case of diabetes, the PI3K-Akt signaling pathway is activated when insulin released after a meal is bound to the insulin receptor substrate-1/2 (Fruman et al. 2017). Glucose utilization and insulin release will increase as the free fatty acid circulation will decrease as this pathway proceeds (Fruman et al. 2017). In the case of a surplus of free fatty acid, there will be a reduction in glucose transport, disruption of beta-cell function, and increase in hepatic glucose production which will start to inhibit the PI3K-Akt signal. Upon the impairment of this pathway is impaired, it will result in insulin resistance and obesity (Fruman et al. 2017). (Table 1.1)

Table 1.1 Polarization of M1- and M2-activated macrophages

	M1	M2
Activation	Classical	Alternative
Produce cytokines to	Create positive immune response to release pro-inflammatory cytokines	Create an anti-inflammatory response
Activated by	IFN-γ	IL-4
Involved in	Protection against bacteria and viruses	Wound healing and tissue repair

Source: Martinez and Gordon (2014)

5 Inflammatory Reactions

Resident macrophages possess the ability to increase the quantity of effector cells that are required in an inflammatory reaction by not only increasing proliferation through self-renewal but also recruiting blood monocytes through monocyte subset trafficking that give way to inflammatory macrophages (Italiani and Boraschi 2014). Research by Steven Jenkins and colleagues regarding type 2 inflammation studied the capacity of macrophages to undergo in situ proliferation (Jenkins et al. 2011). The recruitment of macrophages by exposure to rodent filarial nematode *Litomosoides sigmodontis* was compared to the response given by an intra-thoracic injection of thioglycollate which gives a known classical inflammation response (Jenkins et al. 2011). Results showed that the thioglycollate recruited a large amount of neutrophils differentiating into F4/80 macrophages, but the *Litomosoides sigmodontis* did not and resulted in a substantial total of F4/80 macrophages similar to the resident macrophages (Jenkins et al. 2011). These macrophages were induced by cytokine IL-4 in the pleural and peritoneal cavities lacking bone marrow progenitors or the need to recruit other inflammatory cells (Jenkins et al. 2011).

Further studies suggested that IL-4 promotes proliferation in cells and a lack of IL-4 prevents cells from M2 polarizing and activating (Jenkins et al. 2011). When mice were injected with recombinant IL-4 complexed to IL-4 antibody, there was a rise of F4/80 macrophages without the recruitment of neutrophils during type 2 inflammation (Jenkins et al. 2011). Classical inflammation, or type 1 inflammation, involves the recruitment of macrophages and usually produces lower levels of F4/80 (Jenkins et al. 2011). This method is significantly more rapid than type 2 inflammation and uses more resources and energy (Jenkins et al. 2011). Type 2 inflammation involves the proliferation of macrophages without the recruitment of additional cells while speeding up the repair time (Jenkins et al. 2011). This inflammation process may be slower, but it avoids potential tissue damage and (Jenkins et al. 2011). M2 macrophages are not restricted to an individual precursor though, so within a Th2 environment, both resident and recruited macrophages have the ability to proliferate and activate (Jenkins et al. 2011).

Through emigration, tissue adherence, clodronate via liposomes, radiation, and cell death, there could be a significant decrease in resident tissue macrophages which need to be recovered (Italiani and Boraschi 2015). The renewal of these cells can be done by bone marrow transplantation which leads to long-term macrophage reconstitution and monocyte recruitment and differentiation (Italiani and Boraschi 2015). Monocytes recruited to the spleen will automatically differentiate to red pulp macrophages in an environment that causes them to deplete to maintain homeostasis in the body (Italiani and Boraschi 2015).

6 Phagocytosis

Macrophages are acknowledged for their phagocytic activity as they have the ability to ingest and destroy bacteria, foreign substances, and apoptotic cells (Uribe-Querol and Rosales 2020). Other cells that can perform this include neutrophils, dendritic cells, and osteoclasts (Uribe-Querol and Rosales 2020). They use specific cell surface receptors to recognize the particle larger than 0.5 μm through physical contact to initiate destruction and alert the body of any similar substances (Uribe-Querol and Rosales 2020). Cytokines will signal these cells to travel to the site of the virus or bacteria through a process called chemotaxis (Uribe-Querol and Rosales 2020). The most common types of receptors are opsonin, scavenger, toll-like, and antibodies which signal pseudopods, which are made of the actin cytoskeleton and lipid membrane, to surround the foreign substance and create a pocket to engulf it (Uribe-Querol and Rosales 2020). This pocket-like space encloses becoming a phagosome which prevents the plasma membrane of the cell from being damaged (Uribe-Querol and Rosales 2020). It fuses with a lysosome to convert into a phagolysosome which then starts the mechanism of breaking down the substance by lowering the pH and creating an acidic internal environment called lysis (Uribe-Querol and Rosales 2020).

Oxygen radicals, nitric oxide, and other antimicrobial proteins can be used to neutralize the virus or bacteria inside the phagolysosome (Uribe-Querol and Rosales 2020). Antimicrobial proteins and peptides attack the cell walls, cell membranes, and proteins of bacteria to destroy it and can include defensins, proteases, and lysozyme (Uribe-Querol and Rosales 2020). Binding proteins will aid in this by binding to specific proteins preventing further replication and essentially causing destruction (Uribe-Querol and Rosales 2020). The cell may also release oxygen radicals or nitric oxide that can either cause oxidative stress or reacting with superoxide to target and destroy specific cell structures (Uribe-Querol and Rosales 2020). Foreign substances then became neutralized waste products which are signaled to be removed from the cell (Uribe-Querol and Rosales 2020).

7 Non-opsonic Receptors

Macrophages utilize a combination of non-opsonic and opsonic receptors bound to the plasma membrane to detect foreign substances throughout the body (Uribe-Querol and Rosales 2020). Scavenger receptors, C-type lectins, and lectin-like recognition molecules make up the non-opsonic receptors that recognize patterns presented by foreign molecules (Uribe-Querol and Rosales 2020). C-type lectin receptors utilize carbohydrate recognition domains to bind carbohydrate ligands which they have a very high affinity for (Bermejo-Jambrina et al. 2018). Dectin-1 is a C-type lectin receptor that participates with other phagocytic receptors to recognize yeast polysaccharides and fungal beta-glucan (Bermejo-Jambrina et al. 2018). Trehalose dimycolate (TDM), also called cord factor, is glycolipid in mycobacterium targeted by macrophage-inducible C-type lectin (Mincle) and macrophage C-type lectin, Dectin-3 (MCL) receptors that form heterodimers to enhance their activity (Bermejo-Jambrina et al. 2018). MCL also recognizes α-mannans, which is another component found in the cell wall, but in yeasts instead of bacteria (Bermejo-Jambrina et al. 2018). Dendritic cell-specific ICAM-3-grabbing non-integrin (DC-SIGN) receptors bind to microbial pathogens through the identification of mannose and fructose glycans (Bermejo-Jambrina et al. 2018). This receptor interacts with gp120 in phagocytic cells to internalize the HIV-1 virus and deliver it to lysosomes for degradation, but studies indicate that it may possess the capability to avoid this pathway and reroute for transmission (Bermejo-Jambrina et al. 2018). The Mannose Receptor (MR) recognizes and internalizes components bound to the surfaces of viruses and bacteria including mannose, N-acetylglucosamine, and fucose (Bermejo-Jambrina et al. 2018). Like DC-SIGN, MR is a receptor that may allow HIV and dengue fever virus (DENV) to avoid internalization, allowing it to transmit and affect other cells (Bermejo-Jambrina et al. 2018). Other receptors that aid in the phagocytic process, but will not initiate the action on their own, are MARCO, which recognizes bacteria, CD14 and SR-A for lipopolysaccharide-binding protein on gram-negative bacteria and toll-like receptors (Bermejo-Jambrina et al. 2018).

Innate immune cells contain specific pattern recognition receptors (PRRs) that recognize pathogen-associated molecular patterns (PAMPs) generated by stressors and pathogens to initiate the correct response (Hirayama et al. 2018). Due to research with the *Drosophila* fruit fly, toll-like receptors (TLRs) were discovered and said to engage in detecting specific components including lipopeptides and nucleic acids in bacteria and viruses (Hirayama et al. 2018). Toll-like receptors can recognize patterns and cooperate with non-opsonic receptors but are not considered phagocytic as they prime the cell for phagocytosis instead of initiating (Uribe-Querol and Rosales 2020). Adapter molecules such as toll/interleukin-1 receptor (TIR) will send a signal to transcription factors once these components are recognized to induce gene expression (Weiss and Schaible 2015). Toll-like receptors are found on the cell membrane, but there are other PRRs that found within the cytoplasm (Weiss and Schaible

2015). The inflammasome is responsible for maturing pro-interleukin (IL)-1β to IL-1β and is composed of NOD-like receptors (NLRs), ASC, and caspase-1 (Weiss and Schaible 2015). These respond to stress in the body by removing infection and activating cell death to induce inflammation (Weiss and Schaible 2015). RIG-1-like receptors (RLRs) pair with an interferon-β promoter stimulator to signal an antiviral immune reaction by producing type I interferons and inflammatory cytokines when exposed to viral RNA (Weiss and Schaible 2015). Scavenger receptors identify the extracellular matrix of bacteria by binding to the proteins distinct to that bacteria and not found in eukaryotic cells (Uribe-Querol and Rosales 2020).

8 Opsonic Receptors

Immune cells produce antibodies that help identify substances in the body that need to be removed, and opsonic receptors will recognize substances bound to immunoglobulin G antibodies to mark them for destruction (Uribe-Querol and Rosales 2020). Fcγ receptors, classified into FcγRI, FcγRII, and FcγRIII, are responsible for antigen recognition and bind to the Fc portion of the immunoglobulin G antibody (Uribe-Querol and Rosales 2020). The more Ig-like domains the receptor contains, the higher its affinity for the immunoglobins (IG) molecules is (Uribe-Querol and Rosales 2020). FcγRI has three domains, while FcγRIII has only two permitting it to bind with multimeric immune complexes only (Uribe-Querol and Rosales 2020). To generate a signal for phagocytosis, an antigen–antibody complex is formed and the FcγRI is bound to FcRγ chain associated with tyrosine residues (Uribe-Querol and Rosales 2020). The increase of FcRs becoming stimulated in a close proximity causes the phosphorylation of tyrosine in the cytoplasm containing the immunoreceptor tyrosine-based activation motif (ITAM) (Uribe-Querol and Rosales 2020). FcγRII is only present in the FcγRIIa form in phagocytic cells and has an ITAM, but does not contain FcRγ chain (Uribe-Querol and Rosales 2020). The second form FcγRIIb is found in B lymphocytes and is responsible for negative regulation of phagocytosis (Uribe-Querol and Rosales 2020). FcγRIIIa is the only isoform found in macrophages containing FcRγ chains, while the second isoform FcγRIIIb is found in neutrophils (Uribe-Querol and Rosales 2020).

Like Fcγ receptors, complement receptors are divided into three groups, but the two are located on the surface of macrophages CR1 and CR3 (Law 1988). CR1 is formed by short consensus repeat elements, which are about 60–65 amino acids each, and recognizes C3b, C4b, C1q, and mannan-binding lectin (Law 1988). CR3 is part of the β2 integral family and contains both an alpha and beta subunit that helps bind to iC3b, which is the most efficient of the receptors (Law 1988). Receptors possess the capability to cooperate with each other to enhance their ability (Uribe-Querol and Rosales 2020). Due to the large size of most receptors, there is difficulty with free diffusion and movement across the membrane, even though phagocytic receptors are smaller than other transmembrane glycoproteins (Uribe-Querol and Rosales 2020). Integrins have the ability to remove larger proteins hinder these receptors from moving so they may interact with IgG molecules to initiate phagocytic activity (Uribe-Querol and Rosales 2020). They can be activated by an inside-out signal from phagocytic receptors leading to a production of the diffusion barrier (Uribe-Querol and Rosales 2020). This can also cause a movement of multiple molecules creating micro-clusters to increase the adhesion among the receptor and the foreign substance (Uribe-Querol and Rosales 2020).

9 Apoptotic Cells

In order to maintain homeostasis and remodel tissues, cells are eliminated over time through a process called apoptosis and are recognized by molecules only presented on the membrane when undergoing cell death (Gordon and Pluddemann 2018). During apoptosis, lipids in the plasma membrane bilayer are reorganized and phosphatidylserine ligands will appear on the membrane surface to signal receptors including TIM-1 and

TIM-4, stabilin-2, and BAI-1 for phagocytosis (Gordon and Pluddemann 2018). To prevent other cells that contain phosphatidylserine on their membrane from being destroyed, CD47 is released and recognized by a signal-regulatory protein α receptor that alerts the cells not to destroy them (Gordon and Pluddemann 2018). When activated, B and T lymphocytes contain this ligand (Gordon and Pluddemann 2018). Other receptors that recognized apoptosis in cells are lactadherin and αvβ3 that bind with MFG-E8, CD36 that bind with oxidized lipids, CD14 that may recognize phosphatidylserine as well, and αvβ5 that recognizes other components in apoptotic cells (Gordon and Pluddemann 2018). (Table 1.2)

Table 1.2 Phagocytic receptors and ligands

Receptors	Ligands	References
Non-opsonic receptors		
Dectin-1	Fungal beta-glucan and yeast polysaccharides	Bermejo-Jambrina et al. (2018)
DC-SIGN	Mannose and fructose glycans	Bermejo-Jambrina et al. (2018)
Mincle	Trehalose dimycolate (TDM)	Bermejo-Jambrina et al. (2018)
MCL	Trehalose dimycolate (TDM) and α-mannans	Bermejo-Jambrina et al. (2018)
Mannose receptors	Mannose, N-acetylglucosamine, and fucose	Bermejo-Jambrina et al. (2018)
Opsonic receptors		
FcγRI	IgG1, IgG3, IgG4	Uribe-Querol and Rosales (2020)
FcγRII	IgG3, IgG1, IgG	Uribe-Querol and Rosales (2020)
FcγRIII	IgG	Uribe-Querol and Rosales (2020)
CR1	C3b, C4b, iC3b	Law (1988)
CR2	C3d, C3dg, iC3b	Law (1988)
CR3	iC3b, C3dg, C3d, ICAM-1, ICAM-2, LPS, fibronectin	Law (1988)
CR4	iC3b, C3dg, C3d	Law (1988)
Apoptotic cells		
TIM-1	Phosphatidylserine	Gordon and Pluddemann (2018)
TIM-4	Phosphatidylserine	Gordon and Pluddemann (2018)
Stabilin-2	Phosphatidylserine	Gordon and Pluddemann (2018)
BAI-1	Phosphatidylserine	Gordon and Pluddemann (2018)
Integrin αvβ3	Phosphatidylserine	Gordon and Pluddemann (2018)
CD14	Phosphatidylserine and LPS-binding protein	Gordon and Pluddemann (2018)

10 Phagosome Signaling

When the receptor and ligand are bound by antigen-presenting cells, phagocytosis is initiated by activating many different signaling pathways to maintain homeostasis in the body (Lee et al. 2020). A phagosome starts to form by reorganizing the actin cytoskeleton and lipid composition as it engulfs the foreign body and encloses it inside for destruction (Lee et al. 2020). The opsonic receptors are well studied in these pathways, which include the signaling from Fcγ receptors and complement receptors (Lee et al. 2020). Fcγ receptors are involved in the phosphorylation of linker for activation of T cells by spleen tyrosine kinase leading to recruitment of critical adaptor molecules including Grb2-associated binder 2 that helps regulate growth and differentiation (Gu et al. 2003). Without this scaffolding molecule, there is a significant decrease in the activation of Akt suggesting that it plays a role in increasing PI3K signaling (Gu et al. 2003).

Phosphatidylinositol 3-kinase (PI3K) plays a role in promoting pseudopod extension and phosphorylates phosphatidylinositol-4,5-bisphosphate to lipid phosphatidylinositol-3,4,5-trisphosphate (PIP3) which regulated contractile proteins like myosin and actin (Gu et al. 2003). The GTPase Rac is critical in pseudopod extension as well as it controls actin polymerization and causes the activation of transcription factors such as NF-κB and JNK (Gu et al. 2003). It is regulated by PIP3 and the Guanine nucleotide exchange factor Vav (Gu et al. 2003). The Fcγ receptors are also involved in the activation of phospholipase Cγ (PLCγ), which releases second messengers to activate protein kinase C and extracellular signal-regulated kinases (Gu et al. 2003). One of the second messengers is inositol triphosphate (IP3), which increases the amount of Ca2+ in the cytosol, regulates the reformation of actin, and activates cytokines to control inflammation and the immune system (Lee et al. 2020).

Pseudopodia, the finger-like protrusions formed around the virus to create the phagosome, are formed by the polymerization and depolymerization of F-actin (Lee et al. 2020). Phosphoinositides control the activity of debranching proteins like coronins that deconstruct the actin filaments (Yan et al. 2005). Although there are six isoforms found in humans, coronin-1 is used in pseudopodia formation with two actin-binding sites (Yan et al. 2005). After the action is debranched, cofilin and gelsolin further break down the filaments, while the Arp2/3 protein complex controls the nucleation of new filaments (Lee et al. 2020). In Fcγ-mediated, WASP and N-WASP control the activation of this complex, while in CR3-mediated phagocytosis it is controlled by Rho-kinase and myosin II (Lee et al. 2020). As actin is removed from the phagocytic cup, the pseudopodia engulf the particle and meet at the distal end to enclose it (Lee et al. 2020). PI 3-K produces PI(3,4,5)P3 which stimulates GTPase-activating proteins to stop the polymerization of actin (Lee et al. 2020). When PI(4,5)P2 is removed from the phagocytic cup, it enables the extension of pseudopods for closure (Lee et al. 2020).

The activity of different myosin proteins increases at different stages in the phagosome formation (Swanson et al. 1999). As the phagocytic cup is formed, class II and IXb myosins are present before the phagosome starts to form around the foreign substance as they help signal the reorganization of the cytoskeleton (Swanson et al. 1999). The extension of the pseudopods is initiated, and just as the phagosome starts to enclose myosin, Ic is present around the distal end as well as a ring of actin that is initiated by myosin light-chain kinase (Swanson et al. 1999). Myosin X is also responsible for the extension and spreading of pseudopodia (Swanson et al. 1999). When the phagosome is fully formed and closed, myosin V appears and is involved in the movement of the phagosome (Swanson et al. 1999).

CR3-mediated phagocytosis may be different than Fcγ-mediated as it does not involve the use of pseudopods but still creates a phagosome with microtubule cytoskeletons and not just actin (Dustin 2016). Instead of depending on activation of GTPase Rac, it involves the activation of GTPase Rho, leading to the polymerization of F-actin, that stimulates Rho-kinase-activating myosin II (Dustin 2016). Actin is assembled in the phagocytic cup for the production of the lipid phosphatidylinositol-3,4,5-trisphosphate after the Arp2/3 complex is activated by myosin II (Dustin 2016). Mammalian diaphanous-related formin 1 binds to CLIP-170 and stimulates linear actin polymerization which creates a link to the microtubule skeleton (Dustin 2016). Spleen tyrosine kinase is found in both Fcγ-mediated and CR3-mediated phagocytosis but appears much earlier in the CR3-mediated pathway and is necessary for the completion (Dustin 2016).

11 Phagosome Maturation

The development from a phagosome to a phagolysosome is referred to as phagosome maturation and involves separation from the membrane (Weiss and Schaible 2015). Maturation of the phagosome involves three stages: early phagosome, late phagosome, and phagolysosome (Pauwels et al. 2017). GTPase Rab5 initiates membrane fusion with the help of early endosome antigen 1 to add sorting enzymes (Pauwels et al. 2017). At this point, the environment is only mildly acidic and Rab7 is recruited for the transformation to late phagosome (Pauwels et al. 2017). During late phagosome, V-ATPase molecules move protons into the phagosomal lumen to create an acidic interior, and lysosomal-associated membrane proteins are recruited for the next stage (Pauwels et al. 2017). The phagolysosome is created by the fusion of the late phagosome and lysosomes to degrade microorganisms (Pauwels et al. 2017). Once fully converted, reactive oxygen species is produced by NADPH oxidase complex into the phagolysosome creating H2O2 (Pauwels et al. 2017). This will react with the Cl- and with the assistance of myeloperoxidase, hypochlorous acid is created to destroy microorganisms (Pauwels et al. 2017). The acidic environment also contains antimicrobial peptides, glycosidases, proteases, lysozymes, and lipases to assist in the elimination of foreign bacteria and viruses (Pauwels et al. 2017).

After foreign substances are engulfed and neutralized, antigen-presenting cells deliver the antigens to T cells to create a specific cellular immunity allowing the specific T cells to recognize this foreign substance in the future (Hughes et al. 2016). Macrophages can be involved in the activation of memory CD8+ T lymphocytes and when exposed to specific proteins, macrophages have the ability to produce inflammatory cytokines (Hughes et al. 2016). The neutralized pathogens are then released into the extracellular matrix by the ER–Golgi secretory pathway or lysosome exocytosis and recycled (Hughes et al. 2016)

12 Drug Delivery

The use of macrophages and other nanoparticles have been recently studied in regards to drug delivery due to their mobility (Visser et al. 2019). Antimicrobials may be attached to these macrophages, and when immune cells target a tissue or pathogen, the drugs will be delivered more efficiently (Visser et al. 2019). These methods are currently being researched in hopes of this method requiring less of the drug itself and avoiding adverse effects, but the risk of lysosomal degradation may hinder this method from being successful (Visser et al. 2019). Microparticle encapsulation has been suggested to enhance the protection and maintenance of the drug over time (Visser et al. 2019). A study regarding the use of acetylated β-cyclodextrin nanoparticles, which are pH sensitive, with microbubbles to help with the delivery of antitumor drugs was performed (Lv et al. 2016). The results indicated that the nanoparticles encapsulated in the microbubbles were more effective in suppressing the tumor compared to other methods indicating that the encapsulation may have prevented the pH sensitivity from degrading the drug (Lv et al. 2016).

13 Conclusion

The innate immune system consists of neutrophils, macrophages, and dendritic cells that are the first line of defense to maintain homeostasis by phagocytosis of foreign bacteria and viruses. Initially originating from either hematopoietic stem cells, embryonic yolk sacs (YS), or fetal liver (FL), monocytes differentiate into macrophages as they enter specific tissues that determine their structure, pathogens they target, and the inflammatory cytokines and nitric oxide release. Macrophages can be polarized toward either M1 that promote inflammation through a positive immune response and are activated by interferon-gamma (IFN-γ) or M2 that inhibit inflammation through anti-inflammatory effects and are activated by interleukin-4 (IL-4). Many different factors and pathways can affect the polarization of these macrophages including the Notch pathway, the JAK-STAT signaling pathway, and the phosphatidylinositol 3′-kinase (PI3K) pathway.

Phagocytosis is the ability to destroy bacteria or other foreign substances by cellular digestion and neutralization through lysosomes. A phagosome is formed around the target substance to engulfing and enclosing it to protect the cell membrane while releasing proteins and oxidative species to initiate breakdown. The two types of receptors used are non-opsonic, which include scavenger receptors, C-type lectins, and lectin-like recognition molecules, and opsonic, which include Fcγ receptors and complement receptors. These receptors also have the ability to cooperate and enhance their action to speed up the process. Apoptotic cells are also ingested by phagocytic cells once they undergo cell death for removal from the body.

Once the receptor is bound to the ligand, there is a signal cascade to initiate phagocytosis. The two common pathways are Fcγ-mediated, involving pseudopodia, and CR3-mediated, involving a more sinking-in method. A phagocytic cup is formed to engulf the foreign substance, and the actin/microtubule cytoskeleton is reorganized to create pseudopodia or allow the membrane to start surrounding the particle. Myosin filaments will help initiate the extension and the closer of the phagosome. Phagosome maturation is the process of converting a phagosome to a phagolysosome to destroy and neutralize viruses and bacteria. Macrophages may also act as antigen-presenting cells to activate the adaptive immune system to recognize similar foreign pathogens in the future.

14 Future Trends

Multiple organizations have focused their research on the development of a nanoparticle-based drug delivery system with macrophages to treat tumors or Parkinson's (Nguyen et al. 2020). These macrophages may have the ability to deliver therapeutic treatment through phagocytosis targeting the inside of tumors for more effective treatment (Nguyen et al. 2020). An experiment conducted by V.D. Nguyen and colleges studied the use of small-sized gold nanorods and anticancer-containing nanoliposomes to treat solid tumors (Nguyen et al. 2020). The use of macrophages was shown to increase tumor penetration when paired with other therapy and may effectively target these tumors with more efficiency (Nguyen et al. 2020). Other experiments have been focused around the connection between polarization of macrophages and the different pathways associated with phagosome maturation (Pauwels et al. 2017). These can be influenced by cytokines, PAMPs, exposed phagocytes, and duration of exposure to many factors (Pauwels et al. 2017). Signaling pathways for phagosome development, maturation, and destruction of pathogens still need to be studied further to be fully understood (Pauwels et al. 2017).

References

Bermejo-Jambrina M, Eder J, Helgers LC, et al. C-type lectin receptors in antiviral immunity and viral escape. Front Immunol. 2018;9(590) https://doi.org/10.3389/fimmu.2018.00590.

Cheon H, Yang J, Stark GR. The functions of signal transducers and activators of transcriptions 1 and 3 as cytokine-inducible proteins. J Interf Cytokine Res. 2011;31(1):33–40. https://doi.org/10.1089/jir.2010.0100.

Dustin ML. Complement receptors in myeloid cell adhesion and phagocytosis. Microbiol Spectr. 2016;4(6) https://doi.org/10.1128/microbiolspec.MCHD-0034-2016.

Fruman DA, Chiu H, Hopkins BD, Bagrodia S, Cantley LC, Abraham RT. The PI3K pathway in human disease. Cell. 2017;170(4):605–35. https://doi.org/10.1016/j.cell.2017.07.029.

Gordon S, Pluddemann A. Macrophage clearance of apoptotic cells: a critical assessment. Front Immunol. 2018;9(127) https://doi.org/10.3389/fimmu.2018.00127.

Gu H, Botelho RJ, Yu M, Grinstein S, Neel BG. Critical role for scaffolding adapter Gab2 in fc gamma R-mediated phagocytosis. J Cell Biol. 2003;161:1151–61. https://doi.org/10.1083/jcb.200212158.

Haldar M, Murphy KM. Origin, development, and homeostasis of tissue-resident macrophages. Immunol Rev. 2014;262(1):25–36. https://doi.org/10.1111/imr.12215.

Hirayama D, Iida T, Nakase H. The phagocytic function of macrophage-enforcing innate immunity and tissue homeostasis. Int J Mol Sci. 2018;19(1):92. https://doi.org/10.3390/ijms19010092.

Hughes CE, Benson RA, Bedaj M, Maffia P. Antigen-presenting cells and antigen presentation in tertiary lymphoid organs. Front Immunol. 2016;7:481. https://doi.org/10.3389/fimmu.2016.00481.

Ignacio Saldana J. Macrophages. British Society for Immunology; (n.d.)

Italiani P, Boraschi D. From monocytes to M1/M2 macrophages: phenotypical vs. functional differentiation. Front Immunol. 2014;5:514. https://doi.org/10.3389/fimmu.2014.00514.

Italiani P, Boraschi D. New insights into tissue macrophages: from their origin to the development of memory. Immune Network. 2015;15(4):167–76. https://doi.org/10.4110/in.2015.15.4.167.

Jenkins SJ, Ruckerl D, Cook PC, et al. Local macrophage proliferation, rather than recruitment from the blood, is a signature of TH2 inflammation. Science. 2011;332(6035):1284–8. https://doi.org/10.1126/science.1204351.

Kopan R. Notch signaling. Cold Spring Harb Perspect Biol. 2012;4(10):a011213. https://doi.org/10.1101/cshperspect.a011213.

Law SK. C3 receptors on macrophages. J Cell Sci. 1988;9:67–97. https://doi.org/10.1242/jcs.1988.supplement_9.4.

Lee HJ, Woo Y, Hahn TW, Jung YM, Jung YJ. Formation and maturation of the phagosome: a key mechanism in innate immunity against intracellular bacterial infection. Microorganisms. 2020;8(9):1298. https://doi.org/10.3390/microorganisms8091298.

Lv Y, Hao L, Hu W, et al. Novel multifunctional pH-sensitive nanoparticles loaded into microbubbles as drug delivery vehicles for enhanced tumor targeting. Scientific Report. 2016;6(29321) https://doi.org/10.1038/srep29321.

Martinez F, Gordon S. The M1 and M2 paradigm of macrophage activation: time for reassessment. F1000 Prime Reports. 2014;6:13. https://doi.org/10.12703/P6-13.

Nguyen VD, Min HK, Kim DH, Kim CS, Han J, Park JO, Choi E. Macrophage-mediated delivery of multifunctional nanotherapeutics for synergistic chemo-photothermal therapy of solid tumors. ACS Appl Mater Interface. 2020;12(9):10130–41. https://doi.org/10.1021/acsami.9b23632.

Pauwels AM, Trost M, Beyaert R, Hoffmann E. Patterns, receptors, and signals: regulation of phagosome maturation. Trends Immunol. 2017;38(6):407–22. https://doi.org/10.1016/j.it.2017.03.006.

Rawlings JS, Rosler KM, Harrison DA. The JAK/STAT signaling pathway. J Cell Sci. 2004;117:1281–3. https://doi.org/10.1242/jcs.00963.

Seif F, Khoshmirsafa M, Aazami H, et al. The role of JAK-STAT signaling pathway and its regulators in the fate of T helper cells. Cell Commun Signal. 2017:15(23). https://doi.org/10.1186/s12964-017-0177-y.

Swanson JA, Johnson MT, Beningo K, Post P, Mooseker M, Araki N. A contractile activity that closes phagosomes in macrophages. J Cell Sci. 1999;112:307–16.

Uribe-Querol E, Rosales C. Phagocytosis: our current understanding of a universal biological process. Front Immunol. 2020;11 https://doi.org/10.3389/fimmu.2020.01066.

Visser JG, Van Staden ADP, Smith C. Harnessing macrophages for controlled-release drug delivery: lessons from microbes. Front Pharmacol. 2019;10:22. https://doi.org/10.3389/fphar.2019.00022.

Weiss G, Schaible UE. Macrophage defense mechanisms against intracellular bacteria. Immunol Rev. 2015;264(1):182–203. https://doi.org/10.1111/imr.12266.

Yan M, Collins RF, Grinstein S, Trimble WS. Coronin-1 function is required for phagosome formation. Mol Biol Cell. 2005;16(7) https://doi.org/10.1091/mbc.e04-11-0989.

Macrophage-Associated Disorders: Pathophysiology, Treatment Challenges, and Possible Solutions

Krishna Yadav, Madhulika Pradhan, Deependra Singh, and Manju Rawat Singh

Abstract

Immune disorders are a self-attacking state of our defense mechanism influencing diverse body tissues and organs. One of the most significant obstacles in treatment for immune system ailments is the imprecise targeting of therapeutic molecules to the ideal cells due to a lack of insights into disease pathophysiology prompting deficient therapeutic viability. The activated macrophages are chieftain in immune reaction safeguarding the body from unauthorized invaders. They are ubiquitous in almost all autoimmune conditions finding a link between acquired and innate immunities. In recent days, understanding macrophage structural configuration and functioning has opened a new portal between the molecular aspects and their therapeutic targeting in the path of treatment. These have conceivably revolutionized the treatment pattern for autoimmune disorders from targeting perspectives. Even though macrophages are impending targets for immune disorders, the detailing related to their targeting approaches is still lacking. This chapter is an endeavor to summarize various key aspects of macrophage-associated disorders, pathophysiology, treatment challenges, and possible solutions for safe and compelling targeting of therapeutics in various immune disorders.

Keywords

Macrophage targeting · Nanoparticles · Surface modification · Autoimmune disorder

1 Introduction

Macrophages (MPs) are specialized mononuclear phagocytes that are differentiated from monocytes and that own more intracellular organelles within a large cell. They perform diverse functions ranging from sensing pathogens, phagocytosis, digesting cell debris, and initiation of inflammation by the release of cytokine molecules that subsequently activate other cells (Williams et al. 2018). Monocytes reside in the circulating blood, whereas MPs are present in the tissues. They have a comparatively long life span, and they produce diverse surface receptors and a variety of secretory products contributing to various inflammatory and anti-inflammatory responses. They acclimatize easily to any changes in their environment and support to maintain local and systemic homeostasis (Guilliams et al. 2014; Wynn and Vannella 2016).

K. Yadav · D. Singh · M. R. Singh (✉)
University Institute of Pharmacy, Pt. Ravishankar Shukla University, Raipur, Chhattisgarh, India

M. Pradhan
Rungta College of Pharmaceutical Sciences and Research, Bhilai, Chhattisgarh, India

S. Gupta, Y. V. Pathak (eds.), *Macrophage Targeted Delivery Systems*,
https://doi.org/10.1007/978-3-030-84164-5_4

Cytokines and other inflammatory mediators released by activated helper T cells (T_H cells) may increase the activation of the macrophage during the inflammatory reaction, and active MPs are essential for clearing the pathogens. MPs display superior phagocytic activity, improved production of inflammatory chemicals, and higher expression levels of Class II MHC molecules, which can present antigen molecules to T_H cells. On the other hand, if MPs are incapable to respond sufficiently against an infectious or injurious stimulus, they encourage a chronic inflammatory process which may lead to persistent tissue damage or other harmful diseases (Perdiguero and Geissmann 2016; Jakubzick et al. 2017).

1.1 Polarization of MPs

The immune system consists of many specialized that gives an immediate response to foreign pathogens or inflammatory activators (Juhas et al. 2015). MPs are one of the pivotal cells that perform a prime role in sustaining the inflammatory reaction and encourage the renewal of tissue homeostasis after injury (Mantovani et al. 2013; Mills 2012). Macrophage polarization is the event that encourages MPs to get transformed into different functional phenotypes under the influence of microenvironmental stimuli (Li et al. 2018). MPs could be transformed into M1 (classically activated) and M2 (alternatively activated) MPs (Zhu et al. 2015). The M1 and M2 MPs are developed following two antagonistic pathways of arginine metabolism called iNOS pathway (where citrulline and NO are produced from arginine) and the arginase pathway (where ornithine and urea are produced from arginine), respectively (Gordon and Martinez 2010; Wang et al. 2014; Vergadi et al. 2017).

1.1.1 Pro-inflammatory M1 Macrophages

The MPs are polarized to M1 MPs by activation of LPS and Th1 cytokines (such as IFN-γ and TNF-α). Polarized M1 MPs are characterized by the presence of CD80, CD86, TLR-2, TLR-4, iNOS, and MHC-II surface phenotypes (Zhu et al. 2015). These cells exert a positive feedback reaction on unpolarized MPs by producing a variety of immune mediators such as TNF-α, IL-1α, IL-1β, IL-6, IL-12, CXCL9, and CXCL10 (Chylikova et al. 2018). Subsequently, these mediators encourage the transformation of unpolarized MPs into the M1 state. The expression of M1 genes is controlled by crucial immune mediators, such as STAT1, STAT5, NF-kB, IRF3, and IRF5. NF-κB and STAT1 are supposed to be the foremost pathways involved in polarization of M1 macrophage (Yunna et al. 2020; Shapouri-Moghaddam et al. 2018).

1.1.2 Anti-inflammatory M2 Macrophages

The MPs are polarized to M2 MPs under the influence of signaling molecules such as IL-10, IL-13, IL-33, and TGF-β. Remarkably, only IL-4 and IL-13 cause direct trigger and activation of M2 MPs, while signaling mediators such as IL-33 and IL-25 augment activation of M2 MPs by releasing Th2 signaling molecules (Chistiakov et al. 2018). M2 MPs can be characterized by the presence of surface markers, such as mannitol receptor, CD206, CD163, and CD209. Overproduction of CCL22, TGF-β, CCL1, CCL18, CCL24, and CCL17 also stimulates unpolarized MPs to get transformed into M2 MPs. Further, the STAT6 pathway has also been reported as a crucial pathway to trigger M2 MPs (Juhas et al. 2015; Arora et al. 2018). MPs perform their role in infection prevention, tissue repairing, the formation of new blood vessels, and immunomodulation. M2 MPs are further subdivided into M2a, M2b, M2c, and M2d. These MPs vary in their cell surface markers, biological functions, and released immune mediators (Yunna et al. 2020). The subtypes of M2 MPs and their characteristics have been shown in Table 1.

1.2 Macrophage Function

All through the process of inflammation, there is the presence of inflammatory cytokines and chemokines produced by various infiltrated immune

Table 1 The subtypes of M2 MPs and their characteristics

Subtype of M2	Activated by	Characterization	Functions	References
M2a	IL-4 or IL-13	Increased expression of IL-10, TGF-β, CCL17, CCL18, and CCL22	These MPs enhance the endocytic activity, remove debris, and promote cell growth and tissue repairing	Arora et al. (2018) and Murray et al. (2014)
M2b	Toll-like receptor (TLR) ligands and IL-1β	Increased expression of cytokines, such as TNF-α, IL-1β, IL-6, and IL-10	MPs regulate the breadth and depth of immune responses and inflammatory reactions	Wang et al. (2019c) and Mulder et al. (2014)
M2c	Glucocorticoids, IL-10, and TGF-β	Increased expression of IL-10, TGF-β, CCL18, and CCL16	Play crucial roles in the phagocytosis of apoptotic cells process	Gordon and Plüddemann (2017) and Murray et al. (2014)
M2d	TLR antagonists	IL-10 and VEGF	M2d MPs lead to the release of and promote angiogenesis and tumor progression	Orecchioni et al. (2019)

cells. The communication between immune cells and their inflammatory secretions might stimulate the development and activation of specific MPs phenotype, subsequently leading to alteration in their functions. MPs perform the following functions:

- *Phagocytosis*: Following the event of inflammation, monocytes undergo transformation into MPs and engulf the foreign cell/particle. Subsequently, the cellular enzymes present inside the macrophage destroy the engulfed cell/particle by the process of phagocytosis (Hirayama et al. 2017).
- *Immune functions*: MPs contribute crucially to achieve innate and adaptive immunity, as they own the ability to functions as antigen-presenting cells (APCs), subsequently leading to activation of B and T cells. They also perform pro-inflammatory and anti-inflammatory functions (Gordon and Martinez 2010).
- The MPs actively participate in the elimination of dead cells. Local cell death may result from autophagy, necrosis, caspase-mediated apoptosis, excitotoxicity, and necroptosis, and caspase-mediated apoptosis (Gordon and Plüddemann 2017).
- *Regulatory functions*: MPs perform regulation functions of the immune system by releasing IL-10. When the IL-10 is produced in a higher amount, the immune responses are repressed thereby regulating and limiting the inflammation.
- *Erythropoiesis*: It is the process of formation of red blood cells in the bone marrow. MPs contribute to erythropoiesis by producing ferritin, which is then engulfed by committed erythroblasts (by endocytosis). Further, the acidification and breakdown of ferritin inside the erythroblasts liberate iron from ferritin which is subsequently used for the production of heme (Jacobsen et al. 2015).
- *Wound healing*: MPs perform a crucial function in wound healing. Following any tissue damage, the released cytokine IL-4 provokes the production of MPs that are engaged in the conversion of arginine to ornithine. Further, the ornithine mediated production of collagen and polyamines that are key components engaged in the wound healing process (Krzyszczyk et al. 2018).
- Apart from the above-mentioned functions, MPs also contribute to bone remodeling, synaptic pruning, and tissue repair.

2 Role of MPs in Disease Pathogenesis

The MPs are the leading defender of the body against unwanted invaders and are omnipresent in almost all autoimmune conditions (Fig. 1). A new gateway between molecular and therapeutic dimensions has been launched in recent days to explain the nature of the MPs and their role in various immune-related disorders. The role of MPs in disease pathogenesis is gathered in this section.

2.1 Role of MPs in the Pathogenesis of Acute/Chronic Inflammation

2.1.1 Type 1 Diabetes (TD1)

TD1 is described by the obliteration of β-cells in the pancreas, resulting in stopping insulin formation causing ultimate rise in blood glucose levels. Compared to earlier investigations, there is moderately less information clarifying the part of MPs in the TD1 headway (Carrero et al. 2016). Obesity encourages an insulin-resistant condition in muscles, adipose, and hepatic tissues, thereby increasing the risk of development of type 1/type 2 diabetes. Development of insulin resistance in an obese environment results from malfunctioning of insulin target cells and the accumulation of MPs that secrete pro-inflammatory signaling molecules (Rehman and Akash 2016). It is considered that insulin resistance is promoted by the transformation of macrophage polarization from M2 state to M1 state under influence of STAT6, PPARs, NF-κB, and other signal-dependent transcription factors (Chen et al. 2015).

2.1.2 Inflammatory Bowel Disease (IBD)

IBD is described by inflammation in the gastrointestinal tract. An elevated level of CD68+ MPs is apparent in Crohn's disease (ChD) as well as ulcerative colitis (UC), whereas an augmented level of CD163+ is also evident in ChD (Na et al. 2019; Skytthe et al. 2020). ChD MPs demonstrate minimized retinoic acid production and unusually rapid cytokine lysosomal debasement, while ChD and UC MPs show insufficient expression and functionality of the CM-CSF receptor (Magnusson et al. 2016; Goldstein et al. 2011). In this context, animal models of IBD supported the participation of mal-regulated MPs in the disease pathogenesis (Kamada et al. 2010; Danese 2011). Murine colitis models presented enhanced recruitment of monocytes and undeveloped MPs into the mucosa of gut that sustained subsequent differentiation during inflammation. They also released a huge quantity of pro-inflammatory molecules such as nitric oxide, TNF, and IL-6 (Meroni et al. 2019). These results suggested that MPs encourage the progression of inflammation in the intestine.

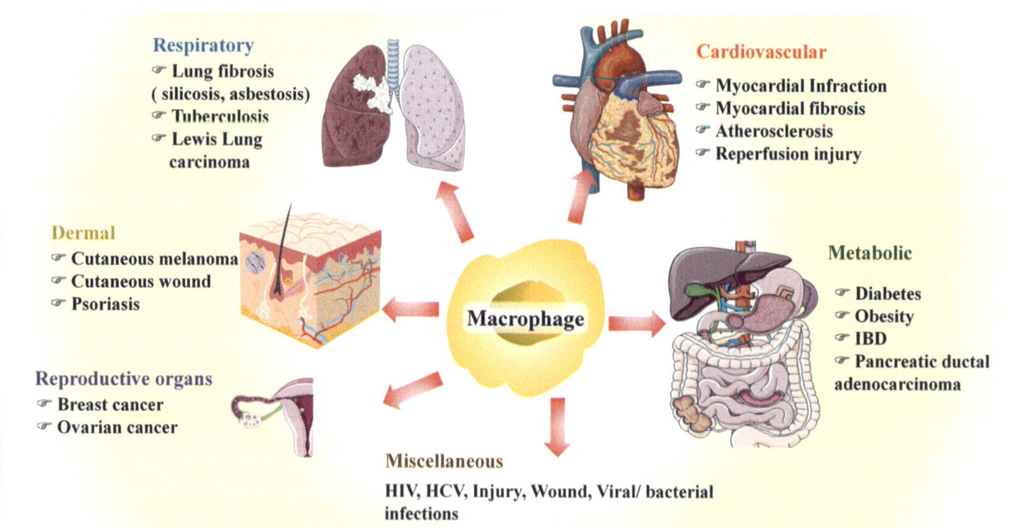

Fig. 1 The illustration showing the involvement of MPs in various organs and related issues

2.1.3 Rheumatoid Arthritis (RA)

RA causes chronic joint inflammation which can cause the annihilation of the cartilage and erosion of the bone. It functions by accumulating synovial tissue MPs (Sack et al. 1994). Elevated synovial MPs penetration triggers more conspicuous IL-6 and TNF-alpha discharge, ensuing in amplified joint disintegration (Paoletti et al. 2019; Udalova et al. 2016). It has been observed that M1 is more prevalent than M2 in RA patients and animal models (You et al. 2014; Yang et al. 2020). The RA synovium contains a variety of cellular infiltrates including MPs which release chemokines and cytokines that interact with other cells and contribute to the pathogenesis of RA. MPs release IL-1β and TNF and stimulate activation and proliferation of fibroblast-like synoviocytes (FLS). Further, activated FLSs release nuclear factor-kappa B ligand and M-CSF, consequently leading to the formation and activation of osteoclast under the influence of IL-1β, IL-6, and TNF secreted by the macrophage. MPs also encourage the recruitment of monocytes in RA synovium by releasing IL-1β and TNF. The differentiation of T cell also takes place under the influence of macrophages secreted, IL-23, IL-12, and TNF, which ultimately contributes to the RA pathogenesis (Udalova et al. 2016; Culemann et al. 2019; van der Vorst et al. 2017).

A high M1/M2 ratio, combining intemperate and pro-inflammatory M1 polarization with insufficient M2 polarization, favors the polarization of MPs in RA (Fukui et al. 2017). In comparison, expanded emissions of CD50 and CD36 are paired with reduced expressions of CD163 and CD209 (Soler Palacios et al. 2015). The effects of polarization over M1 increased TNF-alpha, IL-1, IL-6, and IL-12 levels. In the progression of this disorder, primary MP-inferred arbiters include IL-1, IL-6, IL-12, and TNF-alpha that are released by active MPs and participate in inflammation and cartilage degradation at both systemic and a local level (Soler Palacios et al. 2015; Pope and Shahrara 2013; Semerano et al. 2014; Yamanaka 2015). Synovial membrane occupant MPs and monocyte-inferred MPs have diverse functionality in RA that has a significant concern when design targeted treatments (Sehnert et al. 2020; Menarim et al. 2020).

2.1.4 Systemic Lupus Erythematosus (SLE) Treatment

Widespread inflammation of connective tissues with no cure is distinguished in SLE (Gatto et al. 2019). There are differing phagocytic thresholds for MPs present in SLE, allowing the clearance of apoptotic cells and immune structures to be inhibited (Kavai and Szegedi 2007; Baumann et al. 2002). SLE MPs likewise have greater antigen-presenting capacities and are associated with the regulation of the acquired immunity via an improved functionality to trigger autoreactive B cells and T cells (Lee et al. 2020a). SLE is characterized by a wide range of clinical indications. T lymphocytes and autoreactive B cells, in association with the innate immune system, chiefly participate in the development of SLE (Miron et al. 2013). In SLE, there is disturbed M1/M2 status resulting in excessive release of signaling molecules such as TNF-α, IL-6, IL-10, and type I IFNs (Liu et al. 2013). Considering inflammatory functions, SLE monocytes and MPs deliver self-antigens to autoreactive T cells instead of immune-silent presentation usually linked with constituent of apoptotic cells (Jiang et al. 2014). Overexpression of adhesion molecules, ICAM-1, can likewise prompt compromised MPs activation (Orme and Mohan 2012). SLE MPs additionally overexpress CD40, CD86, and Siglec-1 (Blanco et al. 2001; Decker et al. 2006; Katsiari et al. 2002; Biesen et al. 2008).

2.1.5 Multiple Sclerosis (MS)

MS is a multifaceted, immune-responsive, demyelinating state of the CNS with reformist neurodegenerative promoting the myelin tissue damage followed by MPs upgrade driving the pro-inflammatory reaction of T and B cells (van Langelaar et al. 2020). MS indicates an expanded total mononuclear phagocyte measure like MPs caused by monocytes and microglia tissue (Lee et al. 2020a; Mallucci et al. 2015; Bramow et al. 2010). The tissue injuries and the enhancement of microglial knobs have also been reported in mononuclear phagocytes (van der Valk and De Groot 2000; Moll et al. 2011). In the initial stage, peripheral MPs penetrate the CNS and cause initiation and progression of MS, in association with residential microglia (Sestak et al. 2011). In the initial stage, microglia/MPs are instantly triggered to transform into M1 cells, producing pro-inflammatory mediators and destructing CNS tissue. In the later stages, M2 cells are the major players of CNS inflammation, liberating anti-inflammatory cytokines. The equilibrium between M1 MPs and M2 MPs in the CNS is vital for the progression of MS (Byrne et al. 2012).

In murine models, neuronal limits and reduced regenerative responses were prevented by the degree of mononuclear phagocyte diffusion cohorts with cerebral decay (Moll et al. 2011; He et al. 2019; Planche et al. 2017; Tambalo et al. 2015). The occurrence of CD68+ MPs adds to the structure of MS abrasions (van der Valk and De Groot 2000). Expanded emission of CD68, HLA, CD86, and iNOS also contains functional MPs defects, as are bizarre metabolic modifications from oxidative to glycolytic (Tannahill et al. 2013; Zrzavy et al. 2017). These factors add up to higher antigen penetration thresholds, weakened myelin demolition remediation, and weakened neurotoxicity (Zrzavy et al. 2017; Kigerl et al. 2009). The polarization of MPs in complex MS wounds exposes instantaneous elements, such as CD40 and MR, with co-emanations of M1 and M2 markers (He et al. 2019; Peferoen et al. 2015). NLRP3 inflammatory agent that drives the autoreactive cell-type T migration and release of IL-1beta, IL-6, and IL-23, which subsequently promotes the formation and conservancy of Th17 cells, are the most important mediators derived by the MPs in disease progression (Wang et al. 2018; Holz et al. 2018; Zhang et al. 2017; Inoue et al. 2012). TNF-beta, IL-1beta, IL-6, IL-12, and ROS intercede inflammatory reactions, whereas BAFF and IL-6 intervene in B cell differentiation and endurance (Giacomini et al. 2013).

2.2 Role of MPs in the Pathogenesis of Cardiovascular Disease (CVD)

MP-incited inflammation is associated with practically all CVDs, comprising atherosclerosis (AS), myocarditis, pulmonary arterial hypertension (PAH), and other heart disorders. Consequently, targeting transformed MPs is a commending technique to deal with CVDs. This segment features the various advancements that can target vascular MPs for the treatment of CVD.

2.2.1 Myocardial Fibrosis (MF)

Cardiac/myocardial fibrosis is the formation of scar in the heart muscle which leads to abnormal functioning of the heart muscle. The anomalous amassing of ECM proteins, for example, laminins, collagen, fibronectin, and elastins, induces MF as cardiovascular disease (Afratis et al. 2018). Primarily, the accumulation of ECM is a shielding mechanism and could be helpful in tissue regeneration, though extreme and incessant deposition of ECM leads to abnormal tissue function (Hinderer and Schenke-Layland 2019). Further, selective cardiac infiltration by MPs causes macrophage-specific overexpression of urokinase plasminogen activator (uPA), and enhanced activity of myocardial uPA activity, leads to myocardial fibrosis, probably by unbalanced matrix metabolism (Moriwaki et al. 2004). MF may indeed be categorized into interstitial reactive fibrosis, interstitial invasion, and substitution fibrosis (Hinderer and Schenke-Layland 2019). Two proteases, MMPs and lysyl oxidase (LOX), are essential to the generation of fibrosis when cross-linked with collagen, and MMPs are over-expressed in the MI-incited MF (Afratis et al. 2018).

2.2.2 Myocardial Infarction (MI)

MI is a major coronary condition that causes recurrent death. After MI, MPs are amassed at the grievance site, causing substantial inflammation and uncomfortable cardiovascular remodeling and failure (Duncan et al. 2020). The chief constituent of the cardiac ECM is fibronectin, glycosaminoglycans, and proteoglycans, which offer mechanical strength to the heart (Okuda et al. 2016). Following MI, enhanced inflammation adversely controls enzymes such as proteases, MMP-2 and MMP-9, leading to decomposition of ECM with a subsequent decrease in mechanical strength of cardiac tissues, thereby preventing function of the heart (Dick et al. 2019). Monocytes and MPs are hindrances in cardiovascular tissue regeneration after MI is the key therapeutic approach (Fujiwara et al. 2019).

2.3 Role of MPs in the Pathogenesis of Cancer

The most common inflammatory MPs recruited to the tumor have been termed tumor-associated macrophages (TAMs). They exist during all stages of tumor development, stimulate neoangiogenesis, and promote invasion of tumor cells. Initially, TAMs mainly exhibit an M1 phenotype, but their continuous presence in tumor-rich environment facilitates their transformation to M2 phenotype, which ultimately promotes the endurance of existing tumor and directs further advancement of cancer. The significance of TAMs in progression of malignant cancers has been elaborated in this part.

2.3.1 MPs and Tumor Cell Activation

It has been established by numerous researches that co-culture of numerous kinds of cancer cells, such as glioma cell, breast cancer, colorectal cancer, and liver cancer with MPs, augments its proliferation (Noy and Pollard 2014). MPs release heparin-binding EGF-like growth factor (HB-EGF) and STAT3 such as oncostatin M, IL-6, and IL-10. STAT3 and NF-kB signals have been reported to play very crucial function in the preservation of cancer stem-like cells (CSCs). Further, the activation of NF-kB is maintained by STAT3; therefore, macrophage-mediated activation of STAT3 could be a crucial event for the establishment and advancement of cancer (Saito et al. 2014; West et al. 2018). Additionally, some investigations have established that direct interaction between TAM and tumor cells through intracellular adhesion molecule-1, and ephrin is vital for their direct cell–cell interaction and cell–cell interaction among them contributes to tumor development, relapse, or resistance to cancer treatment (Malfitano et al. 2020).

2.3.2 MPs and Immunosuppression

Delay of intrinsic antitumor immune response attributable to phenotypic TAMs polarization remains a prime issue during recurrent phases of cancer development. TAMs lead to overexpression of B7-H4 by producing IL-10 and IL-6 resulting in inhibition of host antitumor immunity (Costa et al. 2013; Kuang et al. 2009). Moreover, the M2 subtype of TAMs owns the ability to interfere with antigen presentation and cellular immunity by increasing the production of IL-10 and TGF-β. IL-10 reduces the synthesis of GM-CSF which in turn decreases the production of T cells, and consequently the probable antitumor immune response. Also, IL-10 jointly participates with TNF-α in the production of an immune mediator called programmed cell death 1 (PD-L1) which exert harmful effect on the production and function of cytotoxic T cell (Kuang et al. 2009).

2.3.3 MPs and Angiogenesis

Angiogenesis is the process of production of new-fangled blood capillaries from pre-existing. Tumor angiogenesis is essential for providing essential nutrients and oxygen to speedily growing malignant tissues. An angiogenic switch permits the survival and metastatic progression of the disease. MPs recruited and redirected by cancer cells are key players for enhancing the angiogenic shift (Zhu et al. 2017a). Hypoxia in avascular areas of tumors is sensed by MPs, and they produce angiogenic factors such as VEGF-A for arousing chemical-induced movement of endothelial cells (EC) and macrophages. Another key contributor to tumor angiogenesis is MMP that breaks down the proteins essential for maintaining the turgidity of ECM of existing endothelial vessels. This breakdown of protein lets individual EC penetrate tumor stroma and associate with pre-existing VEGF-A to produce new blood capillaries. Since TAMs help the expression of MMPs (MMP-1, MMP-7, and MMP-9), their contribution to angiogenesis and tumor growth becomes very ostensible (Hwang et al. 2020; Singh et al. 2017). Other angiogenic mediators released by TAM are basic fibroblast growth factor (bFGF), thymidine phosphorylase (TP), urokinase-type plasminogen activator (uPA), and adrenomedullin (ADM) (Riabov et al. 2014).

2.3.4 MPs and Tumor Invasion and Metastasis

MPs encourage invasion and metastasis in the tumor site by engaging tumor cells in an autocrine circle that facilitates migration of tumor cells. This signaling results in the production of colony-stimulating factor-1 from the cancer cells that include the MPs to produce EGF. This growth factor assists in the movement of MPs trailed by tumor cells toward tumor blood vessels where macrophage-derived VEGF-A encourages the entry of cancer cells into the blood vessels (Nielsen and Schmid 2017; Lin et al. 2019). Further, ECM acts as a barricade for tumor cell migration, and degradation of ECM is an important happening in metastasis. It is also well recognized that TAMs are proficient in releasing various enzymes, such as cathepsins, MMPs, and serine proteases, which are key factors facilitating the degradation of ECM (Wu et al. 2020).

3 Treatment Challenges in MP-Based Therapy

3.1 Status Quo of MPs

The restricted knowledge of circumstances or context or Status quo is a substantial impediment to the analysis of MP-based therapies. In some cases, M1-like MPs inhibit recovery and facilitate recovery in other conditions (Spiller and Koh 2017; Kumar et al. 2020). It is valid for MPs like M2 which are associated with both the regeneration of tissue and fibrosis. As regards tissue biomaterial reconstruction, all microphages encompassing M1 and M2 are autonomously associated with all vascular or fiber-related regeneration (Spiller and Koh 2017). These dispute records are proposed to be identified with MPs activation, where variations from the standard M1-to-M2 classification contribute to a hindrance. Another possible purpose behind the conflicting reports may be that there is no consistent classification for the representation of the phenotype of MPs, particularly for MPs obtained from the in vivo setting, which is likely to lead to misidentification of entirely different phenotypes (Murray et al. 2014). For instance, in either case, considering the fact that these MPs exhibit distinctive phenotypes with diverse impacts on angiogenesis, fibrosis, and tissue repair, two separate phenotypes of MPs are typically denoted as M2, combining that invigorated in vitro with IL-4 and those activated with IL-10 (Spiller et al. 2016; Lurier et al. 2017; Lolmede et al. 2009; Lu et al. 2013). Moreover, it is generally accepted that the expressions "M1 and M2" are mischaracterized and that MPs often occur as phenotypes of half-breeds just as phenotypes that do not have a direct alikeness to the MPs of M1 and M2 that have been portrayed in vitro (Murray et al. 2014). In conclusion, the assessment of M2 markers relies all the more frequently on which antibodies are useful to use rather than which marker would normally be suitable for the solicitation. This is exclusively troublesome concerning the fact that M1 and M2 markers are not on/off switches but are somewhat up- or down-managed with activation changes, and both M1 and M2 markers do not similarly alter expression when triggered so that MPs are possibly misclassified simply by tallying cells that stain emphatically for one or even a few markers (Spiller and Koh 2017). To facilitate our interpretation of the meaning of subservient components of in vivo MPs, their phenotypic representations must be as detailed as might be required under the circumstances, and when deciphering details from each test, techniques used to generate these definitions must be unmistakably explained and deliberately thought about. For instance, one empirical investigation utilized whole-genome expression signatures to compare the characteristics of MPs retrieved from the damaged spinal cord of mice to 19 currently accessible MPs information indexes polarized with different stimuli (Zhu et al. 2017b). They observed that the MPs of the spinal cord more closely resemble foam cells found in atherosclerotic mice, and subsequent studies reported that silencing the structure of foam cells increased regeneration from spinal cord injury (Zhu et al. 2017b). Although this MPs phenotyping methodology is extraordinarily detailed, in many implementations it may not be possible or monetarily conceivable. Various markers of phenotypes of M1 and M2 can be used at any rate. In addition, while multiple tests measure phenotype markers in tissue homogenates or for populaces of tissue-separated cells, the use of assays that can ascertain various markers on single cells (e.g., flow cytometry, single-cell genomics/proteomics) can provide important evidence on the heterogeneity of MPs in vivo in both physiological and pathological environments.

3.2 Imperative Concerns for Preclinical Models

Since the disparity between human and animal models has long been seen as a foremost obstacle for understanding scientific revelations, it could be exceptionally challenging for immune system therapies to resolve mouse–human differences. There have been substantial differences in the response to injury and infection between the mouse and human inflammation (Seok et al. 2013). A study on MPs and their phenotypes indicates that they are particularly responsive to mouse–human variations since certain main genes in the mouse and human MPs are controlled differently according to normal stimuli including LPS (M1 stimuli) and IL-4 (M2 stimuli) (Spiller et al. 2016; Raes et al. 2005; Schroder et al. 2012; Martinez et al. 2013). The normal murine M1 and M2 markers iNOS and Arg1 do not manifest at all, for example, human MPs, and essential AIF factors TGF-b1and IL-10 are reversed in the mouse and humans M1 and M2 MPs in vitro (Raes et al. 2005; Scheib and Höke 2016; Murray and Wynn 2011). For instance, age and sex are both essential considerations with various variables. In aged rats contrasted with young rats, axon regeneration following peripheral nerve grafting had been prevented and was associated with post-invasion, phagocytosis reduced, and the emission of pro-and AIF cytokines decreased (Scheib and Höke 2016). Essentially, another study has shown that MPs in older mice are smaller than MPs in young mice that cause the recuperation of injuries with disabilities (Swift et al. 1999). Further, limiting the activity of M-CSF receptor kinase decreased MPs counts and caused an extended pattern of bone growth for old but not young mice (Slade Shantz et al. 2014). This knowledge demonstrates that the ability of MPs can change over time and that when preparing macrophagic methodologies, age, and other related elements must be pondered. In addition, MPs may display gender-subordinated behavioral contrasts during tissue recovery, although these sex contrasts have not yet been comprehensively studied (Banerjee et al. 2013).

3.3 Issues Regarding Quality Control

The versatile uniqueness of MPs renders the drugs that target them extremely vulnerable to quality control concerns. For instance, it is not obvious, whether exogenously activated MPs maintain their attributes after in vivo administration. Intriguingly, Cao and co-workers showed that ex vivo-polarized splenic MPs, but not bone marrow-determined MPs, had an M2-like feature and had been shielded against injury for quite a long period after administering to the murine renal injury model (Cao et al. 2014). This significant distinction in the therapeutic effects of MPs arising from the two unique origins was shown to be attributed to the ability of bone marrow-determined MPs to replicate in vivo, as opposed to splenic MPs that are relatively mature and differentiated. In another report, by Lavin and co-workers (Lavin et al. 2014), found that transplantation of peritoneal MPs into the lungs, which are also a moderately mature MPs population, allowed them to take on the epigenetic legacy and pattern of gene expression that is common for lung MPs, regardless of the fact that they had peritoneal MPs characteristics. Correspondingly, it would be necessary to formulate systems that pledge their functional consistency in vivo before MP-based cell therapies can be translated into therapeutic use or if nothing else has the option of predicting changes in the phenotype that may occur after administration. Dosage and dosing frequency would be a major quality control problem for biomaterial-based procedures. Jiang and co-researchers (Jiang et al. 2017) have shown that longstanding effects of just 10–30 days of sustained release of dexamethasone, an AIF glucocorticoid, from synthetic systems capturing pancreatic islet cells for the treatment of diabetes by cell transplantation could be achieved. In this investigation, on the other hand, the release of dexamethasone from the delivery system was described in vitro, which may not be exactly the same as in vivo conditions (Jiang et al. 2017; Sundararaj et al. 2016). The dosage would be another simple quality control issue for drug-delivering implants. Low dosages of delivered dexamethasone enhanced graft durability in the islet transplantation sample, while high doses inappropriately smothered MPs invasion, thereby disrupting vascularization and graft incorporation and decreasing islet cell endurance (Jiang et al. 2017). In order to better balance MPs activity, an overly low dosage will fail the test. The importance of cautiously defining release profiles in vivo is present in these findings and fitting them to the current application.

3.4 Miscellaneous Concerns

The perception of MPs-centric therapies may also be a matter for person-to-person heterogeneity, as the immune cells of a patient can respond distinctively to inflammatory stimuli based on specific clinical records of the patient. In all areas of the progenitors of the bone marrow, chronic, pro-inflammatory conditions will significantly affect the phenotype of MPs to a large extent through epigenetic scripting. In a perpetual report, diabetic db/db mice wound MPs displayed M1-like features in contrast to non-diabetic mice in 4 or 7 days of post-injury (Bannon et al. 2013). Cultured bone marrow-inferred MPs of the diabetic db/db mice have displayed more M1-like traits, in similar cultural settings, than MPs of non-diabetic control mice (Bannon et al. 2013). Related results were observed using prediabetic high-fat diet (HFD) mice including wound MPs, exhibiting more M1-like features in the third and seventh day after the injury, and this distinction was shown by transferring the bone marrow from HFD to the lean benefactor mice (Gallagher et al. 2015). Inborn programming of diabetic MPs will elucidate an epigenetic aspect, which decreases the suppressive histone methylation of the IL-12 promoter of progenitors in bone marrow, resulting in the expansion of IL-12 in MPs with descendants with wounds (Gallagher et al. 2015). Once again, some examinations have shown that MPs in patients with persistent inflammatory disorders can be hypo-receptive to inflammatory stimuli irrespective of higher inflammatory activity levels. For instance, Malaponte and co-investigators (Malaponte et al. 2002) have shown that monocytes from dialyzed patients emit more substantial concentrations of pro-inflammatory cytokines as compared to monocytes from healthy volunteers, but monocytes emit less than those from healthy volunteers when endotoxin has been stimulated. In addition, the inflammatory reactions of the endotoxin were proportional immediately to the dialytic patient's time measurement, recommending that dialytic pain caused a constant one, resulting in a dose-responsive monocyte resistance. In another study, the discharge of inflammatory cytokines causing severe internal or bacterial damage compared with healthy controls was decreased in ChD patients, suggesting that the condition caused a smothering inflammatory reaction to acute stimuli (Marks et al. 2006). The concept of the state of the MPs activity after treatment is particularly relevant in the construction of biomaterials that change the MPs phenotype with consequences on the repair of tissue if MPs behavior controls are a deliberate part of its architecture (Brown et al. 2009). Currently, in stable and disorderly control animal models, the inflammatory response to biomaterials is distinctive (Oliva et al. 2015). A superior understanding of how the disease environment can affect therapies intended to balance MPs phenotype and provide additional objectives to control MPs' functionality, including epigenetic programming. Such an understanding will also assist with studies aimed at assessing the processes through which exogenous phenotypes of MPs are regulated in various disease conditions following ex vivo stimulation and therapy administration. The overall impacts of proliferation, cell death, and migration into/out of the target tissue on MPs accretion and phenotypes are other unexplored regions. Most regenerative medicine studies agree that the collection of MPs is dominated by the persistent mobilization of blood-borne monocytes, but the role of cell proliferation, as well as the work of tissue-occupant MPs, is best appreciated (Spiller and Koh 2017). In addition, after playing their function in the target tissue, the fate of MPs is inadequately interpreted and may have important consequences for both local and systemic inflammatory reactions, the latter likely through interacting with lymph nodes and even inverse migration to the bone marrow (Spiller and Koh 2017). In translational regenerative drug techniques, a superior understanding of these cycles could cause better targeting of MPs reactions.

4 Treatment Approaches as Promising Solution for Dealing with MPs Associated Disorders

A variety of immune systems diseases have represented elevated levels of MPs infiltration (Poltavets et al. 2020). In its pathology, every issue of the immune system is relevant provided that the recruitment and activation of MPs following accredited pathways (Lee et al. 2016). The advocacy of immune system disorders can be influenced by differing cytokine, chemokine, and transduction variables under the influence of MPs, thereby resulting in anomalous intermodulation of dendritic or regulatory T cells (Udalova et al. 2016; He et al. 2019). Therefore, targeting such disorders driven by the MPs could be a breakthrough in treatment. In this section, we have discussed the targeting aspects of MPs in different disorders toward achieving efficient and precise treatment in such debilitating conditions.

4.1 Targeting MPs for Acute/Chronic Inflammation

4.1.1 Type 1 Diabetes (TD1) Treatment

The explorations have indicated that MPs from non-obese diabetic (NOD) mice show faulty phagocytic functioning, bringing about impeded accord of apoptotic cells (He et al. 2019). MPs from NOD mice were likewise exhibited anomalously, interceding the degeneration of islet β-cells by means of cytolytic action plus ROS generation (Calderon et al. 2006; Thayer et al. 2011). Moreover, MPs show interactivity with cells in the adaptive immune system and are associated with the functioning of CD4+ and CD8+ T cells, essentially through type1 IFN signaling (Marro et al. 2019; Vomund et al. 2015). The polarization of TD1-MPs supports pro-inflammatory activity, which is related to more significant levels of CRP, IFN-γ, CXCL10, and CCL2 (Ma et al. 2019), bringing about more prominent emissions of IL-6, IL-1beta, and TNF-alpha. It is recommended that raised levels of IL-1 and IL-6 produce Th17 cells (Ma et al. 2019). Like other diseases, monocyte-inferred MPs and pancreas-inhabitant MPs in TD1 have various activities, which is a significant approach when creating MPs-targeted treatments (Carrero et al. 2016; Albiero et al. 2015; Ferris et al. 2017).

The initial strategy of treatment for TD1 has centered on eliminating or repolarizing the MPs. Targeted diminution of MPs by clodronate liposomes (clodrolip) was appeared to abrogate diabetes in NOD mice, despite the fact that inflammation continued (Calderon et al. 2006). To restrict MPs-inferred TNF-alpha, TNF-alpha blockages have exhibited clinical viability; however, they stay questionable because different cases have indicated the disquieting influence of glycemic control (Tack et al. 2009; Mastrandrea et al. 2009). Endeavors to stifle M1 phenotypes via the receptive exchange of M2 MPs diminished TD1 onset in NOD mice and decreased hyperglycemia, kidney damage, and insulitis in vitro (Parsa et al. 2012; Cao et al. 2013). In a further analysis, the researchers attempted to use the immunomodulatory effects of MSCs by building TGF-beta gene MSCs, which were found to restore antagonistic immune response control and another TD1 feature in the murine model (Daneshmandi et al. 2017). Treatment of NOD mice with CSF-1 antibody monoclonal antibodies decreased the incidence of diabetes and advanced the regulatory mechanism for the advancement of the immune system (Carrero et al. 2017). In addition, indigenous scaffold release of dexamethasone, an anti-inflammatory (AIF) therapeutic agent, sustained phenotypic transition to M2 MPs and accelerated engraftment of islet transplants (Jiang et al. 2017). These medications, however, are in preclinical trials, and concentrating on MPs for the treatment of TD1 is at the naive level, suggesting a comprehensive treatment void.

4.1.2 Inflammatory Bowel Disease (IBD) Treatment

Numerous standard IBD treatments impact macrophage works by restraining inflammatory pathways or inciting polarization of activated macrophages. MPs-exhibited TNF-alpha in ChD intensifies the emission of other pro-inflammatory cytokines and assigns other immune cells to the microenvironment of the neighborhood tissue, encouraging the production of granulomas (Na et al. 2019). The polarization of MPs in IBD is intricate, with M2 polarization being favored by ChD MPs, as shown by the expanded development of IL-13 and expanded expression of CD163 (Demetter et al. 2005; Bailey et al. 2012). In UC, conversely, M1 MPs emit more elevated IL-23 and TNF-alpha levels and lesser IL-10 levels, while M2 MPs emit more CD163 and CD206. IL-1beta, IL-6, IL-23, TNF-a, and TNF-like protein 1A are key MPs-inferred arbiters in disease progression, advancing Th1 and Th17 cell reactions (Demetter et al. 2005; Bouma and Strober 2003; Zhou et al. 2015; Cosín-Roger et al. 2013). Concealment of the pro-inflammatory M1 activity or/and initiation of the tissue-reparation and anti-inflammatory M2 activity has been appeared to lessen IBD. This has been shown through the assenting transference of M2a macrophages, oral administration of pentacyclic triterpene lupeol, TNF-α-simulated gene 6 protein (TSG-6), IL-33, and the probiotic *C. butyricum* as a probiotic. The receptive transference of peritoneal cells likewise has been appeared to control regulatory B cells and macrophages in colitis experiments (He et al. 2019).

4.1.3 Rheumatoid Arthritis (RA) Treatment

The standard care for RA includes disease-modifying anti-rheumatic drugs, which usually incorporate steroids, methotrexate, and NSAIDs. These drugs are not exclusive to MPs, however, rather relieve certain inflammatory symptoms associated with the activation of MPs (Scott et al. 2010). The disposal of inflammatory synovial MPs is one of the MP-based tactics to manage RA. It appears that synovial inflammation in arthritis has recovered (i) after infusion of clodronate liposomes primarily to drain local MPs (Van Lent et al. 1998), (ii) after SIRT1 obstruction of monocyte differentiation into MPs, and (iii) after CD64-regulated immunotoxin (van Vuuren et al. 2006; van Roon et al. 2003). The repolarization of MPs into an AIF phenotype is another method to treat RA. In investigational arthritis, IL-10 was displayed to diminish M1 activity, while initiation of SIRT1/AMPK signaling and THAP-administration, a Notch inhibitor, reduced inflammation of the knee joints. MSCs that have immunomodulatory roles were administered systemically to collagen-induced arthritis mice in a comparable vein and recovered the severity of the disease (He et al. 2019; Lopez-Pedrera et al. 2020). A group of researchers developed HABN as part of another approach, namely, an amphiphilic profile that aims to the inflamed colonic epithelium and pro-inflammatory MPs post-oral administration. The introduction of HABN increased the involvement in the lamina propria of DSS-colitis mice of AIF CD11b + Ly6G- MHCII+ TRMs and lightened the indications of colitis more appropriately than other standard IBD therapies (Lee et al. 2020b). In a mouse model of the metabolic disease, another research investigated a nanogel device that may defeat the dysregulated gut microbiota for immunization applications, which can handle potential undertakings that use nanomaterials for immunomodulation in gut-associated problems (Mosquera et al. 2019).

4.1.4 Systemic Lupus Erythematosus (SLE) Treatment

There is a transcendence of pro-inflammatory MPs in SLE, bringing about expanded degrees of IL-1beta, IFN-γ, CXCL10, CCL2, and GM-CSF. This M1-dominance is combined with lacking M2-polarization, described by more elevated magnitudes of IL-10 (He et al. 2019). IL-10 authoritatively has AIF capacities; however, in SLE, IL-10 exhibits pro-inflammatory highlights, given by type 1 IFNs (Sharif et al. 2004). In specific, IL-1beta, IL-6, TNF-alpha, and IL-10 are the chief cytokine referees in SLE progression, all of which lead to systemic and local inflammation (Sharif et al. 2004; Yang et al. 2015). Intriguingly, monocyte-gathered MPs were more receptive to cytokine impelling than renal MPs in lupus-inclined mice (Sahu et al. 2014).

Acquiescent transferal of M2a MPs, incited by IL-4, in a murine model of SLE generally lessened SLE pursuit (Li et al. 2015). Incitement of M2 polarization in pioglitazone-treated patients correspondingly smothered the formation of incendiary cytokines (Mohammadi et al. 2017). An examination by researchers shows that sodium valproate, an HDAC inhibitor, can trigger MPs in ex vivo and avert M1 inflammatory aggregation in vitro, exhibiting that epigenetic changes can play out a vital function in MPs polarization (Mohammadi et al. 2018a). The AhR-interceded signaling and expression of PPARγ in like manner help to the emission of AIF cytokines and diction of M2 markers from monocyte-surmised MPs of SLE patients (Mohammadi et al. 2017, 2018b).

4.1.5 Multiple Sclerosis (MS) Treatments

The MPs depletion tends to stifle CNS damage by restricting the production of myelin-explicit T cell as a way of reducing clinical signs of EAE (He et al. 2019). While MPs depletion, specifically monocyte-based MPs, has been shown to catch the progression of the disease, it is not sufficient to facilitate recovery (Moreno et al. 2016). Another method that shows optimism, particularly as M2 MPs are required for the differentiation of oligodendrocyte, is therefore prompting AIF M2 properties (Miron et al. 2013; Kigerl et al. 2009). The exploitation of fasudil (selective Rho-kinase blocker), an impeding of succinate emissions, and the supportive transition of all potential AIF M2 features from IL-33-recourse MPs and increased EAE (Liu et al. 2013; Peruzzotti-Jametti et al. 2018; Jiang et al. 2012). The derivation of glatiramer acetic acid as a treatment for MS also strengthened M2 phénotypes and developed IL-1-receptor disputes, while the T-celled IL-1beta was diminished by human monocytes and MS. Glatiramer acetic acid derivatives natalizumab, dimethyl fumarate, and fingolimod are included in other clinically employed medicinal products that promote M2 polarization. The anticipation of microglial stimulation to quell inflammatory impacts has likewise been investigated by means of microglial paralysis, concealment of CXCR7, administration of 18-beta-glycyrrhetinic, and hydroxychloroquine (an antimalarial drug) therapy (Weber et al. 2007; Burger et al. 2009). Nr4a1, a gene encrypting for an orphan receptor, likewise incites M2 topographies in MPs and is a crucial blocker for Th1/Th17 cell differentiation, demonstrating remedial perspective (Wang et al. 2018). There are hardly any compounds under preclinical screening that affect MPs and microglial capability (Wang et al. 2019a). The cause of MPs and microglia in the interior of CNS may be inhibited by ethyl pyruvate to guard against EAE (Djedović et al. 2017). Forskolin illuminates EAE by smearing the CD86 MPs and upgrading the arginase-1 (ARG1) MPs expression (Veremeyko et al. 2018). Bryostatin-1 initiates rapid and powerful AIF effects for MPs through hindrances of IL-12, IL-6, and ARG1 expression orientation, subsequent conversion to Th2, and enhancement of EAE indication for mice (Kornberg et al. 2018). In addition, PADRE-Kv1.3 vaccines, which demonstrate the existence of the therapeutic neutralization and MPs penetration of M2 in CNS, are used for the treatment of EAE (Fan et al. 2018). A further possible target is mineralocorticoid steroid receptors seen on MPs because of the lesser clinical manifestations of EAE in mice clearly defective in terms of steroid receptors (Montes-Cobos et al. 2017a). Glucocorticoids, the present-day standard of care for MS, were formed in inorganic–organic blend NPs and were particularly engulfed by MPs, considering superior remedial feasibility in EAE models (Montes-Cobos et al. 2017b).

4.2 MPs Targeting for Treating Cardiovascular Disease (CVD)

4.2.1 Myocardial Fibrosis (MF) Treatment

The studies have found that the regulated local distribution to the infarct region using MMP-degradable hydrogel, a recombinant metalloproteinase tissue inhibitor (TIMP-3), enhances MF, as affected by MI (Purcell et al. 2014). The MF of the myocardial MPs aggregation inferred from monocytes is identified as interacting with the fibroblasts and as a SPARC that causes the emission of matricellular proteins (Falkenham et al. 2015; McDonald et al. 2018). M1 MPs-interceded inflammation damages tissues in substituted fibrosis causing cardiac damage and enhance fibrotic reactions; for different forms of MF, M2 MPs discharge AIF and pro-inflammatory fibrotic factors, including IL-10, TGF-beta, CTGF, and PDGF, which advance fibroblast multiplication and ECM formation (Heymann et al. 2009). One research group created cytotoxic clodrolip to fight circulating monocytes that break into the myocardial region and lead to MF. Monocyte penetration was fundamentally decreased in angiotensin II-injected mice following intravenous administration of liposomes (Falkenham et al. 2015). Targeting phosphatidylserine (PS)-presenting liposomes after intravenous infusion of infarct-MPs advanced tissue-inhabitant MPsAIF polarization, as evidenced by elevated CD206 expression that assisted angiogenesis and forestalled ventricular dilatation and remodeling (Harel-Adar et al. 2011). Further research indicated that TNF-alpha decreased release and reduced inflammation and fibrosis by the transportation of rhASB through myocardial agents in an oblique model of aortically constricted rat (Zhao et al. 2018). In comparison, the delivery of antioxidants to MPs using NPs has had a similar effect on fibrous reduction (Jin et al. 2020). Another way of coping with MF is to use hydrogels to recuperate heart tissue (He et al. 2019). Recent treatments were shown to decrease ECM deposition, restrict myocardial thinner wall, and strengthen the left ventricle function by injecting hyaluronic acid-based MPs with AIF properties (Le et al. 2018; Wang et al. 2019b). Like MI therapy, a safe anti-fibrotic response by injecting controlled drug delivery locally benefits MF treatment adequacy.

4.2.2 Treatment of Myocardial Infarction (MI)

After MI, in patients, there are typically elevated IL-1beta serum levels that are evident (Chen et al. 2020). Administration of IL-1beta-based antibody promotes the drop of neutrophils and monocytes in blood and cardiac infarction zones, especially in the MI-mice model of AS, which has led to attenuated left-ventricular reshaping (Sager et al. 2015). A patient with a higher hsCRP and MI history favorably foreshadowed antagonistic heart activities for the treatment of canakinumab (FDA-affirmed IL-1beta antibody) (Crossman and Rothman 2017). Another research was conducted to avoid the cardiovascular aggregation of Ly-6C high monocytes through PLGA-based TAK-242 NP transmission. Furthermore, HMGB1 circulating, NF-ŚB activation in the heart, and IL-6 and MCP-1 articulated after MI lessened (Fujiwara et al. 2019). Promoting phenotypic regenerative transmission is a positive approach that can be used in the promotion of myocardial remodeling following MI (He et al. 2019). MiR-155 can be carried out in MPs through cardiac inflammation, hypertrophy, or failure to respond to pressure burden (Heymans et al. 2013). MiR-155 prohibitions have been observed to inhibit NGF production by reducing M1 polarization and cytokine signaling paths 1(SOCS1)/NFk-B (Hu et al. 2019). MiR-21 is yet a nucleic acid alternate capable of modifying MPs polarization. The NP-regulated delivery of miR-21 to cardiac MPs following MI exhibited polarization toward remediating angiogenesis, a decrease in left ventricle reshaping, and lesser apoptotic cell number (Bejerano et al. 2018). Similarly, the transition to the regenerative phenotype (CD206+) of antigen, neuregulin-1, by using microparticles of PLGA, was also seen that advanced cardiovascular regeneration by remedial polarization (Garbayo et al. 2016; Pascual-Gil et al. 2019). Conversely, introduction to dexamethasone repressed the discharge of NO and other pro-inflammatory components from MPs, in this way empowering cardiac reparation (Feiner et al. 2018). Another study demonstrated that berberine encapsulated liposomes can effectively be amassed in infarcted areas and were taken up by MPs, bringing about huge restraint of IL-6 discharge and superior cardiac expulsion fraction (Allijn et al. 2017). A delivery system administrated locally that can control drug discharge has been appeared to encourage cardiac tissue repair after MI, especially through the improved shift from M1 MPs toward reparative phenotype (He et al. 2019). A group of researchers has established NPs that retort to up-regulated MMPs in extreme MI and a mass in infarcts at 28 days of post-injection to sustain the preservation of NPs in infarcts after MI (Nguyen et al. 2015). In the same way, the enzyme-response hydrogel that advances the accretion and safety of drugs in infarctions has been investigated by Carlini and co-researchers. Overall, the different delivery mechanisms found are based on the transfer of biological drugs to MPs for the treatment of MI (Carlini et al. 2019). Specifically, canakinumab, the antibody indicated, enhances compelling cardiovascular repair in MI patients but requires continued dosing. In order to increase treatment adequacy, a locally dosed delivery mechanism with staggered dosage retention and prolonged drug release can be implemented.

4.3 Targeting MPs for Cancer Therapy

Progression of malignancy has sturdily compared with MPs homing and polarization. An assortment of struggles has demonstrated that the provision of solid tumors (STs) can be facilitated by tumor-associated MPs (TAMs), which involve up to half of the tumor mass in certain conditions (Cassetta and Pollard 2018; Liu and Wang 2020). Over 80% of exploration have indicated that STs with greater densities of TAMs identify with denied patient predictions (Qian and Pollard 2010). Regardless, a couple of uncommon cases exist, for instance, colorectal and pancreatic tumors (Qian and Pollard 2010; Edin et al. 2012, 2013). Plus, the incidence of TAMs has been seemed to affect the sufficiency of immunotherapy in STs (Pascual-García et al. 2019). Inflammation has immovably related to the progression of disease in various destructive conditions (Qian and Pollard 2010). While in-depth mechanics are so far hazy, persistent inflammation triggers key transcriptional factors, for instance, STAT3, NF-κB, and HIF1alpha, which control MPs to grasp inflammatory highlights and discharge positive pro-inflammatory adjudicators that are associated with a mutagenic microenvironment (Cassetta and Pollard 2018; Balkwill and Mantovani 2012). This is thought to be the fundamental aim behind why primary inflammatory cells in sicknesses like IBD are identified with a raised peril for cancerous growth (Ponzoni et al. 2018). In the course of chronic inflammation, these arbiters work to deliver tremendous amounts of myeloid cells, for instance, monocytes that changed into MPs upon their advent (Cassetta and Pollard 2018). When perceived, TAMs get AIF activity, outfitting encompassing cells, and tissues with signs to propel tumor evolution (de Visser et al. 2006). In like manner, TAMs are proven to support tumor advancement by propelling tumor commencement, tumor cell development, immune cell intrusion, and metastasis by methods for positive input pathways including CCL2 or possibly CSF, angiogenesis through implication with VEGF nodding, and immune subdual through PD-L1 or B7-H4 among others (He et al. 2019). In precise, TAMs are related to all episodes of metastasis, prepared for setting up ideal regions for metastatic cell development by propelling tumor cell extravasation, founding, and assuring the improvement of metastatic lesions (Qian and

Pollard 2010; Ruffell and Coussens 2015). Also, TAMs incite chemotherapy obstruction by giving endurance factors or potentially enacting hostile to apoptotic programs in cancerous cells. Given their perpetual function in malignant growth initiation and progression, MPs have arisen as a significant target for disease treatment.

4.3.1 MPs Targeting to Aim STs

Ongoing headways in immunotherapy have revised clinical therapeutics in various disorders designed to produce versatile immune reactions (Liu and Wang 2020; Zhang and Chen 2018). Given the essential capacity of MPs and monocytes in malignant growth, it remains the best juncture to target inherent immune cells to destroy ailment. On a very basic level, three methodologies are used to target MPs for disease treatment: first is to curb monocyte intrusion into STs, second to repolarize TAMs to shield STs, and at last wipe out TAMs from the tumor ambiance (Cassetta and Pollard 2018). This part will explore the key improvements, restrictions, and prospects for MPs and monocyte-targeting considering drug transporter for cancer treatment.

Stalling Monocyte Infiltration

The number of proofs has discovered that tissue-resident MPs (TRMs) are gotten from the fetal liver or yolk sac (Hoeffel et al. 2015; Schulz et al. 2012). Furthermore, neighborhood TRMs, TAMs of specific tumors have been set up to make from permeating BMPCs like circulating monocytes (Cassetta and Pollard 2018; Guerriero 2018). Accordingly, conduits to dismay monocyte entrance to aroused tissues identified with neoplastic tumors have occurred as an encouraging way to treat basic tumors and curb the advancement of metastatic states. TAMs archetypally recruit monocytes to STs by the emission of CSF-1, VEGF, CCL2-5, CCL8, and mCSF (Murdoch et al. 2004; Lewis and Pollard 2006). CCL2 explicitly is a viable chemoattractant that has accumulated enormous support as a significant facilitator of myeloid cell recruitment, as CCR2+ monocytes are the inflammatory forerunner of M2-polarized TAMs (Madsen et al. 2017). In this way, approaches pointed toward repressing the CCR2/CCL2 chemokine axle have become fundamental to numerous therapeutic procedures (Peña et al. 2015; Sanford et al. 2013). Two medications that restrain the CCR2/CCL2 chemokine axle are as of now under clinical examination: PF-04136309 and CNTO 888 (carlumab). PF-04136309 is a blocker of CCR2, which is in Phase-Ib clinical preliminary in a fusion with FOLFIRINOX, a therapeutic medicine blend including irinotecan and oxaliplatin with fluorouracil and leucovorin (Nywening et al. 2016). The examination found that the amalgamation was nontoxic and prolific compared to FOLFIRINOX alone. However, carlumab is an immunizer that complexed to CCL2. When conveyed intravenously, carlumab was seemed to impel backslide of prostate cancer growth by reducing monocyte and MPs permeation into the tumor microenvironment (Loberg et al. 2007a, b). Stage I and II clinical preliminaries were accomplished with carlumab; nevertheless, the subdual of CCL2 was transitory, and no therapeutic assets were found (Pienta et al. 2013). One examination establishes that the surface of the CCL2 deterrent achieved vigorous metastasis of breast carcinomas through angiogenesis (Bonapace et al. 2014). Consequently, auxiliary work is required to survey the safety and feasibility of CCL2 blockers as monotherapy. Regardless of this, the usage of CCL2 blockers in combination therapy is encouraging. Obstruction against CCL2 seemed to progress the sufficiency of chemotherapy with paclitaxel and carboplatin in a mouse ovarian cancer model (Moisan et al. 2014). Another looming target to thwart monocyte ingress may be through hindering the angiopoietins receptor TIE-2 on monocytes. Another examination found TIE-2-emitting whacked-out cells in vivo quite diminished angiogenesis in human glioma xenografts and incited broad tumor regress (De Palma et al. 2005). Disregarding these degrees of progress, an extended biological perception of monocyte recruitment to essential or metastatic tumors is vital before this procedure may show up at its greatest limit (He et al. 2019).

Repolarizing TAMs

The restoration of pro-inflammatory properties in TAMs exemplifies the fastest-growing approach to attacking MPs in cancer treatment (Wynn et al. 2013). This technique immediately uses intrinsic cells to have an extreme antitumor reaction that can function synergistically with medications that augment T cell penetration (e.g., checkpoint inhibitors) (Barclay and Van den Berg 2014). CD47 antibodies are valuable assets for tweaking MPs characteristics. During homeostasis, the CD47 function is to prevent phagocytosis by networking with thrombospondin 1 and the SIRP-α that are present in phagocytic myeloid cells like MPs and dendritic cells (Barclay and Van den Berg 2014). On the other hand, CD47 is typically over-expressed by every STs, giving a "don't consume me signal" to the adjacent milieu. Antibodies against CD47 have demonstrated extraordinary advancement in xenograft mouse models via mechanisms associated with phagocytosis pathways' activation in TAMs (He et al. 2019; Edris et al. 2012). A few inventive drug delivery systems have been created to convey anti-CD47. A research investigation explored the applicability of liposome-like particles to concurrently hinder the mCSF-CSF-1R signaling by means of obstructing CSF-1R while repressing the CD47-SPIR alpha signaling through SIRPα-impeding antibodies (Kulkarni et al. 2018). The treatment in combination with repolarized TAMs headed for M1 characteristics to advance anti-metastasis and antitumor impacts in breast and melanoma murine models. One TLR9 ligand (IMO-2055) and two small-particle TLR7 ligands (852A and imiquimod) are presently going through clinical examination (He et al. 2019). Zhang and co-researchers followed a strategy for developing injectable mannose-based PLGA-based NPs that encoded M1-polarized mRNA transcripts into TAMs (Zhang et al. 2019). They find that NP infusions transformed the immunosuppressive, tumor-assistant state of TAMs into phenotypes programmed in the melanoma, ovarian, and glioblastoma malignancy model for antitumor immunity to trigger and curb tumor development (Zhang et al. 2019). Another research reveals that conditions for removing the miR-processing enzyme DICER in MPs promote the programming of M1-like TAMs that are essentially the source of cytotoxicity tumor recruitment and tumors disrupted when utilized in fusions with anti-PD-1 or CD40 agonists (Baer et al. 2016). Different material structures have been tested in preclinical exams to potentiate improvements in TAM to M1 characteristics (He et al. 2019; Ovais et al. 2019). Another likely methodology to repolarize TAMs is by means of fragmentary radiotherapy. Customary ablative radiotherapy is a broadly embraced way to deal with executing tumorous tissues; notwithstanding, their impact at lesser dosages to potentiate antitumor aggregates is as yet going through examination. After a solitary ablative portion of 10Gy, the intrinsic immune system is commonly triggered by pro-fibrotic factors and inflammatory cytokines that solicit MPs with a tissue-reparative characteristic and helping tumor development (Barker et al. 2015). More effort requisite is done to assess the effectiveness and viability of this methodology before clinical analysis. In comparison, low-dose chemotherapy has been appeared to upgrade the impact of immunotherapies to reorient the immune system to receive antitumor characteristics (Pusuluri et al. 2019). Generally, whereas their phenotypic versatility marks MPs as alluring targets for malignancy treatment, it additionally makes the structure of sturdy therapeutic intercessions a key challenge. MPs constantly accustom their polarization because of the encompassing tissue microenvironment. Therefore, delivery vehicles with fleeting delivery profiles regularly bring about negligible potencies. Therefore, it is envisioned that controlled delivery approaches through materials configuration will turn out to be progressively significant (He et al. 2019).

Depleting TAMs

The abolition of TAMs is an approach to carnage MPs at tumor locations that are supposed to conflict with the organization of signals that facilitate the growth and development of tumors (Ponzoni et al. 2018). Strenuous antibodies or clodronate-containing liposomes are the most commonly accepted way to achieve TAM depletion to date.

The signaling pathway of CSF-1R pushes the conscription of TAMs to tumors and advances the differentiation of TAMs against protumor features. CSF-1R is a class III transmembrane tyrosine kinase receptor that is necessary for most MPs to be present. It is related to two ligands, CSF-1 and IL-34, and in both humans and mice, it directs MPs expansion, differentiation, and endurance (Cassetta and Pollard 2018). A barrier against this mechanism has consequently been an important area of focus for preventing TAM mobilization and advancing their depletion. PLX3397 (pexidartinib) has been shown to obstruct glioma formation, smother tumor cell expansion, and decrease tumor grade (Yan et al. 2017). In a preclinical experimental model of glioblastoma, CSF-1R inhibition with PLX3397 additionally made tumor cells extra prone to receptor tyrosine kinase restraint. PLX3397 was tested in clinical preliminary phases I and II, exhibiting antitumor efficacy as monotherapy in 52 percent of patients (Tap et al. 2015). Other small molecules that repress CSF-1R have been examined, comprising PLX7486, which may also interact with TRKS, and ARRY-382, both of which are subject to clinical examination (Cannarile et al. 2017). The antibody blockage was also established for CSF-1R. CSF-1R is related to RG7155 (emactuzumab) and inhibits its dimerization. Preclinical investigations showed that RG7155 caused an augmented proportion of CD8+/CD4+ T cells, which decreased CSF-1R+ CD163+ MPs in STs (Ries et al. 2014). In phase I clinical preliminaries in patients with advanced and metastatic tumors, this counteracting agent has experienced a devastating deal (Cassier et al. 2015). IMC-CS4 and FPA008 (cabiralizumab) are both CSF-1R antibodies now experiencing three sets of autonomous clinical preliminaries (Cannarile et al. 2017; Qiu et al. 2018). Similarly, AMG 820 has proceeded to preclinical studies, but insufficient antitumor activity has been noticed (Papadopoulos et al. 2017). Generally, CSF-1R inhibitors are an encouraging way to deal with deplete TAMs; nonetheless, discoveries from these preliminaries have restricted dose accelerations because of toxic issues that come about because of clearing MPs in healthy tissues. Hence, it envisions that CSF-1R repressor in fusion with different medications might be encouraging. The succeeding way to deal with TAMs depletion is by means of bisphosphonates (e.g., alendronate, clodronate), which can counteract MPs because of their particular interactivity with mononuclear cells. The upgrade of nanotechnology has arisen as an incredible asset to plan and convey bisphosphonates, commonly as liposomes (He et al. 2019). Clodrolip is imperative to specifically trigger MPs self-destruction, since phagocytosis is the characteristic fate of liposomes; hence, lysosomal phospholipases disrupt lipid bilayer. However, clodronate does not readily leave the cell and does not move into new phagocytic cells (Van Rooijen and Sanders 1994). The study found that clodronate-containing liposomes hindered tumor development by 55% compared to untreated controls while maintaining the chemo-attraction of fresh circulating monocytes (Banciu et al. 2008). Drenched angiogenesis, which was discovered recently, is an impending mechanism behind the limited chemo-attraction of new monocytes (Zeisberger et al. 2006). Liposomal clodronate constructed utilizing DOTAP, a positively charged surfactant extensively utilized to frame cationic liposomes, was subsequently used to inhibit multiplication and facilitate apoptosis of MPs in a dose- and time-dependent fashion, resulting in condensed tumor volumes in mouse melanoma models (Piaggio et al. 2016). Another study showed the improved affinity of clodronate to MPs by forming ALN to free glucomannan, a polysaccharide with a high affinity for the mannose receptor, for the effective consumption of TAMs (Zhan et al. 2014). Clodrolip has additionally been utilized against lung carcinomas, colon malignancy, T cell lymphoma, pancreatic disease, and different myeloma, among others (He et al. 2019). Most examinations including clodrolip and different bisphosphonates to deplete TAMs have been acted in mice; nonetheless, the small bunch of clinical preliminaries that have been delivered has demonstrated conflicting outcomes, proposing a need to streamline molecule plan and organization (Van Acker et al. 2016). Different examinations are exploring the impacts of joining clodronate as a novel adjuvant with hormonal treatments and chemotherapeutics (Cassetta and Pollard 2018).

Another moderately less-contemplated way to deplete TAMs is via trabectedin, which is a chemotherapeutic (small molecular size) able for targeting tumor cells and prompting apoptosis of monocytes and MPs (Germano et al. 2013). Trabectedin initiates caspase-8 via TRAIL to exhaust TAMs (Liguori et al. 2016). Trabectedin can likewise restrain aggravation, accordingly repressing tumor development, by stifling pro-inflammatory arbiters, CCL2, and IL-6, which ordinarily enroll monocytes at tumor locales and encourage tumor development, respectively (Allavena et al. 2005; Germano et al. 2010). In general, while encouraging in numerous preclinical and clinical investigations, a more extensive concern of the overall technique to deplete TAMs is that the shielding elements of MPs will be undermined, which can expand the threat for the disease (Purnama et al. 2014). In this way, it is accepted that cautious must be given to therapeutic conveyance methodologies and dosing to improve targeting and restrict undesired impacts. All in all, targeting TAMs is a proficient approach to treat malignancy. Above all the targeting techniques depicted, depletion of TAMs is a generally straightway to battle tumors since TAMs frequently comprise a huge division of the tumor load. Various biological medications and DDS utilizing this pathway are going through clinical analysis, demonstrating the potential for interpretation. Nonetheless, no medications are endorsed for clinical use due to the likely expanded danger for disease, which might be tended to with DDS with multiplexed cargos. For repolarizing TAMs, long haul introduction of boosts to incite solid immune response is needed; appropriately, DDS with continued delivery profile after some time is especially reasonable for TAM repolarization.

5 Nanomedicine: A Novel Approach to Deal with Cancer as a Prototype of MPs Targeting

In the arena of molecular targeting, the utilization of nanotechnology has given a wide scope of nanostructures intended to advance the conveyance of therapeutics toward MPs (Liu et al. 2020). Regardless of their underlined huge potential to expand the viability and to decrease the toxic effect of immunomodulatory drugs, upgradation in the endurance of patients is as yet modest. Diverse nanostructures have been utilized to exemplify pharmacological actives permitting them to beat regular issues of solvency and stability, to diminish their side effects, to expand their circulating half-time, and, sometimes, to empower their controlled delivery toward the objective cell (Andón et al. 2017).

5.1 MPs Captured Nanomedicines: New Live Cell-Intervened Drug Delivery Systems (LCDDS) for Cancer Therapy

An advancement in medication conveyance techniques, purported LCDDS, comprises the utilization of host cells of the patient (for example, stem cells, MPs, or monocytes) (Su et al. 2015), either in general or by utilizing chosen basic fragments of these cells (for example, outer cell layers) (Andón et al. 2017), as "Trojan Horses" stacked with medications. Significant favorable circumstances of live cell imaging attributes, of concern for their solicitation as "Trojan Horses," incorporate extended circulating time in blood, adaptable morphology, the proportion of ligands on the exterior of cells, and biological digestion (Su et al. 2015). Moreover, the ephemeral life expectancy of live cells (for example, leukocytes) must be utilized as an apt period for the conveyance, therapeutic activity, and physiological disposal of the "carrier." The most recent examinations, identified with the cell elements of the tumor microenvironment, show that TAMs are principally originated from circulating monocytes, which are stemmed from the peripheral blood in reply to wide-ranging scope of molecular signals (for example, CCL2/MCP-1 delivered from tumor cells) (Qian et al. 2011). Of interest, however, the greater part of the chemotherapeutic medications utilized in the clinical studies can just arrive at the vascular territory, and monocytes/MPs present a great capacity to infiltrate the tumor tissue profoundly, straight coming to the hypoxic area (Andón et al. 2017). This exciting capacity of myeloid cells to invade ailing tissues and to infiltrate testing biological obstructions, even when they are stacked with a therapeutic load, has begun to be researched. Moreover, it has been illustrated, essentially in vitro, that the phagocytic behavior of monocytes/MPs is valuable for their simple stacking with drug-enclosing nanostructures (Andón et al. 2017). For instance, Qin and co-workers have examined the ex vivo stacking of THP-1 monocytes with cRGD-altered liposomes enclosing anti-depressive macromolecular medication (trefoil factor 3). The intravenous delivery of these "loaded cells" in germane murine models brought about an acceptable execution to cross the BBB and the enhancement of indications (Qin et al. 2015). In another approach, Yuan and the co-investigator tried to target M1 and M2 MPs by dextran-modified polystyrene nanoparticles (DEX-PS-NPs) to the acute peritonitis and 4 T1 breast tumor (Fig. 2). The in vitro and in vivo association for NPs targeting at the specified area of the tumor was achieved (Yuan et al. 2020). The targeting and distribution of NPs at the tumor site were estimated after IV administration by tracing the intensity of loaded Nile red fluorescent dye. The active targeting of NPs to M2 MPs was observed by effectual recognition in in vitro scenario, however, lack to distinguish various subtypes of MPs in in vivo encounter. Eventually, the approach seemed to be effective in tracking the TAMs for targeting both acute peritonitis and tumor. An additionally testing tactic involves the in vivo targeting and stacking of monocytes in the circulation with nano constructs, which should then have the option to arrive at the heft of the tumor. For this, Smith et al. have created RGD (Arg-Gly-Asp) peptide named single-walled carbon nanotubes (SWCNTs) which were explicitly engulfed by Ly-6C high monocytes in the blood and hence conveyed into the tumor, indicating critical tumor entrance abilities (Smith et al. 2014). Even though these RGD-SWCNTs were not stacked with any medication, the positive result of this effort energizes the venture of forthcoming endeavors toward this path. Another intriguing ligand to arrive at monocytes with regard to flow can be β-glucan to focus on the pattern recognition receptor Dectin-1 receptor (Wu et al. 2010). Notwithstanding the focus on conceivable targets, the utilization of nanotechnology offers the possibility to stack the LCDDS with various pharmacological agents and to keep up these medications securely limited, until they arrive at the tumor site. When the LCDDS arrives at the suitable area for its therapeutic activity (for example, the focal point of the tumor), the medication must be delivered. For this, a choice has been the usage of nanotechnology moves toward that prompt the release of the medication upon physical or chemical incitement. However, the passive release of the medication at the targeted tissue might be accomplished via the choice of the apposite fragments of the nanocarrier, which essentially have a decent biodegradation profile when the LCDDS arrives at the target. For instance, notable biodegradable polymers with good clinical profiles, for example, polylactic acid (PLA), are accessible, and their biodegradation profiles have been appropriately examined (Xie et al. 2014; Makadia and Siegel 2011). Then again, to accomplish the active release of the medication, further developed nanotechnological techniques are required, to excavate the nanocarrier and to prompt the release of the medication at the desired time. For this, external stimulus, for example, magnetic field, and temperature, light, or ultrasound methodologies can be used (Su et al. 2015). In this way, a study showed disulfide linkages, L-asparaginase, for medication for intense lymphoblastic leukemia, with a CCP. This composite was captured into RBCs, and when it arrives at the cytoplasm the decrease of disulfide bonds by glutathione brings about the disassociation of the medication from the peptide, drug delivery, and therapeutic activity (He et al. 2014). Fundamentally, the stacking of monocytes/MPs with nanomedicines, afore they arrive at the tumor, is as yet in the beginning phases of examination and exhausting to apply in the facility. However, this procedure as of now exploits the physiological assets of the "carrier" cells, consequently limiting issues of toxicity,

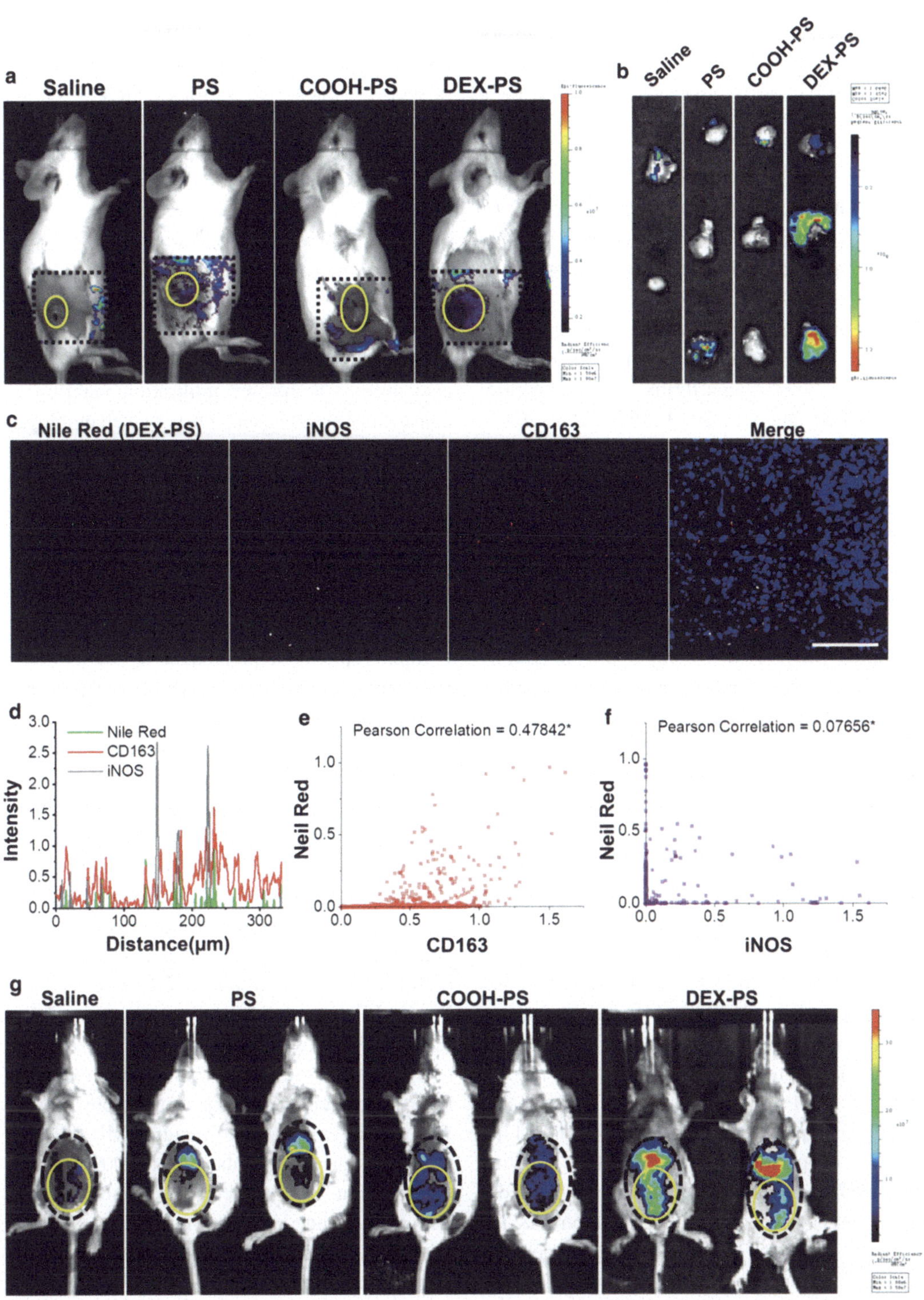

Fig. 2 *The illustration showing the in vivo targeting potential of DEX-PS to the acute peritonitis and tumor.* (**a**) Fluorescent signals emitted by the Nile red in a tumor of mice model (**In vivo**); (**b**). Fluorescent signals emitted by the Nile red in the isolated tumor of mice (**ex vivo**); (**c**). CLSM interpretation for NPs distribution in the tumor; (**d**) NPs localization assessment with CD163 and iNOS; Pearson co-relation assessments of (**e**) CD163 and NPs; and (**f**) iNOS and NPs; (**g**) Fluorescent signals emitted by the Nile red in acute peritonitis (**in vivo**). The black dotted marks are indicative of disease, whereas the solid yellow line represents the area of interest. (The illustration was adopted from Yuan et al. (2020))

and of their inborn ability to enter the tumor microenvironment in any event, coming to the most troublesome hypoxic locales in the focal point of the tumor.

5.2 Nanoparticles for MPs Imaging: For Prognosis and Identification of Cancer

A focal utility for MPs and monocytes has been depicted in malignancy as well as in different pathologies, including arthritis, CVD, tuberculosis, and TD1 (Jin et al. 2020; Andón et al. 2017). In this manner, the explicit targeting and perception of MPs and monocytes signify an exceptional prospect to screen and break down the advancement of these maladies. A broad assortment of symptomatic imaging endeavors have been merged with nanoparticle contrast agents and targeting probes for imaging techniques, for example, MRI, X-ray computed tomography (CT), PET, and single-photon emission computed tomography (SPECT) (Andón et al. 2017). These days, optical imaging shows a legitimate and upfront option in contrast to typical clinical imaging procedures for fundamental examination in animals, where the requirement for more profound tissue infiltration is decreased (Weissleder and Ntziachristos 2003). Besides, the progress of different fluorescent nanomaterials (for example, quantum dots) permits currently the upgraded enactment of preclinical optical imaging strategies (Cormode et al. 2008). MPs-targeted imaging is progressively utilized for the examination of malignant growth (Andón et al. 2017). The imaging of MPs invasion in STs gives applicable prophetic data, permitting the distinguishing proof of neoplastic edges and at times showing the adequacy of antitumor treatments (Weissleder et al. 2014). For instance, MRI empowers the representation of TAM in an assortment of tumors, comprising malignant growth in the breast (Daldrup-Link and Coussens 2012). Superparamagnetic iron oxide NPs (SPIO) are helpful contrast agents for MRI, and they can be fittingly designed, with suitable molecule surface and size properties, to be phagocytosed by MPs in various tissues. While 75–145 nm SPIO are quickly taken up by MPs of the mononuclear phagocyte system (MPS), for example, liver and spleen, conversely, ultra-small SPIO (USPIO) with a more modest width (<50 nm), endure lower retention in MPS-MPs, stay for an extra period in blood circulation, and a mass in tumor locales and inflamed tissues, arriving at TAMs and tissue MPs (Rosenblum et al. 2010; Corot et al. 2006). Ferumoxytol® is a USPIO that has been effectively utilized to measure MPs invasion by MRI in tumors. Ferumoxytol® has been also utilized to foresee infiltration viability of conveyance systems, and it is very well employed to anticipate patients' results in breast malignancy and to observe TAM-targeted treatments in clinical exercise. As of late, TAM-targeted dextran NPs, a fluorescent and cross-linked form of Ferumoxytol®, have been created to picture TAM penetration in a 3D organ imaging archetypal of lung carcinoma (Cuccarese et al. 2017). Lymphotropic superparamagnetic NPs, which are engulfed by MPs, have been utilized in fusion with high-resolution MRI for the non-intrusive identification of trifling lymph node metastasis in prostate cancer (Harisinghani et al. 2003). Another methodology, comprising the formation, by radiolabeling, of dextran NPs (recognized to amass in MPs) with 89Zr, has been employed to measure TAM penetration in a syngenic colon carcinoma mouse model (Keliher et al. 2011). Additionally, 89Zr radiolabeled reconstructed high-thickness lipoprotein (rHDL) has been utilized in PET imaging for the non-obtrusive observation of TAM in a mouse model of breast cancer (Pérez-Medina et al. 2015). In another examination, fluorescent molecular tomography (FMT) was merged with MRI to imagine MPs in mice with sarcoma (Leimgruber et al. 2009). In this investigation, TAMs were specifically marked with surface-tuned magneto-fluorescent NPs permitting their tracking inside the tumor microenvironment by both MRI and FMT. Another methodology for the recognition of pro-angiogenic TAM in tumors was reported by Movahedi and co-workers. They created 99mTc-marked MPs mannose receptor targeting nanobodies (single-domain antigen-binding fragments from *Camelidae* heavy-chain antibodies) with admirable features for the imaging by SPECT of MPs over exhibited the mannose receptor in the tumor stroma (Movahedi et al. 2012).

As an unabridged, these imaging methods, sometimes with appropriate alterations, are pertinent to consider the MPs invasion in STs, and all the more significantly, to assess their explicit localization (for example, central or periphery), phenotypes, and capacities. This exhaustive investigation of MPs in tumors, which can be accomplished utilizing novel nanotechnological imaging methods, can prompt a superior assessment of the malignant growth and for a superior expectation of a response to antitumoral treatments. These identification and prognostic examinations increase expanding importance with the escalating utilization of disease immunotherapeutics (e.g., checkpoint antagonists) which are employed now in the clinical practice to battle the immunosuppressive cells of the tumor microenvironment, among TAM (Allen et al. 2017; Gordon et al. 2017).

6 Future Prospects

MPs are omnivorous in every immune response and form a bridge between innate and acquired immunity in nearly all autoimmune disorders. Further phenotypic variants of MPs and their subsequent polarization are key aspects in aggravating tissue damage or other harmful diseases. MPs have emerged as imminent targets for immune disorders but are still explored to infancy. This chapter highlights macrophage-related diseases, pathophysiology, obstacles for recovery, and potential alternatives to the clinical focus of various immune disorders along with nanotechnological targeting approaches. Although enormous efforts have been specified to target and modify the functionality of MPs, their in vitro and in vivo correlation, preclinical models, and person-to-person diseased variability are still visions to achieve. These opportunities would lay the base for further exploration and targeting of MPs by merging the nanotechnological approaches for achieving the most effective treatment.

Acknowledgments The authors are grateful for the literature access facility required for this research at the UIOP, PRSU, Raipur, India. The authors are also thankful to ICMR (V.25011/286-HRD/2016-HR) for the funding.

Declaration of Interest None.

Acronyms

AMPK	AMP-activated protein kinase
CCL2	The chemokine (C-C motif) ligand 2
CD	Cluster of differentiation
CM-CSF	Granulocyte–macrophage colony-stimulating factor
CNS	Central nervous system
COOH-PS	Carboxyl-functionalized polystyrene
CRP	C-reactive protein
CTGF	Connective tissue growth factor
CXCL10	C-X-C motif chemokine ligand 10
DEX-PS	dextran-functionalized polystyrene
DOTAP	1,2-Dioleoyl-3-trimethyl ammonium-propane
DSS-colitis mice	Dextran sodium sulfate-colitis mice
EAE	Experimental autoimmune encephalomyelitis
ECM	Extracellular matrix
HABN	Hyaluronic acid–bilirubin nanomedicine
HDAC	Histone deacetylases
HIF1alpha	Hypoxia-inducible factor 1-alpha
HLA	Human leukocyte antigens
HMGB1	High mobility group box 1 protein
ICAM-1	Intercellular adhesion molecule 1
IFN	Interferon
IL	Interleukin
iNOS	inducible nitric oxide synthase

LPS	Lipopolysaccharide
Ly6G	Lymphocyte antigen 6 complex locus G6D
MCP-1	Monocyte chemoattractant protein-1
MHC	Major histocompatibility complex
MMPs	Matrix metalloproteinases
MSCs	Mesenchymal stem cells
NF-κB	Nuclear factor-kappa light chain enhancer of activated B cells
NSAIDs	Nonsteroidal anti-inflammatory drugs
PDGF	Platelet-derived growth factor
PI3Kγ	Phosphoinositide 3-kinase gamma
PLGA	Poly Lactic-co-Glycolic Acid
PPARγ	Peroxisome proliferator-activated receptor gamma
PS	Polystyrene
ROS	Reactive oxygen species
SIRT1	Sirtuin1
SPARC	Secreted protein acidic and rich in cysteine
STAT3	Signal transducer and activator of transcription 3
TAK-242	Resatorvid
TGF-beta	Transforming growth factor beta
THAP	Thapsigargin
TNF	Tumor necrosis factor
TRMs	Tissue-resident macrophages
VEGF	Vascular endothelial growth factor

References

Afratis NA, Klepfish M, Karamanos NK, Sagi I. The apparent competitive action of ECM proteases and cross-linking enzymes during fibrosis: applications to drug discovery. Adv Drug Deliv Rev. 2018;129:4–15. https://doi.org/10.1016/j.addr.2018.03.004.

Albiero M, Poncina N, Ciciliot S, et al. Bone marrow macrophages contribute to diabetic stem cell mobilopathy by producing oncostatin M. Diabetes. 2015;64(8):2957–68. https://doi.org/10.2337/db14-1473.

Allavena P, Signorelli M, Chieppa M, et al. Anti-inflammatory properties of the novel antitumor agent yondelis (trabectedin): inhibition of macrophage differentiation and cytokine production. Cancer Res. 2005;65(7):2964–71. https://doi.org/10.1158/0008-5472.CAN-04-4037.

Allen E, Jabouille A, Rivera LB, et al. Combined anti-angiogenic and anti-PD-L1 therapy stimulates tumor immunity through HEV formation. Sci Transl Med. 2017;9(385) https://doi.org/10.1126/scitranslmed.aak9679.

Allijn IE, Czarny BMS, Wang X, et al. Liposome encapsulated berberine treatment attenuates cardiac dysfunction after myocardial infarction. J Control Release. 2017;247:127–33. https://doi.org/10.1016/j.jconrel.2016.12.042.

Andón FT, Digifico E, Maeda A, et al. Targeting tumor associated macrophages: the new challenge for nanomedicine. Semin Immunol. 2017;34:103–13. https://doi.org/10.1016/j.smim.2017.09.004.

Arora S, Dev K, Agarwal B, Das P, Syed MA. Macrophages: their role, activation and polarization in pulmonary diseases. Immunobiology. 2018;223(4-5):383–96. https://doi.org/10.1016/j.imbio.2017.11.001.

Baer C, Squadrito ML, Laoui D, et al. Suppression of microRNA activity amplifies IFN-γ-induced macrophage activation and promotes anti-tumour immunity. Nat Cell Biol. 2016;18(7):790–802. https://doi.org/10.1038/ncb3371.

Bailey JR, Bland PW, Tarlton JF, et al. IL-13 promotes collagen accumulation in Crohn's disease fibrosis by down-regulation of fibroblast MMP synthesis: a role for innate lymphoid cells? PLoS One. 2012;7(12):e52332. https://doi.org/10.1371/journal.pone.0052332.

Balkwill FR, Mantovani A. Cancer-related inflammation: common themes and therapeutic opportunities. Semin Cancer Biol. 2012;22(1):33–40. https://doi.org/10.1016/j.semcancer.2011.12.005.

Banciu M, Metselaar JM, Schiffelers RM, Storm G. Antitumor activity of liposomal prednisolone phosphate depends on the presence of functional tumor-associated macrophages in tumor tissue. Neoplasia. 2008;10(2):108–17. https://doi.org/10.1593/neo.07913.

Banerjee A, Wang J, Bodhankar S, Vandenbark AA, Murphy SJ, Offner H. Phenotypic changes in immune cell subsets reflect increased infarct volume in male vs. female mice. Transl Stroke Res. 2013;4(5):554–63. https://doi.org/10.1007/s12975-013-0268-z.

Bannon P, Wood S, Restivo T, Campbell L, Hardman MJ, Mace KA. Diabetes induces stable intrinsic changes to myeloid cells that contribute to chronic inflammation during wound healing in mice. Dis Model Mech. 2013;6(6):1434–47. https://doi.org/10.1242/dmm.012237.

Barclay AN, Van den Berg TK. The interaction between signal regulatory protein alpha (SIRPα) and CD47:

structure, function, and therapeutic target. Annu Rev Immunol. 2014;32:25–50. https://doi.org/10.1146/annurev-immunol-032713-120142.

Barker HE, Paget JTE, Khan AA, Harrington KJ. The tumour microenvironment after radiotherapy: mechanisms of resistance and recurrence. Nat Rev Cancer. 2015;15(7):409–25. https://doi.org/10.1038/nrc3958.

Baumann I, Kolowos W, Voll RE, et al. Impaired uptake of apoptotic cells into tingible body macrophages in germinal centers of patients with systemic lupus erythematosus. Arthritis Rheum. 2002;46(1):191–201. https://doi.org/10.1002/1529-0131(200201)46:1<191::AID-ART10027>3.0.CO;2-K.

Bejerano T, Etzion S, Elyagon S, Etzion Y, Cohen S. Nanoparticle delivery of miRNA-21 mimic to cardiac macrophages improves myocardial remodeling after myocardial infarction. Nano Lett. 2018;18(9):5885–91. https://doi.org/10.1021/acs.nanolett.8b02578.

Biesen R, Demir C, Barkhudarova F, et al. Sialic acid-binding Ig-like lectin 1 expression in inflammatory and resident monocytes is a potential biomarker for monitoring disease activity and success of therapy in systemic lupus erythematosus. Arthritis Rheum. 2008;58(4):1136–45. https://doi.org/10.1002/art.23404.

Blanco P, Palucka AK, Gill M, Pascual V, Banchereau J. Induction of dendritic cell differentiation by IFN-alpha in systemic lupus erythematosus. Science. 2001;294(5546):1540–3. https://doi.org/10.1126/science.1064890.

Bonapace L, Coissieux M-M, Wyckoff J, et al. Cessation of CCL2 inhibition accelerates breast cancer metastasis by promoting angiogenesis. Nature. 2014;515(7525):130–3. https://doi.org/10.1038/nature13862.

Bouma G, Strober W. The immunological and genetic basis of inflammatory bowel disease. Nat Rev Immunol. 2003;3(7):521–33. https://doi.org/10.1038/nri1132.

Bramow S, Frischer JM, Lassmann H, et al. Demyelination versus remyelination in progressive multiple sclerosis. Brain. 2010;133(10):2983–98. https://doi.org/10.1093/brain/awq250.

Brown BN, Valentin JE, Stewart-Akers AM, McCabe GP, Badylak SF. Macrophage phenotype and remodeling outcomes in response to biologic scaffolds with and without a cellular component. Biomaterials. 2009;30(8):1482–91. https://doi.org/10.1016/j.biomaterials.2008.11.040.

Burger D, Molnarfi N, Weber MS, et al. Glatiramer acetate increases IL-1 receptor antagonist but decreases T cell-induced IL-1beta in human monocytes and multiple sclerosis. Proc Natl Acad Sci U S A. 2009;106(11):4355–9. https://doi.org/10.1073/pnas.0812183106.

Byrne JC, Ní Gabhann J, Lazzari E, et al. Genetics of SLE: functional relevance for monocytes/macrophages in disease. Niewold TB, ed. Clin Dev Immunol. 2012;2012:582352. https://doi.org/10.1155/2012/582352.

Calderon B, Suri A, Unanue ER. In CD4+ T-cell-induced diabetes, macrophages are the final effector cells that mediate islet beta-cell killing: studies from an acute model. Am J Pathol. 2006;169(6):2137–47. https://doi.org/10.2353/ajpath.2006.060539.

Cannarile MA, Weisser M, Jacob W, Jegg A-M, Ries CH, Rüttinger D. Colony-stimulating factor 1 receptor (CSF1R) inhibitors in cancer therapy. J Immunother Cancer. 2017;5(1):53. https://doi.org/10.1186/s40425-017-0257-y.

Cao Q, Wang Y, Harris DCH. Pathogenic and protective role of macrophages in kidney disease. Am J Physiol Renal Physiol. 2013;305(1):F3–F11. https://doi.org/10.1152/ajprenal.00122.2013.

Cao Q, Wang Y, Zheng D, et al. Failed renoprotection by alternatively activated bone marrow macrophages is due to a proliferation-dependent phenotype switch in vivo. Kidney Int. 2014;85(4):794–806. https://doi.org/10.1038/ki.2013.341.

Carlini AS, Gaetani R, Braden RL, Luo C, Christman KL, Gianneschi NC. Enzyme-responsive progelator cyclic peptides for minimally invasive delivery to the heart post-myocardial infarction. Nat Commun. 2019;10(1):1735. https://doi.org/10.1038/s41467-019-09587-y.

Carrero JA, Ferris ST, Unanue ER. Macrophages and dendritic cells in islets of Langerhans in diabetic autoimmunity: a lesson on cell interactions in a mini-organ. Curr Opin Immunol. 2016;43:54–9. https://doi.org/10.1016/j.coi.2016.09.004.

Carrero JA, McCarthy DP, Ferris ST, et al. Resident macrophages of pancreatic islets have a seminal role in the initiation of autoimmune diabetes of NOD mice. Proc Natl Acad Sci U S A. 2017;114(48):E10418–27. https://doi.org/10.1073/pnas.1713543114.

Cassetta L, Pollard JW. Targeting macrophages: therapeutic approaches in cancer. Nat Rev Drug Discov. 2018;17(12):887–904. https://doi.org/10.1038/nrd.2018.169.

Cassier PA, Italiano A, Gomez-Roca CA, et al. CSF1R inhibition with emactuzumab in locally advanced diffuse-type tenosynovial giant cell tumours of the soft tissue: a dose-escalation and dose-expansion phase 1 study. Lancet Oncol. 2015;16(8):949–56. https://doi.org/10.1016/S1470-2045(15)00132-1.

Chen L, Chen R, Wang H, Liang F. Mechanisms linking inflammation to insulin resistance. Int J Endocrinol. 2015;2015:508409. https://doi.org/10.1155/2015/508409.

Chen M, Li X, Wang S, Yu L, Tang J, Zhou S. The role of cardiac macrophage and cytokines on ventricular arrhythmias. Front Physiol. 2020;11:1113. https://doi.org/10.3389/fphys.2020.01113.

Chistiakov DA, Myasoedova VA, Revin VV, Orekhov AN, Bobryshev YV. The impact of interferon-regulatory factors to macrophage differentiation and polarization into M1 and M2. Immunobiology. 2018;223(1):101–11. https://doi.org/10.1016/j.imbio.2017.10.005.

Chylikova J, Dvorackova J, Tauber Z, Kamarad V. M1/M2 macrophage polarization in human obese adipose tissue. Biomed Pap Med Fac Univ Palacky Olomouc Czech Repub. 2018;162(2):79–82. https://doi.org/10.5507/bp.2018.015.

Cormode DP, Skajaa T, van Schooneveld MM, et al. Nanocrystal core high-density lipoproteins: a multimodality contrast agent platform. Nano Lett. 2008;8(11):3715–23. https://doi.org/10.1021/nl801958b.

Corot C, Robert P, Idée J-M, Port M. Recent advances in iron oxide nanocrystal technology for medical imaging. Adv Drug Deliv Rev. 2006;58(14):1471–504. https://doi.org/10.1016/j.addr.2006.09.013.

Cosín-Roger J, Ortiz-Masiá D, Calatayud S, et al. M2 macrophages activate WNT signaling pathway in epithelial cells: relevance in ulcerative colitis. PLoS One. 2013;8(10):e78128. https://doi.org/10.1371/journal.pone.0078128.

Costa NL, Valadares MC, Souza PPC, et al. Tumor-associated macrophages and the profile of inflammatory cytokines in oral squamous cell carcinoma. Oral Oncol. 2013;49(3):216–23. https://doi.org/10.1016/j.oraloncology.2012.09.012.

Crossman D, Rothman A. The Canakinumab Antiinflammatory Thrombosis Outcome Study trial-the starting gun has fired. J Thorac Dis. 2017;9(12):4922–5. https://doi.org/10.21037/jtd.2017.11.96.

Cuccarese MF, Dubach JM, Pfirschke C, et al. Heterogeneity of macrophage infiltration and therapeutic response in lung carcinoma revealed by 3D organ imaging. Nat Commun. 2017;8(1):14293. https://doi.org/10.1038/ncomms14293.

Culemann S, Grüneboom A, Nicolás-Ávila JÁ, et al. Locally renewing resident synovial macrophages provide a protective barrier for the joint. Nature. 2019;572(7771):670–5. https://doi.org/10.1038/s41586-019-1471-1.

Daldrup-Link H, Coussens LM. MR imaging of tumor-associated macrophages. Onco Targets Ther. 2012;1(4):507–9. https://doi.org/10.4161/onci.19456.

Danese S. Immune and nonimmune components orchestrate the pathogenesis of inflammatory bowel disease. Am J Physiol Liver Physiol. 2011;300(5):G716–22. https://doi.org/10.1152/ajpgi.00472.2010.

Daneshmandi S, Karimi MH, Pourfathollah AA. TGF-β engineered mesenchymal stem cells (TGF-β/MSCs) for treatment of type 1 diabetes (T1D) mice model. Int Immunopharmacol. 2017;44:191–6. https://doi.org/10.1016/j.intimp.2017.01.019.

De Palma M, Venneri MA, Galli R, et al. Tie2 identifies a hematopoietic lineage of proangiogenic monocytes required for tumor vessel formation and a mesenchymal population of pericyte progenitors. Cancer Cell. 2005;8(3):211–26. https://doi.org/10.1016/j.ccr.2005.08.002.

de Visser KE, Eichten A, Coussens LM. Paradoxical roles of the immune system during cancer development. Nat Rev Cancer. 2006;6(1):24–37. https://doi.org/10.1038/nrc1782.

Decker P, Kötter I, Klein R, Berner B, Rammensee H-G. Monocyte-derived dendritic cells over-express CD86 in patients with systemic lupus erythematosus. Rheumatology (Oxford). 2006;45(9):1087–95. https://doi.org/10.1093/rheumatology/kel061.

Demetter P, De Vos M, Van Huysse JA, et al. Colon mucosa of patients both with spondyloarthritis and Crohn's disease is enriched with macrophages expressing the scavenger receptor CD163. Ann Rheum Dis. 2005;64(2):321–4. https://doi.org/10.1136/ard.2003.018382.

Dick SA, Macklin JA, Nejat S, et al. Self-renewing resident cardiac macrophages limit adverse remodeling following myocardial infarction. Nat Immunol. 2019;20(1):29–39. https://doi.org/10.1038/s41590-018-0272-2.

Djedović N, Stanisavljevic S, Jevtić B, Momčilović M, Lavrnja I, Miljković D. Anti-encephalitogenic effects of ethyl pyruvate are reflected in the central nervous system and the gut. Biomed Pharmacother. 2017;96:78–85. https://doi.org/10.1016/j.biopha.2017.09.110.

Duncan SE, Gao S, Sarhene M, et al. Macrophage activities in myocardial infarction and heart failure. Cardiol Res Pract. 2020;2020:4375127. https://doi.org/10.1155/2020/4375127.

Edin S, Wikberg ML, Dahlin AM, et al. The distribution of macrophages with a M1 or M2 phenotype in relation to prognosis and the molecular characteristics of colorectal cancer. PLoS One. 2012;7(10):e47045. https://doi.org/10.1371/journal.pone.0047045.

Edin S, Wikberg ML, Oldenborg P-A, Palmqvist R. Macrophages: good guys in colorectal cancer. Onco Targets Ther. 2013;2(2):e23038. https://doi.org/10.4161/onci.23038.

Edris B, Weiskopf K, Volkmer AK, et al. Antibody therapy targeting the CD47 protein is effective in a model of aggressive metastatic leiomyosarcoma. Proc Natl Acad Sci U S A. 2012;109(17):6656–61. https://doi.org/10.1073/pnas.1121629109.

Falkenham A, de Antueno R, Rosin N, et al. Nonclassical resident macrophages are important determinants in the development of myocardial fibrosis. Am J Pathol. 2015;185(4):927–42. https://doi.org/10.1016/j.ajpath.2014.11.027.

Fan C, Long R, You Y, et al. A novel PADRE-Kv1.3 vaccine effectively induces therapeutic antibodies and ameliorates experimental autoimmune encephalomyelitis in rats. Clin Immunol. 2018;193:98–109. https://doi.org/10.1016/j.clim.2018.02.012.

Feiner R, Fleischer S, Shapira A, Kalish O, Dvir T. Multifunctional degradable electronic scaffolds for cardiac tissue engineering. J Control Release. 2018;281:189–95. https://doi.org/10.1016/j.jconrel.2018.05.023.

Ferris ST, Zakharov PN, Wan X, et al. The islet-resident macrophage is in an inflammatory state and senses microbial products in blood. J Exp Med. 2017;214(8):2369–85. https://doi.org/10.1084/jem.20170074.

Fujiwara M, Matoba T, Koga J-I, et al. Nanoparticle incorporating Toll-like receptor 4 inhibitor attenuates myocardial ischaemia-reperfusion injury by inhibiting monocyte-mediated inflammation in mice. Cardiovasc Res. 2019;115(7):1244–55. https://doi.org/10.1093/cvr/cvz066.

Fukui S, Iwamoto N, Takatani A, et al. M1 and M2 monocytes in rheumatoid arthritis: a contribution of imbalance of M1/M2 monocytes to osteoclastogenesis. Front Immunol. 2017;8:1958. https://doi.org/10.3389/fimmu.2017.01958.

Gallagher KA, Joshi A, Carson WF, et al. Epigenetic changes in bone marrow progenitor cells influence the inflammatory phenotype and alter wound healing in type 2 diabetes. Diabetes. 2015;64(4):1420–30. https://doi.org/10.2337/db14-0872.

Garbayo E, Gavira JJ, de Yebenes MG, et al. Catheter-based intramyocardial injection of FGF1 or NRG1-loaded MPs improves cardiac function in a preclinical model of ischemia-reperfusion. Sci Rep. 2016;6:25932. https://doi.org/10.1038/srep25932.

Gatto M, Zen M, Iaccarino L, Doria A. New therapeutic strategies in systemic lupus erythematosus management. Nat Rev Rheumatol. 2019;15(1):30–48. https://doi.org/10.1038/s41584-018-0133-2.

Germano G, Frapolli R, Simone M, et al. Antitumor and anti-inflammatory effects of trabectedin on human myxoid liposarcoma cells. Cancer Res. 2010;70(6):2235–44. https://doi.org/10.1158/0008-5472.CAN-09-2335.

Germano G, Frapolli R, Belgiovine C, et al. Role of macrophage targeting in the antitumor activity of trabectedin. Cancer Cell. 2013;23(2):249–62. https://doi.org/10.1016/j.ccr.2013.01.008.

Giacomini E, Severa M, Rizzo F, et al. IFN-β therapy modulates B-cell and monocyte crosstalk via TLR7 in multiple sclerosis patients. Eur J Immunol. 2013;43(7):1963–72. https://doi.org/10.1002/eji.201243212.

Goldstein JI, Kominsky DJ, Jacobson N, et al. Defective leukocyte GM-CSF receptor (CD116) expression and function in inflammatory bowel disease. Gastroenterology. 2011;141(1):208–16. https://doi.org/10.1053/j.gastro.2011.03.060.

Gordon S, Martinez FO. Alternative activation of macrophages: mechanism and functions. Immunity. 2010;32(5):593–604. https://doi.org/10.1016/j.immuni.2010.05.007.

Gordon S, Plüddemann A. Tissue macrophages: heterogeneity and functions. BMC Biol. 2017;15(1):53. https://doi.org/10.1186/s12915-017-0392-4.

Gordon SR, Maute RL, Dulken BW, et al. PD-1 expression by tumour-associated macrophages inhibits phagocytosis and tumour immunity. Nature. 2017;545(7655):495–9. https://doi.org/10.1038/nature22396.

Guerriero JL. Macrophages: the road less traveled, changing anticancer therapy. Trends Mol Med. 2018;24(5):472–89. https://doi.org/10.1016/j.molmed.2018.03.006.

Guilliams M, Ginhoux F, Jakubzick C, et al. Dendritic cells, monocytes and macrophages: a unified nomenclature based on ontogeny. Nat Rev Immunol. 2014;14(8):571–8. https://doi.org/10.1038/nri3712.

Harel-Adar T, Ben Mordechai T, Amsalem Y, Feinberg MS, Leor J, Cohen S. Modulation of cardiac macrophages by phosphatidylserine-presenting liposomes improves infarct repair. Proc Natl Acad Sci U S A. 2011;108(5):1827–32. https://doi.org/10.1073/pnas.1015623108.

Harisinghani MG, Barentsz J, Hahn PF, et al. Noninvasive detection of clinically occult lymph-node metastases in prostate cancer. N Engl J Med. 2003;348(25):2491–9. https://doi.org/10.1056/NEJMoa022749.

He H, Ye J, Wang Y, et al. Cell-penetrating peptides meditated encapsulation of protein therapeutics into intact red blood cells and its application. J Control Release. 2014;176:123–32. https://doi.org/10.1016/j.jconrel.2013.12.019.

He W, Kapate N, Shields CW, Mitragotri S. Drug delivery to macrophages: A review of targeting drugs and drug carriers to macrophages for inflammatory diseases. Adv Drug Deliv Rev. 2019;(December). https://doi.org/10.1016/j.addr.2019.12.001

Heymann F, Trautwein C, Tacke F. Monocytes and macrophages as cellular targets in liver fibrosis. Inflamm Allergy Drug Targets. 2009;8(4):307–18. https://doi.org/10.2174/187152809789352230.

Heymans S, Corsten MF, Verhesen W, et al. Macrophage microRNA-155 promotes cardiac hypertrophy and failure. Circulation. 2013;128(13):1420–32. https://doi.org/10.1161/CIRCULATIONAHA.112.001357.

Hinderer S, Schenke-Layland K. Cardiac fibrosis – a short review of causes and therapeutic strategies. Adv Drug Deliv Rev. 2019;146:77–82. https://doi.org/10.1016/j.addr.2019.05.011.

Hirayama D, Iida T, Nakase H. The phagocytic function of macrophage-enforcing innate immunity and tissue homeostasis. Int J Mol Sci. 2017;19(1) https://doi.org/10.3390/ijms19010092.

Hoeffel G, Chen J, Lavin Y, et al. C-Myb(+) erythro-myeloid progenitor-derived fetal monocytes give rise to adult tissue-resident macrophages. Immunity. 2015;42(4):665–78. https://doi.org/10.1016/j.immuni.2015.03.011.

Holz K, Prinz M, Brendecke SM, et al. Differing outcome of experimental autoimmune encephalitis in macrophage/neutrophil- and T cell-specific gp130-deficient mice. Front Immunol. 2018;9:836. https://doi.org/10.3389/fimmu.2018.00836.

Hu J, Huang C-X, Rao P-P, et al. Inhibition of microRNA-155 attenuates sympathetic neural remodeling following myocardial infarction via reducing M1 macrophage polarization and inflammatory responses in mice. Eur J Pharmacol. 2019;851:122–32. https://doi.org/10.1016/j.ejphar.2019.02.001.

Hwang I, Kim JW, Ylaya K, et al. Tumor-associated macrophage, angiogenesis and lymphangiogenesis markers predict prognosis of non-small cell lung cancer patients. J Transl Med. 2020;18(1):443. https://doi.org/10.1186/s12967-020-02618-z.

Inoue M, Williams KL, Gunn MD, Shinohara ML. NLRP3 inflammasome induces chemotactic immune cell migration to the CNS in experimental autoimmune encephalomyelitis. Proc Natl Acad Sci U S A. 2012;109(26):10480–5. https://doi.org/10.1073/pnas.1201836109.
Jacobsen RN, Perkins AC, Levesque J-P. Macrophages and regulation of erythropoiesis. Curr Opin Hematol. 2015;22(3):212–9. https://doi.org/10.1097/MOH.0000000000000131.
Jakubzick CV, Randolph GJ, Henson PM. Monocyte differentiation and antigen-presenting functions. Nat Rev Immunol. 2017;17(6):349–62. https://doi.org/10.1038/nri.2017.28.
Jiang H-R, Milovanović M, Allan D, et al. IL-33 attenuates EAE by suppressing IL-17 and IFN-γ production and inducing alternatively activated macrophages. Eur J Immunol. 2012;42(7):1804–14. https://doi.org/10.1002/eji.201141947.
Jiang Z, Jiang JX, Zhang G-X. Macrophages: a double-edged sword in experimental autoimmune encephalomyelitis. Immunol Lett. 2014;160(1):17–22. https://doi.org/10.1016/j.imlet.2014.03.006.
Jiang K, Weaver JD, Li Y, Chen X, Liang J, Stabler CL. Local release of dexamethasone from macroporous scaffolds accelerates islet transplant engraftment by promotion of anti-inflammatory M2 macrophages. Biomaterials. 2017;114:71–81. https://doi.org/10.1016/j.biomaterials.2016.11.004.
Jin C, Wang K, Oppong-Gyebi A, Hu J. Application of nanotechnology in cancer diagnosis and therapy – a mini-review. Int J Med Sci. 2020;17(18):2964–73. https://doi.org/10.7150/ijms.49801.
Juhas U, Ryba-Stanisławowska M, Szargiej P, Myśliwska J. Different pathways of macrophage activation and polarization. Postepy Hig Med Dosw (Online). 2015;69:496–502. https://doi.org/10.5604/17322693.1150133.
Kamada N, Hisamatsu T, Honda H, et al. TL1A produced by lamina propria macrophages induces Th1 and Th17 immune responses in cooperation with IL-23 in patients with Crohn's disease. Inflamm Bowel Dis. 2010;16(4):568–75. https://doi.org/10.1002/ibd.21124.
Katsiari CG, Liossis S-NC, Souliotis VL, Dimopoulos AM, Manoussakis MN, Sfikakis PP. Aberrant expression of the costimulatory molecule CD40 ligand on monocytes from patients with systemic lupus erythematosus. Clin Immunol. 2002;103(1):54–62. https://doi.org/10.1006/clim.2001.5172.
Kavai M, Szegedi G. Immune complex clearance by monocytes and macrophages in systemic lupus erythematosus. Autoimmun Rev. 2007;6(7):497–502. https://doi.org/10.1016/j.autrev.2007.01.017.
Keliher EJ, Yoo J, Nahrendorf M, et al. 89Zr-labeled dextran nanoparticles allow in vivo macrophage imaging. Bioconjug Chem. 2011;22(12):2383–9. https://doi.org/10.1021/bc200405d.
Kigerl KA, Gensel JC, Ankeny DP, Alexander JK, Donnelly DJ, Popovich PG. Identification of two distinct macrophage subsets with divergent effects causing either neurotoxicity or regeneration in the injured mouse spinal cord. J Neurosci. 2009;29(43):13435–44. https://doi.org/10.1523/JNEUROSCI.3257-09.2009.
Kornberg MD, Smith MD, Shirazi HA, Calabresi PA, Snyder SH, Kim PM. Bryostatin-1 alleviates experimental multiple sclerosis. Proc Natl Acad Sci U S A. 2018;115(9):2186–91. https://doi.org/10.1073/pnas.1719902115.
Krzyszczyk P, Schloss R, Palmer A, Berthiaume F. The role of macrophages in acute and chronic wound healing and interventions to promote pro-wound healing phenotypes. Front Physiol. 2018;9:419. https://doi.org/10.3389/fphys.2018.00419.
Kuang D-M, Zhao Q, Peng C, et al. Activated monocytes in peritumoral stroma of hepatocellular carcinoma foster immune privilege and disease progression through PD-L1. J Exp Med. 2009;206(6):1327–37. https://doi.org/10.1084/jem.20082173.
Kulkarni A, Chandrasekar V, Natarajan SK, et al. A designer self-assembled supramolecule amplifies macrophage immune responses against aggressive cancer. Nat Biomed Eng. 2018;2(8):589–99. https://doi.org/10.1038/s41551-018-0254-6.
Kumar S, Ramesh A, Kulkarni A. Targeting macrophages: a novel avenue for cancer drug discovery. Expert Opin Drug Discov. 2020;15(5):561–74. https://doi.org/10.1080/17460441.2020.1733525.
Lavin Y, Winter D, Blecher-Gonen R, et al. Tissue-resident macrophage enhancer landscapes are shaped by the local microenvironment. Cell. 2014;159(6):1312–26. https://doi.org/10.1016/j.cell.2014.11.018.
Le LV, Mohindra P, Fang Q, et al. Injectable hyaluronic acid based microrods provide local micromechanical and biochemical cues to attenuate cardiac fibrosis after myocardial infarction. Biomaterials. 2018;169:11–21. https://doi.org/10.1016/j.biomaterials.2018.03.042.
Lee S, Kivimäe S, Dolor A, Szoka FC. Macrophage-based cell therapies: the long and winding road. J Control Release. 2016;240:527–40. https://doi.org/10.1016/j.jconrel.2016.07.018.
Lee DSW, Rojas OL, Gommerman JL. B cell depletion therapies in autoimmune disease: advances and mechanistic insights. Nat Rev Drug Discov Published online. 2020a; https://doi.org/10.1038/s41573-020-00092-2.
Lee Y, Sugihara K, Gillilland MG 3rd, Jon S, Kamada N, Moon JJ. Hyaluronic acid-bilirubin nanomedicine for targeted modulation of dysregulated intestinal barrier, microbiome and immune responses in colitis. Nat Mater. 2020b;19(1):118–26. https://doi.org/10.1038/s41563-019-0462-9.
Leimgruber A, Berger C, Cortez-Retamozo V, et al. Behavior of endogenous tumor-associated macrophages assessed in vivo using a functionalized nanoparticle. Neoplasia. 2009;11(5):459-468, 2 p following 468. https://doi.org/10.1593/neo.09356.
Lewis CE, Pollard JW. Distinct role of macrophages in different tumor microenvironments. Cancer Res. 2006;66(2):605–12. https://doi.org/10.1158/0008-5472.CAN-05-4005.

Li F, Yang Y, Zhu X, Huang L, Xu J. Macrophage polarization modulates development of systemic lupus erythematosus. Cell Physiol Biochem Int J Exp Cell Physiol Biochem Pharmacol. 2015;37(4):1279–88. https://doi.org/10.1159/000430251.

Li C, Xu MM, Wang K, Adler AJ, Vella AT, Zhou B. Macrophage polarization and meta-inflammation. Transl Res. 2018;191:29–44. https://doi.org/10.1016/j.trsl.2017.10.004.

Liguori M, Buracchi C, Pasqualini F, et al. Functional TRAIL receptors in monocytes and tumor-associated macrophages: a possible targeting pathway in the tumor microenvironment. Oncotarget. 2016;7(27):41662–76. https://doi.org/10.18632/oncotarget.9340.

Lin Y, Xu J, Lan H. Tumor-associated macrophages in tumor metastasis: biological roles and clinical therapeutic applications. J Hematol Oncol. 2019;12(1):76. https://doi.org/10.1186/s13045-019-0760-3.

Liu Y, Wang R. Immunotherapy targeting tumor-associated macrophages. Front Med. 2020;7:583708. https://doi.org/10.3389/fmed.2020.583708.

Liu C, Li Y, Yu J, et al. Targeting the shift from M1 to M2 macrophages in experimental autoimmune encephalomyelitis mice treated with fasudil. PLoS One. 2013;8(2):e54841. https://doi.org/10.1371/journal.pone.0054841.

Liu R, An Y, Jia W, et al. Macrophage-mimic shape changeable nanomedicine retained in tumor for multimodal therapy of breast cancer. J Control Release. 2020;321:589–601. https://doi.org/10.1016/j.jconrel.2020.02.043.

Loberg RD, Ying C, Craig M, et al. Targeting CCL2 with systemic delivery of neutralizing antibodies induces prostate cancer tumor regression in vivo. Cancer Res. 2007a;67(19):9417–24. https://doi.org/10.1158/0008-5472.CAN-07-1286.

Loberg RD, Ying C, Craig M, Yan L, Snyder LA, Pienta KJ. CCL2 as an important mediator of prostate cancer growth in vivo through the regulation of macrophage infiltration. Neoplasia. 2007b;9(7):556–62. https://doi.org/10.1593/neo.07307.

Lolmede K, Campana L, Vezzoli M, et al. Inflammatory and alternatively activated human macrophages attract vessel-associated stem cells, relying on separate HMGB1- and MMP-9-dependent pathways. J Leukoc Biol. 2009;85(5):779–87. https://doi.org/10.1189/jlb.0908579.

Lopez-Pedrera C, Barbarroja N, Patiño-Trives AM, et al. Effects of Biological Therapies on Molecular Features of Rheumatoid Arthritis. Int J Mol Sci. 2020;21(23) https://doi.org/10.3390/ijms21239067.

Lu J, Cao Q, Zheng D, et al. Discrete functions of M2a and M2c macrophage subsets determine their relative efficacy in treating chronic kidney disease. Kidney Int. 2013;84(4):745–55. https://doi.org/10.1038/ki.2013.135.

Lurier EB, Dalton D, Dampier W, et al. Transcriptome analysis of IL-10-stimulated (M2c) macrophages by next-generation sequencing. Immunobiology. 2017;222(7):847–56. https://doi.org/10.1016/j.imbio.2017.02.006.

Ma W-T, Gao F, Gu K, Chen D-K. The role of monocytes and macrophages in autoimmune diseases: a comprehensive review. Front Immunol. 2019;10:1140. https://doi.org/10.3389/fimmu.2019.01140.

Madsen DH, Jürgensen HJ, Siersbæk MS, et al. Tumor-associated macrophages derived from circulating inflammatory monocytes degrade collagen through cellular uptake. Cell Rep. 2017;21(13):3662–71. https://doi.org/10.1016/j.celrep.2017.12.011.

Magnusson MK, Brynjólfsson SF, Dige A, et al. Macrophage and dendritic cell subsets in IBD: ALDH+ cells are reduced in colon tissue of patients with ulcerative colitis regardless of inflammation. Mucosal Immunol. 2016;9(1):171–82. https://doi.org/10.1038/mi.2015.48.

Makadia HK, Siegel SJ. Poly lactic-co-glycolic acid (PLGA) as biodegradable controlled drug delivery carrier. Polymers (Basel). 2011;3(3):1377–97. https://doi.org/10.3390/polym3031377.

Malaponte G, Bevelacqua V, Fatuzzo P, et al. IL-1beta, TNF-alpha and IL-6 release from monocytes in haemodialysis patients in relation to dialytic age. Nephrol Dial Transplant. 2002;17(11):1964–70. https://doi.org/10.1093/ndt/17.11.1964.

Malfitano AM, Pisanti S, Napolitano F, Di Somma S, Martinelli R, Portella G. Tumor-Associated Macrophage Status in Cancer Treatment. Cancers (Basel). 2020;12(7) https://doi.org/10.3390/cancers12071987.

Mallucci G, Peruzzotti-Jametti L, Bernstock JD, Pluchino S. The role of immune cells, glia and neurons in white and gray matter pathology in multiple sclerosis. Prog Neurobiol. 2015;127-128:1–22. https://doi.org/10.1016/j.pneurobio.2015.02.003.

Mantovani A, Biswas SK, Galdiero MR, Sica A, Locati M. Macrophage plasticity and polarization in tissue repair and remodelling. J Pathol. 2013;229(2):176–85. https://doi.org/10.1002/path.4133.

Marks DJB, Harbord MWN, MacAllister R, et al. Defective acute inflammation in Crohn's disease: a clinical investigation. Lancet. 2006;367(9511):668–78. https://doi.org/10.1016/S0140-6736(06)68265-2.

Marro BS, Legrain S, Ware BC, Oldstone MB. Macrophage IFN-I signaling promotes autoreactive T cell infiltration into islets in type 1 diabetes model. JCI Insight. 2019;4(2) https://doi.org/10.1172/jci.insight.125067.

Martinez FO, Helming L, Milde R, et al. Genetic programs expressed in resting and IL-4 alternatively activated mouse and human macrophages: similarities and differences. Blood. 2013;121(9):e57–69. https://doi.org/10.1182/blood-2012-06-436212.

Mastrandrea L, Yu J, Behrens T, et al. Etanercept treatment in children with new-onset type 1 diabetes: pilot randomized, placebo-controlled, double-blind study. Diabetes Care. 2009;32(7):1244–9. https://doi.org/10.2337/dc09-0054.

McDonald LT, Zile MR, Zhang Y, et al. Increased macrophage-derived SPARC precedes collagen depo-

sition in myocardial fibrosis. Am J Physiol Heart Circ Physiol. 2018;315(1):H92–H100. https://doi.org/10.1152/ajpheart.00719.2017.
Menarim BC, Gillis KH, Oliver A, et al. Macrophage activation in the synovium of healthy and osteoarthritic equine joints. Front Vet Sci. 2020;7:568756. https://doi.org/10.3389/fvets.2020.568756.
Meroni E, Stakenborg N, Viola MF, Boeckxstaens GE. Intestinal macrophages and their interaction with the enteric nervous system in health and inflammatory bowel disease. Acta Physiol (Oxf). 2019;225(3):e13163. https://doi.org/10.1111/apha.13163.
Mills CD. M1 and M2 macrophages: oracles of health and disease. Crit Rev Immunol. 2012;32(6):463–88. https://doi.org/10.1615/critrevimmunol.v32.i6.10.
Miron VE, Boyd A, Zhao J-W, et al. M2 microglia and macrophages drive oligodendrocyte differentiation during CNS remyelination. Nat Neurosci. 2013;16(9):1211–8. https://doi.org/10.1038/nn.3469.
Mohammadi S, Saghaeian-Jazi M, Sedighi S, Memarian A. Immunomodulation in systemic lupus erythematosus: induction of M2 population in monocyte-derived macrophages by pioglitazone. Lupus. 2017;26(12):1318–27. https://doi.org/10.1177/0961203317701842.
Mohammadi S, Saghaeian-Jazi M, Sedighi S, Memarian A. Sodium valproate modulates immune response by alternative activation of monocyte-derived macrophages in systemic lupus erythematosus. Clin Rheumatol. 2018a;37(3):719–27. https://doi.org/10.1007/s10067-017-3922-0.
Mohammadi S, Memarian A, Sedighi S, Behnampour N, Yazdani Y. Immunoregulatory effects of indole-3-carbinol on monocyte-derived macrophages in systemic lupus erythematosus: a crucial role for aryl hydrocarbon receptor. Autoimmunity. 2018b;51(5):199–209. https://doi.org/10.1080/08916934.2018.1494161.
Moisan F, Francisco EB, Brozovic A, et al. Enhancement of paclitaxel and carboplatin therapies by CCL2 blockade in ovarian cancers. Mol Oncol. 2014;8(7):1231–9. https://doi.org/10.1016/j.molonc.2014.03.016.
Moll NM, Rietsch AM, Thomas S, et al. Multiple sclerosis normal-appearing white matter: pathology-imaging correlations. Ann Neurol. 2011;70(5):764–73. https://doi.org/10.1002/ana.22521.
Montes-Cobos E, Schweingruber N, Li X, Fischer HJ, Reichardt HM, Lühder F. Deletion of the mineralocorticoid receptor in myeloid cells attenuates central nervous system autoimmunity. Front Immunol. 2017a;8:1319. https://doi.org/10.3389/fimmu.2017.01319.
Montes-Cobos E, Ring S, Fischer HJ, et al. Targeted delivery of glucocorticoids to macrophages in a mouse model of multiple sclerosis using inorganic-organic hybrid nanoparticles. J Control Release. 2017b;245:157–69. https://doi.org/10.1016/j.jconrel.2016.12.003.
Moreno MA, Burns T, Yao P, Miers L, Pleasure D, Soulika AM. Therapeutic depletion of monocyte-derived cells protects from long-term axonal loss in experimental autoimmune encephalomyelitis. J Neuroimmunol. 2016;290:36–46. https://doi.org/10.1016/j.jneuroim.2015.11.004.
Moriwaki H, Stempien-Otero A, Kremen M, Cozen AE, Dichek DA. Overexpression of Urokinase by macrophages or deficiency of plasminogen activator inhibitor type 1 causes cardiac fibrosis in mice. Circ Res. 2004;95(6):637–44. https://doi.org/10.1161/01.RES.0000141427.61023.f4.
Mosquera MJ, Kim S, Zhou H, et al. Immunomodulatory nanogels overcome restricted immunity in a murine model of gut microbiome-mediated metabolic syndrome. Sci Adv. 2019;5(3):eaav9788. https://doi.org/10.1126/sciadv.aav9788.
Movahedi K, Schoonooghe S, Laoui D, et al. Nanobody-based targeting of the macrophage mannose receptor for effective in vivo imaging of tumor-associated macrophages. Cancer Res. 2012;72(16):4165–77. https://doi.org/10.1158/0008-5472.CAN-11-2994.
Mulder R, Banete A, Basta S. Spleen-derived macrophages are readily polarized into classically activated (M1) or alternatively activated (M2) states. Immunobiology. 2014;219(10):737–45. https://doi.org/10.1016/j.imbio.2014.05.005.
Murdoch C, Giannoudis A, Lewis CE. Mechanisms regulating the recruitment of macrophages into hypoxic areas of tumors and other ischemic tissues. Blood. 2004;104(8):2224–34. https://doi.org/10.1182/blood-2004-03-1109.
Murray PJ, Wynn TA. Obstacles and opportunities for understanding macrophage polarization. J Leukoc Biol. 2011;89(4):557–63. https://doi.org/10.1189/jlb.0710409.
Murray PJ, Allen JE, Biswas SK, et al. Macrophage activation and polarization: nomenclature and experimental guidelines. Immunity. 2014;41(1):14–20. https://doi.org/10.1016/j.immuni.2014.06.008.
Na YR, Stakenborg M, Seok SH, Matteoli G. Macrophages in intestinal inflammation and resolution: a potential therapeutic target in IBD. Nat Rev Gastroenterol Hepatol. 2019;16(9):531–43. https://doi.org/10.1038/s41575-019-0172-4.
Nguyen MM, Carlini AS, Chien M-P, et al. Enzyme-responsive nanoparticles for targeted accumulation and prolonged retention in heart tissue after myocardial infarction. Adv Mater. 2015;27(37):5547–52. https://doi.org/10.1002/adma.201502003.
Nielsen SR, Schmid MC. Macrophages as key drivers of cancer progression and metastasis. Mediat Inflamm. 2017;2017:9624760. https://doi.org/10.1155/2017/9624760.
Noy R, Pollard JW. Tumor-associated macrophages: from mechanisms to therapy. Immunity. 2014;41(1):49–61. https://doi.org/10.1016/j.immuni.2014.06.010.
Nywening TM, Wang-Gillam A, Sanford DE, et al. Targeting tumour-associated macrophages with CCR2 inhibition in combination with FOLFIRINOX

in patients with borderline resectable and locally advanced pancreatic cancer: a single-centre, open-label, dose-finding, non-randomised, phase 1b trial. Lancet Oncol. 2016;17(5):651–62. https://doi.org/10.1016/S1470-2045(16)00078-4.

Okuda K, Fu HY, Matsuzaki T, et al. Targeted therapy for acute autoimmune myocarditis with nano-sized liposomal FK506 in rats. PLoS One. 2016;11(8):e0160944. https://doi.org/10.1371/journal.pone.0160944.

Oliva N, Carcole M, Beckerman M, et al. Regulation of dendrimer/dextran material performance by altered tissue microenvironment in inflammation and neoplasia. Sci Transl Med. 2015;7(272):272ra11. https://doi.org/10.1126/scitranslmed.aaa1616.

Orecchioni M, Ghosheh Y, Pramod AB, Ley K. Macrophage polarization: different gene signatures in M1(LPS+) vs. classically and M2(LPS-) vs. alternatively activated macrophages. Front Immunol. 2019;10:1084. https://doi.org/10.3389/fimmu.2019.01084.

Orme J, Mohan C. Macrophages and neutrophils in SLE-An online molecular catalog. Autoimmun Rev. 2012;11(5):365–72. https://doi.org/10.1016/j.autrev.2011.10.010.

Ovais M, Guo M, Chen C. Tailoring nanomaterials for targeting tumor-associated macrophages. Adv Mater. 2019;31(19):e1808303. https://doi.org/10.1002/adma.201808303.

Paoletti A, Rohmer J, Ly B, et al. Monocyte/macrophage abnormalities specific to rheumatoid arthritis are linked to miR-155 and are differentially modulated by different TNF inhibitors. J Immunol. 2019;203(7):1766–75. https://doi.org/10.4049/jimmunol.1900386.

Papadopoulos KP, Gluck L, Martin LP, et al. First-in-human study of AMG 820, a monoclonal anti-colony-stimulating factor 1 receptor antibody, in patients with advanced solid tumors. Clin Cancer Res. 2017;23(19):5703–10. https://doi.org/10.1158/1078-0432.CCR-16-3261.

Parsa R, Andresen P, Gillett A, et al. Adoptive transfer of immunomodulatory M2 macrophages prevents type 1 diabetes in NOD mice. Diabetes. 2012;61(11):2881–92. https://doi.org/10.2337/db11-1635.

Pascual-García M, Bonfill-Teixidor E, Planas-Rigol E, et al. LIF regulates CXCL9 in tumor-associated macrophages and prevents CD8+ T cell tumor-infiltration impairing anti-PD1 therapy. Nat Commun. 2019;10(1):2416. https://doi.org/10.1038/s41467-019-10369-9.

Pascual-Gil S, Abizanda G, Iglesias E, Garbayo E, Prósper F, Blanco-Prieto MJ. NRG1 PLGA MP locally induce macrophage polarisation toward a regenerative phenotype in the heart after acute myocardial infarction. J Drug Target. 2019;27(5-6):573–81. https://doi.org/10.1080/1061186X.2018.1531417.

Peferoen LAN, Vogel DYS, Ummenthum K, et al. Activation status of human microglia is dependent on lesion formation stage and remyelination in multiple sclerosis. J Neuropathol Exp Neurol. 2015;74(1):48–63. https://doi.org/10.1097/NEN.0000000000000149.

Peña CG, Nakada Y, Saatcioglu HD, et al. LKB1 loss promotes endometrial cancer progression via CCL2-dependent macrophage recruitment. J Clin Invest. 2015;125(11):4063–76. https://doi.org/10.1172/JCI82152.

Perdiguero EG, Geissmann F. The development and maintenance of resident macrophages. Nat Immunol. 2016;17(1):2–8. https://doi.org/10.1038/ni.3341.

Pérez-Medina C, Tang J, Abdel-Atti D, et al. PET imaging of tumor-associated macrophages with 89Zr-labeled high-density lipoprotein nanoparticles. J Nucl Med. 2015;56(8):1272–7. https://doi.org/10.2967/jnumed.115.158956.

Peruzzotti-Jametti L, Bernstock JD, Vicario N, et al. Macrophage-derived extracellular succinate licenses neural stem cells to suppress chronic neuroinflammation. Cell Stem Cell. 2018;22(3):355.e13–68.e13. https://doi.org/10.1016/j.stem.2018.01.020.

Piaggio F, Kondylis V, Pastorino F, et al. A novel liposomal clodronate depletes tumor-associated macrophages in primary and metastatic melanoma: anti-angiogenic and anti-tumor effects. J Control Release. 2016;223:165–77. https://doi.org/10.1016/j.jconrel.2015.12.037.

Pienta KJ, Machiels J-P, Schrijvers D, et al. Phase 2 study of carlumab (CNTO 888), a human monoclonal antibody against CC-chemokine ligand 2 (CCL2), in metastatic castration-resistant prostate cancer. Investig New Drugs. 2013;31(3):760–8. https://doi.org/10.1007/s10637-012-9869-8.

Planche V, Panatier A, Hiba B, et al. Selective dentate gyrus disruption causes memory impairment at the early stage of experimental multiple sclerosis. Brain Behav Immun. 2017;60:240–54. https://doi.org/10.1016/j.bbi.2016.11.010.

Poltavets AS, Vishnyakova PA, Elchaninov AV, Sukhikh GT, Fatkhudinov TK. Macrophage Modification Strategies for Efficient Cell Therapy. Cells. 2020;9(6) https://doi.org/10.3390/cells9061535.

Ponzoni M, Pastorino F, Di Paolo D, Perri P, Brignole C. Targeting macrophages as a potential therapeutic intervention: impact on inflammatory diseases and cancer. Int J Mol Sci. 2018;19(7) https://doi.org/10.3390/ijms19071953.

Pope RM, Shahrara S. Possible roles of IL-12-family cytokines in rheumatoid arthritis. Nat Rev Rheumatol. 2013;9(4):252–6. https://doi.org/10.1038/nrrheum.2012.170.

Purcell BP, Lobb D, Charati MB, et al. Injectable and bioresponsive hydrogels for on-demand matrix metalloproteinase inhibition. Nat Mater. 2014;13(6):653–61. https://doi.org/10.1038/nmat3922.

Purnama C, Ng SL, Tetlak P, et al. Transient ablation of alveolar macrophages leads to massive pathology of influenza infection without affecting cellular adaptive immunity. Eur J Immunol. 2014;44(7):2003–12. https://doi.org/10.1002/eji.201344359.

Pusuluri A, Wu D, Mitragotri S. Immunological consequences of chemotherapy: single drugs, combination therapies and nanoparticle-based treatments. J Control

Release. 2019;305:130–54. https://doi.org/10.1016/j.jconrel.2019.04.020.
Qian B-Z, Pollard JW. Macrophage diversity enhances tumor progression and metastasis. Cell. 2010;141(1):39–51. https://doi.org/10.1016/j.cell.2010.03.014.
Qian B-Z, Li J, Zhang H, et al. CCL2 recruits inflammatory monocytes to facilitate breast-tumour metastasis. Nature. 2011;475(7355):222–5. https://doi.org/10.1038/nature10138.
Qin J, Yang X, Zhang R-X, et al. Monocyte mediated brain targeting delivery of macromolecular drug for the therapy of depression. Nanomedicine. 2015;11(2):391–400. https://doi.org/10.1016/j.nano.2014.09.012.
Qiu S-Q, Waaijer SJH, Zwager MC, de Vries EGE, van der Vegt B, Schröder CP. Tumor-associated macrophages in breast cancer: innocent bystander or important player? Cancer Treat Rev. 2018;70:178–89. https://doi.org/10.1016/j.ctrv.2018.08.010.
Raes G, Van den Bergh R, De Baetselier P, et al. Arginase-1 and Ym1 are markers for murine, but not human, alternatively activated myeloid cells. J Immunol. 2005;174(11):6561.; author reply 6561-2. https://doi.org/10.4049/jimmunol.174.11.6561.
Rehman K, Akash MSH. Mechanisms of inflammatory responses and development of insulin resistance: how are they interlinked? J Biomed Sci. 2016;23(1):87. https://doi.org/10.1186/s12929-016-0303-y.
Riabov V, Gudima A, Wang N, Mickley A, Orekhov A, Kzhyshkowska J. Role of tumor associated macrophages in tumor angiogenesis and lymphangiogenesis. Front Physiol. 2014;5:75. https://doi.org/10.3389/fphys.2014.00075.
Ries CH, Cannarile MA, Hoves S, et al. Targeting tumor-associated macrophages with anti-CSF-1R antibody reveals a strategy for cancer therapy. Cancer Cell. 2014;25(6):846–59. https://doi.org/10.1016/j.ccr.2014.05.016.
Rosenblum LT, Kosaka N, Mitsunaga M, Choyke PL, Kobayashi H. In vivo molecular imaging using nanomaterials: general in vivo characteristics of nano-sized reagents and applications for cancer diagnosis. Mol Membr Biol. 2010;27(7):274–85. https://doi.org/10.3109/09687688.2010.481640.
Ruffell B, Coussens LM. Macrophages and therapeutic resistance in cancer. Cancer Cell. 2015;27(4):462–72. https://doi.org/10.1016/j.ccell.2015.02.015.
Sack U, Stiehl P, Geiler G. Distribution of macrophages in rheumatoid synovial membrane and its association with basic activity. Rheumatol Int. 1994;13(5):181–6. https://doi.org/10.1007/BF00390265.
Sager HB, Heidt T, Hulsmans M, et al. Targeting interleukin-1β reduces leukocyte production after acute myocardial infarction. Circulation. 2015;132(20):1880–90. https://doi.org/10.1161/CIRCULATIONAHA.115.016160.
Sahu R, Bethunaickan R, Singh S, Davidson A. Structure and function of renal macrophages and dendritic cells from lupus-prone mice. Arthritis Rheumatol. 2014;66(6):1596–607. https://doi.org/10.1002/art.38410.
Saito Y, Komohara Y, Niino D, et al. Role of CD204-positive tumor-associated macrophages in adult T-cell leukemia/lymphoma. J Clin Exp Hematop. 2014;54(1):59–65. https://doi.org/10.3960/jslrt.54.59.
Sanford DE, Belt BA, Panni RZ, et al. Inflammatory monocyte mobilization decreases patient survival in pancreatic cancer: a role for targeting the CCL2/CCR2 axis. Clin Cancer Res. 2013;19(13):3404–15. https://doi.org/10.1158/1078-0432.CCR-13-0525.
Scheib JL, Höke A. An attenuated immune response by Schwann cells and macrophages inhibits nerve regeneration in aged rats. Neurobiol Aging. 2016;45:1–9. https://doi.org/10.1016/j.neurobiolaging.2016.05.004.
Schroder K, Irvine KM, Taylor MS, et al. Conservation and divergence in Toll-like receptor 4-regulated gene expression in primary human versus mouse macrophages. Proc Natl Acad Sci U S A. 2012;109(16):E944–53. https://doi.org/10.1073/pnas.1110156109.
Schulz C, Gomez Perdiguero E, Chorro L, et al. A lineage of myeloid cells independent of Myb and hematopoietic stem cells. Science. 2012;336(6077):86–90. https://doi.org/10.1126/science.1219179.
Scott DL, Wolfe F, Huizinga TWJ. Rheumatoid arthritis. Lancet (London, England). 2010;376(9746):1094–108. https://doi.org/10.1016/S0140-6736(10)60826-4.
Sehnert B, Burkhardt H, Dübel S, Voll RE. Cell-type targeted NF-kappaB inhibition for the treatment of inflammatory diseases. Cells. 2020;9(7) https://doi.org/10.3390/cells9071627.
Semerano L, Thiolat A, Minichiello E, Clavel G, Bessis N, Boissier M-C. Targeting IL-6 for the treatment of rheumatoid arthritis: phase II investigational drugs. Expert Opin Investig Drugs. 2014;23(7):979–99. https://doi.org/10.1517/13543784.2014.912276.
Seok J, Warren HS, Cuenca AG, et al. Genomic responses in mouse models poorly mimic human inflammatory diseases. Proc Natl Acad Sci U S A. 2013;110(9):3507–12. https://doi.org/10.1073/pnas.1222878110.
Sestak AL, Fürnrohr BG, Harley JB, Merrill JT, Namjou B. The genetics of systemic lupus erythematosus and implications for targeted therapy. Ann Rheum Dis. 2011;70(Suppl 1):i37–43. https://doi.org/10.1136/ard.2010.138057.
Shapouri-Moghaddam A, Mohammadian S, Vazini H, et al. Macrophage plasticity, polarization, and function in health and disease. J Cell Physiol. 2018;233(9):6425–40. https://doi.org/10.1002/jcp.26429.
Sharif MN, Tassiulas I, Hu Y, Mecklenbräuker I, Tarakhovsky A, Ivashkiv LB. IFN-alpha priming results in a gain of proinflammatory function by IL-10: implications for systemic lupus erythematosus pathogenesis. J Immunol. 2004;172(10):6476–81. https://doi.org/10.4049/jimmunol.172.10.6476.
Singh Y, Pawar VK, Meher JG, et al. Targeting tumor associated macrophages (TAMs) via nanocarriers. J Control Release. 2017;254:92–106. https://doi.org/10.1016/j.jconrel.2017.03.395.

Skytthe MK, Graversen JH, Moestrup SK. Targeting of cd163+ macrophages in inflammatory and malignant diseases. Int J Mol Sci. 2020;21(15):1–31. https://doi.org/10.3390/ijms21155497.

Slade Shantz JA, Yu Y-Y, Andres W, Miclau T 3rd, Marcucio R. Modulation of macrophage activity during fracture repair has differential effects in young adult and elderly mice. J Orthop Trauma. 2014;28(Suppl 1(0 1)):S10-4. https://doi.org/10.1097/BOT.0000000000000062.

Smith BR, Ghosn EEB, Rallapalli H, et al. Selective uptake of single-walled carbon nanotubes by circulating monocytes for enhanced tumour delivery. Nat Nanotechnol. 2014;9(6):481–7. https://doi.org/10.1038/nnano.2014.62.

Soler Palacios B, Estrada-Capetillo L, Izquierdo E, et al. Macrophages from the synovium of active rheumatoid arthritis exhibit an activin A-dependent pro-inflammatory profile. J Pathol. 2015;235(3):515–26. https://doi.org/10.1002/path.4466.

Spiller KL, Koh TJ. Macrophage-based therapeutic strategies in regenerative medicine. Adv Drug Deliv Rev. 2017;122:74–83. https://doi.org/10.1016/j.addr.2017.05.010.

Spiller KL, Wrona EA, Romero-Torres S, et al. Differential gene expression in human, murine, and cell line-derived macrophages upon polarization. Exp Cell Res. 2016;347(1):1–13. https://doi.org/10.1016/j.yexcr.2015.10.017.

Su Y, Xie Z, Kim GB, Dong C, Yang J. Design strategies and applications of circulating cell-mediated drug delivery systems. ACS Biomater Sci Eng. 2015;1(4):201–17. https://doi.org/10.1021/ab500179h.

Sundararaj SC, Al-Sabbagh M, Rabek CL, Dziubla TD, Thomas MV, Puleo DA. Comparison of sequential drug release in vitro and in vivo. J Biomed Mater Res B Appl Biomater. 2016;104(7):1302–10. https://doi.org/10.1002/jbm.b.33472.

Swift ME, Kleinman HK, DiPietro LA. Impaired wound repair and delayed angiogenesis in aged mice. Lab Investig. 1999;79(12):1479–87.

Tack CJ, Kleijwegt FS, Van Riel PLCM, Roep BO. Development of type 1 diabetes in a patient treated with anti-TNF-alpha therapy for active rheumatoid arthritis. Diabetologia. 2009;52(7):1442–4. https://doi.org/10.1007/s00125-009-1381-0.

Tambalo S, Peruzzotti-Jametti L, Rigolio R, et al. Functional magnetic resonance imaging of rats with experimental autoimmune encephalomyelitis reveals brain cortex remodeling. J Neurosci. 2015;35(27):10088–100. https://doi.org/10.1523/JNEUROSCI.0540-15.2015.

Tannahill GM, Curtis AM, Adamik J, et al. Succinate is an inflammatory signal that induces IL-1β through HIF-1α. Nature. 2013;496(7444):238–42. https://doi.org/10.1038/nature11986.

Tap WD, Wainberg ZA, Anthony SP, et al. Structure-guided blockade of CSF1R kinase in tenosynovial giant-cell tumor. N Engl J Med. 2015;373(5):428–37. https://doi.org/10.1056/NEJMoa1411366.

Thayer TC, Delano M, Liu C, et al. Superoxide production by macrophages and T cells is critical for the induction of autoreactivity and type 1 diabetes. Diabetes. 2011;60(8):2144–51. https://doi.org/10.2337/db10-1222.

Udalova IA, Mantovani A, Feldmann M. Macrophage heterogeneity in the context of rheumatoid arthritis. Nat Rev Rheumatol. 2016;12(8):472–85. https://doi.org/10.1038/nrrheum.2016.91.

Van Acker HH, Anguille S, Willemen Y, Smits EL, Van Tendeloo VF. Bisphosphonates for cancer treatment: mechanisms of action and lessons from clinical trials. Pharmacol Ther. 2016;158:24–40. https://doi.org/10.1016/j.pharmthera.2015.11.008.

van der Valk P, De Groot CJ. Staging of multiple sclerosis (MS) lesions: pathology of the time frame of MS. Neuropathol Appl Neurobiol. 2000;26(1):2–10. https://doi.org/10.1046/j.1365-2990.2000.00217.x.

van der Vorst EPC, Theodorou K, Wu Y, et al. High-density lipoproteins exert pro-inflammatory effects on macrophages via passive cholesterol depletion and PKC-NF-κB/STAT1-IRF1 signaling. Cell Metab. 2017;25(1):197–207. https://doi.org/10.1016/j.cmet.2016.10.013.

van Langelaar J, Rijvers L, Smolders J, van Luijn MM. B and T cells driving multiple sclerosis: identity. Mechanisms and Potential Triggers Front Immunol. 2020;11:760. https://doi.org/10.3389/fimmu.2020.00760.

Van Lent PL, Holthuysen AE, Van Rooijen N, Van De Putte LB, Van Den Berg WB. Local removal of phagocytic synovial lining cells by clodronate-liposomes decreases cartilage destruction during collagen type II arthritis. Ann Rheum Dis. 1998;57(7):408–13. https://doi.org/10.1136/ard.57.7.408.

Van Rooijen N, Sanders A. Liposome mediated depletion of macrophages: mechanism of action, preparation of liposomes and applications. J Immunol Methods. 1994;174(1-2):83–93. https://doi.org/10.1016/0022-1759(94)90012-4.

van Roon JAG, van Vuuren AJ, Wijngaarden S, et al. Selective elimination of synovial inflammatory macrophages in rheumatoid arthritis by an Fcgamma receptor I-directed immunotoxin. Arthritis Rheum. 2003;48(5):1229–38. https://doi.org/10.1002/art.10940.

van Vuuren AJ, van Roon JAG, Walraven V, et al. CD64-directed immunotoxin inhibits arthritis in a novel CD64 transgenic rat model. J Immunol. 2006;176(10):5833–8. https://doi.org/10.4049/jimmunol.176.10.5833.

Veremeyko T, Yung AWY, Dukhinova M, et al. Cyclic AMP pathway suppress autoimmune neuroinflammation by inhibiting functions of encephalitogenic CD4 T cells and enhancing M2 macrophage polarization at the site of inflammation. Front Immunol. 2018;9:50. https://doi.org/10.3389/fimmu.2018.00050.

Vergadi E, Ieronymaki E, Lyroni K, Vaporidi K, Tsatsanis C. Akt Signaling pathway in macrophage activation and M1/M2 polarization. J Immunol. 2017;198(3):1006–14. https://doi.org/10.4049/jimmunol.1601515.

Vomund AN, Zinselmeyer BH, Hughes J, et al. Beta cells transfer vesicles containing insulin to phagocytes for presentation to T cells. Proc Natl Acad Sci U S A. 2015;112(40):E5496–502. https://doi.org/10.1073/pnas.1515954112.

Wang N, Liang H, Zen K. Molecular mechanisms that influence the macrophage m1-m2 polarization balance. Front Immunol. 2014;5:614. https://doi.org/10.3389/fimmu.2014.00614.

Wang L-M, Zhang Y, Li X, et al. Nr4a1 plays a crucial modulatory role in Th1/Th17 cell responses and CNS autoimmunity. Brain Behav Immun. 2018;68:44–55. https://doi.org/10.1016/j.bbi.2017.09.015.

Wang J, Wang J, Wang J, Yang B, Weng Q, He Q. Targeting microglia and macrophages: a potential treatment strategy for multiple sclerosis. Front Pharmacol. 2019a;10:286. https://doi.org/10.3389/fphar.2019.00286.

Wang N, Liu C, Wang X, et al. Hyaluronic acid oligosaccharides improve myocardial function reconstruction and angiogenesis against myocardial infarction by regulation of macrophages. Theranostics. 2019b;9(7):1980–92. https://doi.org/10.7150/thno.31073.

Wang L-X, Zhang S-X, Wu H-J, Rong X-L, Guo J. M2b macrophage polarization and its roles in diseases. J Leukoc Biol. 2019c;106(2):345–58. https://doi.org/10.1002/JLB.3RU1018-378RR.

Weber MS, Prod'homme T, Youssef S, et al. Type II monocytes modulate T cell-mediated central nervous system autoimmune disease. Nat Med. 2007;13(8):935–43. https://doi.org/10.1038/nm1620.

Weissleder R, Ntziachristos V. Shedding light onto live molecular targets. Nat Med. 2003;9(1):123–8. https://doi.org/10.1038/nm0103-123.

Weissleder R, Nahrendorf M, Pittet MJ. Imaging macrophages with nanoparticles. Nat Mater. 2014;13(2):125–38. https://doi.org/10.1038/nmat3780.

West AJ, Tsui V, Stylli SS, et al. The role of interleukin-6-STAT3 signalling in glioblastoma. Oncol Lett. 2018;16(4):4095–104. https://doi.org/10.3892/ol.2018.9227.

Williams JW, Giannarelli C, Rahman A, Randolph GJ, Kovacic JC. Macrophage biology, classification, and phenotype in cardiovascular disease: JACC macrophage in CVD series (part 1). J Am Coll Cardiol. 2018;72(18):2166–80. https://doi.org/10.1016/j.jacc.2018.08.2148.

Wu M-L, Huang T-C, Hung H-J, Yu J-H, Chung W-B, Chaung H-C. Investigation of beta-glucans binding to human/mouse dectin-1 and associated immunomodulatory effects on two monocyte/macrophage cell lines. Biotechnol Prog. 2010;26(5):1391–9. https://doi.org/10.1002/btpr.429.

Wu H, Zhang X, Han D, Cao J, Tian J. Tumour-associated macrophages mediate the invasion and metastasis of bladder cancer cells through CXCL8. PeerJ. 2020;8:e8721. https://doi.org/10.7717/peerj.8721.

Wynn TA, Vannella KM. Macrophages in tissue repair, regeneration, and fibrosis. Immunity. 2016;44(3):450–62. https://doi.org/10.1016/j.immuni.2016.02.015.

Wynn TA, Chawla A, Pollard JW. Macrophage biology in development, homeostasis and disease. Nature. 2013;496(7446):445–55. https://doi.org/10.1038/nature12034.

Xie Z, Zhang Y, Liu L, et al. Development of intrinsically photoluminescent and photostable polylactones. Adv Mater. 2014;26(26):4491–6. https://doi.org/10.1002/adma.201306070.

Yamanaka H. TNF as a target of inflammation in rheumatoid arthritis. Endocr Metab Immune Disord Drug Targets. 2015;15(2):129–34. https://doi.org/10.2174/1871530315666150316121808.

Yan D, Kowal J, Akkari L, et al. Inhibition of colony stimulating factor-1 receptor abrogates microenvironment-mediated therapeutic resistance in gliomas. Oncogene. 2017;36(43):6049–58. https://doi.org/10.1038/onc.2017.261.

Yang C-A, Huang S-T, Chiang B-L. Sex-dependent differential activation of NLRP3 and AIM2 inflammasomes in SLE macrophages. Rheumatology (Oxford). 2015;54(2):324–31. https://doi.org/10.1093/rheumatology/keu318.

Yang X, Chang Y, Wei W. Emerging role of targeting macrophages in rheumatoid arthritis: focus on polarization, metabolism and apoptosis. Cell Prolif. 2020;53(7):e12854. https://doi.org/10.1111/cpr.12854.

You S, Yoo SA, Choi S, et al. Identification of key regulators for the migration and invasion of rheumatoid synoviocytes through a systems approach. Proc Natl Acad Sci U S A 2014;111(1):550–555. https://doi.org/10.1073/pnas.1311239111.

Yuan Y, Long L, Liu J, et al. The double-edged sword effect of macrophage targeting delivery system in different macrophage subsets related diseases. J Nanobiotechnol. 2020;18(1):1–14. https://doi.org/10.1186/s12951-020-00721-3.

Yunna C, Mengru H, Lei W, Weidong C. Macrophage M1/M2 polarization. Eur J Pharmacol. 2020;877:173090. https://doi.org/10.1016/j.ejphar.2020.173090.

Zeisberger SM, Odermatt B, Marty C, Zehnder-Fjällman AHM, Ballmer-Hofer K, Schwendener RA. Clodronate-liposome-mediated depletion of tumour-associated macrophages: a new and highly effective antiangiogenic therapy approach. Br J Cancer. 2006;95(3):272–81. https://doi.org/10.1038/sj.bjc.6603240.

Zhan X, Jia L, Niu Y, et al. Targeted depletion of tumour-associated macrophages by an alendronate-glucomannan conjugate for cancer immunotherapy. Biomaterials. 2014;35(38):10046–57. https://doi.org/10.1016/j.biomaterials.2014.09.007.

Zhang H, Chen J. Current status and future directions of cancer immunotherapy. J Cancer. 2018;9(10):1773–81. https://doi.org/10.7150/jca.24577.

Zhang H, Bi J, Yi H, et al. Silencing c-Rel in macrophages dampens Th1 and Th17 immune responses and alleviates experimental autoimmune encephalomyelitis in mice. Immunol Cell Biol. 2017;95(7):593–600. https://doi.org/10.1038/icb.2017.11.

Zhang F, Parayath NN, Ene CI, et al. Genetic programming of macrophages to perform anti-tumor functions using targeted mRNA nanocarriers. Nat Commun. 2019;10(1):3974. https://doi.org/10.1038/s41467-019-11911-5.

Zhao R-R, Ackers-Johnson M, Stenzig J, et al. Targeting chondroitin sulfate glycosaminoglycans to treat cardiac fibrosis in pathological remodeling. Circulation. 2018;137(23):2497–513. https://doi.org/10.1161/CIRCULATIONAHA.117.030353.

Zhou J, Cai W, Jin M, et al. 18β-glycyrrhetinic acid suppresses experimental autoimmune encephalomyelitis through inhibition of microglia activation and promotion of remyelination. Sci Rep. 2015;5:13713. https://doi.org/10.1038/srep13713.

Zhu L, Zhao Q, Yang T, Ding W, Zhao Y. Cellular metabolism and macrophage functional polarization. Int Rev Immunol. 2015;34(1):82–100. https://doi.org/10.3109/08830185.2014.969421.

Zhu C, Kros JM, Cheng C, Mustafa D. The contribution of tumor-associated macrophages in glioma neo-angiogenesis and implications for anti-angiogenic strategies. Neuro-Oncology. 2017a;19(11):1435–46. https://doi.org/10.1093/neuonc/nox081.

Zhu Y, Lyapichev K, Lee DH, et al. Macrophage transcriptional profile identifies lipid catabolic pathways that can be therapeutically targeted after spinal cord injury. J Neurosci. 2017b;37(9):2362–76. https://doi.org/10.1523/JNEUROSCI.2751-16.2017.

Zrzavy T, Hametner S, Wimmer I, Butovsky O, Weiner HL, Lassmann H. Loss of "homeostatic" microglia and patterns of their activation in active multiple sclerosis. Brain. 2017;140(7):1900–13. https://doi.org/10.1093/brain/awx113.

Understanding Macrophage-Associated Diseases and Applications of Nanodrug Delivery Systems

Papiya Bigoniya

Abstract

Macrophage phenotypes are associated with the pathogenesis of several diseases involving chronic inflammation, like atherosclerosis, obesity, diabetes, cancer, skin diseases, chronic wound, and neurodegenerative diseases. Macrophages play a crucial role in tissue repair, immunity, and cancer. Macrophages provide immunity by destroying invading pathogens and acting as scavengers, removing dead, necrotic, or apoptotic cells. Macrophages can differentiate into classically activated (M1) kill-type macrophages or alternatively activated (M2) regulatory-type macrophages. An in-depth understanding of macrophages biology and activation states related to different disease progression is crucial for designing nanodrug delivery systems to treat cancer, arthritis, obesity, and diabetes. Many researchers have contributed to the development of macrophage-associated drug delivery systems targeting tumor therapies, such as inhibition of tumor-associated macrophage (TAMs) recruitment, reeducation, and direct killing. Recently, investigations were done on targeting tumor macrophages with therapeutic drugs loaded with nanomedicines called live cell-mediated drug delivery systems (LCDDS) for transporting drugs in the tumor tissue. The major challenge toward tumor macrophage-associated nanodrug delivery system is lack of selectivity, an insufficient antitumor response, and limited uptake of drugs in the solid tumor tissues. Effective nanodrug delivery therapy development can be attained by understanding macrophage diversity and defining the target cells according to anatomical location and functional regulation.

Keywords

Arthritis · Cancer · Chronic inflammation · Diabetes · Macrophage · Nanocarrier · Nanodrug delivery · Phenotype

Abbreviations

AMPK	Adenosine monophosphate kinase
BBB	Blood-brain barrier
CCL	Chemokine ligand
CCR	Chemokine receptor
CNS	Central nervous system
COPD	Chronic obstructive pulmonary disease
CSF 1R	Colony-stimulating factor 1 receptor
CTLA-4	Cytotoxic T lymphocyte-associated antigen

P. Bigoniya (✉)
DSKM College of Pharmacy, RKDF University,
Bhopal, Madhya Pradesh, India

S. Gupta, Y. V. Pathak (eds.), *Macrophage Targeted Delivery Systems*,
https://doi.org/10.1007/978-3-030-84164-5_5

DIO	Diet-induced obese
DOTAP	1,2-dioleoyl-3-trimethylammonium-propane
HIF-1α	Hypoxia-inducible factor 1 alpha
HRG	Histidine-rich glycoprotein
IL	Interleukin
Interferon γ	Interferon gamma
IPCS	Induced pluripotent stem cell
LCDDS	Live cell-mediated drug delivery systems
LDL	Low-density lipoprotein
LPS	Lipopolysaccharide
MAM	Metastasis-associated macrophage
MAP	Mitogen-activated protein
MCP	Monocyte chemoattractant protein
MDSCs	Myeloid-derived suppressor cells
MMP	Metalloproteinases
MnO_2	Manganese dioxide
MPS	Mononuclear phagocytic system
MR or CD	Mannose receptor or cluster of differentiation
NCs	Nanocomplexes
NF-κB	Nuclear factor kappa light-chain enhancer of activated B cells
NPs	Nanoparticles
PEG	Polyethylene glycol
PLA	Polylactic acid
PLGA	Poly lactic-co-glycolic acid
PPAR-γ	Peroxisome proliferator-activated receptor gamma
siRNA	Small interfering RNA
STAT	Signal transducer and activator of transcription
SWCNT	Single-walled carbon nanotube
T2D	Type 2 diabetes
TAM	Tumor-associated macrophage
TAR	Transactive response
TGF-β	Transforming growth factor beta
TLR	Toll-like receptor
TNFα	Tumor necrosis factor alpha
VATM	Visceral adipose tissue macrophage
VCAM	Vascular cell adhesion molecule
VEGF	Vascular endothelial growth factor

1 Introduction

Macrophages play a pivotal role in diseases mostly representing chronic inflammation, like atherosclerosis, obesity, diabetes, cancer, skin diseases, chronic wound, and neurodegenerative diseases. Macrophages have a significant aspect in tissue repair, immunity, chronic inflammatory diseases, and cancer. Macrophages are large phagocytic cells developed by the differentiation of monocyte outside the circulatory system. Based on host tissue, macrophages are named as follows: Langerhans cells (skin), Kupffer cells (liver), microglia (brain), alveolar macrophages (lung), peritoneal macrophages (peritoneum), and osteoclasts (bone). Following inflammation, monocytes depart from blood vessels and undergo a series of changes to transform as macrophages. The current concept suggests that macrophages can be acquired from embryonic progenitors competent to maintain their population unconstrained from blood-borne precursors in adulthood or from hematopoietic stem cell-derived monocytes. In healthy tissues, macrophages are responsible for homeostasis maintenance through the clearance of aged cells and modulation of tissue metabolism.

Macrophages detect, engulf, and destroy pathogens and apoptotic cells. Macrophages guard the body tissue and detect cell categories that do not display the proper protein surface markers. Some macrophages act as scavengers, removing bodies dead, necrotic, or apoptotic cells, while others provide host immunity by engulfing invading microbes. Macrophages can live for several months and kill hundreds of different microbes before they die, providing nonspecific or innate immunity. Macrophages' ability to destroy the invading pathogen is a critical component of host defense. Macrophages engulf intruders via phagocytosis and play a crucial role in alerting the immune system regarding invaders presence. After ingestion of microorganisms, the macrophages express a specific protein on its surface called an antigen, which signals the corresponding T-helper cell about the presence of antigen. Following identification of an antigen, the T-helper cell activates cytotoxic T cells of the immune system to attack the infected B cells to produce antibodies. This "signature" antigen is remembered by antibodies, which can directly target any cells displaying the antigen in the future, termed as adaptive or acquired immunity.

Macrophages are the vital first responders concerning host defense to bacterial infection. Prolonged inflammatory responses are always harmful to the host. Macrophages can differentiate into either classically activated (M1) or alternatively activated (M2) macrophages. The M1 are considered kill-type macrophages that encompass a proinflammatory host defense phenotype that killed the foreign intruder in the course of an infection. M1 macrophages perform this function by metabolizing arginine to nitric oxide, which is cytotoxic to invading cells. M2 macrophages manifest both wound-healing and regulatory phenotypic functions; thus, are considered tissue repair-type macrophages. In the course of an infection, M2 macrophages promote new growth in the surrounding tissue and repair the damage of the infection. M2 macrophages accomplish this cell repair task by metabolizing arginine to ornithine, which is critical in regulating cell proliferation. The advancement from subclinical to clinical stages of the disease results from macrophage phenotype polarization from the M1 host defense type to regulatory M2 phenotype. Macrophage phenotypes are distinguished by cytokine secretion and representation of cell surface markers and receptors. M1 type host defense macrophages express proinflammatory cytokines like interferon gamma (interferon γ), interleukin (IL) 1β, IL-12, IL-23, and tumor necrosis factor (TNF-α). In contrast, resolution/regulatory and repair macrophages are pervaded by expression of the surface markers, mannose receptor, or cluster of differentiation (MR or CD) CD163 and CD206, cytokines IL-1Ra and IL-10, and transforming growth factor (TGF-β).

2 Understanding Macrophages Role in Disease Pathology

Macrophage polarization is plastic and reversible. Polarization between macrophage phenotypes is a dynamic process dominated by the presence of phenotype-dependent stimuli and their persistence. The plasticity of macrophage phenotypes occurs during the process of infection in the presence of heterogeneous populations. M1 polarization occurs at the initial stages of the inflammatory response, while M2 polarization is predominant during inflammation resolution. The sequential occurrence of both M1 and M2 polarization is absolutely required for the appropriate termination of inflammatory responses and initiation of adequate tissue repair response after injury. Alterations in the shifting process between macrophage polarization states result in some chronic inflammatory pathologies, autoimmune diseases, and even metabolic disorders. Macrophage polarization transpires both in physiological and pathological conditions. Polarization stages in a host are the key determinant of disease development and/or regression. Macrophage polarization depends on their metabolic profile. The metabolic character of M1 macrophages is enhanced glycolysis, irregularity in the pentose phosphate pathway (PPP), and accumulation of succinate and citrate due to diminished functioning of the TCA cycle.

In contrast, the metabolic feature of M2 macrophages is defined by oxidative phosphorylation, enhanced fatty acid oxidation, decreased glycolysis, and decline in pentose phosphate pathway activity. Macrophages recognize mitogen-activated protein (MAP) on the pathogens during the infectious process by using pattern recognition receptors. The macrophages' functions and responsiveness can be deranged by MAP through a variety of methods, including preventing macrophage activation, blocking phagosome maturation, preventing macrophage acidification, and delaying macrophage apoptosis. Comprehending macrophages' biology and their activation states in different disease conditions helps in designing therapeutic protocols.

2.1 Atherosclerosis

Atherosclerosis is the persistent chronic inflammation in arteries due to cholesterol plug buildup around the walls causing narrowing and thickening of the arteries. Atherosclerosis is a chronic inflammatory disease characterized by the narrowing and thickening of the arteries due to the buildup of plaque around the artery wall. Atherosclerosis is driven by an imbalance in lipid metabolism and a maladaptive immune response. Accumulation of lipids in large- and medium-sized arteries causes plaque deposition that blocks the flow of the blood. Elevated low-density lipoprotein (LDL) cholesterol, hypertension, obesity, and Type 2 and Type 1 Diabetes are associated with atherosclerosis development. The accumulation of LDL cholesterol promotes the recruitment of monocytes, and the uptake of oxidized LDL induces monocyte differentiation into macrophages that results in foam cell formation with the proliferation of smooth muscle cells. Oxidized phospholipids, oxidized LDL, saturated fatty acids, and lipoprotein(a) can induce apoptosis in the stressed macrophages. Proinflammatory macrophages implicate all pathogenic stages from plaque initiation to progressive deposition, while antiinflammatory macrophages are involved in plaque stabilization. It has been reported that both M1 and M2 macrophage subtypes can be found in atherosclerotic plaques.

Within the atherosclerotic microenvironment, activation of macrophages phenotype-specific transcriptional programs occurs due to accumulation of cholesterol crystals during the early stages of the atherosclerotic process. Cholesterol crystals can promote the caspase-1-activating inflammasome, which results in the cleavage and secretion of IL-1 functioning as an M1-polarizing stimulus. The proinflammatory M1 phenotype can also be promoted by inhibition of the transcription Kruppel-like factor 2 or stimulation of the Toll-like receptor (TLR) 4-mediated pathway that, in turn, leads to the activation of NFκB. In comparison, the antiinflammatory M2 phenotype is activated by 9-oxononanoyl cholesterol, a prominent cholesteryl ester oxidation product that can enhance TGFβ secretion. A third macrophage phenotype has recently been described in atherosclerosis, representing macrophages exposed to oxidized phospholipids termed as Mox. In advanced atherosclerotic lesions, Mox macrophages comprise approximately 30% of the total number of macrophages. Mox macrophages display reduced phagocytic and chemotactic abilities compared to M1 and M2 macrophages. In response to oxidized phospholipids, Mox macrophages lead to an increase in IL-1β and cyclooxygenase-2 expression.

Macrophage polarization is also closely correlated with the clinical course of atherosclerosis. Marked differences are observed between the macrophage subset of symptomatic and asymptomatic plaques. M1 macrophages are abundant in the lipid core of the developed symptomatic plaque but rarely found in the intimal regions of the plaque, while M2 macrophages are higher in asymptomatic atherosclerotic plaques, suggesting the potential protective role of M2 macrophages. Macrophage number gradually decreases in the regressing plaque, and in some cases, enhanced switch to phenotypic characteristics is observed, with enrichment in M2 phenotype.

2.2 Asthma

Allergic asthma is a chronic inflammatory lung disease characterized by airway inflammation, obstruction, hyperresponsiveness, and pathological lung remodeling. The inflammatory response hallmark is the activation of lymphocytes, mast cells, eosinophils, and macrophages in the lung and elevated expression of allergen-specific IgE. Chronic cytokine-mediated airway inflammation is the classic feature of allergic asthma correlated with the persistent presence of macrophage population in the airway lumen. Pulmonary macrophages produce various factors that straightly stimulate airway smooth muscle contractility and contribute to pathological airway remodeling. Airway macrophages release cytokines like IL-4, IL-13, and IL-33, having implications in the pathogenesis of chronic asthma. These macrophages also promote type 2 cytokines production by pulmonary CD4 T lymphocytes (helper T cell) and produce various cytokines and chemokines regulating the enrollment of eosinophils and basophils toward the lung. Macrophages also exacerbate the severity of the allergen-induced disease.

2.3 Obesity and Diabetes

Type 2 diabetes (T2D) is characterized by hyperglycemia due to peripheral resistance to insulin action and the failure of beta cells to compensate. Macrophage polarization has a pivotal role in the pathogenesis of T2D. The polarization of M1/M2 tissue-destructive versus tissue-reparative macrophages has an important effect on β-cell proliferation. Several studies have reported that M1 macrophages cause increased inflammation, obesity, and insulin resistance, while M2 macrophages are associated with reducing both obesity and insulin resistance. Imbalance in the ratio of M1 and M2 adipose macrophages has been directly related to the development of insulin resistance. M2 macrophages also provide a niche for preadipocytes to keep the number and quality of them, thus, maintaining insulin sensitivity. Builder M2-like macrophages regulate β-cell proliferation by releasing a variety of trophic factors such as TGF-β1. M1 macrophage secretes IL-1β, resulting in potent inhibition of insulin secretion, followed by islet destruction. The use of IL-1R antagonists and anti-IL-1β-neutralizing antibodies can abolish these effects on pancreatic islets.

Peroxisome proliferator-activated receptor gamma (PPAR-γ) is essential for macrophage M2 polarization having an antiinflammatory function and being associated with metabolic dysfunction. PPAR-γ is a target gene in regulating macrophage polarization as it interacts with nuclear factor kappa light-chain enhancer in the activated B cells (NF-κB) in the modulation of macrophage polarization. PPAR-γ blocks the proinflammatory pathway of NF-κB and inhibits the expression of relative factors, such as TNF-α. Lately, it has been reported that adenosine monophosphate interleukin β1 plays a significant role in protecting macrophages from inflammation while under high lipid content exposure resulting in a modulation of obesity-induced insulin resistance. Inhibition of proinflammatory cytokines and chemokines, such as TNF-α, IL-1β, IL-6, and chemokine ligand (CCL) 2, may reduce adipose tissue inflammation and insulin resistance.

Obesity is associated with the buildup of proinflammatory cells in visceral adipose tissue. Adipose tissue inflammation is an important underlying cause of insulin resistance and progression to T2D. Obesity escalates the risk of T2D by inducing chronic low-grade inflammation in localized adipose tissue. Tissue macrophages have an important role in inflammatory conditions, including obesity-associated metabolic diseases, insulin resistance, and T2D. Studies have proved that obesity-induced adipocyte hypertrophy results in upregulated surface expression of stress markers. Adipocyte hypertrophy has been reported to create a hypoxic area and activate hypoxia-inducible factor 1. Hypoxia induces inflammatory cytokines and suppresses preadipocyte-related angiogenesis, causing insulin resistance. Macrophages together with other immune cells are accounted to be almost 10% of the normal adipose tissue and play a paramount role in maintaining homeostasis. Diet-induced obesity compromises this homeostasis, resulting in enhanced macrophages' infiltration, causing up to 50% increment of the cells in adipose tissue. Adipose tissue inflammation is aggravated by the secretion of TNF-α, which further increases lipolysis leading to the production of free fatty acids, establishing a vicious circle. Understanding the underlying mechanisms behind obesity-induced visceral adipose tissue inflammation is an essential aspect of diabetes prevention.

Adipose tissue macrophages polarization from M2 to M1 like proinflammatory state results in insulin resistance. Targeting the inflammatory M1/M2 polarization process in obese persons appears to be an encouraging future strategy for prophylaxis against diabetes development. Adipose tissue macrophages from chemokine receptor 2 (CCR2) knockout mice are found to be polarized to the M2 macrophages, even after obesity, while CCR2 knockout mice were observed to be protected from diet-induced insulin resistance. Inhibition of IL-10 secretion by M2-like macrophages enhances the impairment of insulin signaling, confirming its protective role in T2D.

2.4 Cancer

Tumors are abundantly populated by macrophages, which were initially thought to be part of the antitumor tissue response, but data indicate that macrophages foster tumor initiation, progression, and metastasis. In response to persistent infections or chronic irritation, macrophages synthesize inflammatory cytokines (IFNγ, TNFα, and IL6) that engage other immune cells and appear to be causal in tumor initiation and promotion. Cancers particularly adept nonclassical tolerant state macrophages that infiltrate the tumor region, commonly called tumor-associated macrophages (TAMs). In established tumors, the TAMs stimulate tumor cell migration, invasion, intravasation as well as the angiogenic response required for tumor growth. TAMs also remodel the tumor microenvironment via the enhanced expression of proteases such as matrix metalloproteinases, cathepsins, and matrix remodeling enzymes, including lysyl oxidase.

Once tumors are established, the tumor-associated TAMs adopt a trophic immunosuppressive phenotype that promotes tumor progression to malignancy different from the initial immunologically active state. Macrophages pertain an important role in tumor angiogenesis, regulating the angiogenic switch needed to transition to a malignant state. A distinct macrophage population promotes secondary tumor development called metastasis-associated macrophages (MAMs), which differ in origin compared to TAMs. Dissemination of cancer cells from primary tumors to the surrounding tissues is called metastasis, which causes about 90% of cancer-associated deaths. Metastasis includes the involvement of TAMs from the primary tumor following multiple steps to metastatic progression. TAMs activate tumor growth, initiate angiogenesis, tumor cell migration, invasion, and display the unique set of markers. Resident macrophages preferentially engaged in primary tumors generate TAMs, whereas inflammatory macrophages in the metastatic sites generate MAMs. Inflammatory monocytes and MAMs promote tumor cell extravasation through the expression of vascular endothelial growth factor (VEGF) that induces local vascular permeability. MAMs also promote subsequent growth of the metastatic cells and ablate them after the metastases are established, inhibiting metastatic growth.

Cancer cells have the capability to suppress M1 macrophages while supporting the growth of M2 macrophages. The paucity of M1 macrophages results in reduced capability of the immune system to destruct cancer cells. Thus, the abundance of M2 macrophages promotes the unopposed growth and proliferation of the tumor. Cancer patients mostly have a higher ratio of M1 macrophages compared to M2. Studies on cancer proliferation showed that mice with an elevated proportion of M1 macrophages got much slower cancer proliferation than those with a lower ratio of M1 macrophages to M2 macrophages. A promising breakthrough in cancer immunotherapy may be to determine a successful and efficient way of increasing the ratio of M1 macrophages compared to M2. Macrophage populations can be preferentially stimulated to differentiate into M1 macrophages through the introduction of interferon γ. However, a high amount of interferon γ can produce a number of unwanted side effects due to its wide physiological implications.

Alternatively, it may be possible to simply increase interferon γ directed at the tumor site to localize the differentiation of macrophages bypassing the widespread impact of increased interferon γ all over the body. Cytotoxic T cells have the capability to directly kill the tumor cell along with production interferon γ, which increases M1 macrophage differentiation accounting for antitumor immune response. By enhancing the functionality of the T cells, both direct and macrophages-mediated cell killing can be achieved, harnessing potentially ideal immunotherapy.

2.5 Chronic Wound

Macrophages are involved in tissue homeostasis, promotion of inflammatory responses, tissue damage, and also in tissue repair. Inflammation and metabolic stress are associated with cancer, T2D, chronic infection, and sepsis. Burn injury patients have an abundance of free fatty acids and M2-polarized macrophages in the adipose tissue and in the liver, which impairs wound healing and contributes to disease progression. Macrophages are differently reprogrammed in response to lipopolysaccharide (LPS), a "danger signal" associated with gram-negative bacterial infections. Macrophages show classical proinflammatory (M1-polarized) response to LPS but a suppressive, antiinflammatory (M2-polarized) response to chronic palmitate exposure, each associated with specific transcriptional programs and signaling pathway activities.

2.6 Rheumatoid Arthritis

Macrophages have a prominent role in the pathogenesis of a variety of autoimmune diseases, including multiple sclerosis, rheumatoid arthritis, and inflammatory bowel disease. Macrophages are the source of many crucial inflammatory cytokines functioning as important drivers of autoimmune inflammation, that is, IL12, IL18, IL23, and TNFα. Arthritis is a chronic inflammatory autoimmune disorder that affects synovial joints and leads to joint destruction. Arthritis is not curable, and the currently available therapies can only decrease symptom severity and delay progression. Macrophages are the most numerous immune-derived cells found in the arthritis synovium, having implications in the pathogenesis and disease progression. Macrophages produce the predominant proinflammatory cytokines found in the arthritis synovium involved in pathogenesis (TNFα, IL-1β, and IL-6), chemoattractant factors (CCL2 and IL-8), and metalloproteinases (MMP-3 and MMP-12). The enhanced population of sublining macrophages in the synovium is an early hallmark of rheumatic disease activation. The extent of macrophage infiltration in the synovial is strictly related to joint erosion. Macrophages have the potential to act as a crucial target for therapeutic intervention of arthritis. In osteoarthritis, macrophages directly take part in synovial inflammation by producing TNF-α and IL-1β and, indirectly, via the activation of synovial fibroblasts.

2.7 Neurodegenerative Diseases

The central nervous system (CNS) is an "immuno-privileged" organ, hosting high populations of myeloid cells. Defensive barriers such as the meninges, the perivascular space, and the choroid plexus protect CNS. Microglia, meningeal macrophages, perivascular (blood-brain barrier), and choroid plexus (blood-cerebrospinal fluid barrier) macrophages are found within the CNS. These cells maintain homeostasis and are preferentially involved in the progression and resolution of neuroinflammatory processes. Toxins, infections, trauma, and ischemia-like stimuli can elicit rapid activation of the CNS immune system, referred to as acute neuroinflammation, by releasing inflammatory mediators. Following inflammation, circulating monocytes are recruited and enter the CNS, generating brain pathology. These cell populations are crucial players in CNS-related autoimmune diseases (multiple sclerosis) and degenerative diseases (amyotrophic lateral sclerosis and Alzheimer's disease). In uncontrolled cases, it leads to chronic neuroinflammation resulting in neurodegeneration giving rise to neurological disorders. Microglia exerts both neuroprotective and neurotoxic effects. Microglia produces inflammatory cytokines IL-1β, TNFα, IL-6, superoxide, nitric oxide, and excitatory amino acid, as well as neuroprotective factors such as neurotrophins, brain, derived neurotrophic factor, glial cell-derived neurotrophic factor, and nerve growth factor.

Alzheimer's disease is the most frequent cause of dementia with pathological hallmarks of β-amyloid accumulation, neurofibrillary tangles, synaptic loss, and neurodegeneration. Inflammatory components involved in Alzheimer's disease are associated with microglia and astrocytes' neuroinflammation with the activation of complement systems like cytokines and chemokines. Microglia and bone marrow-derived mononuclear phagocytes accumulate around senile plaques in Alzheimer patients. Microglia responds to the β-amyloid peptides and promotes their clearance by releasing cytotoxic factors, which further promotes phagocytosis of these peptides.

Amyotrophic lateral sclerosis is a neurodegenerative disease of motor neurons characterized by moderate and progressive dysfunction and loss of motor neurons. The neuronal injury occurs in motor neurons and microglia involving pathological markers like superoxide dismutase, fused in sarcoma, or TAR (transactive response) DNA-binding protein in motor neurons and oligodendrocytes. Neuroinflammation in amyotrophic lateral sclerosis involves activation and proliferation of microglia and the infiltration of T cells into the brain and spinal cord. Activation of microglia in amyotrophic lateral sclerosis is characterized as a continuum between the neuroprotective M2 and the neurotoxic M1 state phenotype macrophages. Inflammatory monocytes recruitment in the spinal cord also has a pathological significance in amyotrophic lateral sclerosis.

Multiple sclerosis is a demyelinating autoimmune disease of the CNS characterized by progressive axonal damage as a result of the loss of oligodendrocytes and neurodegeneration. Following damage of the blood-brain barrier, peripheral immune cells (T lymphocytes, monocytes, and dendritic cells) invade the CNS and activate the innate immune system. T-helper lymphocytes, cytotoxic T cells, B cells, macrophages, and microglia secrete cytokines that implicate initiation and progression of the deregulated immune response to myelin antigens and subsequent immune-mediated demyelination. In MS, microglia exhibit neuroinflammation and neuroprotective characteristics as following activation, massive immune cell infiltration and demyelination occur, and it finally dominates the remyelination and repair. Various proinflammatory cytokines are produced during the inflammatory process, such as TNFα, interferon γ, IL-1β, IL6, and inducible iNOS, which activates the microglial cells. Activated microglia is also the primary source of interferons in the inflamed CNS, leading to increased phagocytosis of myelin debris.

3 Targeting Macrophages by Nanodrug Delivery System

Due to the wide-ranging involvement of macrophages in the pathogenesis of several types of diseases, they are considered a relevant therapeutic target. Several strategies have been adopted and tested for manipulating macrophage function. Targeting macrophages in a particular diseased tissue ranging from depletion to reprogramming/repolarization has been on prime focus. Macrophage reeducation and depletion are considered the most important methods. Among the different approaches for targeting macrophages, the most revolutionary approach is depletion, which is usually achieved by clodronate-containing liposomal formulations preparation and/or depleting antibodies. Significant depletion of macrophages is associated with immunosuppression, enhanced chances of infection, and delayed wound healing. The latest generation of macrophage-based therapies focuses on repolarizing macrophages instead of eliminating them, like in the case of tumor-associated macrophage-targeted therapies. These strategies aim to inhibit macrophage effector functions or reprogramming of macrophages from achieving an antitumorigenic phenotype. Nanotechnology-based systems, that is, liposomes, dendrimers, gold nanoparticles (NPs), and polymeric NPs, have attracted attention. The advantages are the future perspective of modification according to the targeted disease site and better efficacy than conventional delivery systems with limited side effects. Nanotechnology-based approaches have huge implications as they show noteworthy outcomes in preclinical models.

4 Targeting Cancer

Tumors have a complicated microenvironment representing physicochemical and cellular heterogeneity with the presence of abnormal tumoral stroma and TAM-like immunosuppressive cells. Effective switching of the tumor-promoting immune-suppressive system to antiangiogenic adaptive immune responses is crucial for killing tumor cells. Loading of the blood monocytes/macrophages with nanomedicines enables drug release in the tumor bulk due to myeloid cells' high infiltration ability. TAM-targeted imaging nanostructures can also be employed to study the macrophage content in solid tumors useful for improved diagnosis and prognosis of cancer. A range of nanostructures has been designed to boost the delivery of drug compounds inside cancer cells. Nanostructures encapsulating active pharmacological compounds have the advantages of overcoming common issues of solubility and stability, side-effects reduction, the extension of half-time, and enabling controlled release to the target cell. The challenges associated are the undesirable delivery of nanomedicines in competing organs (liver, lung, or spleen) by the macrophages. Another issue is difficulty penetrating nanomedicines across solid tumors due to the abnormal vasculature presence of excessive extracellular matrix in stromal tumors. The following example provides an overview of the recent investigations targeting tumor macrophages with therapeutic or diagnostic intentions by loading with nanomedicines involving live cell-mediated drug delivery systems (LCDDS) for transporting drugs into the tumor bulk. Following are the examples of various reported nanotechnological approaches.

4.1 Targeting Tumor-Associated Macrophages

The tumor infiltrated TAMs support growth, angiogenesis, invasion, and metastasis. The high density of TAMs is associated with tumor growth progression and development of resistance to therapies making them favorable targets for novel antitumor therapy strategies like:

- Inhibition of TAMs recruitment.
- Reeducate tumor TAMs.
- Directly killing tumor TAMs.

4.1.1 Inhibition of TAMs Recruitment

In response to tumor-induced damage-related immune response, monocytes are recruited from the circulating blood and infiltrate the tumor site, becoming macrophages. Induction of TAM generation impairment in the tumors is possible by targeting to impact the blood, bone marrow, or lymphoid monocytes. Small interfering RNA (siRNA), well known as short interfering RNA or silencing RNA, is a class of double-stranded non-coded RNA molecules with 20–27 base pairs similar to miRNA, the RNA interference pathway. Chemokines play a fundamental role in monocyte recruitment and phenotypic maturation into TAMs linked to fibrogenesis and angiogenesis during chronic liver injury and hepatocarcinogenesis. Among the chemokines' broad family, CCL2 (also known as monocyte chemoattractant protein 1; MCP-1) is secreted by the liver cells upon stress and injury. The CCR2 receptor is expressed in monocytes and liver macrophages. Enhanced signaling of CCL2/CCR2 promotes liver inflammation, fibrosis, and pathologic angiogenesis. An RNA aptamer-based CCL2 inhibitor (CCL2i) was applied in a mouse model of liver fibrosis and hepatocarcinogenesis (diethylnitrosamine and CCl_4 administration). CCL2 inhibition resulted in reduced liver infiltration of M1, pathogenic angiogenesis, and tissue fibrosis with significant inhibition of tumor progression.

Lipid NPs loaded with siRNA were targeted to decrease the expression of the chemokine by the CCR2 receptors, necessitated for the recruitment of monocytes toward the tumor. These intravenous injected NPs accumulate in the spleen and bone marrow and deliver the associated siRNA into the Ly6C high monocytes (TAM precursors), resulting in decreased tumor growth when experimented in two xenografted tumor models. A similar approach has been presented for the CCL5-CCR5 axis by bone marrow-targeted biodegradable mesoporous silicon nanoparticles (MSVs) loaded with liposomal CCL5-

siRNA. The surfaces of NPs were marked with this aptamer targeting the E-selectin expressed on the bone marrow endothelium. When these CCL5-siRNA-loaded MSVs were injected intravenously in mice, they resulted in reprogramming of immunosuppressive myeloid cells in the bone marrow resulting in significantly reduced tumor growth and increased CD8+ T-cell (cytotoxic T cell) infiltration. Enhanced antitumor immune responses were observed when the CCL5-siRNA-loaded MSVs combined with the CCR5 inhibitor maraviroc. For inhibiting TAM recruitment, another potential target is the colony-stimulating factor receptor (CSF-1R). Studies conducted on several CSF-1R inhibitors have demonstrated encouraging antitumoral effects in different murine tumor models and progressed to the level of clinical trials. Further exploratory studies are required for targeting CSF-1R using NPs.

4.1.2 Directly Killing Tumor TAMs

Anticancer nanomedicines have the capability to kill tumor TAM directly. Several molecules and receptors overexpressed in TAM's surface have been marked for the targeting of TAM with NPs decorated with drug delivery nanocarriers. Nanotechnology-based delivery of bisphosphonates (clodronate or zoledronate) into the tumors has been tested 30 years back, which resulted in depletion of TAM alongside potential antitumor effect, impaired angiogenesis, and decreased metastasis. Clodronate is commonly used for the treatment of osteolytic bone disease and postmenopausal osteoporosis as it is able to inhibit osteoclast function. The clodronate liposomes are regularly used in biomedical research for the depletion of macrophages, which act in a nontargeted manner. Van Rooijen and colleagues developed large multilamellar clodronate-containing liposomes in the 1990s as specific tools for the induction of transient depletion in macrophages. Large clodronate liposomes are rapidly identified and taken up by macrophages due to their big size. Following capture and ingestion by macrophages, clodronate liposomes are degraded, releasing the drug, which causes apoptotic cell death via an ATP-dependent mechanism called "liposome-mediated macrophage suicide." Clodronate liposomes effectively deplete tissue-resident macrophages preferentially depending on the route of administration. A new liposomal formulation of clodronate (Clo-Lipo-DOTAP) equipped with stealth features has been developed. The small size of this formulation provides tumor-targeting properties like: long circulating time, enhanced permeability, and longer retention (DOTAP: 1,2-dioleoyl-3-trimethylammonium-propane). This formulation was very specific for macrophages and was effective at 5–10 times lower clodronate dosage levels than usually used. Clo-Lipo-DOTAP was tested to be highly effective on two different xenografted mouse melanoma models, resembling the early-stage primary tumor and advanced-stage lung melanoma metastasis. It showed inhibition of tumor growth, tumor volume reduction, and reduced tumor vascularization. Intratumoral alendronate conjugated with glucomannan was targeted toward TAM mannose receptors on sarcoma-bearing mice resulting in effective depletion. Folate-decorated liposomes loaded with zoledronic acid were targeted to tumor TAM, which did not show any promising results toward TAM load reduction in human nasopharyngeal and mouse colon adenocarcinoma. Targeting highly expressed mannose receptor (MR or CD206) in M2 macrophages with liposomes and poly lactic-co-glycolic acid (PLGA) nanoparticles decorated with mannose especially favors uptake of the mannosylated nanomedicines by tumor TAM and prevents uptake by other macrophages.

Doxorubicin activates antitumor immune responses by depleting myeloid-derived suppressor cells (MDSCs), and inducing immunogenic cell death. Platinum compounds show protumor skewing effects. Docetaxel and gemcitabine also have the potential to abolish the immunosuppressive microenvironment. Trabectedin was registered for the treatment of soft tissue sarcoma and ovarian cancer. Trabectedin was the first marketed drug that showed selective cytotoxic activity toward Ly6C monocytes in circulation and spleen and reduces TAM in the tumor. Many researchers have explored the avenue of cytotoxic drugs loading into nanocarriers to reach immunosuppressive cells in the tumors. Pegylation is the process of both covalent and noncovalent attachment of polyethylene glycol (PEG) polymer chains with drug, therapeutic protein, or vesicle, described as PEGylated. PEGylation of liposomes helps it to reach safely in the tumor sites, avoiding early recognition by the mononuclear phagocytic system. Mannosylated-PLGA NPs shielded with pH-sensitive PEG moieties were developed for the delivery of doxorubicin and siRNA into TAM. PEGylated liposomes conjugated with the peptide LyP-1 loaded with doxorubicin were shown to reach TAM in metastatic lymph nodes, causing lymphatic tumor metastasis inhibition. RNA aptamers have shown an affinity for the murine and human IL-4 receptor resulting in preferential binding to MDSCs in the spleen and TAM in murine tumors. Gemcitabine lipid nanocapsules showed a promising ability toward targeting monocytic MDSCs eliciting antitumor responses by activating T cells in lymphoma and melanoma-bearing mice. In mouse cervical and mammary carcinoma models, the depletion of TAMs by a highly selective CSF 1R inhibitor emanated in the arrest and delay of tumor growth. Moreover, macrophage depletion was parallel with an increase in CD8+ T-cell infiltration in cervical and breast carcinomas. Recently, the safety, pharmacokinetics, pharmacodynamics, and antitumor performance of a human antibody against CSF1R (AMG 820) in its first-ever phase I clinical study were evaluated for advanced solid tumors. This study demonstrated good tolerance of the anti-CSF1R CSF AMG 820.

The current emerging approach for killing TAM is photodynamic or photothermal therapy using engineered nanostructures that was first reported in 2015. Quenched activity-based probes (qABPs) are small molecules that emit fluoresce light upon stimulation. This activity-dependent covalent modification can kill TAM resulting in significant tumor shrinkage in the murine model of breast cancer. Photodynamic therapy was targeted to the mannose receptor with mannose-conjugated chlorin nanoconjugates against the murine colon cancer model. This phthalocyanine dye conjugated to monoclonal anti-CD206 antibody demonstrated potential antitumor activity against lung metastasis. The formulation was also very effective when applied intravenously on breast cancer resistant to sorafenib. All these reports provide encouraging evidence that TAM depletion in tumors resistant to current antitumoral therapies can be promising.

4.1.3 Reeducate Tumor TAMs

The reeducation of TAM aims at reprogramming of macrophages with M2-like protumoral properties to M1-like antitumoral phenotype. The M1 phenotypes can directly kill tumor cells and elicit vascular damage and tissue destruction. Several pharmacological molecules have the ability to reswitch macrophage polarization from M2 immunosuppressive to M1 cytotoxic. Investigations have been done on the use of IFN-γ, immunomodulatory protein histidine-rich glycoprotein (HRG), agonist of antibody CD40, anti-CD47 antibodies, phosphatidylinositide 3-kinase (PI3K)-gamma, or Bruton's tyrosine kinase (BTK) inhibitors. This is also a modified form of vitamin D-binding protein (EF-022) for the antitumoral activation of TAM. The IFN-γ is an exemplary inducer of macrophage M1 polarization inducing cancer cell killing while avoiding problems associated with systemic macrophage activation. It should be administered intraperitoneally (i.p.) for enhanced clinical responses. The CD40 antibody agonist on advanced pancreatic cancer patients and BLZ945 (a highly selective small-molecule inhibitor of CSF-1R) on patients with glioblastoma multiforme lead to functional repolarization of TAM with significant enhancement of the patient's survival.

Long-lasting functional reeducation of TAM can be achieved by nanotechnological approaches. Mannose-decorated polymeric nanocapsules loaded with IκBα siRNA were developed targeting mannose receptors to restore the classical NF-κB activity reeducating TAMs. Mannan-conjugated manganese dioxide (MnO_2) and low molecular weight hyaluronic acid nanoparticles intended to induce proinflammatory signals in macrophages in the breast cancer model. This formulation increases tumor oxygenation, regulates the pH, and downregulates hypoxia-inducible factor 1-alpha (HIF-1α) and VEGF in the tumor microenvironment. This nanosystem also demonstrated interesting performance on tumor imaging and detection and showed enhanced antitumor activity in combination with doxorubicin. Galactosylated cationic dextran nanocomplexes (NCs) are aimed at the galactose-type lectin receptor and deliver anti-IL oligodeoxynucleotide in TAM. These NCs are made of pH-sensitive PEG-histidine-modified alginate that hides the nanocarriers from macrophage uptake, but upon reaching the acidic microenvironment of the tumor, releases the NCs for specific binding with TAM. These NCs administered IV on an allografted hepatoma murine model give rise to the suppression of TAM protumor functions and stimulation of macrophage antitumor activities.

Signal transducer and activator of transcription (STAT3) is a well-known factor for TAMs induction-related tumor progression. Several nanotechnology approaches involving the use of PLGA-PEG NPs and liposomes have been explored for inhibition of STAT3 in TAM. PLGA-PEG NPs loaded with sunitinib, hydrazinocurcumin, and imidazole-loaded legumain-targeted liposomes have shown the excellent capability of tumor microenvironment remodeling. PEG-PLGA and DOTAP-NPs containing IL-12 immunogene plasmids inhibited angiogenesis and tumor growth in colon cancer murine models. Combination delivery of TGF-β inhibitor and IL-2 by nanoscale liposomal polymeric gels enhances tumor immunotherapy. Cytokine-encapsulating biodegradable polymers can codeliver small hydrophobic TGF-β inhibitors and water-soluble IL-2 with noteworthy antitumoral efficacy in melanoma-bearing mice. PEGylated gold nanoparticles with lipid/protamine/hyaluronic acid can deliver biohybrid TGF-β siRNA into tumors. These NPs have the ability to target murine lung cancer cells and TAMs, demonstrating the reeducating effect on the tumor microenvironment, activating a host immune response.

Dextran-coated iron oxide NPs (ferumoxytol), FDA approved for iron deficiency, also possess antitumor effects toward early mammary cancers and lung and liver cancer metastasis. Ferumoxytol accumulates in tumor TAMs and induces reeducation toward M1-like antitumoral macrophages by increasing TNF-α and reactive oxygen species generation through iron oxide Fenton reactions. A combination therapy consisting of PLGA NPs loaded with the photothermally active dye indocyanine green and a TLR7 agonist was administered IV in mice, which was followed by localized near-infrared light application to the tumor. This strategy resulted in tumor ablation by stimulation of antitumoral immune responses, which was significantly enhanced based on photothermal ablation of primary tumors. In combination with cytotoxic T lymphocyte-associated antigen (CTLA-4) checkpoint blockade with anti-CTLA-4 therapy potential, metastasis inhibition was observed (Fig. 1).

5 Arthritis

Liposomal clodronate therapy was assessed first on rheumatoid arthritis like inflammatory disease in the 1990s. Acute administration of clodronate liposomes via intraarticular injection caused reversible depletion of synovial phagocytic cells, accompanied by reduced cartilage destruction in an experimental mouse model of arthritis. Similar results were obtained on rheumatoid arthritis-affected patients scheduled for knee joint replacement when administered intraarticular clodronate liposomes. The formulation was found to be well-tolerated and nontoxic in a study conducted on human volunteers in 2000.

6 Obesity and Diabetes

The depletion of visceral adipose tissue macrophages (VATMs) occurred following an i.p. injection of clodronate liposomes in a high-fat diet animal obesity model, causing blocked weight gain and insulin resistance. Gene expression analysis study revealed that VATMs depletion is connected with the downregulation of lipogenesis and gluconeogenesis genes. Another

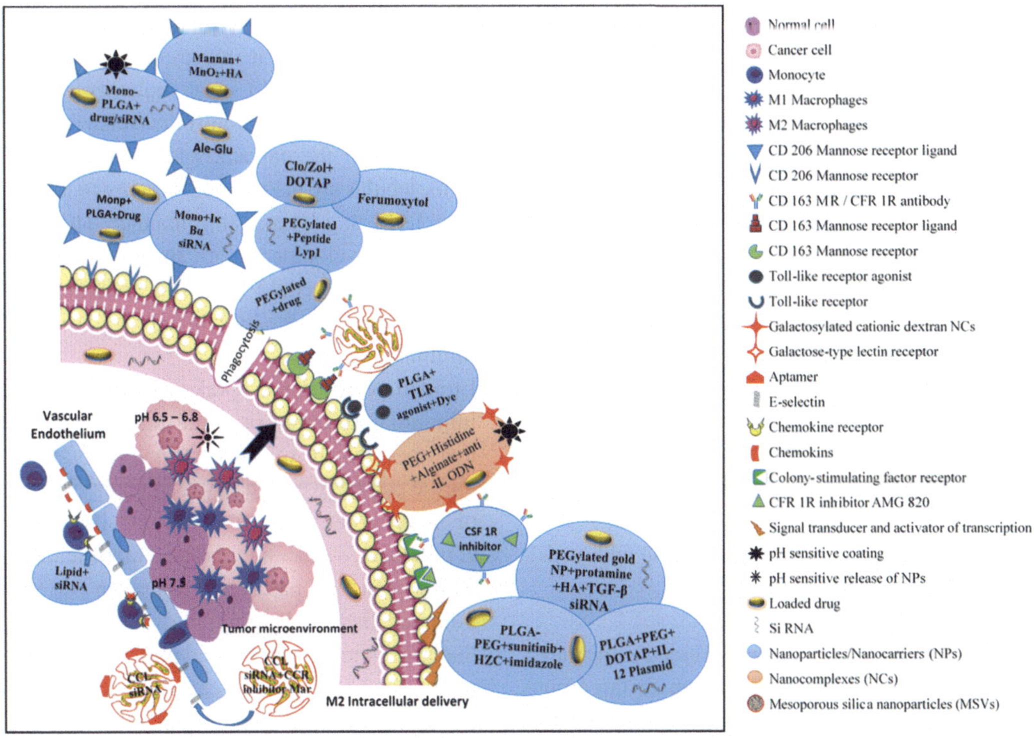

Fig. 1 Tumor macrophages-targeted nanodrug delivery system strategies
Ale Alendronate, *CCL* Chemokine ligand, *CCR* Chemokine receptor, *Clo* Clodronate, *CSF1R* Colony-stimulating factor receptor, *DOTAP* 1,2-dioleoyl-3-trimethylammonium-propane, *Drug* Doxorubicin, *Dye* Indocyanine green, *Ferumoxytol* Dextran-coated iron oxide NPs, *Glu* Glucomannan, *HA* hyaluronic acid, *HZC* hydrazinocurcumin, *Mar* Maraviroc, *Mono* Monosylated, *ODN* oligodeoxynucleotide, *PLGA* Poly lactic-co-glycolic acid, *siRNA* Small interfering RNA, *STAT* Signal transducer and activator of transcription, *TLR* Toll-like receptor, and *Zol* Zoledronate

research has demonstrated a similar effect following i.p. clodronate liposome administration in diet-induced obese (DIO) mice, where it reduces VATMs and improves glucose homeostasis and insulin sensitivity.

7 Inflammatory Diseases

The growth of vascularized endometrial tissue in aberrant locations outside of the uterus is a basic indicator of chronic inflammatory disease endometriosis. Chronic pelvic pain, dysmenorrhea, and reduced fertility are the common symptoms of endometriosis. The endometriosis pathogenesis is associated with altered immune response in the local peritoneal microenvironment due to abundant infiltrating macrophages in endometriotic lesions. They produce high amounts of pro-inflammatory and chemotactic cytokines. In an endometriosis experimental mouse model, clodronate liposomes were injected (i.p.) at different time points of endometrium implantation. Clodronate liposomes potentially reduced the percentage of F4/80- and CD11b-positive cells in the peritoneum, accompanied by a considerable reduction in endometriotic lesions weight.

Inflammatory lung disease, such as granulomatous inflammation, occurs due to *Mycobacterium tuberculosis* infection or chronic obstructive pulmonary disease (COPD). COPD is a life-threatening lung inflammatory disease hallmarked by chronic airway inflammation, mucus hypersecretion, and airway remodeling. In a mice model of cigarette smoke-induced COPD, the pivotal role of macrophages in the pathogenesis of the disease has been established. Intranasal administration of clodronate liposomes induces macrophage depletion causing reduction in smoke-induced nasal epithelial thickening and emphysema development. Depletion of alveolar macrophages induced by intranasal delivery of clodronate liposomes can protect mice from lethality as assessed in a mouse model of pulmonary tuberculosis. This was found to be associated with decreased mycobacteria outgrowth in the lungs and liver and less production of type I cytokines in the lung tissue due to the polarization of macrophages.

8 Major Challenges in Macrophages Targeting

Substantial challenges for the targeting and reeducation of TAM are as follows:

- The crucial concern to be addressed is the lack of selectivity of the formulation in targeting macrophages. The proteins of the target receptors are also present in other immune cells (T or B cells and endothelial cells) and macrophages in other locations. The specificity of ligands is a critical factor while reaching the protumoral M2 macrophages in tumors bypassing other populations of macrophages.
- The resilient character of the immunosuppressive microenvironment Immunosuppressive microenvironment often makes it time-consuming for the immune system to become activated/stimulated/re-educated and acquire the capability to mount a sufficient long-lasting antitumor response causing decisive elimination of cancer. This issue can be addressed by bioengineering new NPs by providing enhanced selectivity in TAM targeting through the synergic response of the ligand-receptor recognition defining a preferential appetite in the macrophages for nanometer-sized particles. Secondly, by combining a unique therapeutic nanostructure with pharmacologically active molecules having the ability to kill TAM.
- Another important challenge is the optimization of NPs "pharmacological" load for the efficacious killing of TAMs. Loading in NPs reduces the amount required to reach the tumor and makes it more manageable to induce an effect with lower quantity due to selective delivery and controlled release characteristics of the nanocarriers. Combinations of molecules are also a good strategy to reeducate TAM in a long-lasting manner. The successful development of these kinds of nanomedicines could lead to a breakthrough in cancer and inflammatory disease immunotherapy.
- Limited or restricted uptake of drugs by the inflammatory organ or solid tumor tissues is due to reduced vascular permeability. This hurdle can be overcome by using nanocarriers of appropriate size (namely 50–100 nm), Z-potential, and membrane modification, which can be passively be taken up by inflamed tissues, helping in accumulating at the target site. This enhanced permeability and retention phenomenon can be particularly exploited on drug-loaded liposomes, using positively for cancer treatment. The unwanted uptake of liposomes by phagocytes of the RES can be prevented by modifying surfaces of the liposomes. Outer membrane modification by PEGylation provides liposomes with stealth features, leading to increased blood circulation time, which is a great characteristic for drug delivery systems.

8.1 Nanomedicines Loaded in Macrophages

Conventionally administered chemotherapeutic drugs are able to reach the vascular area, whereas macrophages have a greater ability to penetrate the solid tumor tissue even in the hypoxic regions. Therapeutic drug-loaded myeloid cells can infiltrate tumor tissues and penetrate challenging biological barriers. The phagocytic nature of monocytes/macrophages is beneficial for easy loading with drug-containing nanostructures. The ex vivo loading of THP-1 (it is a spontaneously immortalized monocyte-like cell line) monocytes was demonstrated with cRGD-modified liposomes containing an antidepressant macromolecule (trefoil factor 3). The "loaded cells" showed satisfactory blood-brain barrier (BBB) transport improving symptoms of depression given IV in murine models. Monocytes or macrophages loaded with nanomedicines are the latest generation of LCDDSs for cancer treatment. It uses the host cells (i.e., monocytes, macrophages, erythrocytes, or stem cells) of patients either as a whole or by modulating a selected vital component of these cells (i.e., external cellular membranes), as "Trojan Horses" loaded with drugs. The paramount advantages of using LCDDS as "Trojan Horses" are long blood circulation time, flexible morphology, high ratio of surface ligands, and physiological metabolism pattern. The transient lifespan of live cells like leukocytes is exploited for extended delivery, therapeutic action, and afterward, physiological elimination of the "carrier."

Although loading of circulating monocytes with nanostructures and further in vivo targeting is a challenging approach. Ly-6C high monocytes in blood circulation can specifically engulf RGD (arginine-glycine-aspartate) peptide-labeled single-walled carbon nanotubes (SWCNTs), and subsequently delivered into the tumor with high penetration capabilities. Another interesting ligand, β-glucan, can also reach monocytes in circulation that could be targeted to the pattern recognition receptor Dectin-1. With nanotechnology's implementation, the LCDDS can be loaded with different pharmacological molecules, and once the LCDDS approaches the appropriate location of its therapeutic action, like the tumor center, the drug is released. A potential approach to achieve this is to use a nanotechnological approach that triggers active drug release upon physical or chemical stimulation. Conversely, passive drug release at the targeted tissue can also be achieved by selecting the appropriate nanocarrier components with a good biodegradation profile. Polylactic acid (PLA) is a conventional biodegradable polymer with a satisfactory therapeutic profile. Highly advanced nanotechnological strategies are required to open up the nanocarrier and trigger the drug release at a chosen time to achieve active drug release. External stimuli, such as temperature, light, magnetic field, or ultrasound approaches, are the options available. L-asparaginase was associated with a cell-penetrating peptide (CCP) through disulfide linkages, and encapsulated into red blood cells (RBC). Glutathione reduces the disulfide bonds resulting in the disassociation of the drug from the peptide, followed by drug release and therapeutic response following reach of this complex to the cytoplasm. The concept of loading live monocytes/macrophages with nanomedicines is still in its initial stages of investigation and onerous to practical implementation. This is a promising concept to be taken over, looking into the "carrier" cells' advantageous physiological properties, minimum chances toxicity, and innate ability to penetrate the tumor microenvironment, even approaching the toughest hypoxic regions of the tumor.

9 Conclusion

Targeting macrophage polarization is leading to the advent of various novel intervention strategies. Macrophages are involved in almost every disease and represent extremely attractive therapeutic targets that either augment or inhibit their responses. Even though a high grade of plasticity characterizes macrophages, they still claim high interest of researchers as potential therapeutic targets. Current approaches using macrophages as targeted therapies are being successfully tested on preclinical mouse models, mainly against cancer. A few drawbacks need to be overcome to ensure successful therapeutic implementation. First, the murine macrophage macrophages do not fully represent the genetic phenotypes of human macrophages and their physiology during homeostasis and disease. Novel human-induced pluripotent stem cell (IPCS)-derived macrophage protocols are reported, which might contribute to a better understanding of macrophage polarization physiology and pathophysiology. Testing of macrophage-targeted NPs on IPCS can give out a similar profile of response expected to be extrapolated on human. Second, the major limitations of macrophage reeducation therapies are the specificity and durability of the treatment.

Targeting a specific subset of macrophages in a particular tissue while avoiding deleterious side effects on adjuvant organs is still difficult to achieve. Discontinuation of reeducation therapies also accompanies major problems of rebound that need to be addressed. The crucial issue related to macrophage recruitment and reeducation therapies is the pathological consequences of repolarization promoting the occurrence of autoimmune or inflammatory diseases. Lack of specific markers is the biggest hurdle for identifying and distinguishing pathological macrophages from the normal counterparts. They can acquire different phenotypes, frequently overlapping each other, which makes therapeutic interventions very challenging.

However, for these therapies to be effective, it is necessary to understand macrophage diversity and define the cells according to anatomical location and function and the regulation of the particular set points that define the particularly recognizable macrophages in microglia, osteoclasts, and Kupffer cells. The recognition of multiple origins may allow unique opportunities to selectively target the recruited ones in the context of the chronic diseases discussed above, thereby inhibiting the pathology without disturbing resident macrophages and thereby regular homeostasis.

Bibliography

Allavena P, Germano G, Belgiovine C, D'Incalci M, Mantovani A. Trabectedin: a drug from the sea that strikes tumor-associated macrophages. OncoImmunology. 2013;2(6):e24614.

Andon FT, Digifico E, Maeda A, Erreni M, Mantovani A, Alonso MJ, Allavena P. Targeting tumor associated macrophages: the new challenge for nanomedicine. Semin Immunol. 2017;34:103–13. https://doi.org/10.1016/j.smim.2017.09.004.

Argyle D, Kitamura T. Targeting macrophage-recruiting chemokines as a novel therapeutic strategy to prevent the progression of solid tumors. Front Immunol. 2018;9:2629. https://doi.org/10.3389/fimmu.2018.02629.

Bacci M, Capobianco A, Monno A, Cottone L, Di Puppo F, Camisa B, et al. Macrophages are alternatively activated in patients with endometriosis and required for growth and vascularization of lesions in a mouse model of disease. Am J Pathol. 2009;175(2):547–56.

Ban Y, Mai J, Li X, Mitchell-Flack M, Zhang T, Zhang L, et al. Targeting autocrine CCL5–CCR5 Axis reprograms immunosuppressive myeloid cells and reinvigorates antitumor immunity. Cancer Res. 2017;77(11):2857–68.

Banciu M, Metselaar JM, Schiffelers RM, Storm G. Antitumor activity of liposomal prednisolone phosphate depends on the presence of functional tumor-associated macrophages in tumor tissue. Neoplasia. 2008;10(2):108–17.

Beatty GL, Chiorean EG, Fishman MP, Saboury B, Teitelbaum UR, Sun W, et al. CD40 agonists alter tumor stroma and show efficacy against pancreatic carcinoma in mice and humans. Science. 2011;331(6024):1612–6.

Beckett EL, Stevens RL, Jarnicki AG, Kim RY, Hanish I, Hansbro NG, et al. A new short-term mouse model of chronic obstructive pulmonary disease identifies a role for mast cell tryptase in pathogenesis. J Allergy Clin Immunol. 2013;131(3):752–62.

Ben-Nun Y, Merquiol E, Brandis A, Turk B, Scherz A, Blum G. Photodynamic quenched Cathepsin activity based probes for cancer detection and macrophage targeted therapy. Theranostics. 2015;5(8):847–62.

Bobryshev YV, Ivanova EA, Chistiakov DA, Nikiforov NG, Orekhov AN. Macrophages and their role in atherosclerosis: pathophysiology and transcriptome analysis. Biomed Res Int. 2016;9582430 https://doi.org/10.1155/2016/9582430.

Bondeson J, Wainwright SD, Lauder S, Amos N, Hughes CE. The role of synovial macrophages and macrophage-produced cytokines in driving aggrecanases, matrix metalloproteinases, and other destructive and inflammatory responses in osteoarthritis. Arthritis Res Ther. 2006;8:R187. https://doi.org/10.1186/ar2099.

Chen Q, Xu L, Liang C, Wang C, Peng R, Liu Z. Photothermal therapy with immune-adjuvant nanoparticles together with checkpoint blockade for effective cancer immunotherapy. Nat Commun. 2016;7:13193. https://doi.org/10.1038/ncomms13193.

Chinetti-Gbaguidi G, Colin S, Staels B. Macrophage subsets in atherosclerosis. Nat Rev Cardiol. 2015;12(1):10–7.

Conde J, Bao C, Tan Y, Cui D, Edelman ER, Azevedo HS, et al. Dual targeted immunotherapy via in vivo delivery of biohybrid RNAi-peptide nanoparticles to tumour-associated macrophages and cancer cells. Adv Funct Mater. 2015;25(27):4183–94.

Davignon JL, Hayder M, Baron M, Boyer JF, Constantin A, Apparailly F, et al. Targeting monocytes/macrophages in the treatment of rheumatoid arthritis. Rheumatol. (United Kingdom). 2013;52:590–8. https://doi.org/10.1093/rheumatology/kes304.

De Palma M, Lewis CE. Macrophage regulation of tumor responses to anticancer therapies. Cancer Cell. 2013;23(3):277–86.

Dijkgraaf EM, Heusinkveld M, Tummers B, Vogelpoel LT, Goedemans R, Jha V, et al. Chemotherapy alters monocyte differentiation to favor generation of cancer-supporting M2 macrophages in the tumor microenvironment. Cancer Res. 2013;73(8):2480–92.

Finucane OM, Reynolds CM, McGillicuddy FC, Roche HM. Insights into the role of macrophage migration inhibitory factor in obesity and insulin resistance. Proc Nut Soc. 2012;71(4):622–33.

Fritz JM, Tennis MA, Orlicky DJ, Lin H, Ju C, Redente EF, et al. Depletion of tumor-associated macrophages slows the growth of chemically induced mouse lung adenocarcinomas. Front Immunol. 2014;5:587. https://doi.org/10.3389/fimmu.2014.00587. eCollection 2014

Geeraerts X, Bolli E, Fendt SM, Van Ginderachter JA. Macrophage metabolism as therapeutic target for cancer, atherosclerosis, and obesity. Front Immunol. 2017;8:289. https://doi.org/10.3389/fimmu.2017.00289.

Ghinoi V, Brandi ML. Clodronate: mechanisms of action on bone remodelling and clinical use in osteometabolic disorders. Expert Opin Pharmacother. 2002;3(11):1643–56.

Ginhoux F, Schultze JL, Murray PJ, Ochando J, Biswas SK. New insights into the multidimensional concept of macrophage ontogeny, activation and function. Nat Immunol. 2016;17:34–40. https://doi.org/10.1038/ni.3324.

Hattori Y, Yamashita J, Sakaida C, Kawano K, Yonemochi E. Evaluation of antitumor effect of zoledronic acid entrapped in folate-linked liposome for targeting to tumor associated macrophages. J Liposome Res. 2015;25(2):131–40.

Hayashi N, Kataoka H, Yano S, Tanaka M, Moriwaki K, Akashi H, et al. A novel photodynamic therapy targeting cancer cells and tumor-associated macrophages. Mol Cancer Ther. 2015;14(2):452–60.

He H, Ye J, Wang Y, Liu V, Chung HS, Kwon YM, et al. Cell-penetrating peptides meditated encapsulation of protein therapeutics into intact red blood cells and its application. J Control Release. 2014;176:123–32.

Huang Z, Zhang Z, Jiang Y, Zhang D, Chen J, Dong L, Zhang J. Targeted delivery of oligonucleotides into tumor-associated macrophages for cancer immunotherapy. J Control Release. 2012;158(2):286–92.

Jenvey CJ, Shircliff AL, Bannantine JP, Stabel JR. Phenotypes of macrophages present in the intestine are impacted by stage of disease in cattle naturally infected with *Mycobacterium avium* subsp. paratuberculosis. PLoS One. 2019;14(5):e0217649. https://doi.org/10.1371/journal.pone.0217649.

Jung U, Choi MS. Obesity and its metabolic complications: the role of adipokines and the relationship between obesity, inflammation, insulin resistance, dyslipidemia and nonalcoholic fatty liver disease. Int J Mol Sci. 2014;15(4):6184–223.

Kadl A, Meher AK, Sharma PR, Lee MY, Doran AC, Johnstone SR, et al. Identification of a novel macrophage phenotype that develops in response to atherogenic phospholipids via Nrf2. Circ Res. 2010;107(6):737–46.

Kodumudi KN, Woan K, Gilvary DL, Sahakian E, Wei S, Djeu JY. A novel chemoimmunomodulating property of docetaxel: suppression of MDSC in tumor bearers. Clin Cancer Res. 2010;16(18):4583–94. https://doi.org/10.1158/1078-0432.CCR-10-0733.

Leemans JC, Juffermans NP, Florquin S, van Rooijen N, Vervoordeldonk MJ, Verbon A. Depletion of alveolar macrophages exerts protective effects in pulmonary tuberculosis in mice. J Immunol. 2001;166(7):4604–11. https://doi.org/10.4049/jimmunol.166.7.4604.

Leuschner F, Dutta P, Gorbatov R, Novobrantseva TI, Donahoe JS, Courties G. Therapeutic siRNA silencing in inflammatory monocytes in mice. Nat Biotechnol. 2011;29(11):1005–10.

Li J, Hsu HC, Mountz JD. Managing macrophages in rheumatoid arthritis by reform or removal. Curr Rheumatol Rep. 2012;14:445–54. https://doi.org/10.1007/s11926-012-0272-4.

Li X, Yao W, Yuan Y, Chen P, Li B, Li J, et al. Targeting of tumour-infiltrating macrophages via CCL2/CCR2 signalling as a therapeutic strategy against hepatocellular carcinoma. Gut. 2017;66(1):157–67.

Liao D, Liu Z, Wrasidlo WJ, Luo Y, Nguyen G, Chen T, et al. Targeted therapeutic remodeling of the tumor microenvironment improves an HER-2 DNA vaccine and prevents recurrence in a murine breast cancer model. Cancer Res. 2011;71(17):5688–96.

Lumeng CN, Bodzin JL, Saltiel AR. Obesity induces a phenotypic switch in adipose tissue macrophage polarization. J Clin Invest. 2007;117(1):175–84.

Mantovani A, Marchesi F, Malesci A, Laghi L, Allavena P. Tumour-associated macrophages as treatment targets in oncology. Nat Rev Clin Oncol. 2017;14(7):399–416.

Mills CD, Lenz LL, Harris RA. A breakthrough: macrophage-directed cancer immunotherapy. Cancer Res. 2016;76(3):513–6.

Molinaro R, Corbo C, Martinez JO, Taraballi F, Evangelopoulos M, Minardi S, et al. Biomimetic proteolipid vesicles for targeting inflamed tissues. Nat Mater. 2016;15(9):1037–46.

Moore KJ, Sheedy FJ, Fisher EA. Macrophages in atherosclerosis: a dynamic balance. Nat Rev Immunol. 2013;13(10):709–21.

Niu M, Naguib YW, Aldayel AM, Shi YC, Hursting SD, Hersh MA, Cui Z. Biodistribution and in vivo activities of tumor-associated macrophage-targeting nanoparticles incorporated with doxorubicin. Mol Pharm. 2014;11(12):4425–36.

Ortega RA, Barham WJ, Kumar B, Tikhomirov O, McFadden ID, Yull FE, Giorgio TD. Biocompatible mannosylated endosomal-escape nanoparticles enhance selective delivery of short nucleotide sequences to tumor associated macrophages. Nanoscale. 2015;7(2):500–10.

Ortega RA, Barham W, Sharman K, Tikhomirov O, Giorgio TD, Yull FE. Manipulating the NF-κB

pathway in macrophages using mannosylated, siRNA-delivering nanoparticles can induce immunostimulatory and tumor cytotoxic functions. Int J Nanomedicine. 2016;11:2163–77.
Palomba R, Parodi A, Evangelopoulos M, Acciardo S, Corbo C, de Rosa E, et al. Biomimetic carriers mimicking leukocyte plasma membrane to increase tumor vasculature permeability. Sci Rep. 2016;6:34422. https://doi.org/10.1038/srep34422.
Papadopoulos KP, Gluck L, Martin LP, Olszanski AJ, Tolcher AW, Ngarmchamnanrith G, et al. First-in-human study of AMG 820, a monoclonal anti-Colony-stimulating factor 1 receptor antibody, in patients with advanced solid tumors. Clin Cancer Res. 2017;23(19):5703–10.
Park J, Wrzesinski SH, Stern E, Look M, Criscione J, Ragheb R, et al. Combination delivery of TGF-β inhibitor and IL-2 by nanoscale liposomal polymeric gels enhances tumour immunotherapy. Nat Mater. 2012;11(10):895–905.
Piaggio F, Kondylis V, Pastorino F, Di Paolo D, Perri P, Cossu I, et al. A novel liposomal Clodronate depletes tumor-associated macrophages in primary and metastatic melanoma: anti-angiogenic and anti-tumor effects. J Control Release. 2016;223:165–77.
Pujade-Lauraine E, Guastalla JP, Colombo N, Devillier P, François E, Fumoleau P, et al. Intraperitoneal recombinant interferon gamma in ovarian cancer patients with residual disease at second-look laparotomy. J Clin Oncol. 1996;14(2):343–50.
Pyonteck SM, Akkari L, Schuhmacher AJ, Bowman RL, Sevenich L, Quail DF, et al. CSF-1R inhibition alters macrophage polarization and blocks glioma progression. Nat Med. 2013;19(10):1264–72.
Qin J, Yang X, Zhang RX, Luo YX, Li JL, Hou J, et al. Monocyte mediated brain targeting delivery of macromolecular drug for the therapy of depression. Nanomedicine. 2015;11(2):391–400.
Rao L, Meng QF, Bu LL, Cai B, Huang Q, Sun ZJ, et al. Erythrocyte membrane-coated Upconversion nanoparticles with minimal protein adsorption for enhanced tumor imaging. ACS Appl Mater Interfaces. 2017;9(3):2159–68.
Ries CH, Cannarile MA, Hoves S, Benz J, Wartha K, Runza V, et al. Targeting tumor-associated macrophages with anti-CSF-1R antibody reveals a strategy for cancer therapy. Cancer Cell. 2014;25(6):846–59.
Roth F, De La Fuente AC, Vella JL, Zoso A, Inverardi L, Serafini P. Aptamer mediated blockade of IL4Rα triggers apoptosis of MDSCs and limits tumor progression. Cancer Res. 2012;72(6):1373–83.
Schultze JL. Reprogramming of macrophages—new opportunities for therapeutic targeting. Curr Opin Pharmacol. 2016;26:10–5. https://doi.org/10.1016/j.coph.2015.09.007.
Shi J, Votruba AR, Farokhzad OC, Langer R. Nanotechnology in drug delivery and tissue engineering: from discovery to applications. Nano Lett. 2010;10:3223–30. https://doi.org/10.1021/nl102184c.
Singh Y, Pawar VK, Meher JG, Raval K, Kumar A, Shrivastava R, et al. Targeting tumor associated macrophages (TAMs) via nanocarriers. J Control Release. 2017;254:92–106. https://doi.org/10.1016/j.jconrel.2017.03.395.
Smith BR, Ghosn EE, Rallapalli H, Prescher JA, Larson T, Herzenberg LA, Gambhir SS. Selective uptake of single-walled carbon nanotubes by circulating monocytes for enhanced tumour delivery. Nat Nanotechnol. 2014;9(6):481–7.
Song M, Liu T, Shi C, Zhang X, Chen X. Bioconjugated manganese dioxide nanoparticles enhance chemotherapy response by priming tumor-associated macrophages toward M1-like phenotype and attenuating tumor hypoxia. ACS Nano. 2016;10(1):633–47.
Strachan DC, Ruffell B, Oei Y, Bissell MJ, Coussens LM, Pryer N, Daniel D. CSF_1R inhibition delays cervical and mammary tumor growth in murine models by attenuating the turnover of tumor-associated macrophages and enhancing infiltration by $CD8^+$ T cells. Onco Targets Ther. 2013;2(12):e26968. https://doi.org/10.4161/onci.26968.
Su Y, Xie Z, Kim GB, Dong C, Yang J. Design strategies and applications of circulating cell-mediated drug delivery systems. ACS Biomater Sci Eng. 2015;1(4):201–17.
Turk MJ, Waters DJ, Low PS. Folate-conjugated liposomes preferentially target macrophages associated with ovarian carcinoma. Cancer Lett. 2004;213(2):165–72.
Van Rooijen N, Sanders A. Liposome mediated depletion of macrophages: mechanism of action, preparation of liposomes and applications. J Immunol Methods. 1994;174(1–2):83–93.
van Rooijen N, van Kesteren-Hendrikx E. Clodronate liposomes: perspectives in research and therapeutics. J Liposome Res. 2002;12(1–2):81–94.
Vilella A, Ruozi B, Belletti D, Pederzoli F, Galliani M, Semeghini V, et al. Endocytosis of nanomedicines: the case of glycopeptide engineered PLGA nanoparticles. Pharmaceutics. 2015;7(2):74–89.
Weisberg SP, Hunter D, Huber R, Lemieux J, Slaymaker S, Vaddi K, et al. CCR2 modulates inflammatory and metabolic effects of high-fat feeding. J Clin Invest. 2006;116(1):115–24.
Wu ML, Huang TC, Hung HJ, Yu JH, Chung WB, Chaung HC. Investigation of beta-glucans binding to human/mouse dectin-1 and associated immunomodulatory effects on two monocyte/macrophage cell lines. Biotechnol Prog. 2010;26(5):1391–9.
Xie Z, Zhang Y, Liu L, Weng H, Mason RP, Tang L, et al. Development of intrinsically Photoluminescent and Photostable Polylactones. Adv Mat. 2014;26(26):4491–6.
Xin H, Zhang C, Herrmann A, Du Y, Figlin R, Yu H. Sunitinib inhibition of Stat3 induces renal cell carcinoma tumor cell apoptosis and reduces immunosuppressive cells. Cancer Res. 2009;69(6):2506–13.
Yan Z, Wang F, Wen Z, Zhan C, Feng L, Liu Y, et al. LyP-1-conjugated PEGylated liposomes: a carrier system

for targeted therapy of lymphatic metastatic tumor. J Control Release. 2012;157(1):118–25.

Yuan A, Hsiao YJ, Chen HY, Chen HW, Ho CC, Chen YY, et al. Opposite effects of M1 and M2 macrophage subtypes on lung cancer progression. Sci Rep. 2015, 5:14273. https://doi.org/10.1038/srep14273.

Zanganeh S, Hutter G, Spitler R, Lenkov O, Mahmoudi M, Shaw A, et al. Iron oxide nanoparticles inhibit tumour growth by inducing pro-inflammatory macrophage polarization in tumour tissues. Nat Nano. 2016;11(11):986–94.

Zaslona Z, Przybranowski S, Wilke C, van Rooijen N, Teitz-Tennenbaum S, Osterholzer JJ, Peters-Golden M, et al. Resident alveolar macrophages suppress, whereas recruited monocytes promote, allergic lung inflammation in murine models of asthma. J Immunol. 2014;193:4245–53. https://doi.org/10.4049/jimmunol.1400580.

Zeisberger SM, Odermatt B, Marty C, Zehnder-Fjällman AH, Ballmer-Hofer K, Schwendener RA. Clodronate-liposome-mediated depletion of tumour-associated macrophages: a new and highly effective antiangiogenic therapy approach. Br J Cancer. 2006;95(3):272–81.

Zhan X, Jia L, Niu Y, Qi H, Chen X, Zhang Q, et al. Targeted depletion of tumour-associated macrophages by an alendronate glucomannan conjugate for cancer immunotherapy. Biomaterials. 2014;35(38):10046–57.

Zhang X, Tian W, Cai X, Wang X, Dang W, Tang H, et al. Hydrazinocurcumin Encapsuled nanoparticles "re-educate" tumor-associated macrophages and exhibit anti-tumor effects on breast cancer following STAT3 suppression. PLoS One. 2013;8(6):e65896. https://doi.org/10.1371/journal.pone.0065896.

Zhang C, Gao L, Cai Y, Liu H, Gao D, Lai J, et al. Inhibition of tumor growth and metastasis by photoimmunotherapy targeting tumor-associated macrophage in a sorafenib-resistant tumor model. Biomaterials. 2016;84:1–12.

Zhu S, Niu M, O'Mary H, Cui Z. Targeting of tumor-associated macrophages made possible by PEG-sheddable, mannose-modified nanoparticles. Mol Pharm. 2013;10(9):3525–30.

Part II

Polymeric Nanoparticles for Macrophage Targeting

Polymeric Nanoparticles-Assisted Macrophage Targeting: Basic Concepts and Therapeutic Goals

Lubna Siddiqui, Asiya Mahtab,
Syed Arman Rabbani, Anita Verma,
and Sushma Talegaonkar

Abstract

Macrophages are the magic cells of innate immune system that are not only capable of fighting foreign bacteria and viruses, but in recent times, our in-depth understanding of their indispensable role in inflammatory diseases, wound healing, cancer, as well as metabolic disorders like atherosclerosis and insulin resistance has opened new gates to achieving better treatment modalities for these fatal diseases. Another field, which has flourished marvelously and is working wonders in the arena of biomedical applications, is the advancement of polymeric nanoparticles as drug carriers for targeting specific receptors and pathways for improved therapeutic efficacy. An amalgamation of these two magic wands has proven to be a boon in combating various fatal disorders including inflammatory diseases and cancer. Polymeric nanoparticles are now extensively used to target drug moieties toward specific macrophage receptors (including activin, CD44, and toll-like receptors) to achieve improved therapeutic effect. Redesigned (genetically modified) and reeducated (tumor-associated macrophages TAMs) macrophages have also been developed to meet the desired clinical effects. This chapter aims to compile details not only on the basics of macrophages and polymeric nanoparticles but also to highlights the bridge between the two streams, which has now been extensively explored to find newer treatments for various diseases and their biomedical translation.

Keywords

Macrophages · Innate immunity · Receptors · Polymeric nanoparticles · Active targeting · Passive targeting

1 Introduction

Macrophages are an important component of innate immunity and play a vital role in the first-line defense against pathogens and modulating

Lubna Siddiqui and Asiya Mahtab contributed equally to this chapter.

L. Siddiqui · A. Mahtab
Department of Pharmaceutics, School of Pharmaceutical Education and Research, Jamia Hamdard, New Delhi, India

S. A. Rabbani
Department of Clinical Pharmacy and Pharmacology, RAK College of Pharmaceutical Sciences, RAK Medical and Health Sciences University, Ras Al Khaimah, UAE

A. Verma
Nanobiotech Lab, Department of Zoology, Kirori Mal College, University of Delhi, Delhi, India

S. Talegaonkar (✉)
Department of Pharmaceutics, Delhi Pharmaceutical Sciences and Research University, New Delhi, India

S. Gupta, Y. V. Pathak (eds.), *Macrophage Targeted Delivery Systems*,
https://doi.org/10.1007/978-3-030-84164-5_6

inflammatory responses (Geissmann et al. 2010). Macrophages comprise of variably mixed populations, including liver Kupffer cells and brain microglial cells, that perform specific functions in the local microenvironment (Ginhoux et al. 2010).

The cells present in immune systems are characterized as lymphocytes, neutrophils, and monocytes or macrophages. These cells are also recognized as white blood cells. Lymphocytes are further categorized into B cells and T cells. B cells are also known as B lymphocytes and are specific cells of immune system whose main task is to produce antibodies. T cells are also known as T lymphocytes and are another type of immune cells. T-cells directly attack cells infected by viruses, and they also function as immune system regulators (The immune system and primary immunodeficiency n.d.).

Macrophages are common phagocytic cell and member of immune cells. It is a white blood cell situated in a tissue derived from monocytes. They are characterized by plasticity and versatility. Macrophages play a crucial role in clearing senescent or apoptotic cells, phagocytosis of immune-related complexes and pathogens, and maintenance of homeostasis.

According to the state of activation, macrophages can be divided into M1 type (classically activated macrophage) and M2 type (alternatively activated macrophage) (Germano et al. 2013). IFN-γ can distinguish macrophages into M1 macrophages that encourage inflammation. Unlike IFN-γ, IL-4 produced by Th2 cells can convert macrophages into M2-type macrophages that inhibit inflammation (Abramson and Gallin 1990). M1 macrophages' role is to secrete proinflammatory cytokines and chemokines, present antigens, and hence, contribute in the positive immune response and function as an immune monitor. The major proinflammatory cytokines it produce are IL-6, IL-12, and TNF-alpha. M2 macrophages mostly secrete arginase-I, IL-10 (Biswas and Mantovani 2010), and TGF-β along with other anti-inflammatory cytokines that have the function of reducing inflammation and contributing to tumor development and immunosuppressive function (Grohmann et al. 2001). It plays a significant role in healing wounds and repairing tissues. M1-type macrophages are not only linked to inflammatory diseases and infectious diseases but also to metabolic diseases including arteriosclerosis and insulin resistance. M2 macrophages are also associated with the development of numerous diseases (Van Ginderachter et al. 2006).

2 Role of Macrophages in Various Diseases Inflammation

Macrophages work in the commencement, maintenance, and determination of inflammation. Macrophages are activated and deactivated in the inflammatory process, macrophages activate and deactivate. Its activation signals comprise of cytokines (interferon γ, TNF-α, and granulocyte-monocyte colony-stimulating factor), extracellular matrix proteins, bacterial lipopolysaccharide (LPS), and various chemical mediators. Deactivation or elimination of mediators and effector cells inhibits inflammation that further allows the host to repair damaged tissues (Wynn et al. 2013a).

The inflammatory response is an arrangement of events which are tightly ordered. The process starts with the release of chemokines as well as soluble mediators from closely residing cells, together with vascular endothelial cells, macrophages, dendritic cells, along with interstitial fibroblasts. When inflammation is triggered by a pathogen, stimulation of resident macrophages occurs via pattern recognition receptors expressed on macrophages as the innate immune response. The significant step in inflammation and functional maturation of macrophages is the alteration from an inactive macrophage to an active macrophage. Activated macrophage specifies that the cell has an increased capacity to kill tumor cells or microbes (Wynn et al. 2013a).

Activated macrophages are large, consist of more pseudopods and noticeable ruffling of the plasma membrane, and they produce a great variation in the biologically lively products that, if unchecked, would lead to damage of the tissue and fibrosis in chronic inflammation. In brief inflammation, if the provoking agent is eliminated, macrophages ultimately vanish. In chronic inflammation, accumulation of macrophage continues. Macrophages play an important role in several inflammatory disorders including rheumatoid arthritis, metabolic homeostasis, myositis, cancer, and liver diseases.

Macrophages have also been concerned in the pathogenesis of a wide variety of autoimmune diseases such as multiple sclerosis, rheumatoid arthritis, and inflammatory bowel disease. In these diseases, macrophages are a major source of many key inflammatory cytokines that have been identified as significant drivers of autoimmune inflammation, including IL12, IL18, IL23, and TNFα (Murray and Wynn 2011). While there is considerable evidence supporting the roles of inflammatory macrophages in autoimmune inflammation, numerous studies have also showed the suppressive roles of macrophages. For instance, macrophages that produce reactive oxygen species can protect mice from arthritis by inhibiting activation of T cell (Gelderman et al. 2007). Proinflammatory cytokines generated by activated macrophages have shown capabilities to defend mice from Crohn's disease by enabling the clearance of pathogenic commensal bacteria from the mucosal lining of the bowel (Smith et al. 2009).

Macrophages seem to play an essential part in RA as they are present in various amounts at the cartilage-pannus junction and in the inflamed synovial membrane. They display clear signs of activation, including proinflammatory or regulatory cytokines and growth factors [IL-1, IL-6, IL-10, IL-13, IL-15, IL-18, TNF-α, and granulocyte-macrophage colony-stimulating factor (GM-CSF)], overexpression of major histocompatibility complex class II molecules, chemokines and chemoattractant [IL-8, macrophage inflammatory protein (MIP)-1, and monocyte chemoattractant protein (MCP)-1], metalloproteinases, and neopterin (Kinne et al. 2000).

The growth in number of sublining macrophages in the synovium is an initial hallmark of active rheumatic disease, and large number of macrophages is a prominent feature of inflammatory lesions. The grade of synovial macrophage infiltration directly relates to the grade of joint erosion, and reduction in these macrophages from inflamed tissue has a deep therapeutic benefit (Udalova et al. 2016).

2.1 Role of Macrophages in Cancer

Macrophages are immune cells which are present plentifully in microenvironment of solid tumors and their presence relates to low survival in most of the cancers. Macrophages are involved in all the phases of tumor development and stimulate angiogenesis, invasion of tumor cell, and intravasation at the primary site. On the metastatic site, macrophages work for the start of dispersed tumor cells and constrain immune-mediated clearance to encourage their extravasation and survival, or engaging directly with tumor cells to stimulate prosurvival signaling pathways. Macrophages also encourage the progress of dispersed tumor cells at the metastatic site by creating the formation of a helpful metastatic (Nielsen and Schmid 2017). In 2009, Qian and colleagues conveyed that macrophages are employed to the extravasation of tumor cells. Removal of these macrophages meaningfully declines the extravasation efficiency, and the following tumor cell survival is remarkably reduced (Qian et al. 2009).

Tissue-resident macrophages act as inborn immune cells in physiological conditions with phagocytic functions. They have enormous heterogeneous features having tissue-specific functions, therefore, playing a role in continuing tissue homeostasis and holding defense contrary to pathogens. Tumor-associated macrophages (TAMs) are also a part of the tumor microenvironment. TAMs can adjust their phenotypes on the basis of signals from the surrounding microenvironment, and can also kill tumor cells or boost tumor cell growth and metastasis. In cancer, TAMs are involved in tumor biology by enabling tumor growth and progress along with contributing to therapy resistance. In breast cancer, TAMs can be abundantly present, and more than half of the number of cells inside TAMs are significant tumor-promoting cells in the breast tumor microenvironment. Breast cancer tumor progression is stimulated by TAMs preclinically comprising of invasion, metastasis, and growth of the tumor cell. TAMs also encourage resistance to different types of treatment in breast cancer models. Multiple studies have reported that increased TAM infiltration of breast tumors is directly correlated with worse patient prognosis. Based on these results, macrophage-targeted treatment approaches have been established and are being evaluated in trials of clinical breast cancer. These approaches include: inhibition of macrophage recruitment, repolarization of TAMs to an antitumor phenotype, and enhancement of macrophage-mediated tumor cell killing or phagocytosis (Qiu et al. 2018).

Chemotherapy is among the most common treatments for breast cancer and it can have a high influence on the breast cancer process. Chemotherapy is a self-defeating process that causes misdirected tissue repair responses orchestrated by TAMs in many examples. TAMs also have the ability to destroy tumor cells when galvanized by bacterial products and cytokines. However, additional evidence suggests that TAMs are drivers of tumor development in recognized tumors, encouraging cancer cell propagation and survival, angiogenesis, and lymphangiogenesis. Polarization of macrophages. Macrophages polarize into M1 and M2 phenotypes. M1 macrophages can be triggered by LPS, TNF-α, and INF-γ and secrete IL1, IL12, CD80, TLR2, and additional cytokines. M2 macrophages can be triggered by IL4, IL10, TGF, as well as some paths like PI3K, and can be further split according to different microenvironmental stimulations. M1 cells have huge bactericidal, proinflammatory, bacteriostatic, and tumor cytotoxic activities. M2 cells contribute to tissue repair, tumor angiogenesis, wound healing, and tumor progression. In consideration of the difficulty of TAMs in breast cancer chemotherapy, additional research is required to investigate the effects and mechanisms of TAMs in breast cancer (Tao et al. 2020). Apart from these functions, it has been established in many studies that M2 macrophages, separately from being intermediaries of chronic inflammation, act as tumor supporters at separate phases of malignant development of gastric, lung, liver, and mammary carcinomas (Dhabekar et al. 2011).

2.2 Role of Macrophages in Metabolic Homeostasis

Pancreas, liver, and adipose tissue are the mammalian metabolic organs that include parenchymal and stromal cells and macrophages, which act collectively to sustain metabolic homeostasis. There is an innate activation of macrophages during bacterial infection that leads to secretion of proinflammatory cytokines (TNFα, IL1β, and IL6) that together encourage peripheral insulin resistance to reduce nutrient storage (Chawla et al. 2011). However, this metabolic adaptive approach of nutrient reallocation progresses to being maladaptive in the setting of diet-induced obesity, a state characterized by chronic low-grade macrophage-mediated inflammation (Olefsky and Glass 2010).

2.3 Role of Macrophages in Liver Diseases

Inflammation is a symbol of almost all liver diseases such as fibrosis, liver cancer, alcoholic liver disease, cholangiopathies, and nonalcoholic steatohepatitis. Liver macrophages mainly contain liver-resident phagocytes, Kupffer cells (KCs) used as substitute for hepatic macrophages, and bone marrow-derived employed monocytes. It is well known that macrophages spotted in the liver ensuing injury are heterogeneous and may begin from different roots including liver-resident macrophages or KCs and two patrolling populations of bone marrow monocyte-derived macrophages (MoMFs) as well as peritoneal macrophages for subcapsular regions of the liver (Guillot and Tacke 2019).

Macrophages play a major role in liver immune homeostasis and in the growth of additional liver diseases that are connected to nonalcoholic fatty liver disease (NAFLD) such as alcoholic liver disease and chronic viral hepatitis.

In several phases of liver disease, resident KC and afresh employed monocyte-derived macrophages play a significant role in the regulation of inflammation, fibrolysis, and fibrogenesis. Macrophages also lead to the development of hepatocellular carcinoma (HCC), which is a rare but authoritative complication of NAFLD and/or NASH. An augmented density of TAMs is strongly related to HCC progression. TAMs have M2-type macrophage properties that overpower tumor-specific T-cell immunity, mainly through PDL1 (Kazankov et al. 2019).

3 Various Receptors Overexpressed in Macrophages During Inflammation

Macrophages play the critical roles in phagocytosis, removal body damage, sterilization and, development of tissue and repair, and transmission of antigen information (Tauber 2003). Macrophages play the critical roles in phagocytosis like removal of body damage, sterilization, and development of tissue and repair, and transmission of antigen information (Wynn et al. 2013b). To achieve protective functions and repair tissues that are damaged tissue, macrophages express an extensive range of surface, cytosolic, and vacuolar receptors for recognition, and uptake of damage signals and foreign particles. These receptors enable phagocytosis, endocytosis, sense parasitic, and viral and bacterial molecules. Following are the receptors overexpressed in macrophage during inflammation (Gordon 2016).

3.1 Activin

Activin A is an antiinflammatory cytokine that plays role in many physiological developments of the body, like cell migration and cell proliferation, and remodeling of bone and embryo development (Sugama et al. 2007). Activin receptor-interacting protein 2 (ARIP2), an upstream signal regulatory protein of Smads, is expressed in a variety of tissues, and prevents production of collagen type IV in mouse Hepa1-6 cells (Matsuzaki et al. 2002). It has been reported that macrophages play a crucial role in inflammatory responses by the release of inflammatory mediators such as TNFα, NO, and interleukin like IL-1β. ARIP2 overexpression lowered the production of IL-1β and TNFα in macrophagic cells. As reported by Wu et al. (2017), *in vitro* activity on RAW 264.7 cells showed that the overexpression of ARIP2 reduced the level of TNFα and IL-1β in RAW264.7 cells treated by LPS. Similarly, *in vivo* ARIP2 overexpression also reduced the production of IL-1β and TNFα in LPS-activated mouse peritoneal macrophages. Therefore, ARIP2 overexpressed in macrophages play an antiinflammatory role by downregulating the secretion of IL-1β and TNFα and phagocytosis (Wu et al. 2017).

3.2 Toll-Like Receptors

Toll-like receptors (TLRs) belong to the class of pattern recognition receptors, which leads to the initiation of innate immune response by identifying the molecular patterns for initial immune recognition of a pathogen. TLRs are overexpressed by various cells such as macrophages, monocytes, and dendritic and various TLR ligands initiate inside the inflamed joints. TLRs allow macrophages to recognize a wide range of microbial ligands, thus, encouraging inflammation (Hoppstädter et al. 2019). The achievement of TLRs to act as key sensors of invading pathogens is their ability to identify a variety of conserved microbial ideas known as "pathogen-associated molecular patterns" (PAMPs) (O'Neill 2006). TLR ligands leads to the activation of TLR that helps to determine proinflammatory cytokines expression via macrophages that further leads to damage of cartilage and bone. TLRs can trigger a cascade of signaling pathways by the recognition of PAMPs that eventually conclude in the initiation of a wide range of immune and inflammatory genes. Innate immune response initiation displayed that TLRs exhibit a role in the functions of macrophage such as phagocytosis, antigen processing, and demonstration and commencement of the adaptive immune response. TLR4 is one of the TLRs that is strongly expressed on monocytes and macrophages. TLR4 mRNA expression increases upon stimulation of LPS (lipopolysaccharide) in human macrophages, whereas TLR4 mRNA is also downregulated in relation to LPS in murine macrophages (Rehli 2002). Apart from TLR 4, TLR9 is also expressed in murine macrophages. More than 10 TLRs are expressed on macrophages which proves how important TLR receptors are for the functioning of an active macrophage. TLRs are not only important for the innate recognition of invasive pathogens – thus, introducing an inflammatory response – but are also substantial for each phase of the phagocytosis (McCoy 2008).

3.3 CD44 Receptor

The CD44 receptor is one of the cell surface glycoproteins that facilitate various physiological and pathological activities. Its main role is to deliver defense contrary to inflammatory reactions through cellular transmigration and cell signaling. CD44 receptor binds to numerous ligands, mainly the hyaluronic acid (HA), chondroitin sulfate (CS), fibronectin, collagen, fibrinogen, laminin, mucosal vascular addressin, osteopontin (OPN), and major histocompatibility complex class II invariant chain (li), and also L-selectin and E-selectin. It is expressed on various cell categories such as hematopoietic cells and endothelial cells. In the immune system, CD44 is expressed on the cells mainly including hematopoietic stem cells, macrophages, monocytes, lymphocytes, neutrophils, and eosinophils (Pandey et al. 2017).

CD44 receptor plays a significant role in regulating cell signaling pathways by enabling signaling protein recruitment and assembly. In the milieu of innate immunity, CD44 receptor was also found to adjust Fc-γ and balance receptor 3–dependent macrophage phagocytosis (Amash et al. 2016). CD44 receptors are overexpressed in both rheumatoid arthritis and osteoarthritis synovial tissues, together with macrophages, fibroblasts, and lining cells. For instance, enhanced expression of CD44 receptor was found on synovial lymphocytes and macrophages of adjuvant arthritis rats. In addition, CD44 expression was also markedly enhanced on lymphocytes from the patient's synovial fluid with rheumatoid arthritis. These disparate findings can be attributed to the fact that CD44 receptors are overexpressed in macrophages. Thus, CD44 is a largely distributed receptor that has been concerned in a number of immunological phenomena, like lymphocyte homing, activation of cell, hemopoiesis, and presentation of growth factor, as well as in such disease conditions as arthritis, tumor metastasis, and allergies (Qadri et al. 2018).

3.4 G Protein-Coupled Receptors (GPCRs)

One of the diverse sentinel cell surface receptors are the G protein-coupled receptors (GPCRs). GPCRs are found in the transmembrane region of the cell surface. They function mainly as transducers of extracellular stimuli into intracellular signals that produce cellular responses (Wang et al. 2019).

GPCRs work as gatekeepers in general and support in particular to modify immune responses of macrophages to extracellular pathogens as well as danger molecules related to injury (Lattin et al. 2008). Mammalian GPCRs feature conspicuously in the regulation of further aspects of macrophage biology, such as development, differentiation, and activation of macrophage. GPCRs are essential for numerous physiological functions and its dysregulated expression and signaling lead to the various chronic inflammatory disorders such as rheumatoid arthritis or diseases having certain inflammatory aspects like cancer. All this make GPCRs significant targets for the development of novel therapeutics (Arakaki et al. 2018).

The class A receptors Mas-related GPCRs (MRGPCRsF) and G protein-coupled receptor family C type 5A (GPCRs5A) are expressed strongly after macrophage differentiation but these have not yet been studied in detail.

3.5 Mannose Receptor

Mannose receptor is a type-1 membrane glycoprotein that consist of a cytoplasmic domain (45 amino acids) and three types of extracellular domains. It is an endocytic receptor having a wide binding specificity for both endogenous and microbial ligands (van Die and Cummings 2017). Mannose receptors also play an important role in cellular activation and signaling. This receptor exhibits high overexpression on polarized macrophages and the overexpression is credited to numerous inflammatory and infection diseases (Hatami et al. 2019). Mannose receptor is mainly expressed on human and mouse macrophages and dendritic cells, but it is also observed on nonvascular endothelial cells. It is able to identify various pathogens and play an important role in the innate immune response by facilitating opsonin-independent phagocytosis. The capability of the mannose receptor to facilitate internalization of soluble glycoconjugates and viruses as well as entire pathogens (bacteria and fungi) recommends that the mannose receptor could lead to the enabling of antigen uptake and processing in the adaptive immune response, as well as intermediating direct uptake of pathogens in the innate immune response (Taylor 2001). There are various mannose receptors, such as MR, CD206, or MRC1, expressed mainly by most tissue macrophages, dendritic cells, and certain lymphatic or liver endothelial cells. Mannose receptors act as a homeostatic receptor by binding and scavenging undesirable high mannose *N*-linked glycoproteins along with pituitary hormones from the circulation. As several pathogenic microbes are coated by mannose-containing structures, the macrophage mannose receptor interrelates with those pathogens in a form of the host molecular mimicry (Medzhitov 2007).

4 Polymeric Nanoparticles

Nanoparticles are solid colloidal particles falling in the nanosize range (preferably <1000 nm) capable of carrying and delivering drug payload to the targeted or desired site of action. A wide range of materials starting from polymers to metals or semiconductors are utilized by the researchers and industries to formulate a plethora of nanoparticles. One such category of greatly appreciated and utilized nanoparticles is polymeric (Bolhassani et al. 2014). The earliest report on use of polymeric nanoparticles dates back to the year 1979, when an anticancer drug (dactinomycin) was adsorbed on the surface of polyalkylcyanoacrylate nanoparticles, and their endocytic uptake and drug release in calf serum were observed (Couvreur et al. 1979).

A great deal of advantages is conferred on polymeric nanoparticles including greater kinetic stability and rigid morphology, easy manipulation of particle size and surface characteristics, sustained/controlled drug delivery, targeting potential and easy to modify organ clearance, and high drug payload carrying capacity. Thus, making these nanoparticles first choice as nanosized colloidal drug carrier for a wide range of applications (Zielińska et al. 2020). Acknowledging the advantages offered by the polymeric nanoparticles, numerous products have been approved by FDA for their commercial application in drugs as well as cosmetics (Table 1).

Polymeric nanoparticles are basically divided into two broad classes on the basis of their morphology and composition (Fig 1a):

i. *Nanospheres*, also known as matrix systems, have a homogenous and uniform structure, that is, a 3D polymeric matrix within which the drug is entrapped or may be adsorbed at the surface of the nanoparticle.
ii. *Nanocapsules*, also known as reservoir system, has a typical core shell structure (core is mainly oily in nature). Drug is usually present in the core (small central cavity), which is surrounded by the polymeric membrane, thus resembling a capsule.

4.1 Formulation of Polymeric Nanoparticles

Polymeric nanoparticles as the name suggest are composed of polymers, which as a general rule

Table 1 FDA-approved marketed polymeric nanoparticles

Trade name	Drug	Company/manufacturer	Polymer used	Route of administration	Indication	FDA approval year
Sublocade®	Buprenorphine	Indivior	PLGA	Subcutaneous injection	Opioid use disorder	2017
Atridox®	Doxycycline hyclate	Tolmar Inc.	PLA	Controlled released gel	Chronic adult periodontitis	2017
Copaxone®	Glatiramer	Teva Pharmaceuticals	Polypeptide (four amino acids)	Subcutaneous injection	Multiple sclerosis	1996/2014
Opaxio®	Paclitaxel	Cell Therapeutics Inc.	Polyglutamate	Intravenous	Glioblastoma	2012
Feridex®	Ferumoxides	AMAG Pharmaceuticals	Dextran	Intravenous	Liver lesions imaging agent	2008
Abraxane®	Paclitaxel	Abraxis Biosciences	Serum albumin	Intravenous	Breast cancer, lung cancer, and pancreatic cancer	2005
Somavert®	Pegvisomant	Pfizer	Polyethylene glycol	Subcutaneous injection	Acromegaly	2003
Eligard®	Leuprolide acetate	Tolmar	PLGA	Intravenous suspension	Prostate cancer, breast cancer, endometriosis, uterine fibroids, and early puberty	2002
Cosmofer®	Iron	Pharmacosmos UK Limited	Dextran	Intravenous injection and infusion	Anemia	2001
Genexol®	Paclitaxel	Samyang Biopharmaceuticals	Copolymer poly(ethylene glycol)-poly(D,L-lactide)	Intravenous	Metastatic breast and prostate cancer	2001
Renagel	Sevelamer	Sanofi	Cross-linked poly-allylamine hydrochloride	Oral	Hyperphosphatemia	2000
Lupron® Depot®	Leuprolide acetate	Abbvie endocrine Inc.	PLGA	Intramuscular suspension depot	Endometriosis	1989

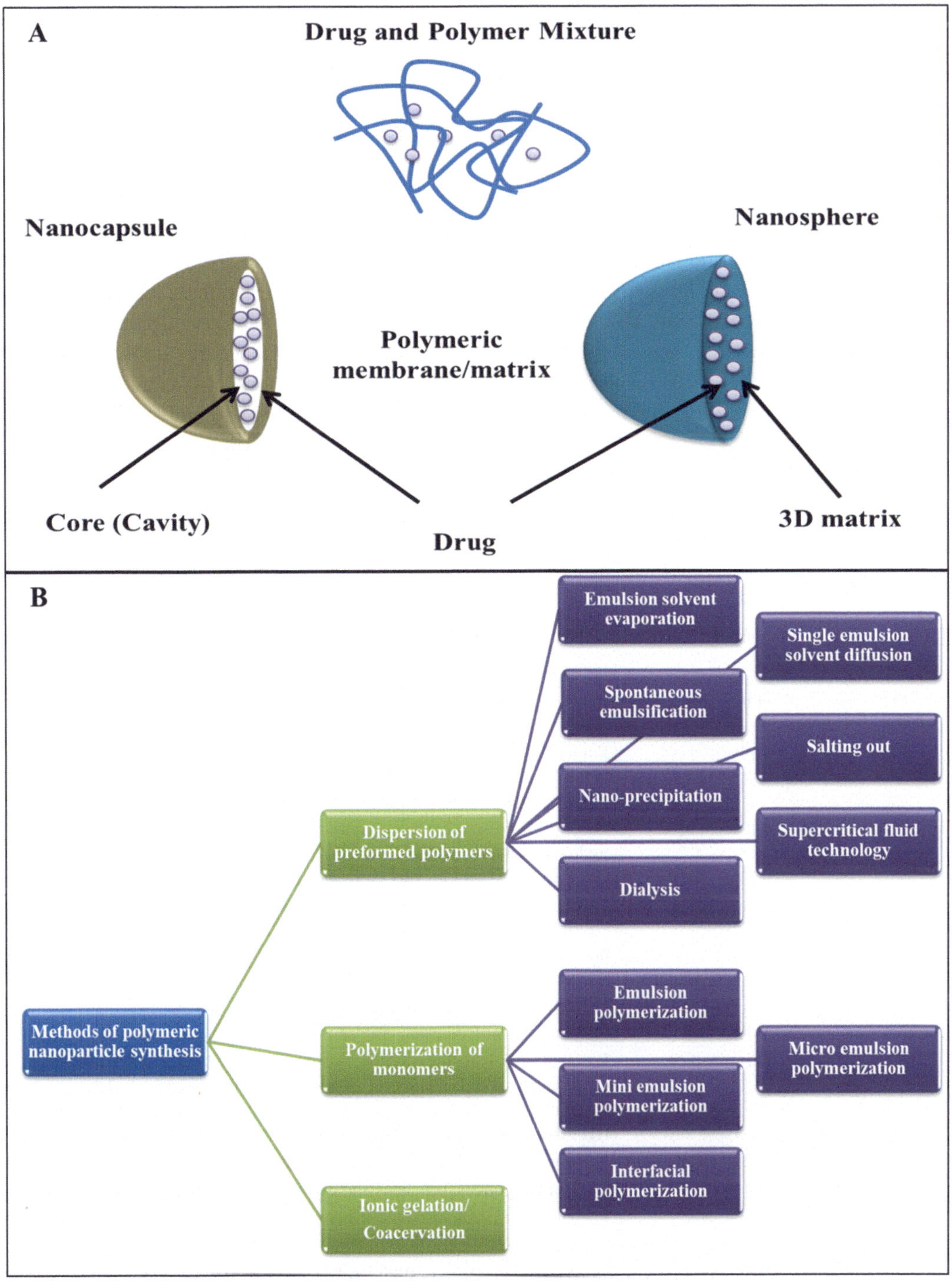

Fig. 1 (**a**) Types of polymeric nanoparticles. (**b**) Various methods for the synthesis of polymeric nanoparticles

should be nontoxic, nonantigenic, biocompatible, and easy to fabricate and manipulate. A wide range of polymers are being employed throughout the globe for formulating nanoparticles, but biodegradable and environment-friendly polymers are getting an edge in the race because not only they afford good tissue compatibility and reduce the apprehension of hypersensitivity reactions but they also reduce bioburden and are usually inexpensive (Elsabahy and Wooley 2012; Siddiqui et al. 2020). This is the main reason why use of polymers like polyacryl amides, polystyrene, polyalkylcyanoacrylate, etc. in nanoparticle synthesis has become history. An immense number of polymers have made into the list of polymers used for nanoparticle synthesis and are majorly classified on the basis of their source, biodegradability, solubility, and stimuli responsiveness (Kumari et al. 2010). Table 2 summarizes the various types of polymers used in nanoparticle synthesis.

After polymer, the next most important constituent is "stabilizers". As the name suggest they are used to stabilize or to make the nanoparticles rigid. Stabilizers range from chemical cross-linkers like formaldehyde and glutaraldehyde to surfactants like anionic (sodium lauryl sulfate, dodecyl sulfate, etc.), cationic (benzalkonium chloride, cetrimide, etc.), and nonionic (Spans/Tweens, poloxomers, etc.) (Liu et al. 2008).

A wide range of methods used for the synthesis of nanoparticles have been developed and perfected by the researchers all over the world (Fig 1b). And the research in this direction is still on the go. However selection of method is based on various factors. Depending upon the type of polymer, required morphology and particle size of nanoparticles, solubility and stability of drug and polymer, desired drug release mechanism, charge and other surface characteristics of the nanoparticles; method for nanoparticle synthesis is selected (Vauthier and Bouchemal 2009).

Table 2 Classification of polymers used for nanoparticle synthesis

Class	Category	Examples
Origin of polymer	Natural	Gelatin, albumin, lignin, dextran, chitosan, alginates, pullulan, pectin and starch
	Synthetic	Poly(glycosides) (PGA), poly(lactides) (PLA), poly(lactide coglycosides) (PLGA), poly-Ɛ-caprolactone, polystyrene, and polyisobutylcyanoacrylates poly(methacrylate) (PMMA)
	Semi-synthetic	Pregelatinized starch, thiolated pectin, ethyl cellulose (EC), hydroxypropyl methylcellulose (HPMC), N,N,N – trimethyl chitosan chloride (TMC), and cellulose acetate phthalate (CAP)
Biodegradability	Biodegradable	Poly(glycosides) (PGA), poly(lactides) (PLA), poly(lactide co-glycosides) (PLGA), poly-Ɛ-caprolactone, gelatin, albumin, lignin, dextran, chitosan, alginates, pullulan, pectin, and starch
	Nonbiodegradable	Polystyrene, acrylates (Eudragit), ethyl cellulose (EC), hydroxypropyl methylcellulose (HPMC), and cellulose acetate butyrate
Solubility	Hydrophilic	Gelatin, albumin, dextran, chitosan, alginates, and pullulan
	Hydrophobic	Lignin, poly(glycosides) (PGA), poly(lactides) (PLA), poly(lactide coglycosides) (PLGA), poly-Ɛ-caprolactone, polystyrene, acrylates (Eudragit), ethyl cellulose (EC), and hydroxy propyl methyl cellulose (HPMC)
Stimuli responsiveness	Photosensitive	N-isopropylacrylamide copolymer gel and cross-linked hyaluronic acid
	Mucoadhesive	Chitosan, alginates, acrylates and cyanoacrylates (superglue)
	pH sensitive	Poly(methacrylic acid-ethylene glycol) copolymer, mixture of acrylic acid, and methacryl-amidopropyl-trimethyl ammonium chloride
	Thermosensitive	N-alkyl acrylamide (N-isopropylacrylamide)
	Inflammation sensitive	Cross-linked hyaluronic acid

5 Polymeric Nanoparticles Assisted Macrophage Targeting

5.1 Passive Targeting of Macrophages

Passive drug targeting assisted by polymeric nanoparticles can be classified mainly in two categories: through mononuclear phagocytic system (MPS) and through enhanced permeability and retention (EPR) effect (Jawahar and Meyyanathan 2012).

Mononuclear phagocytic system (MPS) or reticuloendothelial systems (RESs) comprise of liver, spleen, lungs, and phagocytes including neutrophils, monocytes, macrophages, and dendritic cells. Collectively MPS is responsible for identification, engulfment, and destruction of foreign particles from the bloodstream (dos Santos et al. 2017). Size, surface charge, and hydrophobicity/hydrophilicity of the particle are major parameters that play vital role in uptake of polymeric nanoparticles by macrophages or MPS (Fig 2a). When a foreign particle enters the bloodstream, special blood proteins or opsonins get adsorbed on the surface of the particles, which are recognized by MPS for phagocytosis. There is no definite rule regarding opsonization of nanoparticles, but generally, large-size particles are more readily taken up by the MPS because of available surface area for opsonization. Similarly, neutrally charged nanoparticles usually show slow opsonization when compared to charged nanoparticles. Among charged nanoparticles, negatively charged nanoparticles are usually long circulating as they circumvent phagocytosis. In case of surface properties, that is, hydrophobicity/hydrophilicity, it was observed that hydrophobic nanoparticles are more susceptible to opsonization as blood proteins get easily adsorbed on hydrophobic nanoparticles (Owens and Peppas 2006). MPS-assisted phagocytosis is generally considered as a disadvantage for nanocarrier drug delivery because of smaller blood circulating time. But in case of macrophage targeting, MPS act as a boon and a tool for achieving passive targeting of macrophages and other members of MPS (Storm et al. 1995).

Enhanced permeability and retention (EPR) effect is a phenomenon observed in case of neovasculature of tumor, where uptake and retention of macromolecules, nanoparticles, liposomes, drug, etc. are higher in tumor tissues as compared to normal tissues (Fig 2b). During tumor progression and metastasis, small aggregates of tumor cells are formed and, to fulfill their nutritional requirement, new vessels are formed near these aggregates and this process is called cancer angiogenesis (Bolkestein et al. 2016). Various factors like bradykinins, vascular endothelial growth factor (VEGF), prostaglandins, tumor necrosis factor, etc. are responsible for neovasculature development and EPR effect. These new vessels are architecturally faulty and have leaky walls due to defective endothelial cells, wide lumen, absence of smooth muscles, and effective lymphatic system (Golombek et al. 2018). Role of macrophages in EPR effect is not well defined at present, but various studies suggest that increased infiltration of macrophages within the tumor microenvironment contribute to EPR. Tumor-associated macrophages (TAMs) are the only immune cells that are present abundantly in tumor microenvironment and they play a vital role in developing new blood vessels by secreting proantigenic factors and in remodeling of extracellular matrix of tumor by providing matrix degradation enzymes and by secreting collagen (Park et al. 2019).

MacParland et al. (2017) observed that presence of Kupffer cells or liver-resident macrophages enhances the uptake of hard nanoparticles (like gold) and, on changing the hepatic microenvironment, liver sequestration can also be modified (MacParland et al. 2017). In another study, it was observed that on exposure with cytokines inflammatory macrophages get converted into tumorigenic macrophages and support the formation of neovasculature and, in turn, promote EPR. Thus, efforts were made to use macrophages as drug carriers, and nanomedicines like drug-loaded liposomes were encapsulated within genetically engineered and *ex vivo* generated macrophages of controlled phenotype. These macrophages were then introduced into animal models and an improvement in overall therapeu-

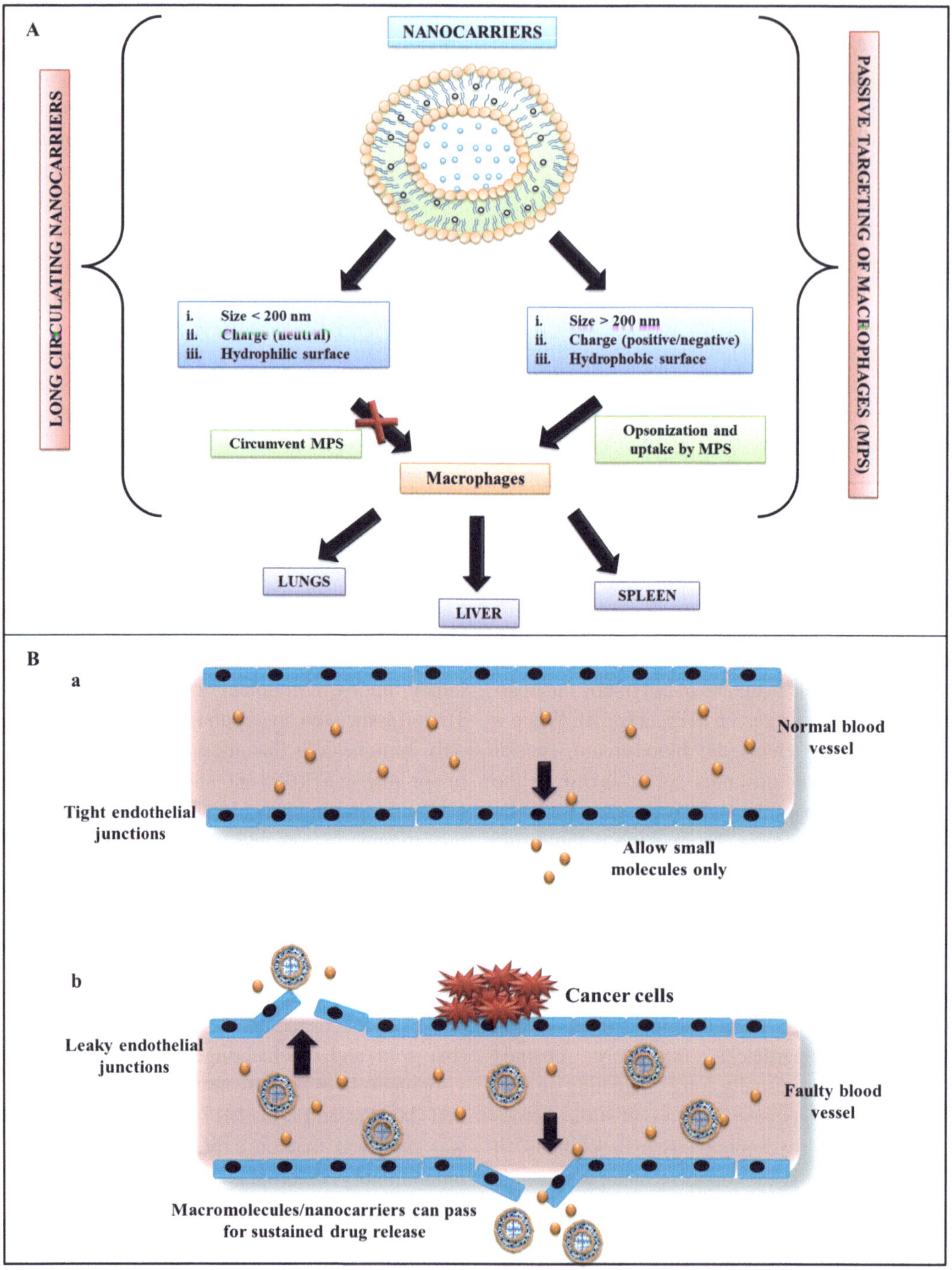

Fig. 2 (**A**). Passive targeting of macrophages can be achieved by modifying the characteristics of nanocarriers. (**B**). EPR effect. (a). Normal blood vessels have tight junctions which allow uptake of only small drug molecules for short duration. (b). Leaky neovasculature of tumor enhances uptake of macromolecules/nanocarriers capable of sustained drug release

tic outcome was achieved by macrophage cells-assisted therapy (Lee et al. 2016). TAMs were also utilized in enhancing the uptake of irinotecan liposomes (Onivyde) within tumor microenvironment and its conversion into active metabolite for improved tumor regression in combination with radiation therapy (Miller et al. 2018). Thus, TAMs can be an attractive passive target for combating solid tumors.

5.2 Active Targeting of Macrophages

Active targeting of the drug delivery system to the macrophages seems to be an attractive intention to advance therapeutic efficacy of encapsulated drug. Therefore, macrophages can be exploited as Trojan horses for target drug delivery of nanoparticles. Nanocarriers can move across the multiple membrane barriers and release their drug at the target (Jain et al. 2013). Active targeting approaches of the delivery systems are based on the surface alteration of the systems with the help of an agent (ligand, antibody, and peptide) that has the abilities to recognize and interact with the specific cell type, tissue, or organ in the body. Active targeting can be attained by employing the presence of numerous receptors and lipid components on the cell plasma membrane. These receptors may be exclusively stated on specific cells or may display a differentially advanced expression on diseased cells compare to the normal cells (Jain and Amiji 2012). A large number of ligands have been recognized and studied for enabling active targeting of NPs (Fig. 3) (Byrne et al. 2008). These ligands frequently bind to specific receptors on the surface of the target cells, and in this way increase cellular uptake of drug-containing NPs and also increase therapeutic efficacy. Compared to singular ligand, an increased density of ligands is advantageous for promoting binding and cellular uptake through the multivalent effect (Montet et al. 2006). Various types of ligands have been employed for this purpose, including proteins, polysaccharides, nucleic acids, peptides, and small molecules.

5.2.1 Hyaluronic Acid

HA, an anionic polysaccharide comprised of N-acetyl-d-glucosamine and d-glucuronic acid, represents one of the prime extracellular matrix components, along with collagen (Yoo et al. 2019). Owing to its biodegradable and biocompatible nature, HA and its derivatives have been widely explored for biomedical applications. HA is a known ligand of various cell surface receptors, including RHAMM, CD44, and lymphatic vessel endothelial hyaluronan receptor 1 (LYVE-1). The catabolization of HA in the human body occurs via HA-degrading enzymes (hyaluronidases), including Hyal1, Hyal2, and PH-20, as well as oxidative stresses (OSs) such as reactive nitrogen species and reactive oxygen species. Notably, the elevated levels of Hyals, CD44, and OSs have been correlated with the development of many inflammatory disorders such as atherosclerosis, traumatic brain injury, and cancer. Due to these multifunctional advantages, HA has been recognized as one of the most popular biomaterials used in tissue engineering and formulation of targeted drug delivery systems during the last two decades (Rao et al. 2020). HA binds to CD44, which is frequently overexpressed on the cancer cell surface and is considered to be a symbolic cancer stem cell marker (Aruffo et al. 1990). Interestingly, CD44 has a specific HA binding affinity, indicating that cells that overexpress CD44 will effectively take up HA-based PNPs. CD44 is a primary HA receptor in human body, and is reported to be expressed in macrophages. Recruitment of macrophages and leukocytes during atherosclerotic plaque formation is likely linked to the HA and CD44 interactions. Kamat et al. designed HA-coated iron oxide magnetic NPs (HA-DESPIONs), which showed efficient uptake by activated macrophages because of biological recognition of HA at the surface of NPs by CD44, which is expressed highly on activated macrophages. In another study, HA-NPs developed by reacting cholanic ester with oligomeric HA functionalized with amine showed significant uptake by proinflammatory macrophages *in vitro* and displayed an increased binding affinity toward atherosclerotic plaque-linked macrophages in mice. Self-assembled HA-NPs can

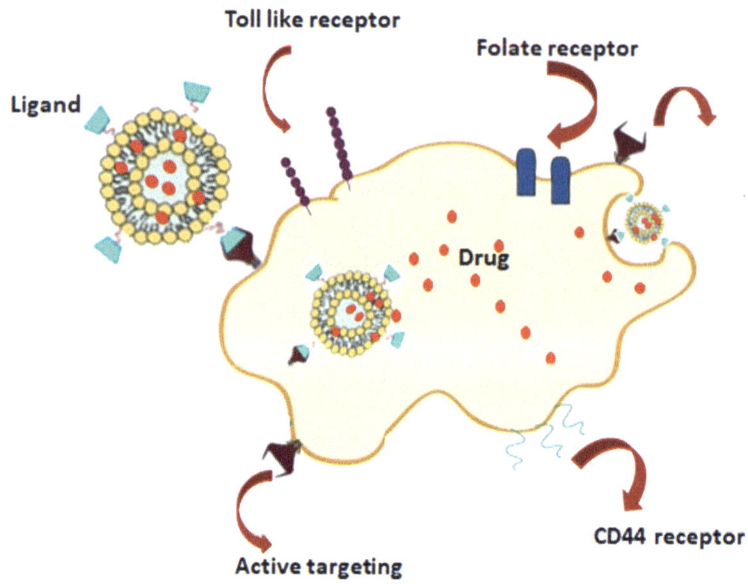

Fig. 3 Active targeting of surface-modified nanoparticle to receptors overexpressed on macrophages

serve as a potential delivery system to target activated macrophages selectively at pathological sites due to the upregulation of HA receptors, including CD44, on these macrophages. This characteristic can be exploited to visualize pathological sites through imaging techniques and/or to cure various inflammatory diseases, including atherosclerosis, arthritis, and T2DM, utilizing placebo HA-NPs themselves, or as delivery system for chemotherapy and immunotherapy (Rao et al. 2020).

5.2.2 Chondroitin Sulfate

Chondroitin sulfate (CS), a glycosaminoglycan with strong CD44 receptor affinity, has a chemical structure identical to that of HA. Recently, CS has drawn great attention in the drug delivery systems as a drug carrier or targeting moiety. CS has good biocompatibility, good biodegradability, and low immunogenicity, which are necessary for the fabrication of drug carriers. Researchers are focusing on designing CS-based NPs for tumor targeting. The CS-drug conjugates of hydrophobic drugs are obtained by conjugating them via linker to hydrophilic CS. CS-drug conjugate formation requires the grafting of low-solubility drugs by chemical reaction to hydrophilic CS. These conjugates self-assemble into micellar NPs containing drugs, as the hydrophobic core, protecting the drug from the hydrophilic environment and blocking leakage before reaching the tumor site, which greatly increases drug solubility and safety (Li et al. 2021).

CS is predominantly present in synovial fluid or cartilage around the joints and lack of it has been associated with the development of RA. In RA, overexpression of various receptors including CD44, leptin, and annexin has been reported in the connective tissue and linked with enhanced chondroitin uptake. Utilizing this process, CS can serve as a targeting ligand for delivering drug-loaded NPs to the joints. Recently, CS has been used as a ligand for the fabrication of targeted NPs of aceclofenac and methotrexate for effective management of RA (Onishi et al. 2013). CS has also been employed as a ligand in the development of immunotherapy-based targeted NPs. For instance, liposomes decorated with derivatives of CS using 1-aminodecane linker conferred pH responsivity and selective delivery to dendritic cells. Moreover, subcutaneous administration of CS liposomes in tumor mice model revealed selective delivery of antigens to dendritic cells, thereby triggering cytokine production and initiating immune response, resulting in tumor inhibition (Okubo et al. 2019).

5.2.3 Folic Acid

Folic acid has appeared as an important ligand for the selective delivery of imaging/therapeutic agents to target cells (cancer cells and activated macrophages) at the inflammatory sites. As a targeting ligand, folic acid has numerous advantages such as great affinity to its target, folate receptors (FRs), and even subsequent conjugation with diagnostic/therapeutic agents; easy conjugation with a variation in imaging as well as therapeutic agents; and very low or untraceable expression of its receptors on normal cells in spite of its high expression on cancer cells and activated macrophages (Low et al. 2008). Folic acid mainly enters non-pathogenic cells via reduced folate carrier to affect its functions, but conjugating agents linked folic acid only enters cells via FR (Yi 2016).

FR is also known for its overexpression in solid tumor cells and macrophages, thus making them attractive targets toward many NPs via receptor-mediated endocytosis. For example, Lv et al. developed mesoporous silica NPs (MSNs) altered with FA for active targeting, and further used the large gas-filled microbubble method for the decoration of NPs (Lv et al. 2017).

Folic acid is also called as folate and vitamin B9. It is crucial for cells to produce DNA, RNA, and metabolic amino acids that are essential for their proliferation and division. Folic acid is delivered into cells either through the concentrated folate carrier, present in all cell types, or the FR, which is expressed in inadequate cells. Although folic acid is delivered into the cells via either system, folate conjugates intended for diagnostics and therapeutics are delivered only via FRs by FR-mediated endocytosis (Yi 2016).

6 Therapeutic Role of Polymeric Nanoparticles in Macrophage Targeting

6.1 Macrophage Targeting in Autoimmune and Autoinflammatory Diseases

Macrophage targeting usually aims at fortifying the immune response but in situations where immunity starts attacking the own body, immunosuppression, or developing tolerance is the way to go (Serra and Santamaria 2015). As discussed, macrophage plays an important role in autoimmune diseases; thus, targeting nanoparticles loaded with immunosuppressant have proven to be a successful mode of treatment. In one such study, polylactide nanoparticles loaded with cyclosporine A were synthesized, which when taken up by lymph nodes through macrophages and dendritic cells led to immunosuppression and reduced T-cell production (Azzi et al. 2010). In another study, poly(amidoamine) dendrimer nanoparticles decorated with folic acid ligand and encapsulating methotrexate were synthesized to actively target macrophages for combating arthritis. *In vitro* cell line studies in human and mouse macrophages showed better cell uptake of nanoparticles as compared to free drug, and *in vivo* studies in rats showed reduction in inflammatory parameters including paw volume, swelling in ankle, bone resorption, cartilage damage, and body weight reduction. Thus, proving that polymeric nanoparticles assisted macrophage targeting is a better approach for treating arthritis (Thomas et al. 2011). Macrophage repolarization from M1 to M2 subtype has also been studied as a means to alleviate arthritis. Plasmid DNA coding cytokine (IL-10) was encapsulated in alginate nanoparticles coated with tuftsin peptide to actively target macrophages. It was observed that after intraperitoneal administration of nanoparticles, an increased number of M2 macrophages (about 66%) were found in arthritic paw of rats as compared to untreated rat. Treated rats were able to retain their mobility as well as reduced inflammation was also observed (Jain et al. 2015).

Getts et al. (2012) studied the effect of encephalitogenic peptides encapsulated in polystyrene or biodegradable poly(lactide-co-glycolide) micro-/nanoparticles on disease progression and observed that these nanoparticles not only prevent the onset of autoimmune encephalomyelitis but also modifies disease progression by targeting macrophages through scavenger MARCO receptors (macrophage receptor with collagenous structure) leading to T-cell suppression (Getts et al. 2012). Colonic macrophages in case of inflammatory bowel disorder (IBD) have also been studied as interesting drug targets. In one such study, Fab' antigen-assisted active targeting on colonic macrophages of TNFα siRNA encapsulated in poly(lactic acid) poly(ethylene glycol) block copolymer (PLA-PEG) nanoparticles was successfully achieved. It was observed as compared to naked nanoparticles, Fab' bearing nanoparticles showed better diminution of colitis and relieved other inflammatory responses like weight loss and myeloperoxidase activity (Laroui et al. 2014).

6.2 Macrophage Targeting in Cancer

Targeting of polymeric nanoparticles both active as well as passive has been employed in treating cancer and to ensure their suitability as drug carrier in combating cancer. Kourtis et al. (2013) synthesized pluronic-stabilized poly(propylene sulfide) nanoparticles and observed that, on intradermal/intramuscular administration in tumor-bearing mice, monocytes and polymorphonuclear myeloid-derived suppressor cells extensively take up these nanoparticles in tumoral stroma, suggesting that they can be used as drug carriers (Kourtis et al. 2013). Wang et al. (2017) synthesized pH-responsive polymeric nanoparticles that release IL-12⊂P1 in the tumor microenvironment leading to conversion of tumor-supporting M2 macrophages into tumor-suppressive M1 macrophages. Also, the cytotoxicity toward normal cells was observed to be negligible. Thus, confirming the suitability of polymeric nanoparticles as immunotherapy payload carriers for defeating cancer with marginal damage to normal cells (Wang et al. 2017).

TAMs are known to play vital role in changing the microenvironment of tumor, and studies have shown that altering the phenotype of TAMs into antitumorigenic macrophage can be beneficial in combating tumors (Mantovani et al. 2017). In one such study, block copolymer-based nanoparticles encapsulating siRNA (responsible for altering phenotype of TAMs) functionalized with mannose moiety were formulated to actively target mannose receptor on TAMs which could alter the phenotype of TAMs to possess immunogenic and anticancer activity. Enhanced uptake and therapeutic effect of nanoparticles was observed in case of lung metastasis (Ortega et al. 2015). In another study, effect of siRNA-loaded glucan-(BG34-10)-based nanoparticles on regression of mammary tumor in balb/c mice by altering macrophage migration inhibitory factor (MIF) was observed, and improvement in MIF manipulation by polymeric nanoparticles was seen as compared to antibodies and antigen-based targeting (Zhang et al. 2015). Hypoxia in tumor region promotes TAMs infiltration and cancer progression as well as causes drug resistance. Manganese dioxide nanoparticles coated with hyaluronic acid were targeted on TAMs to produce O_2 in tumor vicinity and to reprogram them phenotypically to act as anticancer agents (Song et al. 2016).

6.3 Macrophage Targeting to Achieve Metabolic Homeostasis

Resident macrophages like adipose tissues macrophages (ATMs) and Kupffer cells in liver are responsible for fighting local infections and to maintain tissue homeostasis. Thus, up- or down-regulation of resident macrophages plays vital role in various metabolic disorders like obesity-related comorbidities, insulin resistance, or artherosclerosis. These resident macrophages have recently gained popularity as drug targets for overcoming these metabolic disorders (Peterson et al. 2018).

Dextran/glucose-based nanoparticles tagged with fluorophores and antiinflammatory drugs were synthesized and administered peritoneally to lean as well as obese mice to understand biodistribution and localization of these carriers in visceral adipose tissues for establishing their future clinical use in drug delivery to combat visceral diseases. And a reduction in inflammation biomarkers in visceral adipose tissues of obese mice was successfully achieved by these carriers (Ma et al. 2016). Another approach for treating diabetes is gene therapy; however, due to the risk to normal cells this approach faces criticism. CRISPR/Cas9 gene editing in macrophages can be helpful to improve type 2 diabetes, as observed in a study where Cas9 plasmids loaded in cationic lipid-assisted PEG-b-PLGA block copolymer nanoparticles were administered to mice and both *in vitro* and *in vivo* studies confirmed that macrophages can be reprogrammed to alleviate diabetes (Jawahar and Meyyanathan 2012).

Atherosclerotic plaque rupturing and destabilization need to be prevented for combating acute myocardial infarction. Macrophages/monocytes can cause destabilization of plaques, so serum monocyte chemoattractant protein-1 (MCP-1), monocyte colony-stimulating factor, and matrix metalloproteinase-9 need to be kept in check to avoid monocyte infiltration in plaques. When pitavastatin-loaded PLGA nanoparticles were administered to mice on high-fat diet, marked reduction in plaque destabilization was observed in comparison to pure drug (Katsuki et al. 2014). Polymeric nanoparticles can also be used as diagnostic aids for metabolic diseases. Dextran nanoparticles encapsulating desferoxamine for zirconium-89 radiolabeling (^{89}Zr-DNP) were administered to apolipoprotein E knockout mice and it was observed that better noninvasive assessment of arthrosclerosis plaques can be done under a positron emission tomography/magnetic resonance imaging (PET/MRI) scan (Majmudar et al. 2013).

6.4 Macrophage Targeting in Miscellaneous Diseases

Macrophage targeting can be helpful in other diseases also like in case of posttraumatic neurodegeneration in case of spinal cord injury. Macrophages/microglia causes inflammatory responses after spinal injury, which leads to persistent pain and other secondary inflammation-associated nerve damage. *In vitro* and *in vivo* studies suggested that poly(methyl methacrylate) nanoparticles showed improved macrophage uptake and better therapeutic response as compared to pure drug (Papa et al. 2014). Similarly, inflammatory response of macrophages in case of nonalcoholic steatohepatitis (NASH) causes significant tissue damage; thus, PLGA nanoparticles encapsulating small molecule inhibitor R406 to inhibit Fc receptors (spleen tyrosine kinase pathway) that are responsible for macrophage signaling and immune complex-mediated inflammation in NASH were synthesized and it was found that marked attenuation of secondary inflammatory damage was achieved (Kurniawan et al. 2018). Studies have shown that polyalkylcyanoacrylates nanoparticles can be used to target antileishmanial drugs to the macrophages as they not only carry the drug to macrophages but are also responsible for burst release of macrophages in respiratory tract, thus combating leishmaniasis (Bolkestein et al. 2016).

Future Prospects and Conclusion

In-depth understanding of role of macrophages in pathogenesis of various diseases has proven to be of great importance. Drug moieties can be targeted both actively as well as passively on to macrophages for improved site-specific therapy with minimum untoward effects on normal cells and tissues. Combining knowledge of macrophage infiltration time points in various diseases with novel concept of macrophage reeducation and remodeling has shown unanticipated results in combating various inflammatory diseases and cancer. This approach needs to be further exploited for overcoming the adverse events associated with synthetic drug molecules.

Polymeric nanoparticles have proven to be appropriate nanocarriers for transporting drug specifically to macrophages. A wide range of polymers are being employed to synthesize nanoparticles having advantage of ease of fabrication and surface modification with high drug payload. At the same time, functionalization of polymeric nanoparticles with ligands specific to the receptors overexpressed on macrophages is another field which needs to be explored to strike a balance between cost and benefits/risk and benefits. Thus, future holds great inventions in field of macrophage targeting assisted by polymeric nanoparticles.

References

Abramson SL, Gallin JI. IL-4 inhibits superoxide production by human mononuclear phagocytes. J Immunol. 1990;144:625–30.

Amash A, Wang L, Wang Y, et al. CD44 antibody inhibition of macrophage phagocytosis targets Fcγ receptor– and complement receptor 3–dependent mechanisms. J Immunol. 2016;196:3331–40. https://doi.org/10.4049/jimmunol.1502198.

Arakaki A, Pan W-A, Trejo J. GPCRs in cancer: protease-activated receptors, endocytic adaptors and signaling. Int J Mol Sci. 2018;19:1886. https://doi.org/10.3390/ijms19071886.

Aruffo A, Stamenkovic I, Melnick M, et al. CD44 is the principal cell surface receptor for hyaluronate. Cell. 1990;61:1303–13. https://doi.org/10.1016/0092-8674(90)90694-A.

Azzi J, Tang L, Moore R, et al. Polylactide-cyclosporin A nanoparticles for targeted immunosuppression. FASEB J. 2010;24:3927–38. https://doi.org/10.1096/fj.10-154690.

Biswas SK, Mantovani A. Macrophage plasticity and interaction with lymphocyte subsets: cancer as a paradigm. Nat Immunol. 2010;11:889–96. https://doi.org/10.1038/ni.1937.

Bolhassani A, Javanzad S, Saleh T, et al. Polymeric nanoparticles. Hum Vaccin Immunother. 2014;10:321–32. https://doi.org/10.4161/hv.26796.

Bolkestein M, de Blois E, Koelewijn SJ, et al. Investigation of factors determining the enhanced permeability and retention effect in subcutaneous xenografts. J Nucl Med. 2016;57:601–7. https://doi.org/10.2967/jnumed.115.166173.

Byrne JD, Betancourt T, Brannon-Peppas L. Active targeting schemes for nanoparticle systems in cancer therapeutics. Adv Drug Deliv Rev. 2008;60:1615–26. https://doi.org/10.1016/j.addr.2008.08.005.

Chawla A, Nguyen KD, Goh YPS. Macrophage-mediated inflammation in metabolic disease. Nat Rev Immunol. 2011;11:738–49. https://doi.org/10.1038/nri3071.

Couvreur P, Kante B, Roland M, Speiser P. Adsorption of antineoplastic drugs to polyalkylcyanoacrylate nanoparticles and their release in calf serum. J Pharm Sci. 1979;68:1521–4. https://doi.org/10.1002/jps.2600681215.

Dhabekar G, Dandekar R, Kingaonkar A. Role of macrophages in malignancy. Ann Maxillofac Surg. 2011;1:150. https://doi.org/10.4103/2231-0746.92782.

dos Santos SN, Rezende Dos Reis SR, Pires LP, et al. Avoiding the mononuclear phagocyte system using human albumin for mesoporous silica nanoparticle system. Micropor Mesopor Mat. 2017;251:181–9. https://doi.org/10.1016/j.micromeso.2017.06.005.

Elsabahy M, Wooley KL. Design of polymeric nanoparticles for biomedical delivery applications. Chem Soc Rev. 2012;41:2545–61.

Gaspar R, Préat V, Opperdoes FR, Roland M. Macrophage activation by polymeric nanoparticles of Polyalkylcyanoacrylates: activity against intracellular Leishmania Donovani associated with hydrogen peroxide production. Pharm Res. 1992;9:782–7. https://doi.org/10.1023/A:1015807706530.

Geissmann F, Manz MG, Jung S, et al. Development of monocytes, macrophages, and dendritic cells. Science. 2010;327:656–61. https://doi.org/10.1126/science.1178331.

Gelderman KA, Hultqvist M, Pizzolla A, et al. Macrophages suppress T cell responses and arthritis development in mice by producing reactive oxygen species. J Clin Invest. 2007;117:3020–8. https://doi.org/10.1172/JCI31935.

Germano G, Frapolli R, Belgiovine C, et al. Role of macrophage targeting in the antitumor activity of Trabectedin. Cancer Cell. 2013;23:249–62. https://doi.org/10.1016/j.ccr.2013.01.008.

Getts DR, Martin AJ, McCarthy DP, et al. Microparticles bearing encephalitogenic peptides induce T-cell tolerance and ameliorate experimental autoimmune encephalomyelitis. Nat Biotechnol. 2012;30:1217–24. https://doi.org/10.1038/nbt.2434.

Ginhoux F, Greter M, Leboeuf M, et al. Fate mapping analysis reveals that adult microglia derive from primitive macrophages. Science. 2010;330:841–5. https://doi.org/10.1126/science.1194637.

Golombek SK, May J-N, Theek B, et al. Tumor targeting via EPR: strategies to enhance patient responses. Adv Drug Deliv Rev. 2018;130:17–38. https://doi.org/10.1016/j.addr.2018.07.007.

Gordon S. Phagocytosis: an immunobiologic process. Immunity. 2016;44:463–75. https://doi.org/10.1016/j.immuni.2016.02.026.

Grohmann U, Belladonna ML, Vacca C, et al. Positive regulatory role of IL-12 in macrophages and modulation by IFN-γ. J Immunol. 2001;167:221–7. https://doi.org/10.4049/jimmunol.167.1.221.

Guillot A, Tacke F. Liver macrophages: old dogmas and new insights. Hepatol Commun. 2019;3:730–43. https://doi.org/10.1002/hep4.1356.

Hatami E, Mu Y, Shields DN, et al. Mannose-decorated hybrid nanoparticles for enhanced macrophage targeting. Biochem Biophys Rep. 2019;17:197–207. https://doi.org/10.1016/j.bbrep.2019.01.007.

Hoppstädter J, Dembek A, Linnenberger R, et al. Toll-like receptor 2 release by macrophages: an anti-inflammatory program induced by glucocorticoids and lipopolysaccharide. Front Immunol. 2019;10:1634. https://doi.org/10.3389/fimmu.2019.01634.

Jain S, Amiji M. Macrophage-targeted nanoparticle delivery systems. In: Svenson S, Prud'homme RK, editors. Multifunctional nanoparticles for drug delivery applications. Boston: Springer; 2012. p. 47–83.

Jain NK, Mishra V, Mehra NK. Targeted drug delivery to macrophages. Exp Opin Drug Deliv. 2013;10:353–67. https://doi.org/10.1517/17425247.2013.751370.

Jain S, Tran T-H, Amiji M. Macrophage repolarization with targeted alginate nanoparticles containing

IL-10 plasmid DNA for the treatment of experimental arthritis. Biomaterials. 2015;61:162–77. https://doi.org/10.1016/j.biomaterials.2015.05.028.

Jawahar N, Meyyanathan SN. Polymeric nanoparticles for drug delivery and targeting: a comprehensive review. Int J Health Allied Sci. 2012;1:217. https://doi.org/10.4103/2278-344X.107832.

Katsuki S, Matoba T, Nakashiro S, et al. Nanoparticle-mediated delivery of pitavastatin inhibits atherosclerotic plaque destabilization/rupture in mice by regulating the recruitment of inflammatory monocytes. Circulation. 2014;129:896–906. https://doi.org/10.1161/CIRCULATIONAHA.113.002870.

Kazankov K, Jørgensen SMD, Thomsen KL, et al. The role of macrophages in nonalcoholic fatty liver disease and nonalcoholic steatohepatitis. Nat Rev Gastroenterol Hepatol. 2019;16:145–59. https://doi.org/10.1038/s41575-018-0082-x.

Kinne RW, Bräuer R, Stuhlmüller B, et al. Macrophages in rheumatoid arthritis. 2000, 2(3):189-202. https://doi.org/10.1186/ar86.

Kourtis IC, Hirosue S, de Titta A, et al. Peripherally administered nanoparticles target monocytic myeloid cells, secondary lymphoid organs and tumors in mice. PLOS ONE. 2013;8:e61646. https://doi.org/10.1371/journal.pone.0061646.

Kumari A, Yadav SK, Yadav SC. Biodegradable polymeric nanoparticles based drug delivery systems. Colloids Surf B Biointerfaces. 2010;75:1–18.

Kurniawan DW, Jajoriya AK, Dhawan G, et al. Therapeutic inhibition of spleen tyrosine kinase in inflammatory macrophages using PLGA nanoparticles for the treatment of non-alcoholic steatohepatitis. J Control Release. 2018;288:227–38. https://doi.org/10.1016/j.jconrel.2018.09.004.

Laroui H, Viennois E, Xiao B, et al. Fab'-bearing siRNA TNFα-loaded nanoparticles targeted to colonic macrophages offer an effective therapy for experimental colitis. J Control Release. 2014;186:41–53. https://doi.org/10.1016/j.jconrel.2014.04.046.

Lattin JE, Schroder K, Su AI, et al. Expression analysis of G Protein-coupled receptors in mouse macrophages. Immunome Res. 2008;4:5. https://doi.org/10.1186/1745-7580-4-5.

Lee S, Kivimäe S, Dolor A, Szoka FC. Macrophage-based cell therapies: the long and winding road. J Control Release. 2016;240:527–40. https://doi.org/10.1016/j.jconrel.2016.07.018.

Li M, Sun J, Zhang W, et al. Drug delivery systems based on CD44-targeted glycosaminoglycans for cancer therapy. Carbohydr Polym. 2021;251:117103. https://doi.org/10.1016/j.carbpol.2020.117103.

Liu Y, Tong Z, Prud'homme RK. Stabilized polymeric nanoparticles for controlled and efficient release of bifenthrin. Pest Manag Sci. 2008;64:808–12.

Low PS, Henne WA, Doorneweerd DD. Discovery and development of folic-acid-based receptor targeting for imaging and therapy of cancer and inflammatory diseases. Acc Chem Res. 2008;41:120–9. https://doi.org/10.1021/ar7000815.

Luo Y-L, Xu C-F, Li H-J, et al. Macrophage-specific in vivo gene editing using cationic lipid-assisted polymeric nanoparticles. ACS Nano. 2018;12:994–1005. https://doi.org/10.1021/acsnano.7b07874.

Lv Y, Cao Y, Li P, et al. Ultrasound-triggered destruction of folate-functionalized mesoporous silica nanoparticle-loaded microbubble for targeted tumor therapy. Adv Healthcare Mater. 2017;6:1700354. https://doi.org/10.1002/adhm.201700354.

Ma L, Liu T-W, Wallig MA, et al. Efficient targeting of adipose tissue macrophages in obesity with polysaccharide nanocarriers. ACS Nano. 2016;10:6952–62. https://doi.org/10.1021/acsnano.6b02878.

MacParland SA, Tsoi KM, Ouyang B, et al. Phenotype determines nanoparticle uptake by human macrophages from liver and blood. ACS Nano. 2017;11:2428–43.

Majmudar MD, Yoo J, Keliher EJ, et al. Polymeric nanoparticle PET/MR imaging allows macrophage detection in atherosclerotic plaques. Circ Res. 2013;112:755–61. https://doi.org/10.1161/CIRCRESAHA.111.300576.

Mantovani A, Marchesi F, Malesci A, et al. Tumour-associated macrophages as treatment targets in oncology. Nat Rev Clin Oncol. 2017;14:399.

Matsuzaki T, Hanai S, Kishi H, et al. Regulation of endocytosis of activin type II receptors by a novel PDZ protein through Ral/Ral-binding protein 1-dependent pathway. J Biol Chem. 2002;277:19008–18. https://doi.org/10.1074/jbc.M112472200.

McCoy CE. The role of toll-like receptors in macrophages. Front Biosci. 2008;13:62. https://doi.org/10.2741/2660.

Medzhitov R. Recognition of microorganisms and activation of the immune response. Nature. 2007;449:819–26. https://doi.org/10.1038/nature06246.

Miller MA, Mikula H, Luthria G, et al. Modular nanoparticulate prodrug design enables efficient treatment of solid tumors using bioorthogonal activation. ACS Nano. 2018;12:12814–26.

Montet X, Funovics M, Montet-Abou K, et al. Multivalent effects of RGD peptides obtained by nanoparticle display. J Med Chem. 2006;49:6087–93. https://doi.org/10.1021/jm060515m.

Murray PJ, Wynn TA. Obstacles and opportunities for understanding macrophage polarization. J Leukoc Biol. 2011;89:557–63. https://doi.org/10.1189/jlb.0710409.

Nielsen SR, Schmid MC. Macrophages as key drivers of cancer progression and metastasis. Mediators Inflamm. 2017;2017:1–11. https://doi.org/10.1155/2017/9624760.

O'Neill LA. How toll-like receptors signal: what we know and what we don't know. Curr Opin Immunol. 2006;18:3–9. https://doi.org/10.1016/j.coi.2005.11.012.

Okubo M, Miyazaki M, Yuba E, Harada A. Chondroitin sulfate-based pH-sensitive polymer-modified liposomes for intracellular antigen delivery and induction of cancer immunity. Bioconjugate Chem.

2019;30:1518–29. https://doi.org/10.1021/acs.bioconjchem.9b00221.
Olefsky JM, Glass CK. Macrophages, inflammation, and insulin resistance. Annu Rev Physiol. 2010;72:219–46. https://doi.org/10.1146/annurev-physiol-021909-135846.
Onishi H, Isoda Y, Matsuyama M. In vivo evaluation of chondroitin sulfate-glycyl-prednisolone for anti-arthritic effectiveness and pharmacokinetic characteristics. Int J Pharm. 2013;456:113–20. https://doi.org/10.1016/j.ijpharm.2013.08.007.
Ortega RA, Barham WJ, Kumar B, et al. Biocompatible mannosylated endosomal-escape nanoparticles enhance selective delivery of short nucleotide sequences to tumor associated macrophages. Nanoscale. 2015;7:500–10. https://doi.org/10.1039/C4NR03962A.
Owens DE, Peppas NA. Opsonization, biodistribution, and pharmacokinetics of polymeric nanoparticles. Int J Pharm. 2006;307:93–102. https://doi.org/10.1016/j.ijpharm.2005.10.010.
Pandey S, Mahtab A, Rai N, et al. Emerging role of CD44 receptor as a potential target in disease diagnosis: a patent review. IAD. 2017;11 https://doi.org/10.2174/1872213X11666170907111858.
Papa S, Ferrari R, De Paola M, et al. Polymeric nanoparticle system to target activated microglia/macrophages in spinal cord injury. J Control Release. 2014;174:15–26. https://doi.org/10.1016/j.jconrel.2013.11.001.
Park J, Choi Y, Chang H, et al. Alliance with EPR effect: combined strategies to improve the EPR effect in the tumor microenvironment. Theranostics. 2019;9:8073–90. https://doi.org/10.7150/thno.37198.
Peterson KR, Cottam MA, Kennedy AJ, Hasty AH. Macrophage-targeted therapeutics for metabolic disease. Trends Pharmacol Sci. 2018;39:536–46. https://doi.org/10.1016/j.tips.2018.03.001.
Qadri M, Almadani S, Jay GD, Elsaid KA. Role of CD44 in regulating TLR2 activation of human macrophages and downstream expression of proinflammatory cytokines. J Immunol. 2018;200:758–67. https://doi.org/10.4049/jimmunol.1700713.
Qian B, Deng Y, Im JH, et al. A distinct macrophage population mediates metastatic breast cancer cell extravasation, establishment and growth. PLoS ONE. 2009;4:e6562. https://doi.org/10.1371/journal.pone.0006562.
Qiu S-Q, Waaijer SJH, Zwager MC, et al. Tumor-associated macrophages in breast cancer: Innocent bystander or important player? Cancer Treatment Reviews. 2018;70:178–89. https://doi.org/10.1016/j.ctrv.2018.08.010.
Rao NV, Rho JG, Um W, et al. Hyaluronic acid nanoparticles as nanomedicine for treatment of inflammatory diseases. Pharmaceutics. 2020;12:931. https://doi.org/10.3390/pharmaceutics12100931.
Rehli M. Of mice and men: species variations of toll-like receptor expression. Trends Immunol. 2002;23:375–8. https://doi.org/10.1016/S1471-4906(02)02259-7.
Serra P, Santamaria P. Nanoparticle-based autoimmune disease therapy. Clin Immunol. 2015;160:3–13.
Siddiqui L, Bag J, Seetha, et al. Assessing the potential of lignin nanoparticles as drug carrier: synthesis, cytotoxicity and genotoxicity studies. Int J Biol Macromol. 2020;152:786–802. https://doi.org/10.1016/j.ijbiomac.2020.02.311.
Smith AM, Rahman FZ, Hayee B, et al. Disordered macrophage cytokine secretion underlies impaired acute inflammation and bacterial clearance in Crohn's disease. J Exp Med. 2009;206:1883–97. https://doi.org/10.1084/jem.20091233.
Song M, Liu T, Shi C, et al. Bioconjugated manganese dioxide nanoparticles enhance chemotherapy response by priming tumor-associated macrophages toward M1-like phenotype and attenuating tumor hypoxia. ACS Nano. 2016;10:633–47. https://doi.org/10.1021/acsnano.5b06779.
Storm G, Belliot SO, Daemen T, Lasic DD. Surface modification of nanoparticles to oppose uptake by the mononuclear phagocyte system. Adv Drug Deliv Rev. 1995;17:31–48. https://doi.org/10.1016/0169-409X(95)00039-A.
Sugama S, Takenouchi T, Kitani H, et al. Activin as an anti-inflammatory cytokine produced by microglia. J Neuroimmunol. 2007;192:31–9. https://doi.org/10.1016/j.jneuroim.2007.08.016.
Tao S, Zhao Z, Zhang X, et al. The role of macrophages during breast cancer development and response to chemotherapy. Clin Transl Oncol. 2020;22:1938–51. https://doi.org/10.1007/s12094-020-02348-0.
Tauber AI. Metchnikoff and the phagocytosis theory. Nat Rev Mol Cell Biol. 2003;4:897–901. https://doi.org/10.1038/nrm1244.
Taylor ME. Structure and function of the macrophage mannose receptor. In: Mammalian carbohydrate recognition systems, vol. 33. Berlin: Springer; 2001. p. 105–21.
The immune system and primary immunodeficiency. In: The Immune Deficiency Foundation
Thomas TP, Goonewardena SN, Majoros IJ, et al. Folate-targeted nanoparticles show efficacy in the treatment of inflammatory arthritis. Arthritis Rheum. 2011;63:2671–80. https://doi.org/10.1002/art.30459.
Udalova IA, Mantovani A, Feldmann M. Macrophage heterogeneity in the context of rheumatoid arthritis. Nat Rev Rheumatol. 2016;12:472–85. https://doi.org/10.1038/nrrheum.2016.91.
van Die I, Cummings RD. The mannose receptor in regulation of helminth-mediated host immunity. Front Immunol. 2017;8:1677. https://doi.org/10.3389/fimmu.2017.01677.
Van Ginderachter JA, Movahedi K, Hassanzadeh Ghassabeh G, et al. Classical and alternative activation of mononuclear phagocytes: Picking the best of both worlds for tumor promotion. Immunobiology. 2006;211:487–501. https://doi.org/10.1016/j.imbio.2006.06.002.
Vauthier C, Bouchemal K. Methods for the preparation and manufacture of polymeric nanoparticles. Pharm Res. 2009;26:1025–58.
Wang Y, Lin Y-X, Qiao S-L, et al. Polymeric nanoparticles promote macrophage reversal from M2 to M1 pheno-

types in the tumor microenvironment. Biomaterials. 2017;112:153–63. https://doi.org/10.1016/j.biomaterials.2016.09.034.

Wang X, Iyer A, Lyons AB, et al. Emerging roles for G-protein coupled receptors in development and activation of macrophages. Front Immunol. 2019;10:2031. https://doi.org/10.3389/fimmu.2019.02031.

Wu Q, Qi Y, Wu N, et al. Expression and anti-inflammatory role of activin receptor-interacting protein 2 in lipopolysaccharide-activated macrophages. Sci Rep. 2017;7:10306. https://doi.org/10.1038/s41598-017-10855-4.

Wynn TA, Chawla A, Pollard JW. Origins and hallmarks of macrophages: development, homeostasis, and disease. Nature. 2013a;496:445–55. https://doi.org/10.1038/nature12034.

Wynn TA, Chawla A, Pollard JW. Macrophage biology in development, homeostasis and disease. Nature. 2013b;496:445–55. https://doi.org/10.1038/nature12034.

Yi Y-S. Folate receptor-targeted diagnostics and therapeutics for inflammatory diseases. Immune Netw. 2016;16:337. https://doi.org/10.4110/in.2016.16.6.337.

Yoo J, Park C, Yi G, et al. Active targeting strategies using biological ligands for nanoparticle drug delivery systems. Cancers. 2019;11:640. https://doi.org/10.3390/cancers11050640.

Zhang M, Gao Y, Caja K, et al. Non-viral nanoparticle delivers small interfering RNA to macrophages in vitro and in vivo. PLOS ONE. 2015;10:e0118472. https://doi.org/10.1371/journal.pone.0118472.

Zielińska A, Carreiró F, Oliveira AM, et al. Polymeric nanoparticles: production, characterization, toxicology and ecotoxicology. Molecules. 2020;25 https://doi.org/10.3390/molecules25163731.

Polymeric Nanoparticles-Based Drug and Gene Delivery to Macrophages

Fitsum Feleke Sahle

Abstract

Recently drug and gene delivery to macrophages has attracted tremendous attention because studies indicated that chronic inflammatory disorders, tissue degeneration due to secondary injury, and cancer are associated with increased infiltration and/or activity of a specific macrophage phenotype. Conseqently, different types of nanoparticles have been designed for macrophage-targeted drug/gene delivery, and polymeric nanoparticles emerged as one of the most widely explored nanoparticles. Commonly, chronic inflammation and tissue degeneration due to secondary injury are associated with increased infiltration and/or activity of M1-like macrophages into the inflamed tissue, and different polymer-based nanocarriers have been designed to deplete and/or modulate the secretion of M1-like macrophages or to repolarize/reprogram the proinflammatory M1-like macrophages into the antiinflammatory M2-like macrophages. Conversely, cancer is associated with increased infiltration and/or activity of the immunosuppressive and tumor-promoting M2-like macrophages, including tumor-associated macrophages (TAMs), and polymeric nanoparticles have been designed to deplete the M2-like macrophages or repolarize/reprogram them into the antitumoral M1-like macrophages. Retroviruses like HIV and lentiviruses utilize monocytes and macrophages to replicate and hide from immune-surveillance system and polymeric nanoparticles have also been designed to target these affected macrophages. In addition, macrophages inherently infiltrate into inflamed and cancerous tissues and they can be utilized as Trojan horses for polymeric nanoparticles loaded with antiinflamatory and anticancer drugs/genes, respectively. Therefore, in this chapter, the challenges and potentials applications of drug/gene delivery to macrophages using conventional and stimuli-responsive polymeric nanoparticles are systematically reviewed.

Keywords

Cancer · Deplete macrophage · Inflammatory disorder · Macrophage targeted · Polymeric nanoparticle · Repolarize macrophage · Trojan horse

F. F. Sahle (✉)
Department of Pharmaceutical and Biomedical Sciences, College of Pharmacy, California Northstate University, Elk Grove, CA, USA

GlaxoSmithKline, Richmond, VA, USA

S. Gupta, Y. V. Pathak (eds.), *Macrophage Targeted Delivery Systems*,
https://doi.org/10.1007/978-3-030-84164-5_7

1 Introduction

Macrophages are phagocytic cells that are found in different tissues, body cavities, and mucosal surfaces and play a critical role in host defense against infection and maintaining tissue homeostasis. They originate from hematopoietic stem cells or blood monocytes and depending on the extracellular environment they are activated/polarized into mature proinflammatory M1-like or antiinflammatory M2-like macrophages (Reichel et al. 2019; Orecchioni et al. 2019). M2-Like macrophages are further subcategorized into M2a, M2b, M2c, and M2d macrophages based on their gene expression profiles (Arora et al. 2018). In addition, each category or subcategory of macrophage represents a heterogeneous population of macrophages that express and produce different molecules (Atri et al. 2018). Generally, M1-like macrophages phagocytose pathogens and promote secretion of proinflammatory cytokines (e.g., TNF-α, IL-23, IL-1β, IL-6, and IL-12), reactive oxygen species, inducible nitric oxide synthase (iNOS), cyclooxygenase-2, and reactive nitrogen species to combat infection and promote tissue repair (Zhang et al. 2019; Swart and Troeberg 2019). However, uncontrolled overactivation of the immune system causes persistent secretion of proinflammatory cytokines that may result in chronic inflammatory conditions, including rheumatoid arthritis, metabolic syndrome-associated disorders (e.g., type 2 diabetes and atherosclerosis), inflammatory bowel diseases (e.g., Crohn's disease and ulcerative colitis), osteoarthritis, asthma, systemic lupus erythematous, psoriasis, multiple sclerosis, and Alzheimer's disease (Alvarado-Vazquez et al. 2017; Xiao et al. 2013; Gouveia et al. 2019; He et al. 2020). M1-like macrophages may also initiate and promote secondary injury and tissue degeneration following primary injury. For example, following spinal cord injury, M1-like resident macrophages (microglia), neutrophils, and lymphocytes infiltrate into the injured site and cause secondary injury and neurodegeneration (Papa et al. 2014). However, in tumors, M1-like macrophages exhibit antitumoral properties by promoting and amplifying Th1-type responses and secreting a series of cytokines, including TNFs, growth inhibitors, and antiangiogenetic factors that possesses antitumoral properties (Zhang et al. 2019).

Conversely, M2-like macrophages secrete cytokines (e.g., IL-4, IL-10, and IL-13), VEGFs, and TGF-β, which elicit antiinflammatory responses and promote tumor progression, metastasis, and resistance to chemotherapy (Atri et al. 2018; Zhang et al. 2019). The best example of M2-like macrophages is tumor-associated macrophages (TAMs), which are the most abundant nonneoplastic cell types found in discrete cancer and perform immunosuppressive and tumor-promoting functions (Zhang et al. 2019). They secrete angiogenic growth factors including VEGF-A and placental growth factor that increase blood vessel formation by tumor endothelial cells, increase nutrient delivery to cancer cells, and enhance cancer cell growth within dense tumors (Reichel et al. 2019; Yu et al. 2013). TAMs secrete enzymes such as matrix metalloproteinase-9 (MMP-9), serine proteases, and cathepsins that degrade the extracellular matrix, facilitate tumor invasion into adjacent organs, and promote cancerous cells metastasis (Reichel et al. 2019; Yu et al. 2013). They also discourage the infiltration of antitumor lymphocytes into the tumor by secretion of the immunosuppressive cytokines including IL-10 and TGF-β that keep tumor progression unchecked (Yu et al. 2013). In addition, TAMs may promote the survival, migration, and chemoresistance of cancer-associated stem cells that facilitate tumors regrowth following treatments (Reichel et al. 2019). Actually, an interesting correlation is established between TAMs infiltration and cancer growth, tumor metastasis, poor prognosis, and shorter tumor survival in a variety of human carcinomas (Steidl et al. 2010; Zhu et al. 2013; Bingle et al. 2002).

Some retroviruses and lentiviruses use monocytes and macrophages to replicate and hide from immune-surveillance system. For example, macrophages located in the reticuloendothelial system (RES) and the brain act as major reservoirs for HIV because of their long-term survival after HIV infection and ability to spread virus particles to CD4-positive lymphocytes (Kaur et al. 2008).

Some viruses such as, HIV, HHV-6, and Maedi visna virus manipulate macrophages and utilize them as their vessels for dissemination or long-term reservoirs (Nikitina et al. 2018). Apart from exploiting the macrophages as reservoirs, some viruses may alter cytokine production in macrophages and induce chronic inflammation and extensive tissue damage (Kaur et al. 2008). Consequently, targeting macrophages emerged as a novel approach to treat chronic inflammatory diseases, tissue degeneration due to secondary injury, cancer, and infections, and over the last few years various types of nanocarriers have been designed and investigated for targeted drug and gene delivery to macrophages.

Various types of nanocarriers including inorganic nanoparticles (Song et al. 2016; Zanganeh et al. 2016; Kanwar et al. 2012; Zazo et al. 2017), lipid nanoparticles (Kharaji et al. 2016; Vieira et al. 2018; Shrivastava et al. 2021), polymeric nanoparticles (Papa et al. 2014; Shilakari Asthana et al. 2014; Jain et al. 2015), mesoporous silica nanoparticles (Giménez et al. 2015), quantum dots (Chakravarthy et al. 2011), carbon nanotubes (Zhang et al. 2021), and hybrid nanoparticles (He et al. 2017a; Bagalkot et al. 2015) have been investigated for targeted drug/gene delivery to macrophages, each with attributable advantages and limitations. Polymeric nanoparticles are one of the most commonly investigated, if not the most investigated, class of nanoparticles for macrophage-targeted drug and gene delivery mainly due to the recent advancement in polymer science, which enabled easy modification of nanoparticles surface and bulk properties for improved drug loading, longer circulation half-life, macrophage targeting, and controlled drug delivery (Jain et al. 2008). In addition, compared to lipid nanocarriers and many other nanoparticles, polymeric nanoparticles have higher drug/gene encapsulation efficiency, excellent stability upon storage and administration to the body, and are easy and cheap to prepare (Peres et al. 2017).

Various types of natural or synthetic polymers have been used for the preparation of polymeric nanoparticles for macrophage-targeted drug/gene delivery. Some of the most commonly investigated biocompatible and biodegradable natural polymers include dextran (Ma et al. 2016), chitosan (Shilakari Asthana et al. 2014), alginate (Jain et al. 2015), gelatin (Kaur et al. 2008; Jain et al. 2008), and hyaluronic acid (Tran et al. 2016). Certain polysaccharides such as dextran can be efficiently and selectively internalized by macrophages by interacting with dextran-binding C-type lectins and scavenger receptors on the macrophages (Ma et al. 2016). Hyaluronic acid nanoparticles can bind specifically to CD44 receptors that are overexpressed on macrophages (Tran et al. 2016). Natural polymers mimic the extracellular matrix, and have tunable mechanical properties like stimuli responsiveness, degradation, swelling, and cross-linking capabilities. However, the application of natural polymers in drug and gene delivery nanoparticle preparation is often hampered by the presence of contaminants, batch-to-batch variability, low hydrophobicity to encapsulate lipophilic drugs, and rapid drug release from the matrix (Panyam and Labhasetwar 2003). Contrarily, synthetic polymers are stable, can be produced reproducibly, their hydrophilicity and lipophilicity can be tailored, and may control drug release from days up to several weeks (Panyam and Labhasetwar 2003). Polylactic acid (PLA), polyglycolic acid (PGA), and their copolymer poly(lactic-co-glycolic acid) (PLGA) represent the most commonly investigated biodegradable and biocompatible synthetic polymers for drug delivery (Peres et al. 2017; Kiefer et al. 2020; Pang et al. 2018). Especially, various grades of PLGA with desired biodegradation and drug release profiles can be obtained by varying the PLA and PGA ratios (Peres et al. 2017). Many other novel synthetic and natural polymers used for the preparation of drug-loaded conventional and stimuli-responsive polymeric nanoparticles for macrophage-targeted drug delivery have also been discussed under this chapter.

In gene delivery, polyethylenimine (PEI) represents the widely investigated polymer. It is a synthetic cationic polymer that electrostatically condenses nucleic acids and forms a stable complex called polyplex. Every third atom of PEI is a protonatable nitrogen atom that can buffer the acidification within the endosome after endocytosis, which results in endosomal chloride accumulation, osmotic swelling of endosomes, rupture of the endosomes, and endosomal escape of the polyplex, a phenomenon described as the "proton sponge effect." A branched PEI can better condense nucleic acids and effectively transfect cells than a linear PEI (Kim et al. 2012). However, the *in vivo* application of linear and branched-chain PEIs as gene transfection agent is restricted by PEIs cytotoxicity (as a positively charged nanocarrier it interacts with the cell membrane and impair its integrity) and its rapid clearance by the RES (Kim et al. 2012; Tran et al. 2015). It also interferes with few critical intracellular processes, including disruption of protein kinase C activity (Xiao et al. 2013). As a result, many research groups modified PEI in different ways to reduce its cytotoxicity and improve its cell targeting potential (Kim et al. 2012). For example, PEI was conjugated with poly(ethylene glycol) (PEG) to render the polyplex hydrophilic and provide steric stabilization, which reduced polyplex aggregation, increased polyplex circulatory half-life, and reduced the toxicity of the polyplexes (Kim et al. 2012). PEI was also conjugated with hyaluronic acid to decrease its toxicity and clearance (Tran et al. 2016; Tran et al. 2015). The gene delivery and macrophage-targeting potentials of some other positively charged synthetic (e.g., poly(β-amino ester) (Zhang et al. 2019)), semi-synthetic (e.g., 6-Amino-6-deoxy-curdlan (Ganbold and Baigude 2018)), and natural (e.g., chitosan (Shilakari Asthana et al. 2014)) polymers have also been investigated.

2 Challenges of Drug and Gene Delivery to Macrophages Using Polymeric Nanoparticles

Polymeric nanoparticles have shown tremendous potential for targeted drug and gene delivery to macrophages (Table 1). They can encapsulate different drugs, therapeutic proteins, and genes, and prevent their degradation and enhance their permeability across different biological membranes (Cohen et al. 2000). For instance, genes are negatively charged and can hardly permeate the negatively charged cell membranes, including macrophages, and their permeability can only be realized by encapsulating them in viral vectors or appropriate nonviral nanocarriers (Kim et al. 2015). Polymeric nanoparticles can also extend drugs/genes circulation half-life and enable controlled and targeted drug/gene delivery to macrophages (Giménez et al. 2015). However, there are numerous challenges hindering the clinical applications of polymeric nanoparticles as macrophage-targeted drug/gene delivery systems and some of these challenges (Table 1) are summarized under this section.

2.1 Drug/Gene Loading and Release

Advancement in polymer chemistry enabled designing of various polymeric nanoparticles with tunable surface and bulk properties. However, loading and/or release of some therapeutic agents like nucleic acid in polymeric nanocarriers can still be very challenging. Usually, loading of negatively charged nucleic acids on conventional nanoparticles is challenging, and commonly, positively charged polymers (polycations) such as PEI and poly(β-amino ester) (Zhang et al. 2019) have been utilized to electrostatically interact with the nucleic acids and form polyplexes (Kim et al. 2012). However, positively charged polymeric nanoparticles tend to be toxic than their neutral or negatively charged counterparts because the positively charged polymers may interact with biological membranes and affect their integrity. In addition, gene release from positively charged nanocarriers could be a problem due to the strong electrostatic interaction between the positively charged nanocarrier and the negatively charged nucleic acid (Zakeri et al. 2018). As a result, the search for novel and safe polymers that can adequately load genetic materials and release them at the target site at the desired rate is intensified.

Table 1 Opportunities and main challenges associated with polymeric nanocarriers designed for macrophage-targeted drug/gene delivery

Opportunities	Main Challenges
• Improved drug/gene stabiity and permeability • Controlled drug/gene release • Biodegradable nanoparticles prevent nanoparticle accumulation in the body	• Drug laoding and release • Targetting of phenotype-specific macrophages • Inhibition of endosomal/lysosomal degradation of loaded therapeutic agents • Predicting the effect of protein corona on nanoparticle uptake

2.2 Endosomal Uptake of Polymeric Nanoparticles and Macrophage Targeting

Polymeric nanoparticles can considerably enhance the permeability of drugs, biologics, and genes across different types of cells including macrophages. However, the therapeutic success of the nanocarriers is highly dependent on the fate of the nanoparticles in the cells and the ability of the nanoparticles to deliver the specific molecules to the cytosol, nucleus, or another specific intracellular target site (Behzadi et al. 2017). Nanoparticles enter cells mainly through endocytosis, which depending on the type of cell and physicochemical properties of the nanoparticle could be phagocytosis, clathrin-mediated endocytosis, caveolin-mediated endocytosis, clathrin/caveolae-independent endocytosis, and/or micropinocytosis. Commonly, macrophages are professional phagocytes and they tend to take up nanoparticle mainly through phagocytosis (Behzadi et al. 2017). However, after endocytosis, the nanoparticles in the endocytic vesicle fuse with the early endosomal compartment, mature into a late endosome, and accumulate in the lysosomes. During maturation, the pH in the endosome decreases from 7.4, down to ~ 6.6 in the early endosome, ~6.0 in the late endosome, and ~5.0 in lysosomes. In addition, macrophages have highly degradative phagocytic, endosomal, and lysosomal compartments, which can degrade some nanoparticles as well as the loaded therapeutic agents. As a result, for most nanocarriers, escape from the endocytic pathway is a critical step for successful delivery of the loaded therapeutic agents to macrophages (Yu et al. 2013; Smith et al. 2019) and polymeric nanoparticles with specific functional groups such as positively charged groups that induce proton sponge effect and pH-sensitive moieties that facilitate escape of the encapsulated therapeutic agents from degradation by the endocytic pathway (Smith et al. 2019) have been utilized.

2.3 Targeting of Polymeric Nanoparticles into Phenotype-Specific Macrophages

Macrophage targeting is a main challenge of drug and gene delivery to macrophage. Apparently, due to the enhanced permeation and retention (EPR) effect, nanoparticles tend to migrate into inflamed tissues, where macrophages also accumulate and targeting macrophages seems to be less problematic (Binnemars-Postma et al. 2017). However, passive targeting of macrophages by polymeric nanoparticles may not provide the desired level of targeting and active targeting of nanoparticles might be necessary through modification of nanoparticles surfaces using different targeting moieties that can interact with the different receptors that are overexpressed on macrophage surfaces, including mannose receptor (recognize mannose terminal, N-acetylglucosamine, and fucose residues), scavenger receptor (highly recognizes modified low-density lipoprotein), dectin-1 receptor, tuftsin peptide (a tetra-peptide sequence Thr-Lys-Pro-Arg that activates macrophages and enhances their phagocytic ability), cluster of differentiation (CD) 44 receptor (a distinctive receptor for hyaluronic acid and can also interact with other ligands, such as osteopontin, collagens, and matrix metalloproteinases), folate receptor-β (specifically expressed by activated macrophages and shows a high affinity for folic acid), and phosphatidylserine receptor (He et al. 2017b). Accordingly, various nanoparticle-targeting moieties including, mannose, lectin, fibronectin, antibodies, monosaccharides, polysaccharides, and folic acid have been investigated (Kaur et al. 2008; Ma et al. 2016; He et al. 2017b). Besides, the existence of M1-like and M2-like macrophages with contrasting responses also adds another layer of complexity to macrophage-targeted drug and gene delivery and effective macrophage-targeted therapy may demand phenotype-specific targeting of macrophages.

2.4 Protein Corona and Cellular Uptake of Polymeric Nanoparticles

Protein corona is another phenomenon that is not well understood and usually overlooked during development of various nanotherapeutics. Protein corona describes the modification of nanoparticles surfaces in biological fluids by the adsorption of different biomolecules including proteins on the nanoparticle's surfaces. The type, amount, and orientation of biomolecules adsorbed on nanoparticles surfaces are highly dependent on the collective physicochemical properties of the nanoparticles such as size, polydispersity, shape, charge, surface chemistry, and surface hydrophobicity/hydrophilicity, which makes its prediction difficult. Consequently, the interaction of various polymeric nanoparticles with cells will be unpredictable since the protein corona provides a new identity to the nanoparticles (Behzadi et al. 2017).

2.5 Accumulation of Polymeric Nanoparticles in the Body

One of the major challenges of transitioning nanotherapeutics into the clinic is accumulation of nanoparticles in the body and the associated acute and chronic toxicity. To address this issue, various types of biodegradable nanoparticles have been designed and investigated. This can be achieved either by using biodegradable polymers such as PLGA and poly(β-amino ester) (Zhang et al. 2019) or rendering the nanoparticles biodegradable through introduction of biodegradable linkers such as polycaprolactone and PLA (Andrade et al. 2020) into the backbone of the polymeric nanoparticles. Nanoparticle degradation in endosomes and lysosomes also facilitates escape of the loaded active molecules from the lysosome/endosome. However, the nanoparticles should degrade in a controlled manner to prevent premature nanoparticle degradation before the nanocarrier crosses the cell membrane and to achieve the desired controlled drug release (Xiao et al. 2013). To address this, multifunctional nanoparticles, such as surface-functionalized and pH-sensitive polymeric nanoparticles (Zhao et al. 2017), have been designed and investigated.

3 Polymeric Nanoparticles for Drug and Gene Delivery to Macrophages

Various types of conventional and stimuli-responsive polymer-based nanocarriers (Fig. 1), including polymeric nanoparticles (Jain et al. 2008), nanogels (Nuhn et al. 2018), polymeric micelles (Yu et al. 2013), polymersomes (Gouveia et al. 2019), dendrimers (He et al. 2018), and core-shell polymeric nanoparticles (Wang et al.

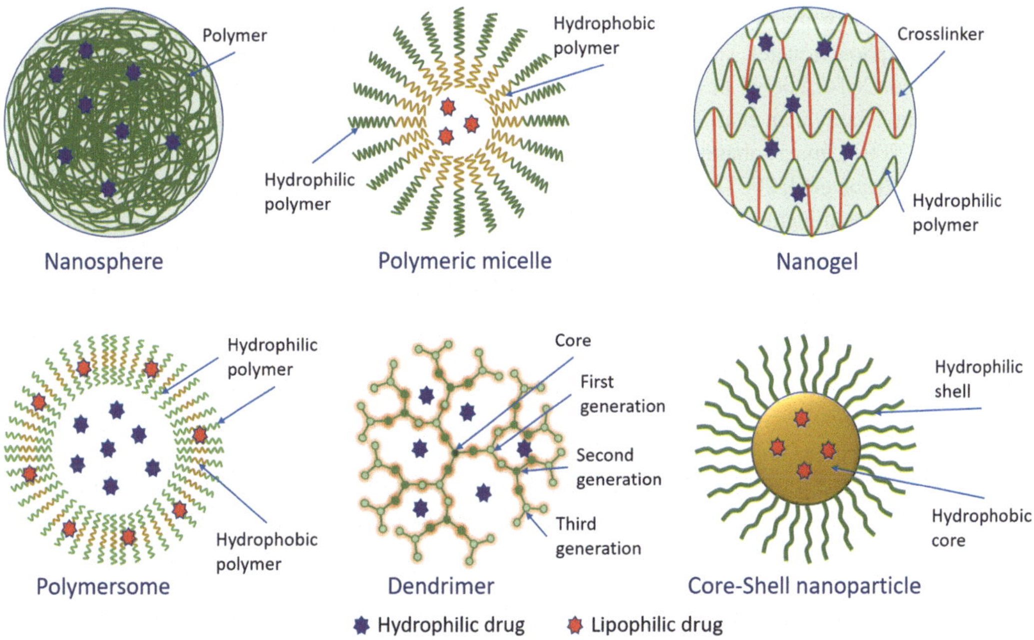

Fig. 1 Schematic representation of different polymer-based nanoparticles designed for macrophage-targeted drug/gene delivery

2019) have been designed to target macrophages for the management of chronic inflammatory disorders, cancer, viral infections, and tissue degeneration due to secondary injury. Chronic inflammation and tissue degeneration are characterized by increased infiltration and activity of M1-like macrophages, and polymer-based nanoparticles are mainly designed to deplete and/or modulate the secretion of M1-like macrophage or repolarize/reprogram the culprit M1-like macrophages into M2-like macrophages. Conversely, in cancer, there is increased infiltration and activity of TAMs and nanoparticles are mainly designed to deplete TAMs or repolarize/reprogram them into M1-like macrophages. In some diseases like HIV, viruses utilize macrophages as their sanctuary and nanoparticles can be used to target the infected macrophages. Macrophages also infiltrate into inflamed tissues and tumors and they can be recruited as Trojan horses to target drugs into inflamed or cancerous tissues. Under this section, the different drug/gene-loaded polymeric nanoparticles designed to (1) deplete macrophages, (2) modulate the secretion of macrophage-derived inflammatory modulators, (3) repolarize/reprogram macrophages, or (4) utilize macrophage as Trojan horses are discussed.

3.1 Polymeric Nanoparticles Designed to Deplete Macrophages

Chronic inflammation is characterized by increased infiltration of monocytes into the inflamed tissue, unbalanced polarization of monocytes into proinflammatory M1-like macrophages, and increased secretions of various pro-inflammatory mediators that induce tissue damage. Different strategies have been explored to deplete M1-like macrophages and one approach is to inhibit monocyte-to-macrophage differentiation. For example, SIRT1 inhibits monocyte differentiation to macrophages by inhibiting the phosphorylation and nuclear translocation of PU.1, and agents that activate SIRT1 or genes that promote SIRT1 secretion (e.g., anti-miR-449a) can be used to deplete M1-like macrophages (Park et al. 2016). Other compounds that deplete M1-like macrophages by apoptosis including bisphosphonates (e.g., clodronate) (Kameka et al. 2014; Niebel et al. 2012), IFN-β (Kumaran Satyanarayanan et al. 2019), rituximab (Toubi et al. 2007), macrophage-targeted immunotoxins (van Roon et al. 2003), and few antiviral drugs have also been identified (Jain et al. 2008). However, their poor bioavailability, short circulation half-life, lack of tissue targeting, and the resulting toxicity restricted their clinical applications and, consequently, various types of polymeric nanoparticles have been designed and investigated to target these drugs/genes to macrophages. For example, Papa et al. (2014) developed To-Pro3-loaded poly(methyl methacrylate) nanoparticles to target activated microglia (macrophage-like phagocytic cells) in the spinal cord to counteract secondary injury and prevent posttraumatic neural degeneration after primary injury. The nanoparticles were effectively taken up by activated microglia obtained from the spinal cord of mouse embryos *in vitro*, but PEGylation slowed down nanoparticle cellular uptake. Following parenchymal delivery *in vivo* in mouse, the nanoparticles were selectively taken up by the microglia. In another study, Niebel et al. (2012) developed clodronate-loaded polymethacrylate (Eudragit RL) nanoparticles for the treatment of ulcerative colitis. Unlike the free clodronate, the nanoparticles significantly decreased TNFα and IL-6 secretions in cultured RAW 264.7 cells *in vitro* but no satisfactory anti-inflammatory effect was obtained *in vivo* due to lack of adequate targeting of the clodronate to the target site.

Surface functionalization of nanoparticles can improve their cellular uptake, and commonly, polymeric nanoparticles designed for macrophage-targeted drug/gene delivery are surface functionalized using different targeting moieties such as mannose, polysaccharides, antibodies, and folic acid (Kaur et al. 2008; Ma et al. 2016; He et al. 2017b), which interact with receptors that are overexpressed on macrophage surfaces. One of the widely investigated surface-functionalized nanoparticles are mannose-functionalized (mannosylated) polymeric nanoparticles (Table 2), which can interact with the mannose receptor – a 175-kDa transmembrane protein of the C-type lectin family that is exclusively expressed on macrophages.

Polysaccharide nanoparticles are another class of nanoparticles investigated for macrophage-targeted drug delivery because they inherently interact with polysaccharide receptors on macrophage surfaces. For example, Ma et al. (2016) designed dextran nanoparticles to ameliorate obesity-associated inflammation by targeting visceral adipose tissue macrophages. In healthy state adipose tissue, macrophages make 10–15% of the total interstitial cell population and are predominantly M2 like. Whereas, in an obese state, macrophages constitute 40–50% of the total interstitial cell population and are predominantly polarized into M1-like phenotype, which secrete proinflammatory cytokines and induce pathological effects both locally and systemically, resulting in glucose intolerance, insulin resistance, and vascular dysfunction (Ma et al. 2016). The dextran nanoparticles selectively interacted with dextran-binding C-type lectins and scavenger receptors on the macrophages and, following intravenous administration, the radiolabeled dextran nanoparticles prepared using 70 KDa and 500 KDa dextran predominantly distributed in the liver of lean and obese C57BL/6J mice models. However, after intraperitoneal administration, a large fraction of the nanoparticles distributed into the visceral adipose tissue of obese mice, while liver remained the dominant distribution site in lean mice. The selective uptake of the nanoparticles by visceral adipose tissue in obese mice was not observed when the receptor targeting was attenuated by PEGylation of the dextran nanoparticles (Ma et al. 2016).

Folate receptor-β is overexpressed on activated macrophages, and folate-functionalized nanoparticles have been investigated for targeted drug and gene delivery to macrophages. For example, Thomas et al. (2011) developed folate-conjugated, generation 5 (G5) poly(amidoamine) dendrimers to target macrophages in inflammatory arthritis. Methotrexate, a drug that inhibits DNA formation and results in reduced cell proliferation and induction of apoptosis, was also conjugated to the dendrimers surfaces. An *in vitro*

Table 2 Mannose-functionalized polymer-based nanoparticles for targeted drug or gene delivery to macrophages

No.	Nanoparticle	Purpose	Result	References
1.	Mannosylated gelatin nanoparticles	Controlled and site-specific delivery of didanosine to HIV-infected macrophages	Mannosylation increased nanoparticle uptake 2.7-fold in alveolar macrophages *in vitro*. It also significantly enhanced didanosine uptake by macrophage-rich organs including the lungs, liver, and lymph nodes in albino rats. The free drug was cleared quickly from the blood.	Jain et al. (2008)
2.	Mannosylated, PEGylated polyamidoamine dendrimers	Targeted delivery of liver X receptor ligands to atherosclerotic plaque-associated macrophages to attenuate plaque burden	After 4 weeks of treatment, the mannosylated dendrimer significantly reduced atherosclerotic plaque size and plaque inflammation in mice aortic arch and aortic root. In addition, unlike the free liver X receptor ligands, the nanoparticle did not cause hepatic lipogenesis.	He et al. (2018)
3.	Mannan-coated gelatin nanoparticles	Targeted delivery of didanosine to HIV-infected macrophages	After subcutaneous administration of the mannan-coated nanoparticles to Sprague-Dawley rats, 1.7, 12.6, and 12.4 times higher concentration of didanosine were obtained in the macrophage-rich organs spleen, lymph nodes, and brain, respectively, compared to the free didanosine.	Kaur et al. (2008)
4.	Mannosylated, PEGylated PEI polyplex	Targeted delivery of siRNA to macrophages	The polyplex was efficiently endocytosed by and inhibited gene expression in RAW264.7 cells (murine macrophage cell line that is known to express mannose receptors and typically hard to transfect) *in vitro*. PEGylation of the polyplex reduced nanoparticle toxicity, without affecting gene knockdown efficiency.	Kim et al. (2012)
5.	Mannosylated 6-amino-6-deoxy-curdlan nanoparticle	Targeted delivery of TNFα-inhibiting siRNA to macrophages	The nanoparticle delivered siTNFα to mouse peritoneal macrophages, liver, and lung and knockdown TNFα expression *in vivo*.	Ganbold and Baigude (2018)
6.	Mannosylated chitosan nanoparticles	Targeted delivery of antisense oligonucleotides to macrophages	The nanoparticle significantly increased gene transfection efficiency in Raw 264.7 cells. The degree of transfection was significantly higher in Raw 264.7 cells that express moderate mannose receptors than HeLa cells that do not express mannose receptors.	Shilakari Asthana et al. (2014)

cellular uptake study showed that, unlike the free dendrimer, the FA-bearing dendrimer was taken up by RAW264.7 cells in a dose-dependent fashion. Following thrice a week intravenous administration in a collagen-induced arthritis model of female Lewis rats, the surface-functionalized dendrimer acted as a potent antiinflammatory agent and reduced arthritis-induced symptoms including ankle swelling, paw volume, cartilage damage, bone resorption, and body weight loss.

Various types of stimuli-responsive polymeric nanoparticles that can sense their environment and control drug/gene release in response to their surrounding have also been designed and investigated for macrophage-targeted drug delivery. pH-sensitive nanoparticles represent the commonly investigated stimuli-responsive polymeric nanoparticles for macrophage-targeted drug delivery. It is well understood that, in inflamed tissues, the pH drops due to infiltration and activation of inflammatory cells and the resulting enhanced energy and oxygen demand accelerated glucose consumption via glycolysis, and increased lactic acid production, which leads to the local acidosis (Xie et al. 2020). In addition, the pH in endosomes is very low and nanoparticles degradation in endosomes facilitates endosomal escape of therapeutics agents. Consequently, different pH-sensitive nanoparticles have been designed for macrophage-targeted drug and gene delivery. For example, Zhao et al. (2017) developed methotrexate-loaded, folic acid-functionalized, pH-responsive polymeric nanoparticles for the treatment of rheumatoid arthritis. The nanoparticles comprise a hydrophobic core made of PLGA and the pH-responsive polymer poly(cyclohexane-1,4-diylacetone dimethylene ketal), which is coated by a hydrophilic shell comprising egg phosphatidylcholine and PEG. The nanoparticle exhibited pH-dependent drug release *in vitro* and drug release increased significantly at low pH (pH 5). The cytotoxicity of free methotrexate and methotrexate loaded in nonfunctionalized and folic acid-functionalized nanoparticles was also determined *in vitro* in RAW264.7 cells and the functionalized nanoparticles showed superior cytotoxic effect than the nonfunctionalized nanoparticles and the free drug. The antiinflammatory effect of the nanoparticles was also investigated *in vivo* in rheumatoid arthritis rat models and the folic acid-functionalized nanoparticle significantly reduced the progression of rheumatoid arthritis compared to the nonfunctionalized nanoparticles. Similarly, Yu et al. (2019) investigated dexamethasone-loaded, hyaluronic acid-coated, pH-sensitive polymeric nanoparticles composed of egg phosphatidylcholine, polyethylenimine, and poly (cyclohexane-1,4 diyl acetone dimethylene ketal) for the treatment of rheumatoid arthritis. The acid-sensitive poly (cyclohexane-1,4-diyl acetone dimethylene ketal) rendered the nanoparticles pH-responsive and dexamethasone release from the nanoparticles increased markedly when the pH decreased from 7.4 to 4.5. The cellular uptake of hyaluronic acid-functionalized and nonfunctionalized nanoparticles was investigated *in vitro* in activated RAW 264.7 cell lines, and surface functionalization of the nanoparticle significantly increased nanoparticle cellular uptake due to the ability of hyaluronic acids to bind to CD44 receptors that are overexpressed on macrophages. The hyaluronic acid coating may also render the nanoparticles hydrophilic and prolong nanoparticle circulation time and accumulation in the inflamed tissue. In adjuvant-induced arthritis rat models, the nanoparticles reduced inflammatory cell infiltration, bone damage, and cartilage damage in the ankle joints and inhibit the progression of rheumatoid arthritis, which was attributed to the targeting and acid-sensitive nature of the nanoparticles. However, the effect of pH responsiveness on drug release and particularly nanoparticle endosomal escape was not supported using non-pH-responsive control nanoparticles.

Redox-responsive nanoparticles are good candidates for targeted drug and gene delivery to macrophages because in chronic inflammation there is production of high level of reactive oxygen species. For example, Gupta et al. (2012) developed redox-responsive micelles using propylene sulfide and N,N-dimethylacrylamide diblock copolymer to target inflamed tissues. Release of model drugs from the micelles was investigated *in vitro* and when the micelles are exposed to oxidants hydrogen peroxide, 3-morpholinosydnonimine, and peroxynitrite, the micelles core oxidized and triggered micelle disassembly and cargo release. In addition, the release of a model drug in the activated RAW 264.7 macrophages was significantly higher than in the nonactivated macrophages because activated macrophages contain high level of endogenous oxidants. Temperature-responsive nanoparticles represent another category of smart nanocarriers for macrophage-targeted drug/gene delivery. Commonly, thermoresponsive polymers with lower critical solution temperature (LCST) such as poly(N-isopropylacrylamide), poly(N-vinylcaprolactam), and poly(ethylene oxide)-poly(propylene oxide) have been used to prepare thermoresponsive nanoparticles (Xie et al. 2020). Thermoresponsive nanocarriers that exhibit LCST are hydrophilic below the LCST and considerably swell in water because of hydrogen bonds formed between water molecules and the polymer functional groups and can load a significant amount of a therapeutic agent. However, above the LCST (usually 38.5–43 °C), the polymers undergo hydrophilic-to-hydrophobic phase transition that results in polymer shrinkage and squeezing out of internal water molecules and release of the encapsulated therapeutic molecule (Xie et al. 2020). For example, Wang et al. (2020) designed simple thermosensitive and biodegradable polymeric nanoparticles containing a diblock copolymer monomethoxy poly(ethylene glycol)-block-poly(trimethylene carbonate) for drug delivery to primary osteoclast precursor cells (bone marrow macrophages). *In vitro* the nanoparticles were safe and effectively taken up by bone marrow-derived macrophages showing great potential to target macrophage *in vivo*. However, despite their potentials, there are not many redox-responsive and thermoresponsive polymeric nanoparticles investigated for macrophage-targeted drug or gene delivery.

Polymeric nanoparticles have also been designed to deplete TAMs for the treatment of cancer. TAMs can be depleted by blocking the CSF-1R signaling pathway – an important signaling pathway that drives the recruitment of TAMs to tumors and promotes macrophage differentiation into protumoral phenotypes. A few drugs (e.g., pexidartinib (Yan et al. 2017), BLZ945 (Strachan et al. 2013), and PLX7486 (Cannarile et al. 2017)) and antibodies (e.g., emactuzumab (Ries et al. 2014) and cabiralizumab (Qiu et al. 2018)) that block CSF-1R signaling pathway have been investigated for the treatment of cancer. Some antibodies (e.g., trabectedin) that activate caspase-8 through TRAIL and deplete TAMs have also been investigated (Germano et al. 2013). However, lack of drug targeting and the associated toxicity and increased risk of infection restricted the clinical application of these drugs and genes. Consequently, various polymeric nanoparticles have been designed to target TAMs depleting agents to TAMs, which in most cases are surface functionalized using different macrophage-specific targeting moieties to achive TAMs targeting. For example, Nuhn et al. (2018) functionalized the surfaces of methoxy triethyeneglycol methacrylate and pentafluorophenyl methacrylae block copolymer-based nanogels using macrophage mannose receptor 1 (MRC1)-specific nanobody to target MRC1-expressing TAMs for cancer therapy. MRC1, also known as CD206, is a type I transmembrane protein that belongs to the C-type lectin family that is expressed by macrophages and dendritic cells. Nanobodies are the smallest available antigen-binding fragment derived from camelid heavy-chain-only antibodies that is about 15 kDa, which is 10 times smaller than a conventional antibody. They lack the Fc portion of antibodies and, unlike antibody-functionalized nanoparticles, nanobody-functionalized nanocarriers avoid Fc-mediated aspecific binding to macrophages. The anti-MRC1 nanobody-functionalized nanogels were efficient *in vitro* in MRC1-expressing TAMs and *in vivo* in Lewis lung carcinoma-bearing mice. Some other common targeting moieties including mannose (Zhu et al. 2013), galactose (Huang et al. 2012), and folate (Meng et al. 2019) were also used to target TAMs. However, mannose, galactose, and folate bind to other cell populations including dendritic cells and epithelial cells (Napoletano et al. 2012; van Dam et al. 2011; Sallusto et al. 1995) making TAM-specific targeting of nanoparticles a challenge.

In addition to surface functionalization, TAM-targeted polymeric nanoparticles can be rendered stimuli responsive to improve macrophage targeting. For example, Zhu et al. (2013) designed mannosylated PLGA nanoparticles for targeted delivery of TAMs in the tumor environment and shielded the nanoparticles using a shedable PEG layer by attaching a long flexible PEG2000 polymer through a hydrazone linker to the PLGA nanoparticles. Under normal physiological condition, the PEG shield remained intact and prevented recognition of the mannose-functionalized nanocarriers by macrophages (Fig. 2). However, under acidic tumor microenvironment, the hydrazone linker was cleaved to shed the PEG layer off the nanoparticles and exposed the mannosylated surface to effect ligand-receptor interaction and nanoparticle uptake by TAMs. The uptake of the nanoparticles was investigated *in vitro* in murine J774A.1 cells and PEGylation of the mannosylated nanoparticle decreased cellular uptake by 75%. However, preincubation of the nanoparticles in a pH 6.8 media prior to cellular uptake increased the cellular uptake of the nanoparticles 3.1-fold due to removal of the PEG layer. Preincubation of the control mannosylated nanoparticles, which were PEGylated without an acid-labile cross-linker, has no effect on the nanoparticles cellular uptake. In addition, after tail vein injection of the stimuli-responsive nanoparticles to healthy C57BL/6 mice, the nanoparticles accumulated in major organs, such as liver and spleen in the order of mannosylated, non-PEGylated nanoparticles > nonmannosylated, non-PEGylated nanoparticles > mannosylated, and PEGylated nanoparticles, indicating that PEGylation significantly decreased the accumulation of the nanoparticles, which is in line with many findings that PEGylation decreases opsonization of nanocarriers by macrophages and their accumulation in the liver and kidney. Increased accumulation of mannosylated nanoparticles over nonmannosylated nanoparticles was attributed to the presence of liver Kupffer cells and splenic macrophages that express high level of mannose receptors. Moreover, 12 hours after injection of the nanoparticles to B16-F10 mouse melanoma tumor-bearing C57BL/6 mice, the accumulation of PEGylated nanoparticles in the tumor site was significantly higher than the non-PEGylated nanoparticles. This is attributed to decreased opsonization and prolonged circula-

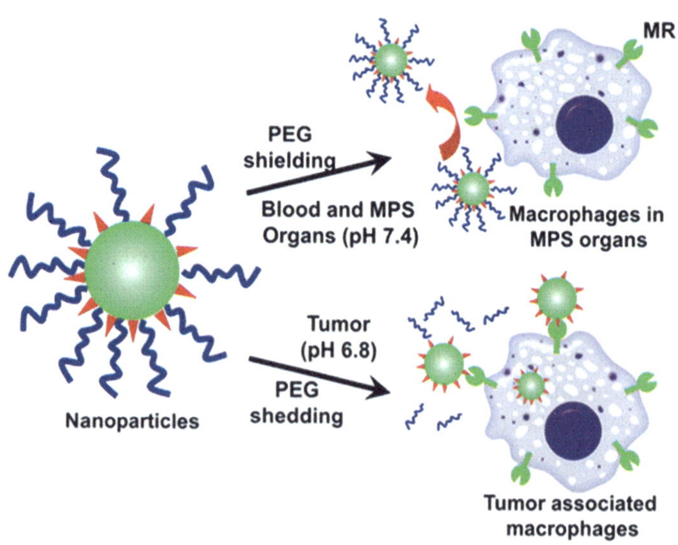

Fig. 2 Mannose-functionalized, PEG-shielded PLGA nanoparticles for targeted drug delivery to TAMs. The PEG moiety is conjugated to the mannose-functionalized nanoparticles through acid cleavable hydrazone linker, which remains intact in healthy tissues but degraded in acidic tumor environment to free the mannose-functionalized nanoparticles. Reprinted with permission from (Zhu et al. 2013). Copyright (2021) American Chemical Society

tion of the PEGylated nanoparticles, which favored tumor accumulation. Nanoparticles with acid cleavable PEG moiety also colocalized into macrophages better than the control mannosylated nanoparticles containing nonacid cleavable PEG shield.

3.2 Polymeric Nanoparticles Designed to Modulate the Secretion of Macrophage-Derived Inflammatory Mediators

Inflammatory stimuli activates macrophages and initiates the nuclear factor-κB (NF-κB) translocation signaling pathway – a common pathological mechanism in several inflammatory diseases –, JAK signaling pathway, and other signaling pathways, and secretes a wide variety of pro- or anti-inflammatory mediators including cytokines, chemokines, inflammatory enzymes, and growth factors (Gouveia et al. 2019). Mostly, the pathogenesis of chronic inflammatory disorders is associated with overexpression of proinflammatory macrophage-derived mediators and different strategies have been applied as an innovative therapeutic approach to block these mediators and treat chronic inflammatory diseases. A common approach involves neutralization of the disease-causing proinflammatory cytokines and their receptors using monoclonal antibodies such as adalimumab (van Schouwenburg et al. 2013), infliximab (van Schouwenburg et al. 2013), golimumab (van Schouwenburg et al. 2013), tocilizumab (Semerano et al. 2014), and anti-IL-15 (Baslund et al. 2005), or receptor fusion proteins such as etanercept (van Schouwenburg et al. 2013). A second approach is to inhibit the different inflammatory signaling pathways using inhibitors such as tofacitinib (Sandborn et al. 2017) and phosphodiesterase-4 (Mazur et al. 2015). A third approach could be gene delivery to modify the secretion of the various proinflammatory modulators. As a result, most of these compounds were subjected to different preclinical and clinical investigations but the desired product stability, safety, bioavailability, or efficacy was not obtained due to lack of adequate macrophage targeting. Consequently, different types of polymeric nanoparticles have been designed to target drugs/genes that modulate macrophage-derived mediators to macrophages. For example, Gouveia et al. (2019) designed pH-responsive polymersomes using the amphiphilic diblock copolymer poly(2-methacryloyloxyethyl phosphorylcholine)-poly(2-(diisopropylamino) ethyl methacrylate) (PMPC-PDPA) for targeted delivery of glucocorticoids (agents that suppress inflammation – activated signaling) to macrophages for the treatment of inflammatory diseases. The PMPC segment of the copolymer binds with the scavenger receptor B type I that is overexpressed in macrophages for targeted macrophage delivery. *In vitro* investigation showed that the polymersomes disassembled below pH 6.2 to release the loaded drug in the endosomes and enabled endosomal escape of the loaded drug molecules in the cytosol.

Laroui et al. (2014) synthesized PLA-PEG block copolymer nanoparticles for targeted delivery of TNFα siRNA to colon macrophages for the treatment of inflammatory bowel diseases (IBD). TNFα siRNA prevents expression of TNFα, a major proinflammatory cytokine mainly secreted by macrophages during IBD. The siRNA was complexed with PEI before it was loaded to the nanoparticles. During nanoencapsulation, the hydrophilic PEG portion of the copolymer faced outward, and the hydrophobic PLA faced inward. The surface of the pegylated nanoparticle was grafted with the Fab' portion of an F4/80 antibody (the Fc portion was removed by pepsin digestion to minimize interaction with immune cells and prevent nonspecific binding) via maleimide/thiol-mediated covalent bonding to target colon macrophages after oral delivery. The Fab in the Fab'-bearing TNFα siRNA-loaded PLA-PEG nanoparticles boosted the phagocytosis of the nanoparticles in RAW264.7 murine macrophage cell lines *in vitro* via direct interaction between the Fab'-bearing and the F4/80 antigens on RAW264.7 cells. In addition, the TNFα expression in the inflamed RAW264.7 cells was determined and the TNFα siRNA-loaded PLA-PEG nanoparticles significantly decreased the level of TNFα secretion unlike the empty nanoparticle, nanoparticles loaded with scrambled siRNA, or TNFα siRNA nanoparticles loaded in lipofectamine. Moreover, the nanoparticles were dispersed in an alginate/chitosan (7:3) hydrogel (the hydrogel prevents nanoparticle release in the upper part of the GIT but collapses and releases the nanoparticles at pH 5 or 6 in the colon) and was administered orally to C57BL/6 mice colitis models. The mice that received Fab'-bearing TNFα siRNA-loaded nanoparticles showed an average weight loss of 6%, which was low compared to the 15% and 25% weight loss in mice that received Fab'-bearing scrambled siRNA-loaded nanoparticles and nongrafted scrambled siRNA-loaded nanoparticles, respectively. Histological investigations showed that mice treated with Fab'-bearing TNFα siRNA-loaded nanoparticles showed near-normal colonic histology with intact tridimensional organization of colonic epithelial cells and the mucosa like the control group. In contrast, mice treated with Fab'-bearing scrambled siRNA-loaded nanoparticles demonstrated multifocal inflammatory cell infiltration into the submucosa, severe denudation of the surface epithelium (erosion), and mucodepletion of glands.

In another study, Xiao et al. (2013) synthesized a bioreducible poly (cysteamine bisacrylamide-PEI) polycation and formed a polyplex with TNF-α siRNA for the treatment of IBD. The polycation was mannosylated using a PEG spacer and, during nanoparticle formation, sodium triphosphate was added to serve as an ionic cross-linker between the polycation and the siRNA and produced a condensed stable nanoparticle (Fig. 3). Sodium triphosphate has a linear structure that carries five negative charges and can cross-link the protonated polycation in an aqueous medium to enhance siRNA condensation capacity of polycations and increases the efficiency of nanoparticle-mediated gene silencing. The TNF-α gene silencing effect of the nanoparticles was investigated in lipopolysaccharide-activated Raw 264.7 cells *in vitro* and cells that were pretreated with siRNA-loaded nanoparticles secreted a significantly less amount of TNF-α. Contrarily, cells treated with scrambled siRNA-loaded nanoparticles exhibited negligible effect on TNF-α gene expression. The gene-silencing effects of the nanoparticle were also comparable with siRNA-loaded Oligofectamine™ that contains double-siRNA concentration. MTT assay showed that, unlike siRNA-loaded Oligofectamine™ or PEI (25 KDa) nanoparticles, the siRNA-loaded reducible mannosylated nanoparticle was not toxic to RAW 264.7 macrophages and Caco2-BBE cells. The nanoparticles also efficiently inhibited TNF-α secretion in 3% (w/v) dextran sodium sulfate-treated FVB male mice colitis models.

Tao et al. (2020) designed siRNA-loaded polymeric nanoparticles to suppress the expression of the macrophage gene Camk2g and improve atherosclerotic plaque stability in mice. Camk2g encodes a calcium-activated kinase called CaMKIIγ (Ca2+/calmodulin-dependent protein kinase γ), which promotes the conversion of a relatively benign, fibrous atherosclerotic lesions into necrotic lesions with thin fibrous caps. The nanoparticle was prepared by complexing the siRNA with a PEG-conjugated cationic molecule and by depositing the formed complex on a PLGA core nanoparticle. The surface of the nanoparticles was further functionalized with S2P peptide (CRTLTVRKC), which recognizes the macrophage receptor stabilin-2 and promotes macrophage targeting. Following intravenous injection in mice, the S2P peptide-functionalized siCaMKIIγ-loaded nanoparticles silenced the expression of the plaque-destabilizing gene in atherosclerotic lesion macrophages and promoted plaque stability.

3.3 Polymeric Nanoparticles Designed to Repolarize/Reprogram Macrophages

The M1/M2 macrophage dichotomy plays a significant role in maintaining tissue homeostasis. Higher M1/M2 ratio is associated with chronic inflammation conditions, whereas lower M1/M2 ratio is associated with cancer (Zhang et al. 2019). As a result, reprogramming/repolarizing of M1-like macrophages into M2-like macro-

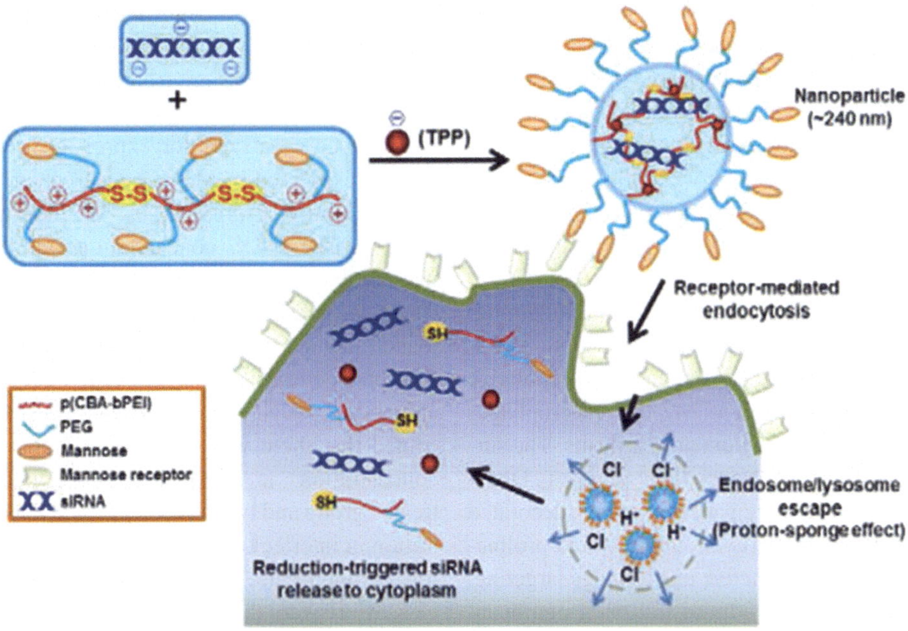

Fig. 3 Schematic illustration showing cell-specific uptake of a mannosylated bioreducible poly(amido amine)-based cationic polymer/siRNA polyplex by macrophages, endosomal uptake of the polyplex, and release of the siRNAs to the cytoplasm by proton sponge effect. Reprinted with permission from (Xiao et al. 2013)

phage and vice versa emerged as a novel strategy for the treatment of chronic inflammatory diseases and cancer, respectively. Naturally, the M1/M2 dichotomy is determined by the tissue environment. For example, in the wound-healing environment, harsh tumor environment, or in the presence of antiinflammatory cytokines, such as IL-4 or IL-10, M1-like macrophages tend to polarize into M2-like macrophages (Tran et al. 2016; Pang et al. 2018). On the other hand, lipopolysaccharide, IFN-γ, TNF-α, IL-12, toll-like receptor agonists, and CD40 agonists induce repolarization of TAMs into M1-like macrophages (Zhang et al. 2019; Alvarado-Vazquez et al. 2017; Tran et al. 2016). As a result, the applications of some immunomodulators; receptor agonist molecules (e.g., Imiquimod (Chi et al. 2017), IMO-2055 (Smith et al. 2014), and TMP195 (Guerriero et al. 2017)), fusion proteins (e.g., TTI-621 (Petrova et al. 2017)), and antibodies (e.g., Hu5F9-G4 (Gholamin et al. 2017), HPA063793 (Georgoudaki et al. 2016), and CC-90002 (Huang et al. 2017)), and genes that modulate expression of target immunomodulators in inducing macrophage phenotype transition were investigated. However, the therapeutic potential of these compounds and genes could not be realized mainly because of their low efficacy associated with their poor bioavailability, short circulation half-life, and poor cell-targeting potential (Shahbazi et al. 2018). As a result, different polymer-based nanoparticles were designed and investigated for targeted delivery of drugs/genes that induce macrophages phenotype transition.

Shahbazi et al. (2018) assembled pure hyaluronic acid nanoparticles for targeted delivery of IL-4 or other immunoregulatory cytokines that can induce M1-to-M2 macrophage phenotype transition to treat inflammatory diseases, regenerate injured tissues, and relief autoimmune disorders. The nanoparticles were capable of targeting macrophages on their right due to the high affinity of hyaluronic acid to CD44 receptors that are overexpressed on macrophage surfaces, and based on the report, the hyaluronic acid nanoparticle alone (unloaded nanoparticle) induced M1-to-M2 macrophage transition *in vitro* in J774A.1 macrophage. However, the IL-4-laoded hyaluronic nanoparticle was superior in inducing M1-to-M2 macrophage phenotype transition than the unloaded nanoparticles. The authors further showed that nanoparticles assembled using high molecular weight hyaluronic acid were more effective than nanoparticles prepared using medium and low molecular weight hyaluronic acid.

Jain et al. (2015) delivered an antiinflammatory cytokine IL-10-encoding plasmid DNA-loaded alginate nanoparticles to repolarize M1-like macrophages to M2-like macrophages for the treatment of rheumatoid arthritis. The surfaces of the IL-10-encoding plasmid DNA-loaded alginate nanoparticles were decorated using tuftsin peptide (L-threonine-lysine-proline-arginine) to achieve active macrophage targeting. After intraperitoneal administration, a significant amount of the tuftsin-modified alginate nanoparticles accumulated in the inflamed paws of arthritic rats unlike the nonmodified nanoparticles. Similarly, other gene-loaded polymeric nanoparticles were designed to reprogram M1-like macrophages into M2-like macrophages (Table 3).

Different types of polymeric nanoparticles have also been designed and investigated to repolarize/reprogram TAMs into M1-like macrophage for the management of cancer. For example, Wang et al. (2017) synthesized pH-responsive poly(β-amino ester) copolymer-based nanoparticles for targeted delivery of IL-12 to TAMs. The nanoparticles were synthesized through Michael addition reaction between the acrylate groups of a hydrophobic monomer 1,6-hexanediol diacrylate and the amine groups of a hydrophilic amino-terminated PEG polymer and the pH-responsive monomer 2-(4-imidazolyl) ethylamine. The formed nanoparticle was responsive to pH (Swart and Troeberg 2019; Alvarado-Vazquez et al. 2017) and swelled and released IL-12 in pH 6.5 media. Following tail vein administration to B16-F10 cell xenografted tumor mice model, unlike the free IL-12, the IL-12-loaded nanoparticles accumulated in the tumor site through EPR effect and released the interleukin in a controlled manner. In addition, tumor was harvested from free IL-12, unloaded nanoparticle, and IL-12-loaded nanoparticle-treated mice, TAMs were isolated by tumor macrophage isolation kits, and the isolated cell lysates were analyzed for expression of Arg-1 (highly expressed in M2-like macrophages) and iNOS (highly expressed in M1-like macrophages). The iNOS concentration significantly increased and the Arg-1 concentration significantly decreased in mice treated with IL-12-loaded nanoparticle signaling TAMs to M1-like macrophage transition. A real-time PCR relative expression of iNOS in the isolated TAMs confirmed that there was three times higher iNOS concentration in IL-12-loaded nanoparticles treated group and two times higher iNOS concentration in interleukin-treated mice group than the mice group treated with the unloaded nanoparticles. Moreover, the isolated TAMs were incubated with free IL-12, unloaded nanoparticles, and IL-12-loaded nanoparticles for 6 h and the phenotype conversion was investigated by immunofluorescence method. During the investigation, CCR7 and CD206 were used as M1-like and M2-like macrophage biomarkers. The relative decrease in fluorescent signals of Cy5.5-CD206 and increased fluorescent signals FITC-CCR7 in IL-12 nanoparticle-treated group also supported the M2-like to M1-like macrophage phenotype transition.

Wang et al. (2019) designed IMD-0354-loaded core-shell nanoparticles for tumor-localized chemoimmunotherapy of hepatocellular carcinoma. IMD 0354 is a selective IKKβ inhibitor that causes repolarization of TAMs into M1-like macrophages. The nanoparticles contain a mannosylated cationic lipid-based core and were coated by a pH-responsive anionic natural polysaccharide derivative O-carboxymethyl chitosan, which undergoes charge reversal in the acidic tumor microenvironment. The researchers also designed similar but nonmannosylated core-shell nanoparticles loaded with sorafenib (an anticancer agent used for the treatment of hepatocellular carcinoma and other types of carcinoma) to target cancerous cells. Upon administration of the two nanocarriers to tumor-bearing mice, the tumor volume decreased, the ratio of M1/M2

Table 3 Gene-loaded polymeric nanoparticles designed to repolarize/reprogram macrophages

No.	Nanoparticle	Nucleic acid (repolarization)	Results	Reference (s)
1.	Tuftsin peptide-functionalized alginate nanoparticles	IL-10 plasmid DNA (M1 like to M2 like)	After intraperitoneal administration in rheumatoid arthritis rats model, the percentage of M2-like macrophages increased from 9 to 62%. The nanoparticles also reduced the expression of systemic and joint tissue proinflammatory cytokines (TNF-α, IL-1β, and IL-6) and prevented the progression of inflammation and joint damage.	Jain et al. (2015)
2.	Hyaluronic acid-PEI nanoparticles	miR-223 duplexes (a critical macrophage polarization regulator that suppresses the proinflammatory pathways and enhances antiinflammatory responses) (M1 like to M2 like)	Increased miR-223 expression in primary peritoneal macrophages isolated from C57BL/6 mice 90-fold, induced M1-to-M2 phenotypic transition, and significantly suppressed inflammation in peritoneal macrophages by decreasing proinflammatory cytokines.	Tran et al. (2016)
3.	Hyaluronic acid-PEI nanoparticles	IL-4 and IL-10 plasmid DNA-expressing genes (the plasmid DNAs were loaded and investigated separately) (M1 like to M2 like)	After intraperitoneal administration of the IL-4- and IL-10-loaded nanoparticles to C57Bl/6 mice, expression of IL-4 mRNA and IL-10 mRNA increased 400- and 1200-fold, respectively. In addition, both IL-4- and IL-10-expressing gene-loaded nanoparticles significantly decreased the M1-like macrophage population.	Tran et al. (2015)
4.	Mannosylated cationic poly(β-amino ester) nanoparticles	IRF5- and IKKβ-encoding mRNAs (M2 like to M1 like)	After intraperitoneal administration of the nanoparticles, the percentage of M1-like macrophage increased and M2-like macrophages dropped. The nanoparticles also cleared 40% of the tumors in a mouse model of ovarian cancer.	Zhang et al. (2019)
5.	Hyaluronic acid-PEI nanoparticles	miR-125b (M2 like to M1 like)	After intraperitoneal administration to naive and nonsmall-cell lung cancer C57BL/6 mice models, the nanoparticles accumulated in the macrophage-ablated lung tissues and the M1/M2 macrophage ratio increased sixfold as compared to the control group.	Parayath et al. (2018)

macrophages increased, and the secretion of IFN-γ and IL-12 immunogenic cytokines secreted by M1-type TAM increased significantly compared to a group treated with only the sorafenib-loaded core-shell nanoparticles.

Yu et al. (2013) reported pH-responsive, endosomolytic, polymeric micelles for targeted delivery of siRNA to TAMs. The micelles were formed upon self-aggregation of a triblock 2-azidoethyl methacrylate, 2-dimethylaminoethyl methacrylate, and 2-dimethylaminoethyl methacrylate-co-butyl methacrylate-co-2-propylacrylic acid terpolymer. In the design, the cationic 2-dimethylaminoethyl methacrylate segment of the polymer was used to condense the anionic siRNA. The micelles formed a stable complex with the siRNA and protected the SiRNA degradation by RNases. The terminal azide group in the 2-azidoethyl methacrylate segment served as an anchor to an alkyne-functionalized mannose through the azide-alkyne "click" chemistry. The micelles significantly increased SiRNA uptake into human macrophages and, relative to the non-targeted diblock micelles, the mannose-functionalized micelles enhanced the SiRNA delivery into the macrophages threefold.

In another study, Zhang et al. (2019) developed biodegradable polymeric nanoparticles to genetically reprogram TAMs into antitumor macrophages through complexation of a positively charged poly(β-amino ester) polymer with the negatively charged interferon regulatory factor 5 (IRF5) and an IKKβ (a kinase that phosphorylates and activates IRF5) encoding mRNAs. Phenotypic, functional, and gene expression studies revealed that coexpression of IRF5/IKKβ favors the polarization of macrophages toward the M1 phenotype. The poly(β-amino ester) polymer contains biodegradable ester bonds in its backbone to undergo intracellular hydrolytic cleavage and release the loaded mRNAs. The surfaces of the nanosized complexes were mannosylated using polyglutamic acid linker, which is added to target macrophages and shield the positive charge of the nanocomplexes. The transfection efficiency of the nanoparticles in murine bone marrow-derived macrophages was investigated *in vitro* after loading green fluorescent protein-encoding mRNA in place of the IRF5/IKKβ-encoding mRNAs and, following a single application, the nanoparticles transfected 31.9 ± 8.5% of the primary macrophages without reducing their viability. The transfection efficiency of nonmannosylated nanoparticles, but with the polyglutamic acid linker, was only 25 ± 2.1%. The effect of the nanoparticles to reprogram M2-like macrophages into the therapeutically desirable anticancer M1 phenotype was also investigated in bone marrow-derived macrophages by gene expression analysis and the nanoparticle significantly downregulated signature M2-like macrophage genes, such as Serpinb2 and Ccl11, but upregulated key M1 differentiation genes, such as Ccl5. The performance of the IRF5/IKKβ-encoding nanoparticles was also investigated *in vivo* in ovarian tumor-bearing C57BL/6 mice model. Following biweekly intraperitoneal administration of the nanoparticles for 3 weeks, the fraction of M1-like macrophages in the rat peritoneum increased from 0.5 ± 0.2% to 10.2 ± 4.1%. In addition, the release of the proinflammatory cytokines IL-12, IFN-γ, and TNF-α increased 3.4, 8.4,1.5-folds and the levels of IL-6 reduced 97-fold. Moreover, the disease regressed and was eventually cleared in 40% of animals and the median mice survival was 142 days versus 60 days in the control group. The distribution of the mRNAs in the ovarian tumor-bearing C57BL/6 mice model was also investigated 24 h after intraperitoneal injection of the IRF5/IKKβ mRNA-loaded nanoparticles and the highest concentrations of the mRNAs were found in organs located in the peritoneum, including liver, spleen, intestine, pancreas, and diaphragm. The distribution was also investigated after intravenous infusion of the nanoparticles and the nanoparticles preferentially delivered their mRNA cargo to organs with high levels of resident macrophages/phagocytes, mostly the spleen, liver, and lungs. The nanocarriers caused only modest increases in the expression levels of inflammatory cytokines. Likewise, a few other polymeric nanoparticles have been designed to delivery genes that reprogram TAMs into M1-like macrophages (Table 3).

3.4 Polymeric Nanoparticles Designed to Utilize Macrophage as Trojan Horses

Macrophages migrate into solid tumors and inflamed tissues and can be utilized as Trojan horses for nanoparticles loaded with anticancer and antiinflammatory agents to the target sites. Particularly, macrophages are nondividing differentiated cells and are resistant to the cytotoxic DNA replication inhibitors such as doxorubicin and daunorubicin (Soto et al. 2012) and have been investigated as Trojan horses to many nanoparticles loaded with lethal dose of anticancer drugs. For example, Pang et al. (2018) prepared doxorubicin-loaded PLGA nanoparticles and loaded the nanoparticles into bone marrow-derived M1-like macrophages by incubating the macrophages with the nanoparticles in a cell culture media for targeted delivery of doxorubicin to glioma (one of the refractory tumors threatening people's lives by rapid development and poor prognosis). M1-like macrophages have stronger capacity to internalize particulate maters more than regular macrophages and the nanoparticles were effectively loaded into the macrophages. Then, the necked nanoparticles and the nanoparticles loaded into the macrophages (Trojan horses) were administered in nude mice-bearing intracranial U87 glioma models and the distribution of the nanoparticles in different tissues was investigated. The free nanoparticles mainly distributed in the liver and spleen, while nanoparticles loaded in the macrophages entered liver, spleen, and the lungs. A significantly higher amount of the nanoparticles was also detected in the brains of the mice group treated with the Trojan horses compared to the mice group treated with the necked nanoparticles, showing that the macrophages were able to carry nanoparticles across the endothelial barrier of brain tissue by the mediation of cell adhesion molecules and further overcome the high interstitial fluid pressure into the core of glioma. In addition, the pure doxorubicin, the necked nanoparticles, and the Trojan horses extended the median mice survival by 21, 26.5, and 38.5 days, respectively.

4 Concluding Remarks and Future Perspectives

Several studies correlated autoimmune diseases, chronic inflammation, and cancer with uncontrolled macrophage infiltration, polarization, and phenotype translation. Several drugs and genes that deplete or repolarize/reprogram the culprit macrophages or modulate their secretion function have also been identified. As a result, targeted delivery of these therapeutic agents to specific macrophage population emerged as a novel therapeutic strategy for the management of chronic inflammatory conditions, cancer, and neural degeneration due to secondary injury and infection. Various types of nanocarriers, including polymeric nanoparticles, lipid nanoparticles, and inorganic nanoparticles, have been designed to target drugs/genes to macrophages, and polymeric nanoparticles represent one of the most investigated nanocarriers, if not the most, because recent advancement in polymer science has enabled designing of spectral of polymer-based nanocarriers with tunable surface and bulk characteristics. Compared to lipid and other nanoparticles, generally polymeric nanoparticles are relatively stable during storage and under physiological conditions, have high drug encapsulation efficiency, and are relatively easy to prepare. Various types of natural and synthetic polymers have been used to prepare polymeric nanoparticles for targeted drug or gene delivery to macrophages and mostly the nanoparticles are mannosylated or functionalized using other targeting moieties including polysaccharides, folic acid, antibodies, and nanobodies. Interestingly, polysaccharides such as dextran can interact with dextran-binding C-type lectins, and scavenger receptors on macrophages surfaces and different polysaccharides were investigated as building blocks or surface modifiers. Similarly, hyaluronic acid specifically interacts with CD44 receptors that are overexpressed on macrophages and various nanoparticles based on hyaluronic acid have also been investigated as both building block and targeting agents. Macrophage-targeted polymeric nanoparticles can also be modified using other surface-modifying agents like PEG to prolong nanoparticle circulation time and prevent non-specific nanoparticle absorption. Many novel stimuli-responsive polymeric nanoparticles with improved drug-loading capacity, bioavailability, longer plasma half-life, and better macrophage targeting potential have also been designed and investigated both *in vitro* and *in vivo* in animals and promising results were obtained. However, the clinical translation of both conventional and stimuli-responsive macrophage-targeted polymeric nanoparticles is not up to expectation due to a few challenges including the need for phenotype-specific macrophage targeting, inefficient escape of the encapsulated therapeutic agents from endocytic pathway, alteration of nanoparticles bulk and surface properties in biological systems due to formation of protein corona, and accumulation of nonbiodegradable nanoparticles in the body. To address these challenges, attention has been given to smart nanoparticles with multiresponsive behaviors. Moreover, macrophages migrate into inflamed or tumorous tissues and recently due attention has been given to exploit them as Trojan horses to drug/gene-loaded nanocarriers.

Glossary of Abbreviations EPR Enhanced permeation and retention

IBD,	Inflammatory bowel disease
iNOS	Inducible nitric oxide synthase
IRF5	Interferon regulatory factor 5
MRC1	Macrophage mannose receptor 1
PEG	Poly(ethylene glycol)
PEI	Polyethylenimine
PGA	Polyglycolic acid
PLA	Polylactic acid
PLGA	Poly(lactic-co-glycolic acid)
RES	Reticuloendothelial system
TAMs	Tumor-associated macrophages

References

Alvarado-Vazquez PA, Bernal L, Paige CA, Grosick RL, Moracho Vilrriales C, Ferreira DW, et al. Macrophage-specific nanotechnology-driven CD163 overexpression in human macrophages results in an M2 phenotype under inflammatory conditions. Immunobiology. 2017;222(8–9):900–12.

Andrade RGD, Reis B, Costas B, Lima SAC, Reis S. Modulation of macrophages M1/M2 polarization using carbohydrate-functionalized polymeric nanoparticles. Polymers (Basel). 2020;13(1)

Arora S, Dev K, Agarwal B, Das P, Syed MA. Macrophages: their role, activation and polarization in pulmonary diseases. Immunobiology. 2018;223(4-5):383–96.

Atri C, Guerfali FZ, Laouini D. Role of human macrophage polarization in inflammation during infectious diseases. Int J Mol Sci. 2018;19(6):1801.

Bagalkot V, Badgeley MA, Kampfrath T, Deiuliis JA, Rajagopalan S, Maiseyeu A. Hybrid nanoparticles improve targeting to inflammatory macrophages through phagocytic signals. J Control Release. 2015;217:243–55.

Baslund B, Tvede N, Danneskiold-Samsoe B, Larsson P, Panayi G, Petersen J, et al. Targeting interleukin-15 in patients with rheumatoid arthritis: a proof-of-concept study. Arthritis Rheum. 2005;52(9):2686–92.

Behzadi S, Serpooshan V, Tao W, Hamaly MA, Alkawareek MY, Dreaden EC, et al. Cellular uptake of nanoparticles: journey inside the cell. Chem Soc Rev. 2017;46(14):4218–44.

Bingle L, Brown NJ, Lewis CE. The role of tumour-associated macrophages in tumour progression: implications for new anticancer therapies. J Pathol. 2002;196(3):254–65.

Binnemars-Postma K, Storm G, Prakash J. Nanomedicine strategies to target tumor-associated macrophages. Int J Mol Sci. 2017;18(5):979.

Cannarile MA, Weisser M, Jacob W, Jegg AM, Ries CH, Rüttinger D. Colony-stimulating factor 1 receptor (CSF1R) inhibitors in cancer therapy. J Immunother Cancer. 2017;5(1):53.

Chakravarthy KV, Davidson BA, Helinski JD, Ding H, Law WC, Yong KT, et al. Doxorubicin-conjugated quantum dots to target alveolar macrophages and inflammation. Nanomedicine. 2011;7(1):88–96.

Chi H, Li C, Zhao FS, Zhang L, Ng TB, Jin G, et al. Anti-tumor activity of toll-like receptor 7 agonists. Front Pharmacol. 2017;8(304)

Cohen H, Levy RJ, Gao J, Fishbein I, Kousaev V, Sosnowski S, et al. Sustained delivery and expression of DNA encapsulated in polymeric nanoparticles. Gene Ther. 2000;7(22):1896–905.

Ganbold T, Baigude H. Design of mannose-functionalized curdlan nanoparticles for macrophage-targeted siRNA delivery. ACS Appl Mater Interfaces. 2018;10(17):14463–74.

Georgoudaki A-M, Prokopec Kajsa E, Boura Vanessa F, Hellqvist E, Sohn S, Östling J, et al. Reprogramming tumor-associated macrophages by antibody targeting inhibits cancer progression and metastasis. Cell Rep. 2016;15(9):2000–11.

Germano G, Frapolli R, Belgiovine C, Anselmo A, Pesce S, Liguori M, et al. Role of macrophage targeting in the antitumor activity of trabectedin. Cancer Cell. 2013;23(2):249–62.

Gholamin S, Mitra SS, Feroze AH, Liu J, Kahn SA, Zhang M, et al. Disrupting the CD47-SIRPα anti-phagocytic axis by a humanized anti-CD47 antibody is an efficacious treatment for malignant pediatric brain tumors. Sci Transl Med. 2017;9(381)

Giménez C, de la Torre C, Gorbe M, Aznar E, Sancenón F, Murguía JR, et al. Gated mesoporous silica nanoparticles for the controlled delivery of drugs in cancer cells. Langmuir. 2015;31(12):3753–62.

Gouveia VM, Rizzello L, Nunes C, Poma A, Ruiz-Perez L, Oliveira A, et al. Macrophage targeting pH responsive polymersomes for glucocorticoid therapy. Pharmaceutics. 2019;11(11):614.

Guerriero JL, Sotayo A, Ponichtera HE, Castrillon JA, Pourzia AL, Schad S, et al. Class IIa HDAC inhibition reduces breast tumours and metastases through anti-tumour macrophages. Nature. 2017;543(7645):428–32.

Gupta MK, Meyer TA, Nelson CE, Duvall CL. Poly(PS-b-DMA) micelles for reactive oxygen species triggered drug release. J Control Release. 2012;162(3):591–8.

He X-Y, Liu B-Y, Ai S-L, Xu L, Zhuo R-X, Cheng S-X. Functional polymer/inorganic hybrid nanoparticles for macrophage targeting delivery of oligodeoxynucleotides in cancer immunotherapy. Mater Today Chem. 2017a;4:106–16.

He H, Ghosh S, Yang H. Nanomedicines for dysfunctional macrophage-associated diseases. J Control Release. 2017b;247:106–26.

He H, Yuan Q, Bie J, Wallace RL, Yannie PJ, Wang J, et al. Development of mannose functionalized dendrimeric nanoparticles for targeted delivery to macrophages: use of this platform to modulate atherosclerosis. Transl Res. 2018;193:13–30.

He W, Kapate N, Shields CW, Mitragotri S. Drug delivery to macrophages: A review of targeting drugs and drug carriers to macrophages for inflammatory diseases. Adv Drug Deliv Rev. 2020;165–166:15–40.

Huang Z, Zhang Z, Jiang Y, Zhang D, Chen J, Dong L, et al. Targeted delivery of oligonucleotides into tumor-associated macrophages for cancer immunotherapy. J Control Release. 2012;158(2):286–92.

Huang Y, Ma Y, Gao P, Yao Z. Targeting CD47: the achievements and concerns of current studies on cancer immunotherapy. J Thorac Dis. 2017;9(2):E168–e74.

Jain SK, Gupta Y, Jain A, Saxena AR, Khare P, Jain A. Mannosylated gelatin nanoparticles bearing an anti-HIV drug didanosine for site-specific delivery. Nanomed Nanotechnol Biol Med. 2008;4(1):41–8.

Jain S, Tran T-H, Amiji M. Macrophage repolarization with targeted alginate nanoparticles containing IL-10 plasmid DNA for the treatment of experimental arthritis. Biomaterials. 2015;61:162–77.

Kameka AM, Haddadi S, Jamaldeen FJ, Moinul P, He XT, Nawazdeen FHP, et al. Clodronate treatment significantly depletes macrophages in chickens. Can J Vet Res. 2014;78(4):274–82.

Kanwar RK, Chaudhary R, Tsuzuki T, Kanwar JR. Emerging engineered magnetic nanoparticulate probes for targeted MRI of atherosclerotic plaque macrophages. Nanomedicine (Lond). 2012;7(5):735–49.

Kaur A, Jain S, Tiwary AK. Mannan-coated gelatin nanoparticles for sustained and targeted delivery of didanosine: in vitro and in vivo evaluation. Acta Pharm. 2008;58(1):61–74.

Kharaji MH, Doroud D, Taheri T, Rafati S. Drug targeting to macrophages with solid lipid nanoparticles harboring paromomycin: an in vitro evaluation against L. major and L. tropica. AAPS PharmSciTech. 2016;17(5):1110–9.

Kiefer R, Jurisic M, Dahlem C, Koch M, Schmitt MJ, Kiemer AK, et al. Targeted delivery of functionalized PLGA nanoparticles to macrophages by complexation with the yeast Saccharomyces cerevisiae. Biotechnol Bioeng. 2020;117(3):776–88.

Kim N, Jiang D, Jacobi AM, Lennox KA, Rose SD, Behlke MA, et al. Synthesis and characterization of mannosylated pegylated polyethylenimine as a carrier for siRNA. Int J Pharm. 2012;427(1):123–33.

Kim J, Wilson DR, Zamboni CG, Green JJ. Targeted polymeric nanoparticles for cancer gene therapy. J Drug Target. 2015;23(7-8):627–41.

Kumaran Satyanarayanan S, El Kebir D, Soboh S, Butenko S, Sekheri M, Saadi J, et al. IFN-β is a macrophage-derived effector cytokine facilitating the resolution of bacterial inflammation. Nat Commun. 2019;10(1):3471.

Laroui H, Viennois E, Xiao B, Canup BSB, Geem D, Denning TL, et al. Fab'-bearing siRNA TNFα-loaded nanoparticles targeted to colonic macrophages offer an effective therapy for experimental colitis. J Control Release. 2014;186:41–53.

Ma L, Liu T-W, Wallig MA, Dobrucki IT, Dobrucki LW, Nelson ER, et al. Efficient targeting of adipose tissue macrophages in obesity with polysaccharide nanocarriers. ACS Nano. 2016;10(7):6952–62.

Mazur M, Karczewski J, Lodyga M, Żaba R, Adamski Z. Inhibitors of phosphodiesterase 4 (PDE 4): a new therapeutic option in the treatment of psoriasis vulgaris and psoriatic arthritis. J Dermatolog Treat. 2015;26(4):326–8.

Meng F, Sun Y, Lee RJ, Wang G, Zheng X, Zhang H, et al. Folate receptor-targeted albumin nanoparticles based on microfluidic technology to deliver cabazitaxel. Cancers. 2019;11(10):1571.

Napoletano C, Zizzari IG, Rughetti A, Rahimi H, Irimura T, Clausen H, et al. Targeting of macrophage galactose-type C-type lectin (MGL) induces DC signaling and activation. Eur J Immunol. 2012;42(4):936–45.

Niebel W, Walkenbach K, Béduneau A, Pellequer Y, Lamprecht A. Nanoparticle-based clodronate delivery mitigates murine experimental colitis. J Control Release. 2012;160(3):659–65.

Nikitina E, Larionova I, Choinzonov E, Kzhyshkowska J. Monocytes and macrophages as viral targets and reservoirs. Int J Mol Sci. 2018;19(9):2821.

Nuhn L, Bolli E, Massa S, Vandenberghe I, Movahedi K, Devreese B, et al. Targeting protumoral tumor-associated macrophages with nanobody-

functionalized nanogels through strain promoted azide alkyne cycloaddition ligation. Bioconjugate Chem. 2018;29(7):2394–405.
Orecchioni M, Ghosheh Y, Pramod AB, Ley K. Macrophage polarization: different gene signatures in M1(LPS+) vs. classically and M2(LPS–) vs. alternatively activated macrophages. Front Immunol. 2019;10(1084)
Pang L, Zhu Y, Qin J, Zhao W, Wang J. Primary M1 macrophages as multifunctional carrier combined with PLGA nanoparticle delivering anticancer drug for efficient glioma therapy. Drug Deliv. 2018;25(1):1922–31.
Panyam J, Labhasetwar V. Biodegradable nanoparticles for drug and gene delivery to cells and tissue. Adv Drug Deliv Rev. 2003;55(3):329–47.
Papa S, Ferrari R, De Paola M, Rossi F, Mariani A, Caron I, et al. Polymeric nanoparticle system to target activated microglia/macrophages in spinal cord injury. J Control Release. 2014;174:15–26.
Parayath NN, Parikh A, Amiji MM. Repolarization of tumor-associated macrophages in a genetically engineered nonsmall cell lung cancer model by intraperitoneal administration of hyaluronic acid-based nanoparticles encapsulating microrna-125b. Nano Lett. 2018;18(6):3571–9.
Park SY, Lee SW, Kim HY, Lee SY, Lee WS, Hong KW, et al. SIRT1 inhibits differentiation of monocytes to macrophages: amelioration of synovial inflammation in rheumatoid arthritis. J Mol Med (Berl). 2016;94(8):921–31.
Peres C, Matos AI, Conniot J, Sainz V, Zupančič E, Silva JM, et al. Poly(lactic acid)-based particulate systems are promising tools for immune modulation. Acta Biomater. 2017;48:41–57.
Petrova PS, Viller NN, Wong M, Pang X, Lin GH, Dodge K, et al. TTI-621 (SIRPαFc): a CD47-blocking innate immune checkpoint inhibitor with broad antitumor activity and minimal erythrocyte binding. Clin Cancer Res. 2017;23(4):1068–79.
Qiu S-Q, Waaijer SJH, Zwager MC, de Vries EGE, van der Vegt B, Schröder CP. Tumor-associated macrophages in breast cancer: innocent bystander or important player? Cancer Treatment Reviews. 2018;70:178–89.
Reichel D, Tripathi M, Perez JM. Biological effects of nanoparticles on macrophage polarization in the tumor microenvironment. Nanotheranostics. 2019;3(1):66–88.
Ries CH, Cannarile MA, Hoves S, Benz J, Wartha K, Runza V, et al. Targeting tumor-associated macrophages with anti-CSF-1R antibody reveals a strategy for cancer therapy. Cancer Cell. 2014;25(6):846–59.
Sallusto F, Cella M, Danieli C, Lanzavecchia A. Dendritic cells use macropinocytosis and the mannose receptor to concentrate macromolecules in the major histocompatibility complex class II compartment: downregulation by cytokines and bacterial products. J Exp Med. 1995;182(2):389–400.
Sandborn WJ, Su C, Sands BE, D'Haens GR, Vermeire S, Schreiber S, et al. Tofacitinib as induction and maintenance therapy for ulcerative colitis. N Engl J Med. 2017;376(18):1723–36.
Semerano L, Thiolat A, Minichiello E, Clavel G, Bessis N, Boissier M-C. Targeting IL-6 for the treatment of rheumatoid arthritis: Phase II investigational drugs. Exp Opin Investig Drugs. 2014;23(7):979–99.
Shahbazi M-A, Sedighi M, Bauleth-Ramos T, Kant K, Correia A, Poursina N, et al. Targeted reinforcement of macrophage reprogramming toward M2 polarization by IL-4-loaded hyaluronic acid particles. ACS Omega. 2018;3(12):18444–55.
Shilakari Asthana G, Asthana A, Kohli DV, Vyas SP. Mannosylated chitosan nanoparticles for delivery of antisense oligonucleotides for macrophage targeting. BioMed Res Int. 2014;2014:526391.
Shrivastava P, Gautam L, Sharma R, Dube D, Vyas S, Vyas SP. Dual antitubercular drug loaded liposomes for macrophage targeting: development, characterisation, ex vivo and in vivo assessment. J Microencapsul. 2021;38(2):108–23.
Smith DA, Conkling P, Richards DA, Nemunaitis JJ, Boyd TE, Mita AC, et al. Antitumor activity and safety of combination therapy with the Toll-like receptor 9 agonist IMO-2055, erlotinib, and bevacizumab in advanced or metastatic non-small cell lung cancer patients who have progressed following chemotherapy. Cancer Immunol Immunother. 2014;63(8):787–96.
Smith SA, Selby LI, Johnston APR, Such GK. The endosomal escape of nanoparticles: toward more efficient cellular delivery. Bioconjugate Chem. 2019;30(2):263–72.
Song M, Liu T, Shi C, Zhang X, Chen X. Bioconjugated manganese dioxide nanoparticles enhance chemotherapy response by priming tumor-associated macrophages toward M1-like phenotype and attenuating tumor hypoxia. ACS Nano. 2016;10(1):633–47.
Soto ER, Caras AC, Kut LC, Castle MK, Ostroff GR. Glucan particles for macrophage targeted delivery of nanoparticles. J Drug Deliv. 2012;2012:143524.
Steidl C, Lee T, Shah SP, Farinha P, Han G, Nayar T, et al. Tumor-associated macrophages and survival in classic Hodgkin's lymphoma. N Engl J Med. 2010;362(10):875–85.
Strachan DC, Ruffell B, Oei Y, Bissell MJ, Coussens LM, Pryer N, et al. CSF1R inhibition delays cervical and mammary tumor growth in murine models by attenuating the turnover of tumor-associated macrophages and enhancing infiltration by CD8(+) T cells. Oncoimmunology. 2013;2(12):e26968-e.
Swart M, Troeberg L. Effect of Polarization and chronic inflammation on macrophage expression of heparan sulfate proteoglycans and biosynthesis enzymes. J Histochem Cytochem. 2019;67(1):9–27.
Tao W, Yurdagul A, Kong N, Li W, Wang X, Doran AC, et al. siRNA nanoparticles targeting CaMKIIγ in lesional macrophages improve atherosclerotic plaque stability in mice. Sci Transl Med. 2020;12(553):eaay1063.
Thomas TP, Goonewardena SN, Majoros IJ, Kotlyar A, Cao Z, Leroueil PR, et al. Folate-targeted nanopar-

ticles show efficacy in the treatment of inflammatory arthritis. Arthritis Rheum. 2011;63(9):2671–80.

Toubi E, Kessel A, Slobodin G, Boulman N, Pavlotzky E, Zisman D, et al. Changes in macrophage function after rituximab treatment in patients with rheumatoid arthritis. Ann Rheum Dis. 2007;66(6):818–20.

Tran T-H, Rastogi R, Shelke J, Amiji MM. Modulation of macrophage functional polarity towards anti-inflammatory phenotype with plasmid DNA delivery in CD44 targeting hyaluronic acid nanoparticles. Sci Rep. 2015;5:16632.

Tran T-H, Krishnan S, Amiji MM. MicroRNA-223 induced repolarization of peritoneal macrophages using CD44 targeting hyaluronic acid nanoparticles for anti-inflammatory effects. PloS one. 2016;11(5):e0152024-e.

van Dam GM, Themelis G, Crane LM, Harlaar NJ, Pleijhuis RG, Kelder W, et al. Intraoperative tumor-specific fluorescence imaging in ovarian cancer by folate receptor-α targeting: first in-human results. Nat Med. 2011;17(10):1315–9.

van Roon JA, van Vuuren AJ, Wijngaarden S, Jacobs KM, Bijlsma JW, Lafeber FP, et al. Selective elimination of synovial inflammatory macrophages in rheumatoid arthritis by an Fcgamma receptor I-directed immunotoxin. Arthritis Rheum. 2003;48(5):1229–38.

van Schouwenburg PA, Rispens T, Wolbink GJ. Immunogenicity of anti-TNF biologic therapies for rheumatoid arthritis. Nat Rev Rheumatol. 2013;9(3):164–72.

Vieira ACC, Chaves LL, Pinheiro M, Lima SAC, Ferreira D, Sarmento B, et al. Mannosylated solid lipid nanoparticles for the selective delivery of rifampicin to macrophages. Artif Cells Nanomed Biotechnol. 2018;46(sup1):653–63.

Wang Y, Lin Y-X, Qiao S-L, An H-W, Ma Y, Qiao Z-Y, et al. Polymeric nanoparticles promote macrophage reversal from M2 to M1 phenotypes in the tumor microenvironment. Biomaterials. 2017;112:153–63.

Wang T, Zhang J, Hou T, Yin X, Zhang N. Selective targeting of tumor cells and tumor associated macrophages separately by twin-like core-shell nanoparticles for enhanced tumor-localized chemoimmunotherapy. Nanoscale. 2019;11(29):13934–46.

Wang P, Yang R, Liu S, Ren Y, Liu X, Wang X, et al. Thermosensitive nanoparticle of mPEG-PTMC for oligopeptide delivery into osteoclast precursors. J Bioact Compat Polym. 2020;35(4-5):426–34.

Xiao B, Laroui H, Ayyadurai S, Viennois E, Charania MA, Zhang Y, et al. Mannosylated bioreducible nanoparticle-mediated macrophage-specific TNF-α RNA interference for IBD therapy. Biomaterials. 2013;34(30):7471–82.

Xie Y, Tuguntaev RG, Mao C, Chen H, Tao Y, Wang S, et al. Stimuli-responsive polymeric nanomaterials for rheumatoid arthritis therapy. Biophys Rep. 2020;6(5):193–210.

Yan D, Kowal J, Akkari L, Schuhmacher AJ, Huse JT, West BL, et al. Inhibition of colony stimulating factor-1 receptor abrogates microenvironment-mediated therapeutic resistance in gliomas. Oncogene. 2017;36(43):6049–58.

Yu SS, Lau CM, Barham WJ, Onishko HM, Nelson CE, Li H, et al. Macrophage-specific RNA interference targeting via "click", mannosylated polymeric micelles. Mol Pharm. 2013;10(3):975–87.

Yu C, Li X, Hou Y, Meng X, Wang D, Liu J, et al. Hyaluronic acid coated acid-sensitive nanoparticles for targeted therapy of adjuvant-induced arthritis in rats. Molecules. 2019;24(1):146.

Zakeri A, Kouhbanani MAJ, Beheshtkhoo N, Beigi V, Mousavi SM, Hashemi SAR, et al. Polyethylenimine-based nanocarriers in co-delivery of drug and gene: a developing horizon. Nano Rev Exp. 2018;9(1):1488497.

Zanganeh S, Hutter G, Spitler R, Lenkov O, Mahmoudi M, Shaw A, et al. Iron oxide nanoparticles inhibit tumour growth by inducing pro-inflammatory macrophage polarization in tumour tissues. Nat Nanotechnol. 2016;11(11):986–94.

Zazo H, Colino CI, Warzecha KT, Hoss M, Gbureck U, Trautwein C, et al. Gold nanocarriers for macrophage-targeted therapy of human immunodeficiency virus. Macromol Biosci. 2017;17(3)

Zhang F, Parayath NN, Ene CI, Stephan SB, Koehne AL, Coon ME, et al. Genetic programming of macrophages to perform anti-tumor functions using targeted mRNA nanocarriers. Nat Commun. 2019;10(1):3974.

Zhang Y, Ye J, Hosseini-Nassab N, Flores A, Kalashnikova I, Paluri SL, et al. Macrophage-targeted single walled carbon nanotubes stimulate phagocytosis via pH-dependent drug release. Nano Res. 2021;14(3):762–9.

Zhao J, Zhao M, Yu C, Zhang X, Liu J, Cheng X, et al. Multifunctional folate receptor-targeting and pH-responsive nanocarriers loaded with methotrexate for treatment of rheumatoid arthritis. Int J Nanomed. 2017;12:6735–46.

Zhu S, Niu M, O'Mary H, Cui Z. Targeting of tumor-associated macrophages made possible by PEG-sheddable, mannose-modified nanoparticles. Mol Pharm. 2013;10(9):3525–30.

Surface Modification of Nanoparticles for Macrophage Targeting

Neelu Singh, Priyanka Maurya, Nidhi Mishra, Samipta Singh, Ravi Raj Pal, Priya Singh, Poonam Parashar, Alka Sonkar, and Shubhini A. Saraf

Abstract

Among various targeting techniques, macrophage targeting has achieved special consideration in the field of biomedicals. Macrophage targeting is facilitated through macrophages and is bestowed with several properties, such as greater payload at the desired site, minimum delivery to off-targets, and stability. Macrophage targeting can be achieved through various strategies such as receptor-mediated phagocytosis, cytoplasmic delivery with the help of endocytosis, and endocytic receptor-mediated active transport by proteins clathrin and caveolin. All the merits of macrophage targeting eventually prove it a suitable and potential target for numerous biomedical and therapeutic applications. In recent years, considerable literature has suggested macrophage targeting as important in various therapies such as cancer, tuberculosis, macular degeneration, and angiogenesis. Macrophage targeting is achieved through numerous nanodelivery systems such as niosomes, carbon nanotubes, liposomes, and dendrimers and has been explored for their drug delivery and therapeutic potential.

Further, these nanosystems are surface modified to improve the delivery of drugs and therapeutic efficacy of targeting. Surface modifiers include CD163, CD204, vascular endothelial growth factor (VEGF), cMAF, folic acid, mannose, etc. This chapter describes several strategies and their targeting potential based on macrophage targeting developed explicitly to deliver drugs and other therapeutic applications.

Keywords

Macrophage · Surface modification · Targeting · TAM · Ligands

N. Singh · P. Maurya · N. Mishra · S. Singh
R. R. Pal · P. Singh · P. Parashar · A. Sonkar
S. A. Saraf (✉)
Department of Pharmaceutical Sciences, Babasaheb Bhimrao Ambedkar University, Vidyavihar, Lucknow, Uttar Pradesh, India

1 Introduction

The macrophage is a type of phagocytic cell that lives in all kinds of tissues and metabolic organs, such as liver and adipose tissue, of the innate type immune system (Peterson et al. 2018). Macrophages are the resultant of the cells known as monocytic lineage precursor cells that are significant for the innate and adaptive-type immune responses. Macrophages have the unprecedented power to ingest cellular waste, foreign particles, and stressed cells, and thus, they act as the body's primary scavenger cells. They can, therefore, maintain cellular homeostasis and immune surveillance. Macrophages also serve as the essential linkers for

S. Gupta, Y. V. Pathak (eds.), *Macrophage Targeted Delivery Systems*,
https://doi.org/10.1007/978-3-030-84164-5_8

adaptive immunity by using antigen processing, presentation, and consequently, T-lymphocyte priming. Macrophages are of two types – M1 cells (classically activated) and M2 cells (alternatively activated) – based on their exposure to certain types of microbial stimuli, such as lipopolysaccharides and cytokines (interferon-γ, interleukin-4, and interleukin-10). The main function of the macrophage phenotypes is their capability to repress inflammation, forage debris, and encourage tissue repair mechanisms with the help of the phagocytosis phenomenon. Various receptors, such as opsonic receptor CD16 and mannose receptor present in macrophages, help target several therapeutic agents (Qie et al. 2016).

2 Strategies for Macrophage Targeting

Receptor-mediated phagocytosis is one of the essential macrophage-targeted therapeutic strategies. Provided surface modifications are done on compounds so that they can target receptors available on macrophages. The receptors expressed on macrophages are F4/80, CD11b, CD68, mannose, lectin, adenosine, and folate. The receptor-focused approach offers direct import into macrophages and, as a result, minimizes off-target effects.

Different methods of internalization can achieve delivery of therapeutic agents:

(i) Activation of therapeutic agents or compounds can be done in the acidic lysosomal compartment. The first step toward this process is the fusion with lysosomes of destructive nature before they are recognized, engulfed, and entrapped into phagosomes. The therapeutic activity can be initiated by utilizing the acidic microenvironment of lysosomes or phagosomes, but it is also dependent on nanoparticle encapsulation approaches. The acidic environment within the lysosome triggers the rupture of the nanoparticle capsule to release its contents.

(ii) Cytoplasmic delivery with the aid of endocytosis and lysosomal escape is the second approach. This helps to circumvent phagocytic destruction. The particles are also prepared in such a manner that they can escape phagosomes following endocytosis but just before fusion with the lysosome (Placha and Jampilek 2021).

(iii) The third strategy involves utilizing endocytic receptors that exploit macrophage-mediated active transport proteins like clathrin and caveolin. Phenotype/function-altering cargo can be delivered in a Trojan horse when the particle enters into the intracellular space (Peterson et al. 2018).

3 Drug Delivery to the Macrophages Via Targeting Approach

Targeted drug delivery can be achieved by following the surface modification approach on the therapeutic agents. Macrophages can be subjugated as Trojan horses meant for drug delivery via targeting methods. Nanocarriers can wander through various membrane barriers and can discharge their drug cargo at targeting sites of infection or wherever targeting is required.

Various advantages of nanotechnology-based delivery systems of the drug are as follows:

(a) Improved solubility
(b) Enhanced bioavailability
(c) Elevated drug payload
(d) Prolonged drug half-life
(e) Enhanced therapeutic index
(f) Controlled release of therapeutic agents
(g) Reduced toxicity and immunogenicity.

On any given tissue, there are several sites for receptor, therefore, targeted delivery is predominantly a smart approach for the bioactive material delivery, which has a narrow therapeutic window or concentration is very low at the targeting site (Jain et al. 2013).

4 Macrophage Targeting Via Surface-Engineered Nanocarriers

Nanocarriers are the drug delivery systems that can deliver the drug at the targeting site and thus can recover the efficacy and diminish the off-target side-effects of drugs (Ng et al. 2020). Therefore, if a drug delivery platform is developed, which is capable of modulating polarized macrophages and targets the sites, then it possesses therapeutic importance (He et al. 2020). (Fig. 1)

4.1 Microparticles/Microspheres-Mediated Macrophage Targeting

Microspheres are solid particles of spherical shape with a size range from 1 to 1000 μm (Swati et al. 2020). Rotman et al. formulated antibiotic microspheres to manage bone infections and targeted the drug delivery to the receptor site. The prepared antibiotic-loaded polycaprolactone (PCL) and poly-D,L-lactic acid (PDLLA) microspheres showed sustained in vitro release of antibiotics with an antimicrobial potential against the OM microbial and *S. aureus* for 5 days (Bai et al. 2019) (Rotman et al. 2020).

4.2 Macrophage Targeting Via Liposomes

Spherical vesicles made up of lipids that can assemble into amphiphilic bilayers similar to the composition and structure of cell membranes are called liposomes. Liposome surface properties are determined by the hydrophilic head, whereas the hydrophobic tail present interior to the bilayer tells about the fluidity of liposomal membranes (Wang et al. 2020a).

Medina et al. formulated alpha-melanocyte-stimulating hormone (α-MSH)-targeted liposomal nanoparticle for imaging in inflammatory bowel disease. NPD-αMSH labeled with Alexa and liposomes were developed and were subsequently identified in the kidneys and inflamed bowel regions, where melanocortin-1 receptor (MC1-R) is available in abundance (Peñate-Medina et al. 2020).

4.3 Nanoparticles-Mediated Macrophage Targeting

Nanoparticles are solid particles or particulate dispersions with a diameter ranging from 10 to 1000 nm. The drug is either entrapped, dissolved, attached, or encapsulated to a nanoparticle matrix (Fernando et al. 2018).

Yang et al. formulated macrophage-targeting silver nanoparticles to treat rheumatoid arthritis. The prepared folic acid-functionalized silver nanoparticles (FA-AgNPs) demonstrated decreased inflammation by synergistic reduction in M1 macrophages and M2 macrophage induction for improved rheumatoid arthritis treatment. The stability and opsonization of nonspecific targets that enable passive accumulation was further improved by PEGylation of FA-AgNPs (Nemeth et al. 2007) (Yang et al. 2020a).

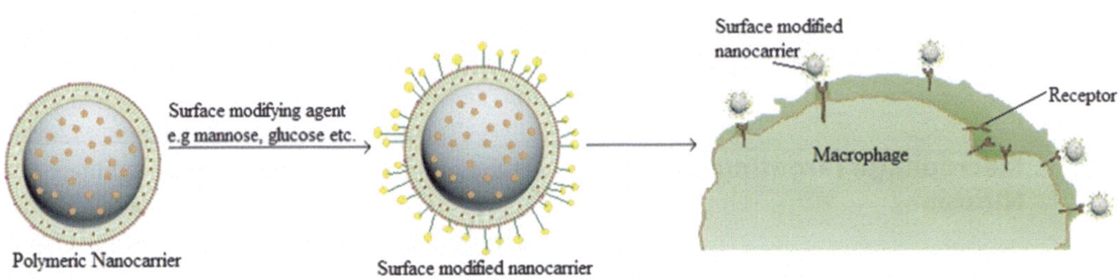

Fig. 1 Diagrammatic representation of nanocarriers and possible receptor-mediated binding of surface-modified nanocarriers with macrophages

4.4 Macrophage Targeting Via Dendrimers

Dendrimers are novel polymeric 3D-structured systems, which are characterized by its hyperbranched structure. Numerous functional groups are present on the surface, enhancing their functional properties, and making them biocompatible and versatile (Sherje et al. 2018).

Sharma et al. formulated and evaluated 5,7-dimethylpyrazolo[1,5-α]pyrimidin-3-ylacetamide (DPA) conjugated dendrimer to target targeting tumour-associated macrophages (TAMs) within the mitochondria. Results of systemic administration have eventually ascertained that dendrimers can achieve TAM-specific targeting in glioblastoma and may be better suited for organelle-specific drug delivery (Sharma et al. 2020).

4.5 Macrophage Targeting Via Carbon Nanotubes (CNTs)

CNTs are the one-dimensional carbon nanomaterials with a huge length-to-diameter ratio with sp^2-hybridized structures considerably superior to that of any other carbon nanomaterial (Wang et al. 2020b).

Zhang et al. prepared single-walled carbon nanotubes that stimulate phagocytosis by the use of pH-dependent release of drug and target the drug delivery to macrophages which showed that single-walled carbon nanotubes (SWNTs) can convey tyrosine phosphatase inhibitor 1 (TPI) intracellularly to macrophages and can stimulate efferocytosis; thus, proving that SWNT can deliver TPI that stimulates macrophage efferocytosis, with the prospective to decrease or prevent atherosclerotic disease (Zhang et al. 2020).

4.6 Macrophage Targeting Via Niosomes

Niosomes are self-assembled nanosized lamellar vesicles made up of nonionic surfactants and cholesterol with hydration in aqueous media (Singh et al. 2017). Singh et al. formulated niosomes for the delivery of isoniazid for efficient management of tuberculosis. Cellular uptake studies of the drug-loaded niosomes showed 61.8% uptake by macrophage cells. The prepared niosome formulation displayed sustained release for up to 30 h and the ability to remain at the treated site for extended periods (Singh et al. 2011).

4.7 Macrophage Targeting Via Polymerosomes

Artificial vesicles enclosing an aqueous cavity, made of self-assembly of amphiphilic copolymers, are known as polymerosomes (Zhang and Zhang 2017).

Gouveia et al. formulated pH-responsive polymerosomes for glucocorticoid therapy using macrophage-targeting approach showing inflammation-activated NF-B signaling pathway was fastened in inflamed macrophages treated with prednisolone disodium phosphate (PDP) loaded within polymerosomes which resulted in reduced expression of proinflammatory genes and proteins. pH-responsive polymerosomes ensure stability and selectivity toward targeted cells and promote inflammation resolution in vitro (Gouveia et al. 2019; Biswas and Mantovani 2014). (Fig. 2; Table 1)

4.8 Surface Modification Strategies for Anticancer Purposes

Tumor stroma contains macrophages as one of their essential components (50% of tumor mass in some cases), involved in the rejection, promotion, and metastasis of the tumor (Nemeth et al. 2007). Tumor-associated macrophages (TAMs) aid the promotion of tumors. They facilitate the suppression of the immune system, angiogenesis, and inflammation. They also influence the relapse of tumor after conventional therapies (Poh and Ernst 2018). TAMs can be differentiated into M1 and M2 macrophages (Fig. 3), and as per tumor

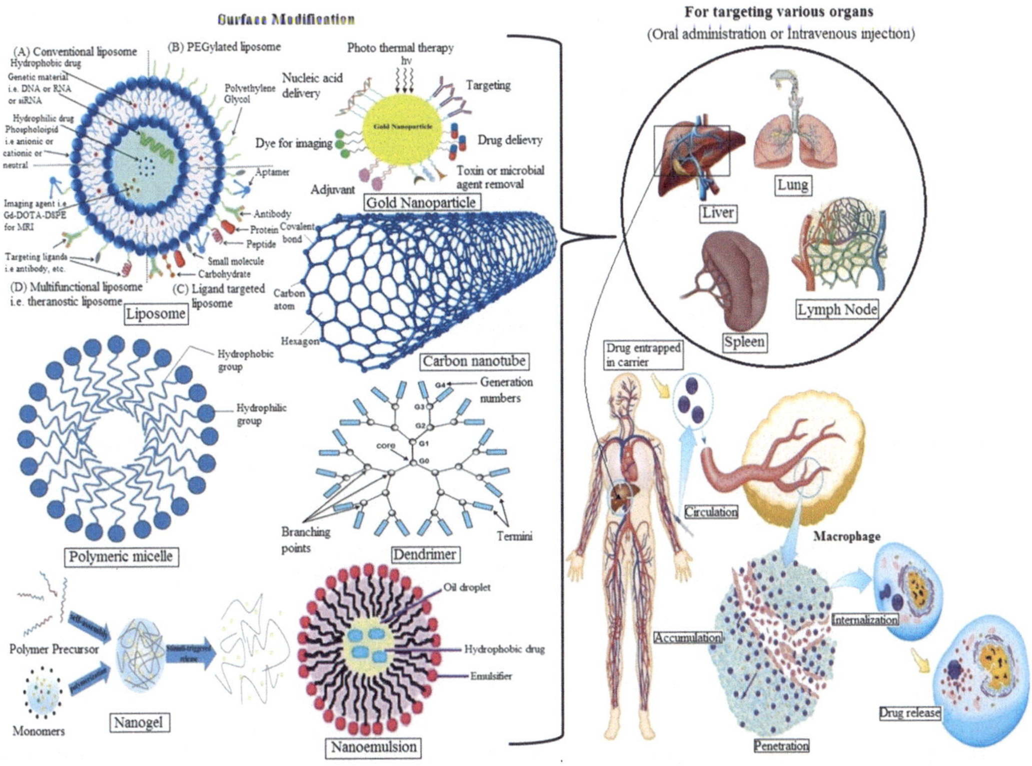

Fig. 2 A graphical outline of treatment-based nanotechnology. Different nanocarriers can target vital organs after the surface alteration. The parasites penetrate the human skin and are absorbed into the bloodstream where they travel via the blood vessels. The administration of a nanotechnology-based drug (oral or intravenous injection) results in the membrane (tegument) of the microbes being broken, releasing the drug to destroy the microbes

microenvironment, relative abundance varies (Kim et al. 2020). M1 TAMs suppress the progression of cancer, whereas M2 TAMs promote it (Zhang et al. 2014).

The surface receptors that TAM overexpresses are exploited (or can be potentially exploited) for the surface-modified nano-based delivery system to target TAM actively. Various such surface receptors are enumerated below (Binnemars-Postma et al. 2017).

4.8.1 CD163

It is a hemoglobin scavenger receptor (Etzerodt et al. 2012) and is considered a marker of alternatively activated macrophages. Postactivation it is cleaved from the macrophage surface, resulting in a soluble CD163 (sCD163) (Matute-Blanch et al. 2018). CD163+ TAMs accumulation is reported to be related to poor survival in breast cancer patients (Ramos et al. 2020; Ishida et al. 2015). It also affects the prognosis or survival in patients with other types of cancers such as oral squamous cell carcinoma (OSCC) (He et al. 2014) and leukemia/lymphoma (Komohara et al. 2013).

4.8.2 CD204

Scavenger receptor class A (SR-A or CD204) is overexpressed on macrophages (Miyasato et al. 2017). Apart from CD163+, CD204 + TAMs also promote the apoptosis of T cell and causes immunosuppression via IL-10 in OSCC patients (Haque et al. 2019), and the aggressiveness of tumor in lung adenocarcinoma (Ohtaki et al. 2010).

4.8.3 CD206

CD206, alternatively called mannose receptor, is a C-type lectin commonly expressed by tissue macrophages, specific lymphatic cells, dendritic

Table 1 Targeting moieties and techniques used for targeting macrophages and their merits, demerits, and applications

Targeting moieties	Techniques used for targeting macrophages	Merits	Demerits	Applications	References
VEGF receptor	Ligands' systematic evolution through exponential enrichment	Possible to develop for any target	Extremely high production cost	Treat age-related macular degeneration of the retina	Stein and Castanotto (2017)
Transferrin receptor	Antibody-mediated targeting	In clinical trials currently	High production cost	Treatment of malignancies via therapeutic agents through receptor-mediated targeting	Luria-Pérez et al. (2016)
Folate receptor	Antibody farletuzumab (anti-FR), folate conjugates	Simple chemistry, low molecular weight, and low production cost	Could reduce circulation time	Recognition of α and β isoforms of folate receptor (FR)	Ledermann et al. (2015)
Human epidermal growth factor receptor 2 (HER2) Cluster of differentiation 4 (CD 4) CD 4 (T4) antigen	Antibody-drug conjugate	High affinity and strong binding, therapeutic potential	High production cost, pharmacokinetics, binding site, barrier effect, and potential immunogenicity	Inhibition of angiogenesis and decreased downstream signaling by phosphatidylinositol 3-kinase and protein kinase activated by mitogens	Kagan et al. (2015) and Barginear et al. (2012)
$\alpha_v\beta_3$ integrins and aminopeptidase N	[^{18}F] Galacto-RGD	High affinity	Reduced circulation half-life	Provides valuable information for antiangiogenic therapy preparation and control aimed at targeting	Gregorc et al. (2009) and Haubner et al. (2005)

cells, or endothelial cells. It plays a crucial role in maintaining immune homeostasis. The elevated expression of CD206 is detected in the microenvironment of the tumor (Haque et al. 2019) in OSCC patients.

4.8.4 VEGF

VEGF-A is an important proangiogenic cytokine released by TAM. The levels of VEGF-A correlate with TAM density in several types of human cancer (Kzhyshkowska et al. 2014).

4.8.5 cMAF

c-Maf is reported to be crucial for the self-renewal of macrophages. Liu et al. demonstrated that c-Maf is an important molecular checkpoint that controls the immunosuppressive macrophage polarization and function in cancer. Furthermore, the c-Maf expression is high in TAMs and governs TAM's immunosuppressive function. Deletion of c-Maf in myeloid cells has decreased tumor load with enhanced immunity of the anti-tumor T cells (Liu et al. 2020).

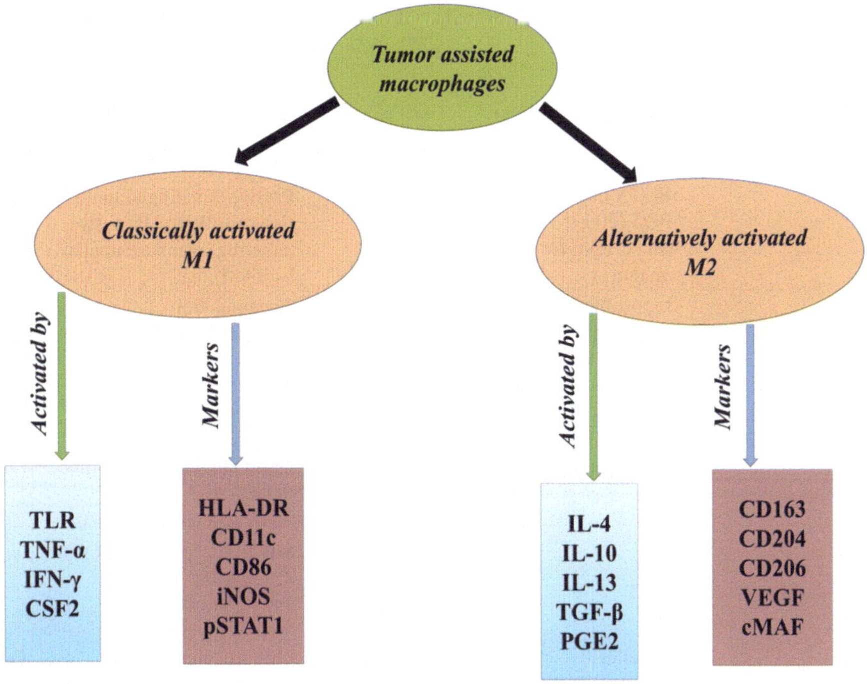

Fig. 3 Common immunohistochemistry markers to distinguish M1 and M2 TAMs and how M1 or M2 macrophages are activated (Jayasingam et al. 2019)

4.9 TAM Targeting Ligands for Nanoparticle Surface Modification

Targeting ligands for nanoparticle surface modification to actively target TAM include small molecules, oligomers, and macromolecules.

4.9.1 Small Molecules

Examples include mannose, folic acid, etc. Specific ligand-receptor interaction aids in preferential delivery of cargos to the TAM via specific small molecules-modified nanoparticles. Mannose receptor (CD206) is among commonly targeted receptors (Zhu et al. 2013). A simple carbohydrate, mannose, can easily be conjugated as a ligand to the nanocarriers (Locke et al. 2012). Folate receptor beta (FRβ) mRNA is also expressed in TAM, which can be targeted via the use of nanoparticle carrying folic acid as a ligand (Ravindra et al. 2018; Hattori et al. 2015). To improve FRβ specificity, various antifolate analogs can be used (Golani et al. 2016). Several other examples are briefed in Table 2.

4.9.2 Oligomers/Saccharide

Legumain substrate, peptides, Lyp-1, mannan, and IL4Rα-targeted aptamer are also used to modify nanoparticles (Ngambenjawong et al. 2017) to target their specific receptors. Legumain's proteolytic activity specific to alanine-alanine-asparagine substrate sequence is a prodrug/smart DDS approach that targets both TAMs and cancer cells (Liu et al. 2014). Other studies are briefed in Table 2.

4.9.3 Macromolecules

Transferrin, antibodies or nanobodies are few common macromolecules used for TAM targeting (Ngambenjawong et al. 2017). Apart from mannose, the anti-CD206 antibody is also exploited to target CD-206+ TAMs (Sun et al. 2015). Modification with anti-FRβ Fv can be helpful in selectively targeting the FRβ + TAMs (Nagai et al. 2009). Nanobodies (15 kDa) are the smallest fragment that binds antigen and is derived from camelid heavy-chain-only antibodies. High affinity, stability, solubility, and increased tissue penetration make them suitable

Table 2 Different ligands used for surface modification to target TAM

TAM targeting agent	Nanoparticle	Targeted	Outcome	References
Small molecules:				
Mannose	Polymeric micelles (to deliver siRNA)	CD206	Enhanced uptake in lung metastasis-associated macrophages	Ortega et al. (2015)
Mannose	PLGA masked with PEG (2000) (FITC used for imaging)	CD206	Passive targeting to tumor tissues followed by active targeting (mannose uptake by TAM)	Zhu et al. (2013)
Mannotriose	Liposomes (^{64}Cu labeled, DOTA encapsulated)	CD206	Accumulation of mannosylated liposomes in TAMs and little accumulation in remote lung areas	Locke et al. (2012)
Mannose	PLGA NPs (drug: Doxorubicin)	CD206	Mannosylated NPs were found to be more effective in mice that were not pretreated with zoledronic acid (ZA; pretreatment with ZA was done to deplete macrophages nonselectively)	Niu et al. (2016)
Mannose	Micelles (drug: siRNA)	CD206	- Modify signaling pathways (using siRNA) - Enhanced uptake	Ortega et al. (2016)
Mannose	Polymer-coated iron oxide NPs	CD206	- Decreased the transverse relaxation time in tumors (as determined by MRI imaging)	Li et al. (2020)
Folic acid	Liposomes (zoledronic acid loaded)	FRβ	- High cytotoxicity against nasopharyngeal KB (tumor) cells and RAW 264.7 (murine macrophage) cells - Limitation: In vivo toxicity of zoledronic acid liposomes	Hattori et al. (2015)
Oligomer/saccharide:				
Ala-Ala-Asp	Liposomes (drug: Doxorubicin)	Legumain	- Tumoricidal effect of doxorubicin increased - Reduction in systemic adverse effects	Liu et al. (2014)
nRGD (Ala-Ala-Asp conjugated to iRGD)	Liposomes (drug: Doxorubicin)	Legumain	- Antitumor efficacy improvement - TAM depletion	Song et al. (2016)
RR-11a	Liposomes (drug: Hydrazinocurcumin)	Legumain	- Reprogramming of TAM to M1 phenotype	Zhang et al. (2013)
Peptide	Gold nanoparticles (drug: siRNA)	M2 (via ligand) VEGF (via iRNA)	- Immune modulation and tumor suppression - Reduced inflammatory TAM in lung tumor tissues	Conde et al. (2015)
β-(1 → 3)-(1 → 4)-glucan	Glucan-based nanoparticle MIF siRNA	CD11b	TAM reprogramming murine 4 T1 breast cancer	Zhang and Kim (2012)

(continued)

Table 2 (continued)

TAM targeting agent	Nanoparticle	Targeted	Outcome	References
Oligonucleotides (CpG, anti-IL-10, and anti-IL-10R), galactosylated dextran in combination	Alginate-based nanoparticles; oligonucleotides	Macrophage galactose-type lectin	- Inhibition of protumoral functions - TAM reprogramming and initiation of the immune response -Stimulation of antitumor activities	Huang et al. (2012)
Macromolecules:				
CD163-specific monoclonal antibody	Liposomes; (fluorescent label, doxorubicin)	CD163	- Increased uptake - Strong cytotoxic effect in CD-163+ monocytes	Etzerodt et al. (2012)
CD163 antibody	Au/SiO2 nanoparticles	CD163	Decreased M2 and increased M1 TAM	Kim et al. (2020)
Polysaccharide from Bletilla striata/MR	Polysaccharide from Bletilla striata-alendronate conjugate	CD206	TAM elimination in murine S180 sarcoma	Zhan et al. (2014)
Rabies virus glycoprotein	PLGA core with a mixed lipid-coating paclitaxel	Nicotinic acetylcholine receptor	TAM depletion in U87 glioma xenograft	Zhan et al. (2014)

to target tumors (Schoonooghe et al. 2012; Movahedi et al. 2012). The avidity of a bivalent nanoformulation for competitively binding to the resident macrophages is higher than that of the monovalent nanoformulation. However, because of its bigger size, its penetration into the tumor is lesser than that of the monovalent nanoformulation, which can diffuse into the tumor (Ngambenjawong et al. 2017). Several other examples are briefed in Table 2.

5 Surface Modification Strategies for Lung Infections Via Surface-Modified Nanoparticles Targeting Macrophages

Lungs carry two distinct macrophage populations, that is, alveolar and interstitial. The alveolar macrophage resides in the alveolar epithelium, whereas the interstitial ones reside in the parenchyma sandwiched between microvascular endothelium and alveolar epithelium (Hu and Christman 2019). The targeting strategies include selecting functionalizing biocompatible agents that possess specificity to the macrophage populations intended to be treated. Various functionalizing agents such as mannose, tuftsin, mannan, Labrafil, hyaluronic acid, and PEGylation are used to stabilize and surface functionalize the formulations for better targeting efficiency (Hu and Christman 2019; Horváti et al. 2018; Leber et al. 2019; Ma et al. 2020a; Marcianes et al. 2019; Muppidi et al. 2011; Parashar et al. 2018; Shen et al. 2015; Yu et al. 2010).

Mannosylation of the formulation is found to be of utmost importance in terms of surface functionalization and can be utilized to deliver drugs to macrophages. Ma et al. mannosylated their formulation with 6-octadecylimino-hexane-1, 2, 3, 4, 5-pentanol and achieved 97.2% uptake compared to 42.4% uptake in unmodified formulations (Ma et al. 2020b). The modified formulation improved the intracellular activity of the drug four times more than that of the unmodified formulation (Ma et al. 2020a). Leber et al. in 2019 had prepared mannosylated nanogel containing siRNA for cancer. The functionalized nanocarriers (20 nm) effectively delivered siRNA to immunocompromised macrophages as evidenced by uptake studies. The nanocarriers were characterized for their efficacy by studying gene knockdown (CSF-1R) in M2 macrophages. They were stable enough to deliver the siRNA payload to the target macrophage and successfully provided combinatorial therapy (Leber et al. 2019).

Cell targeting ability of DNA-incorporated SLNs can be improved by modifying the formulation with mannan. The transfection ability was assessed on the RAW 264.7 cell line. The in vivo activity was checked on delivering the formulation to the lungs and evaluating its transfection ability. The functionalized formulations possessed high gene expression when compared to the unmodified formulation. The targeted modified formulation, therefore, sets prospects for gene delivery (Yu et al. 2010).

Hyaluronic acid-modified polycaprolactone nanoparticles formulated by Parashar et al. through layer-by-layer technique were employed to functionalize the prepared nanoparticles. The cell uptake study done on A549 cells showed enhanced uptake, followed by cell cycle analysis, which revealed better cytotoxic effects and active targeting. The formulation showed promise in delivering anticancer compound naringenin to the lung cancer tissues against urethane-induced carcinoma (Parashar et al. 2018).

Tuftsin, a tetrapeptide present in immunoglobulin G, was employed by Horvati et al. in conjunction with pluronic F127 for functionalizing PLGA nanoparticles. The binding of pluronic to tuftsin is a single-step reaction. The modification efficiently increases the internalization of the formulation, thereby increasing its activity. The strategy can be easily employed to target mycobacterium tuberculosis. Pluronic acts as a surface modifier to prevent aggregation and is also biocompatible (Horváti et al. 2018).

Labrafil, a nonionic surfactant, is also utilized to improve the targeting ability toward macrophages. Gatifloxacin-entrapped PLGA microparticles were prepared by Marcianes et al. The uptake was studied by preparing fluorescein-loaded microparticles, and RAW 264.7 cells were used to study the uptake. The study showed that the microparticles were internalized rapidly within 3 hrs and remained inside for 2 days (Marcianes et al. 2019; Ma et al. 2020b).

PEGylation is another technique to improve the internalization of nanoparticles in macrophages and also improve the circulation time (Shen et al. 2015). Treatment of methicillin-resistant *Staphylococcus aureus* (MRSA) (Hibbitts and O'Leary 2018) related pneumonia is limited by poor drug penetration and the limited ability of the drug to achieve bactericidal effects. To encounter this shortcoming, Muppidi et al. prepared PEGylated liposomes containing vancomycin. PEGylated formulations possessed better uptake in lung tissues as well as increased circulation time. The improved lung targeting was further amplified with its limited renal parenchymal exposure. This improved the MRSA-associated pneumonia and attenuated the risk of renal damage (Muppidi et al. 2011).

6 Surface Modification Strategies for Inflammation

It has been established that macrophages are an indispensable part of the immune system with the pivotal role of removing foreign pathogens. Activated macrophage triggers chronic inflammation. In general, macrophages remain activated in various pathological conditions and display a wide range of surface receptors that have an affinity toward modified polysaccharides and lipoproteins (Singh et al. 2020), like mannose receptor, Fc gamma receptor, glucosamine carbohydrate receptors, europilin-1 receptors, N-acetyl glucosamine receptors, and pattern recognition receptors. Based on these receptors and their opsonization process, the surface of the nanoparticle is often modified. Nanoparticles have a core for drug loading as well as multiple binding sites for surface modification.

Nanoparticle surface modification for macrophage targeting can be done before or after nanoparticle formation using various techniques, such as electrostatic binding, physical adsorption, and covalent coupling. Several coating agents are utilized based on their binding affinity with the receptor on the macrophage surface. The materials like chitosan have a binding affinity with glucosamine carbohydrate receptor, lactoferrin, mannose; O-palmitoyl mannan has an affinity with mannose receptor sodium alginate has binding affinity Fc and europilin-1 receptors;

lectin has a binding affinity with N-acetyl glucosamine) receptors tuftsin has a binding affinity with Fc-receptor and stearylamine has a binding affinity with pattern recognition receptors (Kumar Singh et al. 2020). Macrophage-targeted chitosan-coated PLGA nanoparticles of doxorubicin and amphotericin B were prepared through the nanoprecipitation method against visceral leishmaniasis (Singh et al. 2016). A 1,2-diacyl-sn-glycero-3-phospho-l-serine (PS)-coated gelatin nanoparticles were reported for macrophages targeting (Khatik et al. 2014), and lactoferrin-coated nanoreservoir for macrophage-specific delivery (Asthana et al. 2015).

7 Surface Modification of Nanoformulations for the Effective Delivery of Drugs to the Macrophage Site

Recently, surface-modified material has gained increasing attention from scientists worldwide since surface-modified nanoformulation specifically targets the drug delivery site. These systems are capable of delivering lipophilic and hydrophilic drugs in a controlled manner. The macrophage is a critical element in inflammatory diseases. The mechanism that initiated macrophage-mediated inflammation dramatically influences the development and outcome of inflammatory diseases. Proinflammatory macrophages initiate vascular inflammation and oxidative stress, while activated macrophages accelerate the inflammatory process through various proinflammatory mediators. This nature of macrophage makes it a prime target for drug delivery to ameliorate macrophage-mediated inflammation (Kalyane et al. 2019).

Currently, nanotechnology is providing a robust platform to deliver drug candidates from tissue to the cellular level. Activated macrophages initiate multiple inflammatory conditions, so it becomes a target for drug delivery and imaging application. Multiple nanoformulations, specifically macrophage targeted, were developed and patented (Table 3). Based on inflammatory conditions, the surface architecture of the nanoparticle is modified for effective site targeting, which is often ignored in designing the traditional formulation.

These surface-modified nanoparticles specifically bind to targeted structures by utilizing a physiological microenvironment. In inflammatory conditions, permeability is enhanced due to leaky vasculature. This leaky vasculature favors nanoparticle penetration as well as retention within the target structure[1]. Every pathological condition is associated with a variety of inflammatory mediator which is over- or underexpressed. These overexpressed molecules utilized for surface modification of nanoparticle like the folate receptor is overexpressed in several types of cancers. In this case, the folic acid-coated nanoparticle is utilized for cancer cell-specific delivery (Sandeep et al. 2020).

The surface-modified nanoparticles have precisely delivered the drug at the target site; moreover, they may also utilize for theranostics purposes, including molecular imaging. The unmodified dextran nanometer-sized conjugates [WO2017100697A1WIPO (PCT)] and multicompartmental nanoparticle [US10500156B2 United States] are patented for macrophage-specific therapy and imaging. The acrylyl oxyethyl phosphocholine surface-modified magnetic nanoparticle (CN102552945A China) and superparamagnetic ultrasmall iron oxide nanoparticles were [US20140249413A1 United States] patented and have been reported to detect and kill activated macrophage, respectively. A suture made up of mannose-coated nanoparticles and loaded with the antiinflammatory drug was patented [KR101925442B1 South Korea] as a medical device for macrophage-specific drug delivery and surgical applications. An invention [CA2735318C Canada] related to designing surface-modified nanoparticles was patented for delivering immunotherapeutics to the antigen-presenting cells. Similarly, surface-modified nanoparticles of organic materials, organometallic materials, inorganic materials, and metals and metal oxides were also recorded [US8642787B2 United States]. The patent of polysaccharide microparticles was recorded for alveolar macrophage

Table 3 Patented macrophage-targeted nanoformulations

S. No	Claim product and activity	Patent no.	Inventor	References
1.	This relates to macrophage-specific nanometer-sized nanoparticles made from unmodified dextran (DNPsssleder Edmund J. Keliher Matthias Nahrendorf	(C:\users\G publishing\ downloads\86 - _ENREF_86)		
2.	The patent-related detecting-activated macrophase in a subject through superparamagnetic ultra-small iron oxide nanoparticles	US20140249413A1 United States	Gerald L. Wolf Karl F. Schmidt	Wolf and Schmidt (2014)
3.	A patent related to imaging and targeted drug delivery to macrophage by nanoparticulate	US10500156B2 United States	Mansoor M. Amiji Mayur Kalariya Shardool Jain Husain Attarwala	Amiji et al. (2019)
4.	The invention is related to designing surface-modified nanoparticles for delivering immunotherapeutics to the antigen-presenting cells	US832696B2 United States	Jeffrey A. Hubbell Conlin P. O'Neil Sai T. Reddy Melody A. Swartz Diana Velluto Andre van Der Vlies Eleonora Simeoni	Hubbell et al. (2012)
5.	The patent related to the surface-modified nanoparticle of organic materials, organometallic materials, inorganic materials, metals, metal oxides, and combinations for drug delivery	US8642787B2 United States	Kazuki Fukushima James Lupton Hedrick Alshakim Nelson Daniel Paul Sanders	Fukushima et al. (2014)

targeting [WO2016174573A1WIPO (PCT)]. These surface-modified nanoparticles provide a variety of applications from drug delivery to diagnosis, monitoring, and inflammatory disease management.

8 Macrophage Targeting in Infectious Diseases/ Leishmaniasis Therapy

The macrophages form an essential fragment of the mononuclear phagocyte and are composed of cells closely associated with their bone marrow sources, including blood monocytes and tissue macros. Monocytes migrate from blood to various tissues, transforming macrophages. Macrophages have three key roles in the development of innumerable cytokines and growth factors; antigen presentation, phagocytoses, and immunomodulation. In inducing, maintaining, and resolving inflammation, macrophages play a vital role (Fujiwara and Kobayashi 2005). It is generally recognized that chronic inflammation is linked to a range of diseases, such as diabetes, obesity, atherosclerosis, rheumatoid arthritis, and cancer. Besides, macrophages are known as the critical components of chronic inflammation in most diseases of human beings (Ponzoni et al. 2018). The two main obstacles in close relation with the majority of microbial infectious diseases are drug resistance and drug toxic. The two main obstacles in close relation with the majority of microbial infectious diseases are drug resistance and drug toxicity. The development of many microbial diseases that affect visceral organisms such as the liver and spleen is affected by the macrophages. The intrinsic mechanism can, in theory, interfere and promote various microbial infections by a macrophage (or circulating monocytes), usually called "scavengers" (since it is capable of extracting exogenous materials from the systemic circulation). Visceral leishmaniasis (VL) is an incurable disease caused by *Leishmania donovani* that has fatal effects on the spleen, and lymphatic node expansion due to parasites' build-up (Kunjachan et al. 2011). Zaslona and colleagues have shown that alveolar macrophages show a role in asthma and that alveolar macrophages play a defensive role in the early stage of allergic pulmonary inflammation. Conversely, it has been reported that recruited monocytes, the alveolar macrophages, encourage lung inflammation through asthma pathogenesis (Organization WH 2013).

In the innate as well as adaptive immune responses, macrophages serve a significant role as these cells learn multiple ways to detect the presence of pathogens in any body tissue. Pathogenically related molecular patterns (PAMPs), which are molecules related to pathogenic groups, are engineered to discriminate against the presence of toll-like receptors (TLRs). The molecules in the macrophage and other immune systems cells may be called cytokines. Components of bacterial surfaces invite a response from the innate receptors of TLR and PRR. These components are recognized. Innate receptors allow the microbicide role of the macrophage to be acquired against the ligands of pathogens (Organization WH 2013).

The key feature of drug targeting is designing a device capable of selectively delivering the drug to the target site (Dave et al. 2020). The approach to drug delivery targeted by macrophages can help as a skilful means of overcoming many of the above problems. Localization of the leishmania parasite inside the macrophage phagolysosome limits the bioavailability of certain antileishmania drugs that are potentially useful. For these parasites, macrophages act as host cells that can prevent the growth of phagosomes to live and reproduce inside the macrophages. However, it is difficult to access these comparatively inaccessible sites selectively; therefore, the subject of concern is often macrophage-specific drug delivery. The idea of "magic bullet," a new notion of drug targeting paradigm, exploits the use of surface-engineered vehicles for site-specific delivery, thus becomes imperative to exploit. Also, a targeting system must be configured to hold in mind the ultimate drug site of action, macrophages in the present case (Rittig and Bogdan 2000). Thus, the carrier should be so conditioned in the scenario that it is either passively or otherwise actively guided to the buffer cells (Croft 1986).

Macrophages are inflammatory regulators for many infectious diseases because they isolate many inflammatory mediators and function as the potential pharmaceutical goal for various species and human diseases. While certain microorganisms such as *Toxoplasma gondii, Leishmania sp., Mycobacterium tuberculosis,* and *Listeria monocytogenes* have gained the ability to resist such phagocytosis, carriers may be suitable for the removal of such cellular reservoirs by supplying the antimicrobial agent(s) in pathogen-contained intracellular vacuoles. This can minimize the administered side-effects and the release of pro-inflammatory cytokines, thus decreasing therapeutic concentration in infected macrophage vacuoles (Anne Février et al. 1999). Therefore, nanosystems built using the above approach can be very effective in targeting diseases such as leishmaniasis in macrophages. The increased permeability of blood vessels due to infectious diseases produces leaky blood vasculature that can move from the walls of the blood vessels and intestinal areas by 10–500 nm of diametric particles. The existence of the condition affects the porosity of the vasculature, which allows regulation of the spread of the drug, which in turn will allow the drug to be withdrawn from the blood vessel by choosing proper carrier dimensions. This is called "enhanced permeation and retention" (EPR) (Anne Février et al. 1999).

The function in which the leishmania parasite resides poses drug delivery problems, and the drug needs to reach antiparasitic levels at different sites depending on the person and the region affected. Before they operate on the parasite, the antileishmanial drug has to cross the different membrane barriers. A promising technique to overcome resistance has appeared as a nanoparticle-mediated drug delivery method; to cross cell membrane barriers, the drug goes to internal area of the cell and mainly targets leishmania-infected macrophages (Shahnaz et al. 2017).

Targeting the macrophages with surface-modified nanocarriers through these receptors contributes to the growing up of large amounts of medication in the same niche where the parasite resides within the macrophages. The development of 1, 2-diacyl-sn-glycero-3-phospho-l-serine (PS)-coated gelatin nanoparticles (GNPs) carrying amphotericin B for enhanced antileishmanial efficacy in vitro/in vivo in VL has been recorded in another research. The 1, 2-diacyl-sn-glycero-3-phospho-l-serine (PS-AmB-GNPs)-decorated nanocarriers were more effective in terms of absorption by J774A.1 macrophages analyzed through flow cytometry (Shahnaz et al. 2018).

9 Macrophage Targeting in Atherosclerosis and Rheumatoid Arthritis

In the last two decades, macrophages and their thrombotic complications were established as the key factors in atherogenesis. Recent macrophage metabolism studies also contributed to a growing interest in immune- metabolism and new clinical treatment options for atherosclerosis. As potential beneficial methods to treating atherosclerosis, recruitment, polarization, and extracellular matrix reshaping in the cytokine profile were suggested, and metabolism of cholesterol, oxidative stress, inflammatory activity, and monocyte/macrophage RNAs were proposed (Taghizadeh et al. 2019).

The C-reactive protein helps VCAM-1, ICAM-1, E-selectin, and monocyte chemoattractant protein-1 expression and promotes inflammation, a significant inflammatory marker of atherosclerotic progression. These membrane surface proteins may be added to macrophages for the design of selective drug delivery systems. To attain CD206 (mannose receptor)-targeted siRNA delivery, Yu et al. engineered a pH-responsive polymeric micelle with further mannose modification. The mannosylated nanoparticles boost the delivery of siRNA to primary macrophages by four times compared to transport of the same carrier's nontarget version (Peng et al. 2020; Yu et al. 2013).

In a macrophage drug carrier, the drug is primarily loaded into the macrophage by in vitro incubation or direct injection in vivo. The process of in vitro incubation includes the proper incubation of macrophages with drugs in vitro and then the reinjection of the cargo-loaded macrophage carrier for therapy into the body. The in vivo injection approach refers to a direct injection inside the body with the required particle size of the adapted particular ligands or drug delivery systems to extract the phagocytic macrophages as a therapeutic drug (Hu et al. 2013; Yoo et al. 2011).

A new biomimetic nanoparticles delivery system immune to reactive oxygen (ROS) macrophage membrane enables the treatment of targeted pharmaceutical atherosclerosis in mice and prevents local inflammation by sequestering inflammatory conditions (Fernández-Ruiz 2020).

Rheumatoid arthritis (RA) is a common autoimmune disorder in which macrophages may play a key pathogenic role, while secondary dysfunctions may be present in fibroblasts, T and B lymphocytes, and neutrophils (Ulbrich and Lamprecht 2010). Their high plasticity during development is an evolving notion that defines the role of macrophages in RA (Li et al. 2012).

The potential role of macrophages as a new therapeutic target for autoimmune diseases has been demonstrated in multiple studies. Changes in the number and expression of synovial macrophages suggest the therapeutic effectiveness of RA (Yang et al. 2020b).

Reactive oxygen species (ROS) manufactured from macrophages have a fundamental function in triggering this inflammatory disease known as rheumatoid arthritis. Reactive oxygen species usually cause the oxidation sequence at the cellular level. SOD can prevent the growth of macrophage ROS and is used to treat arthritis and other autoimmune disorders. Conversely, STP has antiinflammatory, proteolytic, and pain-relieving properties that together aid patients with various conditions such as arthritis wound healing (Srivastava et al. 2017).

10 Future Prospects

Nanomedicine will be the forthcoming medicine, and the heart of it lies in nanoparticle-based therapeutics. It is also essential to carefully research the metabolism, pharmacokinetics, and self-assembled nanodrugs therapeutic efficacy to advance the use of self-assembled macromolecule-based treatment. Most importantly, the long-term safety/toxicity of the nanoparticles must be investigated. It is feasible to develop advanced colloidal nanocarriers to strengthen combination therapy and to use combination medicines with target inhibitors with a better perception of resistance mechanisms and improvement in the nature of polymer/lipid excipient. Exhaustive studies are required to recognize and characterize the association between macrophages and receptors expressed on the macrophages for efficient drug delivery. However, potential developments in this field are possibly based on the design of multifunctional nanoplatforms, which combine diagnostic (and imaging) and therapeutic ligands into a single entity.

11 Conclusion

As discussed in this chapter, macrophages play an essential function in mediating an extensive range of diseases that make them a superior target for nanoparticle-mediated therapies. Macrophages possess an ordinary tendency to wander into unhealthy tissues in response to signals, such as cytokines. This ability is subjugated by macrophage modification and reintroducing them back into the bloodstream. The mechanism mentioned above has proven macrophage targeting as a promising targeting agent for delivering a variety of molecules, namely neoplastic agents, antibacterial drugs, small molecule drugs, gene, and many more, resulting in enhanced therapeutic and targeting efficacy. Numerous studies have established the macrophage targeting potential in various fields. However, it is still crucial that better novel systems be developed which can encounter the challenges associated with such strategies. The demand for novel surface engineering approaches can further enhance the specificity and selectivity that can promote improved therapeutic efficacy of the delivery systems, which have clinical trial prospective and can be finally approved for clinical use.

References

Amiji MM, Kalariya M, Jain S, Attarwala H. Multi-compartmental macrophage delivery. Google Patents; 2019.

Anne Février MEF, Fournet A, Gloria Yaluff AI, de Arias AR. Acetogenins and other compounds from Rolliniu emurginutu and their antiprotozoal activities. Planta Med. 1999;65:47–9.

Asthana S, Gupta PK, Jaiswal AK, Dube A, Chourasia MK. Targeted chemotherapy of visceral leishmaniasis by lactoferrin-appended amphotericin B-loaded nanoreservoir: in vitro and in vivo studies. Nanomedicine. 2015;10(7):1093–109.

Bai S-B, Liu D-Z, Cheng Y, Cui H, Liu M, Cui M-X, et al. Osteoclasts and tumor cells dual targeting nanoparticle to treat bone metastases of lung cancer. Nanomed Nanotechnol Biol Med. 2019;21:102054.

Barginear MF, John V, Budman DR. Trastuzumab-DM1: a clinical update of the novel antibody-drug conjugate for HER2-overexpressing breast cancer. Mol Med. 2012;18(11):1473–9.

Binnemars-Postma K, Storm G, Prakash J. Nanomedicine strategies to target tumor-associated macrophages. Int J Mol Sci. 2017;18(5):979.

Biswas SK, Mantovani A. Macrophages: biology and role in the pathology of diseases. Springer; 2014.

Conde J, Bao C, Tan Y, Cui D, Edelman ER, Azevedo HS, et al. Dual targeted immunotherapy via in vivo delivery of biohybrid RNAi-peptide nanoparticles to tumor-associated macrophages and cancer cells. Adv Funct Mater. 2015;25(27):4183–94.

Croft S. In vitro screens in the experimental chemotherapy of leishmaniasis and trypanosomiasis. Parasitol Today. 1986;2(3):64–9.

Dave V, Gupta N, Sur S. Nanopharmaceutical advanced delivery systems. John Wiley & Sons; 2020.

Etzerodt A, Maniecki MB, Graversen JH, Møller HJ, Torchilin VP, Moestrup SK. Efficient intracellular drug-targeting of macrophages using stealth liposomes directed to the hemoglobin scavenger receptor CD163. J Control Release. 2012;160(1):72–80.

Fernández-Ruiz I. Macrophage mimetics to target atherosclerosis. Nat Rev Cardiol. 2020;17(8):454.

Fernando S, Gunasekara T, Holton J. Antimicrobial nanoparticles: applications and mechanisms of action. 2018.

Fujiwara N, Kobayashi K. Macrophages in inflammation. Curr Drug Targets Inflamm Allergy. 2005;4(3):281–6.

Fukushima K, Hedrick JL, Nelson A, Sanders DP. Surface modified nanoparticles, methods of their preparation, and uses thereof for gene and drug delivery. Google Patents; 2014.

Golani LK, Wallace-Povirk A, Deis SM, Wong J, Ke J, Gu X, et al. Tumor targeting with novel 6-substituted pyrrolo [2, 3-d] pyrimidine antifolates with heteroatom bridge substitutions via cellular uptake by folate receptor α and the proton-coupled folate transporter and inhibition of de novo purine nucleotide biosynthesis. J Med Chem. 2016;59(17):7856–76.

Gouveia VM, Rizzello L, Nunes C, Poma A, Ruiz-Perez L, Oliveira A, et al. Macrophage targeting pH responsive polymersomes for glucocorticoid therapy. Pharmaceutics. 2019;11(11):614.

Gregorc V, Santoro A, Bennicelli E, Punt C, Citterio G, Timmer-Bonte J, et al. Phase Ib study of NGR–hTNF, a selective vascular targeting agent, administered at low doses in combination with doxorubicin to patients with advanced solid tumours. Br J Cancer. 2009;101(2):219–24.

Haque AR, Moriyama M, Kubota K, Ishiguro N, Sakamoto M, Chinju A, et al. CD206+ tumor-associated macrophages promote proliferation and invasion in oral squamous cell carcinoma via EGF production. Sci Rep. 2019;9(1):1–10.

Hattori Y, Yamashita J, Sakaida C, Kawano K, Yonemochi E. Evaluation of antitumor effect of zoledronic acid entrapped in folate-linked liposome for targeting to tumor-associated macrophages. J Liposome Res. 2015;25(2):131–40.

Haubner R, Weber WA, Beer AJ, Vabuliene E, Reim D, Sarbia M, et al. Noninvasive visualization of the activated αvβ3 integrin in cancer patients by positron emission tomography and [18 F] Galacto-RGD. PLoS Med. 2005;2(3):e70.

He K-F, Zhang L, Huang C-F, Ma S-R, Wang Y-F, Wang W-M, et al. CD163+ tumor-associated macrophages correlated with poor prognosis and cancer stem cells in oral squamous cell carcinoma. Biomed Res Int. 2014;2014

He W, Kapate N, Shields CW IV, Mitragotri S. Drug delivery to macrophages: a review of targeting drugs and drug carriers to macrophages for inflammatory diseases. Adv Drug Deliv Rev. 2020;165:15–40.

Hibbitts A, O'Leary C. Emerging nanomedicine therapies to counter the rise of methicillin-resistant Staphylococcus aureus. Materials. 2018;11(2):321.

Horváti K, Gyulai G, Csámpai A, Rohonczy J, Kiss É, Bősze S. Surface layer modification of Poly(d,l-lactic-co-glycolic acid) nanoparticles with targeting peptide: a convenient synthetic route for pluronic F127-Tuftsin conjugate. Bioconjug Chem. 2018;29(5):1495–9.

Hu G, Christman JW. Editorial: Alveolar Macrophages in Lung Inflammation and Resolution. Front Immunol. 2019;10(2275)

Hu C-MJ, Fang RH, Luk BT, Chen KN, Carpenter C, Gao W, et al. 'Marker-of-self' functionalization of nanoscale particles through a top-down cellular membrane coating approach. Nanoscale. 2013;5(7):2664–8.

Huang Z, Zhang Z, Jiang Y, Zhang D, Chen J, Dong L, et al. Targeted delivery of oligonucleotides into tumor-associated macrophages for cancer immunotherapy. J Control Release. 2012;158(2):286–92.

Hubbell JA, O'neil CP, Reddy ST, Swartz MA, Velluto D, van Der Vlies A, et al. Nanoparticles for immunotherapy. Google Patents; 2012.

Ishida J, Kurozumi K, Ichikawa T, Otani Y, Onishi M, Fujii K, et al. Evaluation of extracellular matrix protein CCN1 as a prognostic factor for glioblastoma. Brain Tumor Pathol. 2015;32(4):245–52.

Jain NK, Mishra V, Mehra NK. Targeted drug delivery to macrophages. Expert Opin Drug Deliv. 2013;10(3):353–67.

Jayasingam SD, Citartan M, Thang TH, Zin AAM, Ang KC, Ch'ng ES. Evaluating the polarization of tumor-associated macrophages into M1 and M2 phenotypes in human cancer tissue: technicalities and challenges in routine clinical practice. Front Oncol. 2019;9

Kagan JM, Sanchez AM, Landay A, Denny TN. A brief chronicle of CD4 as a biomarker for HIV/AIDS: a tribute to the memory of John L. Fahey. In: Forum on immunopathological diseases and therapeutics. Begel House Inc; 2015.

Kalyane D, Raval N, Maheshwari R, Tambe V, Kalia K, Tekade RK. Employment of enhanced permeability and retention effect (EPR): nanoparticle-based precision tools for targeting of therapeutic and diagnostic agent in cancer. Mater Sci Eng C. 2019;98:1252–76.

Khatik R, Dwivedi P, Khare P, Kansal S, Dube A, Mishra PR, et al. Development of targeted 1,2-diacyl-sn-glycero-3-phospho-l-serine-coated gelatin nanoparticles loaded with amphotericin B for improved in vitro and in vivo effect in leishmaniasis. Expert Opin Drug Deliv. 2014;11(5):633–46.

Kim MS, Lee JS, Kim JE, Kim J-W, Bok S, Keum KC, et al. Enhancement of antitumor effect of radiotherapy via combination with Au@ SiO2 nanoparticles targeted to tumor-associated macrophages. J Ind Eng Chem. 2020;84:349–57.

Komohara Y, Niino D, Saito Y, Ohnishi K, Horlad H, Ohshima K, et al. Clinical significance of CD 163+ tumor-associated macrophages in patients with adult T-cell leukemia/lymphoma. Cancer Sci. 2013;104(7):945–51.

Kumar Singh P, Gorain B, Choudhury H, Kumar Singh S, Whadwa P, Shilpa, et al. Macrophage targeted amphotericin B nanodelivery systems against visceral leishmaniasis. Material Sci & Eng: B. 2020;258:114571.

Kunjachan S, Gupta S, Dwivedi AK, Dube A, Chourasia MK. Chitosan-based macrophage-mediated drug targeting for the treatment of experimental visceral leishmaniasis. J Microencapsul. 2011;28(4):301–10.

Kzhyshkowska J, Riabov V, Gudima A, Wang N, Orekhov A, Mickley A. Role of tumor associated macrophages in tumor angiogenesis and lymphangiogenesis. Front Physiol. 2014;5:75.

Leber N, Kaps L, Yang A, Aslam M, Giardino M, Klefenz A, et al. α-mannosyl-functionalized cationic nanohydrogel particles for targeted gene knockdown in immunosuppressive macrophages. Macromol Biosci. 2019;19(7):e1900162.

Ledermann JA, Canevari S, Thigpen T. Targeting the folate receptor: diagnostic and therapeutic approaches to personalize cancer treatments. Ann Oncol: Official J Eu Soc Med Oncol. 2015;26(10):2034–43.

Li J, Hsu H-C, Mountz JD. Managing macrophages in rheumatoid arthritis by reform or removal. Curr Rheumatol Rep. 2012;14(5):445–54.

Li Y, Wu H, Ji B, Qian W, Xia S, Wang L, et al. Targeted imaging of CD206 expressing tumor-associated M2-like macrophages using mannose-conjugated antibiofouling magnetic iron oxide nanoparticles. ACS Appl Bio Material. 2020;3(7):4335–47.

Liu Z, Xiong M, Gong J, Zhang Y, Bai N, Luo Y, et al. Legumain protease-activated TAT-liposome cargo for targeting tumours and their microenvironment. Nat Commun. 2014;5(1):1–11.

Liu M, Tong Z, Ding C, Luo F, Wu S, Wu C, et al. Transcription factor c-Maf is a checkpoint that programs macrophages in lung cancer. J Clin Invest. 2020;130(4)

Locke LW, Mayo MW, Yoo AD, Williams MB, Berr SS. PET imaging of tumor associated macrophages using mannose coated 64Cu liposomes. Biomaterials. 2012;33(31):7785–93.

Luria-Pérez R, Helguera G, Rodríguez JA. Antibody-mediated targeting of the transferrin receptor in cancer cells. Bol Med Hosp Infant Mex. 2016;73(6):372–9.

Ma C, Wu M, Ye W, Huang Z, Ma X, Wang W, et al. Inhalable solid lipid nanoparticles for intracellular tuberculosis infection therapy: macrophage-targeting and pH-sensitive properties. Drug Deliv Transl Res. 2020a;

Ma C, Wu M, Ye W, Huang Z, Ma X, Wang W, et al. Inhalable solid lipid nanoparticles for intracellular tuberculosis infection therapy: macrophage-targeting and pH-sensitive properties. Drug Deliv Transl Res. 2020b:1–18.

Marcianes P, Negro S, Barcia E, Montejo C, Fernández-Carballido A. Potential active targeting of gatifloxacin to macrophages by means of surface-modified PLGA microparticles destined to treat tuberculosis. AAPS PharmSciTech. 2019;21(1):15.

Matute-Blanch C, Montalban X, Comabella M. Multiple sclerosis, and other demyelinating and autoimmune inflammatory diseases of the central nervous system. Handbook of clinical neurology, vol. 146. Elsevier; 2018. p. 67–84.

Miyasato Y, Shiota T, Ohnishi K, Pan C, Yano H, Horlad H, et al. High density of CD 204-positive macrophages predicts worse clinical prognosis in patients with breast cancer. Cancer Sci. 2017;108(8):1693–700.

Movahedi K, Schoonooghe S, Laoui D, Houbracken I, Waelput W, Breckpot K, et al. Nanobody-based targeting of the macrophage mannose receptor for effective in vivo imaging of tumor-associated macrophages. Cancer Res. 2012;72(16):4165–77.

Muppidi K, Wang J, Betageri G, Pumerantz AS. PEGylated liposome encapsulation increases the lung tissue concentration of vancomycin. Antimicrob Agents Chemother. 2011;55(10):4537–42.

Nagai T, Tanaka M, Tsuneyoshi Y, Xu B, Michie SA, Hasui K, et al. Targeting tumor-associated macrophages in an experimental glioma model with a recombinant immunotoxin to folate receptor β. Cancer Immunol Immunother. 2009;58(10):1577–86.

Nemeth JA, Nakada MT, Trikha M, Lang Z, Gordon MS, Jayson GC, et al. Alpha-v integrins as therapeutic targets in oncology. Cancer Invest. 2007;25(7):632–46.

Ng PQ, Ling LSC, Chellian J, Madheswaran T, Panneerselvam J, Kunnath AP, et al. Applications of nanocarriers as drug delivery vehicles for active phytoconstituents. Curr Pharm Des. 2020;

Ngambenjawong C, Gustafson HH, Pun SH. Progress in tumor-associated macrophage (TAM)-targeted therapeutics. Adv Drug Deliv Rev. 2017;114:206–21.

Niu M, Valdes S, Naguib YW, Hursting SD, Cui Z. Tumor-associated macrophage-mediated targeted therapy of triple-negative breast cancer. Mol Pharm. 2016;13(6):1833–42.

Ohtaki Y, Ishii G, Nagai K, Ashimine S, Kuwata T, Hishida T, et al. Stromal macrophage expressing CD204 is associated with tumor aggressiveness in lung adenocarcinoma. J Thorac Oncol. 2010;5(10):1507–15.

Organization WH. WHO traditional medicine strategy: 2014–2023: World Health Organization; 2013.

Ortega RA, Barham WJ, Kumar B, Tikhomirov O, McFadden ID, Yull FE, et al. Biocompatible mannosylated endosomal-escape nanoparticles enhance selective delivery of short nucleotide sequences to tumor associated macrophages. Nanoscale. 2015;7(2):500–10.

Ortega RA, Barham W, Sharman K, Tikhomirov O, Giorgio TD, Yull FE. Manipulating the NF-κB pathway in macrophages using mannosylated, siRNA-delivering nanoparticles can induce immunostimulatory and tumor cytotoxic functions. Int J Nanomedicine. 2016;11:2163.

Parashar P, Rathor M, Dwivedi M, Saraf SA. Hyaluronic acid decorated naringenin nanoparticles: appraisal of chemopreventive and curative potential for lung cancer. Pharmaceutics. 2018;10(1)

Peñate-Medina T, Damoah C, Benezra M, Will O, Kairemo K, Humbert J, et al. Alpha-MSH targeted liposomal nanoparticle for imaging in inflammatory bowel disease (IBD). Curr Pharm Des. 2020;26(31): 3840–6.

Peng R, Ji H, Jin L, Lin S, Huang Y, Xu K, et al. Macrophage-based therapies for atherosclerosis management. J Immunol Res. 2020;2020

Peterson KR, Cottam MA, Kennedy AJ, Hasty AH. Macrophage-targeted therapeutics for metabolic disease. Trends Pharmacol Sci. 2018;39(6):536–46.

Placha D, Jampilek J. Chronic inflammatory diseases, anti-inflammatory agents and their delivery Nanosystems. Pharmaceutics. 2021;13(1):64.

Poh AR, Ernst M. Targeting macrophages in cancer: from bench to bedside. Front Oncol. 2018;8:49.

Ponzoni M, Pastorino F, Di Paolo D, Perri P, Brignole C. Targeting macrophages as a potential therapeutic intervention: impact on inflammatory diseases and cancer. Int J Mol Sci. 2018;19(7):1953.

Qie Y, Yuan H, Von Roemeling CA, Chen Y, Liu X, Shih KD, et al. Surface modification of nanoparticles enables selective evasion of phagocytic clearance by distinct macrophage phenotypes. Sci Rep. 2016;6(1):1–11.

Ramos RN, Rodriguez C, Hubert M, Ardin M, Treilleux I, Ries CH, et al. CD163+ tumor-associated macrophage accumulation in breast cancer patients reflects both local differentiation signals and systemic skewing of monocytes. Clinic & Translation Immunol. 2020;9(2):e1108.

Ravindra M, Wallace-Povirk A, Karim MA, Wilson MR, O'Connor C, White K, et al. Tumor targeting with novel Pyridyl 6-substituted Pyrrolo [2, 3-d] pyrimidine Antifolates via cellular uptake by folate receptor α and the proton-coupled folate transporter and inhibition of De novo purine nucleotide biosynthesis. J Med Chem. 2018;61(5):2027–40.

Rittig M, Bogdan C. Leishmania–host-cell interaction: complexities and alternative views. Parasitol Today. 2000;16(7):292–7.

Rotman SG, Thompson K, Grijpma DW, Richards RG, Moriarty TF, Eglin D, et al. Development of bone seeker–functionalised microspheres as a targeted local antibiotic delivery system for bone infections. J Orthop Translat. 2020;21:136–45.

Sandeep D, AlSawaftah NM, Husseini G. Immunoliposomes: synthesis, structure, and their potential as drug delivery carriers; 2020.

Schoonooghe S, Laoui D, Van Ginderachter JA, Devoogdt N, Lahoutte T, De Baetselier P, et al. Novel applications of nanobodies for in vivo bio-imaging of inflamed tissues in inflammatory diseases and cancer. Immunobiology. 2012;217(12):1266–72.

Shahnaz G, Edagwa BJ, McMillan J, Akhtar S, Raza A, Qureshi NA, et al. Development of mannose-anchored thiolated amphotericin B nanocarriers for treatment of visceral leishmaniasis. Nanomedicine. 2017;12(2):99–115.

Shahnaz G, Sarwar HS, Yasinzai M. Crossing biological barriers for Leishmaniasis therapy: from Nanomedicinal targeting perspective. Leishmanias Re-emerg Diseas. 2018;199

Sharma A, Liaw K, Sharma R, Thomas AG, Slusher BS, Kannan S, et al. Targeting mitochondria in tumor-associated macrophages using a dendrimer-conjugated TSPO ligand that stimulates antitumor signaling in glioblastoma. Biomacromolecules. 2020;21(9):3909–22.

Shen TW, Fromen CA, Kai MP, Luft JC, Rahhal TB, Robbins GR, et al. Distribution and cellular uptake of PEGylated polymeric particles in the lung towards cell-specific targeted delivery. Pharm Res. 2015;32(10):3248–60.

Sherje AP, Jadhav M, Dravyakar BR, Kadam D. Dendrimers: a versatile nanocarrier for drug delivery and targeting. Int J Pharm. 2018;548(1):707–20.

Singh G, Dwivedi H, Saraf SK, Saraf SA. Niosomal delivery of isoniazid-development and characterization. Tropic J Pharma Res. 2011;10(2)

Singh PK, Sah P, Meher JG, Joshi S, Pawar VK, Raval K, et al. Macrophage-targeted chitosan anchored PLGA nanoparticles bearing doxorubicin and amphotericin B against visceral leishmaniasis. RSC Adv. 2016;6(75):71705–18.

Singh N, Parashar P, Tripathi CB, Kanoujia J, Kaithwas G, Saraf SA. Oral delivery of allopurinol niosomes in treatment of gout in animal model. J Liposome Res. 2017;27(2):130–8.

Singh PK, Gorain B, Choudhury H, Singh SK, Whadwa P, Sahu S, et al. Macrophage targeted amphotericin B nanodelivery systems against visceral leishmaniasis. Mater Sci Eng B. 2020;258:114571.

Song X, Wan Z, Chen T, Fu Y, Jiang K, Yi X, et al. Development of a multi-target peptide for potentiating chemotherapy by modulating tumor microenvironment. Biomaterials. 2016;108:44–56.

Srivastava S, Singh D, Patel S, Singh MR. Treatment of rheumatoid arthritis by targeting macrophages through folic acid tailored superoxide dismutase and serratiopeptidase. J Drug Deliv Sci & Technol. 2017;41:431–5.

Stein CA, Castanotto D. FDA-approved oligonucleotide therapies in 2017. Mol Ther. 2017;25(5):1069–75.

Sun X, Gao D, Gao L, Zhang C, Yu X, Jia B, et al. Molecular imaging of tumor-infiltrating macrophages in a preclinical mouse model of breast cancer. Theranostics. 2015;5(6):597.

Swati S, Batta S, Pandala S, Sravanthi TS, Vineesha S. Formulation and in vitro characterisation of floating microspheres of glipizide. J Pharm Sci Res. 2020;12(5):684–90.

Taghizadeh E, Taheri F, Renani PG, Reiner Ž, Navashenaq JG, Sahebkar A. Macrophage: a key therapeutic target in atherosclerosis? Curr Pharm Des. 2019;25(29):3165–74.

Ulbrich W, Lamprecht A. Targeted drug-delivery approaches by nanoparticulate carriers in the therapy of inflammatory diseases. J Royal Soc Interface. 2010;7(suppl_1):S55–66.

Wang D-Y, Van Der Mei HC, Ren Y, Busscher HJ, Shi L. Lipid-based antimicrobial delivery-systems for the treatment of bacterial infections. Front Chem. 2020a;7:872.

Wang X, Dong A, Hu Y, Qian J, Huang S. A review of recent work on using metal–organic frameworks to grow carbon nanotubes. Chem Commun. 2020b;56(74):10809–23.

Wolf GL, Schmidt KF. Theranosis of macrophage-associated diseases with ultrasmall superparamagnetic iron oxide nanoparticles (uspio). Google Patents; 2014.

Yang Y, Guo L, Wang Z, Liu P, Liu X, Ding J, et al. Targeted silver nanoparticles for rheumatoid arthritis therapy via macrophage apoptosis and re-polarization. Biomaterials. 2020a;264:120390.

Yang X, Chang Y, Wei W. Emerging role of targeting macrophages in rheumatoid arthritis: Focus on polarization, metabolism and apoptosis. Cell Prolif. 2020b:e12854.

Yoo J-W, Irvine DJ, Discher DE, Mitragotri S. Bio-inspired, bioengineered and biomimetic drug delivery carriers. Nat Rev Drug Discov. 2011;10(7):521–35.

Yu W, Liu C, Liu Y, Zhang N, Xu W. Mannan-modified solid lipid nanoparticles for targeted gene delivery to alveolar macrophages. Pharm Res. 2010;27(8):1584–96.

Yu SS, Lau CM, Barham WJ, Onishko HM, Nelson CE, Li H, et al. Macrophage-specific RNA interference targeting via "click", mannosylated polymeric micelles. Mol Pharm. 2013;10(3):975–87.

Zhan X, Jia L, Niu Y, Qi H, Chen X, Zhang Q, et al. Targeted depletion of tumour-associated macrophages by an alendronate–glucomannan conjugate for cancer immunotherapy. Biomaterials. 2014;35(38):10046–57.

Zhang M, Kim JA. Effect of molecular size and modification pattern on the internalization of water soluble β-(1→ 3)-(1→ 4)-glucan by primary murine macrophages. Int J Biochem Cell Biol. 2012;44(6):914–27.

Zhang X-Y, Zhang P-Y. Polymersomes in nanomedicine – a review. Curr Nanosci. 2017;13(2):124–9.

Zhang X, Tian W, Cai X, Wang X, Dang W, Tang H, et al. Hydrazinocurcumin Encapsuled nanoparticles "re-educate" tumor-associated macrophages and exhibit anti-tumor effects on breast cancer following STAT3 suppression. PLoS One. 2013;8(6):e65896.

Zhang M, He Y, Sun X, Li Q, Wang W, Zhao A, et al. A high M1/M2 ratio of tumor-associated macrophages is associated with extended survival in ovarian cancer patients. J Ovarian Res. 2014;7(1):19.

Zhang Y, Ye J, Hosseini-Nassab N, Flores A, Kalashnikova I, Paluri SL, et al. Macrophage-targeted single walled carbon nanotubes stimulate phagocytosis via pH-dependent drug release. Nano Res. 2020:1–8.

Zhu S, Niu M, O'Mary H, Cui Z. Targeting of tumor-associated macrophages made possible by PEG-sheddable, mannose-modified nanoparticles. Mol Pharm. 2013;10(9):3525–30.

Part III

Lipid Nanoparticles for Macrophage Targeting

Liposomal Delivery for Targeting Macrophages

Bahareh Asadi Aghbolagh and Uyen Le

Abstract

Macrophages are cells not only associated with body defense but also involved in the precursor of human diseases. Targeting macrophages has potential not only for the treatment but also for the prevention of the diseases. Different approaches in nanoparticles have been explored in macrophage-targeting macrophages for cancer therapy. Liposomes, a type of nanocarrier, are a promising candidate for the delivery of drug targeted to macrophages for many reasons. Firstly, liposomes are lipid bilayers that are composed of a phospholipid membrane and an aqueous core. The unique amphiphilic structure enables the liposomes to flexibly encapsulate hydrophilic drugs and/or to incorporate hydrophobic drugs. In addition, liposomes are biodegradable and capable of being phagocyted, bringing the drugs inside the macrophages. Furthermore, liposomes have demonstrated to be a good nanocarrier for different drugs to target macrophages for the treatment of infection, inflammation, cancers, cardiovascular diseases, cerebral ischemia and stroke, etc. Phospholipid composition, particle size, surface charge, and modified surface are major factors affecting the efficacy of liposomes as a drug delivery system in targeting macrophages. Commonly used phospholipids are phosphatidic acid, lysophosphatidic acid phosphatidylcholine, lysophosphatidylcholine, phosphatidylethanolamine lysophosphatidylethanolamine, phosphoinositide, etc. The particle size plays a critical factor in the opsonization of drug-carried liposomes. Liposomes with diameter sizes larger than 200 nm have the tendency of opsonizing by protein better than the smaller ones. Neutral, negatively-, or positively charged liposomes can influence macrophages on liposome uptake. Liposome's surface charge majorly results from the phospholipid composition. Commonly, cationic liposomes are made from N-[1-(2,3-dioleyloxy)propyl]-N,N,N-trimethylammonium chloride (DOTMA), 1,2-dioleoyl-3-trimethylammonium-propane (DOTAP), and N(1)-cholesteryloxycarbonyl-3,7-diazanonane-1,9-diamine (CDAN), while anionic liposomes are associated with phosphatidic acid (PA), lysophosphatidic acid (LPA), cyclic lysophosphatidic acid (CPA), 1,2-dioleoyl-sn-glycero-3-phospho-L-serine (DOPS), and dioleoylphosphatidylethanolamine (DOPE). Typically, a mixture of

B. A. Aghbolagh
California Northstate University College of Pharmacy, Elk Grove, CA, USA
e-mail: bahareh.aghbolagh9548@cnsu.edu

U. Le (✉)
Department of Pharmaceutical & Biomedical Sciences, California Northstate University College of Pharmacy, Elk Grove, CA, USA
e-mail: uyen.le@cnsu.edu

S. Gupta, Y. V. Pathak (eds.), *Macrophage Targeted Delivery Systems*,
https://doi.org/10.1007/978-3-030-84164-5_9

charged-associated phospholipid and neutral phospholipid provides more stable liposomes than a single type of phospholipid. Conventional liposomes possess appropriate properties for the macrophage uptake, while the stealth liposomes can escape the entrapment. However, stealth liposomes with added ligand-targeted binding can yield a specific delivery to macrophages.

Keywords

Liposome · Macrophage · Properties · Target · Disease · Treatment · Drug delivery

1 Introduction

First discovered in 1968 by Bangham (Bangham et al. 1965), liposomes have been shown interest by scientists due to their special composition, which is similar to the structure of a cell membrane. Some years later, liposomes have been developed as one of the microencapsulation delivery systems for drugs to treat diseases. So far, liposomes have moved a long way from being pharmaceutical, therapeutic, and personal care delivery systems for numerous practical applications, to becoming a promising object of functional food ingredients' research.

Liposomes are small spherical vesicles encapsulated with phospholipids. Liposomes are composed of phospholipid bilayers, in which the polar groups of phospholipids face the inside to form an outer aqueous phase. Liposomes have amphiphilic properties. Their bilayers are composed of amphiphilic lipids that have a hydrophilic head and two hydrophobic tails. In general, the hydrophobic tails have two fatty acid chains containing 10–24 carbon atoms and 0–6 double bonds in each chain, and the hydrophilic head is the phosphoric acid bound to a water-soluble molecule (Lasic and Papahadjopoulos 1995). When phospholipids are dispersed in aqueous medium, "the hydrophobic effect" makes the hydrophobic tail remain in contact with the polar environment and shields the aqueous medium. In addition, van der Waals interactions and the bond of hydrogen molecules from water to phospholipids also help arrange phospholipids into closed bilayered vesicles, which offer a unique function for carrying both hydrophobic and hydrophilic molecules (Mufamadi et al. 2011). This property of liposome's structure defines the tendency that liposomes can encapsulate hydrophilic drugs in the inner aqueous compartment and incorporate hydrophobic drugs in the lipid bilayers.

The utilization of liposomes in drug delivery has attracted attention for more than 40 years. As a drug delivery system, liposomes have many advantages as follows: delivering both hydrophilic and lipophilic drugs, possessing targeting capacity, controlled release properties, cell affinity, tissue compatibility, and reducing drug toxicity and improving drug stability. Over time, the original structures of the liposomes have changed, which have created a series of new type liposomes, such as long-circulating liposomes, stimuli-responsive liposomes, cationic liposomes, and ligand-targeted liposomes. Liposomes can serve as the carriers of antitumor drugs, antifungal drugs, analgesic drugs, gene therapeutics, and vaccines. One of the aspects of liposome function is its capability to interact with macrophages (Kelly et al. 2015). Therefore, the entrapment of drugs into liposomes can be an effective approach of drug targeting to this system.

Macrophages are cells relating to the phagocytosis and destruction of bacteria and other organisms. Macrophages have an important role in forming the first line of defense against the microbial evasion. They act like "scavengers" in digesting foreign substances that might be potentially harmful to the body. Macrophages are not only associated with inflammatory diseases but can also participate in tissue remodeling. During normal healing, M1 macrophages are predominant in removing debris, infection, and damaged cells. Macrophage transition to M2 phenotype is necessary to promote healing and regeneration. Dysregulation of M1 or M2 expression has been associated with major diseases including chronic wounds (diabetic, pressure, and venous ulcers), chronic obstructive pulmonary disease (COPD), infection, atherosclerosis, cancer, asthma, Parkinson's disease, Alzheimer's disease, and cancer.

Macrophages play crucial roles in the development, homeostatic tissue processes, tissue repair, and immunity (Wynn et al. 2013). At present, a variety of diseases (i.e., diabetes, atherosclerosis, rheumatoid arthritis, obesity, and cancer) are widely recognized to be associated with chronic inflammation (Schultze et al. 2015). Macrophages transform from circulating peripheral blood mononuclear cells which move to tissue in the steady state or in response to inflammation (Gordon and Taylor 2005). These cells are grown from a common myeloid progenitor cell in the bone marrow, which is the precursor of neutrophils, eosinophils, basophils, macrophages, dendritic cells, and mast cells. During monocyte development, myeloid progenitor cells continuously produce monoblasts, promonocytes, and then monocytes, which are released from the bone marrow into the bloodstream (Gordon and Taylor 2005). From the blood, monocytes will move into the tissue to reload long-lived tissue-specific macrophages of the bone, alveoli, central nervous system (CNS), connective tissue, gastrointestinal (GI) tract, liver (Kupffer cells), spleen, and peritoneum.

Since largely associated with the pathogenesis of several types of human diseases, macrophages are considered to be relevant therapeutic targets. The aim for targeting macrophages using liposomes has potential for the treatment of different diseases such as cardiovascular diseases, cancers, HIV, metabolic impairment, infection, etc. The efficiency of targeting is based upon different factors including liposome composition, charge, size, release mechanism, and structure.

2 Physicochemical Properties of Liposomes Affecting Drug Delivery and Macrophage Targeting

2.1 Phospholipids in Liposome Composition

Phospholipid is the major ingredient of liposomes. Due to its phospholipid similarities to the cellular structure besides the capability of membrane modification, liposomes can be used to target cells of the body while avoiding the cells of the immune system. This with their ability to carry drugs makes liposomes strong candidates for cell- or tissue-specific drug delivery. Being the major component in liposome structures, phospholipids can affect the delivery efficiency of drugs (Akbarzadeh et al. 2013). It is important to count on many different factors to identify appropriate phospholipids when formulating liposomes. Phospholipids can be natural or synthetic, which both have their own advantages and disadvantages. Synthetic phospholipids are relatively stable and highly pure, but the drawback is that the price is relatively high (Mufamadi et al. 2011). Natural phospholipid is less pricey. However, the purity is difficult to control, and its nature is relatively unstable because it can be metabolized to lysophospholipids in the process of usage and storage. The lipid profile of a liposome is one of the key factors that determine the chemical properties, which can influence the delivery efficacies, loading capacity, pharmacological and toxicological properties, and overall quality of the liposomes (Bozzuto and Molinari 2015). The profile remarkably influences the uptake of liposomes by macrophages.

In targeting bone marrow macrophages, liposomes made with phosphatidylcholine, phosphatidylserine (PS), or gangliosides provide a high uptake (Allen et al. 1991). However, monosialoganglioside, polyethylene glycol-derived liposomes, or the addition of sphingomyelin and cholesterol demonstrate high rigidity and reduce the uptake by macrophages.

The evaluation of total lipid content in liposome profiles is necessary for the FDA approval. According to Avanti Polar Lipids, a popular provider of phospholipids in the USA, phospholipids can be categorized in different types (Table 1) (Avantilipids.com 2020). The selection of phospholipids in liposome formulation is based mainly upon the drug properties, disease, and purpose of drug delivery.

2.2 Size of Liposomes

The majority of physicochemical properties of a liposomal product rely on the structure, method of preparation, composition, and application (White et al. 2000). Both size and size distribution of liposomes are important in the physical properties of liposomal formulations (Nagayasu et al. 1999). Over more than 20 years, numerous studies have shown that the liposome size can influence drug bioavailability. Most of these studies conclude that even a small change in size distribution can influence the performance of a liposomal formulation altering delivery and circulation efficacy.

The particle size of liposome is extremely important for its application in vivo because it could potentially impact different properties such as stability, encapsulation efficiency, drug release, mucoadhesion, and the cellular uptake of liposome (Juliano and Stamp 1975). Therefore, it is very important to make uniform-sized liposomes when liposomes are prepared. Liposomes with diameters >0.1 μm can be, in comparison to smaller ones, opsonized more rapidly and to a greater extent, which leads to a more rapid removal from the body by the reticuloendothelial system (RES). Also, the biodistribution of liposomes is related to particle size (Liu et al. 2020). Larger liposomes (diameter greater than 200 nm) have been reported to increase protein opsonization (Harashima et al. 1994). Large sizes of liposomes are not preferred in liposomal drug delivery where a long-circulating time of lipo-

Table 1 Common types of phospholipids

Phospholipid types	Properties	References
Isometrically pure mixed-acyl glycerophospholipids	Synthetic mixed-acyl glycerophospholipids. Important to the identification of glycerophospholipids in biological samples	Ekroos et al. (2003); Zacek et al. (2016) and Maccarone et al. (2014)
Phosphatidic acid (PA), lysophosphatidic acid (LPA), And cyclic lysophosphatidic acid (CPA)	PA is an anionic bioactive lipid. LPA has more negative charge than PA. CPA is an analog of LPA. PA and LPA can affect various cellular functions. CPA can affect antimitogenic regulation of cells, induction of stress fiber formation, inhibition of tumor cell invasion, and regulation of survival of neuronal cells	Kooijman et al. (2005) and Fujiwara (2008)
Phosphatidylcholine (PC) and lysophosphatidylcholine (LPC)	PC can be involved in intracellular cholesterol transport and membrane cholesterol homeostasis. LPC can be a factor associated with cardiovascular and neurodegenerative diseases	Lagace (2015) and Law et al. (2019)
Phosphatidylethanolamine (PE) and lysophosphatidylethanolamine (LPE)	PE can affect autophagy, cell division, and protein folding. LPE can affect cell signaling and enzymatic activation	Farine et al. (2015)
Phosphatidylglycerol (PG) and lysophosphatidylglycerol (LPG)	PG is an anionic phospholipid. PG plays an important role in regulating the innate immune response in the lungs. LPG can block the increase of intracellular calcium induced by LPA in ovarian cancer cells	Hagio et al. (2002) and Shim et al. (2009)
Phosphoinositide (PI) and lysophosphatidylinositol (LPI)	PI can regulate the innate immune response. LPI can affect cell growth, differentiation, and mobility in cell types such as cancer cells, endothelial cells, and nerve cells	Falkenburger et al. (2010) and Arifin and Falasca (2016)
Phosphatidylserine (PS) And lysophosphatidylserine (LPS)	PS has a crucial role in exocytosis. LPS has functions in degranulation, calcium mobilization, enhanced efferocytosis, migration, and differentiation	Kim et al. (2014 and Frasch and Bratton (2012)
Cardiolipin (CL)	CL can affect the activation of different signaling pathways	Dudek (2017)
Ether lipids	Ether lipids can relate to membrane trafficking, cell signaling and differentiation, and cellular antioxidants	Dean and Lodhi (2018)

some is desirable. However, they can facilitate the uptake of macrophages, which are the aim of targeting macrophages in the delivery of drug using liposomes (Chono et al. 2007).

The initial heterogeneous suspensions of liposomes can be reduced in size and size distribution by several different methods. Homogenization is one of the methods used to reduce size and distribution, which is also suitable for large-scale production. During homogenization, heterodispersed liposome preparation is cycled pumped under high pressure through a small reaction tank, until a desired average size of liposome is achieved. Sonication or ultrasonic irradiation is another method for reducing the size of the liposome because of the shear forces present during the process. Another size-processing method for uniform liposome preparation is extrusion through uniform pore-size membranes.

2.3 Surface Charge of Liposomes

Liposome's surface can be neutral, negative, or positive. The charge may have the following critical effects on the adjuvant properties of liposomes:

- Cationic liposomes may react to the negatively charged cellular surfaces and produce IL-4, IgG2a, and IgG1 responses (Mehravaran et al. 2019).
- Neutral liposomes can promote a Th1 (e.g., IFNγ and IL-12) immune response even more strongly than positively charged liposomes (Badiee et al. 2009).
- Anionic liposomes may modulate immune via Th2 (e.g., IL-5, IL-6, and IL-10) responses besides inducing platelet aggregation (Zbinden et al. 1989).

Liposome charge is mainly made from the phospholipid composition and possible linkages grated to its surface. The charge can affect the capability of uptake and mechanism of uptake by the macrophages. In a study in vitro, neutral and negatively charged liposomes tend to bind competitively, as compared to the positively charged ones (Dijkstra et al. 1985). However, in a recent in vivo study (Ibaraki et al. 2021), positively charged liposomes demonstrated a high accumulation in hepatic inflammatory sites and were taken up by macrophages better than neutral or negatively charged ones. Therefore, using the influence of liposomal charge on the uptake of liposomes by macrophages needs to be carefully examined based on the delivery purpose and sites of action. Common composition associated with different liposome charges is shown in Table 2.

2.4 Conventional and Modified Liposomes

2.4.1 Conventional Liposomes

The first liposomes were introduced in 1965 by Alec Bangham. Since liposomes are similar to natural cells, they are able to overcome biological barriers. Conventional liposomes mainly consist of phospholipids, which then contain active pharmaceutical ingredients. Hydrophilic drugs can be encapsulated in the aqueous core, while hydrophobic drugs can be incorporated in the phospholipid membrane (Fig. 1).

Conventional liposomes are convenient to prepare but have disadvantages of rapid clearance due to the uptake of phagocytes. Although that would be a disadvantage in drug delivery that desires a long systemic circulation time of the drug, the property makes conventional liposomes a good candidate to target macrophages. Advanced liposome technology has changed from conventional vesicles to "second-generation liposomes," in which liposomes are improved in terms of long-circulation time, composition, size, and charge. The advancement has resulted in new generations of liposomes, which are stealth liposomes and targeted liposomes.

2.4.2 Stealth Liposomes

Stealth liposomes are long-circulating liposomes which are grafted with polymer polyethylene glycol (PEG) in the membrane composition. The schematic structure of stealth liposome is given in Fig. 2. These liposomes exhibit increasing drug stability and solubility, lowering toxicity, increas-

Table 2 Examples of liposome's surface charge and its component

Type	Component resulting in charge	Application	Reference
Cationic liposomes	N-[1-(2,3-dioleyloxy)propyl]-N,N,N-trimethylammonium chloride (DOTMA), 1,2-dioleoyl-3-trimethylammonium-propane (DOTAP), and N(1)-cholesteryloxycarbonyl-3,7-diazanonane-1,9-diamine (CDAN)	Gene delivery and transfection	Elsana et al. (2019) and Spagnou et al. (2004)
Neutral liposomes	Phosphatidylcholine (PC) and lysophosphatidylcholine (LPC). Examples: 1,2-dioleoyl-sn-glycerophosphatidylcholine (DOPC), 2-distearoyl-sn-glycero-3-phosphocholine (DSPC), and dioleoyl-L-alpha-phosphatidylethanolamine (DOPE). Dipalmitoylphosphatidylcholine (DPPC), egg yolk PC, soy PC, etc.	Various drug delivery applications. Mixed with other neutral, negatively, or positively charged phospholipids	Chowdhury et al. (2020) and Tucci et al. (2019)
Anionic liposomes	Phosphatidic acid (PA), lysophosphatidic acid (LPA), cyclic lysophosphatidic acid (CPA), and 1,2-dioleoyl-sn-glycero-3-phospho-L-serine (DOPS); dioleoylphosphatidylethanolamine (DOPE) and dimyristoylphosphatidylcholine (DMPC)	Gene delivery and transfection	Smith et al. (2017); Neves et al. (2016) and Aoki et al. (2015)

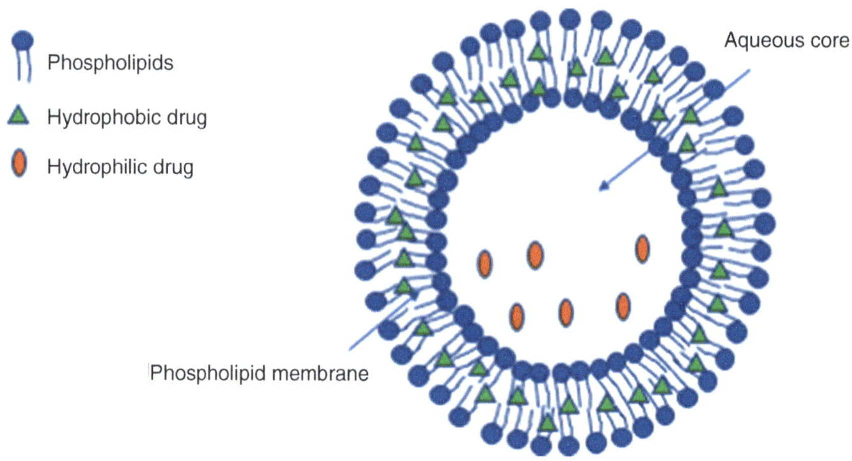

Fig. 1 Schematic structure of conventional liposome

ing half-life, decreasing clearance, and immunogenicity. Consequently, by adding PEG into the liposome composition, liposomes are able to prolong its blood circulation time. Because of the presence of PEG, the mononuclear phagocyte system uptake is decreased and since they are "invisible" to the body's defense system (reticuloendothelial system (RES)), sterically stabilized liposomes are often called stealth liposomes (Nogueira et al. 2013). Liposomes with modified surfaces have been engineered using several molecules such as glycolipids or sialic acid, and modified and unmodified dextrans, which afford stealth liposomal properties. However, stealth liposomes without modified function are not a great candidate for targeting macrophage because the coating leads to reduction of internationalization by macrophages and escape the clearance by macrophages (Lee et al. 2015).

2.4.3 Targeted Liposomes

In the development of targeted formulations, which contain different agents targeted to specific place in the body (e.g., anticancer agent delivery to the tumors), functionalization of liposome with a suitable ligand, (i.e., peptides, antibodies, aptamers, small molecules, etc.) should be prepared with the so-called ligand-targeted liposome or targeted liposome (Fig. 3).

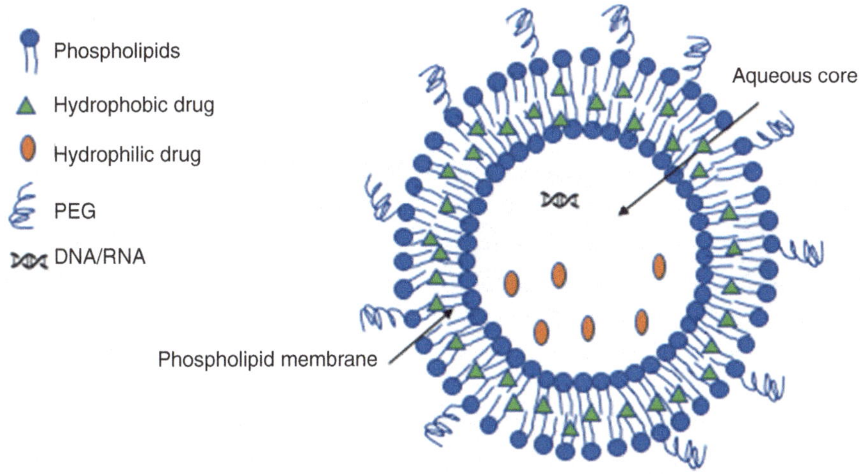

Fig. 2 Schematic structure of stealth liposome

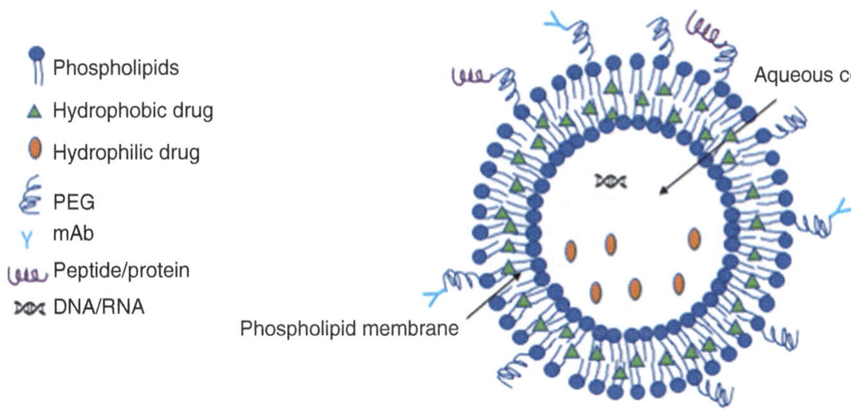

Fig. 3 Schematic structure of targeted liposome

Targeted liposomes help to reduce the impact of drugs on healthy tissues, as compared to conventional liposomes, for cancer therapy (Riaz et al. 2018). The main strategies for the therapeutic agent delivery at the affected site involve two approaches: passive targeting (PEGylated liposome with enhanced permeability and retention effect) and active targeting (functionalized liposome with targeting ligands). In the targeting of macrophages, conventional liposomes could be used besides targeted liposomes. The mechanism of action for macrophage-targeting liposomes is briefly described in Fig. 4. The activity majorly relies on the endocytosis of a drug-encapsulated liposome inside macrophages. Then, phospholipids in liposomes can be digested by lysosomal phospholipases, but the liposome-encapsulated drug still remains in the macrophage and generates its programmed cell death or apoptosis (van Rooijen and Hendrikx 2010).

3 Liposome Targeting Macrophages for the Treatment of Various Diseases

3.1 Diseases Causing Infection

Pathogenic antigens are involved in the process of reacting to phagocytes that are caught and brought to them by mononuclear phagocytes. However, some of the pathogens like *Brucella* species or mycobacteria have the ability to go

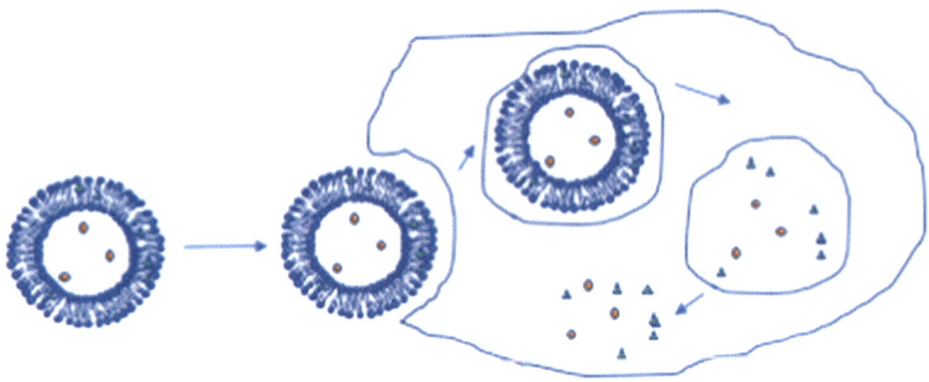

Fig. 4 Schematic mechanism of action of drug-encapsulated liposomes on macrophages

undetected and not be caught by macrophage phagocytosis (Dornand et al. 2002). When these bacteria and viruses escape phagocytosis, it causes them to proliferate inside their cells. In addition, some viruses and bacteria, such as *Brucella* species, have the ability to prolong the survival time of pathogen-infected cells when *Brucella* species are capable of delaying macrophage involvement in the apoptotic process. Many of the undetected pathogen-infected cells can cross tissues like the blood-brain-barrier, when viruses can evade the cells with not much restriction (Carter and Ehrlich 2008). As a result of these kinds of viral and bacterial activities, and their abilities to cause infection and interfere with the phagocytic system, it has become very important to target these pathogens through targeted-therapeutics using mainly negatively charged liposomes. There have been many studies done to assess the effectiveness of liposome target therapies and many of these studies have concluded that liposomes that contain anti-infective agents are very much effective to slow down cell's toxicity and increase the positive outcomes of using anti-infective medications to decrease infection diseases and treating them.

3.2 Cancer and Inflammation Causing Diseases

When there are injuries or cancer, mononuclear phagocytes are the first to appear at the site and eventually lead to a great number of macrophages to be present at the site as well (Ponzoni et al. 2018). Macrophages secrete many proinflammatory cytokines like TNFα) that enhance inflammation. This mechanism has been studied and used by researches in targeted-drug-delivery to treat cancer. At the same time, targeted-drug-therapies use anti-inflammatory agents that are encapsulated in liposomes, which ultimately result in decreases in population of macrophages at the site of injuries with inflammation or cancer. Researchers are currently assessing a liposomal MTP-PE) formulation (L-MTPPE; mifamurtide) that could be used when treating patients with high-risk osteosarcoma (Sousa et al. 2015). Furthermore, many diseases such as endometriosis, lung cancer, and arthritis that involve inflammation have been treated with liposomes that are encapsulated with bisphosphonate. Researchers also have used liposomes encapsulated with propamide to enhance macrophages activity in the apoptotic process (Rao et al. 1995).

3.3 Cardiovascular Diseases

There are a great number of studies done which show that macrophages and monocytes play an important role in atherosclerosis. When endothelial cells are damaged, the chemokines are released, which cause the presence of monocytes at the site. Furthermore, macrophages have a key role in the process of the atherosclerotic plaque formation through the scavenging of oxidized low-density lipoprotein (LDL) and the differentiation into foam cells that make the atherosclerotic plaque core (Bobryshev et al. 2016). In addition, the CD36, a glycoprotein that belongs to scavenger receptor class B, also has a major impact on the plaque formation process. CD36 is expressed in endothelial cells, macrophages, platelets, and monocytes (Park 2014). CD36 deactivation that leads to a decrease in the size of lesion has been greatly studied. Therefore, ligands have been used to target macrophages that express CD36. Hexarelin is a growth peptide with a dual impact and it is used as a ligand with the ability to deliver medication to lesions (Kelly et al. 2011). Hexarelin also has the ability to attach to CD36 receptors because it belongs to the category of hexapeptide-growth-hormone-releasing peptides (GHRPs). More researches about liposome targeting when treating atherosclerotic lesions have led to the discovery of delivering contrast agents that are used in diagnostic imaging. In one study done by Chono and colleagues, researches have studied the liposomal delivery to macrophages when treating atherosclerosis. In that study, researchers used anionic liposomes containing egg yolk phosphatidylcholine (PC), cholesterol, and dicetylphosphate (DCP) at a molar ratio of 7:2:1 and sized to 70, 200, and 500 nm (Chono et al. 2005). In vitro, uptake by macrophages and foam cells was enhanced with increasing particle size when, in vivo, targeted aortic delivery in atherogenic mice was reached using 200-nm liposomes. In addition, dexamethasone encapsulated in liposome has been researched as a possible anti-atherosclerotic liposome-targeting treatment (Wijagkanalan et al. 2008; Tam et al. 2005).

3.4 Cerebral Ischemia and Stroke

There have been a great number of studies done to evaluate the relationship between infiltrating macrophages, the innate immune system, and resident microglia. Studies have shown that one of the factors affecting cerebral ischemic injury is inflammation. When there is inflammation, it causes the upregulation of CD36, which has also been detected in stroke and cerebral ischemia. The upregulation of CD36 would result in excessive inflammatory responses. To prevent such responses, there have been studies done where researches used infiltrating macrophages to deliver a systemically administered gene therapy in stroke (Tanaka et al. 2004). Researches in the studies observed that plasmids expressing enhanced green fluorescent protein (EGFP) and fibroblast growth factor-2 (FGF-2) were complexed with cationic liposomes, administered into the femoral vein, resulting in the expression of EGFP and FGF-2 in infiltrating macrophages and in the cerebral infarction.

4 Conclusions

Over time, roles of monocytes and macrophages in various diseases have been further investigated, which led to the development of different treatment strategies. Improved therapeutic purposes are focused on the effectiveness and safety. The unique characteristics of liposomes have made them an excellent candidate for targeting macrophages in the treatment of diseases, such as infection, cancer, inflammation, etc. This natural targeting capacity can be of great value for drug delivery to target areas. By controlling the liposome's physicochemical properties of size, charge, and lipid composition, natural targeting can be enhanced.

References

Akbarzadeh A, et al. Liposome: classification, preparation, and applications. Nanoscale Res Lett. 2013;8(1):102.

Allen TM, et al. Uptake of liposomes by cultured mouse bone marrow macrophages: influence of liposome composition and size. Biochim Biophys Acta. 1991;1061(1):56–64.

Aoki A, et al. Preparation of pH-sensitive anionic liposomes designed for drug delivery system (DDS) application. J Oleo Sci. 2015;64(2):233–42.

Arifin SA, Falasca M. Lysophosphatidylinositol signalling and metabolic diseases. Metabolites. 2016:**6**(1).

Avantilipids.com. [cited 2020 February 10]; Available from: https://avantilipids.com/product-category/phospholipids.

Badiee A, et al. The role of liposome charge on immune response generated in BALB/c mice immunized with

recombinant major surface glycoprotein of Leishmania (rgp63). Exp Parasitol. 2009;121(4):362–9.
Bangham AD, Standish MM, Watkins JC. Diffusion of univalent ions across the lamellae of swollen phospholipids. J Mol Biol. 1965;13(1):238–52.
Bobryshev YV, et al. Macrophages and their role in atherosclerosis: pathophysiology and transcriptome analysis. Biomed Res Int. 2016;2016:9582430.
Bozzuto G, Molinari A. Liposomes as nanomedical devices. Int J Nanomedicine. 2015;10:975–99.
Carter CA, Ehrlich LS. Cell biology of HIV-1 infection of macrophages. Annu Rev Microbiol. 2008;62:425–43.
Chono S, et al. Efficient drug delivery to atherosclerotic lesions and the antiatherosclerotic effect by dexamethasone incorporated into liposomes in atherogenic mice. J Drug Target. 2005;13(4):267–76.
Chono S, et al. Uptake characteristics of liposomes by rat alveolar macrophages: influence of particle size and surface mannose modification. J Pharm Pharmacol. 2007;59(1):75–80.
Chowdhury N, et al. Targeted delivery of doxorubicin liposomes for Her-2+ breast cancer treatment. AAPS PharmSciTech. 2020;21(6):202.
Dean JM, Lodhi IJ. Structural and functional roles of ether lipids. Protein Cell. 2018;9(2):196–206.
Dijkstra J, van Galen M, Scherphof G. Influence of liposome charge on the association of liposomes with Kupffer cells in vitro. Effects of divalent cations and competition with latex particles. Biochim Biophys Acta. 1985;813(2):287–97.
Dornand J, et al. The innate immune response against Brucella in humans. Vet Microbiol. 2002;90(1–4):383–94.
Dudek J. Role of Cardiolipin in mitochondrial signaling pathways. Front Cell Dev Biol. 2017;5:90.
Ekroos K, et al. Charting molecular composition of phosphatidylcholines by fatty acid scanning and ion trap MS3 fragmentation. J Lipid Res. 2003;44(11):2181–92.
Elsana H, et al. Evaluation of novel cationic gene based liposomes with cyclodextrin prepared by thin film hydration and microfluidic systems. Sci Rep. 2019;9(1):15120.
Falkenburger BH, et al. Phosphoinositides: lipid regulators of membrane proteins. J Physiol. 2010;588(Pt 17):3179–85.
Farine L, et al. Phosphatidylethanolamine and phosphatidylcholine biosynthesis by the Kennedy pathway occurs at different sites in Trypanosoma brucei. Sci Rep. 2015;5:16787.
Frasch SC, Bratton DL. Emerging roles for lysophosphatidylserine in resolution of inflammation. Prog Lipid Res. 2012;51(3):199–207.
Fujiwara Y. Cyclic phosphatidic acid - a unique bioactive phospholipid. Biochim Biophys Acta. 2008;1781(9):519–24.
Gordon S, Taylor PR. Monocyte and macrophage heterogeneity. Nat Rev Immunol. 2005;5(12):953–64.
Hagio M, et al. Phosphatidylglycerol is essential for the development of thylakoid membranes in Arabidopsis thaliana. Plant Cell Physiol. 2002;43(12):1456–64.
Harashima H, et al. Enhanced hepatic uptake of liposomes through complement activation depending on the size of liposomes. Pharm Res. 1994;11(3):402–6.
Ibaraki H, et al. In Vivo fluorescence imaging of passive inflammation site accumulation of liposomes via intravenous administration focused on their surface charge and PEG modification. Pharmaceutics. 2021:**13**(1).
Juliano RL, Stamp D. The effect of particle size and charge on the clearance rates of liposomes and liposome encapsulated drugs. Biochem Biophys Res Commun. 1975;63(3):651–8.
Kelly C, Jefferies C, Cryan SA. Targeted liposomal drug delivery to monocytes and macrophages. J Drug Deliv. 2011;2011:727241.
Kelly C, et al. High-throughput methods for screening liposome-macrophage cell interaction. J Liposome Res. 2015;25(3):211–21.
Kim HY, Huang BX, Spector AA. Phosphatidylserine in the brain: metabolism and function. Prog Lipid Res. 2014;56:1–18.
Kooijman EE, et al. What makes the bioactive lipids phosphatidic acid and lysophosphatidic acid so special? Biochemistry. 2005;44(51):17007–15.
Lagace TA. Phosphatidylcholine: greasing the cholesterol transport machinery. Lipid Insights. 2015;8(Suppl 1):65–73.
Lasic DD, Papahadjopoulos D. Liposomes revisited. Science. 1995;267(5202):1275–6.
Law SH, et al. An updated review of lysophosphatidylcholine metabolism in human diseases. Int J Mol Sci. 2019:**20**(5).
Lee JS, Hwang SY, Lee EK. Imaging-based analysis of liposome internalization to macrophage cells: effects of liposome size and surface modification with PEG moiety. Colloids Surf B Biointerfaces. 2015;136:786–90.
Liu C, et al. Barriers and strategies of cationic liposomes for cancer gene therapy. Mol Ther Methods Clin Dev. 2020;18:751–64.
Maccarone AT, et al. Characterization of acyl chain position in unsaturated phosphatidylcholines using differential mobility-mass spectrometry. J Lipid Res. 2014;55(8):1668–77.
Mehravaran A, et al. Immunogenicity and protection effects of cationic liposome containing imiquimod adjuvant on leishmaniasis in BALB/c mice. Iran J Basic Med Sci. 2019;22(8):922–31.
Mufamadi MS, et al. A review on composite liposomal technologies for specialized drug delivery. J Drug Deliv. 2011;2011:939851.
Nagayasu A, Uchiyama K, Kiwada H. The size of liposomes: a factor which affects their targeting efficiency to tumors and therapeutic activity of liposomal antitumor drugs. Adv Drug Deliv Rev. 1999;40(1–2):75–87.
Neves LF, et al. Preparation and optimisation of anionic liposomes for delivery of small peptides and cDNA

to human corneal epithelial cells. J Microencapsul. 2016;33(4):391–9.
Nogueira E, et al. Liposome and protein based stealth nanoparticles. Faraday Discuss. 2013;166:417–29.
Park YM. CD36, a scavenger receptor implicated in atherosclerosis. Exp Mol Med. 2014;46:e99.
Ponzoni M, et al. Targeting Macrophages as a Potential Therapeutic Intervention: Impact on Inflammatory Diseases and Cancer. Int J Mol Sci. 2018:**19**(7).
Rao M, et al. Intracellular processing of liposome-encapsulated antigens by macrophages depends upon the antigen. Infect Immun. 1995;63(7):2396–402.
Riaz MK, et al. Surface functionalization and targeting strategies of liposomes in solid tumor therapy: a review. Int J Mol Sci. 2018;(1):19.
Schultze JL, Schmieder A, Goerdt S. Macrophage activation in human diseases. Semin Immunol. 2015;27(4):249–56.
Shim JW, et al. Lysophosphatidylglycerol inhibits formyl peptide receptorlike-1-stimulated chemotactic migration and IL-1beta production from human phagocytes. Exp Mol Med. 2009;41(8):584–91.
Smith MC, et al. Zeta potential: a case study of cationic, anionic, and neutral liposomes. Anal Bioanal Chem. 2017;409(24):5779–87.
Sousa S, et al. Liposome encapsulated zoledronate favours M1-like behaviour in murine macrophages cultured with soluble factors from breast cancer cells. BMC Cancer. 2015;15:4.
Spagnou S, Miller AD, Keller M. Lipidic carriers of siRNA: differences in the formulation, cellular uptake, and delivery with plasmid DNA. Biochemistry. 2004;43(42):13348–56.
Tam SP, et al. Peptides derived from serum amyloid a prevent, and reverse, aortic lipid lesions in apoE–/– mice. J Lipid Res. 2005;46(10):2091–101.
Tanaka S, et al. Infiltrating macrophages as in vivo targets for intravenous gene delivery in cerebral infarction. Stroke. 2004;35(8):1968–73.
Tucci ST, et al. Tumor-specific delivery of gemcitabine with activatable liposomes. J Control Release. 2019;309:277–88.
van Rooijen N, Hendrikx E. Liposomes for specific depletion of macrophages from organs and tissues. Methods Mol Biol. 2010;605:189–203.
White GF, et al. Physical properties of liposomes and proteoliposomes prepared from Escherichia coli polar lipids. Biochim Biophys Acta. 2000;1468(1–2):175–86.
Wijagkanalan W, et al. Enhanced anti-inflammation of inhaled dexamethasone palmitate using mannosylated liposomes in an endotoxin-induced lung inflammation model. Mol Pharmacol. 2008;74(5):1183–92.
Wynn TA, Chawla A, Pollard JW. Macrophage biology in development, homeostasis and disease. Nature. 2013;496(7446):445–55.
Zacek P, et al. Quantitation of isobaric phosphatidylcholine species in human plasma using a hybrid quadrupole linear ion-trap mass spectrometer. J Lipid Res. 2016;57(12):2225–34.
Zbinden G, Wunderli-Allenspach H, Grimm L. Assessment of thrombogenic potential of liposomes. Toxicology. 1989;54(3):273–80.

Solid Lipid Nanoparticles-Based Drug and Gene Delivery to Macrophages

Srinivas Reddy Jitta and Lalit Kumar

Abstract

The current chapter discusses targeting macrophages with solid lipid nanoparticles (SLNs) loaded with drugs and genetic material. Macrophages are the key cells and play a crucial role in the maintenance of homeostasis, tissue maintenance, and also the immune system. Targeting macrophages to deliver drugs and genes is an important aspect in the development of delivery systems in the treatment of infections involving macrophages such as leishmaniasis, tuberculosis, and cancers related to lungs. Solid lipid nanoparticles (SLNs) are the emerging nanocarriers to deliver drugs and genes to the target cells, and the application of SLNs to deliver drugs and genetic materials to macrophages is a promising approach to follow. The properties of SLNs and various types of methods of preparation are discussed in this chapter along with the application of SLNs in targeting macrophages. Also, the current scenario of SLNs' application for targeted therapy to enhance its targetability and internalization is discussed in this chapter.

S. R. Jitta · L. Kumar (✉)
Department of Pharmaceutics, Manipal College of Pharmaceutical Sciences, Manipal Academy of Higher Education, Udupi, Karnataka, India
e-mail: lalit.kumar@manipal.edu

Keywords

Solid lipid nanoparticles · Macrophages · Gene delivery · Nanotechnology · Targeted drug delivery · Macrophage receptors

1 Introduction

Macrophages are important cells of not only the immune system but also take part in homeostasis and the maintenance of various kinds of tissues. Macrophages are present in almost all the tissues, especially the liver, lungs, spleen, and lymph nodes. They are actively involved in the detection and destruction of foreign particles and harmful organisms such as bacteria by phagocytosis. Macrophages play a key role in inflammation by presenting antigens to T-cells and by releasing cytokines that activate other cells. Macrophages also play a major role in various chronic diseases, which include asthma, arthritis, atherosclerosis, fibrosis, and various lung cancers. Hence, targeting macrophages to deliver drugs and genetic material that can alter its dysfunction helps in treating various diseases that involve macrophages. Solid lipid-based nanoformulations are the emerging novel drug delivery systems that are promising in addressing key issues associated with conventional carriers. Development of solid lipid nanoparticle-based drug delivery systems to target macrophages to deliver the drugs and genes

S. Gupta, Y. V. Pathak (eds.), *Macrophage Targeted Delivery Systems*,
https://doi.org/10.1007/978-3-030-84164-5_10

is becoming one of the key areas of research. The current chapter discusses the importance of targeting macrophages to deliver drugs and genes using solid lipid nanoparticles (SLNs). The different methods for the preparation of SLNs and various surface modifiers to improve the targeting and cellular uptake are also discussed. Transfection enhancers that are used in the preparation of SLNs-based gene delivery systems to improve transfection and finally, various assays to evaluate transfection agents are discussed in this chapter.

2 Nanotechnology in Gene Therapy

Currently, nanotechnology has become the most widely used technique to improve treatment strategies in the field of cancer and also other diseases (Gong et al. 2020). A variety of nanovectors are being developed by researchers in the currently to deliver drugs and gene (Yu et al. 2010). The delivery of nucleic acids using nonviral transfection agents such as peptides, polymer nanoparticles, lipid nanoparticles (such as liposomes, SLNs, and nanostructured lipid carriers (NLCs)), and nanoemulsions is creating novel opportunities in the field of gene therapy by the applications of nanotechnology in novel drug delivery systems. Targeted delivery of gene is feasible with both viral and nonviral vectors, as both of them can bind with nucleic acids and form a complex known as polyplexes (Limeres et al. 2019; Ye et al. 2008). Vectors such as adenoviruses and retroviruses can be used to deliver genes, though this approach results in higher efficiencies, but nonspecific binding, DNA package size limit, insertional mutagenesis, and immunogenicity are the few disadvantages which make researchers to look for other alternatives (Suñé-Pou et al. 2018). Nonviral transfection agents are safer as they show low cytotoxicity, include easy and cost-effective preparation methods, and the feasibility of scale-up and low immunogenicity make them a preferable choice over viral vectors (Limeres et al. 2019; Ye et al. 2008). Alternatively, physical transfection and mechanical transfection methods can also be used for the delivery of a gene to the targeted tissue or cells. Laser irradiation/sonoporation and electroporation are the physical transfection methods, whereas direct microinjection and gene gun are the mechanical methods for delivering genetic material to the targeted cells. Even though some of these methods are successful in the transfer of genetic material to the target cells with good transgene expression, still stability and degradation of the gene and toxicity are the major issues associated with it. Development of nanotechnology-based nucleic acid delivery systems has gained considerable attention from the researchers. Nanoparticle-based delivery systems are considered to be less toxic, with less oncogenic and immunogenic nature.

3 Drug Delivery Systems for Targeted Therapy

Targeted therapy is one of the successful drug delivery strategies followed by the researchers in the recent past to address challenges in the treatment of various cancers, such as breast cancer, lung cancer, and bladder cancer. A drug delivery system that can achieve specific localization to the tumor with low toxicity is the ideal one for cancer treatment (Liang et al. 2021). Conventional drug delivery systems are associated with several limitations such as nonspecific distribution, lack of targetability, high toxicity, low solubility, and narrow therapeutic index window. Application of nanotechnology to develop drug delivery systems can address all these limitations of conventional delivery systems. Nanoformulations are usually up to the size of 500 nm. Nanoparticles increase the surface area of the drug and hence enhance the solubility of drugs in the aqueous medium. The nanoformulations that are designed with optimal size, morphology, and surface properties enhance solubility, improve biodistribution, and also exhibit low immunogenicity. As most of the raw materials that are used in the preparation of nanoformulations are biocompatible, the risk of toxicity is also less. Nanoformulations are not only efficient in delivering drugs to the targeted sites but also very effective in the delivery of nucleic acids such as DNA/RNA/short genetic sequences. Surface properties of nanoformulations play a key role in the targeting and distribution of therapeutic agents. Nanoformulations are more stable in an in vivo environment in comparison with the traditional delivery systems, and this stability helps in exhibiting a longer half-life in systemic circulation and prolonged therapeutic effect. As therapeutic agents are encapsulated in nanoformulations, they can protect from environment such as low pH conditions of the stomach. Nanoformulations with enhanced permeability and retention (EPR), and mononuclear phagocytosis are considered as the important parameters in designing the types of nanoformulations for a specific type of disease targeting (Jin et al. 2019).

4 Gene Therapy

Gene therapy has become one of the most promising therapeutic approaches to cure diseases or to improve the ability of body to fight against diseases. Control administration of a therapeutic gene or nucleic acids to target cells helps in effective curing of genetic diseases. Gene delivery systems are the novel drug delivery approaches to use nucleic acids for the treatment of various diseases. Gene therapy is effective in a wide range of diseases, such as cancers of different organs (Liang et al. 2021; dos Santos et al. 2020; Vighi et al. 2010). Gene therapy has also become one of the novel approaches in various kinds of drug therapies and vaccination also, where nucleic acids are being administered, which can regulate the suppression or expression of proteins of interest resulting in the correction of genetic alteration or induction of cell-mediated immunity or reversing disorders. Even though a lot of research is going on in this field, still poor clinical outcomes of the application of plasmid DNA in the treatment of diseases emphasize the importance of further efforts to address the key issues. Lower cellular uptake, instability because of rapid degradation by the nucleases, and ineffective delivery to the target calls are some of the critical challenges. Viral vectors are the biological carriers that are derived from naturally evolved viruses and are capable of transferring genetic material to the target cells. But these carriers suffer from drawbacks such as mutations as a safety concern, need for packaging cell lines, immunogenicity, and carcinogenesis. Nonviral vectors are safer than the viral vectors and also offer flexibility in the design of formulation, size, and topology of plasmid vector complex, which can be tailored. Additionally, nonviral vectors are capable of carrying large inserts (Yu et al. 2009, 2010; Limeres et al. 2019; Liang et al. 2021; Vighi et al. 2010). Various kinds of nonviral vector delivery systems have been developed by researchers to transfer genetic material to the target cells. These include cationic peptides, lipids, and polymeric carriers. SLNs serve as a very good carrier to transfer gene, as they offer many technical advantages such as the possibility of sterilization, lyophilization, better storage stability, and also the possibility to manufacture on a large scale (Yu et al. 2009). Enhanced permeability and retention (EPR) is an effective strategy for delivering genes to tumor sites that are large and well vascularized. But this approach has limitations in delivering nanochemotherapeutics to tumors, which are small in size with poor vascularization and high dispersion. Development of a safe vector that can deliver transgene to the specific target site with a long-term expression is the key challenge in the development of successful gene therapy (Suñé-Pou et al. 2018; dos Santos et al. 2020).

5 Macrophages

Macrophages are important cells of the immune system, and through a direct or an indirect mechanism these cells play a multifunctional role in the immune response (Sun et al. 2008). These are the components of innate immunity and are mainly divided into two types based on their function and phenotypes: M1 macrophages and M2 macrophages. M1 macrophages are known as classically activated and M2 macrophages are called as alternatively activated macrophages. M1 macrophages are involved in the production of the pro-inflammatory process that protects against foreign invaders, whereas M2 macrophages express anti-inflammatory activity in inflammatory diseases. The macrophages present in tumor macro environment are called as Tumor-Associated Macrophages (TAMs) and these macrophages suppress the anti-tumor immunity. Mononuclear cells present in the blood transform into TAMs in tumor site, whose phenotype is similar to that of M2 macrophages. As M1 macrophages are crucial in inflammation and M2 macrophages in the tumor, targeting these macrophages for treatment strategies will be a successful approach to design formulations for macrophage targeting (Jin et al. 2019).

Macrophages are monocyte-derived myeloid cells and play key roles in autoimmune diseases related to inflammation and cancers. But uncontrolled activation of myeloid cells results in chronic illness, including autoimmune, neoplastic, metabolic, pulmonary, cardiovascular, and neurodegenerative diseases (Wang et al. 2015). As macrophages are crucial in pathogen control, tissue maintenance, and immune regulation, they can serve as important targets for gene delivery to modulate their function (Mahor et al. 2012). Macrophages influence the physiological process such as the development of tissues, remodeling, and homeostasis. Cancer progression and metastasis are influenced by these physiological processes (Gong et al. 2020). Application of gene therapy to target macrophages in pulmonary conditions such as inflammation, infectious diseases, and lung cancers is an exciting area of research. It is well known that alveolar macrophages are key cells in the first-line host immune response system and also a key component of cancer-inducing inflammatory reactions. Targeting of macrophage-derived mediators with drugs and gene provides a novel strategy against metastasis and tumor invasion (Yu et al. 2010). Figure 1 illustrates the diseases associated with macrophages.

Mycobacterium tuberculosis is one of the most threatening infectious bacteria that transmit mainly through aerosolized particles exhaled by the people who have tuberculosis. It is well known that *Mycobacterium tuberculosis* mainly resides in macrophages, and the cell-mediated immune response induced to clear the pathogens involves the alveolar macrophages, so targeting these macrophages to deliver antitubercular drugs enhances the treatment efficiency. Moreover, targeting alveolar macrophages is

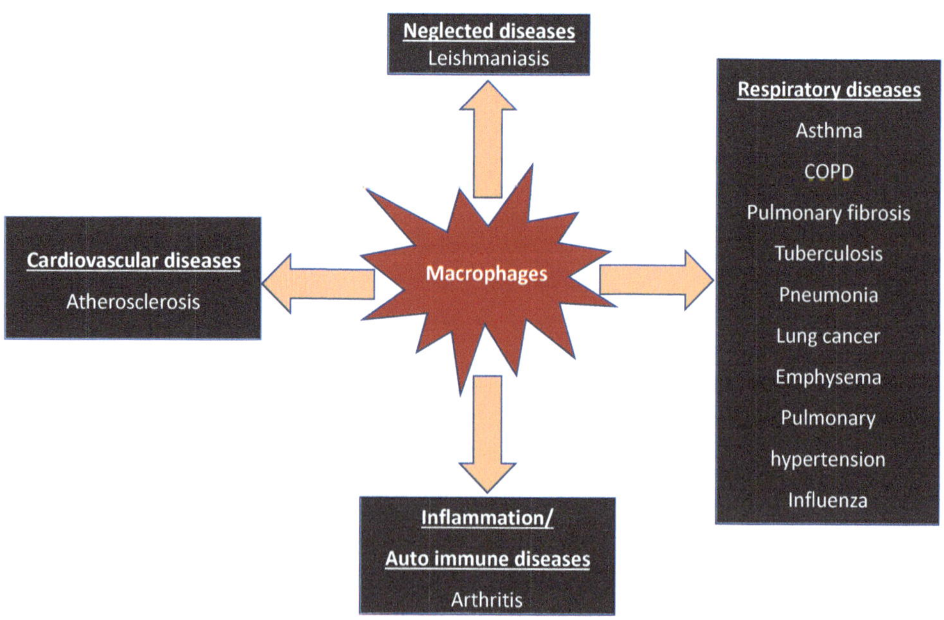

Fig. 1 Various diseases that involve macrophages

advantageous in the case of development of nasal drug delivery systems. Also, as long-term treatment schedules with conventional therapy usually associated with severe adverse effects such as hepatotoxicity, resulting in poor patient compliance to the treatment, targeted drug delivery systems such as SLNs help in enhancing the concentrations in macrophages of affected organs and also a controlled release of drugs is possible. Furthermore, the appearance of multidrug-resistant strains may also happen in the case of infections of *Mycobacterium tuberculosis*. Hence, the development of novel drug delivery systems that can deliver drugs to the specific target cells is required to improve the drug delivery, so that treatment duration can be reduced. Eventually, it helps in reducing adverse effects, improves patient compliance toward the treatment, and most importantly avoids the development of multidrug-resistant strains (Vieira et al. 2018; Ma et al. 2020).

Uncontrolled activation of macrophages results in many kinds of acute and chronic diseases. Pro-inflammatory mediators such as nitric oxide, IL-6, IL-1β, and TNF- α are released by activated macrophages, act as immunomodulation markers, and play a critical role in inflammation. Various kinds of cytokines produced by the macrophages during the process of inflammation play a key role in the initiation, maintenance, and resolution of inflammation by taking part in various functions, such as phagocytosis, immunomodulation, and antigen presentation (Wang et al. 2015). Inflammation in rheumatoid arthritis mainly affects activated macrophages, and the continuous release of cytokines by these macrophages aggravates the arthritic condition. CD44, a pro-inflammatory cell surface glycogen protein, is overexpressed on the surface of macrophages, fibroblasts, and lymphocytes present in the inflamed joints in rheumatoid arthritis. This protein facilitates the activation and migration of various effector cells and thus promotes the progression of arthritis (Zhou et al. 2018). Wang et al. prepared SLNs to encapsulate curcumin to target inflammatory cells and the study reported that the curcumin-loaded SLNs were superior in delivering and internalization of curcumin in RAW264.7 cell lines in comparison with free curcumin. The curcumin-loaded SLNs inhibited release of pro-inflammatory mediators in comparison with free curcumin. Furthermore, researchers were able to improve the internalization of curcumin because of the smaller particle size of SLNs and also the viability of cells improved by inhibiting apoptosis (Wang et al. 2015).

Most of the antibiotics such as beta-lactams that are effective in in vitro assay sometimes fail to show good permeability and intracellular accumulation properties in in vivo. This results in poor efficacy against intracellular bacteria such as *Mycobacterium*, *Salmonella*, and *Staphylococcus* that are present in macrophages. Drugs that cannot penetrate into and accumulate in macrophages allow these bacteria to survive for longer periods. Development of SLNs-loaded antibiotics against these kinds of bacteria helps in improved permeation and also cellular internalization in macrophages. Moreover, the development of SLNs for these drugs is of great interest globally for clinical applications. Long-chain saturated fatty acids are one of the commonly used lipid matrix materials in the preparation of SLNs. The length of the fatty acid chain may have an impact on the in vitro drug release and also the pharmacokinetics of the drug encapsulated in it. In some of the studies, it was reported that SLNs prepared with long-chain fatty acids are effective in enhancing the internalization of drug in RAW 264.7 cells, in comparison with free drug. Long-chain fatty acids also improves retention time of drugs in the body (Meng et al. 2020).

Macrophages along with monocytes play an important role in the development of atherosclerosis by contributing to plaque growth, cholesterol accumulation, and inflammatory response at various stages of the disease development. Atherosclerosis is characterized by the chronic inflammatory response and formation of plaques in arterial walls leading to the narrowing of the affected vessel lumen, which causes a higher risk of thrombosis. Ischemic stroke and cardiac death are the lethal events associated with atherosclerosis. Mitochondrial DNA damage leading to the dysfunction of mitochondria is a well-studied cause of death observed in atherosclerosis. The dysfunctional mitochondrial DNA is a promising target to treat atherosclerosis using lipid-based formulations such as gene-based SLNs. Zakirov et al. summarized current knowledge on gene-loaded lipid-based nanoformulations to target macrophage mitochondria (Zakirov et al. 2020).

6 Macrophages as Therapeutic Targets

Macrophages are important cells to consider for therapeutic targeting in a variety of diseases such as diabetes, cardiovascular dysfunctions, cancers, etc. As macrophages reside in almost of all the tissues, it makes a good choice as a therapeutic target to deliver the drugs and nucleic acids. They are given different names in different tissues, such as Kupffer cells in the liver, adipose tissue macrophages, arterial macrophages in atherosclerotic plaques, red pulp macrophages in red blood cells, tissue-associated macrophages in tumors, and microglia in the brain. Macrophages play a key role in inflammation by acting as pro-inflammatory and inflammatory cells. Macrophages convert their phenotypes in different conditions, for example, the phenotype is relevant to acute infection when they fight against phagocyte dead cells and pathogens but they convert their phenotype to assist in the process of wound healing. Unregulated activation of macrophages results in several abnormal effects in chronic inflammation such as excess cytokine release which can interfere with insulin signaling that leads to insulin resistance in adipose tissue. Nonalcoholic fatty liver diseases occur due to high-fat or carbohydrate consumption for a long time. The chemokines released by the inflamed hepatocytes activate Kupffer cells causing them to secrete cytokines such as TNF-α, IL-6, IL-1β, and nitric oxide synthase 2. Inflamed hepatocytes and Kupffer cells also signal for the infiltration of monocyte-derived macrophages that express phenotype as pro-inflammatory cells. The other critical disorder that involves is atherosclerosis. Healthy artery wall consists of macrophages like other tissues, and recruitment of additional macrophages takes place in hyperlipidemic conditions leading to the formation of atherosclerotic lesion. Myelopoiesis is the process by which recruitment of monocytes takes place in bone marrow, and it is induced by hyperlipidemia and high-fat food intake. These monocytes differentiate into macrophages, and oxidation of lipids by endocytosis takes place because of the express scavenger receptors leading to lipid accumulation. The foamy macrophages resulted from lipid accumulation send chemotactic signals to recruit additional monocytes, which ultimately results in the formation of a lipid-enriched inflammatory milieu. Targeting these macrophages to reduce its inflammatory potential may provide a very good therapeutic target in the treatment of atherosclerosis (Peterson et al. 2018). Hence, to consider macrophages as a therapeutic target, it is important to take note of the pathogenesis of the disease treated and also the phenotype of the macrophages targeted.

The most common therapeutic strategies for targeting macrophages are proliferation, depletion, inflammation, and gene silencing, based on the type of diseases to treat. The condition in which depletion of macrophages is required, apoptosis will be induced following the accumulation of toxic component's release into macrophages. Conditions like atherosclerosis, targeting proliferation, help in reducing the number of macrophages being recruited. Alteration of signal transduction pathways of inflammation is another approach. Modulation of inflammatory cytokine production by using anti-inflammatory agents is one of the most commonly used strategy to target macrophages in disease conditions where inflammation is involved. Involvement of small interference RNA (siRNA) is the most advanced and attractive approach through which inflammatory gene expression can be modulated and also downregulation of multiple genes is possible simultaneously.

7 Targeting of Macrophage Receptors

Currently, most of the macrophage-targeting strategies are based on receptor-mediated phagocytosis. The delivery systems modified with specific surface modifiers that can be recognized by the receptors on macrophages are one of the promising approaches to deliver drugs and nucleic acids to macrophages. Few receptors such as CD68, CD11b, and F4/80 are expressed on almost all the macrophages, but receptors such as mannose, folate, adenosine, and lectin are expressed by some specific macrophages. Receptor-focused approach of targeting offers a direct route of delivering drugs and nucleic acids while minimizing off-targets (Peterson et al. 2018).

Though, several receptors which are considered to be the potential targets to target macrophages in treating diseases are associated with macrophage involvement, mannose receptor, a C-type lectin I transmembrane protein, is one of the most promising targets for nanoparticle-based delivery systems to target macrophages for delivery of drugs and genetic materials. Mannose receptors are overexpressed on APCs (antigen-presenting cells) such as dendritic cells and macrophages. Mannose receptors play a crucial role in inflammatory response, endocytosis, antigen presentation, and can also recognize N-acetylglucosamine, fructose, and mannose. The phenomena of mannose receptor endocytosis because of the abundant expression of mannose on macrophages could be used in designing nanoformulations to target macrophages to deliver the drugs and genetic material. Surface modification of nonviral vectors with mannose improves transfection efficiency and cellular uptake by the macrophages (Mahor et al. 2012; Zakirov et al. 2020). Integrins are the other class of surface proteins that can be targeted to deliver drugs and genes to macrophages. Integrins play a key role in cell adhesion and can recognize phospholipids, proteins, extracellular matrix, and also various amino acid sequences. Conjugation of some specific proteins to the lipid nanocarriers can enhance cellular uptake by macrophages. Pattern recognition receptors and toll-like receptors are the other receptors that can be targeted with the help of a suitable ligand to target macrophages to deliver the drugs and genetic material with the help of nanodelivery systems (Zakirov et al. 2020). Despite a lot of research going on in the development of transfection agents with low toxicity and high efficiency, a critical gap remains in the development of effective targeting techniques. Surface modification of nanovectors with ligands to target and internalize in specific cells is one of the promising approaches in the development of successful gene vectors in the treatment of many diseases such as cancers (Yu et al. 2010).

Mannose receptors are known to recognize several microorganisms, such as *Mycobacterium, Candida*, *Leishmania*, *Pneumocystis*, etc., owing to their mannan coating on their cell wall. The lungs are constantly exposed to the particles and microbes by the airways, alveolar macrophages with a phagocytotic activity which is mainly because of mannose receptors playing a crucial role in the defensive mechanism. The surface expression of mannose receptor during infections is upregulated by the surface protein A present in the lungs. Mannosylated drug delivery systems are efficient in delivering the drug to macrophages, as the surface modification directs the carriers to get phagocytosed by macrophages. This kind of approach is very much useful in infections related to lungs, especially in the treatment of tuberculosis. Maretti et al. developed rifampicin-loaded solid lipid nanoparticles that were surface modified with mannose derivative to target alveolar macrophages through inhalation. The study demonstrates that the mannosylated surface modification of SLNs is effective as a surfactant in terms of improving respirability and also cellular uptake by alveolar macrophages with mannose receptors (Maretti et al. 2019). Few examples of drugs targeting macrophages by various receptors are presented in Table 1.

Table 1 Macrophage-targeting drugs through different receptor targets

S. no.	Drug name	Target	Effect
1	Hu5F9-G4	CD47	Repolarize TAMs
2	Imiquimod	TLR7	Repolarize TAMs
3	Emactuzumab	CSF-1R	Deplete macrophages
4	Cabiralizumab	CSF-1R	Deplete macrophages
5	Bisphosphonates	Phagocytes	Deplete macrophages
6	Clodronate	Phagocytes	Deplete macrophages
7	Tofacitinib	JAK1, JAK2	Inhibit macrophages
8	Infliximab	TNF	M2 polarization
9	Dimethyl fumarate	NRF-2, NF-κB	M2 polarization
10	Fingolimod	Sphingosine-1-phosphate, lymphocytes	M2 polarization
11	Forskolin	CD86, ARG1	M2 polarization

Gholamin et al. (2017); Chi et al. (2017); Ries et al. (2014); Qiu et al. (2018); Cannarile et al. (2017); Zeisberger et al. (2006); Zhan et al. (2014); Van Lent et al. (1998); Calderon et al. (2006); Yarilina et al. (2012); He et al. (2020); Van Schouwenburg et al. (2013); Nally et al. (2019); Veremeyko et al. (2018)

8 Solid Lipid Nanoparticles

Solid lipid nanoparticles were introduced during the 1990s in an attempt to use as an alternative drug delivery system to traditional colloidal carriers, such as polymeric nanoparticles, emulsions, and liposomes (Qi et al. 2012; Olbrich et al. 2001). SLNs are being considered as one of the most promising approaches to drug delivery systems for many diseases. Application of SLNs in the field of clinical medicine is increasing currently because of its potentiality as well-tolerated drug carriers. SLNs are known to have very good biocompatibility, high stability, and a well-established safety profile of the excipients that make them a good choice for developing the drug delivery systems. Solid lipid nanoparticles has the flexibility of drug administration. These can be administered by oral, ocular, topical, nasal and intravenous route. Moreover, the physicochemical properties of lipids used in the preparation of SLNs, such as low melting point, are an added advantage, as it is possible to prepare SLNs by direct emulsification so that toxic solvents can be avoided. Also, the existence of other robust methods and the possibility of sterilization by lyophilization make them a strongly recommended nanodrug delivery system for many diseases. It is also possible to prepare controlled release therapeutic agents by choosing a suitable lipid matrix in the preparation of SLNs. Figure 2 presents the advantages of SLNs (Ye et al. 2008; Mannucci et al. 2020; Fàbregas et al. 2014; Rudolph et al. 2004; Akbaba et al. 2018).

8.1 Preparation of Solid Lipid Nanoparticles

Solid lipid nanoparticles are lipid-based drug delivery systems with distinctive properties, such as high drug loading capacity and capacity to enhance the bioavailability of drugs. There are two models for the drug incorporation in SLNs, solid solution model in which the drug is dispersed in the lipid matrix by the application of a surfactant, where the drug will have distinct interaction with the lipid matrix. The second model is drug-enriched shell model, where the drug is added to the molten lipid. The recrystallization of lipid happens with cooling down resulting in the formation of a solid lipid core. The drug concentrates in the liquid outer shell of SLN because of reduction in dispersion temperature. Cooling of nanoemulsion results in supersaturation of drug that is dissolved in the molten lipid (Dolatabadi et al. 2015).

Preparation of SLNs mainly involves excipients such as solid lipid, emulsifier, and solvent/water. Fatty acids, partial glycerides, triglycerides, waxes, and sterols are the most commonly used solid lipids in the preparation of SLNs. A variety of emulsifiers are being used by researchers to stabilize the solid dispersion and a mixture

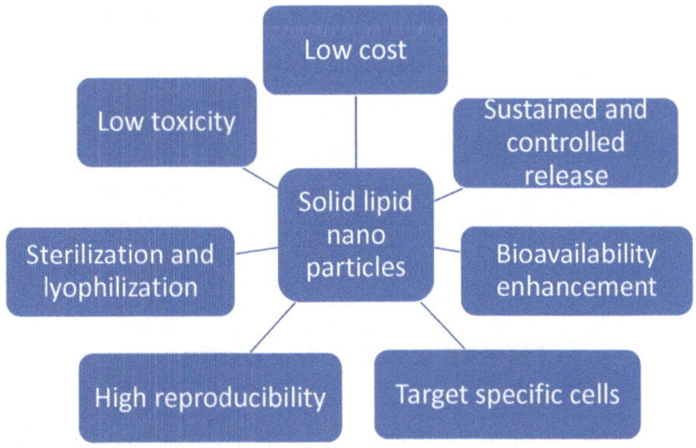

Fig. 2 Characteristics of solid lipid nanoparticles

of emulsifiers helps in avoiding agglomeration of nanoparticles during and after the preparation of SLNs (Dolatabadi et al. 2015). Following are the techniques for the preparation of SLNs.

8.1.1 Solvent Emulsification/Evaporation Method

The main principle of this technique is the precipitation of dispersion in the emulsion which is typical of oil/water (o/w) kind. The lipid is dissolved in an organic solvent that is immiscible with water followed by emulsification in the aqueous phase. The dispersion of SLNs precipitates with the evaporation of organic solvent. The method is capable of producing small-size nanoparticles with high reproducibility. The major disadvantage of this method is that it causes residual solvent after preparation of nanoparticles which results in toxicity (Dolatabadi et al. 2015).

8.1.2 Double Emulsion Method

This technique is useful in the preparation of SLNs of hydrophilic drugs. This method is also based on emulsification and solvent evaporation. The procedure involves the preparation of water/oil/water emulsion and a stabilizer to avoid the partitioning of hydrophilic drug into the outer water phase during the process of evaporation. SLNs can be recovered with centrifugation, once the solvent gets evaporated (Dolatabadi et al. 2015).

8.1.3 Homogenization Method

This method is based on the homogenization of the pre-emulsion of drug incorporated in lipid. Hot homogenization and cold homogenization are two types of techniques that come under this method. Hot homogenization technique involves the preparation of pre-emulsion by adding the drug to the molten lipid followed by high-speed or high-pressure homogenization. For thermolabile/heat-sensitive compounds, cold homogenization technique is suitable. The cold homogenization technique involves the preparation of pre-emulsion by adding the drug to the molten lipid followed by rapid cooling with the help of ice or liquid nitrogen. High-pressure or high-speed homogenization helps to reduce the size of particles and produces the nanoparticles (Dolatabadi et al. 2015).

8.1.4 Hot Emulsion and Ultrasonication

This method is one of the simplest and it involves the melting of lipid by heating above its melting point followed by the emulsification with an aqueous phase. The emulsion is sonicated using a sonicator, resulting in the formation of nanoparticles.

8.1.5 Microemulsion-Based SLN Preparation

The technique involves the dilution of microemulsion, where SLNs are produced by stirring the mixture consisting of low melting fatty acids, emulsifiers, co-emulsifiers, and water, and maintained at optimum temperature. This is followed by the dispersion of hot emulsion in cold water under stirring. The usual dilution ratio of 1:25 or 1:50 of microemulsion to cold water is used.

8.1.6 Supercritical Fluid Technique

This is the most advanced technique used in the preparation of SLNs. The main advantage of this method is, it does not involve any solvent in the process, instead, supercritical carbon dioxide is used as the solvent for the preparation of SLNs.

8.2 SLNs as a Carrier for Gene

The most common nonviral transfection agents include cationic lipids, liposomes, polycationic polymers, cationic peptides, and dendrimers. Cationic SLNs (cSLNs) are considered as one of the most promising nonviral transfection agents or an alternative DNA carrier for the assembly of gene transfer systems. Cationic SLNs are o/w micro-emulsions where the liquid lipid is replaced by a solid lipid dispersed in an aqueous surfactant solution or water (Limeres et al. 2019; Olbrich et al. 2001; Montana et al. 2007). Manipulation of biological processes such as receptor targeting, nuclear translocation of DNA, and endosomal escape limits the efficiency of these nonviral transfection agents (Sun et al. 2008; Olbrich et al. 2001). Apart from the efficiency of nonviral agents, it is also important to have a suitable efficiency/toxicity ratio to become an ideal transfection agent. Type of lipid matrix, particle size, and surface modifier are few key components that decide the efficiency of the systems and also, encapsulation of transfection enhancers helps in the improvement of efficiency of transfection agents. The selection of surfactant in the preparation of nonviral transfection agents, especially in the case of SLNs, plays a key role in the binding of DNA. Differences in DNA binding may result because of the insufficient affinity between hydrophobic domain of the surface modifier and lipid matrix, as it may lead to dissociation of surfactant from particle surface during the interaction with DNA. The ratio of lipid matrix to DNA influences the binding behavior of DNA with the SLN. Olbrich et al. observed an increase in the binding behavior of DNA with SLN when the ratio of lipid matrix to DNA increases from 10 to 60 equivalents. The researchers also observed that higher amounts of transfection agent were required in the absence of enhancer to promote the transfection, which emphasizes the importance of an enhancer in gene transfer by SLNs (Olbrich et al. 2001). Even though nonviral delivery systems are advantageous for gene therapy in aspects of high reproducibility, low-cost production, and safety, the lower transfection efficiency is its major limitation. In the case of cationic SLNs, the positive charge of the particles causes aggregation and hemagglutination that sometimes result in toxicity. Development and optimization of a method that can improve the transfection efficiency of nonviral vectors to deliver genetic material with the least toxicity is one of the most important aspects (Delgado et al. 2012).

Administration of naked genetic material such as DNA/siRNA, especially through intravenous administration, is associated with several problems such as elimination by hepatic clearance and digestion by nucleases in biological fluids. This emphasizes the importance of gene delivery systems development for safe and effective delivery of genetic material to the specific target cells. Conventional drug delivery systems are not so capable of treating intracellular infections where parasites are localized intracellularly. Conventional nanocarriers are prone to clear from the systemic circulation by reticuloendothelial system (RES) and macrophages in various organs such as liver, spleen, and lungs. SLNs are capable of enhancing cellular uptake of drugs or genomic material and are an important treatment strategy for intracellular infections. SLNs are efficient in delivering, from simple lipophilic molecules to complex biomolecules such as genetic material (DNA/siRNA). Lipid nanoparticle – a nucleic acid complex that contains cationic lipids with pKa value less than 7 that exhibit positive charge in biological systems are considered to be the most ideal delivery systems for the delivery of nucleic acids. The positive charge on the surface is advantageous in forming an association with the macrophage membrane which is negatively charged. This helps in increasing cellular uptake but at the same time, there is a chance of an increase in association with serum protein that leads to a quick clearance from the systemic circulation by reticuloendothelial system (RES) causing reduced delivery to the target cells (Soni et al. 2014; Lin et al. 2013).

Association of lipids with cell-penetrating peptides such as sweet arrow peptide (SAP) or transcriptional activator protein (TAT) is one of the approaches to improve the transection efficiency (Delgado et al. 2012). Gene transfer efficiency mediated by nanoparticles remains lower in the presence of endosomolytic agents such as chloroquine. Standard transfection agents such as polyethyleneimine are superior in enhancing transfection efficiency in comparison to chloroquine. Pre-compaction of genomic material with peptides during the formulation of gene vector complex formation is the alternative methodology followed for the improvement of gene transfer efficiency. Rudolph et al. demonstrated up to two orders of magnitude enhancement in the transfection efficiency of the gene vector complex using the HIV-1 TAT peptide for the pre-compaction of DNA (Rudolph et al. 2004).

Many strategies have been followed by the researchers to improve the transfection efficiency of nonviral gene delivery systems, such as employing the various compositions of constituents used in the formulation and using transfection enhancing agents, which are some of the successful approaches. Tabatt et al. worked on the enhancement of transfection activity of SLNs by optimizing cationic lipid and matrix lipid composition. From this study, researchers concluded that the transfection activity depends on both matrix lipid and cationic lipid used in the preparation of formulation and that the cytotoxicity of the formulation depends on the cationic lipid used in the preparation. A combination of cetyl palmitate and N-[1-(2,3-dioleoyloxy) propyl]-N,N,N-trimethylammonium chloride (DOTMA) showed a superior transfection efficiency in comparison with other combinations that include compritol ATO 888, N,N-di-(beta-stearoylethyl)-N,N-dimethylammonium chloride, benzalkonium chloride, cetylpyridinium chloride, cetrimide, chloroquine phosphate, and polyethyleneimine. The detergents used in the preparation of SLNs will have a great impact on the cytotoxicity of the formulations. Detergents with one tail exhibit higher toxicity in comparison with two-tailed cationic lipids when cytotoxicity assay was performed on COS-1 cell lines. Cytotoxicity of formulations depends not only on the aliphatic tails of the cationic lipids but also on its biodegradable nature. Cytotoxicity will be less with cationic lipids which can metabolize easily and it is more with detergents with lower biodegradability. The activity of liposomal transfection agents is influenced by the presence of serum and to some extent SLNs also. The effect of serum on the activity can be determined by supplementing the medium with serum. Though the binding of the plasmid is important, insufficient dissociation could occur after internalization because of too-tight binding. But there is no correlation between the effective transfection and ability of plasmid immobilization (Tabatt et al. 2004).

Cetyltrimethylammonium, a single-tailed cationic lipid, is one of the most commonly used lipids in the development of gene delivery systems and it is known to have a good loading capacity and to promote gene transfection. Though it may exhibit higher toxicity than its double-tailed counterparts, such as dimethyl dioctadecyl ammonium bromide (DDAB) and N-[1-(2,3-dioleoyloxy)propyl]-N,N,N-trimethylammonium chloride (DOTMA), cytotoxicity can be attributed to its biocompatibility than the number of aliphatic chains. Yu et al. prepared cationic solid nanoparticles to deliver the gene. In this study, researchers synthesized a novel single-tailed cationic lipid and prepared SLNs by the nanoprecipitation method. The novel cationic lipid was evaluated for its potential as a gene transfection agent. The toxicity studies by the in vitro method demonstrated lesser toxicity than cetyltrimethylammonium bromide (CTAB) and Lipofectamine. The prepared nanoparticles exhibited a good transfection efficiency in A549 and HeLa cells than CTAB (Yu et al. 2009). In another study by Choi et al., cationic SLNs were prepared by the hot melt homogenization method where no organic solvent was required to be used. The nanoparticles were evaluated to confirm its potential as a nonviral vector to transfer p53 gene to lung cancer cells. It was reported in this study that, the cationic SLNs prepared were showing higher transfection efficiency than the commercially available Lipofectin when they performed in vitro by Fluorescence activated cell sorting (FACS) analysis. The gene expression studies conducted by Reverse transcription - Polymerase chain reaction (RT-PCR) and Western blot analysis revealed high protein expression of the gene (Choi et al. 2008).

Slow, sustained, and controlled release of drugs or genetic materials from drug delivery systems helps in reducing dosing frequency that leads to effective therapy. Apart from crossing biological barriers, frequent dosing for long-term therapeutic effect is the other major issue faced by the researchers in the development of siRNA-based therapeutics. In most of the cells, uptake of unmodified siRNA is very less because of its larger size and also the negative charge is another limitation for effective gene delivery in in vivo. Development of SLNs is one of the approaches to overcome all these problems associated with the development of controlled and sustained release delivery systems for siRNA. One such kind of study was performed by Lobovkina et al., in which researchers demonstrated the sustained release of siRNA from cationic SLNs in in vivo studies where drug release was up to 13 days followed by intradermal injection. As siRNA is anionic, it is difficult to form a complex with cationic SLNs. To overcome this issue, researchers followed hydrophobic ion pairing (HIP) approach which involves the formation of hydrophilic drug and surfactant complex. As this complex increases the lipophilicity of the drug, it thus enhances the loading into SLNs lipid core (Lobovkina et al. 2011).

The balance between the capacity of the carrier to condense DNA, to protect from endonucleases and the ability to release DNA is very important in the development of a successful carrier for transferring the gene to the target cells. The most critical phase of transfection is the entry of genetic material into the nucleus. The favorable time to cross cell membrane is during the cell division and in nondividing cells, it is problematic to cross the nuclear membrane by a transfection agent (Apaolaza et al. 2014). Few examples of SLNs-based gene delivery systems and their method of preparation are given in Table 2.

Table 2 Examples of solid lipid nanoparticles-based gene delivery systems

S. no.	Formulation type	Transfection agent	Method of preparation
1	SLN	pCMS–EGFP plasmid	Solvent emulsification method
2	Cationic SLN	pEGFP-N1 plasmid	Solvent displacement technique
3	Cationic SLN	pEFBOS-ECFP-T7-TCERG1 plasmid	Hot microemulsion technique
4	Cationic SLN	pMS2-TCERG1 plasmid	Microemulsion technique
5	SLN	Kinesin spindle protein siRNA	Thin film hydration method

Yu et al. (2010); Fàbregas et al. (2014); del Pozo-Rodríguez et al. (2010); Carrillo et al. (2013); Ying and Campbell (2014)

8.3 SLNs as a Drug Carrier

Delivery of drugs to macrophages is a very promising approach of therapeutic strategy in the treatment of respiratory diseases such as tuberculosis, pneumonia, pulmonary hypertension, and other diseases which involve alveolar macrophages as the major target for the treatment. Makled et al. prepared sildenafil citrate-loaded SLNs to deliver the drug into the lungs through the nasal route of administration in nebulized form. Sildenafil citrate belongs to phosphodiesterase type 5 inhibitors that are used in the management of pulmonary hypertension. As the normal route of administration of this drug is oral or intravenous which is not effective to deliver drugs to lungs, nebulization or inhalation could be a promising approach to localize the drug in lungs. In this study, researchers were able to achieve a smaller size of SLNs using triglycerides with melt emulsification method, which also helps in avoiding organic solvents used in the preparation of nanoparticles. The spherical shape and smaller particle size help in controlling the release of drug from the particles. Even though the drug is hydrophilic and the method of preparation enhances the partition of more amount of drug into an aqueous medium, researchers were able to control drug leakage by adjusting pH during the preparation considering the pH-dependent solubility of the drug (Makled et al. 2017).

SLNs play a key role as the drug delivery systems through the inhalation route of administration, especially in infections associated with lungs such as tuberculosis. Delivery of drugs through inhalation increases the local delivery of therapeutic agent to the lungs. Though the concept of inhalation delivery of drugs is not new, currently many researchers are exploring SLNs for respiratory delivery of drugs in infections such as tuberculosis. Gasper et al. prepared rifabutin-loaded SLNs for pulmonary release. The study demonstrates the preparation of stable SLNs and the nanoparticles can release the drug completely in in vitro assays. Cell line studies conducted using THP1 cells differentiated in macrophages showed approximately 50% of cellular uptake. Though the toxicity is less, lower uptake is the limitation. Surface modification of the nanoparticles would help to overcome the issues (Gaspar et al. 2016). SLNs are gaining very much attention by researchers in the development of drug delivery systems for neglected diseases and infections. Cellular internalization of drugs is very important in disease conditions such as Leishmaniasis, where the protozoan parasites reside inside the cell. Conventional drug delivery systems face many barriers to enter the target cells. For instance, 17-N-allylamino-17-demethoxygeldanamycin (17-AAG) is one of the Hsp90 inhibitors that is being used in the treatment of Leishmaniasis, and it is known to have low aqueous solubility, short half-life, and also low stability. To address these issues, Pires et al. formulated 17-AAG-loaded SLNs to target Leishmaniasis. To minimize the toxicity of the formulation, glyceryl palmitostearate was used as the lipid matrix in the preparation which can solubilize 17-AAG. As cellular uptake of the drug by macrophages is a crucial assay in the development of drug delivery systems that are targeted for infections such as Leishmaniasis, and the developed SLNs showed a very good internalization that was confirmed by determining the accumulation of fluorescent markers in the cellular cytoplasm (Pires et al. 2020). SLNs-based drug delivery systems designed for various drugs are given in Table 3.

8.4 Surface Modifications of SLNs to Target Macrophages

Macrophages tend to swallow the particles that are in the size range of 200 nm–10 μm. However, for development of formulations to enhance cellular uptake by the macrophages, only controlling particle size is not sufficient. The smaller size of nanoparticles along with surface modification enhances their internalization and also uptake by the macrophages. Following are few surface modifiers that help to target macrophages by various mechanisms and facilitate nanoparticles to deliver the drug and genetic materials effectively.

2-Hydroxypropyl-β-cyclodextrin (HPCD) enhances the oral bioavailability and cellular uptake by inhibiting the p-glycoprotein (p-gp) efflux pump. The cellular internalization of SLNs in macrophages can be increased by surface modification of SLNs with HPCD. Visceral leishmaniasis, a neglected tropical disease, is caused by a protozoan parasite and the Leishmania amastigote resides in macrophages. The drugs used for the treatment are associated with adverse effects such as toxicity of the drugs and potential drug resistance developed by the parasites. Moreover, high cost and prolonged treatment regime are the other key issues. Additionally, the drugs like amphotericin B and AmBisome are usually administered through IV infusion which causes chills, infusion-related fever, and rigor. Parvez et al. prepared SLNs for the oral delivery of amphotericin B and paromomycin against visceral leishmaniasis. Surface modification of optimized SLNs with HPCD (2-hydroxypropyl-β-cyclodextrin) enhanced oral bioavailability of the drugs loaded in the formulation. Surface modification of SLNs with HPCD enhances cellular uptake in macrophages where

Table 3 Examples of solid lipid nanoparticles-based drug delivery systems

S. no.	Formulation type	Drug/gene	Disease type	Method of preparation
1	SLN	Diclofenac sodium	Chronic pain associated with cancer	Hot homogenization method
2	SLN	Quinine hydrochloride	Cerebral malaria	Ethanol injection method
3	SLN	Noscapine	Brain cancer	Hot homogenization method
4	SLN	Paclitaxel	Brain cancer	Microemulsion technique
5	SLN	Camptothecin	Glioblastoma	High shear homogenization and ultrasonication technique

Blasi et al. (2013); Gupta et al. (2007); Vaghasiya et al. (2013); Koziara et al. (2004); Martins et al. (2013)

Leishmania amastigote resides. The in vitro anti-leishmanial study results of this experiment revealed the enhancement in efficacy of HPCD-modified SLNs on the macrophage intracellular amastigote growth inhibition. This emphasizes SLNs to become a promising approach to target macrophages in neglected and infectious diseases (Parvez et al. 2020). Even though many novel drug delivery systems are developed by researchers such as nanoparticles in the recent past, one of the major drawbacks of these delivery systems is lower drug-loading capacities. SLNs generally show higher drug-loading capacities and also have several other advantages; hence, SLNs have the potential to become an alternative drug delivery system in targeting specific cells with drugs/genetic material (Vieira et al. 2018).

Genes can be transferred to alveolar macrophages to inactivate oncogenes or to correct genetic lesions (Yu et al. 2010). From the literature, it is evident that DNA activates macrophages and there are delivery systems developed by the researchers to deliver the drug to macrophages. Many researchers worked on nonviral gene delivery systems to improve the transfection efficiency. One of such delivery systems developed by researchers is liposomes, which were studied intensively as drug carriers to target macrophages. Cationic liposomes were used as nonviral vectors to deliver the siRNA, because of their high efficiency to condense DNA. But cationic liposomes, treatment to RAW 264.7 cells causes reduction of mitochondrial membrane potential and release of cytochrome c. Hence, it may lead to cell apoptosis by mitochondrial pathway. To overcome the limitations associated with cationic liposomes in delivering plasmid DNA, Sun et al. investigated the anionic lipid/peptide/DNA complex consisting of anionic pH-sensitive liposomes and positively charged protamine to transfer plasmid DNA to RAW264.7 cell lines (Sun et al. 2008).

cSLNs gained much attention as an alternative to the conventional drug delivery systems because of several advantages such as low cost, ease of preparation, the safety of the formulation, capability of transfecting nucleic acids, etc. (Limeres et al. 2019; Kharaji et al. 2016). Kharaji et al. prepared SLNs loaded with paromomycin sulfate, one of the promising anti-leishmaniasis drugs. The drug-loaded SLNs demonstrated an increase in efficacy and also efficient delivery of drug to macrophages through nanoparticle utilization (Kharaji et al. 2016). Although cationic vectors are considered as safer transfection agents, there are still few drawbacks associated with them. The inefficiency of these cationic vectors to condense DNA completely results in the formation of a half-condensed complex that leads to the formation of structures with condensed DNA strands. Apart from this, it is also difficult to control the interaction between plasmid DNA and positively charged SLNs resulting in the formation of large aggregates. Though the toxicity of these transfection agents is lower in comparison to other forms of vectors, but still, cytotoxicity remains a concern, as these cationic lipids cause shrinkage of cells, cytoplasmic vacuolization, and mitosis reduction (Ye et al. 2008).

8.5 Enhancement of Transfection Efficiency

Protamine is a polycationic amine with a molecular weight of 4–6 KDa. It is a basic protein derived from salmon fish, and the major amino acid consists of a high amount of arginine, which contributes to its alkaline nature. As protamine is a positively charged protein, it binds to the backbone of DNA in a nonspecific manner using its amino acid domain. As nonviral vectors such as cationic SLNs suffer from limitations such as low transfection efficiency, the addition of transfection promotor such as protamine helps in enhancing the transfection efficiency of the carrier (Limeres et al. 2019; Ye et al. 2008; Vighi et al. 2010). It was demonstrated by the researchers that the addition of protamine to sulfate in the preparation of transfection agents augments the efficiency and stability of the transfection agents (Limeres et al. 2019; Ye et al. 2008). Ye et al. prepared negatively charged SLNs loaded with protamine-plasmid DNA binary complex. Blank SLNs were prepared by film dispersion and ultrasonication methods. The prepared SLNs were adsorbed on to the protamine-plasmid DNA binary complex to form ternary nanoparticles (Ye et al. 2008). In another study, Vighi et al. prepared cationic SLNs by adding protamine to the matrix of SLN formulation. Along with protamine, esterquat1 was also added to the matrix, but as such there was no effect observed with the addition of esterquat1 (Vighi et al. 2010).

Cholesteryl oleate, a derivative of cholesterol, was used in the preparation of SLN-nucleic acid complex to improve cellular uptake and to reduce toxicity. The cell membrane consists of steroid lipid cholesterol as an important component which has multiple functions. Incorporation of cholesteryl oleate in the formulation of transfection agents such as lipid-based formulation SLNs helps in increasing transfection efficiency by promoting the fusion of lipoplex with the cellular membrane and it also increases biocompatibility of the formulation. Moreover, the lower melting point of cholesteryl oleate (44–47 °C) is an added advantage to use it in the preparation of SLNs where lower temperatures are used during the manufacturing of formulations. The concentration of cholesteryl oleate needs to be optimized according to the formulation design protocol, as a higher concentration may result in instability of the nanoparticles. Suñé-Pou et al. studied the effect of cholesteryl oleate on the cytotoxicity and transfection efficiency of cationic SLNs for siRNA delivery. It was reported that cholesteryl oleate can achieve 45% of cell incorporation and minimal degradation by the endonucleases (Suñé-Pou et al. 2018).

Hyaluronic acid, a naturally occurring polysaccharide with a carboxylic acid group, is capable of binding to CD44 surface protein that plays a crucial role in the progression of rheumatoid arthritis. It also shows very good selectivity for targeting drugs to inflammatory cells, apart from its great biodegradability and biocompatibility. Surface modification of SLNs with polysaccharide helps in delivering drugs and genes to the macrophages and other inflammatory cells in inflamed joints of arthritis. Zhou et al. prepared glucocorticoid-loaded SLNs surface modified with hyaluronic acid to target surface protein on macrophages, fibroblasts, and lymphocytes of inflammatory joints. The in vivo study conducted in mice revealed that hyaluronic acid-modified SLNs are therapeutically superior to uncoated SLNs. Furthermore, the enhanced and selective accumulation of these hyaluronic acid-coated SLNs in inflammatory tissues demonstrated improved clinical signs of rheumatoid arthritis in mice (Zhou et al. 2018). In another study, Apaoloza et al. prepared SLNs composing protamine and hyaluronic acid for gene therapy. Stable complexes with net positive charge were obtained with the interaction of positively charged SLN and protamine with negatively charged hyaluronic acid and DNA (Apaolaza et al. 2014). A list of few transfection enhancers and their applications are given in Table 4.

8.6 Characterization of Transfection Agents

The transfection efficiency of the nanovectors depends on some specific properties such as

Table 4 Transfection enhancers used in the preparation of nanoformulations to enhance the transfection efficiency of a gene carrier

S. no.	Transfection enhancer	Drug/genetic material	Type of formulation	Target
1	Protamine	Plasmid DNA	Anionic LPD complex	Macrophage nucleus
2	Protamine	DNA	Anionic SLNs	A549 cells
3	Protamine	pCMS-EGFP plasmid	SLN	HEK-293
4	Peptide	Protein drugs	SLN	Caco-2/HT29-MTX
5	TAT peptide	DNA	SLN	Human broncho-epithelial cells
6	Hyaluronic acid	siRNA	Nanogold carrier	Mesenchymal cells
7	Chloroquine/ cetylpyridinium	Plasmid DNA	cSLN	Cos-1 cells
8	Cholesteryl oleate	Nucleic acid	cSLN	nHEK293T cells
9	Cholesteryl oleate/ protamine	DNA	SLN	HEK293T

Limeres et al. (2019); Ye et al. (2008); Suñé-Pou et al. (2018); Sun et al. (2008); Olbrich et al. (2001); Rudolph et al. (2004); Delgado et al. (2012); Fan et al. (2014); Hsu S-h et al. (2019)

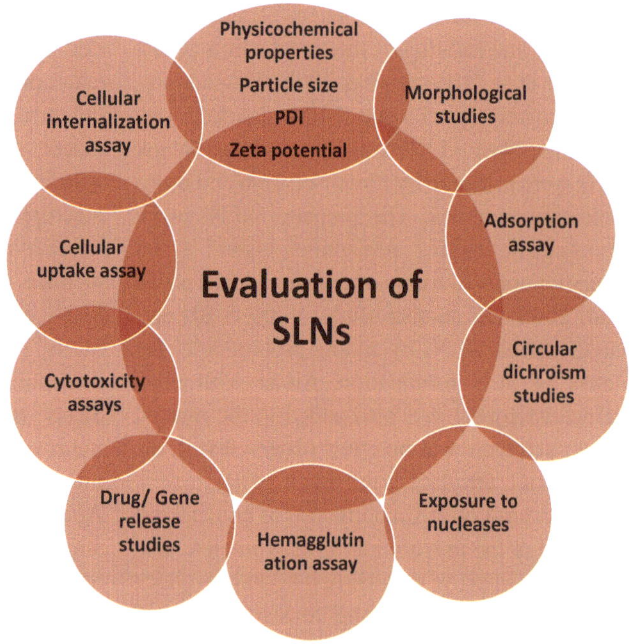

Fig. 3 Evaluation of SLNs by various assays

particle size, shape of the particles, charge on the surface, and the components used in the preparation. Adsorption efficiency, in vitro release, cytotoxicity assay plays a major role to decide the appropriate combination of components for the development of nanovectors. Various parameters for the evaluation of SLNs are explained below in Fig. 3 (Suñé-Pou et al. 2018).

8.6.1 Physicochemical Properties: Particle Size, Zeta Potential, and Morphology

Physicochemical properties of nanoparticles influence the various properties such as cellular uptake, stability, and the release of the entrapped drug from the core of the matrix (Wang et al. 2015). Particle size and zeta potential of the transfection agents play a very crucial role in its efficacy and stability. Particle size is analyzed using photon correlation spectroscopy or laser diffraction spectroscopy. Photon correlation spectroscopy gives information about the average diameter of the population (i.e., z-average) and width of the distribution as polydispersity index (PDI) (Ye et al. 2008; Sun et al. 2008). The decrease in size of particles is attributed to surfactant concentration, as the lower surface tension enables the formation of smaller droplets. Lower size may also result from lipids such as glyceryl monostearate because of its hydrophilic nature (Soni et al. 2014).

Surface charge of the transfection agents is a key property that can influence the stability, transfection efficiency, and also cell adhesion. As cell surfaces are with a negative charge, transfection agents with a net positive charge could help in the attachment to the cell surface (Mahor et al. 2012). Surface charge of the particles is measured by laser Doppler anemometry or laser Doppler microelectrophoresis. The electrophoretic mobility between electrodes of the sample holder containing the sample gives the zeta potential values using the software (Ye et al. 2008; Sun et al. 2008). The size and surface charge of SLNs play a key role in cellular uptake in in vivo. Transfection efficiency can be increased by modifying the preparation strategies to alter the size and electrostatic interaction between the cell surface and vector (Akbaba et al. 2018). The lipid nanoparticles with lower surface charge generally exhibit lesser toxicity and enhance the circulation time, especially in the case of the intravenous route of administration (Lin et al. 2013).

Morphological studies help in revealing the shape of the particles. The scanning electron microscopy (SEM) and transmission electron microscopy (TEM) are the commonly used techniques to determine the shape and texture of particles. Typically, the sample suspension is dropped on to the metallic grid of the instrument and air-dried. To enhance the contrast, staining with reagents such as sodium phosphotungstate is useful. Samples will be observed under the vacuum chamber of the instrument under low vacuum conditions (Ye et al. 2008; Suñé-Pou et al. 2018). The morphology of the nanoparticles varies before and after freeze-drying the product. The procedure involves two steps, freezing of the product followed by drying. Addition of a suitable cryoprotectant or stabilizer is suggested, so that the nanoparticles are protected during the process of freeze-drying. The final product's particle size should be comparable with an initial size. Selection of a suitable cryoprotectant and adequate concentration are crucial to get the final product with the desired morphology (Soni et al. 2014; Kumar et al. 2016).

8.6.2 Adsorption Efficiency

Determination of adsorption efficiency is important to estimate the binding capacity of nanoparticles with nucleic acid and transfection enhancing agents. Adsorption efficiency can be determined in terms of the amount of free nucleic acid content present in the formulation using quantitative spectroscopic methods (Ye et al. 2008).

8.6.3 Circular Dichroism Studies

Conformational changes of proteins in the formulation can be studied with the help of circular dichroism (CD). The structural alterations because of changes in environmental conditions, such as ionic strength, temperature and pH, can be evaluated by circular dichroism studies (Ye et al. 2008).

8.6.4 Exposure of Transfection Agents to Nucleases

This method is useful to test whether the prepared nanovectors could protect plasmid nucleic acids from nuclease enzymatic degradation using enzymes such as DNase, RNase, etc. (Ye et al. 2008). Protection of plasmid against endonucleases is a crucial parameter in the designing of an effective gene transfer by the transfection agent (Mahor et al. 2012).

8.6.5 In Vitro Release Studies

The release of nucleic acids from the transfection agents at the target site or cells is critical to characterize the prepared nanovectors (Ye et al. 2008). In vitro studies using dialysis membrane is one of the most commonly used methods where formulations are loaded into the dialysis bags. This is followed by the incubation in dialysis medium at controlled temperature and stirring. The samples are collected at specific time points and analyzed. The same amount of fresh dialysis medium is replaced to maintain the sink conditions (Zhou et al. 2018; Akbaba et al. 2018; Fernandes et al. 2020).

A biphasic pattern of in vitro release is commonly seen with lipid matrix-based nanoformulations. Initial burst release occurs due to the drug present on the surface of nanoparticles followed by the extended and slow release. The extended release happens because of the biodegradation of matrix and spreading of medium across the lipid. The slow release of hydrophilic drugs is attributed to the strong hydrophilic interaction between drug and surfactant or surface modifiers that is used in the preparation of formulation (Parvez et al. 2020).

8.6.6 Cytotoxicity Assays

Toxicity testing of the nanovectors is very crucial during early development of delivery systems for targeting macrophages. The biocompatibility of lipids and surface modifiers used during the development of nanoparticles plays a key role in the cytotoxicity of formulations. Various kinds of cells are being used by the researchers for cytotoxicity assays. Human THP1 monocytes are the cell lines which can be differentiated into macrophages with the addition of supplements. The differentiated macrophages can be used for the toxicity testing of nanovectors developed for targeting macrophages with drugs or genetic materials (Vieira et al. 2018). In vitro viability test with RAW 264.7 cell lines is one of the most common methods. Most of the researchers revealed the cytotoxicity of the designed transfection agents (Wang et al. 2015; Mahor et al. 2012).

8.6.7 Cellular Uptake Assays and Internalization Pathways

Cellular internalization of nanocarriers is very important to deliver drug or gene of interest to the target cells. Finding out the internalization pathway helps in understanding the mechanism of intracellular delivery of drugs or genes. Differing stabilities, differing abilities to escape from endocytotic pathways and nonuniform uptake by cells result in differences in transfection potency of the delivery systems. Cellular uptake, intracellular distribution, and endocytosis mechanism help in understanding the mechanism by which transfection enhancers work (Apaolaza et al. 2014).

There are three methods by which internalization of delivery of drugs happens. The first method is the activation of therapeutic agents in the lysosomal compartment, nanoparticles engulfed by macrophages are packaged into a phagosome before its destruction by lysosomes, where acidic lysosomes break down the nanocapsule to release its contents. The second method is cytoplasmic delivery via lysosomal escape and endocytosis in which formulation is designed, so that nanoparticles can escape from phagosomes after endocytosis before fusion with the lysosome. The last method is by utilization of endocytic receptors, where the active transport of a therapeutic agent is mediated by coat proteins such as caveolin and clathrin (Peterson et al. 2018).

Endocytosis is the most common mode of mechanism by which nonviral vectors enter cells. The cellular entry of nanoparticles through the endocytic route mainly depends on key physicochemical properties such as particle size, shape, surface charge, material composition, and flexibility. Nanoparticles with size less than 200 nm are known to have extended circulating period than those with higher particle size, as larger particle size is prone to being engulfed by macrophages. Nanoparticles that are smaller than 30 nm have also shown less circulation, as they can be easily eliminated by the renal excretion system. Smaller nanoparticles enter cells by the process called pinocytosis and macroparticles enter through micropinocytosis. Phagocytosis, caveolate-mediated endocytosis, clathrin-mediated endocytosis, clathrin- and caveolate-independent pathways are the other modes of internalization of nanoparticles. Even though the particle size plays an important role in the internalization of nanoparticles, the chemical composition of the formulation is crucial in defining the entry pathway. Nanoparticles enter into cancer tissues by "Enhanced Permeability and Retention (EPR) effect," where nanoparticles accumulate in the interstitial space of tumor tissues (Martins et al. 2012). To evaluate the role of energy in the internalization of nanoparticles, inhibition of active transport can be achieved either by reducing the temperature or by using active transport inhibitors such as sodium azide. Reducing the temperature to 4 °C results in inhibiting cellular active transport and also diffusion, whereas sodium azide only inhibits active transport. By incubation of cell lines with nanoparticles after treating with inhibitors or under low-temperature conditions, it is possible to find out the extent of cellular internalization (Vieira et al. 2018). Specific inhibitors such as sucrose for clathrin and filipin for caveolate-dependent endocytosis can be used during the assay to determine its effect on the uptake of nanoparticles (Martins et al. 2012). For molecules such as curcumin which are fluorescent, the extent of cellular uptake can be demonstrated using the confocal image analysis using confocal laser scanning microscope (Wang et al. 2015).

8.6.8 In Vitro Transfection Assay

Apart from the lower toxicity, transfection efficiency is the other most important factor that is considered during the development of the transfection agent for the delivery of genetic material to the target cells. Transfection assay using fluorescent biomarkers to determine the expression of the genes is one of the most convenient and easy methods. A customized protocol needs to be developed and optimized, according to the type of genetic material intended for the delivery and also the type of target cells (Akbaba et al. 2018).

8.6.9 Hemagglutination Assay

Apart from all these evaluations, the hemolytic potential of the formulations intended for the intravenous administration is very important, especially when surfactants are used in the preparation of formulations. Smaller particle size and other unique physicochemical properties may enhance the interaction of nanoparticles with blood components. In general, the hemolytic potential between 8 and 25% is considered as safe in in vitro studies. The experimental procedure involves the separation of erythrocytes by centrifugation and appropriate dilution using buffer typically phosphate-buffered saline (PBS) solution. This is followed by the incubation of naked plasmid and plasmid-DNA complex with the diluted erythrocyte suspension for around 15 min at room temperature. A microscopical observation of this suspension on a glass slide reveals the hemagglutination of the sample (Delgado et al. 2012; Soni et al. 2014).

9 Conclusion

Macrophages are the cells of an innate immune system that play a key role in homeostasis, tissue maintenance, and are, most importantly, involved in immunoreactions. Targeting macrophages to deliver drugs and genetic material such as RNA/DNA is a promising approach in addressing therapeutic issues associated with macrophages. Conventional drug delivery systems show lack of targetability, display high internalization, are not effective at lower doses, and are toxic at higher doses. Furthermore, as regards drugs with poor aqueous solubility, it is difficult to achieve high bioavailability. Nanotechnology in the development of drug delivery systems is the key area of interest shown by researchers currently. SLNs are the lipid-based nanoformulations considered as the most promising approach in the development of drug delivery systems, especially in targeting specific tissues and cells. SLNs are known to have low toxicity, as the preparation involves biocompatible excipients and their small particle size helps in improving the permeability and solubility, which enhances the bioavailability of drugs. Furthermore, the surface modification of SLNs with specific modifiers such as mannose and hyaluronic acid helps in targeting macrophages of various tissues. By complexing SLNs with nucleic acids, it is possible to transfer a gene of interest to the target cells. The permeability of SLNs carrying the gene of interest can be enhanced using protamine and/or chloroquine. The selection of excipients used in the preparation of SLNs, such as types of lipids and stabilizers, has a great impact not only on the physicochemical properties but also on their efficiency to reach target cells, cellular uptake, and internalization.

References

Akbaba H, Akbaba GE, Kantarcı AG. Development and evaluation of antisense shRNA-encoding plasmid loaded solid lipid nanoparticles against 5-α reductase activity. J Drug Deliver Sci Tech. 2018;44:270–7. https://doi.org/10.1016/j.jddst.2018.01.001.

Apaolaza PS, Delgado D, del Pozo-Rodriguez A, Gascón AR, Solinís MÁ. A novel gene therapy vector based on hyaluronic acid and solid lipid nanoparticles for ocular diseases. Int J Pharm. 2014;465(1–2):413–26. https://doi.org/10.1016/j.ijpharm.2014.02.038.

Blasi P, Schoubben A, Traina G, et al. Lipid nanoparticles for brain targeting III. Long-term stability and *in vivo* toxicity. Int J Pharm. 2013;454(1):316–23. https://doi.org/10.1016/j.ijpharm.2013.06.037.

Calderon B, Suri A, Unanue ER. In CD4+ T-cell-induced diabetes, macrophages are the final effector cells that mediate islet β-cell killing: studies from an acute model. Am J Path. 2006;169(6):2137–47. https://doi.org/10.2353/ajpath.2006.060539.

Cannarile MA, Weisser M, Jacob W, Jegg A-M, Ries CH, Rüttinger D. Colony-stimulating factor 1 receptor (CSF1R) inhibitors in cancer therapy. J Imm Ther Cancer. 2017;5(1):53. https://doi.org/10.1186/s40425-017-0257-y.

Carrillo C, Sánchez-Hernández N, García-Montoya E, et al. DNA delivery via cationic solid lipid nanoparticles (SLNs). Eur J Pharm Sci. 2013;49(2):157–65. https://doi.org/10.1016/j.ejps.2013.02.011.

Chi H, Li C, Zhao FS, et al. Anti-tumor activity of toll-like receptor 7 agonists. Front Pharmacol. 2017;8:304. https://doi.org/10.3389/fphar.2017.00304.

Choi SH, Jin S-E, Lee M-K, et al. Novel cationic solid lipid nanoparticles enhanced p53 gene transfer to lung cancer cells. Eur J Pharm Biopharm. 2008;68(3):545–54. https://doi.org/10.1016/j.ejpb.2007.07.011.

del Pozo-Rodríguez A, Delgado D, Solinís MÁ et al (2010) Solid lipid nanoparticles as potential tools for gene therapy: *in vivo* protein expression after intravenous administration. Int J Pharm 385(1–2):157–162. https://doi.org/10.1016/j.ijpharm.2009.10.020.

Delgado D, Gascón AR, del Pozo-Rodríguez A, et al. Dextran–protamine–solid lipid nanoparticles as a non-viral vector for gene therapy: *in vitro* characterization and *in vivo* transfection after intravenous administration to mice. Int J Pharm. 2012;425(1–2):35–43. https://doi.org/10.1016/j.ijpharm.2011.12.052.

Dolatabadi JEN, Valizadeh H, Hamishehkar H. Solid lipid nanoparticles as efficient drug and gene delivery systems: recent breakthroughs. Adv Pharm Bullet. 2015;5(2):151. https://doi.org/10.15171/apb.2015.022.

dos Santos RB, Kanekiyo T, Singh J. *In vitro* and *in vivo* characterization of CPP and transferrin modified liposomes encapsulating pDNA. Nano Med Nano Tech Biomed. 2020;102225 https://doi.org/10.1016/j.nano.2020.102225.

Fàbregas A, Sánchez-Hernández N, Ticó JR, et al. A new optimized formulation of cationic solid lipid nanoparticles intended for gene delivery: development, characterization and DNA binding efficiency of TCERG1 expression plasmid. Int J Pharm. 2014;473(1–2):270–9. https://doi.org/10.1016/j.ijpharm.2014.06.022.

Fan T, Chen C, Guo H, et al. Design and evaluation of solid lipid nanoparticles modified with peptide ligand for oral delivery of protein drugs. Eur J Pharm Biopharm. 2014;88(2):518–28. https://doi.org/10.1016/j.ejpb.2014.06.011.

Fernandes AV, Pydi CR, Verma R, Jose J, Kumar L. Design, preparation and *in vitro* characterizations of fluconazole loaded nanostructured lipid carriers. Brazilian J Pharm Sci. 2020;56:e18069. https://doi.org/10.1590/s2175-97902019000318069.

Gaspar DP, Faria V, Gonçalves LM, Taboada P, Remuñán-López C, Almeida AJ. Rifabutin-loaded solid lipid nanoparticles for inhaled antitubercular therapy: physicochemical and *in vitro* studies. Int J Pharm. 2016;497(1–2):199–209. https://doi.org/10.1016/j.ijpharm.2015.11.050.

Gholamin S, Mitra SS, Feroze AH, et al. Disrupting the CD47-SIRPα anti-phagocytic axis by a humanized anti-CD47 antibody is an efficacious treatment for malignant pediatric brain tumors. Sci Trans Med. 2017;9(381) https://doi.org/10.1126/scitranslmed.aaf2968.

Gong C, Yu X, You B, et al. Macrophage-cancer hybrid membrane-coated nanoparticles for targeting lung metastasis in breast cancer therapy. J Nano Biotech. 2020;18(1):1–17. https://doi.org/10.1186/s12951-020-00649-8.

Gupta Y, Jain A, Jain SK. Transferrin-conjugated solid lipid nanoparticles for enhanced delivery of quinine dihydrochloride to the brain. J Pharm Pharmacol. 2007;59(7):935–40. https://doi.org/10.1211/jpp.59.7.0004.

He W, Kapate N, Shields CW IV, Mitragotri S. Drug delivery to macrophages: a review of targeting drugs and drug carriers to macrophages for inflammatory diseases. Adv Drug Deliver Rev. 2020;165–166:15–40. https://doi.org/10.1016/j.addr.2019.12.001.

Hsu S-h YA, Yeh C-A, et al. Biocompatible nanogold carrier coated with hyaluronic acid for efficient delivery of plasmid or siRNA to mesenchymal stem cells. ACS App Biomaterial. 2019;2(3):1017–30. https://doi.org/10.1021/acsabm.8b00540.

Jin Y, Hu G, Guo M, et al. Nanoparticles targeting macrophages as potential clinical therapeutic agents against cancer and inflammation. Front Immun. 2019;10:1998. https://doi.org/10.3389/fimmu.2019.01998.

Kharaji MH, Doroud D, Taheri T, Rafati S. Drug targeting to macrophages with solid lipid nanoparticles harboring paromomycin: an *in vitro* evaluation against L. major and L. tropica. AAPS PharmSciTech. 2016;17(5):1110–9. https://doi.org/10.1208/s12249-015-0439-1.

Koziara JM, Lockman PR, Allen DD, Mumper RJ. Paclitaxel nanoparticles for the potential treatment of brain tumors. J Control Release. 2004;99(2):259–69. https://doi.org/10.1016/j.jconrel.2004.07.006.

Kumar L, Reddy MS, Verma R, Koteshwara KB. Selection of cryoprotective agent for freeze drying of valsartan solid lipid nanoparticles. Latin Am J Pharm. 2016;35(2):284–90.

Liang Y, Wang Y, Wang L, et al. Self-crosslinkable chitosan-hyaluronic acid dialdehyde nanoparticles for CD44-targeted siRNA delivery to treat bladder cancer. Bioact Material. 2021;6(2):433–46. https://doi.org/10.1016/j.bioactmat.2020.08.019.

Limeres MJ, Suñé-Pou M, Prieto-Sánchez S, et al. Development and characterization of an improved formulation of cholesteryl oleate-loaded cationic solid-lipid nanoparticles as an efficient non-viral gene delivery system. Colloid Surfac B. 2019;184:110533. https://doi.org/10.1016/j.colsurfb.2019.110533.

Lin PJ, Tam YYC, Hafez I, et al. Influence of cationic lipid composition on uptake and intracellular processing of lipid nanoparticle formulations of siRNA. Nanomedicine: Nano Tech, Biol Med. 2013;9(2):233–46. https://doi.org/10.1016/j.nano.2012.05.019.

Lobovkina T, Jacobson GB, Gonzalez-Gonzalez E, et al. *In vivo* sustained release of siRNA from solid lipid nanoparticles. ACS Nano. 2011;5(12):9977–83. https://doi.org/10.1021/nn203745n.

Ma C, Wu M, Ye W, et al. Inhalable solid lipid nanoparticles for intracellular tuberculosis infection therapy: macrophage-targeting and pH-sensitive properties. Drug Deliver Trans Res. 2020:1–18. https://doi.org/10.1007/s13346-020-00849-7.

Mahor S, Dash BC, O'Connor S, Pandit A. Mannosylated polyethyleneimine–hyaluronan nanohybrids for targeted gene delivery to macrophage-like cell lines. Bioconjug Chem. 2012;23(6):1138–48. https://doi.org/10.1021/bc200599k.

Makled S, Nafee N, Boraie N. Nebulized solid lipid nanoparticles for the potential treatment of pulmonary hypertension via targeted delivery of phosphodiesterase-5-inhibitor. Int J Pharm. 2017;517(1–2):312–21. https://doi.org/10.1016/j.ijpharm.2016.12.026.

Mannucci S, Boschi F, Cisterna B, et al. A correlative imaging study of *in vivo* and *ex vivo* biodistribution of solid lipid nanoparticles. Int J Nanomedicine. 2020;15:1745. https://doi.org/10.2147/IJN.S236968.

Maretti E, Costantino L, Buttini F, et al. Newly synthesized surfactants for surface mannosylation of respirable SLN assemblies to target macrophages in tuberculosis therapy. Drug Deliver Trans Res. 2019;9(1):298–310. https://doi.org/10.1007/s13346-018-00607-w.

Martins S, Costa-Lima S, Carneiro T, Cordeiro-da-Silva A, Souto E, Ferreira D. Solid lipid nanoparticles as intracellular drug transporters: an investigation of the uptake mechanism and pathway. Int J Pharm. 2012;430(1–2):216–27. https://doi.org/10.1016/j.ijpharm.2012.03.032.

Martins SM, Sarmento B, Nunes C, Lúcio M, Reis S, Ferreira DC. Brain targeting effect of camptothecin-loaded solid lipid nanoparticles in rat after intravenous administration. Eur J Pharm Biopharm. 2013;85(3):488–502. https://doi.org/10.1016/j.ejpb.2013.08.011.

Meng K, Chen D, Yang F, et al. Intracellular delivery, accumulation, and discrepancy in antibacterial activity of four enrofloxacin-loaded fatty acid solid lipid nanoparticles. Colloid Surfac B. 2020;194:111196. https://doi.org/10.1016/j.colsurfb.2020.111196.

Montana G, Bondì ML, Carrotta R, et al. Employment of cationic solid-lipid nanoparticles as RNA carriers. Bioconjug Chem. 2007;18(2):302–8. https://doi.org/10.1021/bc0601166.

Nally FK, De Santi C, McCoy CE. Nanomodulation of macrophages in multiple sclerosis. Cell. 2019;8(6):543. https://doi.org/10.3390/cells8060543.

Olbrich C, Bakowsky U, Lehr C-M, Müller RH, Kneuer C. Cationic solid-lipid nanoparticles can efficiently bind and transfect plasmid DNA. J Control Release. 2001;77(3):345–55. https://doi.org/10.1016/S0168-3659(01)00506-5.

Parvez S, Yadagiri G, Gedda MR, et al. Modified solid lipid nanoparticles encapsulated with Amphotericin B and Paromomycin: an effective oral combination against experimental murine visceral leishmaniasis. Sci Report. 2020;10(1):1–14. https://doi.org/10.1038/s41598-020-69276-5.

Peterson KR, Cottam MA, Kennedy AJ, Hasty AH. Macrophage-targeted therapeutics for metabolic disease. Trend Pharmcol Sci. 2018;39(6):536–46. https://doi.org/10.1016/j.tips.2018.03.001.

Pires VC, Magalhães CP, Ferrante M, et al. Solid lipid nanoparticles as a novel formulation approach for tanespimycin (17-AAG) against leishmania infections: preparation, characterization and macrophage uptake. Acta Trop. 2020;211:105595. https://doi.org/10.1016/j.actatropica.2020.105595.

Qi J, Lu Y, Wu W. Absorption, disposition and pharmacokinetics of solid lipid nanoparticles. Cur Drug Met. 2012;13(4):418–28. https://doi.org/10.2174/138920012800166526.

Qiu S-Q, Waaijer SJ, Zwager MC, de Vries EG, van der Vegt B, Schröder CP. Tumor-associated macrophages in breast cancer: innocent bystander or important player? Cancer Treatment Review. 2018;70:178–89. https://doi.org/10.1016/j.ctrv.2018.08.010.

Ries CH, Cannarile MA, Hoves S, et al. Targeting tumor-associated macrophages with anti-CSF-1R antibody reveals a strategy for cancer therapy. Cancer

Cell. 2014;25(6):846–59. https://doi.org/10.1016/j.ccr.2014.05.016.

Rudolph C, Schillinger U, Ortiz A, et al. Application of novel solid lipid nanoparticle (SLN)-gene vector formulations based on a dimeric HIV-1 TAT-peptide *in vitro* and *in vivo*. Pharm Res. 2004;21(9):1662–9. https://doi.org/10.1023/B:PHAM.0000041463.56768.ec.

Soni MP, Shelkar N, Gaikwad RV, Vanage GR, Samad A, Devarajan PV. Buparvaquone loaded solid lipid nanoparticles for targeted delivery in theleriosis. J Pharm Bioallied Sci. 2014;6(1):22. https://doi.org/10.4103/0975-7406.124309.

Sun P, Zhong M, Shi X, Li Z. Anionic LPD complexes for gene delivery to macrophage: preparation, characterization and transfection *in vitro*. J Drug Target. 2008;16(9):668–78. https://doi.org/10.1080/10611860802201381.

Suñé-Pou M, Prieto-Sánchez S, El Yousfi Y, et al. Cholesteryl oleate-loaded cationic solid lipid nanoparticles as carriers for efficient gene-silencing therapy. Int J Nano Med. 2018;13:3223. https://doi.org/10.2147/IJN.S158884.

Tabatt K, Sameti M, Olbrich C, Müller RH, Lehr C-M. Effect of cationic lipid and matrix lipid composition on solid lipid nanoparticle-mediated gene transfer. Eur J Pharm Biopharm. 2004;57(2):155–62. https://doi.org/10.1016/j.ejpb.2003.10.015.

Vaghasiya H, Kumar A, Sawant K. Development of solid lipid nanoparticles based controlled release system for topical delivery of terbinafine hydrochloride. Eur J Pharm Sci. 2013;49(2):311–22. https://doi.org/10.1016/j.ejps.2013.03.013.

Van Lent P, Holthuysen A, Van Rooijen N, Van De Putte L, Van Den Berg W. Local removal of phagocytic synovial lining cells by clodronate-liposomes decreases cartilage destruction during collagen type II arthritis. Ann Rheum Dis. 1998;57(7):408–13. https://doi.org/10.1136/ard.57.7.408.

Van Schouwenburg PA, Rispens T, Wolbink GJ. Immunogenicity of anti-TNF biologic therapies for rheumatoid arthritis. Nat Rev Rheum. 2013;9(3):164. https://doi.org/10.1038/nrrheum.2013.4.

Veremeyko T, Yung AW, Dukhinova M, et al. Cyclic AMP pathway suppress autoimmune neuroinflammation by inhibiting functions of encephalitogenic CD4 T cells and enhancing M2 macrophage polarization at the site of inflammation. Front Immun. 2018;9:50. https://doi.org/10.3389/fimmu.2018.00050.

Vieira AC, Chaves LL, Pinheiro M, et al. Mannosylated solid lipid nanoparticles for the selective delivery of rifampicin to macrophages. Art Cells Nano Med Biotech. 2018;46(1):653–63. https://doi.org/10.1080/21691401.2018.1434186.

Vighi E, Montanari M, Ruozi B, Tosi G, Magli A, Leo E. Nuclear localization of cationic solid lipid nanoparticles containing protamine as transfection promoter. Eur J Pharm Biopharm. 2010;76(3):384–93. https://doi.org/10.1016/j.ejpb.2010.07.012.

Wang J, Zhu R, Sun D, et al. Intracellular uptake of curcumin-loaded solid lipid nanoparticles exhibit anti-inflammatory activities superior to those of curcumin through the NF-κB signaling pathway. J Biomed Nan Tech. 2015;11(3):403–15. https://doi.org/10.1166/jbn.2015.1925.

Yarilina A, Xu K, Chan C, Ivashkiv LB. Regulation of inflammatory responses in tumor necrosis factor–activated and rheumatoid arthritis synovial macrophages by JAK inhibitors. Arth Rheum. 2012;64(12):3856–66. https://doi.org/10.1002/art.37691.

Ye J, Wang A, Liu C, Chen Z, Zhang N. Anionic solid lipid nanoparticles supported on protamine/DNA complexes. Nano Tech. 2008;19(28):285708. https://doi.org/10.1088/0957-4484/19/28/285708.

Ying B, Campbell RB. Delivery of kinesin spindle protein targeting siRNA in solid lipid nanoparticles to cellular models of tumor vasculature. Biochem Biophys Res Comm. 2014;446(2):441–7. https://doi.org/10.1016/j.bbrc.2014.02.120.

Yu W, Liu C, Ye J, Zou W, Zhang N, Xu W. Novel cationic SLN containing a synthesized single-tailed lipid as a modifier for gene delivery. Nano Tech. 2009;20(21):215102. https://doi.org/10.1088/0957-4484/20/21/215102.

Yu W, Liu C, Liu Y, Zhang N, Xu W. Mannan-modified solid lipid nanoparticles for targeted gene delivery to alveolar macrophages. Pharm Res. 2010;27(8):1584–96. https://doi.org/10.1007/s11095-010-0149-z.

Zakirov FH, Zhang D, Grechko AV, Wu WK, Poznyak AV, Orekhov AN. Lipid-based gene delivery to macrophage mitochondria for atherosclerosis therapy. Pharmacol Res Perspect. 2020;8(2):e00584. https://doi.org/10.1002/prp2.584.

Zeisberger S, Odermatt B, Marty C, Zehnder-Fjällman A, Ballmer-Hofer K, Schwendener R. Clodronate-liposome-mediated depletion of tumour-associated macrophages: a new and highly effective antiangiogenic therapy approach. Br J Cancer. 2006;95(3):272–81. https://doi.org/10.1038/sj.bjc.6603240.

Zhan X, Jia L, Niu Y, et al. Targeted depletion of tumour-associated macrophages by an alendronate–glucomannan conjugate for cancer immunotherapy. Biomaterials. 2014;35(38):10046–57. https://doi.org/10.1016/j.biomaterials.2014.09.007.

Zhou M, Hou J, Zhong Z, Hao N, Lin Y, Li C. Targeted delivery of hyaluronic acid-coated solid lipid nanoparticles for rheumatoid arthritis therapy. Drug Deliver. 2018;25(1):716–22. https://doi.org/10.1080/10717544.2018.1447050.

In Vivo Fate of Nanoparticles Undergoing Macrophage Targeting

Anushka Tyagi, Atul Pathak, Yashwant V. Pathak, and Swati Gupta

Abstract

Nanotechnology is an ever-growing field in the healthcare sector due to its effectivity in solubilizing poorly water-soluble drugs, targeting the drug to a particular site/organ which has been a huge challenge for the pharmaceutical industry so far. Due to the small size of the nanoparticle, it has the capability to overcome the barriers present in various sites of the body and this can be employed in the diagnosis of various diseases. Nanoparticles can be administered via various routes such as oral, parenteral, transdermal, and pulmonary. Nanoparticles show high therapeutic action as compared to other conventional dosage forms. To utilize the superiority of nanoparticles one needs to optimize them. Macrophages play a key role in tissue regeneration, apoptosis, inflammation, and various other physiological activities of the body. These activities play a key role in various diseases like Alzheimer's, leishmaniasis, tuberculosis, and other viral and bacterial diseases. Thus, it becomes important to target the macrophages by various nanosystems. For better optimization of the delivery systems one needs to understand the fate of the nanoparticles inside the body. In this chapter, we have discussed the mechanisms behind uptake of nanoparticles by macrophages and summarized the interaction of physiochemical properties like size, charge, surface topography, and how the alteration in the properties can change the uptake drastically. This chapter will be helpful in expanding the understanding of interaction of nanoparticles with cell and cellular environment and how nanotechnology can be translated into medicine.

Keywords

Nanoparticle · In vivo fate · Endocytosis · Macrophages · Physiochemical property · Opsonization · Adverse effects

A. Tyagi · S. Gupta (✉)
Department of Pharmaceutics, Amity Institute of Pharmacy, Amity University Uttar Pradesh, Noida, India

A. Pathak
BioMedica Diagnostics Inc., London, Ontario, Canada

Y. V. Pathak
University of South Florida, Taneja College of Pharmacy, Tampa, FL, USA

Adjunct Professor University of Airlangga, Surabaya, Indonesia

1 Introduction

Nanoparticles pose a tremendous clinical potential as they are capable of recognition and can be processed via various pathways. Nanoparticles can be used to treat viral, bacterial, autoimmune,

S. Gupta, Y. V. Pathak (eds.), *Macrophage Targeted Delivery Systems*,
https://doi.org/10.1007/978-3-030-84164-5_11

and other diseases like cancer (Gupta and Gupta 2017; Yadav and Gupta 2015). On the other hand, certain proteins, nucleic acid, and hydrophobic drugs can be easily encapsulated in the form of a nanoparticle. The biggest advantage nanoformulations have over other conventional formulation is the ease of target delivery to a particular site or organ (Gupta et al. 2019). In the past couple of years, extensive research has been done on the efficient delivery of the nanoparticles inside the cells. Some multivesicular particles are also formed in the cellular compartment called exosomes, which can be utilized to deliver small-sized particles like proteins and nucleic acids like RNA/DNA. Exosomes were recognized as powerful drug delivery agents by loading them with nanoparticles. It is a well-known fact that nanoparticles undergo endocytosis to enter inside the cells, and after entry, they are taken up by lysosomes which can degrade the cargo. On the other hand, the efficiency of nanoparticles decreases as there is very little endosomal escape which causes them to reside inside the lysosomal system, and hence, nanoparticles exit the cellular compartment without causing any significant action. One can enhance the bioavailability of the drug by targeting the endosomal pathway and working efficiently on intracellular barriers. It has been observed that the physiochemical properties of nanoparticles have a major impact on their interaction with the barriers. In this chapter, the effect of the physicochemical property concerning delivery of nanocarrier system has been discussed in brief. This chapter also discusses the toxicity associated with nanoparticles and what we can do to minimize the toxicity. Lastly, some future perspective regarding the improvement of drug delivery of nanoparticle concerning its fate inside the cell has been discussed.

2 What Is a Macrophage?

Macrophages are the cells that take part in providing innate immunity to the body against the foreign substance and were first described in 1882 inside starfish's larvae by Elia Metchnikoff. These were also studied in water flea where they were performing phagocytosis on spores of fungi. It was previously a part of the reticuloendothelial system (RAS) but was later placed into the mononuclear phagocyte system (MPS) in 1968 by Cohn, Furth, and colleagues (Biswas and Mantovani 2014) so that they can be deemed closest to the parent cell, that is, monocytes. The MPS is composed of macrophages, monocytes, and mediators which are responsible for mediating inflammation as well as maintaining homeostasis in the body and maintaining the damaged tissues (Biswas and Mantovani 2010, 2012; Wynn et al. 2013). They are differentiated into different types based on different physiological conditions inside the body and can promote effector activities like anticancer and antimicrobial activities. In a body in any state, four types of macrophages can be found (Williams et al. 2018). Figure 1 illustrates the types of macrophages which are present in various sites of the body.

2.1 Origin of Macrophages

The origin of macrophages has been described in fetal cells (Samokhvalov 2014) where macrophage development occurs from the yolk sac in the primitive hematopoiesis at very early days of embryogenesis, that is, during embryonic days 6.5–8.5. The process of hematopoiesis moves toward the bone marrow from the liver around day 10.5 to 17.5. Around 4 weeks of gestational development, the embryonic yolk sac of a human

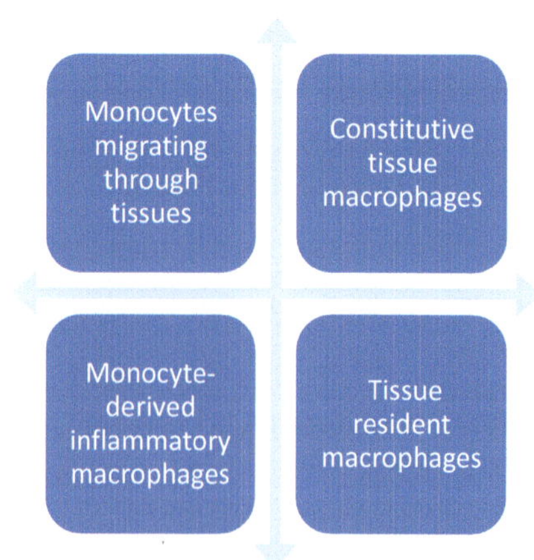

Fig. 1 Four different types of macrophages found in the body

embryo consists of around 70% macrophages. The stem cells are formed by aorta-gonad-mesonephros from ED 8.5 to 10.5, which paves the path for the generation of all other immune cells by hematopoiesis. Cells that are destined to evolve into macrophages undergo development via common myeloid progenitors (CMP), which is also common for granulocytes into promonocytes and monocytes and is governed by expression through various lineage-determining transcription factors (TF). The development of promonocytes to Ly6C+ classical monocytes and Ly6C+ nonclassical monocytes occur in bone marrow after encountering vasculature of peripheral blood circulation. Later, classical macrophages are carried to various organs, where these cells evolve into organ-specific macrophages (Williams et al. 2018).

2.2 Polarization of Macrophages

2.2.1 Polarization of M1 Macrophages into Acquired and Innate Forms

Macrophage polarization into classThe list of transcription factors thatcally activated (M1) or alternatively activated (M2) macrophages is regulated during transduction and transcription. M1 promotes inflammation by releasing cytokines that halts cell division, resulting in further tissue damage. Disruption in the polarization toward M1/M2 macrophage is responsible for inflammatory diseases and cancer (Wang et al. 2014). Classically activated M1 macrophages undergo activation via immune cells like NK-cells and APC, and activate IFN-γ on binding with IFN-γ receptor (IFNGR). The IFNGR releases a set of signals causing activation of M1 macrophages.

After its activation by IFN-γ, M1 macrophages acquire a certain set of functions like increased production of inflammation-promoting cytokines especially IL-12, increased function by APC, and the release of intermediates of oxygen and nitrogen in high amount. These functions result in further polarization of CD4+ lymphocytes. IFN-γ along with GM-CSF (granulocyte-macrophage colony-stimulating factor) and TNF-α function as a cytokine (Martinez and Gordon 2014; Murray 2014).

In innate activation, LPS acts as a trigger that causes the lesser release of IL-12 without phagocytosis. This is observed in another set of cells which are known as M1b (acquired) type cells. Cells that are activated by pathways other than LPS activation form M2a (classical) type of cells.

2.2.2 Polarization of M2 Macrophages

It has been suggested that M2 macrophages assist in the progression of tumors as well as in promoting the formation of newer blood vessels, clearing the metabolic waste out of the body by conveying SPARC toward the stabilizing-1 receptor. M2 activity results in providing adaptive immunity. M2 also inhibits inflammation by releasing cytokines promoting cell division, which results in the repair of damaged tissue. These can be characterized by low IL-12 and IL-23 levels in the blood. M2 macrophages can be classified into three categories based on their gene expression profile which are mentioned in Table 1.

An additional subset of M2 macrophages (M2d macrophage) was identified, and they were actively involved in the reduction of IL-12 and increase of IL-10. They play a key role in slowing the tumor progression and perform functions like tumor-associated macrophages (TAMs).

2.2.3 Regulation of Macrophage Polarization

M1 macrophages can transform into M2 macrophages under a very special condition in specific tissues. Transcription factors play a key role in polarization and these are described by: Lawrence and Natoli (2011), Odegaard (2007), Ruffell (2009), Schonthaler et al. (2011). The list of transcription factors that are involved in polarization is given in Fig. 2.

Signaling pathways play a key role in macrophage polarization. They help in regulating and coordinating the external stimuli given to the body. JAK/STAT pathway is one of the most common pathway which pops up unexpectedly when IFN-γ binds to its receptor present over the surface, which results in the initiation of tran-

Table 1 Activation as well as roles of three subtypes of M2 macrophages

M2a subtype	M2b subtype	M2c subtype
Activated by signals secreted by Th2 immune response and IL4 and IL-13 secreted by mast cells The proteins expressed by these subtype plays a key role in proliferation and repair of tissues as well as in killing of foreign parasites They also downregulate proinflammatory mediators like TNF-α, IFN-γ, IL-6, IL-12, and IL-1β	Activated by signals secreted by Th2, simulation of IL-1R ligand and with complex combination with LPS Here the cells take part in regulating the immunity of body	These cells are activated by TGF-β or IL-10. Glucocorticoids can play a key role as well It is involved in downregulating proinflammatory cytokines and helps in removal of cell debris along with tissue repair

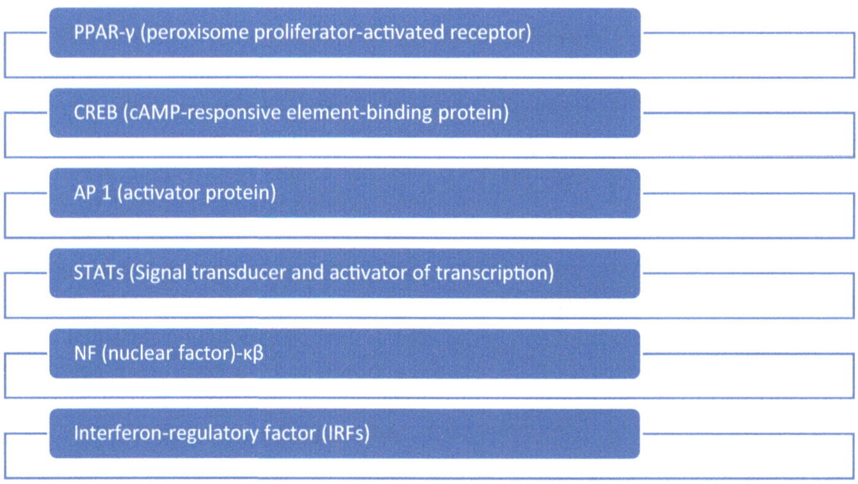

Fig. 2 Transcription factors which are involved in transforming M1 macrophages to M2 macrophages

scription genes associated with M1 macrophage due to dimerization as well as translocation of STAT-1 toward the nucleus. Hence, in M1 macrophages, proinflammatory signals are produced.

Due to the presence of IL-4 and IL-13, the STAT-6 becomes active, which further causes activation of PPARγ and PPARδ. It is to be noted that c-Jun N-terminal kinase and MAPK play an important role in making STAT-6 active. For activation of M2 macrophages by STAT-6, NF-κβ inactivation is much needed, which is done by KLF-2 and KLF-4. The mechanism involves poor attachment of NF-κβ transcriptional coactivator complex PCAF/p300 toward its promoter. The absence of KLF4 can help in the polarization of M1 macrophages and inhibits the polarization of M2 macrophages. On the other hand, activation of Akt2 via PI3K signaling pathway favors polarization of M1 macrophages, and Akt-1 activation promotes polarization of M2 macrophages. Besides, IRF5 (interferon regulatory factor) is responsible for polarizing M1 macrophages. In contrast, M2 polarization is carried out by IRF-4. There are various functions of macrophages which have been mentioned in Table 2.

3 Methods to Study Uptake of Various Delivery Systems by Macrophages

Macrophages play a key role in the regulation of various diseases. Various delivery systems have been targeted toward the macrophages. To completely understand the mechanism of uptake by macrophages, various techniques have been implemented by various scientists. The methods have been mentioned in schematic illustration in Fig. 3.

Table 2 Role of macrophages in various diseases

Serial number	Name of diseases	Observation	References
1.	In autoimmune diseases	Macrophages control the growth and activity of T cells by releasing the transforming growth factor (TGF), contributing toward an increase in inflammation for patients suffering from rheumatoid arthritis	Laria et al. (2016) and Yin et al. (2017)
		In type 1 diabetes mellitus the generation of Th1 cells occurs which then stimulates CD8+ T cells destroying pancreatic cells. Here both M1 and M2 macrophages play a key role in causing type 1 diabetes	Navegantes et al. (2017)
		It has been observed that M1 and M2 cells play a crucial role in the pathogenesis of autoimmune elephantiasis as they start entering inside the brain due to the accumulation of CCL2,3,4 and CCL22	
2.	In tissue regeneration and inflammation	Macrophages play a key role in tissue and organ regeneration by targeting and managing the inflammation of cells and regeneration of cells to replace the structures that were lost. Complete tissue is not regenerated completely and then a scar is formed	Vannella and Wynn (2017) and Das et al. (2015)
		In skeletal muscle, the macrophages play a key role in inflammation that are present in the space between myofibers, in connective tissue between muscle fibers called perimysium, in the perivascular space, and in the connective tissue around the muscle which is known as epimysium	Oishi and Manabe (2018)
		When there is an inflammation in the skeletal muscle of the body, macrophage migrates from the perivascular space into the site of inflammation where they transform into Ly6Chi (inflammatory) and Ly6Clow (regenerative or repair) macrophages	Wang et al. (2014)
		These inflammatory and regenerative macrophages are not relying on NR4A1 (nuclear receptor subfamily 4 group A member 1) or NUR77 or nerve growth factor IB (NGFIB)	Varga et al. (2013)
		The Ly6Clow macrophages are responsible for functional regeneration of skeletal muscle which can be recognized by the display of proresolving signature composed of lipid mediators (SPMs), it also includes resolvins (like RvD1, RvD2, RvE1)	Giannakis et al. (2019)
		Ly6Chi differentiates into M1 and M2 macrophages secreting cytokines promoting inflammation. IFN-γ, TNF-α, IL-1β, and IL-6 are also crucial parts of myogenic precursor cells (MPCs) or myoblasts. MPCs grow and differentiate into other cells by M2 macrophages	
3.	In angiogenesis	Macrophages line themselves up for synthesis and release of various factors like vascular endothelial growth factor-A (VEGF-A), angiopoietin 1, and angiopoietin 2. Later, the same macrophage aligns themselves to act as angiogenic professional cells and arteriogenic professional cells (APC). ANG-1 supports angiogenesis and formation of APC by suppressing PHD2 via (ANG) TIE2 During tumor formation as well as inflammation, the hypoxia is induced due to factors like HIF-1α and HIF-2α. Cytokine and angiogenesis-promoting factors like (CXCR4, GLUT1 [glucose transporter 1], VEGF A, IL-1β, IL-8, adrenomedullin, and ANG 2) are also produced during inflammation and it is induced by TIE2 signaling which depends on ANG2	

3.1 Electron Microscopy

Electron microscopy is one of the most used imaging techniques for studying cell interaction at the vesicular level. It uses an electron beam along with electromagnetic focusing to obtain resolutions of images. Both SEM (scanning electron microscope) and TEM (transmission electron microscope) have been used to understand uptake, particle size, transport, and characteristics of nanoparticle formulation (Patel et al. 2017; Thakor et al. 2016; Echarri and Del Pozo 2015; Hinde et al. 2017; Hernandez et al. 2017) with resolution closer to the atomic level for a long time (Klang et al. 2013; Kulkarni et al. 2018). Aside from nanoparticles, both SEM and TEM have been used to understand uptake and distribution of virus in different tissues. Now, with the advancement in microscopy due to better electron detectors and improved software for resolution, scientists were able to study morphology, conjugates, uptake, transport, and delivery of nanoparticles in the resolution closest to the atomic level (Kulkarni et al. 2018). Some scientists can elucidate the stabilization of membrane proteins in response to saponin nanoparticles.

3.2 Super-Resolution Microscopy

Green fluorescent protein (GFP) from jellyfish was isolated by using super-resolution microscopy in 1962. Currently it is used for observing a wide range of biological phenomena aside from the endocytosis of a drug and its delivery. In an optical microscope, the biggest drawback is their inability to peer inside the cell due to diffraction which was eliminated by fluorescent microscopy. Photoactivation localization microscopy (PALM) which is well-known as stochastic optical reconstruction microscopy (STORM) is a part of super-resolution microscopy that can detect the location of nanoparticles as small as 20 nm by using organic fluorophores. The image is reconstructed using mapping and switching of different types of fluorophores. Now scientists can detect characteristics and mechanism of uptake of nanoparticles more efficiently than the optical microscope. In more advanced forms, 3D images along with the localization of nanoparticles can be obtained easily by using 3D fluorescent probes (Clemments et al. 2017; van der Zwaag et al. 2016) of dendritic cells. Different cell lines have been studied by this method to understand the fate of nanoparticles (van der Zwaag et al. 2016).

3.3 Single-Particle Tracking

Our understanding of the fate of nanoparticles relies on previously studied types of nanoparticles. The diffusion of particles across the environment involves interaction with other biomolecules like lipid, protein, and ergastic material, needs to be studied along with the exosomes the passive or active translocation of a nanoparticle. Those studies require a timescale of seconds to hours. The effective delivery of a

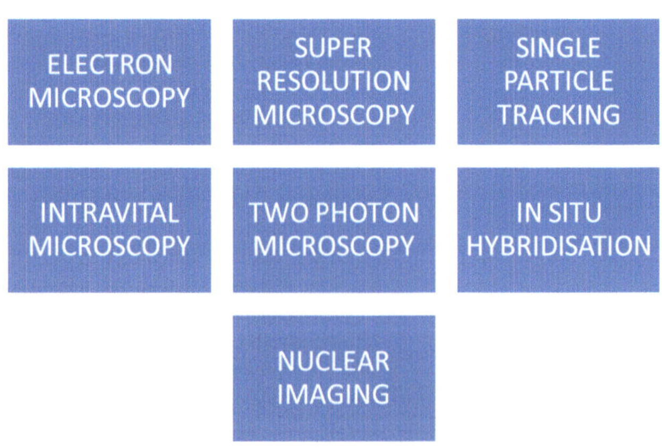

Fig. 3 Various techniques employed to study uptake of various drug delivery systems by macrophages

nanoparticle must transverse this highly dynamic route. Usually single-particle tracking is used to detect the diffusion of quantum dots having cationic, anionic, and zwitterionic nature. It has been observed that charge over quantum dots plays little to no role in transport across the cytosol, but it greatly affects the uptake of molecules across the membrane (Zahid et al. 2018). Welsher and Yang (2014) studied the uptake of modified quantum dot nanoparticles using multiresolution techniques in three dimensions.

3.4 Intravital Microscopy

In fluorescent microscopy, the dynamic environment for cellular uptake by loading fluorophores with various types of materials and fluorescent transgenic animals can be performed easily. The pharmacokinetics and pharmacodynamics of nanoparticles have been studied by this method (Naumenko et al. 2018). This method is different than electron microscopy where one must take a section of tissue for understanding the movement. However, the 3D structures can be assessed by combining intravital microscopy with 3D microscopy (Patel et al. 2019).

3.5 Two-Photon Microscopy

Two or more photons having low energy are utilized to excite a fluorophore unlike high-energy photons used in traditional microscopy. Excitation of the photons can be controlled as low-energy photons are being used. It is to be noted that the lower the energy of a photon, the better is its resolution as low-energy photons have the better capacity to penetrate the tissue due to a lesser amount of diffraction. Firstly used in 1990, the method has been used for imaging various types of particles today.

3.6 In Situ Hybridization

The fluorescence in situ hybridization (FISH) technique was developed in 1982 to identify and locate the hybridized DNA molecules in chromosomes of *Drosophila* by Langer-Safer. In this technique, a single molecule of the probe which is fluorescent is made to bind with a complementary region in the chromosome. The high complementary region in the chromosome leads to more binding of the probe at that region. In single-molecule FISH (smFISH), the individual cells having mRNA are made to appear fluorescent. This process can be used to detect the movement of virus particles inside the cell which can be used for gene delivery. The structure-function relationship between protein and lipid moiety of nanoparticles concerning removal by endosome was discovered recently (Sabnis et al. 2018).

3.7 Nuclear Imaging

While fluorescent techniques have always been the priority for studying the trafficking of nanoparticles, radioisotopic techniques never left the interest of researchers. Nanoparticle distribution inside the tissues has been performed by incorporation of isotopes and scanning via positron emission computed tomography (PET). In the PET technique, radioisotopes having long half-life are tagged to a molecule. It is a highly sensitive technique. Nanoparticles of Zr-89 and Cu-64 are being used as a probe in blood because of their small size and long half-life. Nanoparticles provide the theranostic potential to those radioisotopes (Goel et al. 2017). The large surface area allows more conjugation with the ligand making it possible for a better description of the cellular environment.

4 How Do Macrophages Identify Nanoparticles?

Whenever nanoparticles enter the blood, they encounter various proteins circulating in the blood which are responsible for creating opsonization of protein by adsorbing on the surface of nanoparticles. Opsonization quickly redirects the nanoparticles for recognition by MPS as well as removal out of the body. By controlling the process of identifying the nanoparticles one can regulate the clearance as well as reduce the toxic side effects of nanoparticles. For regulating the fate of nanomaterials by phagocytes it becomes important to understand the conformation as well as surface chemistry of protein molecules attached with nanoparticles (Dobrovolskaia and McNeil 2007). Confirmation of proteins can influence the phagocytosis and inflammatory response out of the nanoparticle (Aggarwal et al. 2009). The charge on the protein can influence the uptake and metabolism of the nanoparticle. For example, polyanionic charged biomaterials are taken up by positively charge domains of scavenger receptors (Chao et al. 2013).

Macrophages as described by Gordon and Boller have adapted their unique recognition mechanism for identifying the foreign substance inside the body (Boller and Felix 2009). These mechanisms can be used for the recognition of nanoparticles as well. Hence, it becomes important to study the uptake and method of recognition of nanoparticles so that their toxicity caused by inflammation can be avoided and phagocytosis can be reduced easily.

Usually, the regulation, uptake, as well as handling of a nanomaterial remains the same as that for any other foreign pathogen which involves a bunch of receptors called pathogen-associated molecular patterns. These receptors can identify epitope from either the damaged tissue or the surface of the causative agent. It is to be noted that the patterns observed on the surface of a pathogen are unique for each one of them which is also known as pattern-associated recognition receptor (PAMPs). According to Kumagai et al., these patterns are stored in the body of the host for a specific pathogen and the patterns can also identify whenever specific damage occurs in the body of a host due to the death of the tissue. The signals associated with the death of host tissues are known as damage-associated molecular patterns (DAMPs) (Kumagai and Akira 2010).

There are four main receptors present on the surface of the macrophage which are as follows.

4.1 Toll-Like Receptors (TLR)

These are glycoproteins which consist of a realm involved in signaling across the cell-like interleukin (Collier et al. 2013). Macrophages and other dendritic cells have an abundant amount of TLR which plays a key role in the activation of cascades responsible for reducing the inflammation. These are present in different parts of the macrophage like on the surface and between the cells. These receptors are also responsible for initiating autophagy of foreign particles (Anand et al. 2011).

4.2 Mannose/Lectin Receptors

These receptors are pluripotent and are responsible for the recognition of sugar/glycoprotein moieties present over the surface of foreign particles. In C-type lectin receptors (CLRs), regulation of APCs like macrophages occur and the delivery of nanoparticle across a specific compartment of macrophages can be possible (Chavez-Santoscoy et al. 2012).

4.3 Scavenger Receptors

These receptors were known to identify lipoproteins. Now scientists have determined its other functions like transport the lipid, removing the pathogen out of the body, and removal of bacteria, which cause contamination of nanoparticles (Canton et al. 2013). They assist in recognition by phagocytes and in the internalization of the pathogen as well as cellular debris. Orr et al. (2011) demonstrated how inhibition of scavenger receptor A can potentially reduce inflammation done by uptake of silica NP. MARCO also known as a macrophage receptor with the collagenous structure reported ingestion of nanoparticles of iron and tin oxide present in the atmosphere.

4.4 Fc Receptors

IgG is adsorbed over the surface of nanoparticles for an immunological response (Bee et al. 2009). These receptors are involved in transporting immunoglobulins across the barriers. FcR, which are involved in phagocytosis in human macrophages, include FcγRI, FcγRIIA, and FcγRIII.

While all the abovementioned receptors play a key role in the uptake of nanoparticles, Fc and mannose receptors allow quicker internalization of nanoparticles and their activity depends on the activation states of the cells. It is to be noted that these receptors work together in a parallel fashion to determine opsonized nanoparticles.

5 Opsonization of Nanoparticles

The adsorption of proteins by the nanoparticle forms the so-called protein corona changes its presentability to macrophages, and then macrophage identifies and expediates its uptake by implementing phagocytosis. The factors affecting the process of opsonization are complex and one cannot predict the fate of an individual type of nanoparticle. Hence, it becomes important to make some model systems and study extensively to see whether the phagocytes are causing events like bioinvisibility or leading toward toxicity or causing immune response like complement activation.

5.1 Factors Affecting Opsonization

5.1.1 Surface Property of NP

Introducing hydrophobicity or hydrophilicity can change the rate of uptake of nanoparticles. Negatively charged silica is readily taken up by macrophages. In some instances, immobilizing polymers like PEG has reduced protein adsorption and they remained in blood circulation for a longer duration of time. (Walkey and chan 2012). Dextran was incorporated on an MRI agent iron oxide nanoparticle so that it stays in circulation for long.

5.1.2 Surface Curvature

Vertegel et al. (2004) observed that with the decrease in size and increase in curvature energy of silica, NP binding with lysozymes at its native conformation was easier. On the other hand, fibrinogen did not bind with its native conformation. Kurylowicz et al. (2014) observed that surface curvature of polystyrene NP can be changed easily by altering the concentration of both monomeric and dimeric units of lactalbumin and lactoglobulin. These observations hint toward the potential effect of curvature over conformation, however, its effect on various types of plasma proteins has not been studied well yet.

5.1.3 Surface Topography

It has been observed that proteins respond differently to different types of nanoparticles. It has been observed that bovine serum albumin responds differently toward gold and platinum NP. BSA attaches itself more on rough surfaces of platinum NP and the same effect was found in gold nanoparticles where cellular uptake was reduced in a rough gold nanoparticle. Recently, Galmarini et al. (2018) observed that only a small fraction of proteins are absorbed preferentially by the presence of surface groups on nanoparticle.

5.1.4 Size

Many people have supported the notion that reduction in the size of nanoparticles leads to the binding of specific type of protein, which has been challenged by Marichal et al. recently. In their study, it was found that 64% of proteins are absorbed commonly on all types of sizes of silica NP, though there remains a tiny minority of specific proteins which adsorbs specifically on a nanoparticle having a certain size (Marichal et al. 2020).

6 Endocytosis of Nanoparticle

When nanoparticles are made to enter inside the body, they encounter various types of cells hence it becomes important to understand the path these nanoparticles travel inside the cell. It has been established by many scientists that the uptake efficiency depends on the size, shape, and surface chemistry of the nanoparticles. Thus, the various mechanisms of intake along with the factors affecting the uptake of drug and interactions are mentioned in Fig. 4.

When nanoparticles travel through the bloodstream it encounters various opsonin proteins which get latched over its surface. Later the mononuclear phagocytic system (MPS) easily identifies the opsonin proteins over the NP surface and binds to them. This in turn results in uneven plasma distribution at various sites of the body causing toxicity.

Nanoparticles face many kinds of force while moving inside the body. When these small particles get clump together due to the presence of strong chemical forces they are known as aggregates. When the same nanoparticles are held together loosely by a weak force and/or packets of aggregates are clump together loosely by a weak force, they are known as agglomerates. By aggregation/agglomeration of particles the effective surface area reduces resulting in the decline of mechanical properties of NP. The agglomeration/aggregation of NP causes a change in its uptake. Nanoparticles interact with cells after reaching a certain site in two ways – either by interacting with a certain type of ligand-based receptors or by interacting with nonspecific interactions like hydrophobic and electrostatic interaction for its internalization. The cell enfolds its plasma membrane along with the extracellular fluid to form a sac called an endosome (Patel et al. 2019). The acquisition of nanoparticles occurs by various mechanisms which have been classified based on the size of nanoparticle containing endosome into phagocytosis and pinocytosis where pinocytosis has been further classified based on effectors responsible for vesicle formation. It has been illustrated schematically in Fig. 4 (Patel et al. 2019).

6.1 Pinocytosis

Pinocytosis includes clathrin-mediated endocytosis, caveolin-mediated endocytosis, macropinocytosis and clathrin, and caveolin-independent endocytosis. By pinocy-

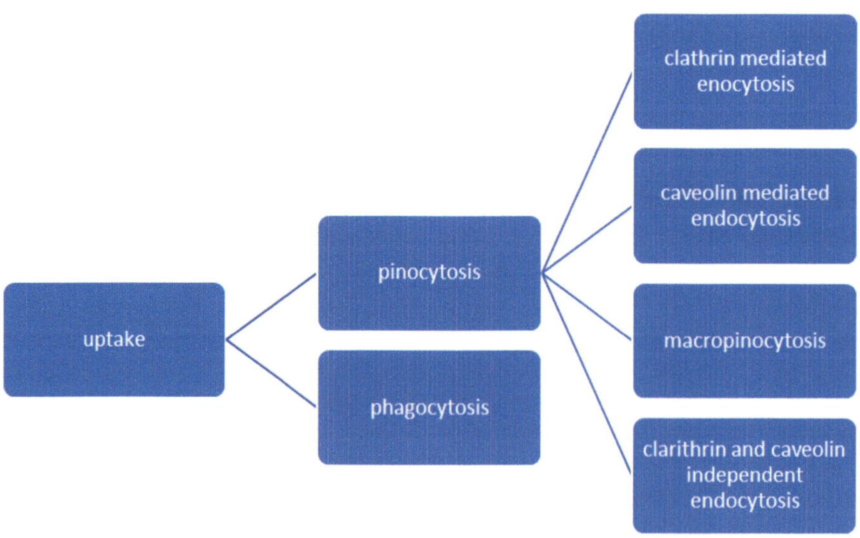

Fig. 4 Various ways of nanoparticle uptake. The process of uptake is broadly classified into pinocytosis and phagocytosis. Pinocytosis has various subtypes under which four different types of endocytosis can take place

tosis, the cell can engulf liquid droplets and small-size particles, whereas larger size particles and cell debris can be engulfed by phagocytosis forming phagosomes. It is to be noted that both phagocytosis and pinocytosis act in the presence of actin protein. Hence, the cells which lack actin-myosin complex (like red blood cells) do not get internalized by these mechanisms (Kuhn et al. 2014).

6.1.1 Clathrin-Dependent Endocytosis

In clathrin-dependent endocytosis, the receptors play a key role in transporting molecules and nanoparticles. This type of endocytosis occurs where lipid rafts are not present. Its mechanism involves clathrin-coated pits which consist of clathrin protein, a transmembrane protein, and other cytosolic proteins along with the AP2 receptor complex (Kuhn et al. 2014).

6.1.2 Caveolin-Dependent Endocytosis

In caveolin-dependent endocytosis, infolding of the plasma membrane into a flask-shaped vesicle occurs. This type of endocytosis occurs due to the presence of lipid rafts where dimeric caveolin-1 protein is attached with cholesterol. Other integral proteins like follitin-1 and follitin-2 are present (Kuhn et al. 2014).

6.1.3 Macropinocytosis

In macropinocytosis, the ruffles are formed by the action of growth hormones and actin filament polymerization which engulfs the NP and macropinosomes may form through them. Macropinocytosis is said to be dependent on the concentration of cholesterol. A large area of macropinosome is said to be composed of lipid rafts from where they have said to be grown (El-Sayed and Harashima 2013).

Lipid rafts are said to be composed of cholesterol, sphingolipids, and unsaturated hydrocarbons. They are also composed of transmembrane proteins whose outer domain is facing toward the surface of the cell. Other protein moieties are either present on the outer leaflet held by glycosylphosphatidylinositol anchors (GPI-AP) or those proteins are attached to the inner leaflet of the membrane (El-Sayed and Harashima 2013).

6.2 Phagocytosis

In phagocytosis, the receptor like the high-affinity IgG receptor FcγRI (CD64) 43, the C-type lectin receptor Dectin-1, and toll receptors present in lipid raft are responsible for carrying out invagination. The lipid raft assists in the formation of the cup-like structure around the nanoparticles. Hence, the concentration of cholesterol plays a key role in the formation of phagosomes. In the later stage, cholesterol depletion occurs and the phagosome acquires an abundant amount of sphingomyelin and ceramide (El-Sayed and Harashima 2013).

After internalization of NP, the phagosome fuses itself to an early endosome so that it can mature itself to form a late endosome to undergo fusion with a lysosome to form a phagolysosome.

The whole process of uptake of nanoparticles can be understood by the illustration given in Fig. 5. In eukaryotic organisms, entry of nanoparticle can be facilitated by various endocytotic mechanisms like clathrin-dependent endocytosis, caveolin-dependent endocytosis, micropinocytosis, etc., after which it gets organized in early endosomes. At later stages, it can be either trafficked toward ER or moved out of cell via late endosomes. Nanoparticles can be degraded only after it is taken up by lysosomes. Exosomes are formed by folding of endosomal membrane, which play a key role in removing the nanoparticles and its debris.

7 Processing of Nanoparticles

Particle opsonization and surface recognition play a key role in the uptake of nanoparticles. Once they are taken up by the cells, they undergo processing, which can be of the following types:

1. General cellular response for antimicrobials is generated.

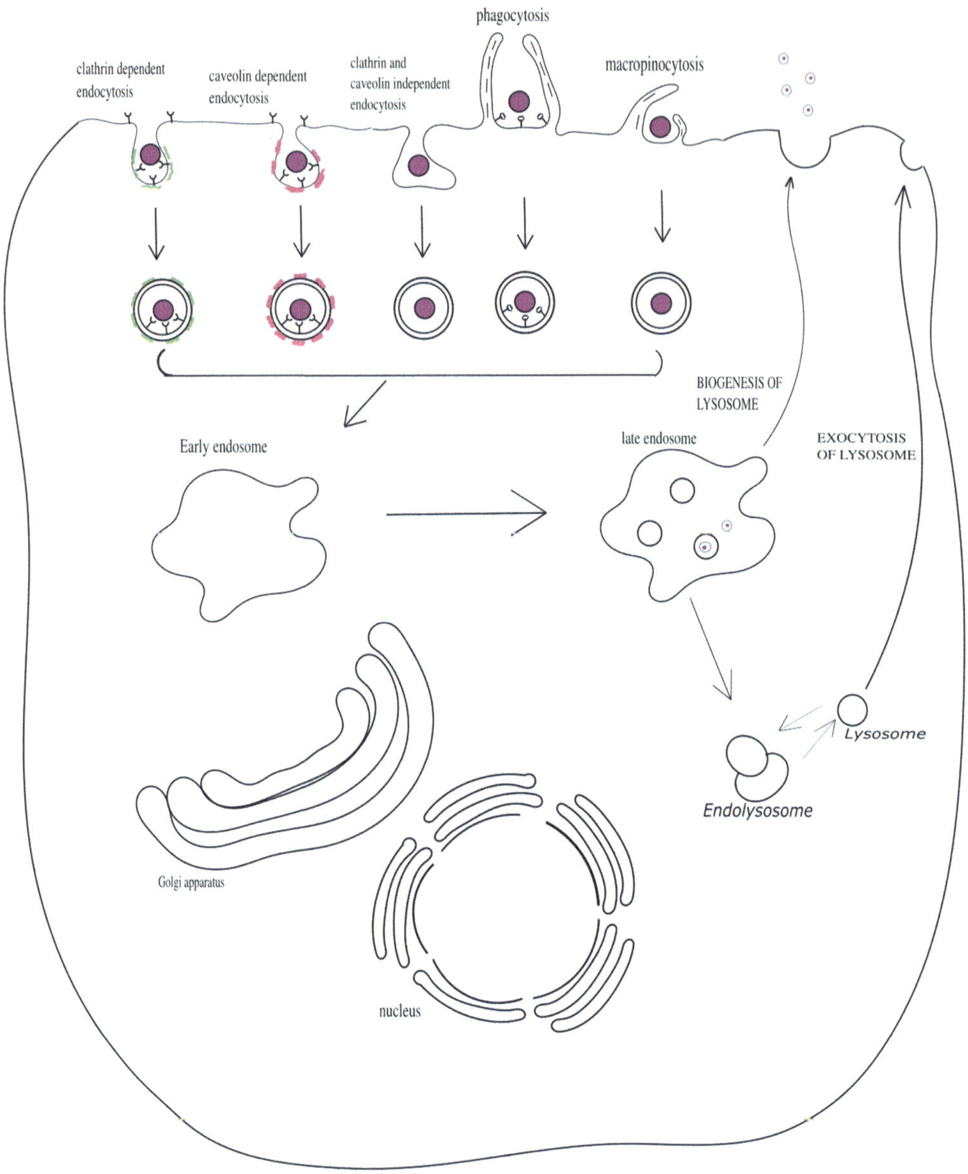

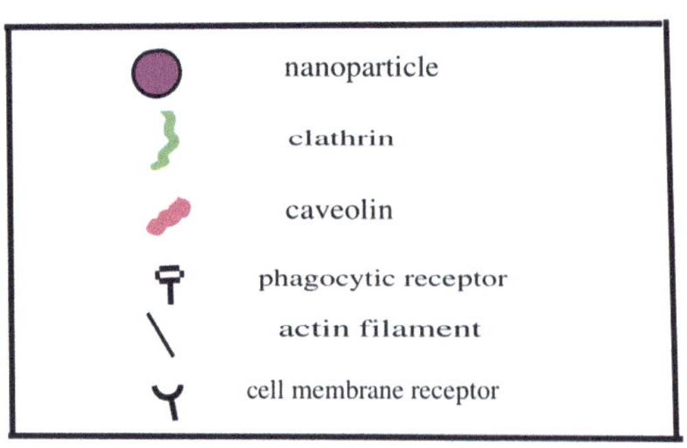

Fig. 5 Uptake and transport of nanoparticles

2. General cellular response for the foreign pathogen is generated.
3. Recognition of corona by specific receptors leading to cellular response for foreign material.

Nanomaterials process themselves in the following two ways:

(A) Induction of Autophagy

It is a form of the innate cellular immune response which involves the engulfment of microbes, poorly folded proteins, and other organelles which are not functioning well inside the body. When the receptor recognizes a nanoparticle, it pushes the nanoparticle for uptake and induces autophagy after which polarization of macrophages and migration of leukocytes occur leading to secretion of factors that are involved in inflammation. After administration of nanoparticle, dysregulation of mitochondria has been observed which was responsible for promoting cell death. Aside from mitochondrial disruption, accumulation of vesicles and dysfunction in autophagy have been observed (Stern et al. 2012). Autophagy serves to provide various advantages like reduction in toxicity as it compartmentalizes the nanoparticle under the stressful condition into autophagosomes. On the other hand, in methods like necrosis as well as apoptosis, the toxicity of nanomaterials increases under stress as the debris is incorporated directly into the cytoplasm causing a change in pH, generating reactive oxygen species leading to inflammation as well as toxicity. Most of the positively charged nanoparticles undergo autophagy. Autophagy is incredibly useful in treating diseases like TB where nanoparticle-induced autophagy can lead to the elimination of *Mycobacterium* from autophagosomes (Gupta et al. 2014) Even though these mechanisms seemingly reduce the toxicity, it can change the transport of cargo to the site that is not specific for eliciting a therapeutic response. For example, autophagy of nucleic acid polyplexes will not allow the delivery of cargo to the nucleus or mitochondria, causing a negative effect on its efficacy. To combat this issue, organic molecules are linked to its nucleic acid nanotube.

(B) Delivery of Nanoparticles Across an Organelle

The delivery of nanomaterials can be manipulated to target specific cellular organelles for a reduction in toxicity. After uptake of foreign particles by phagosomes, there occurs digestion by enzymes resulting in infusion into lysosomes causing the reduction in pH. This helps in the inactivation of foreign particles. On the other hand, the presence of receptors on the surface of vesicles can help in redirecting the particles into specific compartments inside the cell. The pattern over these nanoparticles can also be utilized to cause a burst of the lysosome. After "sponging" out the particles, they can be further carried to other cellular organelles for effective delivery of drugs. Nanoparticles can be engineered to direct the cargo toward mitochondria and can cause programmed cell death necessary to eliminate cancer, proper regulation of potassium pump to cure heart diseases, and cause a reduction in oxidative stress for prolonging the life and reducing aging. Nanoparticles can be attached to mitochondria through the positively charged molecules present on its membranes for better. Such strategies can be used to avoid contact with MPS as well as reduce the nonspecific uptake of nanoparticles resulting in reduced cellular toxicity and better effectivity. Targeting of the nucleus and endoplasmic reticulum has been done by employing peptides as a carrier. A family of proteins known as importin is used for transport across the nucleus as these proteins can assist in identifying nuclear signals by recognizing the attachment of NLS.

8 Interaction of Nanoparticles

There are various routes for the entry of nanoparticles inside the human body. Nanoparticles suspended in air can enter inside the body via the pulmonary system, which is composed of the lungs. The lungs are covered with surfactant molecules to reduce the pleural friction and nanoparticles interact with them (Kapralov et al. 2012). Many nanoparticles are present in the form of parenteral, which is

- Nuclear Magnetic Resonance (NMR)
- Mass spectroscopy (MS)

  } For structure elucidation

- Thin-layer chromatography
- Dynamic light scattering (DLS)
- Transmission electron microscopy (TEM)

  } For elucidation of surface

- Scanning Electron Microscopy (SEM)
- ζ-potential analysis → Change in charge due to interaction with surfactant.
- UV Vis spectroscopy → Change in Surface plasma resonance (SPR) by binding of surfactant.

injected inside the body. When an injection comes in contact with blood it encounters proteins as well as other components of plasma and cells surrounding the tissue and they can absorb amino acid, folic acid, and all other nutrients. Some nanoparticles can penetrate the nuclear membrane and can interact with base pairs of DNA (Prado-Gotor and Grueso 2011). The deficiency of small metabolites of cells can play a key role in causing toxicity to nanoparticles. In this subunit, we have discussed various ways of interaction between nanoparticles and various types of biomolecules that are present inside the body.

8.1 Interaction with Phospholipids

8.1.1 Inside the Lungs

The pulmonary system is surrounded by surfactant which is composed of proteins and phospholipids. When nanoparticles enter inside the lungs, they bound themselves with surfactant and it results in its interaction with lipids. The type of interaction between pulmonary surfactant and nanoparticle plays a key role in the determination of the probable toxicity inside the lungs. Thus, it becomes immensely important to study the interaction between nanoparticle and pulmonary systems so that better therapeutics can be implemented. To create the environment for ex vivo studies of interaction, surfactant solution is prepared, which closely resembles the surfactant itself. The surface property of surfactant solution can also be measured by looking at various parameters like surface tension (Beck-Broichsitter et al. 2011). Nature of interaction between nanoparticles with surfactant can be studied by isolating and screening the bound complex with various techniques which are mentioned next.

8.1.2 With Bilayer Membranes

When any kind of nanoparticle crosses the lipid bilayer there are two main types of interaction that occur in the bilayer – hydrophobic and electrostatic. These kinds of interactions change the permeability as well as the property of membranes, which can be detrimental in nature. Albumin-coated carbon nanotube acquires protein covering causes expulsion of material. Interaction between membranes can be studied taking various models (lipid mono and bilayer, liposomes) by the following means:

- π-A isotherm
- AFM
- Differential scanning calorimetry (DSC)
- isothermal titration calorimetry (ITC)
- quartz crystal microbalance (QCM)
- Fluorescence microscopy
- fluorescence correlation spectroscopy (FCS)
- Foster resonance energy transfer
- NMR
- liposomal leakage assay

Study morphological changes in bilayer due to NP

nanoparticle-induced thermodynamic changes

Study the nanoparticle-lipid membrane interactions.

biophysical changes and disruptions of model lipid membranes

The nanoparticle and lipid membrane interaction can be influenced by various factors like surface property, chemistry, charge on nanoparticles – polyorganic siloxane NP affects the surfactant of lungs at high concentration (Sachan et al. 2012).

8.2 Interaction with the Proteins

After administration of the drug, the nanoparticles distribute itself freely to other sites in the body where it encounters various other types of proteins present over cells. The proteins present over cells bind to the NP and the effect is clearance from various organs as well as its biodistribution. Protein binding with the nanoparticle can affect the conformation of the protein itself, which might result in cell signaling. The interaction between proteins and nanoparticles is very complex and is regulated by both nanoparticle and cellular proteins. The physicochemical changes in nanoparticles or proteins also allow the characterization of the thermodynamic and kinetic processes of their interactions. The interaction between nanoparticles and proteins can be determined by the affinity of binding between nanoparticle and protein, kinetic properties, and stoichiometry. The mechanism between various types of nanoparticles and proteins interaction depends on the type of nanoformulation.

- For carbon nanotube, the protein bonds in layers over the surface of the tube, and here the binding depends on pH and the ionic strength. Therefore, nanoparticle-protein binding events depend on the carbon nanoparticles.
- For gold nanoparticles, the binding of proteins with nanoparticles can be either electrostatic or hydrophobic depending on the size of GNP. For smaller size range the interaction is hydrophobic, whereas for larger-sized particles the interaction is usually hydrophilic.
- In quantum dots, the electrostatic effects as well as the formation of sulfur bond can induce protein binding.
- For silica nanoparticles, the protein binding occurs in a progressive manner where a rapid complex is formed between proteins and

nanoparticles first, followed by conformational change which is usually irreversible in nature. Here, conformation as well as ionic strength plays a key role.

8.3 Interaction with the DNA

Once the NP enters inside the nucleus, interaction can happen between the NP and DNA directly. The interaction itself depends on the properties of DNA as well as the nanoparticle itself. Physiochemical properties of NP change after binding. Here, the DNA warps itself around the nanotube by π-π stacking or by inserting itself into the central cavity. Following methods can be used to determine the NP-DNA interactions:

- IR spectroscopy – monitoring the binding process.
- XPS – determine the binding between DNA and nanoparticle.
- Fluorescence spectroscopy – binding affinity between DNA and nanoparticle.
- RT-PCR – determine the amount of binding to DNA.
- Electrochemical approach – determine desorption/sorption of nanoparticles.
- Circular dichroism – detect conformational changes in DNA.
- Molecular simulations – conformational changes as well as binding pattern.
- UV-vis spectroscopy – determine the type of binding between DNA and NP.

8.4 Interaction with Small Biomolecules

Nanoparticles due to their small particle size possess high surface energy which helps attract tiny molecules present in biological systems which are useful for cell signaling and maintaining biological functions of the cell. These molecules are absorbed via electrostatic interactions aside from π-π stacking. Their removal from the body can cause irregular functioning of cells accompanied by toxicity (Fig. 6).

9 Functional Changes Induced by Nanoparticles

When nanoparticles encounter the cytoplasmic organelles, they cause various types of functional changes in cells, which have been summarized in Table 3.

9.1 Changes in Plasma Membrane

Silica nanoparticles consist of a silanol group which causes lipid peroxidation of membranes by the generation of ROS, resulting in changes in RBC (Slowing et al. 2009). It is to be noted that the needle-shaped nanoparticles can cause more disruption of the cell membrane than their spherical counterparts (Doshi and Mitragotri 2010).

9.2 Changes in Ion Channel

Nanoparticles disrupt ion channels which are responsible for facilitating the development of a gradient across the individual cells in a cellular network. The nanoparticles interact with extracellular realms present in proteins involved in ion channel transport. Single-walled nanotubes can cause irreversible blockage of potassium ion channels due to chemical bond formation. On the other hand, a multiwalled nanotube can suppress potassium ion channel. Contaminations on nanotubes can also cause cellular toxicity, hence the type of disruption caused by different types of nanoformulation must be studied.

9.3 Changes in Cytoskeleton

The cytoskeleton plays a key role in transport, cell division, as well as providing motility. When nanoparticles are taken up by the cell, they cause indirect changes in the property of the cytoskeleton which further leads to a loss in the functions mentioned earlier. Nanotube can cause the formation of bundles of actin fibers in HeLa cells, which can lead to inhibition of cell division (Holt et al. 2010). FeO and gelatin nanoparticles can cause tubulin rearrangement in human fibro-

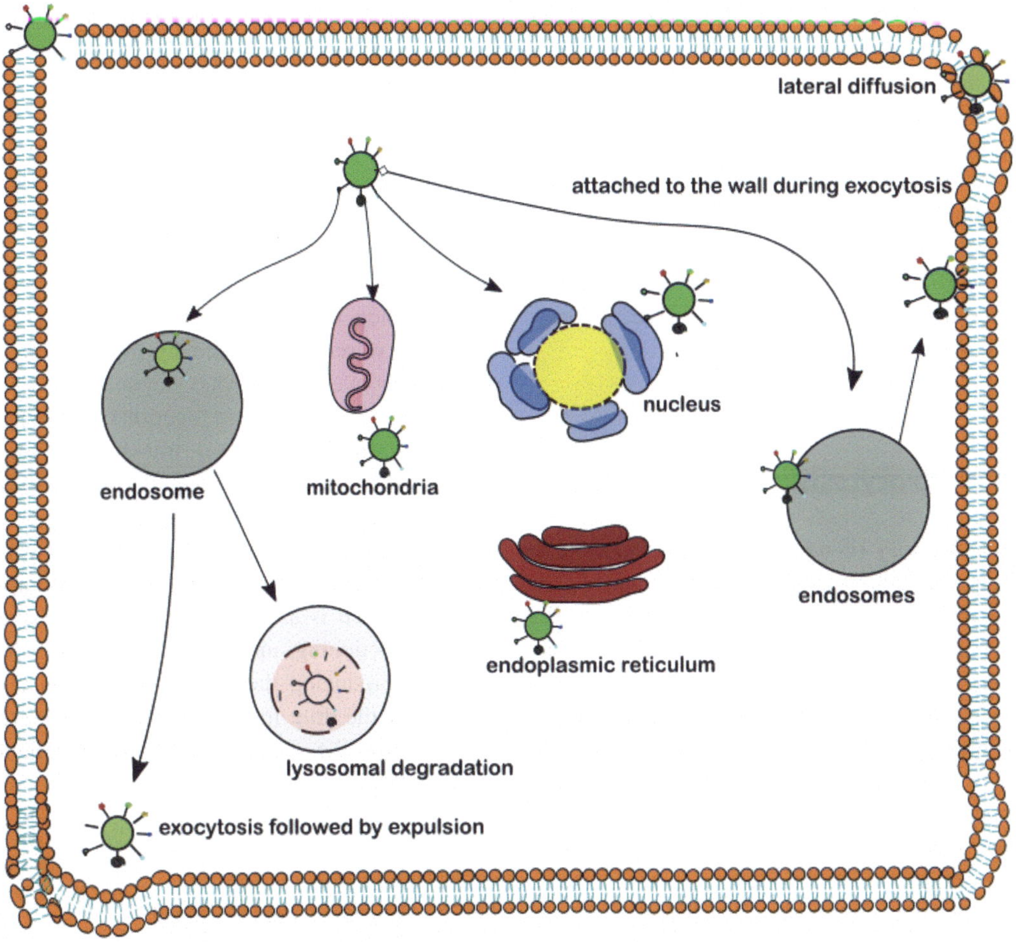

Fig. 6 Lateral diffusion of nanoparticle, from there it attaches itself to the wall during exocytosis, and how it interacts with various other organelles. It can also expel itself from the cell compartment after exocytosis

Table 3 Functional changes encountered by cellular organelle after interacting with nanoparticles

Cellular organelle	Functional changes
1. Plasma membrane	Membrane depolarization, hole formation and resealing, inhibition of cellular proliferation
2. Ion channels	Stimulation or blocking of potassium ion channels
3. Changes in cytoskeleton	Retards cell proliferation and motility
4. Changes in mitochondria	Inhibits signaling and promotes oxidative stress
5. Changes in nucleus	Binds with histone proteins and chromatin fibers

blasts. Coating nanorods with PEG assists in reducing the changes (Tarantola et al. 2009).

9.4 Changes in Mitochondria

Mitochondria serve as the powerhouse of the cells by providing ATP to the cell. Aside from providing energy, these organelles are involved in cell signaling and other processes that maintain homeostasis. Gold nanoparticles bind to the biomolecules present inside the cell by entering via voltage-dependent channels (Karataş et al. 2009). Gold nanoparticles as small as 1.4 nm can cause oxidative stress resulting in the change of permeability inside the cell.

9.5 Changes in the Nucleus

The nucleus is involved in signaling and control of hereditary characteristics of the cell. The nucleus consists of nucleolus, chromosomes, RNA, DNA, and histone proteins. Quantum dots can bind to histone and histone-rich organelle because of electrostatic interactions. Gold nanoparticles were shown to cause hepatotoxicity in the mouse as they can cause decondensation in the chromatin fibers.

10 Exocytosis of Nanoparticles

After delivery of nanoparticles to the cellular target, the nanoparticles are removed out of the body by organs that form a part of MPS like the liver and spleen. It has been observed that many nondegradable nanomaterials reside in clearance organs even after 2 weeks, which remains the biggest contributor to toxicity (Kumar et al. 2010; Hirst et al. 2013). Thus, it becomes important to study the removal of nanoparticles from various types of cells, particularly from macrophages.

In a study, gold nanoparticles were coated with transferrin protein to understand its mechanism of uptake and removal. It has been observed that the larger surface has a small wrapping time which attributes to a slower rate of exocytosis, whereas small particles were able to perform exocytosis quickly due to their larger wrapping time (Zhu et al. 2013).

10.1 Factors Affecting Exocytosis of Nanoparticles

Various factors affect the exocytosis of nanoparticles. It is the exocytosis which decides the in vivo fate of nanoparticles. They have been elucidated in Fig. 7 (Gustafson et al. 2015).

11 Macrophage Targeting by Nanoparticles in Various Diseases

11.1 For Human Immunodeficiency Virus

Macrophages are targeted for the treatment of AIDS since these cells are heavily involved in the immunopathogenesis of HIV-AIDS. Drugs like AZT were incorporated by human serum albumin as well as polyhexylcynoacrylate to form nanoparticles that can function by preventing infection during blood transfusion from infected donors. To determine the efficacy of nanoparticles, the AZT was labeled and the drug was injected via the intravenous route. It was found that the concentration was 18 times higher in RES, which is associated with macrophage and its functioning, and the accumulation was helpful in reducing the dosage as well as the toxicity of the drug itself. Similar results were obtained for the drug given orally (Chellata et al. 2005).

11.2 For Diseases Related to Inflammation

Nanoparticles have been used to treat chronic inflammatory diseases like arthritis (Mishra and Gupta 2020). Since inflammation plays a key role in flaring up and maintaining the inflammation, some diseases like atherosclerosis where macrophages attach themselves to the endothelial wall, secretion of MMP, cytokines, etc. occur as an inflammatory response. To combat it, nanoparticles containing coenzyme Q10 (CoQ10) and apolipoprotein-A1 were used in gene therapy. In rheumatoid arthritis, the level of TNF-α is increased significantly which can be reduced by administering loaded ODN nanoparticles (Bayik et al. 2016) as well as PEI derivatives.

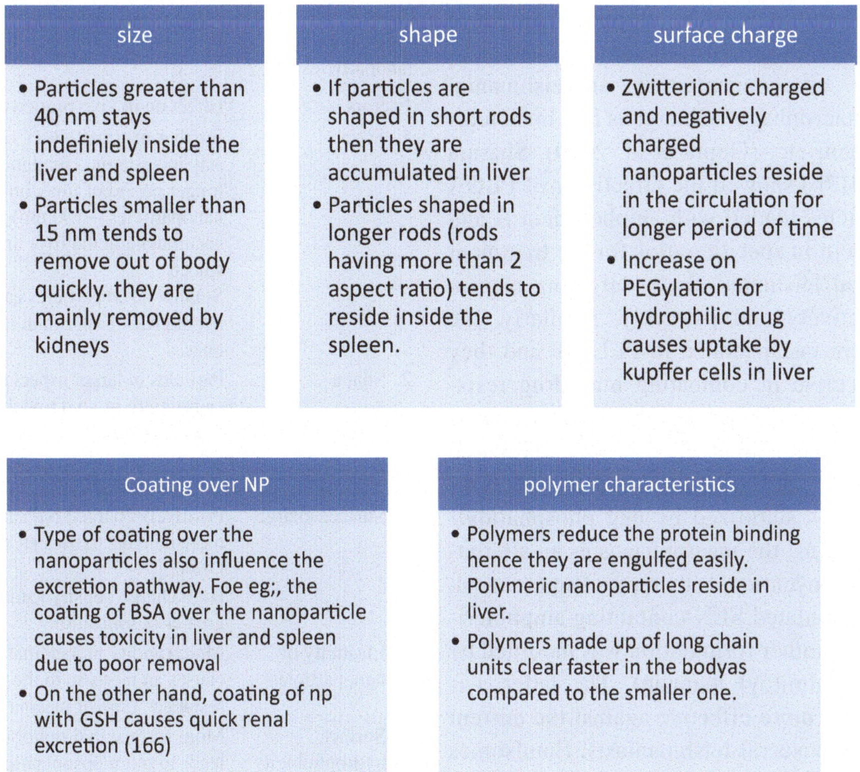

Fig. 7 Factors involved in exocytosis of nanoparticles

11.3 For Diseases Related to Neuroinflammation

In diseases like multiple sclerosis where the neuroinflammation leads to an increase in production of NO, which itself can cross BBB and affect the surrounding neurons, nanoparticles loaded with dichloromethylene diphosphonate were able to eliminate macrophages. After depletion of macrophages, the level of microglial cells declined leading to a lessening of severity. Recently, tolerogenic nanoparticles have been shown to elicit immunomodulatory effects by working on eliminating antigen-presenting cells like macrophages (Nally et al. 2019).

11.4 For Diseases Related to Bacterial Infection

In many diseases, the bacteria reside inside the macrophages for maturation and development which can lead to outbreaks for many diseases. In bacterial diseases like leishmaniasis, the parasite resides inside the phagosomes of macrophages lying inside the reticuloendothelial systems, and it negatively affects the action of antileishmanial drugs (Pal et al. 2012). *Leishmania* if left untreated can reach visceral organs causing visceral leishmaniasis or it can reach mucous membranes causing mucocutaneous leishmaniasis (MCL) (Gupta et al.

2016). Hence, it becomes important to target macrophage for effective therapy against leishmaniasis by concentrating antileishmanial drug in macrophage-rich organs like bone marrow, spleen, etc. (Gupta et al. 2010). Sharma et al. (2017b) showed the effectivity of PLGA nanoparticles loaded with amphotericin B and doxorubicin in specific ratios for the treatment of visceral leishmaniasis by targeting macrophage actively and passively. Similarly, the drugs were encapsulated in PCL-NP and they were effective in combating multidrug resistance against visceral leishmaniasis (Sharma et al. 2017a). Kumar et al. (2016) prepared plain as well as ligand-anchored formulations of emulsions stabilized by egg phosphatidylcholine against the macrophages as an alternative for conventional therapy. Gupta et al. (2013) formulated SLN-containing amphotericin B and another formulation was modified by OPM (O-palmitoyl mannan). The latter was found to be more effective against the current regimen of visceral leishmaniasis. Emulsomes containing amphotericin B that were modified with OPM showed more efficiency than emulsomes containing plain amphotericin B against the treatment of visceral leishmaniasis (Gupta et al. 2007) (Gupta and vyas 2007). To reduce the toxicity of amphotericin B, the drug can be complexed with lipid as the uptake of lipid molecule is more pronounced in RES of macrophages (Gupta et al. 2007). Ligand-anchored aerosolized liposomes for delivery of amphotericin B against the treatment of visceral leishmaniasis were formulated and were found to be effective in achieving high concentration even after 24 hours (Vyas et al. 2005).

12 Effect of Physicochemical Properties on In Vivo Fate of Nanomaterials

It is to be noted that the properties of nanoformulations can be altered easily according to the factors which has been mentioned in Table 4.

Table 4 Factors like size, shape, charge, elasticity, and hydrophobicity and its effect on in vivo property of nanoparticles

Factors	Effect on in vivo property
1. Size	Smaller nanoparticles (~20 nm) stay in systemic circulation for a longer period of time, larger nanoparticles (~100 nm) accumulate in the liver and spleen Smaller sized particles can bypass RES easier than larger ones
2. Shape	Particles in larger aspect ratio cause more in vivo toxicity Nonspherical shapes cause delayed macrophage uptake (Wibroe et al. 2017)
3. Surface charge	Positively charged NP causes more damage to the cell and the DNA High charge density leads to greater opsonization
4. Elasticity of nanoparticles	More rigidity in nanomaterials causes an increase in the residence time of plasma
5. Surface hydrophobicity	More surface hydrophobicity leads to more opsonization To prevent RES clearance the surfaces are linked to polymers like PEG

12.1 Effect of Size

When nanoparticles enter inside the body, the corona is formed on the surface of nanoparticles due to proteins present in the plasma. It greatly affects the in vivo fate of nanoparticles (Lundqvist et al. 2008). Corona formation around the nanoparticles is a size-dependent property. Larger-sized nanoparticles having a diameter of 100 nm have a coating of proteins related to the complement system, on the other hand, smaller-sized nanoparticles having a diameter of 20 nm shows more affinity toward lipoproteins. Clearance, as well as biodistribution of any nanoformulation, depends on the clearance of the drug-loaded carrier in various types of tissues. When a drug passes through a particular tissue it encounters various types of barriers through which only particles of a specific size can cross. Particles of less than 6 nm can cross the kidney and particles above 200 nm can cross the liver and spleen. On the other hand, nanoparticles of size greater than 200 nm get easily recognized by RES, which leads to their decline in half-life inside the body (Kulkarni and Feng 2013).

12.2 Effect of Shape

Some studies have shown that toxicity is attributed due to the shape of nanoparticles (Kinnear et al. 2017). The influence of shape is measured by changing the aspect ratios. An increase in shape can also promote the formation of proinflammatory cells. By changing the shape of nanoparticles into a nonspherical form can be useful in increasing their concentration in cancerous tissue which can be helpful in cancer therapy (Truong et al. 2014). Nonspherical nanocarriers have an extended rate of transport across the circulation as they possess different types of hydrodynamic properties due to their shape. Nonspherical nanocarriers attach themselves to the wall of endothelium strongly as they have a higher surface area and more resistance to flow than their spherical counterparts.

12.3 Charge on Nanoparticles

Surface charge plays a key role in increasing the uptake. While increased intracellular uptake is useful for increased circulation and uniform distribution, it can also contribute to toxicity. Positively charged nanoparticles interact with negatively charged cell membranes and hence cause more toxicity. The positively charged nanoparticle binds with negatively charged histone protein causing damage in the DNA and extending G0/G1 phase (Hühn et al. 2013). These structural and functional changes are responsible for inducing apoptosis. A positively charged nanoparticle can promote opsonization resulting in reduced clearance and hence reduced therapeutic efficacy occurs. More charge density causes more adsorption of proteins on their surface. By cross-linking the surface of nanoparticles with various polymers or enclosing them with lipids or enclosing them with pH-sensitive polymers can reduce the toxicity. Kupffer cells phagocytose highly positive and highly negative charges the most in comparison to neutral charge molecules. However, negatively charged nanoparticle is taken up and accumulated in the liver on a slightly lesser amount. Positively charged nanoparticles exhibit more penetration and damage to tumor tissue despite having a shorter duration of systemic circulation than negative and neutral nanoparticles (Wang et al. 2016).

12.4 Elasticity of Nanoparticles

The ability of a nanoparticle to undergo deformation is called the elasticity of nanoparticles. Different nanomaterial has their elasticity modulus which determines their ability to respond to various types of materials. The elasticity affects the uptake as well as endocytosis of nanoparticles (Hartmann et al. 2014). Nanocarriers can be classified into hard nanoparticles and soft nanoparticles. The former includes metal NP, quantum dots, etc., whereas the latter has liposomes, conjugates, etc. (Nag et al. 2016). Metal oxide NPs are known to trigger inflammatory responses from the body as well as disrupt homeostasis by harming the functioning of organs. On the other hand, metal NP can leak out of the cell causing damage to the cellular organelle (Chen et al. 2013). Müllner et al. (2015) showed that by increasing the rigidity of cylindrical polymer brushes (CPB) clearance rate can be increased and MPS response can be altered which as well decreases its residence time in plasma.

12.5 Surface Hydrophobicity

Surface hydrophobicity determines the opsonization as well as biodistribution of nanocarriers. Nanoparticles with hydrophobic moiety on their surface opsonize more resulting in greater clearance after in vivo administration. To reduce the number of nanoparticles being captured by RES, polymers like chitosan, PEG, etc. were attached to the surface of nanocarriers to mask their hydrophobicity. By changing the density as well as its degree of freedom and length of the polymer one can reduce the accumulation of NP into RES organs. The density of PEG plays a key role in opsonization as the NP having a high density of PEG tend to absorb a less amount of albumin proteins (Jokerst et al. 2011).

13 Adverse Reactions

Interaction between macrophages and nanoparticles can cause activation of the complement system which can be responsible for creating toxicity via the production of cytokines, reactive oxygen, and nitrogen species. Cytokines are responsible for producing an immune response toward the inflammation by sending signals to macrophages for differentiating into effector units and their transport. On the other hand, the complement system can cause the production of factors responsible for thrombosis, anaphylactic reactions which mainly depend on the surface properties of the nanoparticle that is interacting with the environment (Neun and Dobrovolskaia 2011; Thomas et al. 2011). Recently, supermagnetic dextran-coated iron oxide nanoparticles have been discontinued as a supplement because of their complement system response due to their surface properties (Banda et al. 2014). Schematic representation on type of adverse reactions observed due to nanoparticles has been mentioned in Fig. 8.

13.1 Complement Activation

After administration of liposomes, micelle CARPA (complement activation-related pseudoallergy) occurs inside the body. By this reaction, the activation of complement proteins occurs, which is responsible for the activation of basophil via generation of C3a and C5a. After activation of basophils, leukotrienes, thromboxane, prostaglandins, etc. are secreted by them causing anaphylactic symptoms like bronchoconstriction, change in blood pressure, dizziness, etc.; hypersensitivity reactions have been observed in patients after administering the liposomes of doxorubicin. Complement activation can be seen in nanoparticles having hydroxyl surface modifications. Similarly, FeO nanoparticles can cause inflammation by C3 activation. Splenic clearance and binding of nanoparticles with erythrocyte have been observed after activation of iC3b by nanoparticles. To reduce the degree of complement activation one can change the surface properties of nanoparticles. For example, polysaccharide-based dextrans as an excipient in nanoparticles helped increase the half-life of the product as it reduces the complement activity as well as protein adsorption. Sera used in cell cul-

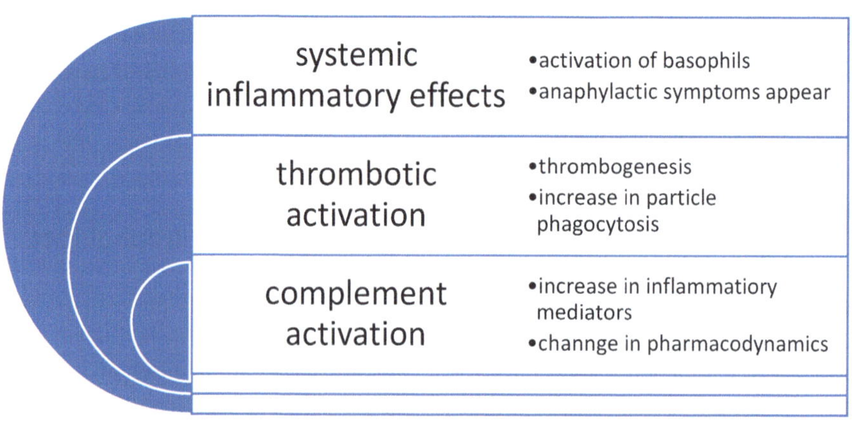

Fig. 8 Types of adverse reactions incited by nanoparticles

ture lacks a complement system because it is exposed to heating at a much earlier stage. Similarly, many systems can be used to facilitate dysopsonin binding in vivo, which helps in reducing the identification of foreign particles by the immune system.

13.2 Thrombotic Activation

In this type of activation aggregation of platelet leads to an increase in phagocytosis of particles. It has been observed that a small change in density and type of charge on the surface of the nanoparticle can induce less phagocytosis (Greish et al. 2012); surface modification of silica with amines can reduce the platelet aggregation. Dendrimers carrying a positive charge on their surface causes aggregation of platelets and the effect is increased with increasing positive charge density (Ilinskaya and Dobrovolskaia 2013); however, the mechanism of the aggregation of cationic dendrimer remains the same.

13.3 Systemic Inflammatory Effects

Usually the inflammatory response produced by the body on exposure to nanoparticles is due to systems that originate from the blood. However, there are some instances where inflammatory effects are observed due to the release of mediators. For example, iron oxide nanoparticles having similar charge and different surface chemistry can contribute to different levels of interleukin and TNF-α inside the body (Dobrovolskaia and McNeil 2013). One study showed that silver nanoparticles caused NF-κB and ROS signaling triggering the release of TNF as well as IL-6, on the other hand, when the residence time of gold nanoparticles inside the body is increased, the inflammatory response inside the body increases. It was also observed that the level of interleukin is increased by more levels of (4 nm) gold nanoparticles. Some other factors contribute to more inflammation like the ability to get oxidized, surface area, and size of nanoparticles. It has been observed that the silver nanoparticles of smaller size cause more toxicity than the larger counterparts (Lim et al. 2012). Oxidation may attribute to leaching and further toxicity of particles in the blood. Larger surface areas contribute more toward toxicity.

14 Toxicity of Nanomaterials

Usually when a nanoparticle is administered to the body it comes in contact with the phagocytic cells of MPS and this contact is responsible for causing toxicity. This contact is independent of the type of groups attached to the surface of nanoparticles. The toxicity is due to nonspecific targeting by macrophages which is very common for all the foreign bodies and this can be measured by the change in levels of cell biomarkers like levels of cytokines, fluctuation in the level of ROS, etc. The levels of monocytes and dendritic cells can be used to determine the level of nanoparticle clearance (Caron et al. 2013). Hence, it becomes imperative to determine the processing of nanoparticles to understand the response of biomarkers so that they can be used to reduce the general toxicity. In some cases, the nanoparticle is responsible for generating apoptosis in cells by disrupting the electron transport chain present over the surface of mitochondria resulting in the generation of cytochrome C and inflammation. Silica-based nanoparticles can also cause necrosis as well as apoptosis following the toxicity. It was determined by JC-1 mitochondrial assay that smaller nanoparticles were responsible for causing more toxicity than the larger ones as they cause more change in the mitochondrial membrane potential. It is to be noted that these nanoparticles can cause hyperpolarization resulting in cell death as well as depolarization leading to necrosis. Such events lead toward nanoparticles induced cellular disorder (Stern et al. 2012). Nanoparticles can damage mitochondria of a macrophage in other ways like by producing reactive oxygen species leading to inflammatory responses as well as upregulation of genres which cause a stress-induced response in the body. Fenton reaction is initiated by interaction between the hydroxyl group and iron in the surrounding environment resulting in the formation of free radicals. Silica nanoparticles were observed to cause a change in the levels of TNF-α in cells responsible for hematopoiesis and autophagy due to an increase in the level of ROS. These effects were more pronounced by nanoparticles possessing more charge on their surface (Shahbazi et al. 2013).

15 Conclusion and Future Perspectives

As discussed earlier, we can conclude that the nanocarrier-mediated drug delivery system offers a reduction in potential toxicity by the virtue of limiting its effects on tissues surrounding the target. Yet, the full potential of nanoparticles has not been utilized completely due to the barriers present in intracellular delivery as well as due to the presence of host-based immune response. One can minimize the engulfment of nanoparticles by immune cells via altering the physicochemical properties of nanoparticles.

Another factor to be taken into consideration is the toxicity induced by nanoparticles. To reduce the effects, one needs to change the size, shape, surface charge, and hydrophilicity of the nanoparticles aside from the type of nanomaterials that are being used. To optimize the delivery of nanoparticles, one needs to focus on the aspect of drug delivery that they are trying to improve. Worm or filament-like morphologies can be incorporated in nanoparticle design as next-generation drug carriers. One can avoid MPS and filtration by kidney, spleen, and liver if they control the size of nanoparticles. Similarly, surface chemistry and elasticity need to be modified to achieve localization only in specific tissues and organs. It is to be noted that more than one property plays a key role in delivering the nanoparticles to a specific target. One needs to go through data that has already been established to modify the release parameter of the formulation. QBD can be a good approach to establish the platform for modifying and implementing those approaches. Other methods can also be employed to study the data of physiochemical properties and associate it with the in vivo fate of various types of nanoparticles. Some methods like HCA (hierarchical cluster analysis) can be used to screen nanoparticles having similar in vivo fate without specifying much about their physiochemical properties.

There are various kinds of toxicities associated with various types of physicochemical properties. Some nanoparticles are known to cause unintended and unfavorable effects. Thus, it becomes necessary to modulate and identify parameters that one needs to modulate to optimize the in vivo fate of nanoparticles.

References

Aggarwal P, Hall JB, McLeland CB, Dobrovolskaia MA, McNeil SE. Nanoparticle interaction with plasma proteins as it relates to particle biodistribution, biocompatibility and therapeutic efficacy. Adv Drug Deliv Rev. 2009;61(6):428–37. https://doi.org/10.1016/j.addr.2009.03.009.

Anand PK, Tait SW, Lamkanfi M, et al. TLR2 and RIP2 pathways mediate autophagy of *Listeria monocytogenes* via extracellular signal-regulated kinase (ERK) activation. J Biol Chem. 2011;286(50):42981–91. https://doi.org/10.1074/jbc.M111.310599.

Banda NK, Mehta G, Chao Y, et al. Mechanisms of complement activation by dextran-coated superparamagnetic iron oxide (SPIO) nanoworms in mouse versus human serum. Part Fibre Toxicol. 2014;11:64. Published 2014 Nov 26. https://doi.org/10.1186/s12989-014-0064-2.

Bayik D, Gursel I, Klinman DM. Structure, mechanism and therapeutic utility of immunosuppressive oligonucleotides. Pharmacol Res. 2016;105:216–25. https://doi.org/10.1016/j.phrs.2015.11.010.

Beck-Broichsitter M, Ruppert C, Schmehl T, et al. Biophysical investigation of pulmonary surfactant surface properties upon contact with polymeric nanoparticles in vitro. Nanomedicine. 2011;7(3):341–50. https://doi.org/10.1016/j.nano.2010.10.007.

Bee JS, Chiu D, Sawicki S, et al. Monoclonal antibody interactions with micro- and nanoparticles: adsorption, aggregation, and accelerated stress studies. J Pharm Sci. 2009;98(9):3218–38. https://doi.org/10.1002/jps.21768.

Biswas SK, Mantovani A. Macrophage plasticity and interaction with lymphocyte subsets: cancer as a paradigm. Nat Immunol. 2010;11(10):889–96.

Biswas SK, Mantovani A. Orchestration of metabolism by macrophages. Cell Metab. 2012;15(4):432–7.

Biswas S, Mantovani A. Macrophages: biology and role in the pathology of diseases. New York: Springer; 2014.

Boller T, Felix G. A renaissance of elicitors: perception of microbe-associated molecular patterns and danger signals by pattern-recognition receptors. Annu Rev Plant Biol. 2009;60:379–406. https://doi.org/10.1146/annurev.arplant.57.032905.105346.

Canton J, Neculai D, Grinstein S. Scavenger receptors in homeostasis and immunity. Nat Rev Immunol. 2013;13(9):621–34. https://doi.org/10.1038/nri3515.

Caron WP, Lay JC, Fong AM, et al. Translational studies of phenotypic probes for the mononuclear phagocyte system and liposomal pharmacology. J Pharmacol Exp Ther. 2013;347(3):599–606. https://doi.org/10.1124/jpet.113.208801.

Chao Y, Karmali PP, Mukthavaram R, Kesari S, Kouznetsova VL, Tsigelny IF, Simberg D. Direct recognition of superparamagnetic nanocrystals by macrophage scavenger receptor SR-AI. ACS Nano. 2013 May 28;7(5):4289–98. https://doi.org/10.1021/nn400769e. Epub 20s13 Apr 30.

Chavez-Santoscoy AV, Roychoudhury R, Pohl NL, Wannemuehler MJ, Narasimhan B, Ramer-Tait AE. Tailoring the immune response by targeting C-type lectin receptors on alveolar macrophages using "pathogen-like" amphiphilic polyanhydride nanoparticles. Biomaterials. 2012;33(18):4762–72. https://doi.org/10.1016/j.biomaterials.2012.03.027.

Chellata F, Merhib Y, Moreauc A, Yahiaa L'H. Therapeutic potential of nanoparticulate systems for macrophage targeting. Biomaterials. 2005;26:7260–75.

Chen J, Guo Z, Wang H, et al. Multifunctional Fe3O4, Cu, Ag hybrid nanoparticles as dual modal imaging probes and near-infrared light-responsive drug delivery platform. Biomaterials. 2013;34(2):571–81. https://doi.org/10.1016/j.biomaterials.2012.10.002.

Clemments AM, Botella P, Landry CC. Spatial mapping of protein adsorption on mesoporous silica nanoparticles by stochastic optical reconstruction microscopy. J Am Chem Soc. 2017;139:3978–81.

Collier MA, Gallovic MD, Peine KJ, et al. Delivery of host cell-directed therapeutics for intracellular pathogen clearance. Expert Rev Anti-Infect Ther. 2013;11(11):1225–35. https://doi.org/10.1586/14787210.2013.845524.

Das A, Sinha M, Datta S, Abas M, Chaffee S, Sen CK, et al. Monocyte and macrophage plasticity in tissue repair and regeneration. Am J Pathol. 2015;185:2596–606.

Dobrovolskaia MA, McNeil SE. Immunological properties of engineered nanomaterials. Nat Nanotechnol. 2007;2(8):469–78. https://doi.org/10.1038/nnano.2007.223.

Dobrovolskaia MA, McNeil SE. Understanding the correlation between in vitro and in vivo immunotoxicity tests for nanomedicines. J Control Release. 2013;172:456–66.

Doshi N, Mitragotri S. Needle-shaped polymeric particles induce transient disruption of cell membranes. J R Soc Interface. 2010;7(Suppl 4):S403–10. https://doi.org/10.1098/rsif.2010.0134.focus.

Echarri A, Del Pozo MA. Caveolae – mechanosensitive membrane invaginations linked to actin filaments. J Cell Sci. 2015;128(15):2747–58.

El-Sayed A, Harashima H. Endocytosis of gene delivery vectors: from clathrin-dependent to lipid raft-mediated endocytosis. Mol Ther. 2013;21(6):1118–30.

Galmarini S, Hanusch U, Giraud M, et al. Beyond unpredictability: the importance of reproducibility in understanding the protein corona of nanoparticles [published correction appears in Bioconjug Chem. 2019 May 15]. Bioconjug Chem. 2018;29(10):3385–93. https://doi.org/10.1021/acs.bioconjchem.8b00554.

Giannakis N, Sansbury BE, Patsalos A, Hays TT, Riley CO, Han X, et al. Dynamic changes to lipid mediators support transitions among macrophage subtypes during muscle regeneration. Nat Immunol. 2019;20(5):626–36.

Goel S, England CG, Chen F, Cai W. Positron emission tomography and nanotechnology: a dynamic duo for cancer theranostics. Adv Drug Deliv Rev. 2017;113:157–76.

Greish K, Thiagarajan G, Herd H, Price R, Bauer H, Hubbard D, Burckle A, Sadekar S, Yu T, Anwar A. Nanotoxicology. 2012;6:713–23.

Gupta S, Gupta M. Possible role of nanocarriers in drug delivery against cervical cancer. Nano Rev Exp. 2017;8(1):1335567. https://doi.org/10.1080/20022727.2017.1335567.

Gupta S, Vyas SP. Development and characterization of amphotericin B bearing emulsomes for passive and active macrophage targeting. J Drug Target. 2007;15:206–17. (Impact factor: 3.277)

Gupta S, Dube A, Vyas SP. Antileishmanial efficacy of amphotericin B bearing emulsomes against experimental visceral leishmaniasis. J Drug Target. 2007;15:437–44. (Impact factor: 3.277)

Gupta S, Pal A, Vyas SP. Drug delivery strategies for therapy of visceral leishmaniasis. Expert Opin Drug Deliv. 2010;7(3):371–402. (Impact factor: 5.40)

Gupta S, Dube A, Vyas SP. Development and characterization of amphotericin B loaded solid lipid nanoparticles against experimental visceral leishmaniasis. Pharm Nanotechnol. 2013;1(1):54–67.

Gupta A, Pant G, Mitra K, Madan J, Chourasia MK, Misra A. Inhalable particles containing rapamycin for induction of autophagy in macrophages infected with Mycobacterium tuberculosis. Mol Pharm. 2014;11(4):1201–7. https://doi.org/10.1021/mp4006563.

Gupta S, Sharma P, Gupta MK. Leishmaniasis – drugs, nanotechnology based delivery systems and recent patents survey. Curr Nanomed. 2016;6:21–42.

Gupta S, Pathak Y, Gupta MK, Vyas SP. Nanoscale drug delivery strategies for therapy of ovarian cancer: conventional vs targeted. Artif Cells Nanomed Biotechnol. 2019;47(1):4066–88. https://doi.org/10.1080/21691401.2019.1677680.

Gustafson HH, Holt-Casper D, Grainger DW, Ghandehari H. Nanoparticle uptake: the phagocyte problem. Nano Today. 2015;10(4):487–510.

Hartmann R, Weidenbach M, Neubauer M, Fery A, Parak W. Stiffness-dependent in vitro uptake and lysosomal acidification of colloidal particles. Angew Chem Int Ed. 2014;54(4):1365–8. https://doi.org/10.1002/anie.201409693.

Hernandez C, Gulati S, Fioravanti G, Stewart PL, Exner AA. Cryo-EM visualization of lipid and polymer-stabilized perfluorocarbon gas nanobubbles – a step towards nanobubble mediated drug delivery. Sci Rep. 2017;7:13517.

Hinde E, Thammasiraphop K, Duong HTT, Yeow J, Karagoz B, Boyer C, Gooding JJ, Gaus K. Pair correlation microscopy reveals the role of nanoparticle shape in intracellular transport and site of drug release. Nat Nanotechnol. 2017;12:81–9.

Hirst SM, Karakoti A, Singh S, Self W, Tyler R, Seal S, Reilly CM.Biodistribution and in-vivo antioxidant effects of cerium oxide nanoparticles in mice.Environ Toxicol. 2013;28:107–18.

Holt BD, Short PA, Rape AD, Wang YL, Islam MF, Dahl KN. Carbon nanotubes reorganize actin structures in cells and ex vivo. ACS Nano. 2010;4(8):4872–8. https://doi.org/10.1021/nn101151x.

Hühn D, Kantner K, Geidel C, et al. Polymer-coated nanoparticles interacting with proteins and cells: focusing on the sign of the net charge. ACS Nano. 2013;7(4):3253–63. https://doi.org/10.1021/nn3059295.

Ilinskaya AN, Dobrovolskaia MA.Nanoparticles and the blood coagulation system.Part II: safety concerns. Nanomedicine (Lond). 2013;8:969–81.

Jokerst J, Lobovkina T, Zare R, Gambhir S. Nanoparticle PEGylation for imaging and therapy. Nanomedicine. 2011;6(4):715–28. https://doi.org/10.2217/nnm.11.19.

Kapralov AA, Feng WH, Amoscato AA, et al. Adsorption of surfactant lipids by single-walled carbon nanotubes in mouse lung upon pharyngeal aspiration. ACS Nano. 2012;6(5):4147–56. https://doi.org/10.1021/nn300626q.

Karataş OF, Sezgin E, Aydin O, Culha M. Interaction of gold nanoparticles with mitochondria. Colloids Surf B Biointerfaces. 2009;71(2):315–8. https://doi.org/10.1016/j.colsurfb.2009.02.020.

Kinnear C, Moore TL, Rodriguez-Lorenzo L, Rothen-Rutishauser B, Petri-Fink A. Form follows function: nanoparticle shape and its implications for nanomedicine. Chem Rev. 2017;117(17):11476–521. https://doi.org/10.1021/acs.chemrev.7b00194.

Klang V, Valenta C, Matsko NB. Electron microscopy of pharmaceutical systems. Micron. 2013;44:45–74.

Kuhn DA, Vanhecke D, Michen B, Blank F, Gehr P, Petri-Fink A, Rothen-Rutishauser B. Different endocytotic uptake mechanisms for nanoparticles in epithelial cells and macrophages. J Nanotechnol. 2014;5:1625–36.

Kulkarni SA, Feng SS. Effects of particle size and surface modification on cellular uptake and biodistribution of polymeric nanoparticles for drug delivery. Pharm Res. 2013;30:2512–22. https://doi.org/10.1007/s11095-012-0958-3.

Kulkarni JA, Darjuan MM, Mercer JE, Chen S, van der Meel R, Thewalt JL, Tam YYC, Cullis PR. On the formation and morphology of lipid nanoparticles containing ionizable cationic lipids and siRNA. ACS Nano. 2018;12:4787–95.

Kumagai Y, Akira S. Identification and functions of pattern-recognition receptors. J Allergy Clin Immunol. 2010;125(5):985–92. https://doi.org/10.1016/j.jaci.2010.01.058.

Kumar R, Roy I, Ohulchanskky TY, Vathy LA, Bergey EJ, Sajjad M, Prasad PN. In vivo biodistribution and clearance studies using multimodal organically modified silica particles. ACS Nano. 2010;4:699–708.

Kumar N, Sharma P, Jaiswal A, Dube A, Gupta S. Development and evaluation of p-aminophenylmannopyranoside anchored emulsomes for treatment of experimental visceral leishmaniasis. Ann Clin Cytol Pathol. 2016;2(6):1042.

Kurylowicz M, Paulin H, Mogyoros J, Giuliani M, Dutcher JR. The effect of nanoscale surface curvature on the oligomerization of surface-bound proteins. J R Soc Interf. 2014;11(94):20130818. Published Feb 26. https://doi.org/10.1098/rsif.2013.0818.

Laria A, Lurati A, Marrazza M, Mazzocchi D, Re KA, Scarpellini M. The macrophages in rheumatic diseases. J Inflamm Res. 2016;9:1–11.

Lawrence T, Natoli G. Transcriptional regulation of macrophage polarization: enabling diversity with identity. Nat Rev Immunol. 2011;11(11):750–61.

Lim DH, Jang J, Kim S, Kang T, Lee K, Choi IH. The effects of sub-lethal concentrations of silver nanoparticles on inflammatory and stress genes in human macrophages using cDNA microarray analysis. Biomaterials. 2012;33:4690–9.

Lundqvist M, Stigler J, Elia G, Lynch I, Cederva T, Dawson KA. Nanoparticle size and surface properties determine the protein corona with possible implications for biological impacts. Proc Natl Acad Sci. 2008 Sep;105(38):14265–70. https://doi.org/10.1073/pnas.0805135105.

Marichal L, Klein G, Armengaud J, et al. Protein corona composition of silica nanoparticles in complex media: nanoparticle size does not matter. Nanomaterials (Basel). 2020;10(2):240. Published 2020 Jan 29. https://doi.org/10.3390/nano10020024.

Martinez FO, Gordon S. The M1 and M2 paradigm of macrophage activation: time for reassessment. F1000Prime Rep. 2014;6:13.

Mishra R, Gupta S. Novel nano carriers for the treatment of progressive auto immune disease rheumatoid arthritis. Curr Pharm Des. 2020;26 https://doi.org/10.2174/1381612826666201021130146.

Müllner M, Dodds S, Nguyen T, et al. Size and rigidity of cylindrical polymer brushes dictate long circulating properties in vivo. ACS Nano. 2015;9(2):1294–304. https://doi.org/10.1021/nn505125f.

Murray PJ. Macrophage activation and polarization: nomenclature and experimental guidelines. Immunity. 2014;41(1):14–20.

Nag O, Field L, Chen Y, Sangtani A, Breger J, Delehanty J. Controlled actuation of therapeutic nanoparticles: an update on recent progress. Ther Deliv. 2016;7(5):335–52. https://doi.org/10.4155/tde-2016-0003.

Nally FK, De Santi C, McCoy CE. Nanomodulation of macrophages in multiple sclerosis. Cells. 2019;8(6):543. Published 2019 Jun 5. https://doi.org/10.3390/cells8060543.

Naumenko V, Van S, Dastidar H, Kim D-S, Kim S-J, Zeng Z, Deniset J, Lau A, Zhang C, Macia N, Heyne B, Jenne CN, Mahoney DJ (2018)

Navegantes KC, de Souza GR, Pereira PAT, Czaikoski PG, Azevedo CHM, Monteiro MC. Immune modulation of some autoimmune diseases: the critical role of macrophages and neutrophils in the innate and adaptive immunity. J Transl Med. 2017;15:36.

Neun BW, Dobrovolskaia MA. Qualitative analysis of total complement activation by nanoparticles . Methods Mol Biol. 2011;697:237–45.

Odegaard JI. Macrophage-specific PPARgamma controls alternative activation and improves insulin resistance. Nature. 2007;447(7148):1116–20.

Oishi Y, Manabe I. Macrophages in inflammation, repair and regeneration. Int Immunol. 2018;30:511–28.

Orr GA, Chrisler WB, Cassens KJ, et al. Cellular recognition and trafficking of amorphous silica nanoparticles by macrophage scavenger receptor A. Nanotoxicology. 2011;5(3):296–311. https://doi.org/10.3109/17435390.2010.513836.

Pal A, Gupta S, Jaiswal A, Dube A, Vyas SP. Development and evaluation of tripalmitin emulsomes for the treatment of experimental visceral leishmaniasis. J Liposome Res. 2012;22(1):62–71.

Patel S, Ashwanikumar N, Robinson E, DuRoss A, Sun C, Murphy-Benenato K, Mihai C, Almarsson Ö, Sahay G. Boosting intracellular delivery of lipid nanoparticle-encapsulated mRNA. Nano Lett. 2017;17:5711–8. https://doi.org/10.1021/acs.nanolett.7b02664.

Patel S, Kim J, Herrera M, et al. Brief update on endocytosis of nanomedicines. Adv Drug Deliv Rev. 2019 April;144:90–111.

Prado-Gotor R, Grueso E. A kinetic study of the interaction of DNA with gold nanoparticles: mechanistic aspects of the interaction. Phys Chem Chem Phys. 2011;13(4):1479–89. https://doi.org/10.1039/c0cp00901f.

Ruffell D. A CREB-C/EBPbeta cascade induces M2 macrophage-specific gene expression and promotes muscle injury repair. Proc Natl Acad Sci U S A. 2009;106(41):17475–80.

Sabnis S, Kumarasinghe ES, Salerno T, Mihai C, Ketova T, Senn JJ, Lynn A, Bulychev A, McFadyen I, Chan J, Almarsson Ö, Stanton MG, Benenato KE. A novel amino lipid series for mRNA delivery: improved endosomal escape and sustained pharmacology and safety in non-human primates. Mol Ther. 2018;26:1509–19.

Sachan AK, Harishchandra RK, Bantz C, Maskos M, Reichelt R, Galla HJ. High-resolution investigation of nanoparticle interaction with a model pulmonary surfactant monolayer. ACS Nano. 2012;6(2):1677–87. https://doi.org/10.1021/nn204657n.

Samokhvalov IM. Deconvoluting the ontogeny of hematopoietic stem cells. Cell Mol Life Sci. 2014;71:957–78.

Schonthaler HB, Guinea-Viniegra J, Wagner EF. Targeting inflammation by modulating the Jun/AP-1 pathway. Ann Rheum Dis. 2011;70(1):109–12.

Shahbazi MA, Hamidi M, Mäkilä EM, et al. The mechanisms of surface chemistry effects of mesoporous silicon nanoparticles on immunotoxicity and biocompatibility. Biomaterials. 2013;34(31):7776–89. https://doi.org/10.1016/j.biomaterials.2013.06.052.

Sharma P, Gupta S. Development and evaluation of macrophage targeted multidrug therapy against visceral leishmaniasis. Pharm Biomed Res. 2017a;3(1):9–18.

Sharma P, Gupta S. PLGA-based macrophage-mediated drug targeting for the treatment of visceral leishmaniasis. Pharm Biomed Res. 2017b;3(1):50–6.

Slowing II, Wu CW, Vivero-Escoto JL, Lin VS. Mesoporous silica nanoparticles for reducing hemolytic activity towards mammalian red blood cells. Small. 2009;5(1):57–62. https://doi.org/10.1002/smll.200800926.

Stern ST, Adiseshaiah PP, Crist RM. Autophagy and lysosomal dysfunction as emerging mechanisms of nanomaterial toxicity. Part Fibre Toxicol. 2012;9:20. Published Jun 14. https://doi.org/10.1186/1743-8977-9-20.

Tarantola M, Schneider D, Sunnick E, et al. Cytotoxicity of metal and semiconductor nanoparticles indicated by cellular micromotility. ACS Nano. 2009;3(1):213–22. https://doi.org/10.1021/nn800721j.

Thakor A, Jokerst J, Ghanouni P, Campbell J, Mittra E, Gambhir S. Clinically approved nanoparticle imaging agents. J Nucl Med. 2016;57:1833–7. https://doi.org/10.2967/jnumed.116.181362.

Thomas SN, van der Vlies AJ, O'Neil CP, Reddy ST, Yu SS, Giorgio TD, Swartz MA, Hubbell JA. Engineering complement activation on polypropylene sulfide vaccine nanoparticle. Biomaterials. 2011;32:2194–203.

Truong N, Whittaker M, Mak C, Davis T. The importance of nanoparticle shape in cancer drug delivery. Expert Opin Drug Deliv. 2014;12(1):129–42. https://doi.org/10.1517/17425247.2014.950564.

van der Zwaag D, Vanparijs N, Wijnands S, De Rycke R, De Geest BG, Albertazzi L. Super resolution imaging of nanoparticles cellular uptake and trafficking. ACS Appl Mater Interfaces. 2016;8:6391–9.

Vannella KM, Wynn TA. Mechanisms of organ injury and repair by macrophages. Annu Rev Physiol. 2017;79:593–617.

Varga T, Mounier R, Gogolak P, Poliska S, Chazaud B, Nagy L. Tissue LyC6-macrophages are generated in the absence of circulating LyC6-monocytes and Nur77 in a model of muscle regeneration. J Immunol. 2013;191:5695–701.

Vertegel AA, Siegel RW, Dordick JS. Silica nanoparticle size influences the structure and enzymatic activity of adsorbed lysozyme. Langmuir. 2004;20(16):6800–7. https://doi.org/10.1021/la0497200.

Vyas SP, Quraishi S, Gupta S, Jaganathan KS. Aerosolized liposome-based delivery of amphotericin B to alveolar macrophages. Int J Pharm. 2005;296:12–25. (Impact factor: 4.213)

Walkey CD, Chan WC. Understanding and controlling the interaction of nanomaterials with proteins in a physiological environment. Chem Soc Rev. 2012;41(7):2780–99. https://doi.org/10.1039/c1cs15233e.

Wang N, Liang H, Zen K. Molecular mechanisms that influence the macrophage m1–m2 polarization balance. Front Immunol. 2014;5:614.

Wang H, Zuo Z, Du J, et al. Surface charge critically affects tumor penetration and therapeutic efficacy of cancer nanomedicines. Nano Today. 2016;11(2):133–44. https://doi.org/10.1016/j.nantod.2016.04.008.

Welsher K, Yang H. Multi-resolution 3D visualization of the early stages of cellular uptake of peptide-coated nanoparticles. Nat Nanotechnol. 2014;9:198–203.

Wibroe P, Anselmo A, Nilsson P, et al. Bypassing adverse injection reactions to nanoparticles through shape modification and attachment to erythrocytes. Nat Nanotechnol. 2017;12:589–94. https://doi.org/10.1038/nnano.2017.47.

Williams J, Giannarelli C, Rahman A, Randolph G, Kovacic J. Macrophage biology, classification, and phenotype in cardiovascular disease. J Am Coll Cardiol. 2018;72(18):2166–80.

Wynn TA, Chawla A, Pollard JW. Macrophage biology in development, homeostasis, and disease. Nature. 2013;496(7446):445–55.

Yadav S, Gupta S. Development and in vitro characterization of docetaxel-loaded ligand appended solid fat nanoemulsions for potential use in breast cancer therapy. Artif Cells Nanomed Biotechnol. 2015;43(2):93–102. https://doi.org/10.3109/21691401.2013.845569.

Yin J, Valin KL, Dixon ML, Leavenworth JW. The role of microglia and macrophages in CNS homeostasis, autoimmunity, and cancer. J Immunol Res. 2017;2017:5150678.

Zahid MU, Ma L, Lim SJ, Smith AM. Single quantum dot tracking reveals the impact of nanoparticle surface on intracellular state. Nat Commun. 2018;9:1830.

Zhu M, Nie G, Meng H, Xia T, Nel A, Zhao Y. Physiochemical properties determine nanomaterial cellular uptake, transport and fate. Acc Chem Res. 2013;46:622–31.

Nanoparticles-Based Theranostics for Macrophage Targeting

Amisha Chauhan, Mahima Gupta, Anushka Tyagi, Yashwant V. Pathak, and Swati Gupta

Abstract

Theranostic nanomedicine is rapidly gaining popularity as a promising therapeutic approach. It takes advantages of nanoplatforms for better imaging and therapeutic functions. The resulting nanosystems, capable of diagnosis, delivery of therapeutics, and monitoring of therapeutic response, are expected to play a significant role in the dawning era of personalized medicine, and much research effort has been devoted toward this goal. Their well-developed surface chemistry makes it simple to stack them with pharmaceutics and elevate them to be theranostic nanosystems. Iron oxide nanoparticles, quantum dots, carbon nanotubes, gold nanoparticles, and silica nanoparticles have been previously well investigated in the imaging setting and are candidate nanoplatforms for building up nanoparticle-based theranostics. Nanoparticle advances are altogether affecting the improvement of both therapeutic and diagnostic effects. At the convergence between treatment and diagnosis, interest has filled in consolidating the two standards into clinically viable definitions. The concept of theranostics is particularly relevant to agents that target atomic biomarkers of infections and is expected to be included in tailored therapy. Highest quality nanoparticles from both a beneficial and symptomatic aspect, as well as the challenges of bringing these fields together, are analyzed. Significant classes of nanoparticles incorporate dendrimers, vesicles, micelles, center shell particles, microbubbles, and carbon nanotubes. Most of these formulations have been described as carriers of either drugs or contrast agents. To observe these formulations and their interactions with disease, a variety of contrast agents have been used, including optically active small molecules, metals and metal oxides, ultrasonic contrast agents, and radionuclides. The ability to rapidly assess and adapt therapy to the needs of the individual has the potential to accelerate the development of theranostic agents. The progress along this line has been outlined in the current chapter. Construction strategies have been focused

A. Chauhan · M. Gupta · A. Tyagi · S. Gupta (✉)
Department of Pharmaceutics, Amity Institute of Pharmacy, Amity University Uttar Pradesh, Noida, India

Y. V. Pathak
University of South Florida, Taneja College of Pharmacy, Tampa, FL, USA

Adjunct Professor University of Airlangga, Surabaya, Indonesia

S. Gupta, Y. V. Pathak (eds.), *Macrophage Targeted Delivery Systems*,
https://doi.org/10.1007/978-3-030-84164-5_12

upon and the challenges and opportunities associated with this emerging technology have been discussed.

Keywords

Nanoparticles · Theranostics · Macrophage · Targeting · Imaging

1 Introduction

The expression "theranostics" has begun to characterize progressing endeavors in facilities to grow more explicit, individualized treatments for different sicknesses and to consolidate analytic and remedial abilities into a single agent (Xie et al. 2010). The rationale emerged from the way that infections, for example, diseases, are colossally heterogeneous, and all current medicines are effective for just restricted patient subpopulations and at specific phases of infection advancement. The development of nanotechnology has offered instances to bring diagnosis and treatment nearer. Nanoparticle (NP)-based imaging and treatment have been examined independently, and their understanding has now advanced to a point empowering the introduction of NP-based theranostics, which can be characterized as nanoplatforms that can co-convey treatment and imaging capacities. This is a supplement to conventional theranostics, with an emphasis on "co-delivery." (Del Vecchio et al. 2007).

It adds to the previous framework for permitting imaging to be conducted not only before or after, but also during a treatment plan. It is helpful that numerous nanomaterials are as of now imaging agents and can be promptly "updated" to theranostic agents by mounting remedial capacities on them. One basic driving power of such a blend is that imaging and treatment both require adequate gathering of agents in infected regions. This usual targeting prerequisite brings the two examination areas closer and, at last, will blur the distinction between them, since numerous methods to improve imaging can, from a certain perspective, be promptly moved to the remedial space and the other way around (Nie et al. 2007). Targeting systems can be changed tremendously to suit the ideal targets. On account of malignancy, it is a typical way to deal with a distinguished biomarker that is unusually communicated on the outside of malignant cells and then to load its related restricting vector onto tests/transporters to accomplish recognition and tumor homing. For nanoplatforms, the extraordinary size of the particles empowers accomplishment of an improved enhanced permeability and retention (EPR) effect in tumor targeting. However, caution must be taken with the particle surfaces to prevent innate immune system detection and to ensure that the agents have long enough circulation half-lives to meet their targets.

Nanoparticle-based imaging and treatment are each attempting to advance into clinical preliminaries and, as relatives of the two, nanoparticle-based theranostics are as yet in their beginning phases of improvement. Nonetheless, the push given by propels in nanotechnology and the call for customized medication have just made nanoparticle-based theranostics an exploration hotspot (Liu et al. 2007).

Clinical nanotechnology has been evolving for quite a long time, and imaginative applications are happening as intended. Nanoparticle details, for example, DoxilTM and AbraxaneTM, have shown clinical importance by expanding drug viability and diminishing toxicity, and various targeting details are under clinical assessment. Some encouraging applications use nanometer-scale particles for synchronous medication conveyance and molecular imaging. There are several instances where multifunctional information may be used for both symptomatic and therapeutic purposes.

The designing of multifunctional theranostic NPs is not direct; besides, informative exercises can be gathered from almost 50 years of research on nanoparticulate drug transporters. Expected impediments to fruitful theranostic NPs incorporate the disclosure and focusing of new biomarkers, the natural toxicity of the nanoparticle segments, production cost, and control of licensed innovation. As these new formulations are developed, disparities regarding the safety of nanoparticulates also emerge. Furthermore, ideal therapeutics and diagnostics are two altogether different substances. Diagnostic agents serve to improve perceivability of determined tissues by increasing the signal-to-noise ratio relative to surrounding tissues and are generally optimized to provide a quick snapshot of the living system. Remedial nanoparticle details that have been utilized clinically are generally long coursing. To develop clinically useful theranostic NPs, this possible difference between fast and slow clearance rates must be resolved (Cai and Chen 2008). Theranostics is turning out to be famous in light of the fact that they are focused on therapeutics and can be utilized with no or insignificant changes for demonstrative imaging to help in precision medicine. Hence, there is a close connection among theranostics and impaired glucose tolerance (IGT), and theranostics are a subclass of IGT in which both remedial and imaging functionalities are ascribed to a solitary stage. A significant theranostic system is naturally pretargeting. In pretargeted IGT, first, the target is distinguished by an objective-specific common or engineered bioligand followed by a nano-scale or subatomic drug delivery fragment, which integrates therapeutic groups by in situ formation responses. If pretargeted drug conveyance stages are marked with multimodal imaging tests, they can be utilized as theranostics for both analytical imaging and treatment. Optical and atomic imaging methods have generally been utilized in confirmation of idea concentrates with pretargeted theranostics.

Theranostics is a field that joins the remarkable open doors offered by nanotechnology with customized medication to give altogether improved treatment viability with diminished impacts through the particular conveyance of treatment to focused tissues. Theranostic methods combine imaging using one of the nonobtrusive imaging modalities with specific delivery of therapeutic components that can be based on a variety of biophysical and natural criteria. Theranostics can be designed to have ideal transportation properties, low renal clearance rate, decreased immunogenicity, and antigenicity (Hapuarachchige and Artemov 2020).

Nanomedicine is characterized as "the utilization of nanotechnology to medication, including the utilization of nanometer-sized transporter materials for facilitating illness determination, illness therapy and treatment checking." Examples of transporter materials regularly utilized for nanomedicine applications are liposomes, polymers, micelles, dendrimers, nanoparticles, and antibodies. Nanomedicines have a few focal points over normal low molecular weight agents. They are, for example, capable (I) to shield the payload from untimely delivery, enzymatic degradation, and additional introduction to conceivably unsafe physiological conditions; (II) to improve the biodistribution and aggregation of medications and imaging agents at target site; (III) to improve the in vivo efficacy of diagnostic and therapeutic interventions; (IV) to constrict aggregation of drug and imaging agent in solid, nontarget tissues; and (V) to reduce the incidence and intensity of side effects (Kunjachan et al. 2014; Roesch 2012). It is critical to quantify various aspects of the drug delivery mechanism, such as pharmacokinetics, biodistribution, target site aggregation, local dissemination at the target site, localization in healthy tissues, drug release kinetics, and therapeutic efficacy, in order to better understand and improve drug delivery to pathological sites.

In this manner, as of late, there has been an increasing focus on the utilization of noninvasive imaging methods, for example, positron emission tomography (PET), single photon emission computerized tomography (SPECT), computed tomography (CT), magnetic resonance imaging (MRI), optical imaging (OI), and ultrasound (US), for observing medication conveyance, drug discharge, and drug adequacy (Muller et al. 2014). Among these strategies, CT, MRI, and US can be utilized both with and without contrast agents. In case of the former, that is, at the point when contrast agents are utilized, these modalities require prescans, to decide the background level of CT, MRI, and US signal preceding contrast agent administration. Such estimations are expected to measure the functional or molecular imaging data. On the other hand, on account of "hot-spot" techniques, for example, PET and SPECT, no background signals are detected in the absence of contrast agents, and prescans are not required. Macrophages are the specific cells which are associated with essential identification, phagocytosis, further destruction of microbes, and other unsafe life forms. Macrophages play a significant role in the inflammatory infections.

Theranostics is characterized as a combination of therapeutic or imaging functionalities on a solitary stage that can analyze, deliver drugs, and screen the treatment prompting customized medicine. By combining imaging and remedial functionalities together, it is conceivable but not exclusive to picture and track the biodistribution of the theranostic progressively yet additionally to anticipate viability and toxicity levels dependent on the tissue accumulation. This data can be utilized to change or alter the treatment methodology (Kunjachan et al. 2012).

Nanotechnology has become the dominant focal point in the advancement of theranostics. Nanoparticles have promising highlights to be used as theranostics, in particular high surface-zone to volume proportions that yield good therapeutic and imaging agent loading, surface functionalization with targeting ligands, and little size for extravasation to broken vasculature. They can be functionalized to balance the delivery dependent on natural stimuli, for example, pH, temperature, chemicals. Blood flow phases of nanoparticles can be upgraded by surface functionalization with a hydrophilic polymer, polyethylene glycol (PEG). High therapeutic and imaging payload, joined with focused conveyance, can build the remedial and imaging viability while lessening off-site toxicity. Nanoparticles, for example, gold nanoparticles and carbon nanotubes, likewise have inalienable theranostic functionalities due to their photo thermal and optical properties (Lammers 2010). Because of the heterogeneity of malignancy, much theranostic research is centered around oncology. Like malignancy, inflammatory disease possesses a challenge due to complexity of inflammation and patient variability. This necessitates the use of patient-specific therapeutics. Malignancy, cardiovascular, neurodegenerative, and immune system illnesses have a functioning provocative component. Inflammatory illnesses are a significant supporter of worldwide wellbeing costs assessed at $57.8 billion in 2010. Therefore, theranostics are presently being explored for incendiary infections (Janib et al. 2010).

Inflammation can be described as a host's attempt to re-establish homeostasis in response to infection, injury, or metabolic irregularity. Irritation that is delayed can cause essential endogenous tissue damage, which can lead to a variety of diseases. The inflammatory mediators released by macrophages cause extensive tissue damage, which leads to the development, progression, and spread of a variety of diseases (Patel and Janjic 2015). Due to their pathogenic roles, macrophages have been focused for remedial and imaging purposes. For instance, perception of macrophage invasion by focused imaging agents has been utilized to survey illness seriousness and therapy efficacy. Treatments targeting macrophages by specific removal, obstruction of their penetration, and decrease in release of inflammatory mediators have demonstrated adequacy in rheumatoid joint inflammation (RA), atherosclerosis, vascular injury, and disease. In either case, mitigating treatments have indicated different viability results across the patient population. In certain cases, significant uptake of macrophages has been related with immunosuppression, infection, and diminished injury healing. This mix of helpful and hurtful impacts can be ascribed to the distinctive initiated conditions of macrophages in infection conditions (Rai et al. 2010).

Macrophages are effector cells of a resistance system that phagocytose bacteria and secrete both pro-inflammatory and antimicrobial mediators. Additionally, macrophages play a big role in eliminating diseased and damaged cells through their programmed death. Macrophages are needed for homeostasis. Macrophages sense response on the pathogen and environmental factors which participate in tissue repair process. Macrophages are present in every tissue and play an awfully important role in the tissue survival process. Macrophages are associated with both commencement and end of inflammatory reaction. They're crucial in RA, where they produce cytokines that enhance inflammation and increase destruction of ligament and bone. Macrophages play a central role in the pathogenesis of atherosclerosis. They actively participate in LDL uptake and lipid accumulation within the arterial wall becoming foam cells. They are heavily involved in cancer-related inflammation. Various formulations and their interaction with diseases and various contrasting agents like metal oxides, ultrasonic agents, optical imaging, radionuclides, etc. will be discussed here. The opportunity to rapidly assess and adjust treatment to the needs of the individual offers potential advantages that will accelerate the development of theranostic agents.

2 Imaging Modalities for Nanotheranostics

The imaging modalities employed in biology and medicine are supported by a range of energy sources, including light, electrons, lasers, X-rays, ultrasound, and nuclear resonance. For understanding the role of macrophages, the imaging modalities techniques are used in various diseases like atherosclerosis, tumor, rheumatoid arthritis. Using these techniques, the severity of diseases and therapeutic efficacy can be determined. These techniques differ in resolution, depth of penetration, acquisition of time, safety, cost, applicability. EMR radiations are used to determine the anatomical and functional changes in various diseases. For this, ultrasound radiation is used in which high frequency waves are required to determine the functioning of the body. Molecular imaging permits the portrayal of organic cycles at the cell and subcellular levels in intact living beings. By exploiting molecular tests or contrasting agents, this incredible strategy can identify and describe diseases in early stages and give a fast technique for assessing treatment. Right now, utilized subatomic imaging modalities incorporate X-ray, CT, US, optical imaging (bioluminescence and fluorescence), single photon emission computerized tomography (SPECT), and positron emission tomography (PET). These modalities require a specific amount of reporter groups to accumulate in the area of interest. Because of the distinctive synthetic nature of correspondents and inborn affectability of each innovation, the tissue concentration needed to achieve signal shifts extensively between the modalities. As a platform, nanoparticles are appropriate for creating focused-on contrast agents, since:

1. They have a surface, which can be functionalized with one or additional targeting ligands at a wide scope of densities.
2. Their plasma dissemination time can be tuned more than a few significant degree dependent on their physico-chemical properties.
3. Contrast agents and medications can be incorporated at predetermined proportions either inside or adsorbed on the surfaces.

The various techniques that have been employed to study imaging modulation have been given in the figure below (Fig. 1):

2.1 Optical Imaging

Optical imaging is one of the most widely used testing modalities. Optical imaging uses photons transmitted from bioluminescent or fluorescent

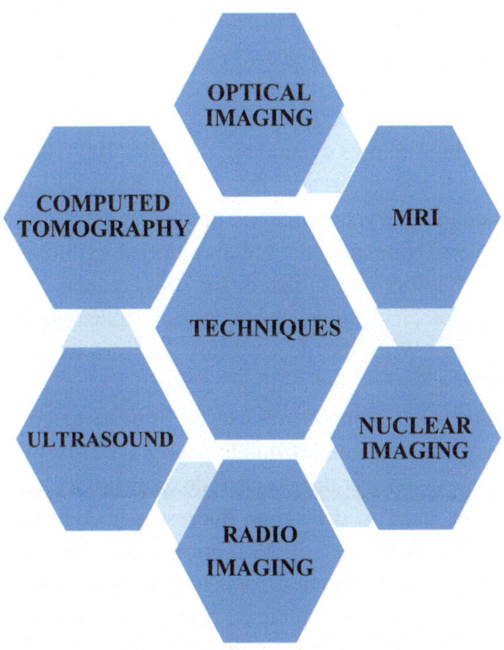

Fig. 1 Various techniques related to imaging modalities

probes. It has preferences over other imaging modalities in that the discovery of low energy photons is moderately low; moreover, the range from visible to near infrared (NIR) light gives great spatial goal, without presentation to ionizing radiation. However, this methodology experiences poor tissue penetration (0–2 cm) and fluorescent imaging is exceptionally vulnerable to disturbance because of the surface dispersing of photons in the visible light locale (395–600 nm). Optical imaging additionally experiences huge background due to tissue autofluorescence as well, absorption of light by proteins (257–280 nm), heme groups (max absorbance at 560 nm), and also water (over 900 nm). In spite of these difficulties, the NIR (700–900 nm) window has the benefits of decreased autofluorescence, diminished tissue scattering, and more prominent depth of penetration, which is generally appropriate for in vivo imaging. The optical imaging method used in UV, IR, and EMR radiations is used to research the cell, subcell, and whole creature because it requires minimal effort and has a high target capacity. An endoscope comprises an adaptable cylinder with a framework to convey light to enlighten an organ or tissue. Clinical optical imaging is the utilization of light as an investigational imaging method for clinical applications. Fluorescence and bioluminescence is utilized to distinguish the light (Malam et al. 2009).

Fluorophores dynamic in the NIR (700–1000) locale can decrease the tissue impacts and increase entry significantly up to very few centimeters. Numerous natural and inorganic fluorescent colors are utilized for this reason, for example, cyanine subsidiaries (indocyanine green). With the accessibility of various frequency lasers and outflow channels, a mix of fluorescent components can be utilized to contemplate the complex inflammatory systems. pH, catalyst, temperature, and redox sensitivity can all be used to determine fluorescence dynamic agents (Park et al. 2009a). For instance, in vivo fluorescence-mediated tomography (FMT) was utilized to envision protease protein movement in macrophages marked with a protease dynamic fluorescent component in a mouse atherosclerosis model. Clinically, fluorescence imaging is utilized for surface tissue imaging (bosom), intravital microscopy, and constant image guided surgery. However, it is restricted by extinguishing and image bleaching impacts of fluorophores. Fluorescence molecular tomography is utilized to identify the protein action in macrophages. Metabolic cofactors NADH and FAD have been utilized to recognize tumor-associated macrophages (TAMs) in creature malignancy models.

2.2 Magnetic Resonance Imaging (MRI)

The origin for MRI signal is the precession of water hydrogen cores inside an applied magnetic field. After application of radiofrequency pulses, the relaxation process through which the nuclei return to the original aligned state can be exploited to produce an image. To improve the separation between tissues, contrast agents are utilized to shorten the relaxation parameters (T1 and T2) of water. Paramagnetic particles, for example, gadolinium (Gd) and manganese (Mn), can be labeled onto little atoms, macromolecules, or nanoparticles. On the other hand, attractive iron-oxide nanoparticles are inherently superparamagnetic and can be incorporated with cargo at their surfaces. Contrasted with radionuclide or optical imaging, MRI has great spatial resolution; however, it experiences low sensibility. To balance this, generally high concentration of contrast agents is needed to produce a signal. The administration of high doses of contrast agents leads to concerns over aggregation and toxicity, which have become a critical issue for Gd (III) models. While Gd (III) gives better tumor and vascular imaging differentiation, issues with moderate discharge and toxicity because of long-term aggregation may have a negative impact on its potential development.

MRI is a clinical imaging strategy utilized in radiology to frame photos of the living structures and the physiological cycles of the body. X-ray can be utilized to recognize cerebrum tumors, awful mind injury, numerous sclerosis, stroke, dementia, contamination, and the reasons for migraine. The MRI procedures for macrophages help in characterizing the part of amassed macrophages and furthermore impact different treatments focused on macrophages aggregation (Richard et al. 2008a).

2.2.1 H MRI

Most contrast agents are produced from paramagnetic gadolinium or superparamagnetic iron oxide (SPIO). They work by modifying the unwinding state of the encompassing protons, consequently delivering contrast. 1H contrast agents are not recognized straightforwardly; however, their impact on unwinding of encompassing water particles produces contrast (Beer and Schwaiger 2008). Contrast agent-based MRI has been utilized clinically for gastrointestinal and blood pool contrast just as to picture numerous neurotic conditions. To get optimal target fixations, contrast agents are fused in nanoparticles and can be targeted to a particular tissue. The affectability of MRI contrast agents is normally in 10^{3} M range. Nanoparticles combining iron oxide and gadolinium agents have been utilized to view macrophages in pathologies, for example, atherosclerosis, RA 61, and numerous sclerosis.

2.2.2 F MRI

F MRI identifies natural fluorine, which is brought exogenously into the host. Like 1H, the fluorine (19F) core has a half turn, practically identical MRI affectability (83%) and reverberation (contrasts by 6%) to 1H. Therefore, 1H MRI machines can recognize 19F cores by tuning to a suitable recurrence. To acquire 19F images, nonharmful perfluorocarbons (PFCs) with countless fluorine are brought into the body in biocompatible formulations. 1H MRI can be enrolled during a similar meeting to get anatomical setting for the 19F sign. Like 1H MRI, 19F MRI has 10^{3} M recognition sensitivity. To expand signal-to-noise ratio slope reverberations, ultrafast outspread arrangements and paramagnetic particles in closeness to PFCs can be used. Nanoparticles, liposomes, and nanoemulsions fusing PFCs for 19F MRI have been broadly reported. PFC nanoemulsions have been used to contemplate cell-cell interactions in an entire creature, identify macrophages for in vivo aggregation measurement, and track the biodistribution of medication stacked nanoparticles. Successful in vivo macrophage representation utilizing 19F MRI has likewise been accounted for in disease models such as diabetes, RA, respiratory irritation, and transplantation models. Unlike 1H contrast agents, 19F cores can be evaluated in vivo due to their near-zero basis and the lack of need for imaging prior to PFC organization. While clinical trials of 19F MRI in cell-based immunotherapy are still in the early stages, clinical trials of 19F MRI in cell-based immunotherapy have begun recently (Beer and Schwaiger 2008).

2.3 Nuclear Imaging

Nuclear drug imaging is a technique for creating imaging by identifying radiation from various pieces of the body after administration of a radioactive tracer to the patient. X-ray method is broadly utilized for different infections like rheumatoid joint inflammation and numerous more. It permits the perception of bone and delicate tissues in three measurements, for example, utilizing multiplanar methods and particularly appropriate for rheumatoid joint pain. Molecular resonance, for example, positron emission tomography (PET) and single-photon emission computerized tomography (SPECT), distinguishes β- or γ-radiation from radionuclides restricted inside the body.

Positron emission tomography is a practical imaging strategy that utilizes radioactive substances known as radiotracers to view and gauge changes in metabolic cycles and in other physiological exercises including blood stream and assimilation. PET and SPECT are 3D-tomographic methods with nanomolar to picomolar recognition affectability, yet with low spatial (1–4 mm) goal. Computed tomography (CT) methods are additionally utilized for this procedure. CT alludes a mechanized X-ray imaging method in which a thin light emission ray is focused on a patient and immediately turned around the body, delivering signals that are prepared by the machine's PC to create cross-sectional pictures or "cuts" of the body.

2.4 Multimodal Imaging

There is no such imaging modality technique that serves for research and medical needs. Therefore, multimodal imaging is used for nanoparticles with complementary imaging agents that offer the high sensitivity detection, anatomical localization, and data validation for different modalities. Nanoparticles incorporate the combination of MRI and FRI imaging agents. There are various techniques like MRI, CT, optical imaging which can be helpful in molecular imaging. The combination of multimodal technique and MRI, PET, and ultrasound gives the better results for molecular imaging, for example, combination of FMRI and fluorescence technique (Haglund et al. 2009).

2.5 Radionuclide Imaging Technique

At the other end of the electromagnetic range, γ-ray emissions are the basis for SPECT and PET. For the two modalities, radiopharmaceuticals are administered that can be detected by a camera. Dissimilar to MRI, SPECT and PET images are acquired over a low background signal and require minimal sign enhancement since the gamma rays have energies in the megavolt range. In contrast with PET, SPECT is roughly multiple times less sensitive; in any case, SPECT is invaluable on the grounds that it enables concurrent imaging of different radionuclides. Also, SPECT is more broadly accessible than PET, and SPECT radionuclides are easy to plan and normally have a more extended half-life than PET radionuclides. SPECT and PET are quantitative procedures, which is an advantage over different modalities, for example, MRI and optical imaging. PET experience poor spatial goal; this can be overcome by hybrid imaging, where PET is utilized to follow molecular regions and high power CT is utilized to limit events (Zalutsky et al. 2007).

2.6 Computed Tomography

While radionuclide-based imaging is capable of giving data about physiological measures, modalities like CT can give integral anatomical data. CT measures the retention of X-rays as they pass through tissues. The capacity of CT to recognize tissues depends on the way that various tissues give distinct degrees of X-ray attenuation, where the attenuation coefficient depends upon the atomic number and electron density of the tissues. Differences in assimilation among bone, fat, air, and water produce high contrast images of anatomical structures.

Presently, CT contrast agents are regularly low molecular weight and are portrayed by quick extravasation and clearance. Then again, macromolecular and nanoparticulate agents have shown predominant prolonged presence in the blood pool, which may make them appropriate agents for vascular CT. Nanoparticles containing electron thick components with high nuclear number, for example, iodine, bismuth, or gold, have been proposed as CT contrast agents. CT contrast requires high convergences of these components; subsequently, most exploration depends on strong nanoparticles or liposomes containing iodinated atoms. Gold nanoparticles have picked up fame as CT contrast agents, despite the fact that it is unclear if gold is an ideal material for physiological use (Cormode et al. 2009).

2.7 Ultrasound

US is one of the most widely recognized clinical imaging modalities because of its ease, speed, straight forwardness, and security. In this methodology, a transducer that discharges high recurrence sound waves (>20 kHz) is put against the skin and US pictures are acquired dependent on the sound wave reflected back from the inner organs. US contrast agents can improve imaging by administration of intravenous contrast agents having microbubbles/nanobubbles of gas nanoparticulate-based contrast agents for the imaging modalities are in different phases of advancement. Commonly, the kind of nanoparticle utilized relies upon the imaging methodology.

3 Different Types of Nanoparticles Used in Imaging

Theranostic NPs are multifunctional because of infusion of both therapeutic and imaging agents. Moreover, theranostic NPs may have components for targeted accumulation, drug activation, or enhancement of contrast. An ideal TNP has molecular targeting that can be imaged during its flow in the body. After arriving at its objective, it might have targeting moieties that partner with cell-surface receptors, internalize into the cytosol, target to the intracellular target if necessary, and release the active therapeutic. Nanoparticles can be produced using various materials including proteins, peptides, polymers, lipids, metals and metal oxides, and carbon. While different materials can frame nanoparticles, these are the predominant materials under development (Haglund et al. 2009). The most important nanoparticle structures incorporating medication are dendrimers, vesicles, micelles, core shell structures, microbubbles, and carbon nanotubes, which would all be able to be functionalized with targeting moiety, drug, and contrast agent.

3.1 Drug Conjugates and Complexes

As opposed to self-assembled structures like vesicles and micelles, complexation and covalent conjugation are other direct routes to design nanoparticles. Drug conjugates depend on reversible interactions among transporter and drug, while drug forms use covalent bond. Drug conjugates can be made through a variety of chemical pathways, which are often dependent on the drug's and carrier's chemistry. The two significant classes of drug conjugates at present being worked on for theranostic use incorporate protein and peptide-associated and polymer-related medications. Likewise, with vesicular structures, the viability of a drug form is identified with its capacity to improve therapeutic index comparative with free drug, for the most part, by lessening harmfulness as well as improving adequacy. Proteins and peptides are flexible materials that can shape nanoparticles variety of ways, including complexation. The best illustration of an effective protein nanoparticle in clinical use is Abraxane™, a formulation of paclitaxel reversibly bound to 130–150 nm albumin nanoparticles by means of high pressure homogenization. Abraxane™ beat regular paclitaxel at an equitoxic dose while diminishing toxicity, because of longer course time and diminished off target action (Gaitanis and Staal 2010).

Two schemes for functionalizing HPMA copolymers with symptomatic and helpful moieties are by means of copolymerization and chemical conjugation. While the last technique permits direct formation of these contrasting agent and therapeutics to the copolymer, it experiences low conjugation productivity and trouble in controlling the level of formation or conjugation. Even so, two chemotherapeutic agents (Dox and gemcitabine) alongside I-131 have been effectively conjugated to a HPMA copolymer utilizing this technique. On the other hand, copolymerization might be the favored formation technique as it offers the adaptability to form both single and multimodal symptomatic agents and chemotherapeutic medications. Peptides that consist the arginine–glycine–aspartate (RGD) motif have been employed as the polymer. It has been marked with In-111(indium) and HPMA build with Gd were set up by copolymerization (Lammers et al. 2009; Hawkins et al. 2008). Dextran-modified polystyrene nanoparticles (DEX-PS) demonstrated that modification of nanoparticles by dextran could specifically enhance their recognition by M2 macrophages in vitro, but it was obstructed by monocytes in peripheral blood according to in vivo assays. DEX-PS not only targeted and became distributed in tumors, M2 macrophage-related disease, but was also highly distributed in M1 macrophage-related disease, namely, acute peritonitis. DEX-PS were synthesized by COOH-PS and amine-dextran under condensation reaction. Compared with COOH-PS, both dextran-NH2 and DEX-PS showed an obvious hydroxyl peak in dextran at 3307 cm^{-1} indicating that dextran was successfully modified on COOH-PS to form DEX-PS.

3.2 Dendrimers

Polymeric dendrimers are hyper-branched nanostructures that can be controlled in size by controlling the quantity of polymerization generations. As polymerization advances, a little, planar molecular structure changes into a round nanostructure, with pits where therapeutics and contrasting agent can be joined with extraordinary loading productivity. Dendrimer polymerization and synthesis can be managed to unequivocally control the atomic weight and chemical synthesis of the end result (Surendiran et al. 2009). The only dendrimer-based formulation to enter clinical preliminaries so far is VivaGel™, which utilizes the dendrimer as a therapeutic agent instead of a carrier – it is at present being assessed for safety and efficacy as a microbicide (Lee et al. 2005). Dendrimers have been developed for quite a long time and have demonstrated to be fruitful medication and imaging agent transporters in various preclinical studies, including tumor regression, gene delivery, and molecular imaging. Barriers to address prior to the theranostic utilization of dendrimers incorporate harmfulness of the nanoparticle, its polymeric part, just as the off-target impacts of the contrasting and restorative agent (Singh et al. 2008).

3.3 Vesicles

Vesicular structures have been widely studied in clinical therapeutics as well as in diagnostic therapeutics. The two significant classes of vesicles, liposomes and polymer vesicles, can be encapsulated within a certain limit covalently and noncovalently for both hydrophobic and hydrophilic payloads. Hence, it becomes easy to deliver diagnostic moiety at specific sites by modifying its physiochemical property. Polymeric vesicle structures, or polymersomes, have shown extraordinary potential for co-encapsulation of drug and contrasting agent (Levine et al. 2008). As contrasting agent, liposomes have been marked with quantum dots (QD) Mn and Gd (which can exist in chelated and aqueous form) radionuclides, for example, Ga-67, In-111 and Tc-99 m, iodine-based agents, or even gas (Table 1).

Vesicle can emerge from the self-assembly of bilayers created only from nonlipid polymers. Polymer vesicles or polymersomes have been made of different polymeric materials including polylactic acid (PLA), polyglycolic acid (PGA), poly lactic-co-glycolic acid (PLGA), polycaprolactone (PCL), chitosan, and PEG.

3.4 Micelle

Micellar nanoparticles are appealing carrier of drug and contrast agent since they can have uniformity in their size and can be prepared by chemicals which are hydrophilic and hydrophobic in nature. It consists hydrophobic moiety which can be modified to increase the solubility. It can consolidate various functionalities into a solitary structure. Without a watery or aqueous core, the drug as well as contrast agent should be bound to the polymer before development, attached to an anchor atom, or captured inside the thick, hydrophobic center of the micelle. Micellar structures, including polymeric micelles, have been broadly studied as drug transporters (Torchilin 2007).

Also, micelles loaded with contrasting agent can be utilized for imaging various organs, tissues, and disease destinations. Membrane tropic chelating agents, for example, DTPA stearyl amine (DTPA-SA) or DTPA-phosphatidylethanolamine (DTPA-PE), have been created whereby the lipid moiety of the particle can be secured in the micelle's hydrophobic center while a more hydrophilic chelate is confined on the hydrophilic crown. Additionally, poly chelating amphiphilic polymers (PAP) have numerous substituted functional groups for chelation that can be secured to the micellar surface. Various blends of chelators and hydrophobic anchors have been tried in the preparing In-111, Tc-99 m, and Gd liposomes. Chelates with high strength under physiological conditions are likely to evade toxicity. Because of variety of connections, PAP can improve image quality at a low micellar concentration.

Table 1 Various formulations along with their theranostic effect

S. no	Formulation	Delivery route	Description	Model used	Theranostic effect	References
1.	Dextranated and DTPA-modified magneto fluorescent nanoparticle	I.v. injection route	Size 20 nm	Mice model	Accumulated dose in apo$E^{-/-}$ aortas determined by gamma counting was 260% and in carotids 392% of respective wild type organs ($p < 0.05$ both). Macrophages (Mø) participate centrally in atherosclerosis and Mø markers (e.g., CD68, MAC-3) correlate well with lesion severity and therapeutic modulation	Nahrendorf et al. (2008)
2.	Macrophage-targeted nanoparticles labeled with fluorine-18 (^{18}F)	Systemic administration intravenous route	Size above 5.5 cm	Mouse model	Validation with scintillation counting, autoradiography, fluorescence, and immunoreactive histology and flow cytometry demonstrated that nanoparticles localized predominantly to monocytes and macrophages within the aneurysmatic wall	Nahrendorf et al. (2011)
3.	Polysulfated gold nanorods (AuNRs)	Intravenous route	Size 780 nm	Mouse model	It shows a robust method to obtain covalently functionalized polyanionic gold nanorods, which are suitable for biological applications as well as a low-cost, actively targeted, and high contrast imaging agent for the diagnosis of rheumatoid arthritis	Vonnemann et al. (2014)
4.	Dextran-modified polystyrene nanoparticles (DEX-PS)	Intravenous injection route	Size 500 nm particle-to-cell ratios in the range of 6.25–100	Murine monocyte/ macrophage cell line)	DEX-PS acts as a double-edged sword in these two different, that is, tumor and inflammatory diseases by reeducating macrophages to a pro-inflammatory phenotype	Yuan et al. (2020)
5.	QDs- nanoparticle	Intravenous injection route	Size 200 or 3.7 nm	Mice model	The multifunctional biomimetic super particle, termed as DOX-QDs-lip@M, which can specifically deliver drugs to tumor and synergistically monitor their therapeutic effects, was fabricated. The integrated nanostructure can greatly increase the fluorescence intensity of the signal unit and tremendously improve the diagnostic sensitivity	Lianga et al. (2020)
6.	^{18}F modified trimodal nanoparticle (^{18}F-CLIO)	Intravenous route	Size 30 nm	Mice model	The presence of ^{18}F dramatically lowers the detection threshold of the nanoparticles, while the facile conjugation chemistry provides a simple platform for rapid and efficient nanoparticle labeling	Devaraj et al. (2009)

(continued)

Table 1 (continued)

S. no	Formulation	Delivery route	Description	Model used	Theranostic effect	References
7.	^{64}Cu-labeled quantum dots nanoparticles	Intravenous route	Size 21–20 nm	Mice model	Knowledge of quantitative biodistribution is of paramount importance in any diagnostic or therapeutic application of nanomaterials in living animals. In nude mice, both well counting and micro-PET show rapid uptake of radiolabeled QD by liver and spleen. Size of the particles has no influence on biodistribution within the range that was tested here	Meike et al. (2007)
8.	Mannose-coated ^{64}Cu nanoparticle	Intravenous route	Size 1.5 mm	Mouse model	Mannose liposome nanoparticles accumulated in TAMs and exhibited little accumulation in remote lung areas. Mannose liposome nanoparticles are a promising new vehicle for the delivery of imaging agents to lung TAMs. In addition to imaging, mannose liposome nanoparticle hold the potential for delivery of therapeutic agents to the tumor microenvironment	Locke et al. (2012)
9.	18F fluorodeoxyglucose (FDG) nanoparticle	Intravenous route	Size NA	Rabbit	It suggests that macrophages are responsible for the accumulation of ^{18}F-FDG in atherosclerotic lesions. Because vulnerable plaques are rich in macrophages, ^{18}F-FDG imaging should be useful for the selective detection of such plaques	Ogawa et al. (2012)
10.	Docetaxel or zoledronic acid nanoparticle	Intravenous or subcutaneous route	Size 10–20 mm	Mice model	The experiments presented here demonstrate the use of multimodality imaging techniques for detection and quantification of multiple interrelated biologic processes affected by therapeutic intervention in a model of metastatic bone disease	Hoff et al. (2012)

3.5 Core Shell

Core shell–organized nanoparticles incorporate particles that are made with a wide assortment of materials like certain metals and their oxides as well as certain polymers. In a core shell is a center encircled by a hydrophilic shell which is quite similar to a micelle where center shell particles are surrounded by covalent or ionic bonds. Core shell nanoparticles offer durability, and surface properties and can even create contrast contingent upon their structure, size, and shape (Keren et al. 2008). Those nanomaterials can be modified easily to control the release and stability of drug. Metal nanoparticles are alluring choices for co-conveyance of medication and contrasting agent as they can accomplish adequate biodistributions, while restricting leeway by the reticuloendothelial system (RES).

3.6 Microbubbles

Microbubbles and nanobubbles are round cavities loaded up with gas and are generally <10 μM in size. These microbubbles can be prompted to grow and contract (reverberate) within the sight of the US conveyed at the resonance frequency of the microbubbles and by changing its composition. Some gases can be incorporated, which serves as contrast agents, for example, octafluoropropane (C3F8), decafluorobutane (C4F10), and sulfur hexafluoride (SF6). The shell that encloses the gas molecules can be made out of phospholipids different polymers of various types of proteins like albumin. One can incorporate some more functional changes like denaturation of protein for targeted therapy (Krupka et al. 2010). Microbubbles are utilized as contrasting agent for imaging inflammation and can be further used for detecting clots.

3.7 Carbon Nanotubes

Carbon nanotubes (CNTs) are tube-shaped cylinders made exclusively out of carbon and can either be shaped single-walled or multiwalled, for greater strength (Richard et al. 2008b). These CNTs are being researched as theranostic nanoparticles as they pose tremendous capacity to consolidate contrasting agent. Amine-functionalized single-walled CNTs were formed through amide linkages to a derivative of cisplatin-folic acid as targeted therapeutics. Likewise, with QDs and gold nanoparticles, a wide excitation profile and coefficient of absorption can be obtained by tuning CNT over a wide scope of frequencies. Moderately few studies have been conducted using CNTs in optical imaging. Nonetheless, NIR and Raman signals from CNTs have been distinguished from inside the subcutaneous tumors in mice (Keren et al. 2008). Examples like ^{64}Cu-Labeled 800- or 525-nm emission wavelength QD (21- or 12-nm diameter), with or without 2000 MW (molecular weight) polyethylene glycol (PEG), were injected intravenously into mice (5.55 MBq/25 pmol QD) and studied using well counting or by serial micro-PET and region-of-interest analysis. CNTs showed good efficacy as a contrast agent as well as a drug delivery agent but further studies to reduce toxicity and override primary barriers need to be done in future.

4 Various Types of Theranostic Agents

Theranostics, as the word indicates, combines therapy and diagnosis of a particular disease into a singular therapeutic agent. This arises from a general consensus between scientists who believe that the existing treatments are only effective to a small amount of patients and treatment needs to become more personalized. Nanotheranostics aims to achieve uniform bioavailability, targeted therapy, as well as minimum side effects. The various types of nanoparticle-based theranostics have been elucidated below (Fig. 2).

4.1 Iron Oxide Nanoparticle–Based Theranostic Agents

IONPs with proper coatings can be effortlessly combined with drugs like methotrexate (MTX), doxorubicin (DOX). The drug is covalently bonded to the carrier material or co-encapsulated

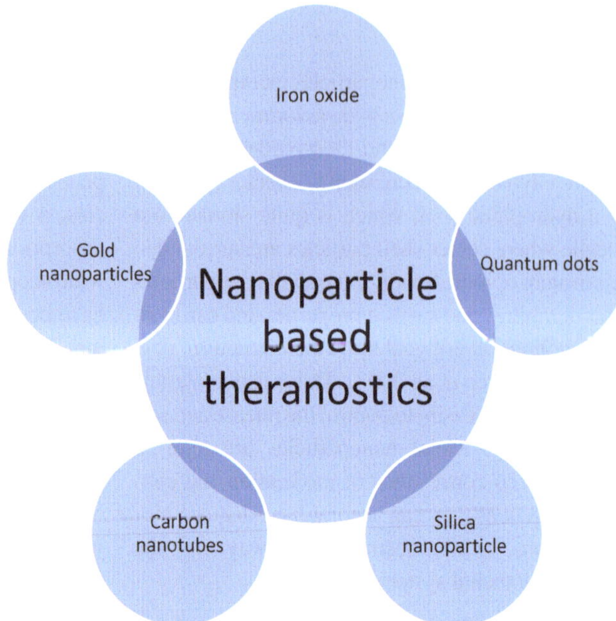

Fig. 2 Various nanoparticles-based theranostic agents

in the polymer matrix of IONP. These nanoparticles have magnetic property which let them accumulate in a particular sit on the application of magnetic field. On the other hand, IONP can receive of magnetic field and can enable the conversion of electromagnetic energy into heat energy which can be further utilized as a theranostic.

Nie et al. modified QDs by using a triblock polymer which has ethylacrylate, butyl acrylate, and methacrylic acid moiety in its chain. They also conjugated the QD with prostate-specific membrane antigen (PSMA). They found that the QD accumulated in tumor area due to EPR effect (Nie et al. 2007).

4.2 Quantum Dots-Based Theranostics Agents

QD-based drug delivery is moderately less explored due to the inborn toxicity of QDs. This issue is more conspicuous with the original QDs, where toxic Cd and Pb are utilized in QD preparation process. Recent advancement in QD synthesis has prompted the development of QDs that do not have cadmium, for example, InAs/ZnSe and InAs/InP/ZnSe. Park et al. did co-encapsulation of hydrophobic QDs and IONPs, alongside DOX, into micelles and they were shaped with PEGylated phospholipid. Such forms were additionally combined with a tumor-homing peptide F3. The mix was infused into a MDA-MB-435 xenograft model. Effective tumor targeting was seen by both optical and MRI imaging modalities (Park et al. 2008).

Yuan et al. stacked MTX onto QD surfaces to prompt photoluminescence quenching (Yuan et al. 2009). The loading was accomplished by reversible physical adsorption, which could have been turned around when uncovered to species with higher affinity, for example, DNA. This coating prompted a rebuilding of the photoluminescence, which could have been possibly useful to check the delivery of drug molecules. QDs may likewise work as gene drug delivery vehicle when changed with lipofectamine or other charged polymers. QDs additionally have extraordinary potential in photodynamic treatment, where they go about as either photosensitizers or carriers. It is believed that QDs can be enacted by light and move the triplet state energy and interact with oxygen molecules to cause cell injury.

4.3 Gold Nanoparticles–Based Theranostic Agent

Bhumkar et al. utilized chitosan as a diminishing agent and coating material to make Au nanoparticles. The chitosan–Au nanoparticles were positively charged and were efficient in stacking insulin through electrostatic association or interaction (53%). Such forms were concentrated in a diabetic model to control postprandial hyperglycemia. Two hours in the wake of administering this insulin stacked Au NPs to diabetic rats, a decrease in their blood glucose level was noticed, of 30.41% and 20.27% for oral (50 IU/kg) and nasal (10 IU/kg) administration, individually. Prabaharan et al. used an amphiphilic-block-copolymer-covered Au NP equation for tumor targeting and drug delivery. Such a nanostructure consisted of an Au NP core shell, a hydrophobic PASP inner shell, and a hydrophilic, folate-formed PEG external shell (PEG-OH/FA). DOX was covalently formed onto the hydrophobic inward shell by acid cleavable hydrazone linkage, with a stacking level of 17 wt.%. Such a nanosystem is fascinating for having both a tumor focusing on an instrument (folate on the external layer) and an intracellular drug release method (hydrazone linkage of DOX on the inward layer). (Bhumkar et al. 2007).

Au NPs have additionally been changed over to polyelectrolytes for studying gene drug delivery. Hence, Au NPs have been functionalized with alkylated quaternary ammonium to stack plasmid DNA. In vitro, such nanocarriers indicated advantages, for example, ensuring DNA from enzymatic processing and GSH release. Afterward, Thomas and Klibanov utilized expanded PEI to present gene stacking ability to Au NPs. Under advanced conditions, transfection power can be expanded by multiple times, contrasted with the parent polymer. Therapeutic gene can be stacked onto Au nanoparticles through covalent bond. DNA oligos can be directly loaded onto Au nanoparticles with high efficiency (Thomas and Klibanov 2003).

4.4 Carbon Nanotube-Based Theranostic Agents

Carbon nanotubes are crucial in the field of nanomedicine because of their unique physical and surface properties. CNTs have the ability to be taken up by cells, which has prompted a surge of research into their possible function in delivery of drug. Truly, CNTs can be disguised by cells by means of various routes if surface coatings are extraordinary, as both endocytosis and passive diffusion have been behind to be liable for carbon nanotube take-up. Previously, Prato et al. coupled MTX onto 1,3-dipolar cycloaddition functionalized CNTs. Comparable CNT equations with various amine ends were utilized to stack and convey DNA plasmid. In any case, in any event in the MTX case, it was discovered that the covalent amide bond was unfavorable to evoking intracellular drug release, since the forms demonstrated no upgrade in helpful viability when contrasted with MTX alone. It applied phospholipid-CNT forms in both imaging and treatment. For instance, they coupled siRNA to CNTs by means of a disulfide bond, which cannot be cleaved in the endolysosome. This CNT carrier indicated high transfection productivity, outperforming lipofectamine in actuating RNAi. Afterwards, they utilized the equivalent nanostructure and effectively transported siRNA into human T cells and primary cells, which are for the most part viewed as hard to transfect with traditional cationic liposome-based transfection agents (Singh et al. 2005).

A similar group also reported the coupling of either Pt(IV) prodrug or PTX onto PEGylated CNTs to improve the pharmacokinetics and therapeutic impacts. On account of PTX, a branched PEG was utilized for phospholipid PEGylation, which was discovered to be invaluable over single-chain PEG in carrying additional strength to CNTs. PTX was coupled through a cleavable ester bond attached to the nanotube surface; furthermore, the build was then tried in a murine 4 T1 bosom malignant growth model. The forms indicated a 10-times increase in tumor homing than PTX alone, which was ascribed to prolong circulation of the nanoformulation. Thus, formula preferred better tumor suppression result over clinically utilized Taxol. Aromatic stacking can likewise serve as a drug loading mechanism. DOX, for example, was loaded onto SWNT by means of this course at a surprisingly high productivity of 4 g DOX/g nanotube. The interaction is pH dependent, proposing a method of emptying the payloads in acidic endolysosome and tumor microconditions. All the more as of late, such DOX stacked nanotubes have been assessed in SCID mice bearing Raji lymphoma xenografts, which indicated more noteworthy helpful viability and less toxicity contrasted with an equimolar measure of free DOX (Liu et al. 2009).

4.5 Silica Nanoparticles–Based Nanotheranostic Agents

It was observed that silica nanoparticle can be made mesoporous by precise pore size control. Such mesoporous nanostructures, consisting of several unfilled channels and a huge surface area (>900 m2/g), are incredible reservoir for small particles and they hold incredible potential in drug delivery. Numerous techniques have been accounted to accomplish such nanostructures. In chemical synthesis process, n-alkyl trialkoxysilane or different surfactants are blended with different precursors and are consolidated into the lattices during particle formation. These surfactants can be removed from the nanostructure by means of postsynthesis dissolvable extraction or calcinations to give the mesoporous structure. With the molecular sieve structure, such mesoporous silica nanoparticles can stack some small drugs through simple interaction. Besides, innovations have been created to cover the mesopores after drug stacking to restrain untimely drug release. For example, mesoporous silica nanoparticles were stacked with PTX, and the mesopores were in this way covered with Au NPs. Such Au NP covering was intended to be photo-labile and can be uncapped to release particles when photo-irradiated (Park et al. 2009b). Some formulations along with their theranostic properties have been further discussed in the table given below.

5 Targeting Theranostics Agent Against Various Diseases

5.1 Targeting Theranostics Agent Against Cancer

Collecting proof from clinical and researcher investigation shows that macrophages in tumor tissues, specifically tumor-associated macrophages (TAMs), play a significant role in metastasis of strong tumors. The initiated macrophages produce numerous tumor-promoting cytokines and various growth factors, for example, endothelial development factor, angiopoietin, metalloproteinase, and tumor necrosis factor alpha (TNF-α), and encourage tumor advancement, angiogenesis, movement, and metastasis. The various nanotheranostic products which have been used in clinical treatment have been mentioned in the table below (Table 2).

5.1.1 Targeting Theranostics Agent Against Breast Cancer

Breast cancer is one of the most common diseases in women and is mostly curable in early stages. After progression of breast cancer, metastasis occurs which is considered incurable with respect to currently available therapy. It occurs due to mutation in BRCA, overactivation of progesterone or estrogen receptor, activation of human epidermal growth factor receptor 2 (HER2). Treatment strategy can be mainly divided into systemic as well as local therapy. There are also other ways to classify the therapy which depends on molecular subtype.

Tao deng et al. demonstrated a theranostic system for challenging issues related with cancers and fabricated the multifunctional biomimetic superparticle, named as DOX-QDs-Lip@M, which can specifically deliver drug to tumor and synergistically screen their therapeutic effect. The anticancer drug doxorubicin hydrochloride (DOX) and imaging agent quaternary quantum dots (QDs) were stacked into the hydrophilic and hydrophobic chamber of liposome. The incorporated nanostructure can enormously expand the fluorescence intensity of signal and increase the

Table 2 Various polymeric and metal-based nanoplatforms under study for clinical applications

Product	Clinical phase	Therapeutic modality	Diagnostic modality	Indication
CriPec® docetaxel	Phase 1	Docetaxel	PET (Zirconium-89)	Tumor
AGuIX®	Phase 1	Radiation therapy	MRI (gadolinium chelates)	Brain (metastases)
AGuIX®	Phase1	Chemotherapy radiation therapy	MRI (gadolinium chelates)	Cancer
SPIONs	Early phase 1	–	Ferumoxytol- based MRI	HNSCC
NBTXR3	Nanobiotix	Phase1,2	Radiation technology	Multiple types of cancer

Abbreviations: *HNSCC* head and neck squamous cell carcinoma, *MRI* magnetic resonance imaging, *PET* positron emission tomography, *SPION* superparamagnetic iron oxide nanoparticles

diagnostic and therapeutic effect. In this manner, the biomimetic DOX-QDs-Lip@M was built by fusing and coating the macrophage layers on the outside of liposome, which can subsequently expand the flow of the entire blood and adequately target on the tumor locales. These incorporated properties endow the biomimetic DOX-QDs-Lip@M with improved tumor imaging and hostile to metastasis treatment in living systems (Liang et al. 2020).

5.1.2 Targeting Theranostics Agent Against Ovarian Cancer

Ovarian malignant growth is the deadliest gynecological disease in women; subsequently, the early diagnose and treatment becomes essentially significant (Gupta et al. 2019). Efforts have been taken to develop drug to treat ovarian malignant growths overexpressing estrogen receptors (ER) and HER2. However, constant use of novel treatments causes drug resistance. So to overcome the chemo-resistance and to treat the ovarian cancer, radioimmunotherapy was used (Honarvar et al. 2016)..

Satpathy M et al. showed an amphiphilic polymer-coated magnetic iron oxide nanoparticle was conjugated with near infrared dye labeled HER2 affibody and chemotherapy drug cisplatin. In this investigation, we have developed HER-2-targeted magnetic iron oxide nanoparticles (IONPs) by conjugating a high affinity and small size HER-2 affibody that is labeled with a unique near infrared dye (NIR-830) to the nanoparticles. Using a clinically relevant orthotopic human ovarian tumor xenograft model, we have shown that HER-2-targeted IONPs are selectively delivered into both primary and disseminated ovarian tumors, enabling noninvasive optical and MR imaging of the tumors as small as 1 mm in the peritoneal cavity. We have determined that HER-2-targeted delivery of the IONPs is essential for specific and sensitive imaging of the HER-2-positive tumor since we are unable to detect the imaging signal in the tumors following systemic delivery of nontargeted IONPs into the mice bearing HER-2-positive SKOV3 tumors (Gupta and Gupta 2017).

5.2 Targeting Theranostics Agent Against Rheumatoid Arthritis

Rheumatoid arthritis is an early advancement of exacerbation of the bone joints. In this disease, synovial fluid erodes away and its progression occurs due to macrophages. Since bone erosions are typically irreversible, the development of imaging modalities is currently focused on new technologies that can be used to detect biological and physiological changes that are associated with the disease (Satpathy et al. 2014). Structural imaging techniques were used to diagnose RA. These techniques were used for the detection of biological and physiological changes that are associated with the diseases. These structural imaging techniques include MRI, fluid radiography, ultrasonography which are used for the detection of bone erosion (Mountz et al. 2012). Persistent synovial inflammation leads to disease progress and also development of bone erosions. At time of onset of RA, several pro-inflammatory cytokines and signaling pathways can be associated with the common outcome of synovial inflammation. However, pro-inflammatory responses differ between different patients with changes in disease progress. The altered pro-inflammatory activity and specific molecular events that cause synovitis can be cured by optical imaging methods, like thermography, near-infrared (NIR) imaging, and also by imaging techniques like PET and single photon emission, CT, etc. As number of RA therapies increases, the use of molecular imaging techniques also increases (Davila and Ranganathan 2011; Kokkonen 2010).

Kalashnikova et al. synthesized (Kalashnikova et al. 2020) albumin-cerium oxide nanoparticles by the biomineralization by indocyanine dye. Their efficacy was determined in collagen-induced arthritis (CIA) mouse via optical or tomographic imaging. It was found that these nanoparticles were able to eliminate reactive oxygen species and convert proinflammatory phenotype to anti-inflammatory phenotype.

5.3 Targeting Theranostics Agent Against Atherosclerosis

Cardiovascular disease causes more than 30 percent of deaths globally. It becomes important to diagnose and treat atherosclerosis as it results in myocardial infarction. The disease is prevented by looking at cardiovascular risk score which tells how much chances are there for occurrence of cardiovascular disease. However, one needs more accurate imaging techniques detection of atherosclerosis. Traditionally the diagnosis revolves around evaluating the stenosis for managing the symptoms in patients suffering from angina.

MA et al. used a magnetic nanoparticle-based approach (Ma et al. 2009) in which SPION that were coated with gold were clustered together. The formed nanoclusters were then surrounded with dextran. A strong NIR contrast was seen in rabbit model of atherosclerosis indicating increase in the uptake by macrophage.

More work has been done on SPION by McCarthy et al., where they modified (McCarthy et al. 2010) the nanoparticle with near infrared fluorophores and with moieties that work on photo activation to clear the plaques. It was observed that the SPION localized in the atherosclerotic region rich with macrophages and phototoxic activation of therapeutic moiety caused extermination of those inflamed macrophages.

5.4 Targeting Theranostics Agent Against AIDS

HIV is a dangerous infection that is spread around the world. It has high death rate. Studies on the pathophysiologic treatment of HIV contamination revealed the pivotal role that macrophages play in the different periods of HIV-1 disease. Macrophages, like CD4+ T aid cells, express elevated levels of the HIV-1 receptor CD4 on their surface and are exceptionally tolerant to HIV-1 contamination. In the beginning stage, macrophages are fundamentally engaged with the beginning and administration of the versatile cell and humoral invulnerable reaction to decrease viral contamination. Macrophages in lymph nodes express popular protein negative regulatory factor (Nef), which is additionally moved to B cells. Upon landing in B lymphocytes, Nef harms defensive elements of B cell. The current HIV-1 treatment is predominantly founded on antiretroviral treatment (ART). However, with long-term treatment with ART, cells actually can't be totally wiped out from the body. Infection levels quickly bounce back once ART treatment is halted. (Torigian 2011).

Recently, Singh et al. introduced the technique of laser ablation (Singh et al. 2020) to encapsulate many anti-HIV drugs like ritonavir, atazanavir, curcumin in a single formulation which act as a nanotheranostic. They produced it in an aqueous media containing Pluronic® F127. They confirmed the increased cellular uptake in microglia via fluorescence signal.

5.5 Targeting Theranostics Agent Against Diabetes

Insulin resistance is a medical condition where insulin becomes ineffective in blood glucose level leading to hyperglycemia and is generally associated with obesity, aging, and physical inactivity among many other potential causes. Islet cells of pancreas increase their insulin secretion to certain level but at some point they may get fail and cause type 2 diabetes. Low-grade systemic inflammation is associated with obesity, with excessive adipocyte necrosis where adipose tissue macrophages (ATMs) contribute to the production of pro-inflammatory cytokines that results in insulin resistance and eventually causes type 2 diabetes. Macrophages accumulate in adipose tissue for over time and produce cytokines like TNF and IL-6 along with chemokines (Aouadi et al. 2013a).

Wang et al. used SPIO particles to evaluate graft survival. They isolated human islets (Wang et al. 2012) and prelabelled them with the theranostic moiety for 48 hours in culture. These human islets were then transferred inside the kidney of immunodeficient B2M-deficient NOD/SCID mice; it is to be noted that the mice didn't have endogenous CD8+ T cells. The mice were made diabetic by streptozotocin. The B2M mRNA expression declined by 46%. By reducing the Beta2-microglobulin (B2M), the functional response of CD8+ T cells lessened, leading to lesser graft rejection.

5.6 Targeting Theranostics Agent Against IBD and Other Inflammatory Disease

Inflammatory bowel disease is a general medical condition used to describe chronic and inflammatory condition of the digestive tract and majorly includes ulcerative colitis (UC) and Crohn's disease (CD). In IBD, there is an increase in autoimmune response of body for food, normal microflora of GIT, and other material. Factor such as diet, consumption of tobacco, and exposure to infection have also been found to be contributing to the etiology of the disease. CD and UD play significant role but in innate immunity CD play a major role where they initiate the inflammatory process as well as adaptive immune response. Macrophages are main part of the normal intestinal tissues and their location is well established in lamina propria and Peyer's patch where they function as immune effector cells against any pathogenic attack. IBD in immune response is a complex cascade of events involving a coordinated activation of immune cells through an array of cytokines and chemokines and is predominated by activation of CD4+ cells (Tillack et al. 2014).

Dexamethasone polymeric nanoshells tagged with Cy5 have been prepared by Lee et al., for treating inflammatory bowel disease. Colitis has been induced (Lee et al. 2017) by dextran sulfate and then accumulation of nanoshells was observed by near infrared fluorescence. It was observed that these nanocarriers reduced the permeation of macrophages and reduced the expression of interleukin-1β and interleukin-6 and assisted in weight loss. Various nanoformulations have been used to deliver therapeutic agent in many inflammatory diseases. They have been discussed in the table below (Table 3).

6 Challenges in Nanotheranostics

However late advances in the field of nanotheranostics, there is no marketed product available. It is due to low affectability of diagnostic agents (e.g., X-ray used to detect iron oxide NPs), cost of production (gold NPs), complex blend, long-term toxicity of nanomaterials utilized, and an absence of human illness models. By far most of studies have been directed in lab animal models that have essentially lot less complex anatomy and physiology. The uptake and processing of nanoparticles get affected by its surface charge, polymer, and shape itself. Accordingly, broad, reasonable, toxicological, and pharmacological examinations are important prior to considering clinical interpretation of NPs (Gobbo et al. 2015). Interpretation from the clinical investigation

Table 3 Macrophages targeted against inflammatory diseases

Disease	Target	Delivery vector	References
IBD	IL-10	Polymer	Bhavsar and Amiji (2008)
	TNF-α	Liposomes	
	Cyclin –D1	Mannose targeted	
	CD-98	Lipid	
RA	IL-10	Polymer	Attarwala (2012)
	TNF-α	Liposomes	
		Lipid	
Atherosclerosis	AIF-1	Lipid	Leuschner et al. (2011)
	CCR-2	Lipid	
Diabetes	CCR-2	Glucan particles	Aouadi et al. (2013b)
	TNF-α	Lipid	
Cancer	CCR-2	Lipid	Tabata et al. (2011)
	IFN-γ	Adenovirus	
	IL-12		

might be quickened by including the novel carriers such as, peptides and nucleic acids to theranostics and additional use of ex-vivo human models. Collective endeavors between nanomaterial engineers, researchers, and clinicians are needed to accomplish positive clinical results. Nanotechnology is one of the most useful procedures for both clinical imaging and treatment of malignant cancer (Shi et al. 2016).

Nanotoxicity is one of the biggest challenges in overcoming the development of nanotheranostics. Silica-based nanoparticles can accumulate over a longer period of time which results in toxicity of tissues. Some nanoparticles stay in the body even after phagocytosing the inflamed macrophages and tissues then cause harm to the nearby healthy cells. They also affect the lungs resulting in fibrosis. In many cases, they promote the ongoing illness. For example, overaccumulation of gold and silica NP also promotes formation of plaques and tangles (Meyers et al. 2013) Hence, it becomes imperative to modify their properties so as they can cause less to no toxicity.

7 Conclusion

In the current chapter, some nanoplatforms that are presently under serious examination for the development of theranostic agents have been featured. The sum of the nanoplatforms examined here have gone through long stretches of improvement and permit effortless and solid capacity. These nanoparticles can possess unique optical or magnetic properties and have been previously studied in the imaging agent and have achieved many successes. It has been shown that therapeutics of various forms, including those that are small molecule, protein, and nucleotide-based, can be conveniently tethered onto nanoplatforms. In addition to achieving and validating the nanoscale integration of imaging and therapeutic functions, it is of significant importance to demonstrate the benefits and synergy of such a combined approach.

Various molecular focuses of macrophages that are shown in a few illnesses were examined and the capability of theranostics in synchronous imaging and treatment were illustrated. Studies additionally demonstrated the utility of imaging of macrophage take-up, biodistribution, representation of target aggregation, and treatment. Macrophages focused on theranostics have exhibited adequacy in a few serious illnesses either by decrease of macrophage numbers or hindrance of inflammation. In view of these outcomes, it has to be presumed that the theranostics focusing on macrophages holds incredible guarantee for diagnosis and treatment of the inflammatory diseases and other illnesses like HIV. Nanotheranostics has quickly advanced to incorporate biocompatible and biodegradable, multifunctional and coordinated nanoplatforms that exemplify drugs and symptomatic agents. Nanotheranostics may convey chemo-, radio-, biologic-, immuno-treatments, or a few blends of these agents. In an ideal world, these nanotheranostics would permit clinicians to analyze tumor growth, begin the treatment, and assess treatment reaction while permitting them to follow the nanoparticles' pharmacokinetics and arrival of the payload following use of suitable triggers.

References

Aouadi M, Tencerova M, Vangala P, Yawe JC, Nicolouro SM, Amano SU, Cohen JL, Czech MP. Gene silencing in adipose tissue macrophages regulates whole-body metabolism in obese mice. Proc Natl Acad Sci U S A. 2013a;110:8278–83.

Aouadi M, Tencerova M, Vangala P, Yawe JC, Nicoloro SM, Amano SU, Cohen JL, Czech MP. Gene silencing in adipose tissue macrophages regulates whole-body metabolism in obese mice. Proc Natl Acad Sci U S A. 2013b;110:8278–83.

Attarwala M. Amiji, multi-compartmental nanoparticles-in-emulsion formulation for macrophage-specific anti-inflammatory gene delivery. Pharm Res. 2012;29:1637–49.

Beer AJ, Schwaiger M. Imaging of integrin alphavbeta3 expression. Cancer Metastasis Rev. 2008;27:631–44. [PubMed: 18523730]

Bhavsar MD, Amiji MM. Oral IL-10 gene delivery in a microsphere-based formulation for local transfection and therapeutic efficacy in inflammatory bowel disease. Gene Ther. 2008;15:1200–9.

Bhumkar DR, Joshi HM, Sastry M, Pokharkar VB. Chitosan reduced gold nanoparticles as novel carriers for transmucosal delivery of insulin. Pharm Res. 2007;24:1415–26.

Cai W, Chen X. Multimodality molecular imaging of tumour angiogenesis. J Nucl Med. 2008;49(Suppl 2):113S–28S.

Cormode DP, Skajaa T, Fayad ZA, Mulder WJ. Nanotechnology in medical imaging: probe design and applications. Arterioscler Thromb Vasc Biol. 2009;29:992–1000.

Davila L, Ranganathan P. Pharmacogenetics: implications for therapy in rheumatic diseases. Nat Rev Rheumatol. 2011;7:537–50.

Del Vecchio S, Zannetti A, Fonti R, Pace L, Salvatore M. Nuclear imaging in cancer theranostics. Q J Nucl Med Mol Imaging. 2007;51:152–63.

Devaraj NK, Keliher EJ, Weissledе R. ^{18}F Labeled Nanoparticles for *in Vivo* PET-CT Imaging. 2009;20(2):397–401.

Gaitanis A, Staal S. Liposomal doxorubicin and nab-paclitaxel: nanoparticle cancer chemotherapy in current clinical use. Methods Mol Biol. 2010;624:385–92. [PubMed: 20217610]

Gobbo OL, Sjaastad K, Radomski MW, Volkov Y, Prina-Mello A. Magnetic nanoparticles in cancer theranostics. Theranostics. 2015;5:1249–63.

Gupta S*, Gupta MK. 2017. Possible role of nanocarriers in drug delivery against cervical cancer. Nano Rev Exp 8 (1):1–25.

Gupta S*, Gupta MK, Pathak Y, Vyas SP. 2019. Nanoscale drug delivery strategies for therapy of ovarian cancer: conventional vs targeted. Artif Cells Nanomed Biotech 47: 4066–4088.

Haglund E, Seale-Goldsmith MM, Leary JF. Design of multifunctional nanomedical systems. Ann Biomed Eng. 2009;37:2048–63. [PubMed: 19169822]

Hapuarachchige S, Artemov D. Theranostic pretargeting drug delivery and imaging platforms in cancer precision medicine. Front. Oncol. 2020; https://doi.org/10.3389/fonc.2020.01131.

Hawkins MJ, Soon-Shiong P, Desai N. Protein nanoparticles as drug carriers in clinical medicine. Adv Drug Deliv Rev. 2008;60:876–85. [PubMed: 18423779]

Hoff BA, Chughtai K, Jeon YH, Kozloff K, Galbán S, Rehemtulla A, Ross BD, Galbán CJ. Multimodality imaging of tumor and bone response in a mouse model of bony metastasis. Transl Oncol. 2012;5(6):415–21.

Honarvar H, Westerlund K, Altai M, Sandstrom M, Orlova A, Tolmachev V, et al. Feasibility of affibody molecule-based pna-mediated radionuclide pretargeting of malignant tumors. Theranostics. 2016;6:93–103. https://doi.org/10.7150/thno.12766.

Janib SM, Moses AS, MacKay JA. Imaging and drug delivery using theranostic nanoparticles. Adv Drug Deliv Rev. 2010;62(11):1052–63. https://doi.org/10.1016/j.addr.2010.08.004.

Kalashnikova I, Chung SJ, Nafiujjaman M, et al. Ceria-based nanotheranostic agent for rheumatoid arthritis. Theranostics. 2020;10(26):11863–11880. Published 2020 Oct 25. https://doi.org/10.7150/thno.49069

Keren S, Zavaleta C, Cheng Z, de la Zerda A, Gheysens O, Gambhir SS. Noninvasive molecular imaging of small living subjects using Raman spectroscopy. Proc Natl Acad Sci U S A. 2008;105:5844–9. [PubMed: 18378895]

Kokkonen H. Up-regulation of cytokines and chemokines predates the onset of rheumatoid arthritis. Arthritis Rheum. 2010;62:383–91.

Krupka TM, Solorio L, Wilson RE, Wu H, Azar N, Exner AA. Formulation and characterization of echogenic lipid-pluronic nanobubbles. Mol Pharm. 2010;7:49–59. [PubMed: 19957968]

Kunjachan S, Jayapaul J, Mertens ME, Storm G, Kiessling F, Lammers T. Theranostic systems and strategies for monitoring nanomedicine-mediated drug targeting. Curr Pharm Biotechnol. 2012;13:609–22.

Kunjachan S, Ehling J, Storm G. Fabian Kiessling, and Twan Lammers noninvasive imaging of nanomedicines and Nanotheranostics: principles. Chem Rev. 2014;5:10–199. https://doi.org/10.1021/cr500314d.

Lammers T. Improving the efficacy of combined modality anticancer therapy using HPMA copolymer-based nanomedicine formulations. Adv Drug Deliv Rev. 2010;62:203–30.

Lammers T, Subr V, Ulbrich K, Peschke P, Huber PE, Hennink WE, Storm G. Simultaneous delivery of doxorubicin and gemcitabine to tumors in vivo using prototypic polymeric drug carriers. Biomaterials. 2009;30:3466–75. [PubMed: 1930432]

Lee CC, MacKay JA, Frechet JM, Szoka FC. Designing dendrimers for biological applications. Nat Biotechnol. 2005;23:1517–26. [PubMed: 16333296]

Lee A, De Mei C, Fereira M, et al. Dexamethasone-loaded polymeric Nanoconstructs for monitoring and treating inflammatory bowel disease. Theranostics. 2017;7(15):3653–66. https://doi.org/10.7150/thno.18183.

Leuschner F, Dutta P, Gorbatov R, Novobrantseva TI, Donahoe JS, Courties G, Lee KM, Kim JI, Markmann JF, Marinelli B, Panizzi P, Lee WW, Iwamoto Y, Milstein S, Epstein-Barash H, Cantley W, Wong J, Cortez-Retamozo V, Newton A, Love K, Libby P, Pittet MJ, Swirski FK, Koteliansky V, Langer R, Weissleder R, Anderson DG, Nahrendorf M. Therapeutic siRNA silencing in inflammatory monocytes in mice. Nat Biotechnol. 2011;29:1005–10.

Levine DH, Ghoroghchian PP, Freudenberg J, Zhang G, Therien MJ, Greene MI, Hammer DA, Murali R. Polymersomes: a new multi-functional tool for cancer diagnosis and therapy. Methods. 2008;46:25–32. [PubMed: 18572025]

Liang B, Deng T, Li J, Ouyang X, Na W, Deng D. Biomimetic theranostic strategy for anti-metastasis therapy of breast cancer via the macrophage membrane camouflaged superparticles. Mater Sci Eng C. 2020;115:111097. https://doi.org/10.1016/j.msec.2020.111097.

Binbin Lianga, Tao Dengb, Jinping Lia, Xueliang Ouyanga, Weidan Naa*, Dawei Denga* Biomimetic theranostic strategy for anti-metastasis therapy of breast cancer via the macrophage membrane camouflaged superparticles 2020. 2020 be found online at https://doi.org/10.1016/j.msec.2020.111097

Liu Y, Miyoshi H, Nakamura M. Nanomedicine for drug delivery and imaging: a promising avenue for cancer therapy and diagnosis using targeted functional nanoparticles. Int J Cancer. 2007;120:2527–37.

Liu Z, Fan AC, Rakhra K, Sherlock S, Goodwin A, Chen X, Yang Q, Felsher DW, Dai H. Supramolecular stacking of doxorubicin on carbon nanotubes for in vivo cancer therapy. Angew Chem Int Ed Engl. 2009;48:7668–72.

Locke LW, Mayo MW, Yoo AD, Williams MB, Berr SS. PET imaging of tumor associated macrophages using mannose coated ^{64}Cu liposomes. 2012;33(2):7785–93.

Ma LL, Feldman MD, Tam JM, et al. Small multifunctional nanoclusters (nanoroses) for targeted cellular imaging and therapy. ACS Nano. 2009;3(9):2686–96. https://doi.org/10.1021/nn900440e.

Malam Y, Loizidou M, Seifalian AM. Liposomes and nanoparticles: nanosized vehicles for drug delivery in cancer. Trends Pharmacol Sci. 2009;30:592–9. [PubMed: 19837467]

McCarthy JR, Korngold E, Weissleder R, Jaffer FA. A light-activated theranostic nanoagent for targeted macrophage ablation in inflammatory atherosclerosis. Small. 2010;6(18):2041–9. https://doi.org/10.1002/smll.201000596.

Meike L, Schipper ML, Cheng Z, Lee SW, Bentolila LA, Iyer G, Rao J, , Chen X, Wu Amweiss S, Gambhir SS. Micropet-based biodistribution of quantum dots in living mice. J Nucl Med. 2007; 48:1511–1518. [Pubmed: 17704240].

Meyers JD, Doane T, Burda C, Basilion JP. Nanoparticles for imaging and treating brain cancer. Nanomedicine (Lond). 2013;8:123–43.

Mountz JM, Alavi A, Mountz JD. Emerging optical and nuclear medicine imaging methods in rheumatoid arthritis. Nat Rev Rheumatol. 2012;8:719–28. published online 25 September 2012; 10.10.38/nrrheum.2012.148

Muller C, Bunka M, Haller S, Koster U, Groehn V, Bernhardt P, van der Meulen N, Turler A, Schibli R. Promising prospects for 44Sc-/47Sc-based Theragnostics: application of 47Sc for radionuclide tumour therapy in mice. J Nucl Med. 2014;55:1658–64.

Nahrendorf M, Zhang H, Hembrador S, Panizzi P, Sosnovik DE, Aikawa E, Libby P, Swirski FK, Weissleder R. Nanoparticle PET-CT imaging of macrophages in inflammatory atherosclerosis. Circulation. 2008;117:379–87. [PubMed: 18158358]

Nahrendorf M, Keliher E, Marinelli B, Leuschner F, Robbins CS, Gerszten RE, Pittet MJ, Swirski FK, Weissleder R. Detection of macrophages in aortic aneurysms by nanoparticle positron emission tomography-computed tomography. Arterioscler Thromb Vasc Biol. 2011;31:750–7. [PubMed:21252070]

Nie S, Xing Y, Kim GJ, Simons JW. Nanotechnology applications in cancer. Annu Rev Biomed Eng. 2007;9:257–88.

Ogawa M, Nakamura S, Saito Y, Kosugi M, Magata Y. What can be seen by 18F-FDG PET in atherosclerosis imaging? The effect of foam cell formation on 18F-FDG uptake to macrophages in vitro. J Nucl Med. 2012;53(1):55–8. https://doi.org/10.2967/jnumed.111.092866. Epub 2011 Nov 29

Park JH, von Maltzahn G, Ruoslahti E, Bhatia SN, Sailor MJ. Micellar hybrid nanoparticles for simultaneous magnetofluorescent imaging and drug delivery. Angew Chem Int Ed Engl. 2008;47:7284–8.

Park K, Lee S, Kang E, Kim K, Choi K, Kwon IC. New generation of multifunctional nanoparticles for cancer imaging and therapy. Adv Funct Mater. 2009a;19:1553–66.

Park JH, Gu L, von Maltzahn G, Ruoslahti E, Bhatia SN, Sailor MJ. Biodegradable luminescent porous silicon nanoparticles for in vivo applications. Nat Mater. 2009b;8:331–6.

Patel SK, Janjic JM. Macrophage targeted theranostics as personalized nanomedicine strategies for inflammatory diseases. Ivyspring International Publisher Theranostics. 2015;5(2):150–72. https://doi.org/10.7150/thno.9476.

Rai P, Mallidi S, Zheng X, Rahmanzadeh R, Mir Y, Elrington S. Development and applications of photo-triggered theranostic agents. Adv Drug Deliv Rev. 2010;62:1094–124.

Richard C, de Chermont QM, Scherman D. Nanoparticles for imaging and tumour gene delivery. Tumouri. 2008a;94:264–70. [PubMed: 18564615]

Richard C, Doan BT, Beloeil JC, Bessodes M, Toth E, Scherman D. Noncovalent functionalization of carbon nanotubes with amphiphilic Gd3+ chelates: toward powerful T1 and T2 MRI contrast agents. Nano Lett. 2008b;8:232–6. [PubMed: 18088153]

Roesch F. Scandium-44: benefits of a long-lived PET radionuclide available from the 44Ti/44Sc generator system. Curr Radiopharm. 2012;5:187–201.

Satpathy M, Wang L, Zielinski R, Qian W, Lipowska M, Capala J, Lee GY, Xu H, Wang YA, Mao H, Yang L. Active targeting using HER-2-affibody-conjugated nanoparticles enabled sensitive and specific imaging of orthotopic HER-2 positive ovarian tumors. Small. 2014;10(3):544–55.

Shi J, Kantoff PW, Wooster R, Farokhzad OC. Cancer nanomedicine: progress, challenges and opportunities. Nat Publ Gr. 2016;17:20–37.

Singh R, Pantarotto D, McCarthy D, Chaloin O, Hoebeke J, Partidos CD, Briand JP, Prato M, Bianco A, Kostarelos K. Binding and condensation of plasmid DNA onto functionalized carbon nanotubes: toward the construction of nanotube-based gene delivery vectors. J Am Chem Soc. 2005;127:4388–96.

Singh P, Gupta U, Asthana A, Jain NK. Folate and folate-PEG-PAMAM dendrimers: synthesis, characterization, and targeted anticancer drug delivery potential in tumor bearing mice. Bioconjug Chem. 2008;19:2239–52. [PubMed: 18950215]

Singh A, Kutscher H, Bulmahn J, Mahajan S, He G, Prasad P. Laser ablation for pharmaceutical nanoformulations: Multi-drug nanoencapsulation and theranostics for HIV. Nanomed Nanotechnol Biol Med. 2020;25:102172. https://doi.org/10.1016/j.nano.2020.102172.

Surendiran A, Sandhiya S, Pradhan SC, Adithan C. Novel applications of nanotechnology in medicine. Indian J Med Res. 2009;130:689–701. [PubMed: 20090129]

Tabata K, Kurosaka S, Watanabe M, Edamura K, Satoh T, Yang G, Abdelfattah E, Wang J, Goltsov A, Floryk D, Thompson TC. Tumor growth and metastasis suppression by Glipr1 gene-modified macrophages in a metastatic prostate cancer model. Gene Ther. 2011;18:969–78.

Thomas M, Klibanov AM. Conjugation to gold nanoparticles enhances polyethylenimine's transfer of plasmid DNA into mammalian cells. Proc Natl Acad Sci U S A. 2003;100:9138–43.

Tillack C, Ehmann LM, Friedrich M, Laubender RP, Papay P, Vogelsang H, Stallhofer J, Beigel F, Bedynek A, Wetzke M, Maier H, Koburger M, Wagner J, Glas J, Diegelmann J, Koglin S, Dombrowski Y, Schauber J, Wollenberg A, Brand S. Gut. 2014;63(4):567–77.

Torchilin VP. Micellar nanocarriers: pharmaceutical perspectives. Pharm Res. 2007;24:1–16. [PubMed: 17109211]

Torigian DA. Feasibility and performance of novel software to quantify metabolically active volumes and 3D partial volume corrected SUV and metabolic volumetric products of spinal bone marrow metastases on 18F-FDG-PET/CT. Hell J Nucl Med. 2011;14:8–14.

Vonnemann J, Beziere N, Böttcher C, Riese SB, Kuehne C, Dernedde J, Licha K, von Schacky C, Kosanke Y, Kimm M, Meier R, Ntziachristos V, Haag R. Polyglycerolsulfate functionalized gold nanorods as optoacoustic signal nanoamplifiers for in vivo bioimaging of rheumatoid arthritis. 2014;4(6):629–41.

Wang P, Yigit MV, Ran C, et al. A theranostic small interfering RNA nanoprobe protects pancreatic islet grafts from adoptively transferred immune rejection. Diabetes. 2012;61:3247–54.

Xie J, Lee S, Chen X. Nanoparticle-based theranostic agents. Adv Drug Deliv Rev. 2010;62:1064–79.

Yuan J, Guo W, Yang X, Wang E. Anticancer drug-DNA interactions measured using a photoinduced electron-transfer mechanism based on luminescent quantum dots. Anal Chem. 2009;81:362–8.

Yuan Y, Long L, Liu J, Lin Y, Peng C, Tang Y, Zhou X, Li S, Zhang C, Li X, Zhou X. The double-edged sword effect of macrophage targeting delivery system in different macrophage subsets related diseases. 2020;18:168.

Zalutsky MR, Reardon DA, Pozzi OR, Vaidyanathan G, Bigner DD. Targeted alpha-particle radiotherapy with 211at-labeled monoclonal antibodies. Nucl Med Biol. 2007;34:779–85. [PubMed: 17921029]

Part IV

Targeting Macrophages for Treatment of Various Diseases

Tumor-Associated Macrophages: Therapeutic Targets of Cancer

Yubin Li, Xuyao Zhang, Xian Zeng, Shaofei Wang, and Hongbin Wang

Abstract

It has been widely known that macrophages play critical roles during infection and immune responses. While as tumor-infiltrating myeloid cells, macrophages accumulated in the tumor microenvironment (TME) could either suppress tumor proliferation through producing pro-inflammatory cytokines or promote tumor growth via activating tumor proliferation, metastasis, and angiogenesis, tumor-associated macrophages (TAMs), the most widely infiltrating macrophages in the TME contributing to immunosuppressive microenvironment establishment, could promote tumor invasion and weaken conventional therapeutic effects. Thus, targeting TAMs is becoming a promising strategy for cancer therapy. Here we will discuss the role of TAMs and their crosstalk with tumor in TME, including immune suppression, tumor progression, tumor metastasis, and angiogenesis. In addition, the role of TAMs in cancer therapy, including chemotherapy, radiotherapy, and immunotherapy, will be presented. Next, TAMs as a promising therapeutic target for cancer treatment will be illustrated, including targeting recruitment and localization of TAMs, targeting TAMs inhibition, and targeting TAMs reprogramming. Finally, the potential of TAMs as drug delivery systems will also be summarized. The chapter highlighted TAMs as promising therapeutic targets of can-

Yubin Li, Xuyao Zhang and Xian Zeng contributed equally with all other contributors.

Y. Li
Department of Dermatology, Perelman School of Medicine, University of Pennsylvania, Philadelphia, PA, USA

Corporal Michael J. Crescenz VA Medical Center, Philadelphia, PA, USA

Department of Neurology, Xinqiao Hospital, Third Military Medical University, Chongqing, People's Republic of China

X. Zhang · X. Zeng
Department of Biological Medicine & Shanghai Engineering Research Center of Immunotherapeutics, School of Pharmacy, Fudan University, Shanghai, People's Republic of China

S. Wang
Department of Cellular and Genetic Medicine, School of Basic Medical Sciences, Fudan University, Shanghai, People's Republic of China

H. Wang (✉)
Department of Pharmaceutical and Biomedical Sciences College of Pharmacy, California Northstate University, Elk Grove, CA, USA

Master of Pharmaceutical Sciences College of Graduate Studies, California Northstate University, Elk Grove, CA, USA

Department of Basic Science College of Medicine, California Northstate University, Elk Grove, CA, USA
e-mail: hongbin.wang@cnsu.edu

S. Gupta, Y. V. Pathak (eds.), *Macrophage Targeted Delivery Systems*,
https://doi.org/10.1007/978-3-030-84164-5_13

cer, which might be useful for drug development and evaluation of targeting TAMs for cancer therapy.

Keywords

Tumor-associated macrophages · Cancer therapy · Drug delivery systems

1 Introduction

Macrophages are white blood cells of the innate immune system that can reside in all tissues of the body and contribute to the development, tissue homeostasis, diseases, and tumor microenvironment (TME). They are given different names based upon their locations, for example, microglia cell in brain, Langerhans cells in skin, Osteoclasts in bone, alveolar macrophages in lung, and Kupffer cells in liver (Epelman et al. 2014; Ginhoux and Guilliams 2016). Tissue-resident macrophages are derived from at least three distinct sources: the yolk sac, the fetal liver, and the bone marrow (Guerriero 2018). Macrophages-mediated innate immunity plays a crucial role in maintaining tissue homeostasis by engulfing and digesting invading pathogens, bacteria, damaged tissue, and malignant cells (Haniffa et al. 2015; Yona and Gordon 2015; Guerriero 2019). They also contribute to specific adaptive immune responses of T lymphocytes activation by antigen presentation. Besides, macrophages can modulate immune system through the secretion of various cytokines, chemokines, and the activation of complement system (Guerriero 2019). Macrophages can be characterized in vitro into two types: classically activated M1 macrophages and alternatively activated M2 macrophages. While M1 type macrophages are induced by T_H1 cytokines such as INF-γ, M2 type macrophages are induced by the T_H2 cytokines such as IL-4/IL-13. Macrophages represent up to 50% of leukocytes in the TME. Clinical data revealed that the poor prognosis is associated with the abundance of tumor-associated macrophages (TAMs) in 80% of human cancer. It was originally thought the macrophages were recruited to malignant sites to kill the cancer cells. However, accumulating evidence demonstrated that tumor cells recruit macrophages to promote tumor malignancy and progression, like macrophages are recruited to a wound site to assist in healing. In this book chapter, we will discuss the role of TAMs and their interplay and crosstalk with T effector cells and tumor cells in TME, including immune suppression, tumor progression, tumor metastasis, and angiogenesis. In addition, the role of TAMs in cancer therapy, including chemotherapy, radiotherapy, and immunotherapy, will be presented. Moreover, TAMs as a therapeutic target for cancer treatment will be illustrated, including the recruitment of TAMs, the signaling pathways involved in TAMs activation, and the production of cytokines and chemokines that are involved in tumor angiogenesis. Then current strategies targeting TAMs for cancer therapy including targeting TAMs inhibition, recruitment, localization, and reprogram will be discussed. Lastly, the potential of TAMs as drug delivery systems will also be summarized.

2 Macrophages and Tumor-Associated Macrophages (TAMs)

Macrophages, the major phagocytic cells in immune system, are derived and matured from monocytes that leave from the circulation system and settle in spleen, lymph nodes, alveoli, and tonsils. Macrophages are also existing in the brain as microglia, in the skin as Langerhans cells, in bone as osteoclasts, and in the liver as Kupffer cells. These tissue-resident macrophages originate from at least three embryonic sources: erythro-myeloid progenitors in the yolk sac and in the fetal liver and macrophage/dendritic cell progenitor cells in the bone marrow that give rise to monocytes (Davies et al. 2013). Tissue-resident macrophages are primarily derived from both the yolk sac and the fetal liver. They reach their tissues during embryonic development, progressing to adulthood, and sustain through proliferation independent of hematopoietic stem cells. Borrow marrow-derived macrophages have a short life span and are only generated in the tissue under inflammatory conditions (Guerriero 2018). In cases of infection, macrophages, through their receptors on the surface, are able to undergo phagocytosis, the process of ingesting of foreign antigens by lysosomes, such as small particles, whole cells, and bacteria. In addition, macrophages in the peripheral lymphoid tissues serve as the major scavengers of abnormal cells and cellular debris. Importantly, macrophages present a vital role in processing antigen to T lymphocytes to activate specific adaptive immune responses. Activated macrophages can release many inflammatory mediators, such as cytokines, chemokines, enzymes, reactive oxidative species, coagulation factors, and growth factors (Ginhoux and Jung 2014; Olingy et al. 2019; Varol et al. 2015). All macrophages express colony-stimulating factor-1 receptor (CSF-1R), which binds to CSF-1 or alternatively to the interleukin IL-34 and regulates macrophage differentiation, proliferation, and survival. The presence and functional states of macrophages are mainly regulated through by CSF-1, granulocyte macrophage CSF (CSF2/GM-CSF), and chemokines.

The activation status of macrophages can be classified into two subpopulations: M1 and M2 macrophages. M1 (also known as classically activated) macrophages, induced by TH_1-type cytokines such as INF-γ and GM-CSF and through Toll-like receptor 4 (TLR4) engagement of lipopolysaccharides (LPS) from gram negative bacteria, give rise to pro-inflammatory, antiviral, antibacterial, antitumoral phenotypes with powerful killing effects on invading pathogens and at the same time destructive effects on normal tissues (Fleetwood et al. 2007; Arnold et al. 2014). Activated macrophages are potent effector cells that can kill tumor cells and microorganisms, trigger massive proinflammatory cytokines production, and activate cytotoxic T lymphocytes. M2 (also known as alternatively activated) macrophages are stimulated by the TH_2 cytokines IL-4/IL-13, IL-10, CSF-1 and play important roles in anti-inflammatory humoral responses and pro-repair, pro-tumoral, and antiparasitic phenotypes (Davies et al. 2013; Murray et al. 2014). Subgroups of M2 macrophages were divided into M2a (IL-4), M2b (INFγ + complexed immunoglobulin (Ig)), and M2c (dexamethasone) (Szulzewsky et al. 2015). The nomenclature of M1 or M2 macrophages describes two extremes states of macrophage activation and functions that have direct in vitro relevance during the infection of bacteria or parasites. In vivo, macrophages are not clearly divided into M1 and M2 classification. Indeed, a study demonstrated that over 60 percent of the upregulated genes in TAMs from brain tumors had no overlap with that from M1 or M2 (M2a, M2b, or M2c) ex vivo macrophages phenotypes, suggesting macrophages activated in TME may not be reflected by ex vivo stimulation of monocytes (Hambardzumyan et al. 2016). Recent studies revealed that there are populations of CD169+ and TCR+ macrophages in vivo, which cannot simply be described as M1 or M2 term and seem to play roles in maintaining homeostasis, immune regulation, and tolerance (Chavez-Galan et al. 2015; Crocker and Gordon 1986; Martinez-Pomares et al. 1996; Martinez-Pomares and Gordon 2012). TCRb gene was discovered rearrangement in the early stage of bone marrow macrophages differentiation. TCR+ macrophages express chemokine (C-C motif) ligand 2 (CCL2) with strong phagocytic functions, which are different from traditional macrophages (Kaminski et al. 2013). It was suggested to move away from the ambiguity of the M1-M2 characterization of TAMs and to define TAMs based on functional, transcriptional, or epigenetic status, which will demonstrate more clear characterization of TAMs (Guerriero 2018).

TAMs are usually referred to as macrophages recruited from circulating monocytes to tumors and influenced by the presence of cancer to promote tumor malignancy, survival, proliferation, angiogenesis, metastatic dissemination, and chemoresistance (Solinas et al. 2009; Qian and Pollard 2010; Kitamura et al. 2017). Rather than a homogenous population, TAMs actually can originate from different sources and exhibit either pro-tumoral or sometimes antitumoral roles. Each population has a unique transcriptional landscape based on the type, the stage, and the immune composition of the tumors they infiltrate. M2 macrophages are defined as TAMs in a narrow sense because active TAMs have various properties similar to M2 macrophages (Murray et al. 2014; Chavez-Galan et al. 2015). The TME consists of variouse cell types, such as tumor cells, granulocytes, macrophages (~50%), mast cells, fibroblast and epithelial cells. Secreted cytokines and chemokines, such as CCL2, CCL11, CCL16, and CCL21, are major determinants for the infiltration of macrophages. Increasing evidence demonstrated that tumor cells recruit macrophages to support tumor growth. Clinical data indicated that poor prognosis in over 80% of human tumor is associated with increased TAMs, which not only lack the phagocytic functions of tumor cells, also promote tumor cells dissemination. High density of TAMs in tumor has been associated with increased vascular density, chemotherapy resistance, and worse outcome in various cancer types, such as colorectal, breast, ovarian, nonsmall cell lung cancer, melanoma, Hodgkin's lymphoma, and multiple myeloma (Guerriero 2018).

3 TAMs and Cancers

3.1 Crosstalk of TAMs and Cancer Cells in TME

Various molecular mechanisms have been documented to play roles in mediating cancer cells that escape form the attacks of macrophages. Programmed cell death protein (PD-1) belongs to CD28 superfamily and plays a significant role in immunosuppression on T effector cells and TAMs. PD-L1 on the surface of tumor cells binding to PD-1 on effector T-cells and macrophages helps tumor cells escape from the attacks of T effector cells as well as TAMs by inhibiting cytokine expression, activation, and proliferation of effector T-cells, and macrophage phagocytosis (Boussiotis et al. 2014; Yu et al. 2015; Gordon et al. 2017; Katsuya et al. 2016). The cluster of differentiation 47 (CD47) molecule, the "do-not-eat-me" signal, is recently characterized as a self-molecule that protects host cells from destruction by macrophages (Jaiswal et al. 2009; Zhao et al. 2016). CD47 expressed on the membrane of tumor cells binding to signal regulatory protein alpha1 (SIRP1α) on macrophages will inhibit phagocytosis by blocking accumulation of myosin IIA at the phagocytic synapse (Okazawa et al. 2005; Barclay and Van den Berg 2014). Antibody-mediated inhibition of human CD47 enhances macrophage-mediated phagocytosis (Tseng et al. 2013). Targeting CD47/SIRP1α signaling pathway is currently being tested in clinical trials (Pathria et al. 2019; Chao et al. 2012). In breast and ovarian cancer, CD24 was found as a dominant innate immune checkpoint, which could enhance tumor escape by providing "do-not-eat-me" signal through the interaction with inhibitory receptor sialic-acid-binding Ig-like lectin 10 (Siglec-10) expressed on TAMs. Suppressing the crosstalk between CD24 and Siglec-10 with respective antibodies or ablating the genes of CD24 or Siglec-10 can obviously promote phagocytic function of TAMs to tumors with CD24 expression (Barkal et al. 2019). Recent study revealed another recognition mechanism that can deter phagocytotic effects of macrophages on tumor cells, which is the interaction between leukocyte immunoglobulin like receptor subfamily B member 1 (LILRB1) on macrophages and major histocompatibility complex (MHC) class I component β2-microglobulin on tumor cells. Either blocking β2-microglobulin or anti-LILRB1 antibody can significantly enhance macrophage phagocytosis and inhibit tumor growth (Barkal et al. 2018). In TME, exosomes were recently discovered as important carriers for proteins, nucleic acid, which can affect survival, growth, and metastasis of cancer cells (Milane et al. 2015; Syn et al. 2016). A recent study found that TAMs characterized as M2-polarized phenotype can secrete exosomes and promote the metastasis of gastric cancer. Apolipoprotein E (ApoE) was identified richly in those exosomes, which can activate phosphatidylinositol 3-kinases (PI3k)-Akt signaling pathway to drive epithelial mesenchymal transition (EMT) and cytoskeleton rearrangement of cancer cells (Zheng et al. 2018). Another study found that TAMs-derived exosomes promote the resistance of PDAC to gemcitabine by transferring miR-365 into cancer cells (Binenbaum et al. 2018). Exploring the roles of macrophage-derived exosomes and their components in cancer development and metastasis will open a new door for the discovery of cancer treatment. Recent study revealed that complement-mediated inflammation could promote tumorigenic progression (Medler et al. 2018). In a mouse squamous cell carcinomas (SCC) model, TAMs-triggered plasminogen produces C5a that is independent of C3 activation. Activation of C5aR by C5a in macrophages and mast cells leads to immunosuppressive macrophages that can inhibit CD8+ T-cell activation (Medler et al. 2018). Combined therapy of C5aR antagonist PMX-53 and paclitaxel significantly inhibits tumor growth as compared to PMX-53 or paclitaxel alone. Moreover, inhibition of C5aR signaling with PMA-53 significantly enhances antitumor efficacy of checkpoint inhibitor anti-PD-L1 antibody, indicating anticomplement therapy might be beneficial for cancer immune therapy (Affara et al. 2014; Zha et al. 2017).

3.2 TAMs and Tumor Progression

It was originally considered that macrophages were recruited to tumor sites to kill the cancer cells. However, accumulating evidence has revealed that TAMs can cause tumor progression and development through secretions of various chemokines and cytokines, such as interleukin-6 (IL-6), interleukin-8 (IL-8), and interleukin-10 (IL-10) (Zhou et al. 2020). IL-6, secreted from tumor-associated endothelial cells and TAMs, has been involved in carcinogenesis and progression of tumors by regulating local inflammation, cell cycle, tumor angiogenesis, and stem cell self-renewal. IL-6 mediates the regulation of signal transducer and activator of transcription 3 (STAT3) phosphorylation, which leads to EMT, the overexpression of snail and Bcl-2, and anti-apoptotic effect in cancers (Gao et al. 2018; Yadav et al. 2011). IL-8, highly secreted by TAMs, is involved in angiogenesis, tumor metastasis, and suppression of immunity (Williams et al. 2016). Monitoring IL-8 level can predict the outcome of immunotherapy of checkpoint blockade (Sanmamed et al. 2017). Exogenous IL-8 causes the activation of JAK2/STAT3/snail signaling pathway by enhancing the expression of p-JAK2 and p-STAT3, further promoting EMT and carcinogenesis (Deng et al. 2013). IL-8 and chemokines, strongly expressed in inflammatory breast cancer (IBC), promote macrophages recruitment and transforming to M2 macrophages, which can secrete high level of IL-8 and chemokines, leading to a feed-forward cytokine/chemokine loop that drives the EMT of IBC (Valeta-Magara et al. 2019). In addition, TAMs also secrete IL-10, transforming growth factor-β (TGF-β), and inflammatory mediators, such as prostaglandin E2 (PGE2) and matrix metalloprotease-7 (MMP-7), which can inhibit antigen-presenting process for T-cell activation. During chronic inflammation, toll-like receptor 4 (TLR4) stimulates M2 to secrete cytokine IL-10 and activation of TLR4 by LPS significantly enhanced the EMT of pancreatic cancer cells (Deng et al. 2013). IL-10 also increases the expression of inhibitor of CIP2A via the PI3K signaling pathway and promotes tumor aggressiveness in lung adenocarcinoma (Sung et al. 2013). IL-10 plasma level showed significant correlation with tumor progression, indicating IL-10 plays a critical role in tumor progression (Sato et al. 2011).

3.3 TAMs and Tumor Metastasis

The underlying mechanisms and processes of cancer metastasis are very complex. The cancer cells start from local invasion, intravasation, and ultimate extravasation at peripheral sites. The disseminated cancer cells need to escape the attack from immune system, survive in the blood or lymphatic circulation, settle down at a distant site, and proliferate in the new hostile environment (Ruffell and Coussens 2015; Cassetta and Pollard 2018). Macrophages have been implicated in all stages of the metastatic processes. High infiltration of macrophages, overexpression of CSF-1, and their chemoattractant CCL2 in human solid tumors are linked to poor clinical prognosis (Cassetta and Pollard 2020). Pollard and his colleagues have demonstrated that in a mouse mammary gland, tumor model genetic depletion of CSF-1 (similar to the depletion of macrophages) can slow the tumor progression and significantly inhibit metastasis, suggesting TAMs play crucial roles in cancer metastasis (Joyce and Pollard 2009). The phenomenon was recapitulated in other preclinical mouse model studies of the development of other cancer types, such as lung cancer, pancreatic cancer, and glioblastoma (Zabuawala et al. 2010; Welm et al. 2007; Rolny et al. 2011; Gocheva et al. 2010; Qian et al. 2011). Metastatic cancer cells recruit bone marrow-derived classical monocytes through CCL2-CCR2 signal mechanism (Qian et al. 2011). These extravasated monocytes differentiate into metastasis-associated macrophages (MAMs) through involvement of chemokine CCL3-CCR1 autocrine signaling (Mummalaneni et al. 2014; Kitamura et al. 2015). MAMs were characterized to express the markers CD11b, VEGF receptor 1, CXCR3, and CCR2, which are different from resident macrophages and MAM precursor cells. MAMs mediate cancer cells metastasis through promoting extravasation and tumor growth, as well as inhibiting cytotoxic T-cells function. Macrophages open up a gate for the cancer cells to escape when arriving at the circulating vessels. In addition, macrophages produce other molecules that help cancer cell invasion (Sangaletti et al. 2008). Cathepsin proteases can remodel the matrix and release sequestered growth factors and TGFβ generated from MAMs that drive epithelial to EMT of cancer cells (Laoui et al. 2011; Quail and Joyce 2013; Bonde et al. 2012). Macrophages in the lung interact with metastatic cancer cells and support their survival through vascular cell adhesion protein 1 (VCAM1)-dependent and Akt-dependent mechanism. Moreover, MAMs crosstalk to metastatic cancer cells to retain MAMs in the metastatic foci, which can further support cancer growth (Kitamura et al. 2015). TAMs inhibit the function of cytotoxic T-cells, which might be one of the mechanisms that help cancer cells escape from attack of T-cells and promote cancer metastasis (Kitamura et al. 2017). TAMs can directly inhibit T-cell cytotoxicity through the depletion of L-arginine via release of arginase 1, which is essential for the re-expression of the T-cell receptor (TCR) after antigen engagement on T-cells. In addition, TAMs generate cytokines such as IL-10 and TGFβ that lead to immunosuppressive microenvironment by inhibiting CD4+ and CD8+ T-cells and inducing Treg cells expansion in the TME. TAM-mediated release of chemokines, like CCL2, CCL3, CCL4, CCL5, and CCL20, contributes to the recruitment of Treg cells in the TME (Noy and Pollard 2014). Moreover, TAM-induced immune suppression is caused by the expression of inhibitory receptors, such as nonclassical major histocompatibility complex class I (MHC-I) molecules such as HLA-E and HLA-G, which can inhibit activation of NK cells and T-cells through the interaction with CD94 and leukocyte immunoglobulin-like receptor subfamily B member 1 (LIR1), respectively (Morandi and Pistoia 2014). The expression of T-cell immune checkpoint ligands by TAMs, such as PDL1, PDL2, B7–1, and B7–2, can directly inhibit T-cell functions (Santarpia and Karachaliou 2015; Buchbinder and Desai 2016). It is of great interest to target TAMs for anticancer metastasis therapy since both preclinical and clinical data have established the proof of principle.

3.4 TAMs and Angiogenesis

Angiogenesis is a key event in tumor growth and progression. Cancerous tissues are characterized by hypoxia and unequal vascularization, which actually can modify macrophage distribution and function. TAMs accumulate preferentially in the inadequately vascularization areas of tumors, which have low oxygen (Mantovani et al. 2002). Under hypoxic condition, macrophage migration is suppressed and TAMs are immobilized in non-vascular, necrotic, hypoxic regions of tumor, from which macrophages cooperate with cancer cells to promote angiogenesis (Grimshaw and Balkwill 2001; Leek et al. 1996, 1999; Lewis et al. 2000). In breast cancer, TAMs overexpress hypoxia-inducible factor-2α (HIF-2α) and HIF-1, which are transcription factors to induce the secretion of pro-angiogenic factors, such as endothelial growth factor VEGF, bFGF, and chemokines CXCL8 and CXCL12 to promote angiogenesis (Talks et al. 2000; Crowther et al. 2001; Lin and Pollard 2007; Hughes et al. 2015). Other factors such as TGFβ, WNT7B, TNF, and thymidine phosphorylase also contribute to angiogenic process by recruitment and activation of endothelial cells or other cells like fibroblast or pericytes that support the generation of vascular networks in microenvironment (Cassetta and Pollard 2018; Yeo et al. 2014; Hirano et al. 2001). A subpopulation of TAMs with the expression of the angiopoietin 1 receptor (TIE2) was identified to play essential roles in tumor angiogenesis. Depletion of these TAMs suppresses tumor growth and metastasis (De Palma et al. 2005; Mazzieri et al. 2011) (Fig. 1).

4 TAMs in Cancer Therapy

TAMs are the dominant cell populations recruited into the TME and are shaped by cancer cells to present M2 properties. The considerable functional plasticity enables TAMs to rapidly adapt to microenvironmental perturbations from various cancer treatment modalities (ranging from radiotherapy, chemotherapy, targeted therapy to immunotherapy) in TME, and clinical consequences of such interactions between TAMs and cancer therapies vary across different cancer types and different therapeutic categories (Vitale et al. 2019; De Palma and Lewis 2013; Larionova et al. 2019; Neophytou et al. 2020). In some cases, therapeutic agents may achieve their pharmacological effects via the function of TAMs. For example, accumulating evidence indicated that metformin suppresses tumorigenesis, cancer cell proliferation, and tumor metastasis by modulating functional states of TAMs. Metformin induced the expression of M1-related cytokines IL-12 and TNF-α and reduced M2-related cytokines IL-8, IL-10, and TGF-β, to revert the polarization of TAMs to M2 phenotype by cancer cells and thus may recover tumor immune surveillance (Ding et al. 2015; Chiang et al. 2017; Ma et al. 2020; de Oliveira et al. 2019). However, in other cases, TAMs may hamper efficacy of drug treatment by forming an immunosuppressive TME.

4.1 TAMs in Radiotherapy

Radiotherapy is the most commonly used modality for clinical treatment of solid tumors. In addition to its direct cell-killing effects, radiotherapy profoundly modified TME, and especially, TAMs are usually reshaped by radiations to exhibit bidirectional consequences (pro-inflammatory or pro-tumorigenic phenotypes) depending on the radiation doses and cancer types (Mantovani and Allavena 2015). On the one hand, immunogenic cell death induced by radiation could activate innate and adaptive immune systems to trigger antitumor immune responses where TAMs are closely involved in (Golden et al. 2014). For example, local low-dose gamma irradiation programs TAMs to iNOS+ M1 state, which subsequently promotes the recruitment of tumor-specific T-cells to TME and then evokes T-cell immune response (Klug et al. 2013; Prakash et al. 2016). On the other hand, accumulating evidence has demonstrated that TAMs mediated radio-resistance in multiple cancers (Wu et al. 2017). Radiotherapy may polarize TAMs to their M2 state to allow immune escape and compromised antitumor effects. For instance, an in vivo study revealed that radiation at 3 Gy in prostate cancer upregulated CSF1 expression on cancer cells, which in turn recruited TAMs (express CSF-1R) to TME and then promoted tumor regrowth (Xu et al. 2013). In glioblastoma, irradiation increased the number of M2 TAMs by upregulating CSF-1R expression, which finally lowers the tumor sensitivity to radiotherapy (Stafford et al. 2016). Bone marrow-derived macrophages (BMDMφ) promoted tumor growth after ionizing radiation at 20 Gy. The underlying mechanism involved in the upregulation of VEGF expression resulted from autocrine or paracrine of TNFα in BMDMφ cells, and removing TNFα could re-sensitize tumors to ionizing radiation (Meng et al. 2010). Recent data demonstrated that fibroblast growth factor 2 (FGF2) is critical for the transition of TAMs toward M2 phenotype

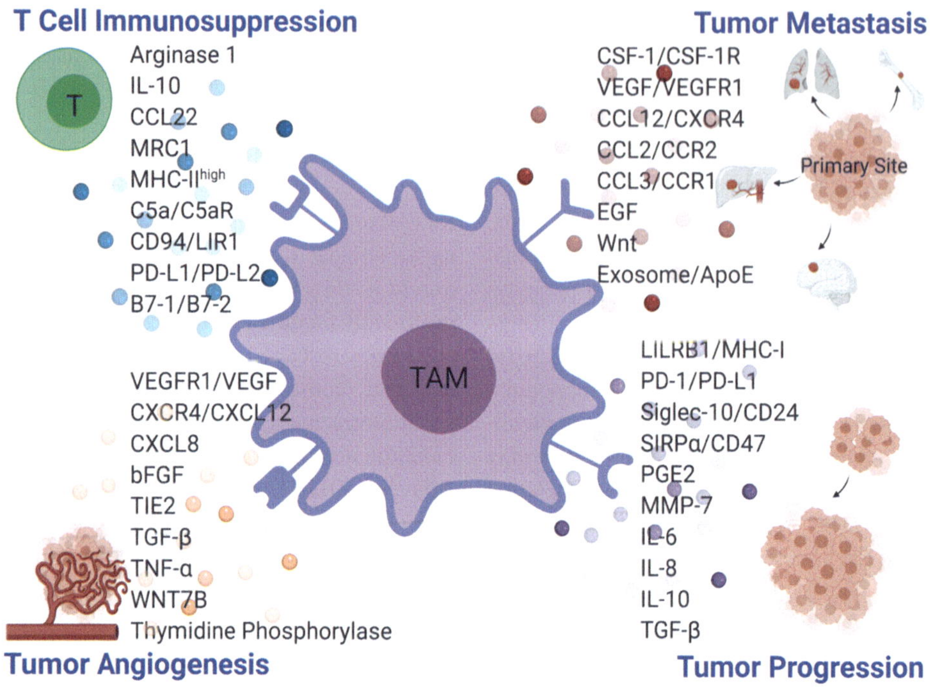

Fig. 1 The role of tumor-associated macrophages (TAMs) in promoting tumor progression, tumor angiogenesis, tumor metastasis, and T-cell immunosuppression. Multiple mechanisms of molecular interactions (CSF-1/CSF-1R, VEGF/VEGFR1, CCL12/CXCR4, CCL2/CCR2, CCL3/CCR1, PD-1/PD-L1, LILRB1/MHC-I, Siglec-10/CD24, SIRPα/CD47, C5a/C5aR, and CD94/LIR1) and secreted growth factors (VEGF, EGF, and bFGF), cytokines (CSF-1, IL-6, IL-8, IL-10, TNFα, TGFβ), chemokines (CXCL8, CXCL12, and CCL22), and other mediators (TIE2, PGE2, MMP-7, exosome, thymidine phosphorylase) are involved in TAMs-mediated different stages of tumor progression and T-cell immunosuppression

after radiotherapy, and blocking FGF2 increased percentage of M1-like TAMs, extended tumor growth delay, and prolonged long-term survival in mouse model (Im et al. 2020). In addition, the FGF2 receptor (FGF2R) has been reported to mediate radio-resistance in glioblastoma (Gouaze-Andersson et al. 2016). Those evidence indicated that reverting the M2 polarization either by direct depletion of TAMs or by targeting abovementioned key players (CSF-1/CSF-1R, TNFα, FGF2/FGF2R) represents a promising strategy for combinatory radiotherapy to improve clinical efficacy.

4.2 TAMs in Chemotherapy

The interactions between TAMs and chemotherapy have been widely cataloged to be involved in variations of clinical efficacy of a wide range of cancer therapeutics among patients (Larionova et al. 2019). Cancer cells can reshape cellular metabolism and thus shift the phenotype of TAMs to protect themselves from drug exposure. For example, TAMs in M2 state have been well documented to be involved in chemoresistance in the treatment of pancreatic ductal adenocarcinoma (PDA) and CCR2 and CSF-1R played key roles in driving chemoresistance (Mitchem et al. 2013). Metabolomics profiling of TAMs and PDA cells revealed the underlying mechanism that TAMs promote pyrimidine biosynthesis from glucose and release deoxycytidine molecules to interact with PDA cells and finally drive the gemcitabine resistance, and blocking deoxycytidine release by simply limiting glucose supply could sharply sensitize PDA cells to gemcitabine exposure, providing novel mechanism of gemcitabine resistance from an intratumoral metabolic crosstalk perspective

(Halbrook et al. 2019). Further retrospective analysis in this study indicated that lower TAMs in microenvironment is a favorable factor for better drug response and improved disease-specific survival outcome of PDA patients. As deoxycytidine release was also observed in a panel of cancer cell lines and patient-derived cancer cells, release of pyrimidine molecules by TAMs may drive chemoresistance in other cancers.

Moreover, TAMs were revealed in vitro and in vivo to mediate chemotherapy resistance by enhancing survival factor activities or activating antiapoptotic programs in cancer cells (Correia and Bissell 2012; Castells et al. 2012; DeNardo et al. 2011). For instance, TAMs suppressed cytotoxic effects of Taxol, a widely used antimitotic agent, on multiple cancer cell lines and on in vivo model models (Olson et al. 2017). TAMs in M2 state are able to release nitric oxide (NO) to protect cancer cell from apoptosis induced by cisplatin, leading to resistance in several cancer cell lines (Perrotta et al. 2018). Coculture studies in vitro using macrophages derived from bone marrow (BM) with mammary carcinoma cell lines demonstrated that macrophages confer to chemotherapy resistance to paclitaxel, doxorubicin, and etoposide and to gemcitabine in PDA cells (Mitchem et al. 2013; Shree et al. 2011). The activation of signal transducer and activator of transcription 3 (STAT3) in TAMs has been shown to promote PDAC cancer cells proliferation and survival, resistance to carboplatin in vivo, and tumor cell growth when synergized with IL-6 and other macrophage-derived factors such as milk fat globule-epidermal growth factor VIII (Jinushi et al. 2011; Taniguchi and Karin 2014; Yu et al. 2014). IL-6 can be generated by co-culture of BM-derived macrophages with neoplastic cells as well as from TAMs in vivo (DeNardo et al. 2009; Movahedi et al. 2010; Song et al. 2009). IL-1β can also induce IL-6 production in monocytes and osteoblasts (Mori et al. 2011; Tosato and Jones 1990). In vivo evidence of IL-6 being chemoresistance was shown in a murine lymphoma model (Gilbert and Hemann 2010). Nevertheless, the source and relevance of IL-6 during chemotherapy for solid tumor remains partly described. In a subcutaneous MCF-7 breast cancer xenografts animal model, anti-CSF-1 neutralizing antibody can increase of chemosensitivity (Paulus et al. 2006). Soluble chemoprotective factors generated from TAMs depend on protease activity of cathepsin B and S. In vivo inhibition of cathepsin protease activity enhances the response of mammary carcinomas to paclitaxel (Shree et al. 2011). Macrophages are main source of tumor necrosis factor (TNF), which may be one of crucial factors mediating chemoresistance either directly through NF-κB activation of indirectly through induced IL-6 expression (Mori et al. 2011; Li and Sethi 2010). TNFα has been reported to mediate chemoresistance to MAPK inhibitors in melanoma through NF-κB-dependent expression of microphthalmia transcription factor (Smith et al. 2014). Taken together, it has a strong rationale to target TAMs that contribute chemoresistance for anticancer therapy.

4.3 TAMs in Cancer Immunotherapy

T-cell-based immunotherapies, such as adoptive T-cell therapy and immune checkpoint blockages (ICBs), are changing the traditional paradigm of cancer treatment in clinical practice (Galon and Bruni 2019). However, only a small fraction of cancer patients can benefit from those therapies. As the dominant immune cell population in TME, TAMs actively interact with other immune cells to interfere antitumor immunity derived from T-cell based immunotherapies, accounting for an important mechanism of immunotherapy failure in primary resistant patients (Li et al. 2019a). T-cell-based immunotherapies exert antitumor activity by boosting T-cell immunity in TME, while T-cells are closely influenced by TAMs. TAMs could not only inhibit T-cell function by suppressing naïve T-cell proliferation, but also inhibit cytotoxic T-cell responses via releasing various anti-inflammatory cytokines (e.g., IL-10, TGF-β, PGE2) or leveraging inhibitory receptors (checkpoint molecules on TAMs such as B7-H4 and VISTA), which finally established an immunosuppressive TME and therefore led to compromised clinical outcomes (De Palma and Lewis 2013; Ruffell et al. 2014).

4.4 TAMs Aid Precision Cancer Treatment

As discussed above, TAMs profoundly influence antitumor activities of cancer therapies. This provides a series of insights to develop predictive or prognostic biomarkers for customized therapeutic regimes and to dig into detailed mechanisms for the discovery of novel TAM-related therapeutic targets, which will finally facilitate precision cancer medicine.

High-throughput omics profiling studies proposed the possibility that using the functional landscape of TAMs as predictive biomarker to guide precise patient stratification for improved therapeutic outcomes. In the TRIBE and FIRE3 clinical trial cohorts, gene mutation sequencing analysis demonstrated that mutation features of TAMs regulating genes could robustly predict progression-free survival of patients with metastatic colorectal cancer treated with bevacizumab plus FOLFIRI (Sunakawa et al. 2015). A recent analysis of multiomics profiles of bulk tumors from 348 patients reported that the infiltration level of M1 macrophages significantly correlated with drug response of ICBs in metastatic urothelial cancer (Zeng et al. 2020), indicating that genomics profile that encoded TAMs states may be an effective biomarker for patient selection in ICB therapy. In a phase IV clinical trial study of NSCLC, DNA methylation microarray profiling revealed that nonresponders to ICBs were characterized as those patients who exhibit high enrichment of TAMs, neutrophils, and cancer-associated fibroblasts in TME, highlighting profiles of TAMs as complementary biomarkers in addition to PD-L1 staining score and tumor mutation burden (Duruisseaux et al. 2018). Transcriptomic profiling of monocytes isolated from breast cancer patients and healthy controls identified TAM signatures, which are highly associated with aggressive subtypes and poor survival, and elucidated that SIGLEC1 and CCL8 were key regulators to govern the interactions between TAMs and cancer cells in this scenario, suggesting novel biomarkers and potential therapeutic targets for breast cancer treatment (Cassetta et al. 2019).

Moreover, TAM transcriptomic profiling at single-cell level depicts more precise TAMs landscape and reveals more clues for personalized cancer therapy. Recent years, single cell sequencing technology is reforming our understanding of tumor heterogeneity. Transcriptomic landscape of tumor at single cell resolution is opening new revenue for personalized cancer treatment by providing individualized immune interaction profiles as novel evidence for tailored therapeutics design (Weissleder and Pittet 2020; Qian et al. 2020). Single-cell RNA-seq profiling of patients with lung cancer demonstrated that TAMs can be functionally categorized into up to 10 subtypes in lung cancer and especially, certain subtypes express both M1 and M2 markers (Weissleder and Pittet 2020), recapitulating that TAMs employ a spectrum of functional states rather than the conventional notion that TAMs simply consist of M1 and M2 subtypes to cope with the complexity of TME.

Interactions of TAMs and cancer therapeutics result in multifaceted effects on clinical outcomes of cancer therapies via various elucidated and unexplored molecular mechanisms. This provides promising opportunities to vanish immunosuppressive TME, sensitize resistant cancer cells to cancer treatment modalities, stratify patients to match optimal treatment strategies, and finally benefit cancer patients in clinical practice, by harnessing those interactions either via depleting tumor-promoting TAMs or re-directing them to their tumor-inhibiting phenotype.

5 Target TAMs for Cancer Therapy

TAMs play essential roles during tumor proliferation, angiogenesis, invasion, metastasis, and immunosuppression and interfere with most standard cancer therapy (Pathria et al. 2019). Elevated number of circulating monocytes from blood and massive macrophage infiltration into tumor tissues, especially accumulation of anti-inflammatory TAMs, is tightly associated with worse clinical outcome and resistance to therapy in various cancer types (La Fleur et al. 2020). Thus, manipulating TAMs in conjunction with chemotherapy, radiotherapy, or immunotherapy might improve standard cancer treatment response (Cheng et al. 2020). Moreover, the potential role of macrophages during cancer development has driven novel antitumor therapy development through targeting macrophages (Mantovani et al. 2017; Sawa-Wejksza and Kandefer-Szerszen 2018). Currently, several pharmacological strategies have been proposed either in preclinical study or in clinical trials through inhibiting TAMs recruitment, depleting TAMs, and reprogramming M2 TAMs to M1 TAMs in laboratory tumor models or clinical trials (Fig. 2) (Anfray et al. 2019). Besides, due to their intrinsic homing property, M1 macrophages or their derived components could also be used as drug delivery systems to actively carry payloads to the tumor sites (Guerra et al. 2017). Here the progress regarding TAMs as drug delivery systems, which mainly focusing on M1 macrophages, their derived exosomes, and membrane-coated nanoparticles will be summarized (Miller et al. 2015).

5.1 Target TAMs Recruitment and Localization for Cancer Therapy

Numerous evidences reveal that recruitment and localization of macrophage in tumors is driven by continuous accumulating monocytes from circulation due to some tumor-derived factors (Qian and Pollard 2010). These factors including colony-stimulating factor-1 (CSF-1), C-C motif chemokine ligand 2 (CCL-2), C-X-C motif chemokine ligand 12 (CXCL-12), and vascular endothelial growth factor (VEGF) have mediated interaction between monocytes and tumors cells; thus, suppressing the monocyte-chemotactic chemokines, cytokines, and their receptors could block the recruitment and localization of TAMs (Argyle and Kitamura 2018; Owen and Mohamadzadeh 2013; Zhang et al. 2020a).

5.1.1 Targeting CSF-1/CSF-1R Signaling

CSF-1/CSF-1 receptor (CSF-1R) signaling is one critical pathway related to TAMs recruitment, proliferation, differentiation, and survival and is also essential for stimulating chemotactic activity of monocytes and macrophages including the transition from M1 TAMs into M2 TAMs (Laoui et al. 2014; Dwyer et al. 2017). Studies have shown that CSF-1 enhances progression of mammary tumors to malignancy, high expression of CSF-1 in tumor tissues, or high expression of CSF-1R in TAMs which are associated with worse prognosis of tumor (Lin et al. 2001; Zhu et al. 2008; Koh et al. 2014). The absence of CSF-1R caused almost all macrophages to be dramatically depleted in mice (Erblich et al. 2011). Besides, G-CSF regulated macrophage phenotype, suppression of G-CSF enhanced circulating monocytes, and TAMs mutation and remarkably limited lung metastasis of triple-negative breast cancer, but if G-CSF was kept in high level, anti-CSF-1R therapy could strength anti-inflammatory TAMs and promote lung metastasis (Hollmen et al. 2016). Inhibition of M-CSF by both antibody and chemical inhibitor significantly suppressed tumor angiogenesis and lymphangiogenesis in mouse osteosarcoma model (Kubota et al. 2009). Antibody against CSF1-R depleted the resident subset of monocytes and TAMs without inhibiting inflammation, and macrophage blockade using a CSF-1R inhibitor also resulted in reduced infiltration of protumorigenic (M2) macrophages (MacDonald et al. 2010). Moreover, the mAb inhibiting the CSF-1R (RG7155) was proved both in vitro and in vivo to decrease F4/80+ TAMs accompanied by an increase in the CD8+/CD4+ T-cell ratio (Ries et al. 2014). Administration of RG7155 to patients led to striking reductions of CSF-1R + CD163+ macrophages in tumor tissues. Some clinical trials of RG7155 are underway in solid tumor treatment as monotherapy or combined with other therapies. Besides, targeting TAMs by CSF-1R blockade in breast cancer mice model could stimulate intratumoral type I interferon signaling, thus enhance the anticancer efficacy of platinum-based chemotherapeutics

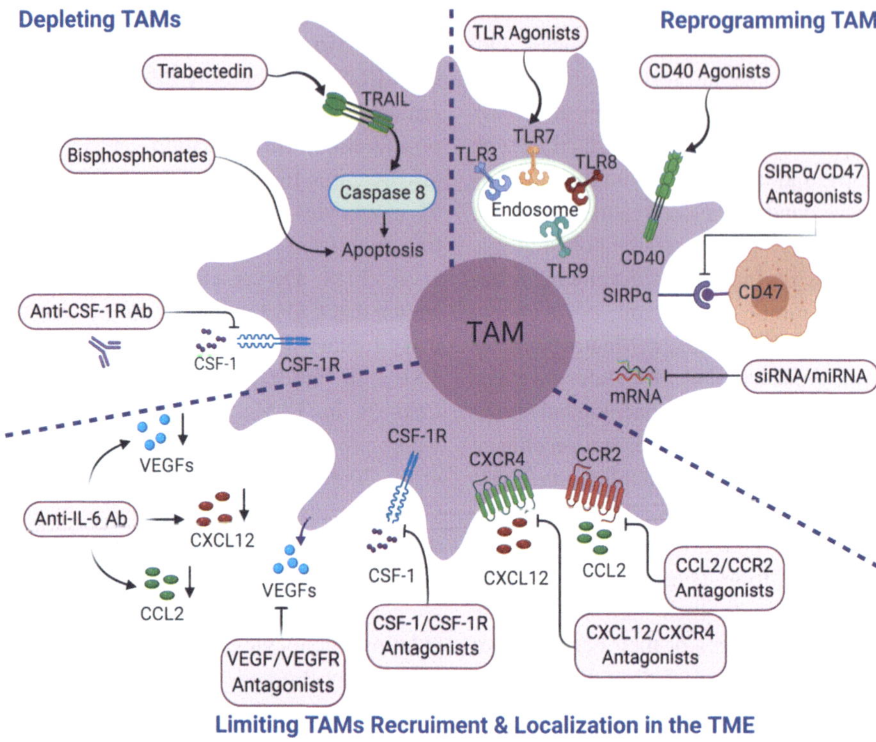

Fig. 2 Targeting TAMs for cancer therapy. Currently, three strategies, including 1) targeting TAMs recruitment and localization, 2) targeting TAMs inhibition, and 3) targeting TAMs reprogram, have been proposed in different stage of studies. For targeting TAMs recruitment and localization, four signaling axis, that is, CSF-1/CSF-1R signaling axis, CCL-2/CCR-2 axis, CXCL-12/CXCR-4 axis, and VEGF/VEGFR signaling axis, were highlighted. For targeting TAMs inhibition, two chemical agents Bisphosphonates and Trabectedin and other targets for TAMs-specific surface markers were discussed. Targeting TLR, SIRPα/CD47 axis, CD40, and other strategies were involved in TAMs reprogramming

(Salvagno et al. 2019). Local targeting of lung TAMs with pulmonary delivery of a CSF-1R inhibitor PLX-3397 could treat lung metastases of breast tumor by decreasing M2 macrophages and increasing M1 macrophages in the tumor microenvironment (Alhudaithi et al. 2020). Researchers also reported a potent, highly selective, and orally bioavailable CSF-1R inhibitor, IACS-9439, which could lead to a dose-dependent reduction in macrophages and lead to tumor growth inhibition in MC38 and PANC02 syngeneic tumor models (Czako et al. 2020). In some other studies, CSF-1R inhibitors could reduce TAMs number in tumor tissue, inhibit tumor growth, and obviously decrease tumor angiogenesis and metastasis. JNJ-28312141, an FMS-related receptor tyrosine kinase-3 (FLT3) inhibitor, suppressed human nonsmall cell lung carcinoma growth and reduced tumor vasculature in a dose-dependent manner; these effects were highly correlated with marked reductions in F4/80+ TAMs. Another CSF-1R inhibitor, BLZ945, showed potentials as a potent antitumor drug in breast cancer, glioblastoma, and cervical carcinoma in preclinical studies. It was confirmed existing potent ability converting tumorigenic M2 macrophages into the antitumor M1 macrophages. Numerous studies revealed that targeting of TAMs through inhibiting CSF-1R could suppress tumor growth and metastasis, and currently massive clinical trials are launched to evaluate bio-safety and bio-activity of antibodies/small molecules targeting CSF-1/CSF-1R signaling alone or combined with other therapies in different types of tumors (Table 1). While researchers also found that anti-CSF-1R and anti-CSF-1 antibodies or CSF-1R small-molecule inhibitors treatment increased breast cancer metastasis, thus the CSF-1/CSF-1R signaling serving as the therapeutic targets for breast cancer still needs further investigation (Swierczak et al. 2014).

5.1.2 Targeting CCL-2/CCR-2 Axis

CCL-2, also widely known as monocyte chemoattractant protein-1(MCP-1), is a C-C motif chemokine overexpressed in solid tumors and tightly associated with increased tumor adhesion, migration, invasion, progression, and angiogenesis (Hao et al. 2020). Besides, CCL-2 could also recruit monocytes into an immunosuppressive tumor microenvironment: Tumor cells recruited inflammatory monocytes to tumors sites through CCL-2/CCR-2 axis, and the recruited monocytes were polarized to TAMs to promote tumor survival and facilitate tumor metastasis (Qian et al. 2011; McClellan et al. 2012). As its main receptor, the upregulation of CCR-2 was also reported correlated with cancer metastasis and relapse (Nagarsheth et al. 2017) and recruitment of CCR-2 positive TAMs to sites of liver metastasis conferred a poor prognosis in human colorectal cancer (Grossman et al. 2018). Recent work suggested that during esophageal carcinogenesis, CCL-2/CCR-2 axis-mediated TAMs recruitment could induce immune evasion through PD-1 signaling (Yang et al. 2020). Blocking of the CCL-2/CCR-2 axis decreased macrophage infiltration and reduced tumor growth (Lim et al. 2016). Besides, disruption of CCL-2/CCR-2 axis could obviously reduce TAMs number in tumors, thereby suppressing tumor growth and dissemination in some certain cancer types (Fujimoto et al. 2009).

Anti-CCL-2 antibody or CCL-2 inhibitors were proved to block glioma, colon, prostate, and melanoma cancers proliferation in animal models (Loberg et al. 2007). Li et al. reported that blockade of CCL-2/CCR-2 axis by using CCR-2 antagonist or CCR-2 knockout could inhibit liver tumor malignant growth and metastasis, reduce postsurgical recurrence, and enhance survival through suppressing inflammatory monocytes recruitment, infiltration, and TAMs M2 polarization (Li et al. 2017). In a preclinical study, a CCL-2 neutralizing antibody led to a remarkable tumor growth suppression in hepatocellular cancer (Teng et al. 2017). Apart from small molecule inhibitors and antibodies, RNA aptamer CCL-2 inhibitor could also reduce M1 TAMs liver infiltration and pathogenic angiogenesis (Bartneck et al. 2019). In clinical trial, Carlumab (CNTO888), an anti-CCL-2 human monoclonal antibody, is being used either alone or in combination with standard chemotherapies for solid tumors patients' treatment (Sandhu et al. 2013). The other approach to decrease the action of CCL-2 is to block CCR-2. Pharmacological CCR-2 inhibitors and humanized antibody that recognizes CCR-2 were examined. BMS-813160, a potent and selective CCR-2/CCR-5 antagonist, alone or combined with chemotherapy or Nivolumab in patients with advanced solid tumors such as colorectal cancer and pancreatic cancer, is launched in clinical trials (NCT03184870). A CCR-2 inhibitor RS 504393 could inhibit xenograft prostate tumor growth in SCID mice fed with high fat diet (Hao et al. 2020). An orally active, high-affinity CCR-2 antagonist CCX872 could also enhance anti-PD-1 effect to slow progression of gliomas and improve survival and is currently in the clinical trial for patients with advanced and metastatic pancreatic cancer (Flores-Toro et al. 2020). PF-04136309, small-molecule inhibitors of CCR-2 showed efficacy in preclinical study: In an orthotopic model of murine pancreatic cancer, CCR-2 blockade by PF-04136309 depleted inflammatory monocytes and macrophages from the primary tumor and premetastatic liver resulting in tumor growth inhibition, antitumor immunity enhancement, and metastasis reduction (Sanford et al. 2013); Currently, PF-04136309 has moved forward to clinical trials, and the inhibitor combined with chemotherapeutic regimen FOLRIRINOX presented favorable response in pancreatic cancer patients during phase 1b clinical trial (Nywening et al. 2016), while another clinical trial in PF-04136309 combined with nab-paclitaxel and gemcitabine for metastatic pancreatic patients was terminated in May

Table 1 List of clinical trials targeting CSF-1/CSF-1R signaling

Identifier	Agent	Description	Phase	Condition
NCT02371369	PLX-3397	Small molecule	Phase III	Pigmented villonodular synovitis or giant cell tumor of the tendon sheath
NCT02071940	PLX-3397	Small molecule	Phase II	Melanoma
NCT02390752	PLX-3397	Small molecule	Phase I/II	Refractory leukemias and refractory solid tumors
NCT01349049	PLX-3397	Small molecule	Phase I/II	Acute myeloid leukemia
NCT02452424	PLX-3397	Small molecule	Phase I/II	Advanced melanoma and other solid tumors
NCT01004861	PLX-3397	Small molecule	Phase I	Solid tumors
NCT01525602	PLX-3397	Small molecule	Phase I	Advanced solid tumors
NCT01042379	PLX-3397	Small molecule	Phase I	Breast cancer
NCT02777710	PLX-3397	Small molecule	Phase I/II	Glioblastoma and gliosarcoma
NCT02584647	PLX-3397	Small molecule	Phase I/II	Unresectable sarcoma and malignant peripheral nerve sheath tumors
NCT01596751	PLX-3397	Small molecule	Phase Ib/II	Metastatic breast cancer
NCT02472275	PLX-3397	Small molecule	Phase I	Intermediate- or high-risk prostate cancer
NCT01790503	PLX-3397	Small molecule	Phase Ib/II	Newly diagnosed glioblastoma
NCT01804530	PLX-7486	Small molecule	Phase I	Advanced solid tumors
NCT02880371	ARRY-382	Small molecule	Phase Ib/II	Advanced solid tumors
NCT02829723	BLZ-945	Small molecule	Phase I/II	Advanced solid tumors
NCT03069469	DCC-3014	Small molecule	Phase I	Hematological tumors and solid tumors
NCT03238027	SNDX-6352	Small molecule	Phase I	Solid tumors
NCT03502330	Cabiralizumab	mAb	Phase II	Advanced pancreatic cancer
NCT03697564	Cabiralizumab	mAb	Phase II	Stage IC pancreatic cancer
NCT03768531	Cabiralizumab	mAb	Phase II	Resectable biliary tract cancer
NCT02526017	Cabiralizumab	mAb	Phase I	Selected advanced tumors
NCT03502330	Cabiralizumab	mAb	Phase I	Advanced melanoma, nonsmall cell lung cancer, and renal cell carcinoma
NCT02471716	Cabiralizumab	mAb	Phase I/II	Tenosynovial giant cell tumor
NCT02265536	LY-3022855	mAb	Phase I	Breast or prostate cancer
NCT01346358	IMC-CS4 (LY-3022855)	mAb	Phase I	Advanced solid tumors
NCT03101254	LY-3022855	mAb	Phase I/II	Melanoma
NCT03153410	LY-3022855	mAb	Phase I	Pancreatic cancer
NCT02923739	Emactuzumab	mAb	Phase II	Platinum-resistant ovarian cancer
NCT02323191	Emactuzumab	mAb	Phase I	Advanced solid tumors

Identifier	Agent	Description	Phase	Condition
NCT01494688	Emactuzumab	mAb	Phase I	Advanced solid tumors
NCT01444404	AMG-820	mAb	Phase I	Advanced solid tumors
NCT02713529	AMG-820	mAb	Phase Ib/II	Pancreatic cancer, colorectal cancer, and nonsmall cell lung cancer
NCT02807844	Lacnotuzumab	mAb	Phase Ib/II	Advanced malignancies
NCT02948101	PD-0360324	mAb	Phase II	Platinum-resistant epithelial ovarian cancer

2017 (Jahchan et al. 2019). Plozalizumab (MLN1202), a humanized mAb to CCR-2, was conducted in phase 2 clinical trial for bone metastasis of unspecified tumors, while Plozalizumab combined with nivolumab (immune-checkpoint inhibitor) for patients with advanced melanoma in phase 1 clinical trial was terminated in May 2018 because of serious adverse events (Vela et al. 2015; Yumimoto et al. 2019). Thus, it should be noted that durable CCL2/CCR2 axis inhibition might lead to the compensation of other chemokine-dependent pathways, which facilitates the recruitment of TAMs in the tumor microenvironment.

5.1.3 Targeting CXCL-12/CXCR-4 Axis

CXCL-12, also known as stromal cell-derived factor-1 (SDF-1), is widely expressed in many different types of tissues. CXCL-12 is the ligand for the chemokine receptor CXCR-4 (Liekens et al. 2010). Recent work also demonstrated that CXCL-12 could bind to another seven-transmembrane span receptor CXCR-7 with high affinity (Sun et al. 2010). Both CXCR-4 and CXCR-7 played critical roles on tumor growth and metastasis, and a lot researches have suggested that the CXCR-4 and its ligand CXCL-12 are potential targets for cancer therapy (Zhou et al. 2019). CXCL-12/CXCR-4 axis could directly or indirectly activate massive signaling pathways including PI3 kinase, stress-activated protein kinase, c-Jun N-terminal kinase, and mitogen-activated protein kinase (MAPK) to promote tumor growth, adhesion, migration, and survival (Dewan et al. 2006). Because of expressing CXCR-4, macrophages could migrate along CXCL-12 gradient. It was shown that CXCL-12 production by breast cancer cells results in increased macrophage number in breast tumor (Bonapace et al. 2014). Similar responses were also showed in monocytes: CXCL-12 released by multiple myeloma cells could trigger monocyte migration and anti-CXCR-4 antibodies could obviously suppress monocyte recruitment. BMS-936564/MDX-1338, a human mAb recognizing human CXCR-4, could significantly induce apoptosis in vitro and inhibit tumor growth in several established experimental tumor models (Kuhne et al. 2013). Plerixafor (AMD3100), CXCR-4 inhibitor, could suppress the postsepsis-induced melanoma progression, TAMs accumulation, and TAMs in situ proliferation in mice model. Plerixafor could also selectively reduce the number of M2 TAMs and suppress tumor re-vascularization and re-growth (Zhou et al. 2018). More specifically, cancer-associated fibroblast-derived CXCL12 attracted CXCR-4 expressing M2-like macrophages to infiltrate into tumors, which promoted cancer stem cells-like transition, proliferation, and migration of cancer cells in oral squamous cell carcinoma (Li et al. 2019b). Progranulin presents carcinogenic roles in breast cancer. Recent studies revealed that Progranulin KO TAMs inhibited invasion, migration, and EMT of breast cancer cells through their exosomes. miR-5100 in Progranulin KO TAMs-derived exosomes was upregulated, which might contribute to CXCL-12 regulation, thereby suppressing CXCL-12/CXCR-4 axis and finally inhibiting the invasion, migration, and EMT of breast cancer cells (Yue et al. 2021). Similar results were also shown in liver metastasis of colorectal cancer: Several miRNAs upregulated in colorectal cancer cells by suppressing CXCL-12/CXCR-4 axis activation could be transferred to TAMs via exosomes and mediate crosstalk between CXCR-4 overexpressing cancer cells and TAMs (Wang et al. 2020). Apart from antibodies, small-molecules, and microRNAs, small modified peptides designed as CXCR-4 antagonists were also widely tested: A novel CXCR-4 antagonist BKT140 showed potent therapeutic efficacy against human nonsmall cell lung cancer in vitro and in vivo. BKT140 could also augment therapeutic effects of chemotherapy and radiotherapy (Fahham et al. 2012). Similarly, a peptidic CXCR-4 inhibitor TF14016 also suppressed metastases of small cell lung cancer cells in mice (Otani et al. 2012). Similarly, as CXCR-4 antagonists, T140 analogs were also known as antimetastatic agents for breast cancer therapy (Tamamura et al. 2003). Targeting CXCL-12/CXCR-4 axis could also be in combination with standard cancer therapies: CXCL-12 upregulation in hepatocellular carcinoma models triggered hypoxia, immunosuppressive cells recruitment,

T-regulatory cells, and M2 TAMs accumulation after sorafenib treatment. CXCR-4 inhibitor AMD3100 and Sorafenib enhanced PD-1 blockade-induced hepatocellular carcinoma suppression through inhibiting the polarization toward an immunosuppressive microenvironment (Chen et al. 2015).

5.1.4 Targeting VEGF/VEGFR Signaling

Tumor angiogenesis is a critical process for necessary nutrients and oxygen supply to support quickly tumor tissues growing (Nishida et al. 2006). Among them, the family of vascular endothelial growth factors (VEGFs) plays critical roles as regulators of angiogenesis. VEGF/VEGFR signaling pathway is upregulated in numerous cancer types and contributes to out-of-controlled angiogenesis and metastatic spreading (Ceci et al. 2020). VEGFs promote tumor angiogenesis via regulating endothelial cells proliferation and survival (Lee et al. 2015) and might promote tumor cell proliferation through activating VEGFR1 signaling (Bhattacharya et al. 2016). Tumor cells could also recruit and reprogram macrophages derived from circulating monocytes; these recruited and reprogrammed macrophages could also work as a main source of angiogenic factors (Cassetta et al. 2019). For example, angiopoietin-2 produced by tumor endothelium could enhance TIE2 receptor expressing monocytes recruiting in tumor cells. TAMs could also release angiogenic factors such as VEGFA as the response of hypoxia in tumor avascular areas; the released VEGFA could further stimulate macrophages and endothelial cells conversely (Mazzone and Bergers 2019). TAMs were able to release cytokines that indirectly contribute to tumor angiogenesis by the induction of a pro-angiogenic program in tumor cells. Tumor cells and recruited TAMs cooperated in the TME to amplify the production of pro-angiogenic factors resulting in an angiogenic switch. Apart from pro-angiogenesis, TAMs also tightly involved in lymph angiogenesis under inflammatory conditions or carcinogenetic conditions (Yang et al. 2018).

VEGFs can recruit macrophages to the tumor and promote TAMs development (Lapeyre-Prost et al. 2017; Linde et al. 2012). Thus, targeting VEGF/VEGFR has potential suppressing macrophage recruit and localize to the tumor (Yang et al. 2018). The potent proangiogenic activities of TAMs in the TME and the existing of angiogenesis-relevant macrophage subpopulation Tie2 expressing monocytes (TEMs) demonstrate novel antitumor strategies through targeting angiogenesis (Johansson-Percival et al. 2018; Chen et al. 2016). Dampening TME recruitment might be an effective method to block tumor angiogenesis. Inhibiting Ang2-Tie2 crosstalk by using Ang2-CovX-Bodies prevented TEMs infiltration (Huang et al. 2011). Anti-ANG2 monoclonal antibody (mAb) could also disrupt connection between TEMs and endothelial cells, leading to suppressed angiogenesis in pancreatic insulinomas and mammary carcinomas (Mazzieri et al. 2011). As ANG2 overexpression could induce anti-VEGF therapy resistance, antibodies targeting both ANG2 and VEGF could abrogate angiogenesis and promote antitumor efficacy (Kloepper et al. 2016; Scheuer et al. 2016). YKL-40, pro-angiogenic factor, could mediate angiogenesis both in vitro and in mice tumor models. Neutralizing mAbs targeting YKL-40 could suppress angiogenesis and progression of tumor (Faibish et al. 2011).

Besides, macrophages could mediate anti-VEGF therapy resistance through changing VEGFR expression level (Dalton et al. 2017), and TAMs might play critical roles during these processes (Itatani et al. 2018): TAMs have been shown to help escape from antiangiogenic therapy of glioblastoma both in preclinical models and in clinic and could work as a potential biomarker and therapeutic target (Lu-Emerson et al. 2013). Similarly, macrophages could be activated and recruited to the TME and further contribute to anti-VEGF resistance in regular ovarian cancer mice model, but there is no resistance anti-VEGF antibody when macrophages were depleted in a macrophage-deficient mouse model, while this resistance could still be triggered by macrophages injection (Dalton et al. 2017). Targeting TAMs might restore or enhance antiangiogenic therapeutic effects. As one of bisphosphonate drugs, Zoledronic was widely used for treating osteoporosis and bone metastases in clinic, while Zoledronic also showed ability to deplete macrophages. Researchers have investigated the role of macrophages depletion in anti-VEGF antibody therapy resistance in mice bearing ovarian tumor models: Macrophages inhibited by Zoledronic could overcome anti-VEGF antibody therapy resistance, enhance tumor growth suppression, and obviously extend survival when compared with vehicle and only anti-VEGF therapy groups (Dalton et al. 2017). Recent study also presented that expression of CCL-2 could promote antiangiogenic therapy resistance; mNOX-E36, a CCL-2 inhibitor inhibiting TAMs recruitment and angiogenesis, could enhance the efficacy of bevacizumab in glioma models (Cho et al. 2019). Similar strategies were also used by targeting CXCL-12/CXCR-4 axis: NOX-A12, a novel CXCL-12 inhibitor, could reverse TAMs recruitment and potentiate the antitumor effects of anti-VEGF therapy (Deng et al. 2017). Combination of CXCR-4 inhibitors and anti-VEGF therapy was also reported to slow GBM xenografts progression both in preclinic and in clinical trial: CXCR-4 antagonist PRX177561 increased the antitumor effects of bevacizumab in preclinical models of human glioblastoma (Gravina et al. 2017); another CXCR-4 antagonist POL5551 combined with VEGF inhibition could also improve survival in an intracranial mouse model of glioblastoma (Barone et al. 2014). AMD3100 against CXCR-4 was applied with a combination of bevacizumab in patients with recurrent high-grade glioma in clinical trial. Thus, suppressing VEGF/VEGFR signaling combined with other inhibiting TAMs strategies will bring more benefits for cancer therapies.

5.2 Target TAMs Inhibition for Cancer Therapy

There are two approaches to decrease the population of TAMs for cancer therapy: One is to reduce the monocytes number in the circulating system and the other is to suppress the population of macrophages already accumulated in the tumor tissues (Sawa-Wejksza and Kandefer-Szerszen 2018).

Bisphosphonates, including alendronate, ibandronate, risedronate, and zoledronic acid, are primary agents in the current pharmacological arsenal to prevent the loss of bone density and treat osteoporosis and related diseases (Drake et al. 2008). The use of bisphosphonates has been related to inhibit proliferation, migration, and invasion of macrophages, which share the same lineage with osteoclasts, causing apoptosis (Rogers and Holen 2011). As chemotherapeutics, no matter traditional free bisphosphonates or nanoparticles-captured bisphosphonates, both showed cytotoxicity to monocytes/macrophages and have been used to decrease their population in tumors (Farrell et al. 2018). In preclinical models, bisphosphonates could effectively inhibit tumor growth through suppressing TAMs infiltration in breast tumors (Holen and Coleman 2010). Some research also presented that Trabectedin, a registered antineoplastic agent that suppresses cancer cell growth, could also trigger specific death of circulating monocytes/macrophages through TNF-related apoptosis-inducing ligand (TRAIL)-dependent apoptosis, as circulating monocytes/macrophages expressed functional TRAIL receptors 1 and 2, which are susceptible to the cytotoxicity of trabectedin (Germano et al. 2013). Trabectedin-treated pancreatic ductal adenocarcinoma showed reduced TAMs, decreased angiogenesis, activated antitumoral CTLs, with promising clinical outcomes (Borgoni et al. 2018). Similarly, in cutaneous melanoma mice model, trabectedin showed significant activity of reducing TAMs and tumor blood vessel density and inhibited lung metastasis of melanoma (Carminati et al. 2019). Besides, a combination of checkpoint blockade and angiogenesis inhibitors could be an effective strategy to promote the curative effect of Trabectedin (Seliger 2019). Trabectedin also revealed a strategy of immunomodulation in chronic lymphocytic leukemia: Trabectedin depleted myeloid-derived suppressor cells and TAMs and increased memory T-cells in xenograft and immunocompetent chronic lymphocytic leukemia mouse models and could also suppress PD-1/PD-L1 axis through targeting PD-L1-positive chronic lymphocytic leukemia cells, monocytes/macrophages, and T-cells (Banerjee et al. 2019). Similar results were also shown in skeletal prostate cancer models: Trabectedin reduced prostate cancer bone resident tumor size, which was in associated with trabectedin-reduced M2 macrophages (Jones et al. 2019). Depletion of TAMs by trabectedin could also switch the epigenetic profile of pancreatic cancer infiltrating T-cells and restore their antitumor phenotype (Borgoni et al. 2018).

There also have other strategies to depleting TAMs through targeting their specific surface markers. Sialic acid receptors in the Siglec family are highly expressed on the surface of TAMs and most have immunosuppressive effects. Targeted delivery of zoledronic acid through the sialic acid-Siglec axis could kill and reversal M2-like TAMs and inhibit S180 tumor growth (Tang et al. 2020). Similarly, Folate receptor beta (FR beta) was found expressed on macrophages in human glioma and rat C6 glioma. A recombinant immunotoxin consisting of anti-FR beta mAbs and Pseudomonas exotoxin A could significantly reduce TAMs numbers and suppress tumor growth (Nagai et al. 2009). Apart from sialic acid receptors and FR beta, activated macrophages also expressed the pattern recognition receptor scavenger receptor A (SR-A), a small peptide SR-A ligand could compete with physiological SR-A ligand in vitro, and deficiency of SR-A suppressed progression and metastasis of ovarian and pancreatic cancer in vivo (Neyen et al. 2013). M2 macrophage-targeting peptide (M2pep), which could bind to murine M2 macrophages and M2-like TAMs, was fused with pro-apoptotic peptide KLA. This fusion peptide could also reduce TAM population in vivo but need high concentrations (Ngambenjawong et al. 2016).

Although targeting CSF-1/CSF-1R axis or CXCL-12/CXCR-4 axis could suppress TAMs through blocking the recruitment and localization of TAM, researchers also find that in some other studies, the use of these axis inhibitors decreased TAMs number in tumor tissue, inhibited tumor proliferation, and significantly decreased angiogenesis and metastasis (Anfray et al. 2019). Besides, the use of the anti-IL-6 antibody had a strong anticancer effect by decreasing CCL-2, VEGF, and CXCL-12. There was a significant decline in CCL-2, CXCL-12, and VEGF as well as the number of TAMs in tumor tissue in patients treated with anti-IL-6 antibody for 6 months (Chen and Chen 2015).

5.3 Reprogramming TAMs for Cancer Therapy

As the most abundant tumor-infiltrating immune cells in TME, TAMs play a crucial role in tumor development, invasion, and metastasis (Pathria et al. 2019). Emerging evidences demonstrate that the plasticity and diversity of TAMs allow them to be classified along the M1-M2 polarization axis (Genard et al. 2017). M2 macrophages, also known as alternatively activated macrophages, can be induced by the TH_2 cytokines such as IL-4, IL-10, and IL-13 and promote tumor angiogenesis, immunosuppression, and drug resistance (Gordon and Taylor 2005). On the contrary, the TH_1 cytokines such as IL12, IL-18, or activated TLRs stimulate macrophages to M1 macrophages, also known as classically activated macrophages (Biswas and Mantovani 2010). M1 macrophages present the ability to eliminate tumor cells by releasing reactive oxygen/nitrogen intermediates, together with pro-inflammatory cytokines such as TNF, IFN-γ. However, in TME of solid tumors, most TAMs are devoid of cytotoxic activity and closely related to the M2 phenotype with poor clinical outcome of patients. Thus, reprogramming the TAMs through chemotherapy, radiotherapy, and immunotherapy could be a promising strategy to eradicate tumors.

5.3.1 TLR Agonists to Reprogram TAMs

Currently, several options including TLRs agonists and compounds or mAb targeting inhibitory proteins of M1 phenotype are used to select M1 phenotype or reprogram TAMs from M2 to M1 phenotype (Mantovani et al. 2017). As pattern-recognition receptors in innate immunity, TLRs stimulate and activate M1-phenotype polarization via their engagement with the ligands. Thus, various agonists targeting TLRs, such as TLR3, TLR7, and TLR9, have been developed to evaluate their capacity to reprogram TAMs into antitumor effectors (Anfray et al. 2019).

Stimulation of TLR3 by poly I:C upregulated costimulatory molecules CD40, CD80, CD86, and M1-specific markers MHC-II on M2 macrophage. Simultaneously, the expression of pro-inflammatory cytokines and M2-specific markers Tim-3, CD206 was reduced (Shime et al. 2012). In MC38 tumor model, poly I:C reverted the M2 phenotype to M1 macrophages and eradicated the tumor in an IFN-α/β-dependent signaling pathway. Recently, Liu et al. developed novel polypeptide micelles to target TAM. Poly I:C-loaded polypeptide micelles showed property for targeting TAM, efficiently reeducated TAMs into M1 macrophages, substantially activated T lymphocytes and NK cells in melanoma models, indicating repolarizing TAMs into M1 phenotype in situ for effective immunotherapy of cancer (Liu et al. 2018). Ferumoxytol, a clinically approved nanoparticle, in combination with TLR3 agonist poly I:C induced pro-inflammatory macrophage polarization for regression of primary and metastatic murine melanoma (Zhao et al. 2018a; Zanganeh et al. 2016). In 2018, a phase II trial has been performed to evaluate the safety and therapeutic effect of combination of Poly I:C and immune checkpoint blockade in unresectable hepatocellular carcinoma, while its analog, poly-ICLC, has been tested as a tumor vaccine to boost antitumor immunity (Anfray et al. 2019).

TLR7 agonist was demonstrated to induce the nuclear translocation of NF-κB in macrophages with the production of pro-inflammatory cytokines (De Meyer et al. 2012). Topical administration of TLR7 agonist imiquimod with radiotherapy has been approved to synergistically inhibit tumor growth and cyclophosphamide further increased the therapeutic effect and induced immunologic memory in cutaneous breast cancer metastases. Currently, imiquimod is the only TLR agonist approved by Food and Drug Administration for topical injection in intraepidermal carcinoma and basal cell carcinoma.

In the past years, resiquimod has attracted more attention for its role to reeducate TAMs into M1 macrophages (Thauvin et al. 2019). Resiquimod, also known as R848, is a dual agonist of the toll-like receptors TLR7/8 (Chi et al. 2017). Similar to imiquimod, R488 showed more powerful as a TLR agonist, with more ability to elicit antitumor immune immunity. Despite several studies have showed promising results based on the topical injection in melanoma patients, there are no ongoing clinical trials due to that systemic administration of R848 is burdened with toxicities: anemia, inflammation, and flu-like symptoms (Perkins et al. 2012; Hasham et al. 2017). To overcome this issue, various methods have been introduced. MEDI9197, a formulation of R848 was designed with a lipid tail to increase lipid solubility and be retained at the injection sites, thus limiting systemic toxicity. As an alternative, R848 was covalently linked with vitamin E and modified by hyaluronic acid into a predrug nanopreparation (CDNP-R848), providing a sustained release of R848 by subcutaneous injection (Rodell et al. 2018). The predrug nanopreparation successfully delivered R848 to TAMs in vivo and altered the functional orientation of the TAMs toward M1 macrophage, leading to inhibited tumor growth and protecting the mice against tumor re-challenge. When CDNP-R848 was combined with anti-PD-1 blockade, improved immunotherapy response rates were observed, including an anti-PD-1 therapy-resistant tumor model. In addition, Rodel et al. identified R848 as a potent driver of M1 macrophage polarization and developed R848-loaded β-cyclodextrin nanoparticles to efficiently deliver R848 to TAM in MC38 colorectal tumors (Rodell et al. 2018). These strategies were able to trigger the IL-12 production of TAMs in TME, demonstrating that engineered R848-nanoparticle combinations could efficiently modulate TAMs for cancer immunotherapy.

As to the TLR9, preclinical data of the agonists singly or in combination with other agents showed efficacy, which led to clinical trials in patients with advanced malignancy (Karapetyan et al. 2020). In the preclinical models, intratumoral, subcutaneous, and intravenous routes of injection have been performed with potent antitumor responses in treated and metastatic sites with the repolarization of TAMs (Hofmann et al. 2008). In the clinic, despite TLR9 agonist monotherapy or combined with chemotherapy and targeted therapy showed no obvious efficacy in advanced solid tumors, especially with an intravenous injection, the greatest excitement has been reserved for TLR9 agonists in combination with immune checkpoint inhibitors. Currently, a clinical trial has been registered to evaluate the antitumor efficacy of TLR9 agonist CpG in combination with nivolumab in metastatic pancreatic cancer patients (NCT04612530).

Despite these in vitro data are promising, there is limited evidence for TLRs agonists as adjuvants in the field of cancer vaccination. In this regard, nanoparticles are considered as a potential strategy to load TLRs agonists to the target tumor sites. For example, nanoparticles loaded with poly I:C or R848 presented the power to elicit potent innate immune responses in lymph nodes, resulting in therapeutic effect without systemic release of inflammatory cytokines (Bocanegra Gondan et al. 2018; Da Silva et al. 2019).

5.3.2 Reprogramming TAMs by mAb and Fusion Protein

In the TME, reprogramming M2 macrophages into M1 proinflammatory phenotype is a potential antitumor immunotherapy which is involved in increased activity of macrophage phagocytosis. The macrophage phagocytosis is balanced by activating and inhibitory signals. CD47 is an immune checkpoint that confers a "do-not-eat-me" signal to host immune surveillance via binding to its ligand, SIRPα on the surface of phagocytic cells, such as macrophages and dendritic cells (Zhang et al. 2017). SIRPα binds to CD47 and transmits intracellular signals through its cytoplasmic domain. The cytoplasmic tail of SIRPα contains four immunoreceptor tyrosine-based inhibition motifs (ITIMs) and becomes phosphorylated after binding of CD47, leading to prevention of myosin-IIA accumulation at the phagocytic synapse and inhibition of phagocytosis (Tsai and Discher 2008; Barclay and Brown 2006; van Beek et al. 2005). The "do-not-eat-me" signals play key roles for tissue homeostasis. However, CD47 overexpression has been adopted by various cancer cells to escape the innate immune surveillance. In this sense, pharmacological blockade of CD47-SIRPα axis by monoclonal antibody or fusion protein could reprogram macrophage phagocytosis in multiple preclinical models (Galli et al. 2015; Weiskopf et al. 2016; Zhang et al. 2018a). For instance, reprogramming of TAMs into M1 macrophages has been achieved by the combination of anti-SIRPα antibody with CSF-1R inhibitor (Kulkarni et al. 2018). TTI-621, a CD47-blocking SIRPαFc fusion protein, triggers tumor cell phagocytosis by M1 macrophages in vitro and exhibits antitumor activity in vivo. Combinational therapies have the potential to increase the effect (Petrova et al. 2017). Data from a phase 1b trial showed that the macrophage checkpoint inhibitor Hu5F9-G4 combined with anti-CD20 antibody rituximab demonstrated promising activity in advanced lymphoma patients (Advani et al. 2018). Based on these promising evidences, several clinical trials are performed to evaluate the safety and activity of anti-CD47/SIRPα monotherapy, also in combination with other agents (Table 2).

As an alternative target to favor cytotoxic functions of TAMs, CD40 is a costimulatory protein found on antigen-presenting cells including macrophages and dendritic cells. CD40L on TH cells binding to CD40 activates antigen presenting cells and triggers various downstream effects, including upregulation of the MHC molecule expression, secretion of proinflammatory cytokines, and finally cytotoxic T-cell activation (Zippelius et al. 2015; Zhang et al. 2018b). Re-educating macrophages using CD40 agonistic antibodies could recover tumor immune surveillance and reprogram TAMs toward M1-polarization in various tumor models (Vonderheide 2020). However, the antitumor activities have been moderate. Since anti-CD40 therapy induced PD-L1 upregulation in TAM, combining CD40 agonists with anti-PD-1 ther-

Table 2 List of clinical trials reprogramming TAMs

Identifier	Agent	Description	Strategy	Phase	Condition
NCT03717103	IBI188	Humanized anti-CD47 mAb	Single agent; combination with rituximab	Phase I	Advanced malignancies
NCT03763149	IBI188	Humanized anti-CD47 mAb	Single agent	Phase I	Advanced malignancies
NCT03834948	AO-176	Humanized anti-CD47 mAb	Single agent; combination with paclitaxel	Phase I/II	Solid tumor
NCT04257617	ZL1201	Humanized anti-CD47 antibody	Single agent	Phase I	Locally advanced solid tumor
NCT04349969	AK117	Humanized anti-CD47 antibody	Single agent	Phase I	Neoplasms malignant
NCT02367196	CC-90002	Humanized CD47-blocking antibody	Single agent; combination with rituximab	Phase I	Hematologic neoplasms
NCT02663518	TTI-621	SIRPα-fc fusion protein, human IgG1 subclass	Single agent; combination with rituximab	Phase I	Hematologic malignancies, solid tumor
NCT03248479	Magrolimab	Humanized anti-CD47 mAb	Single agent; combination with azacitidine	Phase I	Hematological malignancies
NCT04599634	Magrolimab	Humanized anti-CD47 mAb	Combination with obinutuzumab and venetoclax	Phase I	Relapsed and refractory indolent B-cell malignancies
NCT04445701	AO-176	Humanized anti-CD47 mAb	Single agent; combination with bortezomib and/or dexamethasone	Phase I/II	Relapsed/refractory multiple myeloma
NCT03013218	ALX148	High-affinity SIRPα variant	Single agent; combination with pembrolizumab; combination with trastuzumab; combination with rituximab	Phase I	Advanced solid tumors, lymphoma
NCT04306224	IMC-002	Humanized anti-CD47 mAb	Single agent	Phase I	Solid tumor, lymphoma
NCT04417517	ALX148	High-affinity SIRPα variant	Combination with azacitidine	Phase I/II	Higher risk myelodysplastic syndrome
NCT03527147	Hu5F9-G4	Humanized anti-CD47 mAb	Combination with acalabrutinib	Phase I	Relapsed or refractory aggressive non-Hodgkin's lymphoma
NCT04406623	SL-172154	Fusion protein consisting of human SIRPα and CD40L	Single agent	Phase I	Ovarian cancer

apy resulted in significantly lower tumor burden than either monotherapy alone (Diggs et al. 2020). Concurrent CSF-1R blockade and CD40 agonists lead to significant changes in the composition of immune infiltrates, causing an increased differentiation of proinflammatory M1 macrophages and also priming potent cytotoxic T-cell response in the draining lymph nodes of "cold" tumor models (Wiehagen et al. 2017).

5.3.3 Other Strategies to Reprogram TAMs

With the progress of oligonucleotide delivery, gene therapies, including microRNA (miRNA), small interfering RNA (siRNA), and messenger RNA (mRNA), are promising strategies to re-educate macrophages for cancer treatment. MiRNAs are tiny, single-stranded noncoding RNA molecule containing 19–24 nucleotides. Substantial research data depict that miRNAs are capable of silencing the gene expression either by degrading mRNA or repressing the gene transcription (Chatterjee et al. 2020). By performing miRNA profiling studies on macrophages, Graff and colleagues identified the miRNAs were associated with macrophage responses to inflammatory stimuli and demonstrated that miR-125a-3p and miR-26a-2 promote M1-like phenotype (Graff et al. 2012). Cai et al. presented the first evidence that miRNA-155 was a key molecule re-educating TAMs into pro-inflammatory M1 macrophages. In a mouse sarcoma model, delivering miRNA-155 by lipid-coated calcium phosphonate nanoparticles to TAMs successfully increased IL-12 expression with controlled tumor growth and extended survival (Cai et al. 2012). SiRNA has been employed to block the expression of immunosuppressive genes in TAMs. Shi et al. developed a robust nanoparticle platform for efficient siRNA delivery and gene silencing in macrophages (Liang et al. 2018). Silencing CCL-18 in macrophages remarkably suppressed breast cancer cells migration via regulation of the TAMs behavior.

In addition, researchers also exploited new strategies and carriers to increase the target gene expression in TAMs, such as charge-altering releasable transporters. This method has been used to deliver a combination of CD80, CD86, and OX40L mRNAs in several tumor models and induced systemic antitumor responses in vivo (Haabeth et al. 2019). As far as we know, no RNA delivery systems have been approved to reprogram TAMs, and more potent and promising strategies are still needed to initiate RNA delivery technology for TAMs reprogramming.

5.4 TAM as Drug Delivery Systems

As a novel strategy to deliver payloads, the biomimetic delivery system recently showed its superiority on cancer targeting drug delivery (Yoo et al. 2011). Neutrophils, red blood cells, and macrophages are three common drug carriers (Fang et al. 2018). Through binding to IL-6 and TNF-α, surface adhesion molecules expressed on neutrophils lead to the intrinsic target to inflammation. Red blood cells can increase the long-circulation time of cargo nanoparticles in blood circulation without tumor-targeting ability. Macrophages not only circulate in the blood circulation like neutrophils and red blood cells, but also specifically target tumors through integrin on macrophages binding to VCAM-1 of tumor cells. Herein, macrophage-based drug deliveries are more versatile (Xia et al. 2020).

Given the intrinsic homing property, M1 macrophages can be employed as a biomimetic delivery system to actively carry payloads to the tumor sites (Guerra et al. 2017). Up to now, researchers have mainly designed strategies to utilize M1 macrophage and M1 macrophage-derived exosomes as tumor-targeted drug carriers. Furthermore, membrane fragments have also been extracted to coat nanoparticles to increase the targeting ability. Here, these three strategies will be detailed analyzed: M1 macrophages, M1 macrophages-derived exosomes, and M1 macrophage membrane-coated nanoparticles (Fig. 3).

5.4.1 M1 Macrophages as Drug Carriers

M1 macrophages including RAW264.7 cells and bone-marrow-derived macrophages (BMDMs) are extensively used for tumor-targeted delivery

(Xia et al. 2020). Up to now, there are three M1 macrophage-based drug delivery systems. The first system is using macrophage as a direct antitumor drug carrier. Drug-loaded M1 macrophages were developed by simply incubation. For example, Fu and colleagues constructed a doxorubicin-loaded RAW264.7 cells to treat 4 T1 tumors. Data showed that doxorubicin did not significantly affect and alter the tumor-tropic capacity of M1 macrophages to tumor cells in vitro and in vivo. Importantly, the doxorubicin-loaded M1 macrophages potently inhibited the tumor metastasis and prolonged the life span, with reduced toxicity (Fu et al. 2015). The second one is using macrophage as an indirect antitumor drug carrier. Instead of directly loading drug, macrophages were employed to load nanoparticles containing drugs. This strategy could reduce the cytotoxicity of drugs on macrophages and further increase the drug load. Tao et al. used BMDMs to load nano/paclitaxel (paclitaxel-loaded nano-formulations) and found that nano/paclitaxel-loaded M1 macrophages showed better efficacy in malignant glioma (Tao et al. 2013). The third one is to modify macrophage as a drug carrier. To further potentiate the property of these carriers, macrophages can be further modified. M1 macrophages can be genetically or chemically modified to enhance immunological activity. Zhang et al. developed induced pluripotent stem cells-derived engineered chimeric antigen receptor-expressing macrophage and conferred the macrophage antigen-dependent functions such as secretion of cytokines, polarization toward the pro-inflammatory state (Zhang et al. 2020b).

5.4.2 M1 Macrophages-Derived Exosomes as Drug Carriers

Exosomes are membrane-bound extracellular vesicles that originated from the endosomal compartment of cells (Bunggulawa et al. 2018). Compared to M1 macrophage, M1 macrophage-derived exosomes present similar membrane

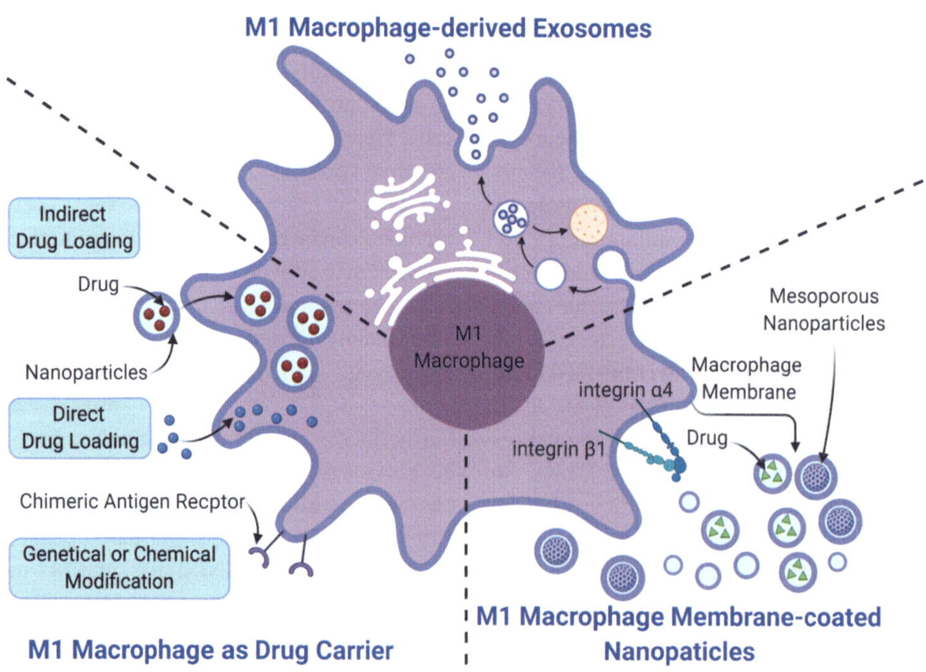

Fig. 3 TAMs as drug delivery systems. Three strategies were highlighted for TAMs as drug delivery systems: 1) M1 macrophages as drug carriers through direct drug loading, indirect drug loading, or genetical/chemical modification; 2) M1 macrophages-derived exosomes as drug carriers; and 3) M1 macrophage membrane-coated nanoparticles as drug carriers

properties and can be utilized to load drugs for cancer treatment. Kim and colleagues used M1 macrophage-derived exosomes to load paclitaxel via a sonication method. This delivery system increased the cytotoxicity more than 50 times in multidrug-resistant cancer cells (Kim et al. 2016). M1 macrophage-derived exosomes also could be a cancer vaccine adjuvant owing to the specific proinflammatory function (Wang et al. 2019). M1 macrophage-derived exosomes not only potentiated tumor-targeted drug delivery, but also could be engineered with liposomes including different components (Rayamajhi et al. 2019). In short, M1 macrophages and M1 macrophage-derived exosomes retained the targeting ability and proinflammatory function. However, the yield of M1 macrophage-derived exosomes is still low, and proper engineering of M1 macrophages and constructing exosome-mimetic vesicles may be the potential strategies to overcome this issue.

5.4.3 Nanoparticles Coated with M1 Macrophage Membrane as Drug Carriers

Considering the excellent tumor-targeting abilities and tumor infiltration of M1 macrophages, cell membranes of M1 macrophages can be extracted to coat nanoparticles and increase the targeting ability of nanoparticles in tumors. M1 macrophage membrane-coated nanoparticles could be constructed by coextruding the isolated macrophage membranes with nanoparticles and exhibited more persistent circulation and higher tumor accumulation (Cao et al. 2016; Zhao et al. 2018b). Cao et al. reported drug-loaded liposomes with isolated macrophage membranes to increase the drug delivery to metastasis sites. The macrophage membranes potentiated the cellular uptake of liposomes in cancer cells and presented potent suppression on metastatic breast cancer (Cao et al. 2016). This is due to the high expression of integrin α4 and β1 on macrophages, which could bind to VCAM-1 on breast cancer cells. Herein, M1 macrophage-derived membrane has the targeting property for tumors. Xuan and colleagues engineered macrophage cell membrane-coated mesoporous silica nano-capsules (MSNCs) by a biomimetic drug delivery system. The macrophage membrane reduced the uptake percentage of MSNCs by immune cells and tissues, effectively prolong the half-life in blood circulation. Compared to the totally cleared non-coated MSNCs, the macrophage membrane-coated MSNCs showed 32% retention in blood circulation and the targeting accumulation of macrophage membrane-coated MSNCs led to complete tumor eradication (Xuan et al. 2015). In summary, membranes from M1 macrophage could also be used to modify various nanoparticles for drug delivery.

6 Conclusions

Pro-tumoral functions of TAMs make them attractive targets for cancer immunotherapy. As described in this book chapter, several preclinical and clinical studies revealed that depletion of TAMs significantly augments the response of immunotherapy and chemo−/radiotherapy and suppresses the cancer metastasis. However, as we know, TAMs represent a heterogeneous population with distinct functions that vary according to cancer types and the stages of cancer progression. One of the future works need to be done to define the subsets of TAMs involving tumorigenesis, which can be targeted for effective therapies, while saving the macrophages with antitumor functions (Guerriero 2018; Cassetta and Pollard 2018). One big challenge is to investigate metastasis-associated macrophage (MAM) phenotypes involved in cancer metastasis, which is major cause of death of cancer patients. To define those subsets of macrophages, new technologies, such as single cell sequencing, digital spatial profiling, multiplex immunofluorescence, and metabolomics, can be applied (Pathria et al. 2019). Despite multiple early stages of clinical trials are going on, scarce knowledge about TAMs in clinic human cancers is available since much of these data were generated from in vitro cellular models or mouse models. TAMs may possess more potential for cancer target therapy than previously thought. More studies should emphasis on the regulation of TAMs, such as reprogramming TAMs, targeting inhibitory molecules, which will help improve the efficacy of current cancer therapeutics, such as immunotherapy, chemotherapy, and radiotherapy, and ultimately open a new door for targeting TAMs immunotherapy (Cassetta and Pollard 2018).

Acknowledgements The work was supported by grants from the National Natural Science Foundation of China (81901232 (to Y.L.), 32000479 (to X.Z.), 82002915 (to S.W.)).

Conflict of Interests The authors declared that there is no conflict of interests.

References

Advani R, Flinn I, Popplewell L, Forero A, Bartlett NL, Ghosh N, et al. CD47 blockade by Hu5F9-G4 and rituximab in non-Hodgkin's lymphoma. N Engl J Med. 2018;379(18):1711–21. https://doi.org/10.1056/NEJMoa1807315.

Affara NI, Ruffell B, Medler TR, Gunderson AJ, Johansson M, Bornstein S, et al. B cells regulate macrophage phenotype and response to chemotherapy in squamous carcinomas. Cancer Cell. 2014;25(6):809–21. https://doi.org/10.1016/j.ccr.2014.04.026.

Alhudaithi SS, Almuqbil RM, Zhang H, Bielski ER, Du W, Sunbul FS, et al. Local targeting of lung-tumor-associated macrophages with pulmonary delivery of a CSF-1R inhibitor for the treatment of breast cancer lung metastases. Mol Pharm. 2020;17(12):4691–703. https://doi.org/10.1021/acs.molpharmaceut.0c00983.

Anfray C, Ummarino A, Andon FT, Allavena P. Current Strategies to Target Tumor-Associated-Macrophages to Improve Anti-Tumor Immune Responses. Cells. 2019;9(1). doi: https://doi.org/10.3390/cells9010046.

Argyle D, Kitamura T. Targeting macrophage-recruiting chemokines as a novel therapeutic strategy to prevent the progression of solid tumors. Front Immunol. 2018;9:2629. https://doi.org/10.3389/fimmu.2018.02629.

Arnold CE, Whyte CS, Gordon P, Barker RN, Rees AJ, Wilson HM. A critical role for suppressor of cytokine signalling 3 in promoting M1 macrophage activation and function in vitro and in vivo. Immunology. 2014;141(1):96–110. https://doi.org/10.1111/imm.12173.

Banerjee P, Zhang R, Ivan C, Galletti G, Clise-Dwyer K, Barbaglio F, et al. Trabectedin reveals a strategy of immunomodulation in chronic lymphocytic leukemia. Cancer Immunol Res. 2019;7(12):2036–51. https://doi.org/10.1158/2326-6066.CIR-19-0152.

Barclay AN, Brown MH. The SIRP family of receptors and immune regulation. Nat Rev Immunol. 2006;6(6):457–64. https://doi.org/10.1038/nri1859.

Barclay AN, Van den Berg TK. The interaction between signal regulatory protein alpha (SIRPalpha) and CD47: structure, function, and therapeutic target. Annu Rev Immunol. 2014;32:25–50. https://doi.org/10.1146/annurev-immunol-032713-120142.

Barkal AA, Weiskopf K, Kao KS, Gordon SR, Rosental B, Yiu YY, et al. Engagement of MHC class I by the inhibitory receptor LILRB1 suppresses macrophages and is a target of cancer immunotherapy. Nat Immunol. 2018;19(1):76–84. https://doi.org/10.1038/s41590-017-0004-z.

Barkal AA, Brewer RE, Markovic M, Kowarsky M, Barkal SA, Zaro BW, et al. CD24 signalling through macrophage Siglec-10 is a target for cancer immunotherapy. Nature. 2019;572(7769):392–6. https://doi.org/10.1038/s41586-019-1456-0.

Barone A, Sengupta R, Warrington NM, Smith E, Wen PY, Brekken RA, et al. Combined VEGF and CXCR4 antagonism targets the GBM stem cell population and synergistically improves survival in an intracranial mouse model of glioblastoma. Oncotarget. 2014;5(20):9811–22. doi: https://doi.org/10.18632/oncotarget.2443.

Bartneck M, Schrammen PL, Mockel D, Govaere O, Liepelt A, Krenkel O, et al. The CCR2(+) macrophage subset promotes pathogenic angiogenesis for tumor vascularization in fibrotic livers. Cell Mol Gastroenterol Hepatol. 2019;7(2):371–90. https://doi.org/10.1016/j.jcmgh.2018.10.007.

Bhattacharya R, Ye XC, Wang R, Ling X, McManus M, Fan F, et al. Intracrine VEGF signaling mediates the activity of Prosurvival pathways in human colorectal cancer cells. Cancer Res. 2016;76(10):3014–24. https://doi.org/10.1158/0008-5472.CAN-15-1605.

Binenbaum Y, Fridman E, Yaari Z, Milman N, Schroeder A, Ben David G, et al. Transfer of miRNA in macrophage-derived exosomes induces drug resistance in pancreatic adenocarcinoma. Cancer Res. 2018;78(18):5287–99. https://doi.org/10.1158/0008-5472.CAN-18-0124.

Biswas SK, Mantovani A. Macrophage plasticity and interaction with lymphocyte subsets: cancer as a paradigm. Nat Immunol. 2010;11(10):889–96. https://doi.org/10.1038/ni.1937.

Bocanegra Gondan AI, Ruiz-de-Angulo A, Zabaleta A, Gomez Blanco N, Cobaleda-Siles BM, Garcia-Granda MJ, et al. Effective cancer immunotherapy in mice by polyIC-imiquimod complexes and engineered magnetic nanoparticles. Biomaterials. 2018;170:95–115. https://doi.org/10.1016/j.biomaterials.2018.04.003.

Bonapace L, Coissieux MM, Wyckoff J, Mertz KD, Varga Z, Junt T, et al. Cessation of CCL2 inhibition accelerates breast cancer metastasis by promoting angiogenesis. Nature. 2014;515(7525):130–3. https://doi.org/10.1038/nature13862.

Bonde AK, Tischler V, Kumar S, Soltermann A, Schwendener RA. Intratumoral macrophages con-

tribute to epithelial-mesenchymal transition in solid tumors. BMC Cancer. 2012;12:35. https://doi.org/10.1186/1471-2407-12-35.
Borgoni S, Iannello A, Cutrupi S, Allavena P, D'Incalci M, Novelli F, et al. Depletion of tumor-associated macrophages switches the epigenetic profile of pancreatic cancer infiltrating T cells and restores their anti-tumor phenotype. Onco Targets Ther. 2018;7(2):e1393596. https://doi.org/10.1080/2162402X.2017.1393596.
Boussiotis VA, Chatterjee P, Li L. Biochemical signaling of PD-1 on T cells and its functional implications. Cancer J. 2014;20(4):265–71. https://doi.org/10.1097/PPO.0000000000000059.
Buchbinder EI, Desai A. CTLA-4 and PD-1 pathways: similarities, differences, and implications of their inhibition. Am J Clin Oncol. 2016;39(1):98–106. https://doi.org/10.1097/COC.0000000000000239.
Bunggulawa EJ, Wang W, Yin T, Wang N, Durkan C, Wang Y, et al. Recent advancements in the use of exosomes as drug delivery systems. J Nanobiotechnology. 2018;16(1):81. https://doi.org/10.1186/s12951-018-0403-9.
Cai X, Yin Y, Li N, Zhu D, Zhang J, Zhang CY, et al. Re-polarization of tumor-associated macrophages to pro-inflammatory M1 macrophages by microRNA-155. J Mol Cell Biol. 2012;4(5):341–3. https://doi.org/10.1093/jmcb/mjs044.
Cao H, Dan Z, He X, Zhang Z, Yu H, Yin Q, et al. Liposomes coated with isolated macrophage membrane can target lung metastasis of breast cancer. ACS Nano. 2016;10(8):7738–48. https://doi.org/10.1021/acsnano.6b03148.
Carminati L, Pinessi D, Borsotti P, Minoli L, Giavazzi R, D'Incalci M, et al. Antimetastatic and antiangiogenic activity of trabectedin in cutaneous melanoma. Carcinogenesis. 2019;40(2):303–12. https://doi.org/10.1093/carcin/bgy177.
Cassetta L, Pollard JW. Targeting macrophages: therapeutic approaches in cancer. Nat Rev Drug Discov. 2018;17(12):887–904. https://doi.org/10.1038/nrd.2018.169.
Cassetta L, Pollard JW. Tumor-associated macrophages. Curr Biol. 2020;30(6):R246–R8. https://doi.org/10.1016/j.cub.2020.01.031.
Cassetta L, Fragkogianni S, Sims AH, Swierczak A, Forrester LM, Zhang H, et al. Human tumor-associated macrophage and monocyte transcriptional landscapes reveal cancer-specific reprogramming, biomarkers, and therapeutic targets. Cancer Cell. 2019;35(4):588–602. e10. https://doi.org/10.1016/j.ccell.2019.02.009.
Castells M, Thibault B, Delord JP, Couderc B. Implication of tumor microenvironment in chemoresistance: tumor-associated stromal cells protect tumor cells from cell death. Int J Mol Sci. 2012;13(8):9545–71. https://doi.org/10.3390/ijms13089545.
Ceci C, Atzori MG, Lacal PM, Graziani G. Role of VEGFs/VEGFR-1 Signaling and its Inhibition in Modulating Tumor Invasion: Experimental Evidence in Different Metastatic Cancer Models. Int J Mol Sci. 2020;21(4). doi: https://doi.org/10.3390/ijms21041388.
Chao MP, Weissman IL, Majeti R. The CD47-SIRPalpha pathway in cancer immune evasion and potential therapeutic implications. Curr Opin Immunol. 2012;24(2):225–32. https://doi.org/10.1016/j.coi.2012.01.010.
Chatterjee B, Saha P, Bose S, Shukla D, Chatterjee N, Kumar S, et al. MicroRNAs: As Critical Regulators of Tumor- Associated Macrophages. Int J Mol Sci. 2020;21(19). doi: https://doi.org/10.3390/ijms21197117.
Chavez-Galan L, Olleros ML, Vesin D, Garcia I. Much more than M1 and M2 macrophages, there are also CD169(+) and TCR(+) macrophages. Front Immunol. 2015;6:263. https://doi.org/10.3389/fimmu.2015.00263.
Chen R, Chen B. Siltuximab (CNTO 328): a promising option for human malignancies. Drug Des Devel Ther. 2015;9:3455–8. https://doi.org/10.2147/DDDT.S86438.
Chen Y, Ramjiawan RR, Reiberger T, Ng MR, Hato T, Huang Y, et al. CXCR4 inhibition in tumor microenvironment facilitates anti-programmed death receptor-1 immunotherapy in sorafenib-treated hepatocellular carcinoma in mice. Hepatology. 2015;61(5):1591–602. https://doi.org/10.1002/hep.27665.
Chen L, Li J, Wang F, Dai C, Wu F, Liu X, et al. Tie2 expression on macrophages is required for blood vessel reconstruction and tumor relapse after chemotherapy. Cancer Res. 2016;76(23):6828–38. https://doi.org/10.1158/0008-5472.CAN-16-1114.
Cheng N, Bai X, Shu Y, Ahmad O, Shen P. Targeting tumor-associated macrophages as an antitumor strategy. Biochem Pharmacol. 2020;183:114354. https://doi.org/10.1016/j.bcp.2020.114354.
Chi H, Li C, Zhao FS, Zhang L, Ng TB, Jin G, et al. Anti-tumor activity of toll-like receptor 7 agonists. Front Pharmacol. 2017;8:304. https://doi.org/10.3389/fphar.2017.00304.
Chiang CF, Chao TT, Su YF, Hsu CC, Chien CY, Chiu KC, et al. Metformin-treated cancer cells modulate macrophage polarization through AMPK-NF-kappaB signaling. Oncotarget. 2017;8(13):20706–18. doi: https://doi.org/10.18632/oncotarget.14982.
Cho HR, Kumari N, Thi Vu H, Kim H, Park CK, Choi SH. Increased antiangiogenic effect by blocking CCL2-dependent macrophages in a rodent glioblastoma model: correlation study with dynamic susceptibility contrast perfusion MRI. Sci Rep. 2019;9(1):11085. https://doi.org/10.1038/s41598-019-47438-4.
Correia AL, Bissell MJ. The tumor microenvironment is a dominant force in multidrug resistance. Drug Resist Updat. 2012;15(1–2):39–49. https://doi.org/10.1016/j.drup.2012.01.006.
Crocker PR, Gordon S. Properties and distribution of a lectin-like hemagglutinin differentially expressed by murine stromal tissue macrophages. J Exp Med. 1986;164(6):1862–75. https://doi.org/10.1084/jem.164.6.1862.
Crowther M, Brown NJ, Bishop ET, Lewis CE. Microenvironmental influence on macrophage

regulation of angiogenesis in wounds and malignant tumors. J Leukoc Biol. 2001;70(4):478–90.

Czako B, Marszalek JR, Burke JP, Mandal P, Leonard PG, Cross JB, et al. Discovery of IACS-9439, a potent, exquisitely selective, and orally bioavailable inhibitor of CSF1R. J Med Chem. 2020;63(17):9888–911. https://doi.org/10.1021/acs.jmedchem.0c00936.

Da Silva CG, Camps MGM, Li T, Chan AB, Ossendorp F, Cruz LJ. Co-delivery of immunomodulators in biodegradable nanoparticles improves therapeutic efficacy of cancer vaccines. Biomaterials. 2019;220:119417. https://doi.org/10.1016/j.biomaterials.2019.119417.

Dalton HJ, Pradeep S, McGuire M, Hailemichael Y, Ma S, Lyons Y, et al. Macrophages facilitate resistance to anti-VEGF therapy by altered VEGFR expression. Clin Cancer Res. 2017;23(22):7034–46. https://doi.org/10.1158/1078-0432.CCR-17-0647.

Davies LC, Jenkins SJ, Allen JE, Taylor PR. Tissue-resident macrophages. Nat Immunol. 2013;14(10):986–95. https://doi.org/10.1038/ni.2705.

De Meyer I, Martinet W, Schrijvers DM, Timmermans JP, Bult H, De Meyer GR. Toll-like receptor 7 stimulation by imiquimod induces macrophage autophagy and inflammation in atherosclerotic plaques. Basic Res Cardiol. 2012;107(3):269. https://doi.org/10.1007/s00395-012-0269-1.

de Oliveira S, Houseright RA, Graves AL, Golenberg N, Korte BG, Miskolci V, et al. Metformin modulates innate immune-mediated inflammation and early progression of NAFLD-associated hepatocellular carcinoma in zebrafish. J Hepatol. 2019;70(4):710–21. https://doi.org/10.1016/j.jhep.2018.11.034.

De Palma M, Lewis CE. Macrophage regulation of tumor responses to anticancer therapies. Cancer Cell. 2013;23(3):277–86. https://doi.org/10.1016/j.ccr.2013.02.013.

De Palma M, Venneri MA, Galli R, Sergi L, Politi LS, Sampaolesi M, et al. Tie2 identifies a hematopoietic lineage of proangiogenic monocytes required for tumor vessel formation and a mesenchymal population of pericyte progenitors. Cancer Cell. 2005;8(3):211–26. https://doi.org/10.1016/j.ccr.2005.08.002.

DeNardo DG, Barreto JB, Andreu P, Vasquez L, Tawfik D, Kolhatkar N, et al. CD4(+) T cells regulate pulmonary metastasis of mammary carcinomas by enhancing protumor properties of macrophages. Cancer Cell. 2009;16(2):91–102. https://doi.org/10.1016/j.ccr.2009.06.018.

DeNardo DG, Brennan DJ, Rexhepaj E, Ruffell B, Shiao SL, Madden SF, et al. Leukocyte complexity predicts breast cancer survival and functionally regulates response to chemotherapy. Cancer Discov. 2011;1(1):54–67. https://doi.org/10.1158/2159-8274.CD-10-0028.

Deng J, Liang H, Zhang R, Sun D, Pan Y, Liu Y, et al. STAT3 is associated with lymph node metastasis in gastric cancer. Tumour Biol. 2013;34(5):2791–800. https://doi.org/10.1007/s13277-013-0837-5.

Deng L, Stafford JH, Liu SC, Chernikova SB, Merchant M, Recht L, et al. SDF-1 blockade enhances anti-VEGF therapy of glioblastoma and can be monitored by MRI. Neoplasia. 2017;19(1):1–7. https://doi.org/10.1016/j.neo.2016.11.010.

Dewan MZ, Ahmed S, Iwasaki Y, Ohba K, Toi M, Yamamoto N. Stromal cell-derived factor-1 and CXCR4 receptor interaction in tumor growth and metastasis of breast cancer. Biomed Pharmacother. 2006;60(6):273–6. https://doi.org/10.1016/j.biopha.2006.06.004.

Diggs LP, Ruf B, Ma C, Heinrich B, Cui L, Zhang Q, et al. CD40-mediated immune cell activation enhances response to anti-PD1 in murine intrahepatic cholangiocarcinoma. J Hepatol. 2020; https://doi.org/10.1016/j.jhep.2020.11.037.

Ding L, Liang G, Yao Z, Zhang J, Liu R, Chen H, et al. Metformin prevents cancer metastasis by inhibiting M2-like polarization of tumor associated macrophages. Oncotarget. 2015;6(34):36441–55. doi: https://doi.org/10.18632/oncotarget.5541.

Drake MT, Clarke BL, Khosla S. Bisphosphonates: mechanism of action and role in clinical practice. Mayo Clin Proc. 2008;83(9):1032–45. https://doi.org/10.4065/83.9.1032.

Duruisseaux M, Martinez-Cardus A, Calleja-Cervantes ME, Moran S. Castro de Moura M, Davalos V, et al. epigenetic prediction of response to anti-PD-1 treatment in non-small-cell lung cancer: a multicentre, retrospective analysis. Lancet Respir Med. 2018;6(10):771–81. https://doi.org/10.1016/S2213-2600(18)30284-4.

Dwyer AR, Greenland EL, Pixley FJ. Promotion of Tumor Invasion by Tumor-Associated Macrophages: The Role of CSF-1-Activated Phosphatidylinositol 3 Kinase and Src Family Kinase Motility Signaling. Cancers (Basel). 2017;9(6). doi: https://doi.org/10.3390/cancers9060068.

Epelman S, Lavine KJ, Randolph GJ. Origin and functions of tissue macrophages. Immunity. 2014;41(1):21–35. https://doi.org/10.1016/j.immuni.2014.06.013.

Erblich B, Zhu L, Etgen AM, Dobrenis K, Pollard JW. Absence of colony stimulation factor-1 receptor results in loss of microglia, disrupted brain development and olfactory deficits. PLoS One. 2011;6(10):e26317. https://doi.org/10.1371/journal.pone.0026317.

Fahham D, Weiss ID, Abraham M, Beider K, Hanna W, Shlomai Z, et al. In vitro and in vivo therapeutic efficacy of CXCR4 antagonist BKT140 against human non-small cell lung cancer. J Thorac Cardiovasc Surg. 2012;144(5):1167–75. e1. https://doi.org/10.1016/j.jtcvs.2012.07.031.

Faibish M, Francescone R, Bentley B, Yan W, Shao R. A YKL-40-neutralizing antibody blocks tumor angiogenesis and progression: a potential therapeutic agent in cancers. Mol Cancer Ther. 2011;10(5):742–51. https://doi.org/10.1158/1535-7163.MCT-10-0868.

Fang RH, Kroll AV, Gao W, Zhang L. Cell membrane coating nanotechnology. Adv Mater. 2018;30(23):e1706759. https://doi.org/10.1002/adma.201706759.

Farrell KB, Karpeisky A, Thamm DH, Zinnen S. Bisphosphonate conjugation for bone specific drug targeting. Bone Rep. 2018;9:47–60. https://doi.org/10.1016/j.bonr.2018.06.007.

Fleetwood AJ, Lawrence T, Hamilton JA, Cook AD. Granulocyte-macrophage colony-stimulating factor (CSF) and macrophage CSF-dependent macrophage phenotypes display differences in cytokine profiles and transcription factor activities: implications for CSF blockade in inflammation. J Immunol. 2007;178(8):5245–52. https://doi.org/10.4049/jimmunol.178.8.5245.

Flores-Toro JA, Luo D, Gopinath A, Sarkisian MR, Campbell JJ, Charo IF, et al. CCR2 inhibition reduces tumor myeloid cells and unmasks a checkpoint inhibitor effect to slow progression of resistant murine gliomas. Proc Natl Acad Sci U S A. 2020;117(2):1129–38. https://doi.org/10.1073/pnas.1910856117.

Fu J, Wang D, Mei D, Zhang H, Wang Z, He B, et al. Macrophage mediated biomimetic delivery system for the treatment of lung metastasis of breast cancer. J Control Release. 2015;204:11–9. https://doi.org/10.1016/j.jconrel.2015.01.039.

Fujimoto H, Sangai T, Ishii G, Ikehara A, Nagashima T, Miyazaki M, et al. Stromal MCP-1 in mammary tumors induces tumor-associated macrophage infiltration and contributes to tumor progression. Int J Cancer. 2009;125(6):1276–84. https://doi.org/10.1002/ijc.24378.

Galli S, Zlobec I, Schurch C, Perren A, Ochsenbein AF, Banz Y. CD47 protein expression in acute myeloid leukemia: a tissue microarray-based analysis. Leuk Res. 2015;39(7):749–56. https://doi.org/10.1016/j.leukres.2015.04.007.

Galon J, Bruni D. Approaches to treat immune hot, altered and cold tumours with combination immunotherapies. Nat Rev Drug Discov. 2019;18(3):197–218. https://doi.org/10.1038/s41573-018-0007-y.

Gao S, Hu J, Wu X, Liang Z. PMA treated THP-1-derived-IL-6 promotes EMT of SW48 through STAT3/ERK-dependent activation of Wnt/beta-catenin signaling pathway. Biomed Pharmacother. 2018;108:618–24. https://doi.org/10.1016/j.biopha.2018.09.067.

Genard G, Lucas S, Michiels C. Reprogramming of tumor-associated macrophages with anticancer therapies: radiotherapy versus chemo- and immunotherapies. Front Immunol. 2017;8:828. https://doi.org/10.3389/fimmu.2017.00828.

Germano G, Frapolli R, Belgiovine C, Anselmo A, Pesce S, Liguori M, et al. Role of macrophage targeting in the antitumor activity of trabectedin. Cancer Cell. 2013;23(2):249–62. https://doi.org/10.1016/j.ccr.2013.01.008.

Gilbert LA, Hemann MT. DNA damage-mediated induction of a chemoresistant niche. Cell. 2010;143(3):355–66. https://doi.org/10.1016/j.cell.2010.09.043.

Ginhoux F, Guilliams M. Tissue-resident macrophage ontogeny and homeostasis. Immunity. 2016;44(3):439–49. https://doi.org/10.1016/j.immuni.2016.02.024.

Ginhoux F, Jung S. Monocytes and macrophages: developmental pathways and tissue homeostasis. Nat Rev Immunol. 2014;14(6):392–404. https://doi.org/10.1038/nri3671.

Gocheva V, Wang HW, Gadea BB, Shree T, Hunter KE, Garfall AL, et al. IL-4 induces cathepsin protease activity in tumor-associated macrophages to promote cancer growth and invasion. Genes Dev. 2010;24(3):241–55. https://doi.org/10.1101/gad.1874010.

Golden EB, Frances D, Pellicciotta I, Demaria S, Helen Barcellos-Hoff M, Formenti SC. Radiation fosters dose-dependent and chemotherapy-induced immunogenic cell death. Onco Targets Ther. 2014;3:e28518. https://doi.org/10.4161/onci.28518.

Gordon S, Taylor PR. Monocyte and macrophage heterogeneity. Nat Rev Immunol. 2005;5(12):953–64. https://doi.org/10.1038/nri1733.

Gordon SR, Maute RL, Dulken BW, Hutter G, George BM, McCracken MN, et al. PD-1 expression by tumour-associated macrophages inhibits phagocytosis and tumour immunity. Nature. 2017;545(7655):495–9. https://doi.org/10.1038/nature22396.

Gouaze-Andersson V, Delmas C, Taurand M, Martinez-Gala J, Evrard S, Mazoyer S, et al. FGFR1 induces glioblastoma Radioresistance through the PLCgamma/Hif1alpha pathway. Cancer Res. 2016;76(10):3036–44. https://doi.org/10.1158/0008-5472.CAN-15-2058.

Graff JW, Dickson AM, Clay G, McCaffrey AP, Wilson ME. Identifying functional microRNAs in macrophages with polarized phenotypes. J Biol Chem. 2012;287(26):21816–25. https://doi.org/10.1074/jbc.M111.327031.

Gravina GL, Mancini A, Marampon F, Colapietro A, Delle Monache S, Sferra R, et al. The brain-penetrating CXCR4 antagonist, PRX177561, increases the antitumor effects of bevacizumab and sunitinib in preclinical models of human glioblastoma. J Hematol Oncol. 2017;10(1):5. https://doi.org/10.1186/s13045-016-0377-8.

Grimshaw MJ, Balkwill FR. Inhibition of monocyte and macrophage chemotaxis by hypoxia and inflammation--a potential mechanism. Eur J Immunol. 2001;31(2):480–9. https://doi.org/10.1002/1521--4141(200102)31:2<480::aid-immu480>3.0.co;2-l.

Grossman JG, Nywening TM, Belt BA, Panni RZ, Krasnick BA, DeNardo DG, et al. Recruitment of CCR2(+) tumor associated macrophage to sites of liver metastasis confers a poor prognosis in human colorectal cancer. Onco Targets Ther. 2018;7(9):e1470729. https://doi.org/10.1080/2162402X.2018.1470729.

Guerra AD, Yeung OWH, Qi X, Kao WJ, Man K. The anti-tumor effects of M1 macrophage-loaded poly (ethylene glycol) and gelatin-based hydrogels on hepatocellular carcinoma. Theranostics. 2017;7(15):3732–44. https://doi.org/10.7150/thno.20251.

Guerriero JL. Macrophages: the road less traveled. Changing Anticancer Therapy Trends Mol Med. 2018;24(5):472–89. https://doi.org/10.1016/j.molmed.2018.03.006.

Guerriero JL. Macrophages: their untold story in T cell activation and function. Int Rev Cell Mol Biol. 2019;342:73–93. https://doi.org/10.1016/bs.ircmb.2018.07.001.

Haabeth OAW, Blake TR, McKinlay CJ, Tveita AA, Sallets A, Waymouth RM, et al. Local delivery of Ox40l, Cd80, and Cd86 mRNA kindles global anticancer immunity. Cancer Res. 2019;79(7):1624–34. https://doi.org/10.1158/0008-5472.CAN-18-2867.

Halbrook CJ, Pontious C, Kovalenko I, Lapienyte L, Dreyer S, Lee HJ, et al. Macrophage-released pyrimidines inhibit gemcitabine therapy in pancreatic cancer. Cell Metab. 2019;29(6):1390–9. e6. https://doi.org/10.1016/j.cmet.2019.02.001.

Hambardzumyan D, Gutmann DH, Kettenmann H. The role of microglia and macrophages in glioma maintenance and progression. Nat Neurosci. 2016;19(1):20–7. https://doi.org/10.1038/nn.4185.

Haniffa M, Bigley V, Collin M. Human mononuclear phagocyte system reunited. Semin Cell Dev Biol. 2015;41:59–69. https://doi.org/10.1016/j.semcdb.2015.05.004.

Hao Q, Vadgama JV, Wang P. CCL2/CCR2 signaling in cancer pathogenesis. Cell Commun Signal. 2020;18(1):82. https://doi.org/10.1186/s12964-020-00589-8.

Hasham MG, Baxan N, Stuckey DJ, Branca J, Perkins B, Dent O, et al. Systemic autoimmunity induced by the TLR7/8 agonist Resiquimod causes myocarditis and dilated cardiomyopathy in a new mouse model of autoimmune heart disease. Dis Model Mech. 2017;10(3):259–70. https://doi.org/10.1242/dmm.027409.

Hirano H, Tanioka K, Yokoyama S, Akiyama S, Kuratsu J. Angiogenic effect of thymidine phosphorylase on macrophages in glioblastoma multiforme. J Neurosurg. 2001;95(1):89–95. https://doi.org/10.3171/jns.2001.95.1.0089.

Hofmann MA, Kors C, Audring H, Walden P, Sterry W, Trefzer U. Phase 1 evaluation of intralesionally injected TLR9-agonist PF-3512676 in patients with basal cell carcinoma or metastatic melanoma. J Immunother. 2008;31(5):520–7. https://doi.org/10.1097/CJI.0b013e318174a4df.

Holen I, Coleman RE. Anti-tumour activity of bisphosphonates in preclinical models of breast cancer. Breast Cancer Res. 2010;12(6):214. https://doi.org/10.1186/bcr2769.

Hollmen M, Karaman S, Schwager S, Lisibach A, Christiansen AJ, Maksimow M, et al. G-CSF regulates macrophage phenotype and associates with poor overall survival in human triple-negative breast cancer. Onco Targets Ther. 2016;5(3):e1115177. https://doi.org/10.1080/2162402X.2015.1115177.

Huang H, Lai JY, Do J, Liu D, Li L, Del Rosario J, et al. Specifically targeting angiopoietin-2 inhibits angiogenesis, Tie2-expressing monocyte infiltration, and tumor growth. Clin Cancer Res. 2011;17(5):1001–11. https://doi.org/10.1158/1078-0432.CCR-10-2317.

Hughes R, Qian BZ, Rowan C, Muthana M, Keklikoglou I, Olson OC, et al. Perivascular M2 macrophages stimulate tumor relapse after chemotherapy. Cancer Res. 2015;75(17):3479–91. https://doi.org/10.1158/0008-5472.CAN-14-3587.

Im JH, Buzzelli JN, Jones K, Franchini F, Gordon-Weeks A, Markelc B, et al. FGF2 alters macrophage polarization, tumour immunity and growth and can be targeted during radiotherapy. Nat Commun. 2020;11(1):4064. https://doi.org/10.1038/s41467-020-17914-x.

Itatani Y, Kawada K, Yamamoto T, Sakai Y. Resistance to Anti-Angiogenic Therapy in Cancer-Alterations to Anti-VEGF Pathway. Int J Mol Sci. 2018;19(4). doi: https://doi.org/10.3390/ijms19041232.

Jahchan NS, Mujal AM, Pollack JL, Binnewies M, Sriram V, Reyno L, et al. Tuning the tumor myeloid microenvironment to fight cancer. Front Immunol. 2019;10:1611. https://doi.org/10.3389/fimmu.2019.01611.

Jaiswal S, Jamieson CH, Pang WW, Park CY, Chao MP, Majeti R, et al. CD47 is upregulated on circulating hematopoietic stem cells and leukemia cells to avoid phagocytosis. Cell. 2009;138(2):271–85. https://doi.org/10.1016/j.cell.2009.05.046.

Jinushi M, Chiba S, Yoshiyama H, Masutomi K, Kinoshita I, Dosaka-Akita H, et al. Tumor-associated macrophages regulate tumorigenicity and anticancer drug responses of cancer stem/initiating cells. Proc Natl Acad Sci U S A. 2011;108(30):12425–30. https://doi.org/10.1073/pnas.1106645108.

Johansson-Percival A, He B, Ganss R. Immunomodulation of tumor vessels: it takes two to tango. Trends Immunol. 2018;39(10):801–14. https://doi.org/10.1016/j.it.2018.08.001.

Jones JD, Sinder BP, Paige D, Soki FN, Koh AJ, Thiele S, et al. Trabectedin reduces skeletal prostate cancer tumor size in association with effects on M2 macrophages and Efferocytosis. Neoplasia. 2019;21(2):172–84. https://doi.org/10.1016/j.neo.2018.11.003.

Joyce JA, Pollard JW. Microenvironmental regulation of metastasis. Nat Rev Cancer. 2009;9(4):239–52. https://doi.org/10.1038/nrc2618.

Kaminski WE, Beham AW, Kzhyshkowska J, Gratchev A, Puellmann K. On the horizon: flexible immune recognition outside lymphocytes. Immunobiology. 2013;218(3):418–26. https://doi.org/10.1016/j.imbio.2012.05.024.

Karapetyan L, Luke JJ, Davar D. Toll-like receptor 9 agonists in cancer. Onco Targets Ther. 2020;13:10039–60. https://doi.org/10.2147/OTT.S247050.

Katsuya Y, Horinouchi H, Asao T, Kitahara S, Goto Y, Kanda S, et al. Expression of programmed death 1 (PD-1) and its ligand (PD-L1) in thymic epithelial tumors: impact on treatment efficacy and alteration in expression after chemotherapy. Lung Cancer. 2016;99:4–10. https://doi.org/10.1016/j.lungcan.2016.05.007.

Kim MS, Haney MJ, Zhao Y, Mahajan V, Deygen I, Klyachko NL, et al. Development of exosome-encapsulated paclitaxel to overcome MDR in cancer cells. Nanomedicine. 2016;12(3):655–64. https://doi.org/10.1016/j.nano.2015.10.012.

Kitamura T, Qian BZ, Soong D, Cassetta L, Noy R, Sugano G, et al. CCL2-induced chemokine cascade promotes breast cancer metastasis by enhancing retention of metastasis-associated macrophages. J Exp Med. 2015;212(7):1043–59. https://doi.org/10.1084/jem.20141836.

Kitamura T, Doughty-Shenton D, Cassetta L, Fragkogianni S, Brownlie D, Kato Y, et al. Monocytes differentiate to immune suppressive precursors of metastasis-associated macrophages in mouse models of metastatic breast cancer. Front Immunol. 2017;8:2004. https://doi.org/10.3389/fimmu.2017.02004.

Kloepper J, Riedemann L, Amoozgar Z, Seano G, Susek K, Yu V, et al. Ang-2/VEGF bispecific antibody reprograms macrophages and resident microglia to anti-tumor phenotype and prolongs glioblastoma survival. Proc Natl Acad Sci U S A. 2016;113(16):4476–81. https://doi.org/10.1073/pnas.1525360113.

Klug F, Prakash H, Huber PE, Seibel T, Bender N, Halama N, et al. Low-dose irradiation programs macrophage differentiation to an iNOS(+)/M1 phenotype that orchestrates effective T cell immunotherapy. Cancer Cell. 2013;24(5):589–602. https://doi.org/10.1016/j.ccr.2013.09.014.

Koh YW, Park C, Yoon DH, Suh C, Huh J. CSF-1R expression in tumor-associated macrophages is associated with worse prognosis in classical Hodgkin lymphoma. Am J Clin Pathol. 2014;141(4):573–83. https://doi.org/10.1309/AJCPR92TDDFARISU.

Kubota Y, Takubo K, Shimizu T, Ohno H, Kishi K, Shibuya M, et al. M-CSF inhibition selectively targets pathological angiogenesis and lymphangiogenesis. J Exp Med. 2009;206(5):1089–102. https://doi.org/10.1084/jem.20081605.

Kuhne MR, Mulvey T, Belanger B, Chen S, Pan C, Chong C, et al. BMS-936564/MDX-1338: a fully human anti-CXCR4 antibody induces apoptosis in vitro and shows antitumor activity in vivo in hematologic malignancies. Clin Cancer Res. 2013;19(2):357–66. https://doi.org/10.1158/1078-0432.CCR-12-2333.

Kulkarni A, Chandrasekar V, Natarajan SK, Ramesh A, Pandey P, Nirgud J, et al. A designer self-assembled supramolecule amplifies macrophage immune responses against aggressive cancer. Nat Biomed Eng. 2018;2(8):589–99. https://doi.org/10.1038/s41551-018-0254-6.

La Fleur L, Botling J, He F, Pelicano C, Zhou C, He C, et al. Targeting MARCO and IL-37R on immunosuppressive macrophages in lung cancer blocks regulatory T cells and supports cytotoxic lymphocyte function. Cancer Res. 2020; https://doi.org/10.1158/0008-5472.CAN-20-1885.

Laoui D, Movahedi K, Van Overmeire E, Van den Bossche J, Schouppe E, Mommer C, et al. Tumor-associated macrophages in breast cancer: distinct subsets, distinct functions. Int J Dev Biol. 2011;55(7–9):861–7. https://doi.org/10.1387/ijdb.113371dl.

Laoui D, Van Overmeire E, De Baetselier P, Van Ginderachter JA, Raes G. Functional relationship between tumor-associated macrophages and macrophage Colony stimulating factor as contributors to cancer progression. Front Immunol. 2014;5:489. https://doi.org/10.3389/fimmu.2014.00489.

Lapeyre-Prost A, Terme M, Pernot S, Pointet AL, Voron T, Tartour E, et al. Immunomodulatory activity of VEGF in cancer. Int Rev Cell Mol Biol. 2017;330:295–342. https://doi.org/10.1016/bs.ircmb.2016.09.007.

Larionova I, Cherdyntseva N, Liu T, Patysheva M, Rakina M, Kzhyshkowska J. Interaction of tumor-associated macrophages and cancer chemotherapy. Onco Targets Ther. 2019;8(7):1596004. https://doi.org/10.1080/2162402X.2019.1596004.

Lee SH, Jeong D, Han YS, Baek MJ. Pivotal role of vascular endothelial growth factor pathway in tumor angiogenesis. Ann Surg Treat Res. 2015;89(1):1–8. https://doi.org/10.4174/astr.2015.89.1.1.

Leek RD, Lewis CE, Whitehouse R, Greenall M, Clarke J, Harris AL. Association of macrophage infiltration with angiogenesis and prognosis in invasive breast carcinoma. Cancer Res. 1996;56(20):4625–9.

Leek RD, Landers RJ, Harris AL, Lewis CE. Necrosis correlates with high vascular density and focal macrophage infiltration in invasive carcinoma of the breast. Br J Cancer. 1999;79(5–6):991–5. https://doi.org/10.1038/sj.bjc.6690158.

Lewis JS, Landers RJ, Underwood JC, Harris AL, Lewis CE. Expression of vascular endothelial growth factor by macrophages is up-regulated in poorly vascularized areas of breast carcinomas. J Pathol. 2000;192(2):150–8. https://doi.org/10.1002/1096-9896(2000)9999:9999<::AID-PATH687>3.0.CO;2-G.

Li F, Sethi G. Targeting transcription factor NF-kappaB to overcome chemoresistance and radioresistance in cancer therapy. Biochim Biophys Acta. 2010;1805(2):167–80. https://doi.org/10.1016/j.bbcan.2010.01.002.

Li X, Yao W, Yuan Y, Chen P, Li B, Li J, et al. Targeting of tumour-infiltrating macrophages via CCL2/CCR2 signalling as a therapeutic strategy against hepatocellular carcinoma. Gut. 2017;66(1):157–67. https://doi.org/10.1136/gutjnl-2015-310514.

Li X, Liu R, Su X, Pan Y, Han X, Shao C, et al. Harnessing tumor-associated macrophages as aids for cancer immunotherapy. Mol Cancer. 2019a;18(1):177. https://doi.org/10.1186/s12943-019-1102-3.

Li X, Bu W, Meng L, Liu X, Wang S, Jiang L, et al. CXCL12/CXCR4 pathway orchestrates CSC-like properties by CAF recruited tumor associated macrophage in OSCC. Exp Cell Res. 2019b;378(2):131–8. https://doi.org/10.1016/j.yexcr.2019.03.013.

Liang S, Zheng J, Wu W, Li Q, Saw PE, Chen J, et al. A robust nanoparticle platform for RNA interference in macrophages to suppress tumor cell migration. Front Pharmacol. 2018;9:1465. https://doi.org/10.3389/fphar.2018.01465.

Liekens S, Schols D, Hatse S. CXCL12-CXCR4 axis in angiogenesis, metastasis and stem cell mobilization. Curr Pharm Des. 2010;16(35):3903–20. https://doi.org/10.2174/138161210794455003.

Lim SY, Yuzhalin AE, Gordon-Weeks AN, Muschel RJ. Targeting the CCL2-CCR2 signaling axis in cancer metastasis. Oncotarget. 2016;7(19):28697–710. doi: https://doi.org/10.18632/oncotarget.7376.

Lin EY, Pollard JW. Tumor-associated macrophages press the angiogenic switch in breast cancer. Cancer Res. 2007;67(11):5064–6. https://doi.org/10.1158/0008-5472.CAN-07-0912.

Lin EY, Nguyen AV, Russell RG, Pollard JW. Colony-stimulating factor 1 promotes progression of mammary tumors to malignancy. J Exp Med. 2001;193(6):727–40. https://doi.org/10.1084/jem.193.6.727.

Linde N, Lederle W, Depner S, van Rooijen N, Gutschalk CM, Mueller MM. Vascular endothelial growth factor-induced skin carcinogenesis depends on recruitment and alternative activation of macrophages. J Pathol. 2012;227(1):17–28. https://doi.org/10.1002/path.3989.

Liu L, He H, Liang R, Yi H, Meng X, Chen Z, et al. ROS-inducing micelles sensitize tumor-associated macrophages to TLR3 stimulation for potent immunotherapy. Biomacromolecules. 2018;19(6):2146–55. https://doi.org/10.1021/acs.biomac.8b00239.

Loberg RD, Ying C, Craig M, Day LL, Sargent E, Neeley C, et al. Targeting CCL2 with systemic delivery of neutralizing antibodies induces prostate cancer tumor regression in vivo. Cancer Res. 2007;67(19):9417–24. https://doi.org/10.1158/0008-5472.CAN-07-1286.

Lu-Emerson C, Snuderl M, Kirkpatrick ND, Goveia J, Davidson C, Huang Y, et al. Increase in tumor-associated macrophages after antiangiogenic therapy is associated with poor survival among patients with recurrent glioblastoma. Neuro-Oncology. 2013;15(8):1079–87. https://doi.org/10.1093/neuonc/not082.

Ma R, Yi B, Riker AI, Xi Y. Metformin and cancer immunity. Acta Pharmacol Sin. 2020;41(11):1403–9. https://doi.org/10.1038/s41401-020-00508-0.

MacDonald KP, Palmer JS, Cronau S, Seppanen E, Olver S, Raffelt NC, et al. An antibody against the colony-stimulating factor 1 receptor depletes the resident subset of monocytes and tissue- and tumor-associated macrophages but does not inhibit inflammation. Blood. 2010;116(19):3955–63. https://doi.org/10.1182/blood-2010-02-266296.

Mantovani A, Allavena P. The interaction of anticancer therapies with tumor-associated macrophages. J Exp Med. 2015;212(4):435–45. https://doi.org/10.1084/jem.20150295.

Mantovani A, Sozzani S, Locati M, Allavena P, Sica A. Macrophage polarization: tumor-associated macrophages as a paradigm for polarized M2 mononuclear phagocytes. Trends Immunol. 2002;23(11):549–55. https://doi.org/10.1016/s1471-4906(02)02302-5.

Mantovani A, Marchesi F, Malesci A, Laghi L, Allavena P. Tumour-associated macrophages as treatment targets in oncology. Nat Rev Clin Oncol. 2017;14(7):399–416. https://doi.org/10.1038/nrclinonc.2016.217.

Martinez-Pomares L, Gordon S. CD169+ macrophages at the crossroads of antigen presentation. Trends Immunol. 2012;33(2):66–70. https://doi.org/10.1016/j.it.2011.11.001.

Martinez-Pomares L, Kosco-Vilbois M, Darley E, Tree P, Herren S, Bonnefoy JY, et al. Fc chimeric protein containing the cysteine-rich domain of the murine mannose receptor binds to macrophages from splenic marginal zone and lymph node subcapsular sinus and to germinal centers. J Exp Med. 1996;184(5):1927–37. https://doi.org/10.1084/jem.184.5.1927.

Mazzieri R, Pucci F, Moi D, Zonari E, Ranghetti A, Berti A, et al. Targeting the ANG2/TIE2 axis inhibits tumor growth and metastasis by impairing angiogenesis and disabling rebounds of proangiogenic myeloid cells. Cancer Cell. 2011;19(4):512–26. https://doi.org/10.1016/j.ccr.2011.02.005.

Mazzone M, Bergers G. Regulation of blood and lymphatic vessels by immune cells in tumors and metastasis. Annu Rev Physiol. 2019;81:535–60. https://doi.org/10.1146/annurev-physiol-020518-114721.

McClellan JL, Davis JM, Steiner JL, Enos RT, Jung SH, Carson JA, et al. Linking tumor-associated macrophages, inflammation, and intestinal tumorigenesis: role of MCP-1. Am J Physiol Gastrointest Liver Physiol. 2012;303(10):G1087–95. https://doi.org/10.1152/ajpgi.00252.2012.

Medler TR, Murugan D, Horton W, Kumar S, Cotechini T, Forsyth AM, et al. Complement C5a fosters squamous carcinogenesis and limits T cell response to chemotherapy. Cancer Cell. 2018;34(4):561–78. e6. https://doi.org/10.1016/j.ccell.2018.09.003.

Meng Y, Beckett MA, Liang H, Mauceri HJ, van Rooijen N, Cohen KS, et al. Blockade of tumor necrosis factor alpha signaling in tumor-associated macrophages as a radiosensitizing strategy. Cancer Res. 2010;70(4):1534–43. https://doi.org/10.1158/0008-5472.CAN-09-2995.

Milane L, Singh A, Mattheolabakis G, Suresh M, Amiji MM. Exosome mediated communication within the tumor microenvironment. J Control Release. 2015;219:278–94. https://doi.org/10.1016/j.jconrel.2015.06.029.

Miller MA, Zheng YR, Gadde S, Pfirschke C, Zope H, Engblom C, et al. Tumour-associated macrophages act as a slow-release reservoir of nano-therapeutic Pt(IV) pro-drug. Nat Commun. 2015;6:8692. https://doi.org/10.1038/ncomms9692.

Mitchem JB, Brennan DJ, Knolhoff BL, Belt BA, Zhu Y, Sanford DE, et al. Targeting tumor-infiltrating macrophages decreases tumor-initiating cells, relieves immunosuppression, and improves chemotherapeutic responses. Cancer Res. 2013;73(3):1128–41. https://doi.org/10.1158/0008-5472.CAN-12-2731.

Morandi F, Pistoia V. Interactions between HLA-G and HLA-E in physiological and pathological conditions. Front Immunol. 2014;5:394. https://doi.org/10.3389/fimmu.2014.00394.

Mori T, Miyamoto T, Yoshida H, Asakawa M, Kawasumi M, Kobayashi T, et al. IL-1beta and TNFalpha-initiated IL-6-STAT3 pathway is critical in mediating inflammatory cytokines and RANKL expression in inflam-

matory arthritis. Int Immunol. 2011;23(11):701–12. https://doi.org/10.1093/intimm/dxr077.
Movahedi K, Laoui D, Gysemans C, Baeten M, Stange G, Van den Bossche J, et al. Different tumor microenvironments contain functionally distinct subsets of macrophages derived from Ly6C(high) monocytes. Cancer Res. 2010;70(14):5728–39. https://doi.org/10.1158/0008-5472.CAN-09-4672.
Mummalaneni S, Qian J, Phan TH, Rhyu MR, Heck GL, DeSimone JA, et al. Effect of ENaC modulators on rat neural responses to NaCl. PLoS One. 2014;9(5):e98049. https://doi.org/10.1371/journal.pone.0098049.
Murray PJ, Allen JE, Biswas SK, Fisher EA, Gilroy DW, Goerdt S, et al. Macrophage activation and polarization: nomenclature and experimental guidelines. Immunity. 2014;41(1):14–20. https://doi.org/10.1016/j.immuni.2014.06.008.
Nagai T, Tanaka M, Tsuneyoshi Y, Xu B, Michie SA, Hasui K, et al. Targeting tumor-associated macrophages in an experimental glioma model with a recombinant immunotoxin to folate receptor beta. Cancer Immunol Immunother. 2009;58(10):1577–86. https://doi.org/10.1007/s00262-009-0667-x.
Nagarsheth N, Wicha MS, Zou W. Chemokines in the cancer microenvironment and their relevance in cancer immunotherapy. Nat Rev Immunol. 2017;17(9):559–72. https://doi.org/10.1038/nri.2017.49.
Neophytou CM, Pierides C, Christodoulou MI, Costeas P, Kyriakou TC, Papageorgis P. The role of tumor-associated myeloid cells in modulating cancer therapy. Front Oncol. 2020;10:899. https://doi.org/10.3389/fonc.2020.00899.
Neyen C, Pluddemann A, Mukhopadhyay S, Maniati E, Bossard M, Gordon S, et al. Macrophage scavenger receptor a promotes tumor progression in murine models of ovarian and pancreatic cancer. J Immunol. 2013;190(7):3798–805. https://doi.org/10.4049/jimmunol.1203194.
Ngambenjawong C, Cieslewicz M, Schellinger JG, Pun SH. Synthesis and evaluation of multivalent M2pep peptides for targeting alternatively activated M2 macrophages. J Control Release. 2016;224:103–11. https://doi.org/10.1016/j.jconrel.2015.12.057.
Nishida N, Yano H, Nishida T, Kamura T, Kojiro M. Angiogenesis in cancer. Vasc Health Risk Manag. 2006;2(3):213–9. https://doi.org/10.2147/vhrm.2006.2.3.213.
Noy R, Pollard JW. Tumor-associated macrophages: from mechanisms to therapy. Immunity. 2014;41(1):49–61. https://doi.org/10.1016/j.immuni.2014.06.010.
Nywening TM, Wang-Gillam A, Sanford DE, Belt BA, Panni RZ, Cusworth BM, et al. Targeting tumour-associated macrophages with CCR2 inhibition in combination with FOLFIRINOX in patients with borderline resectable and locally advanced pancreatic cancer: a single-Centre, open-label, dose-finding, non-randomised, phase 1b trial. Lancet Oncol. 2016;17(5):651–62. https://doi.org/10.1016/S1470-2045(16)00078-4.
Okazawa H, Motegi S, Ohyama N, Ohnishi H, Tomizawa T, Kaneko Y, et al. Negative regulation of phagocytosis in macrophages by the CD47-SHPS-1 system. J Immunol. 2005;174(4):2004–11. https://doi.org/10.4049/jimmunol.174.4.2004.
Olingy CE, Dinh HQ, Hedrick CC. Monocyte heterogeneity and functions in cancer. J Leukoc Biol. 2019;106(2):309–22. https://doi.org/10.1002/JLB.4RI0818-311R.
Olson OC, Kim H, Quail DF, Foley EA, Joyce JA. Tumor-associated macrophages suppress the cytotoxic activity of antimitotic agents. Cell Rep. 2017;19(1):101–13. https://doi.org/10.1016/j.celrep.2017.03.038.
Otani Y, Kijima T, Kohmo S, Oishi S, Minami T, Nagatomo I, et al. Suppression of metastases of small cell lung cancer cells in mice by a peptidic CXCR4 inhibitor TF14016. FEBS Lett. 2012;586(20):3639–44. https://doi.org/10.1016/j.febslet.2012.08.011.
Owen JL, Mohamadzadeh M. Macrophages and chemokines as mediators of angiogenesis. Front Physiol. 2013;4:159. https://doi.org/10.3389/fphys.2013.00159.
Pathria P, Louis TL, Varner JA. Targeting tumor-associated macrophages in cancer. Trends Immunol. 2019;40(4):310–27. https://doi.org/10.1016/j.it.2019.02.003.
Paulus P, Stanley ER, Schafer R, Abraham D, Aharinejad S. Colony-stimulating factor-1 antibody reverses chemoresistance in human MCF-7 breast cancer xenografts. Cancer Res. 2006;66(8):4349–56. https://doi.org/10.1158/0008-5472.CAN-05-3523.
Perkins H, Khodai T, Mechiche H, Colman P, Burden F, Laxton C, et al. Therapy with TLR7 agonists induces lymphopenia: correlating pharmacology to mechanism in a mouse model. J Clin Immunol. 2012;32(5):1082–92. https://doi.org/10.1007/s10875-012-9687-y.
Perrotta C, Cervia D, Di Renzo I, Moscheni C, Bassi MT, Campana L, et al. Nitric oxide generated by tumor-associated macrophages is responsible for cancer resistance to cisplatin and correlated with Syntaxin 4 and acid sphingomyelinase inhibition. Front Immunol. 2018;9:1186. https://doi.org/10.3389/fimmu.2018.01186.
Petrova PS, Viller NN, Wong M, Pang X, Lin GH, Dodge K, et al. TTI-621 (SIRPalphaFc): a CD47-blocking innate immune checkpoint inhibitor with broad antitumor activity and minimal erythrocyte binding. Clin Cancer Res. 2017;23(4):1068–79. https://doi.org/10.1158/1078-0432.CCR-16-1700.
Prakash H, Klug F, Nadella V, Mazumdar V, Schmitz-Winnenthal H, Umansky L. Low doses of gamma irradiation potentially modifies immunosuppressive tumor microenvironment by retuning tumor-associated macrophages: lesson from insulinoma. Carcinogenesis. 2016;37(3):301–13. https://doi.org/10.1093/carcin/bgw007.
Qian BZ, Pollard JW. Macrophage diversity enhances tumor progression and metastasis. Cell. 2010;141(1):39–51. https://doi.org/10.1016/j.cell.2010.03.014.

Qian BZ, Li J, Zhang H, Kitamura T, Zhang J, Campion LR, et al. CCL2 recruits inflammatory monocytes to facilitate breast-tumour metastasis. Nature. 2011;475(7355):222–5. https://doi.org/10.1038/nature10138.

Qian J, Olbrecht S, Boeckx B, Vos H, Laoui D, Etlioglu E, et al. A pan-cancer blueprint of the heterogeneous tumor microenvironment revealed by single-cell profiling. Cell Res. 2020;30(9):745–62. https://doi.org/10.1038/s41422-020-0355-0.

Quail DF, Joyce JA. Microenvironmental regulation of tumor progression and metastasis. Nat Med. 2013;19(11):1423–37. https://doi.org/10.1038/nm.3394.

Rayamajhi S, Nguyen TDT, Marasini R, Aryal S. Macrophage-derived exosome-mimetic hybrid vesicles for tumor targeted drug delivery. Acta Biomater. 2019;94:482–94. https://doi.org/10.1016/j.actbio.2019.05.054.

Ries CH, Cannarile MA, Hoves S, Benz J, Wartha K, Runza V, et al. Targeting tumor-associated macrophages with anti-CSF-1R antibody reveals a strategy for cancer therapy. Cancer Cell. 2014;25(6):846–59. https://doi.org/10.1016/j.ccr.2014.05.016.

Rodell CB, Arlauckas SP, Cuccarese MF, Garris CS, Li R, Ahmed MS, et al. TLR7/8-agonist-loaded nanoparticles promote the polarization of tumour-associated macrophages to enhance cancer immunotherapy. Nat Biomed Eng. 2018;2(8):578–88. https://doi.org/10.1038/s41551-018-0236-8.

Rogers TL, Holen I. Tumour macrophages as potential targets of bisphosphonates. J Transl Med. 2011;9:177. https://doi.org/10.1186/1479-5876-9-177.

Rolny C, Mazzone M, Tugues S, Laoui D, Johansson I, Coulon C, et al. HRG inhibits tumor growth and metastasis by inducing macrophage polarization and vessel normalization through downregulation of PlGF. Cancer Cell. 2011;19(1):31–44. https://doi.org/10.1016/j.ccr.2010.11.009.

Ruffell B, Coussens LM. Macrophages and therapeutic resistance in cancer. Cancer Cell. 2015;27(4):462–72. https://doi.org/10.1016/j.ccell.2015.02.015.

Ruffell B, Chang-Strachan D, Chan V, Rosenbusch A, Ho CM, Pryer N, et al. Macrophage IL-10 blocks CD8+ T cell-dependent responses to chemotherapy by suppressing IL-12 expression in intratumoral dendritic cells. Cancer Cell. 2014;26(5):623–37. https://doi.org/10.1016/j.ccell.2014.09.006.

Salvagno C, Ciampricotti M, Tuit S, Hau CS, van Weverwijk A, Coffelt SB, et al. Therapeutic targeting of macrophages enhances chemotherapy efficacy by unleashing type I interferon response. Nat Cell Biol. 2019;21(4):511–21. https://doi.org/10.1038/s41556-019-0298-1.

Sandhu SK, Papadopoulos K, Fong PC, Patnaik A, Messiou C, Olmos D, et al. A first-in-human, first-in-class, phase I study of carlumab (CNTO 888), a human monoclonal antibody against CC-chemokine ligand 2 in patients with solid tumors. Cancer Chemother Pharmacol. 2013;71(4):1041–50. https://doi.org/10.1007/s00280-013-2099-8.

Sanford DE, Belt BA, Panni RZ, Mayer A, Deshpande AD, Carpenter D, et al. Inflammatory monocyte mobilization decreases patient survival in pancreatic cancer: a role for targeting the CCL2/CCR2 axis. Clin Cancer Res. 2013;19(13):3404–15. https://doi.org/10.1158/1078-0432.CCR-13-0525.

Sangaletti S, Di Carlo E, Gariboldi S, Miotti S, Cappetti B, Parenza M, et al. Macrophage-derived SPARC bridges tumor cell-extracellular matrix interactions toward metastasis. Cancer Res. 2008;68(21):9050–9. https://doi.org/10.1158/0008-5472.CAN-08-1327.

Sanmamed MF, Perez-Gracia JL, Schalper KA, Fusco JP, Gonzalez A, Rodriguez-Ruiz ME, et al. Changes in serum interleukin-8 (IL-8) levels reflect and predict response to anti-PD-1 treatment in melanoma and non-small-cell lung cancer patients. Ann Oncol. 2017;28(8):1988–95. https://doi.org/10.1093/annonc/mdx190.

Santarpia M, Karachaliou N. Tumor immune microenvironment characterization and response to anti-PD-1 therapy. Cancer Biol Med. 2015;12(2):74–8. https://doi.org/10.7497/j.issn.2095-3941.2015.0022.

Sato T, Terai M, Tamura Y, Alexeev V, Mastrangelo MJ, Selvan SR. Interleukin 10 in the tumor microenvironment: a target for anticancer immunotherapy. Immunol Res. 2011;51(2–3):170–82. https://doi.org/10.1007/s12026-011-8262-6.

Sawa-Wejksza K, Kandefer-Szerszen M. Tumor-associated macrophages as target for antitumor therapy. Arch Immunol Ther Exp. 2018;66(2):97–111. https://doi.org/10.1007/s00005-017-0480-8.

Scheuer W, Thomas M, Hanke P, Sam J, Osl F, Weininger D, et al. Anti-tumoral, anti-angiogenic and anti-metastatic efficacy of a tetravalent bispecific antibody (TAvi6) targeting VEGF-A and angiopoietin-2. MAbs. 2016;8(3):562–73. https://doi.org/10.1080/19420862.2016.1147640.

Seliger B. Combinatorial approaches with checkpoint inhibitors to enhance anti-tumor immunity. Front Immunol. 2019;10:999. https://doi.org/10.3389/fimmu.2019.00999.

Shime H, Matsumoto M, Oshiumi H, Tanaka S, Nakane A, Iwakura Y, et al. Toll-like receptor 3 signaling converts tumor-supporting myeloid cells to tumoricidal effectors. Proc Natl Acad Sci U S A. 2012;109(6):2066–71. https://doi.org/10.1073/pnas.1113099109.

Shree T, Olson OC, Elie BT, Kester JC, Garfall AL, Simpson K, et al. Macrophages and cathepsin proteases blunt chemotherapeutic response in breast cancer. Genes Dev. 2011;25(23):2465–79. https://doi.org/10.1101/gad.180331.111.

Smith MP, Sanchez-Laorden B, O'Brien K, Brunton H, Ferguson J, Young H, et al. The immune microenvironment confers resistance to MAPK pathway inhibitors through macrophage-derived TNFalpha. Cancer Discov. 2014;4(10):1214–29. https://doi.org/10.1158/2159-8290.CD-13-1007.

Solinas G, Germano G, Mantovani A, Allavena P. Tumor-associated macrophages (TAM) as major players of the cancer-related inflammation. J Leukoc Biol. 2009;86(5):1065–73. https://doi.org/10.1189/jlb.0609385.

Song L, Asgharzadeh S, Salo J, Engell K, Wu HW, Sposto R, et al. Valpha24-invariant NKT cells mediate antitumor activity via killing of tumor-associated macrophages. J Clin Invest. 2009;119(6):1524–36. https://doi.org/10.1172/JCI37869.

Stafford JH, Hirai T, Deng L, Chernikova SB, Urata K, West BL, et al. Colony stimulating factor 1 receptor inhibition delays recurrence of glioblastoma after radiation by altering myeloid cell recruitment and polarization. Neuro-Oncology. 2016;18(6):797–806. https://doi.org/10.1093/neuonc/nov272.

Sun X, Cheng G, Hao M, Zheng J, Zhou X, Zhang J, et al. CXCL12 / CXCR4 / CXCR7 chemokine axis and cancer progression. Cancer Metastasis Rev. 2010;29(4):709–22. https://doi.org/10.1007/s10555-010-9256-x.

Sunakawa Y, Stintzing S, Cao S, Heinemann V, Cremolini C, Falcone A, et al. Variations in genes regulating tumor-associated macrophages (TAMs) to predict outcomes of bevacizumab-based treatment in patients with metastatic colorectal cancer: results from TRIBE and FIRE3 trials. Ann Oncol. 2015;26(12):2450–6. https://doi.org/10.1093/annonc/mdv474.

Sung WW, Wang YC, Lin PL, Cheng YW, Chen CY, Wu TC, et al. IL-10 promotes tumor aggressiveness via upregulation of CIP2A transcription in lung adenocarcinoma. Clin Cancer Res. 2013;19(15):4092–103. https://doi.org/10.1158/1078-0432.CCR-12-3439.

Swierczak A, Cook AD, Lenzo JC, Restall CM, Doherty JP, Anderson RL, et al. The promotion of breast cancer metastasis caused by inhibition of CSF-1R/CSF-1 signaling is blocked by targeting the G-CSF receptor. Cancer Immunol Res. 2014;2(8):765–76. https://doi.org/10.1158/2326-6066.CIR-13-0190.

Syn N, Wang L, Sethi G, Thiery JP, Goh BC. Exosome-mediated metastasis: from epithelial-mesenchymal transition to escape from Immunosurveillance. Trends Pharmacol Sci. 2016;37(7):606–17. https://doi.org/10.1016/j.tips.2016.04.006.

Szulzewsky F, Pelz A, Feng X, Synowitz M, Markovic D, Langmann T, et al. Glioma-associated microglia/macrophages display an expression profile different from M1 and M2 polarization and highly express Gpnmb and Spp1. PLoS One. 2015;10(2):e0116644. https://doi.org/10.1371/journal.pone.0116644.

Talks KL, Turley H, Gatter KC, Maxwell PH, Pugh CW, Ratcliffe PJ, et al. The expression and distribution of the hypoxia-inducible factors HIF-1alpha and HIF-2alpha in normal human tissues, cancers, and tumor-associated macrophages. Am J Pathol. 2000;157(2):411–21. https://doi.org/10.1016/s0002-9440(10)64554-3.

Tamamura H, Hori A, Kanzaki N, Hiramatsu K, Mizumoto M, Nakashima H, et al. T140 analogs as CXCR4 antagonists identified as anti-metastatic agents in the treatment of breast cancer. FEBS Lett. 2003;550(1–3):79–83. https://doi.org/10.1016/s0014-5793(03)00824-x.

Tang X, Sui D, Liu M, Zhang H, Liu M, Wang S, et al. Targeted delivery of zoledronic acid through the sialic acid - Siglec axis for killing and reversal of M2 phenotypic tumor-associated macrophages - a promising cancer immunotherapy. Int J Pharm. 2020;590:119929. https://doi.org/10.1016/j.ijpharm.2020.119929.

Taniguchi K, Karin M. IL-6 and related cytokines as the critical lynchpins between inflammation and cancer. Semin Immunol. 2014;26(1):54–74. https://doi.org/10.1016/j.smim.2014.01.001.

Tao Y, Ning M, Dou H. A novel therapeutic system for malignant glioma: nanoformulation, pharmacokinetic, and anticancer properties of cell-nano-drug delivery. Nanomedicine. 2013;9(2):222–32. https://doi.org/10.1016/j.nano.2012.10.006.

Teng KY, Han J, Zhang X, Hsu SH, He S, Wani NA, et al. Blocking the CCL2-CCR2 Axis using CCL2-neutralizing antibody is an effective therapy for hepatocellular cancer in a mouse model. Mol Cancer Ther. 2017;16(2):312–22. https://doi.org/10.1158/1535-7163.MCT-16-0124.

Thauvin C, Widmer J, Mottas I, Hocevar S, Allemann E, Bourquin C, et al. Development of resiquimod-loaded modified PLA-based nanoparticles for cancer immunotherapy: a kinetic study. Eur J Pharm Biopharm. 2019;139:253–61. https://doi.org/10.1016/j.ejpb.2019.04.007.

Tosato G, Jones KD. Interleukin-1 induces interleukin-6 production in peripheral blood monocytes. Blood. 1990;75(6):1305–10.

Tsai RK, Discher DE. Inhibition of "self" engulfment through deactivation of myosin-II at the phagocytic synapse between human cells. J Cell Biol. 2008;180(5):989–1003. https://doi.org/10.1083/jcb.200708043.

Tseng D, Volkmer JP, Willingham SB, Contreras-Trujillo H, Fathman JW, Fernhoff NB, et al. Anti-CD47 antibody-mediated phagocytosis of cancer by macrophages primes an effective antitumor T-cell response. Proc Natl Acad Sci U S A. 2013;110(27):11103–8. https://doi.org/10.1073/pnas.1305569110.

Valeta-Magara A, Gadi A, Volta V, Walters B, Arju R, Giashuddin S, et al. Inflammatory breast cancer promotes development of M2 tumor-associated macrophages and cancer mesenchymal cells through a complex chemokine network. Cancer Res. 2019;79(13):3360–71. https://doi.org/10.1158/0008-5472.CAN-17-2158.

van Beek EM, Cochrane F, Barclay AN, van den Berg TK. Signal regulatory proteins in the immune system. J Immunol. 2005;175(12):7781–7. https://doi.org/10.4049/jimmunol.175.12.7781.

Varol C, Mildner A, Jung S. Macrophages: development and tissue specialization. Annu Rev Immunol. 2015;33:643–75. https://doi.org/10.1146/annurev-immunol-032414-112220.

Vela M, Aris M, Llorente M, Garcia-Sanz JA, Kremer L. Chemokine receptor-specific antibodies in cancer immunotherapy: achievements and challenges. Front Immunol. 2015;6:12. https://doi.org/10.3389/fimmu.2015.00012.

Vitale I, Manic G, Coussens LM, Kroemer G, Galluzzi L. Macrophages and metabolism in the tumor microenvironment. Cell Metab. 2019;30(1):36–50. https://doi.org/10.1016/j.cmet.2019.06.001.

Vonderheide RH. CD40 agonist antibodies in cancer immunotherapy. Annu Rev Med. 2020;71:47–58. https://doi.org/10.1146/annurev-med-062518-045435.

Wang P, Wang H, Huang Q, Peng C, Yao L, Chen H, et al. Exosomes from M1-polarized macrophages enhance paclitaxel antitumor activity by activating macrophages-mediated inflammation. Theranostics. 2019;9(6):1714–27. https://doi.org/10.7150/thno.30716.

Wang D, Wang X, Si M, Yang J, Sun S, Wu H, et al. Exosome-encapsulated miRNAs contribute to CXCL12/CXCR4-induced liver metastasis of colorectal cancer by enhancing M2 polarization of macrophages. Cancer Lett. 2020;474:36–52. https://doi.org/10.1016/j.canlet.2020.01.005.

Weiskopf K, Jahchan NS, Schnorr PJ, Cristea S, Ring AM, Maute RL, et al. CD47-blocking immunotherapies stimulate macrophage-mediated destruction of small-cell lung cancer. J Clin Invest. 2016;126(7):2610–20. https://doi.org/10.1172/JCI81603.

Weissleder R, Pittet MJ. The expanding landscape of inflammatory cells affecting cancer therapy. Nat Biomed Eng. 2020;4(5):489–98. https://doi.org/10.1038/s41551-020-0524-y.

Welm AL, Sneddon JB, Taylor C, Nuyten DS, van de Vijver MJ, Hasegawa BH, et al. The macrophage-stimulating protein pathway promotes metastasis in a mouse model for breast cancer and predicts poor prognosis in humans. Proc Natl Acad Sci U S A. 2007;104(18):7570–5. https://doi.org/10.1073/pnas.0702095104.

Wiehagen KR, Girgis NM, Yamada DH, Smith AA, Chan SR, Grewal IS, et al. Combination of CD40 Agonism and CSF-1R blockade reconditions tumor-associated macrophages and drives potent antitumor immunity. Cancer Immunol Res. 2017;5(12):1109–21. https://doi.org/10.1158/2326-6066.CIR-17-0258.

Williams CB, Yeh ES, Soloff AC. Tumor-associated macrophages: unwitting accomplices in breast cancer malignancy. NPJ Breast Cancer. 2016;2 https://doi.org/10.1038/npjbcancer.2015.25.

Wu Q, Allouch A, Martins I, Modjtahedi N, Deutsch E, Perfettini JL. Macrophage biology plays a central role during ionizing radiation-elicited tumor response. Biom J. 2017;40(4):200–11. https://doi.org/10.1016/j.bj.2017.06.003.

Xia Y, Rao L, Yao H, Wang Z, Ning P, Chen X. Engineering macrophages for cancer immunotherapy and drug delivery. Adv Mater. 2020;32(40):e2002054. https://doi.org/10.1002/adma.202002054.

Xu J, Escamilla J, Mok S, David J, Priceman S, West B, et al. CSF1R signaling blockade stanches tumor-infiltrating myeloid cells and improves the efficacy of radiotherapy in prostate cancer. Cancer Res. 2013;73(9):2782–94. https://doi.org/10.1158/0008-5472.CAN-12-3981.

Xuan M, Shao J, Dai L, He Q, Li J. Macrophage cell membrane camouflaged mesoporous silica nanocapsules for in vivo cancer therapy. Adv Healthc Mater. 2015;4(11):1645–52. https://doi.org/10.1002/adhm.201500129.

Yadav A, Kumar B, Datta J, Teknos TN, Kumar P. IL-6 promotes head and neck tumor metastasis by inducing epithelial-mesenchymal transition via the JAK-STAT3-SNAIL signaling pathway. Mol Cancer Res. 2011;9(12):1658–67. https://doi.org/10.1158/1541-7786.MCR-11-0271.

Yang J, Yan J, Liu B. Targeting VEGF/VEGFR to modulate antitumor immunity. Front Immunol. 2018;9:978. https://doi.org/10.3389/fimmu.2018.00978.

Yang H, Zhang Q, Xu M, Wang L, Chen X, Feng Y, et al. CCL2-CCR2 axis recruits tumor associated macrophages to induce immune evasion through PD-1 signaling in esophageal carcinogenesis. Mol Cancer. 2020;19(1):41. https://doi.org/10.1186/s12943-020-01165-x.

Yeo EJ, Cassetta L, Qian BZ, Lewkowich I, Li JF, Stefater JA 3rd, et al. Myeloid WNT7b mediates the angiogenic switch and metastasis in breast cancer. Cancer Res. 2014;74(11):2962–73. https://doi.org/10.1158/0008-5472.CAN-13-2421.

Yona S, Gordon S. From the reticuloendothelial to mononuclear phagocyte system - the unaccounted years. Front Immunol. 2015;6:328. https://doi.org/10.3389/fimmu.2015.00328.

Yoo JW, Irvine DJ, Discher DE, Mitragotri S. Bio-inspired, bioengineered and biomimetic drug delivery carriers. Nat Rev Drug Discov. 2011;10(7):521–35. https://doi.org/10.1038/nrd3499.

Yu H, Lee H, Herrmann A, Buettner R, Jove R. Revisiting STAT3 signalling in cancer: new and unexpected biological functions. Nat Rev Cancer. 2014;14(11):736–46. https://doi.org/10.1038/nrc3818.

Yu GT, Bu LL, Huang CF, Zhang WF, Chen WJ, Gutkind JS, et al. PD-1 blockade attenuates immunosuppressive myeloid cells due to inhibition of CD47/SIRPalpha axis in HPV negative head and neck squamous cell carcinoma. Oncotarget. 2015;6(39):42067–80. doi: https://doi.org/10.18632/oncotarget.5955.

Yue S, Ye X, Zhou T, Gan D, Qian H, Fang W, et al. PGRN(−/−) TAMs-derived exosomes inhibit breast cancer cell invasion and migration and its mechanism exploration. Life Sci. 2021;264:118687. https://doi.org/10.1016/j.lfs.2020.118687.

Yumimoto K, Sugiyama S, Mimori K, Nakayama KI. Potentials of C-C motif chemokine 2-C-C chemokine receptor type 2 blockers including propagermanium as anticancer agents. Cancer Sci. 2019;110(7):2090–9. https://doi.org/10.1111/cas.14075.

Zabuawala T, Taffany DA, Sharma SM, Merchant A, Adair B, Srinivasan R, et al. An ets2-driven transcriptional program in tumor-associated macrophages promotes tumor metastasis. Cancer Res. 2010;70(4):1323–33. https://doi.org/10.1158/0008-5472.CAN-09-1474.

Zanganeh S, Hutter G, Spitler R, Lenkov O, Mahmoudi M, Shaw A, et al. Iron oxide nanoparticles inhibit tumour growth by inducing pro-inflammatory macrophage polarization in tumour tissues. Nat Nanotechnol. 2016;11(11):986–94. https://doi.org/10.1038/nnano.2016.168.

Zeng D, Ye Z, Wu J, Zhou R, Fan X, Wang G, et al. Macrophage correlates with immunophenotype and predicts anti-PD-L1 response of urothelial cancer. Theranostics. 2020;10(15):7002–14. https://doi.org/10.7150/thno.46176.

Zha H, Han X, Zhu Y, Yang F, Li Y, Li Q, et al. Blocking C5aR signaling promotes the anti-tumor efficacy of PD-1/PD-L1 blockade. Onco Targets Ther. 2017;6(10):e1349587. https://doi.org/10.1080/2162402X.2017.1349587.

Zhang X, Fan J, Wang S, Li Y, Wang Y, Li S, et al. Targeting CD47 and autophagy elicited enhanced antitumor effects in non-small cell lung cancer. Cancer Immunol Res. 2017;5(5):363–75. https://doi.org/10.1158/2326-6066.CIR-16-0398.

Zhang X, Chen W, Fan J, Wang S, Xian Z, Luan J, et al. Disrupting CD47-SIRPalpha axis alone or combined with autophagy depletion for the therapy of glioblastoma. Carcinogenesis. 2018a;39(5):689–99. https://doi.org/10.1093/carcin/bgy041.

Zhang JQ, Zeng S, Vitiello GA, Seifert AM, Medina BD, Beckman MJ, et al. Macrophages and CD8(+) T cells mediate the antitumor efficacy of combined CD40 ligation and Imatinib therapy in gastrointestinal stromal tumors. Cancer Immunol Res. 2018b;6(4):434–47. https://doi.org/10.1158/2326-6066.CIR-17-0345.

Zhang SY, Song XY, Li Y, Ye LL, Zhou Q, Yang WB. Tumor-associated macrophages: a promising target for a cancer immunotherapeutic strategy. Pharmacol Res. 2020a;161:105111. https://doi.org/10.1016/j.phrs.2020.105111.

Zhang L, Tian L, Dai X, Yu H, Wang J, Lei A, et al. Pluripotent stem cell-derived CAR-macrophage cells with antigen-dependent anti-cancer cell functions. J Hematol Oncol. 2020b;13(1):153. https://doi.org/10.1186/s13045-020-00983-2.

Zhao H, Wang J, Kong X, Li E, Liu Y, Du X, et al. CD47 promotes tumor invasion and metastasis in non-small cell lung cancer. Sci Rep. 2016;6:29719. https://doi.org/10.1038/srep29719.

Zhao J, Zhang Z, Xue Y, Wang G, Cheng Y, Pan Y, et al. Anti-tumor macrophages activated by ferumoxytol combined or surface-functionalized with the TLR3 agonist poly (I : C) promote melanoma regression. Theranostics. 2018a;8(22):6307–21. https://doi.org/10.7150/thno.29746.

Zhao H, Li L, Zhang J, Zheng C, Ding K, Xiao H, et al. C-C chemokine ligand 2 (CCL2) recruits macrophage-membrane-camouflaged hollow bismuth selenide nanoparticles to facilitate photothermal sensitivity and inhibit lung metastasis of breast cancer. ACS Appl Mater Interfaces. 2018b;10(37):31124–35. https://doi.org/10.1021/acsami.8b11645.

Zheng P, Luo Q, Wang W, Li J, Wang T, Wang P, et al. Tumor-associated macrophages-derived exosomes promote the migration of gastric cancer cells by transfer of functional apolipoprotein E. Cell Death Dis. 2018;9(4):434. https://doi.org/10.1038/s41419-018-0465-5.

Zhou Y, Cao HB, Li WJ, Zhao L. The CXCL12 (SDF-1)/CXCR4 chemokine axis: oncogenic properties, molecular targeting, and synthetic and natural product CXCR4 inhibitors for cancer therapy. Chin J Nat Med. 2018;16(11):801–10. https://doi.org/10.1016/S1875-5364(18)30122-5.

Zhou W, Guo S, Liu M, Burow ME, Wang G. Targeting CXCL12/CXCR4 Axis in tumor immunotherapy. Curr Med Chem. 2019;26(17):3026–41. https://doi.org/10.2174/0929867324666170830111531.

Zhou J, Tang Z, Gao S, Li C, Feng Y, Zhou X. Tumor-associated macrophages: recent insights and therapies. Front Oncol. 2020;10:188. https://doi.org/10.3389/fonc.2020.00188.

Zhu XD, Zhang JB, Zhuang PY, Zhu HG, Zhang W, Xiong YQ, et al. High expression of macrophage colony-stimulating factor in peritumoral liver tissue is associated with poor survival after curative resection of hepatocellular carcinoma. J Clin Oncol. 2008;26(16):2707–16. https://doi.org/10.1200/JCO.2007.15.6521.

Zippelius A, Schreiner J, Herzig P, Muller P. Induced PD-L1 expression mediates acquired resistance to agonistic anti-CD40 treatment. Cancer Immunol Res. 2015;3(3):236–44. https://doi.org/10.1158/2326-6066.CIR-14-0226.

Macrophage Targeting by Nanocarriers for Therapy of Autoimmune Diseases

Largee Biswas, Monika Yadav, Priyanka Singh, Sushma Talegaonkar, and Anita Kamra Verma

Abstract

Macrophages and monocytes are essentially the cells of the innate immune system residing in all tissues that include liver and adipose tissue. Owing to their phenotypic flexibility, they play critical roles in tissue homeostasis, but they may also contribute to the initiation and progression of metabolic disorders. Targeted drug delivery to the macrophages appears to be an attractive proposition to improve therapeutic efficacy of enclosed drug. Thus, they are ideal therapeutic targets for diseases such as insulin resistance, nonalcoholic fatty liver disease, and atherosclerosis. Currently, development in the field of drug delivery has enabled phenotype-specific targeting of macrophages. Macrophages can be exploited as Trojan horses for targeted drug delivery. Nanocarriers can migrate across the different membrane barriers and release their drug cargo at sites of infection. In this review, we deliberate on the advances in therapeutics for autoimmune disorders via macrophage-specific delivery.

We highlight microspheres/microparticles, liposomes, nanoparticles, dendrimers, metallic lipid nanoparticles for targeted delivery to the macrophages and how they can be optimized to alter the macrophage phenotype to exploit its therapeutic potential to combat metabolically favorable tissue environment.

Keywords

Macrophage polarization · Nanocarriers · Macrophage receptors · Autoimmune disorders

1 Introduction

Macrophages are a population of assorted immune cells that exhibit multiple functions to help maintain homeostasis of immune responses. The wide-ranging spectrum of macrophage functions depends on both diversity and plasticity of these cells that are tremendously sensitive to the microenvironment and modulate their properties. Immuno-metabolism is an imminent field that allows understanding of the metabolic pathways of immune cells and regulates how the metabolic phenotype may be modulated to alter the immune cell effector functions. It has been reported that

Authors Largee Biswas and Monika Yadav have equally contributed to this chapter.

L. Biswas · M. Yadav · P. Singh · A. K. Verma (✉)
Nanobiotech Lab, Department of Zoology, Kirori Mal College, University of Delhi, Delhi, India
e-mail: akverma@kmc.du.ac.in

S. Talegaonkar
Department of Pharmaceutics, Delhi Pharmaceutical Sciences and Research University, New Delhi, India

S. Gupta, Y. V. Pathak (eds.), *Macrophage Targeted Delivery Systems*,
https://doi.org/10.1007/978-3-030-84164-5_14

the microenvironment determines cell metabolism thereby contributing to their functionality. Microenvironmental indicators may be involved in metabolic regulation of cytokines, growth factors, oxygen levels, and availability of nutrients. There are innumerable reports signifying that specific microenvironment can modulate the phenotype of immune cells, for example, tumors or inflammation of tissues may reprogram the metabolic phenotype of the immune cells to achieve the cellular necessities that include survival, growth, proliferation, and suppression, or even to enable certain other effector functions, such as cytokine release and phagocytosis. By modulating the immune-metabolism of cells, especially macrophages, it will empower variation of their activity that has advantageous in combating different diseases that necessitate enhanced macrophage commitment. Both monocytes and macrophages possess extensive inflammatory, immuno-modulatory, as well as tissue-healing abilities affecting many autoimmune diseases (Laria et al. 2016). These cells secrete several cytokines and chemokines that induce and recruit additional immune cells to the diseased site (Navegantes et al. 2017). Presence of autoantibodies and autoreactive B and T cells is indicative of the critical role of adaptive immune system in pathogenesis, although it cannot completely justify the onset and progression of autoimmune diseases, as the innate immune response initially plays a pivotal and irreplaceable role in initiation of many autoimmune diseases (Laria et al. 2016; Ma et al. 2017). Actually, infiltration of monocytes or macrophages is usually observed in several autoimmune diseases (Misharin et al. 2014; Bramow et al. 2010). Furthermore, alteration in the frequency and count of monocyte/macrophages is a characteristic feature in various autoimmune disorders such as rheumatoid arthritis (RA), systemic sclerosis (SSc), primary biliary cholangitis (PBC), inflammatory bowel disease (IBD), and Sjögren's syndrome (SS) (Misharin et al. 2014; Wildenberg et al. 2009; Yilmaz et al. 2018; Ma et al. 2019). Hence, understanding the role of macrophages in autoimmune disorders along with its regulation will be quintessential to exploit the macrophage subsets as a therapeutic tool to target disease outcomes (Ma et al. 2019).

2 Understanding the Macrophage Lineage

Although known as monocytes when in circulation, macrophages are present in tissues as resident cells traversing their environs and eliminating the invading pathogens, the apoptotic debris of cells thereby preserving the integrity of tissues. Initially, it was hypothesized that the tissue macrophages differentiate from those monocytes that depart from the bloodstream during inflammatory conditions. But now it is established that the monocyte-derived macrophages originate in the bone marrow during hematopoiesis and the tissue macrophage progenitors are derived from the yolk sac and fetal liver during the primitive and definitive hematopoiesis (Epelman et al. 2014). Remarkably, the embryo-derived macrophages still retain their self-renewal potential, although the monocyte-derived cells are often terminally differentiated (Röszer 2018).

Based on the tissue they occupy, macrophages have been termed as Kupffer cells in liver, alveolar in lungs, histiocytes in bone marrow, microglia in nervous system, peritoneal in the peritoneum, and synovial macrophages in the bone (Fig. 1) (Arango Duque and Descoteaux 2014; Ta et al. 2013). Macrophages are the key players of the innate immunity as their main function is phagocytosis, to engulf any foreign substances, digest the cellular debris, and invade microbes and pathogens. They secrete cytokines and chemokines that regulate the activities of a variety of immune cells in the inflammatory lacerations. Antigen presentation to T cells is another important function of macrophages in the innate immune system which is a link between the innate and the acquired immunity as well as the dendritic cell (Shi and Pamer 2011). Macrophages enable recovery of injured tissue by triggering angiogenesis or fibrosis (Peter and Thomas 2011).

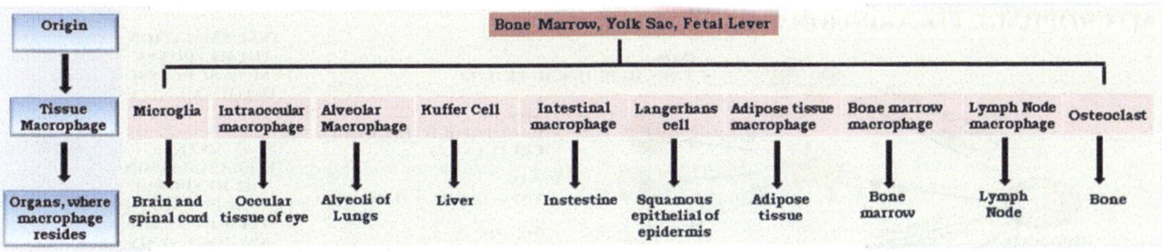

Fig. 1 Macrophage origin and their plasticity

Typically activated (M1) macrophages secrete pro-inflammatory cytokines, like interleukin (IL)-1β, interferon (IFN)-γ, and tumor necrosis factor (TNF)-α that elicit numerous inflammatory responses. On the other hand, activated (M2) macrophages secrete anti-inflammatory cytokines, like IL-4 and IL-10. Generally, three subsets of macrophages, namely M2a, b, and c, have been identified based on their characteristic trigger or specific cytokine profiles (Arango Duque and Descoteaux 2014; Biswas and Mantovani 2010). When naive monocytes come in contact with polarization indicators like hormones, lipids, cytokines, chemokines, and bacterial products, they differentiate into typically activated (M1) or alternatively activated (M2) macrophages (Fig. 2).

On exposure to TNF-α, IFN-γ, or lipopolysaccharide (LPS), differentiation of monocytes to M1 macrophage development is triggered. M1 macrophages in turn secrete pro-inflammatory cytokines, including IL-6, IL-1β, IL-12, IL-23, and TNF-α, that further promote differentiation of Th1 cells (Arango Duque and Descoteaux 2014; Peter and Thomas 2011). On the other hand, M2 macrophages are classified into three subsets, based on the varied responses to several stimuli. Further, when naive monocytes are exposed to IL-4 and IL-13, it stimulates the development of M2a macrophages that express arginase-I and triggers secretion of TGF-β, IL-10, IL-1Ra, and chemokines such as CCL17, CCL22, and CCL24 to help develop Th2 cells, basophils, and eosinophils. The other subset M2b macrophage differentiation is triggered by LPS, apoptotic cells, or even immune complexes. Sequentially, the M2b macrophages produce inducible nitric oxide synthase and secrete high level of TNF-α, IL-1β, IL-6, IL-10, and CCL1, leading to the increased recruitment of Treg cells and eosinophils. Lastly, M2c subsets of macrophages are elicited by exposure to TGF-β, IL-10 or glucocorticoids and they too express arginaseI. M2c macrophages wield an immune-suppressive purpose by endorsing the development of Treg cells and Th2 cells (Arango Duque and Descoteaux 2014; Biswas and Mantovani 2010).

2.1 Role of Macrophages in Autoimmune Diseases

Autoimmune disorders develop through a multifaceted, interlinked immune responses by diverse immune cells in the lymphoid organs as well as the target organs (Cho and Feldman 2015). Quite a few autoimmune disorders involve T cell-mediated autoimmune responses at the onset or progression of the disease. Autoreactive T cells originate in thymus and peripheral lymphoid organs where they develop in an environment that comprises stromal cells, other immune cells, and several epithelial cells (Mueller 2010; Waldmann 2010). Since autoreactive T cells are controlled by an intricate crosslink with regulatory T (Treg) cells, macrophages, dendritic cells, and B cells, the immune-pathogenesis of any autoimmune disease should never be considered merely as T-cell-mediated only (Liston and Gray 2014). Plethora of reports discussing the interactions of T cells with other immune cells in the pathogenesis of autoimmune disease are published (Hogquist and Jameson 2014). Even though the regulatory mechanism of monocytes and macrophages in the progression of autoimmune diseases has yet not been completely elucidated,

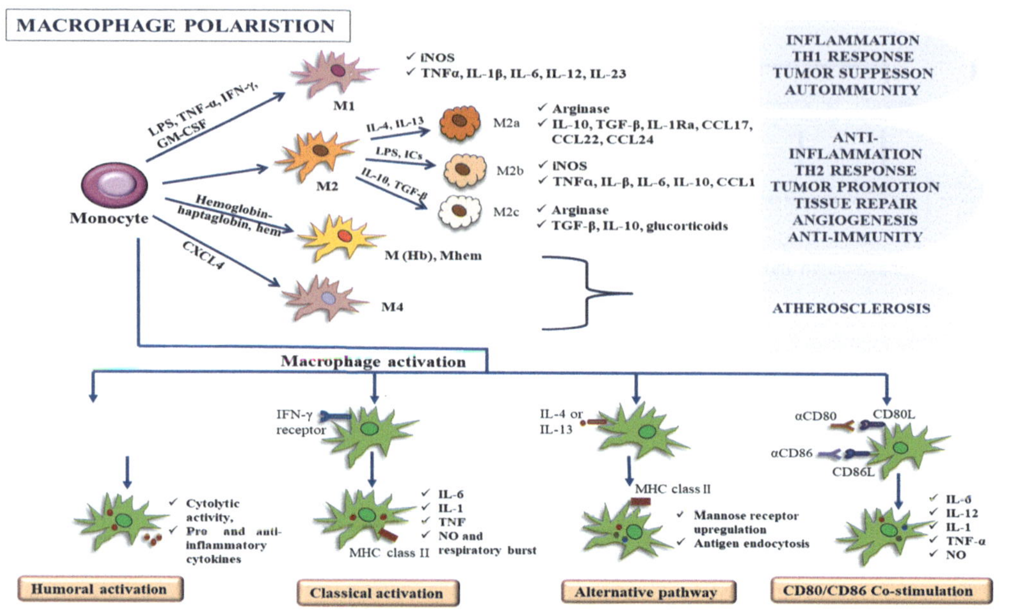

Fig. 2 *Activation and differentiation of macrophages.* Monocytes differentiate into M1, M2, and atherosclerosis-related macrophages [M (Hb), M4] by various stimuli. M1 elevates inflammation by release of TNF-α, IL-1β, IL-6, IL-12, and IL-23. M2 polarize into three subsets: M2a, M2b, and M2c by cytokines or ICs to suppress inflammation and repair tissues through producing regulatory cytokines, such as IL-10, TGF-β, IL-6. *ICs* immune complexes, *IFN* interferon, *TNF* tumor necrosis factor, *IL* interleukin, *TGF-β* transforming growth factor-β, *LPS* lipopolysaccharide, *iNOS* inducible nitric oxide synthase, *GM-CSF* granulocyte-macrophage colony-stimulating factor

general agreement suggests that their aberrant activity plays a pivotal role.

Generally, M1-polarized macrophages are pro-inflammatory and promote release of tumor necrosis factor (TNF)-α and interleukin (IL)-12 to cause local inflammation. Alternately, M2-polarized macrophages secrete IL-4 and IL-10 that generate wound healing, tissue remodeling, and immunomodulatory functions (Funes et al. 2018). Nevertheless, simply using the M1/M2 dichotomy may not justify the complexity of macrophage activation and function. In reality, certain autoimmune diseases show enhanced M1- and M2-polarization of macrophages as both populations are detected simultaneously, and increased cytokine secretion is present after stimulation of both M1- and M2-macrophages (Godsell et al. 2016; Mellor-Pita et al. 2009). Furthermore, macrophages also exhibit a state of intermediate activation by co-expressing both M1- and M2-specific biomarkers in some autoimmune diseases (Soldano et al. 2018; Vogel et al. 2013). Also, polarization in macrophages is a dynamic and reversible occurrence totally governed by the local environs and the stage of disease (Piccolo et al. 2017).

2.1.1 Rheumatoid Arthritis (RA)

Infiltration of macrophages in synovia is the key indicator and a reliable biomarker of RA progression. There is sufficient evidence that total number and the frequency of macrophages are distinctly enhanced in the synovial tissues of RA patients (Sack et al. 1994; Janossy et al. 1981). Mulherin et al. observed that the number of synovial macrophages correlated directly with the articular destruction in RA (Mulherin et al. 1996). Plethora of reports suggest the fundamental role of macrophages activation in RA pathogenesis. To be precise, unrestricted pro-inflammatory M1 polarization along with incomplete M2 polarization often leads to very severe joint pathology, and therefore, modulation of macrophage polarization alters the outcome in experimental arthritis. In a collagen II-induced arthritis rat model, it was observed that a potent pro-arthritic protein, cyclophilin A, intensified the arthritic severity via the initiation of polarization of pro-inflammatory M1 macrophages coupled with cytokine production in the knee joint (Dongsheng et al. 2017). Alternatively, competently suppressed M1 polarization or enhanced anti-inflammatory M2 polarization repressed the synovial inflammation resulting in efficient targeted therapy for RA. Similarly, efficient amelioration of collagen-induced arthritis was observed postadministration of mesenchymal stem cells (MSCs) that have strong immunomodulatory abilities (Shin et al. 2016; Li and Hua 2017). Additionally, IL-10 was capable of suppressing the observed effects of pro-inflammatory M1 macrophages in experimental arthritis, probably due to the inhibition of NF-κB (inflammation-linked nuclear factor kappa-light-chain-enhancer of the activated B cells) signaling pathway or secretion of pro-inflammatory cytokines by macrophages (Chen et al. 2017; Ye et al. 2014). Misharin et al. reported that $Ly6C^{-}$macrophages were recruited to the synovium and differentiated into pro-inflammatory M1 macrophages in the effector phase of arthritis, thereby promoting initiation and development of inflammation in the joints. In fact, the tissue-resident synovial macrophages maintain their anti-inflammatory polarization all through the course of arthritis and hinder joint inflammation in the initiation phase (Misharin et al. 2014).

The essential mediators of effector macrophages in the development of RA are activated macrophages that are an effective source of innumerable pro-inflammatory cytokines (Littlewood-Evans et al. 2016). TNF-α is a significant cytokine secreted by the synovial macrophages and is crucial in the pathogenesis of RA (Sun et al. 2017). This cytokine is expressed in most arthritis biopsies, and its overexpression promotes spontaneous inflammatory arthritis, whereas its inhibition suppresses various rodent arthritis models (Lin 2013). Accordingly, therapeutic targeting of TNF-α signaling has yielded clinical efficacy in patients with established RA, which has also been corroborated by a number of mouse model-based results (Yarilina et al. 2012; McInnes and Schett 2007). Other macrophage-derived cytokines such as IL-1, IL-6, and IL-12 are also abundantly present in the arthritic synovium of patients with RA (Lin 2013) (Fig. 3a).

2.1.2 Systemic Lupus Erythematosus (SLE)

Role for macrophages in the SLE pathogenesis was initially proposed in the early 1980s post the discovery of defect in the ability of SLE macrophages to clear apoptotic cell debris thereby prolonging the exposure of potential autoantigens to the adaptive immune response (Kawane et al. 2001). The innate immune function of monocytes and macrophages in initiation and development of SLE may be understood with respect to their role in (1) formation of immune complex (IC) recognition and clearance, (2) nucleic acid recognition via toll-like receptors (TLRs) and downstream signaling, and (3) interferon signaling (Kawane et al. 2001). Weakened engulfment of cell debris by macrophage causes stimulation of autoimmune responses resulting in acute autoimmune anemia and chronic arthritis (Kawane et al. 2006; Malkiel et al. 2018). Apoptotic cells release apoptotic bodies that work as "find me" indicators, like lysophosphatidylcholine, to recruit phagocytes and depict "eat me" signals on the cellular surface that trigger enhanced engulfment. The other signal is phosphatidylserine, which is overexpressed on cells during apoptosis. Macrophages identify the phosphatidylserine and phagocytose the apoptotic cells that are transported to the lysosomes subjected to lysosomal degradation. The clearance of apoptotic cells in the body is controlled by the multifaceted cellular and molecular interaction with macrophages. If apoptotic cells are not eliminated, they endure secondary necrosis, wherein the plasma membrane disintegrates with burst release of the cellular contents. These later bind to the immunoglobulins and the plasma complement proteins that further activate B cells and macrophages. B cell receptor, Fc receptors, and TLR are capable of recognizing the necrotic compo-

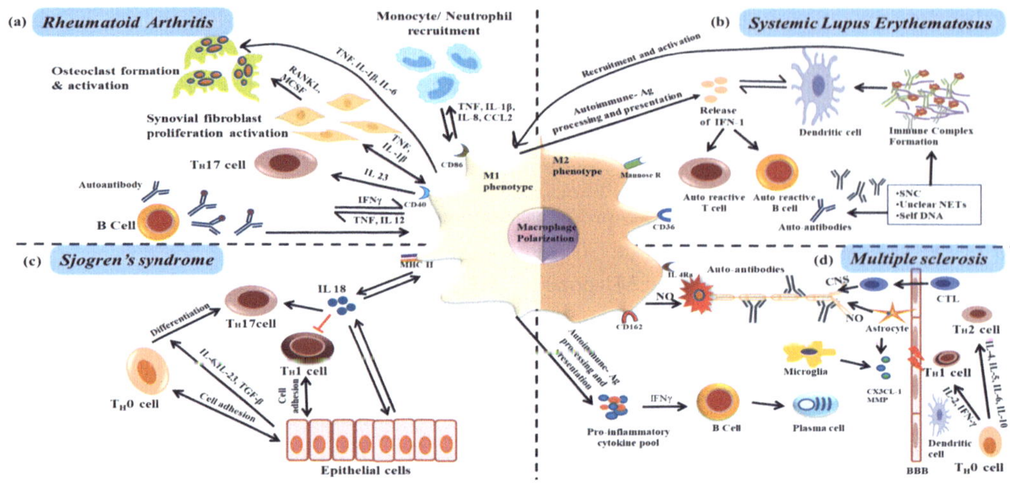

Fig. 3 *Overview of macrophage functions that are capable to initiate, regulate, and perpetuate autoimmune pathology in autoimmune disorders.* Macrophages reveal diverse functional phenotypes with two extreme M1 (pro-inflammatory) and M2 (anti-inflammatory) polarization states in response to various stimuli like injury, repair, and fibrosis. Macrophages secrete cytokines that promote inflammation by recruitment of T cell polarization and fibroblast activation. Activated fibroblasts then secrete receptor activator of nuclear factor κB ligand (RANKL) and macrophage colony-stimulating factor 1 (M-CSF), which induce differentiation of osteoclast, which is further enhanced by cytokines and macrophage-derived TNF. Autoantibodies form immune complexes with antigens and activated macrophages. Moreover, cell–cell contact and cytokines secreted by T-cells influence macrophages, fibroblasts, and the innate immune cells. *CCL2* monocyte chemo-attractant protein 1, *TH1* type 1 T helper cells, *TH17* type 17 T helper cells. (**a**) represents the events occurring in RA patient; (**b**) represents similar events in SLE; (**c**) represents similar events in SS and (**d**) represents similar events in MS

nents to stimulate the macrophages and B cells. These activated macrophages then secrete cytokines that trigger B cells to generate autoantibodies that cause the pathophysiological condition such as systemic erythematosus (Malkiel et al. 2018; Orme and Mohan 2012). Also, if the lysosomal degradation process is impaired, the cell debris accumulates in the lysosomes that cause activation of intracellular pro-inflammatory cytokines like TNF-α and IFN-β by the innate immune system (Malkiel et al. 2018). Therefore, both extrinsic stimulation of the immune responses by the nonphagocytized apoptotic cells and the intracellular activation of the macrophages due to defective processing of apoptotic cells are the key contributors to the onset and progression of autoimmunity.

The main contribution of monocyte/macrophage to SLE pathogenesis is the modulation of adaptive immune system. The binding of co-stimulatory molecule CD40 to its ligand CD40L is necessary for activation of the humoral responses that include activation of B cell, differentiation of plasma cell, secretion of antibodies, and isotype-switching. Overexpression of adhesion molecules may further lead to inconsistent macrophage activation and migration. Macrophage from active SLE patients overexpress intercellular adhesion molecule (ICAM)-1, which is associated with tissue recruitment and inflammatory cytokine production, and this is partially equipoised by corticosteroid treatment (Banwell et al. 2011) (Fig. 3b).

2.1.3 Sjögren's Syndrome (SS)

SS is a systemic autoimmune disease that targets the exocrine glands, such as lacrimal and salivary glands, which causes chronic autoimmune lesions (Katsifis et al. 2007). The mononuclear cell populations infiltrating the salivary gland tissues of SS patients include macrophages, CD4^{+} T cells (Th2), CD8^{+} T cells (Th1), Treg cells, B cells, NK cells, and DCs (Greenwell-Wild et al. 2011). The severity of lesions is correlated to the infiltration of macrophages, DCs, B cells, $CD4^+$ T cells, and Treg cells (Ravetch 2008). Although SS is stimulated by T cell-mediated autoimmune responses, other immune cells like macrophages contribute to the onset or progression of SS. Macrophages are present in the autoimmune lesions in SS patients. Undeniably, an enhanced expression of macrophage-derived molecules, such as chitinase-3-like protein-1 and chitinase-1, is associated with increased severity of SS lesions, suggesting that macrophages are involved in the pathogenesis of SS (Carroll 2004). Additionally, macrophages secrete pro-inflammatory cytokines that include IL-6, TNF-α, IL-12, and IL-1β and have also been linked with the stimulation of autoimmune lesions. Moreover, autoreactive T cells trigger macrophage infiltration in the target organs of macrophage-associated autoimmune lesions (Fig. 3c).

2.1.4 Multiple Sclerosis (MS)

Multiple sclerosis (MS) is an incapacitating neurological disease affecting the central nervous system (CNS). It is a T-cell-mediated autoimmune disorder with characteristic weakness of muscle followed by progressive paralysis (Ousman and Kubes 2012). Th1 and Th17 cells induce autoreactive responses within CNS via the pro-inflammatory cytokines that include IL-12, IL-17, IL-23, and IFN-γ. Th2 and Treg cells regulate the inflammatory processes during the later phases of MS.

Under physiological conditions, few infiltrating macrophages are found in the CNS. But during initiation and progression, macrophages start infiltrating the meninges surrounding the CNS, the peri-vascular space, and the choroid plexus (Jiang et al. 2012). Further, infiltrating macrophages reduce in the CNS during reduction in disease postmedication in parallel with the decrease in lymphocyte infiltrates (Jiang et al. 2012). The chemokines secretion and expression of chemokine receptors in CNS macrophages contribute to the induction and progression of MS (Pyonteck et al. 2013).

Together M1 and M2 macrophages play a pivotal role in developing and controlling the pathogenesis of MS. Activated M1 macrophages secrete molecules like nitric oxide; cytokines, such as IL-12, IL-1β, and TNF-α, stimulate inflammatory responses causing tissue lesions in CNS (Denney et al. 2012; Liu et al. 2013). Reduced M2 subsets when compared to M1 macrophages present in the CNS are the key factors in MS patients. Enhanced M2 macrophage subset stimulates an anti-inflammatory response associated with heightened immune response due to the secretion of TGF- β, IL-4, IL-10, and IL-13 (Ushio et al. 2017) (Table 1). Furthermore, M2 macrophages are the key regulators of the pathogenesis of disease. Along with tissue repair, M2 macrophages promote the differentiation and enrolment of Th2 response and Treg cells to suppress the autoimmune response (Fox et al. 1984) (Fig. 3d).

3 Role of Macrophage Receptor in Autoimmune Diseases

Macrophages express a variety of surface receptors that recognize both pathogenic and host-derived ligands. These are opsonic Fc receptors (Zani et al. 2015), complement receptors (Fu et al. 2012), nonopsonic receptors, like lectins (e.g., DC-SIGN, Mannose Receptor, Dectin-1), and scavenger receptors (e.g., SR-A I/II, MARCO) (Plüddemann et al. 2007). The nonopsonic receptors have a promising role in autoimmune disease and are often ignored and less appreciated, when compared to the importance of Fc and complement receptors.

3.1 Complement Receptor: It is complex and heterogeneous in structure, expression, and function by either activating or suppressing macrophage responses, based on the ligand and receptor (Zani et al. 2015; Fu et al. 2012). Opsonins that promote uptake by these receptors include immune complexes: IgG complexed with antigens, antibody, complement cleavage products, fibronectin, or milk fat globulins. Macrophages overexpress a range of receptors recognizing complement molecules, such as CR3, that play a significant role in apoptotic cell clearance.

Table 1 Triggers for monocytes/or macrophages activation, recruitment, development, molecular metabolic mechanisms, functional abnormalities, as well as macrophage-derived mediators for disease progression in few autoimmune diseases

Disease	Triggers for monocyte/macrophage recruitment and activation	Molecular mechanisms of monocyte/macrophage function	Functional abnormalities	Monocyte/macrophage-derived mediators in disease progression
RA	**Neutrophil micro-vesicles:** inhibiting inflammatory activation of macrophage **GM-CSF, osteopontin, CCL2:** monocyte migration and recruitment **Activin A:** pro-inflammatory macrophage generation	**Liver X receptor pathway:** potentiating TLR-driven cytokine production from macrophage **Succinate/GPR91 signaling:** production of IL-1β from macrophage **NFAT5:** activated macrophage survival	Enhanced monocyte CD276, CD80, and Siglec-1 expression	**TNF-α, IL-1, IL-6 and IL-12**: mediating systemic and local inflammation **TNF-α, IL-1 and IL-6:** mediating cartilage degradation
SLE	**IFN-α:** B-lymphocyte stimulator expression in monocyte **microparticle associated immune-complexes:** Activation of pro-inflammatory monocyte **TNF-α:** monocyte NF-κB inflammatory response	**Increased IRF1 expression:** enhanced inflammasome activity **Decreased PPAR-γ expression:** pro-inflammatory functions **Decreased KLF2, KLF4 expressions:** defective phagocytosis	Enhanced expression of ICAM-1, CD40, Siglec-1, CD86; defective phagocytic ability	**TNF-α, IL-1β, IL-6, and IL-10:** mediating systemic and local inflammation
SS	**Extranuclear accumulation of DNA:** activation of NLRP3 inflammasome **MIF:** local infiltration of macrophage	**Activated NF-κB pathway:** amplifying cytokine production and inflammatory response **MicroRNAs:** targeting the canonical TGF-β signaling pathway	Reduced phagocytic ability	**CCL22:** enhancing autoreactive T-cell response and recruitment **IL-6, IL-18, BAFF and type-I IFN:** mediating pro-inflammatory immune responses
MS	**GM-CSF:** monocyte migration across the blood-brain barrier **IFN-γ**: activation of microglia/macrophage of MS-affected brain tissue **CCL2:** recruitment of M1 macrophage	**KLF2:** negative regulation of macrophage activation **Decreased SHP1 signaling:** enhanced inflammatory activity of monocyte	Enhanced expression of CD68, HLA. Abnormal metabolic changes (more glycolysis)	**GM-CSF:** migration of monocyte across the blood brain barrier **KLF2:** negative regulation of macrophage activation **TNF-α, IL-1β, IL-6, IL-12, reactive oxygen, and nitrogen species:** mediating inflammatory responses

Resident macrophages in pancreas and liver express a newly reported CRIg that has a protective function in autoimmune diabetes (Berwin et al. 2004).

3.2 Fc Receptors: They represent the immunoglobulin superfamily possessing the immunoreceptor tyrosine-based activation or inhibitory motifs (ITAMs and ITIMs) that interplay with other membrane molecules and cytosolic kinases to control complex signaling cascades.

3.3 The macrophage scavenger receptors are a large family of structurally diverse receptors that identify a series of ligands that include altered low-density lipoproteins (LDL), designated polyanionic ligands, and microbial structures (Shimaoka et al. 2004). The Class A scavenger receptor I (SR-AI) was initially identified by Brown and Goldstein. Other Class A scavenger receptors expressed on macrophages include two alternative splice variants of the SR-AI gene, SR-AII, and SR-AIII and the distinct macrophage receptor with collagenous structure (MARCO). Interestingly, all receptors are trimeric, type II trans-membrane glycoproteins usually comprising of numerous domains that include an α-helical coiled domain, a collagenous domain, as well as C-terminal cysteine-rich domain; although the coiled domain is absent in MARCO, it has elongated collagenous domain. For Class B scavenger receptors, CD36 and SR-B, ligand binding is probably mediated through the central domain of the extracellular loop assembly and these receptors identify modified LDL, but the native lipoproteins (VLDL, LDL, and HDL) too, and have a significant role in transport of cholesterol, metabolism, and homeostasis. The other scavenger receptors overexpressed by macrophages are CD68 and FEEL-1 (primarily expressed intracellularly), SREC, and SR-PSOX. SREC receptors bind to the molecular chaperones (e.g., gp96, calreticulin, and heat shock protein 70) that facilitate transport of ligands to the major histocompatibility complex (MHC) Class-I antigen presentation pathway (Gazi and Martinez-Pomares 2009). SR-PSOX is the chemokine ligand for a G protein-coupled CXC chemokine receptor 6 (CXCR6) expressed on activated T cells and NKT cells, supporting adhesion of these cells to dendritic cells (DCs) (Kerrigan and Brown 2011).

3.4 C-type Lectin Receptors: C-type lectin receptors expressed on macrophages are **mannose receptor (MR), DC-SIGN, SIGNR1, Dectin-1, and Dectin-2** (Drummond et al. 2011). These receptors usually have one or more carbohydrate recognition domains that bind ligands in a Ca21-dependent manner that recognize mannose- or galactose-type ligands (Fiete et al. 1998). In most cases, recognition is independent of the carbohydrate recognition domain or Ca21. β-glucan receptor is Dectin-1 that bears a cytoplasmic hemi-ITAM, that is coupled to Syk-kinase and signals through CARD9, which results in stimulation of NF-κB, thereby inducing both innate and adaptive immunity (Ariizumi et al. 2000). All together, these receptors play a role in pathogen recognition (fungi, bacteria, and viruses) and help recognize endogenous ligands. Sulfated carbohydrates are present on endogenous glycoproteins bind to MR (e.g., CD45 and sialoadhesin) through a cysteine-rich domain (Upchurch et al. 2016) and play a vital role in clearance of self-antigens. Dectin-1 further interacts with an endogenous ligand on activated T cells, triggering T cell proliferation, and may stimulate CD41 and CD81 T-cells (Rothlin et al. 2015). Dectin-1 agonists can trigger DCs, further promoting Th1/Th17 differentiation and initiating expression of IL-17 by Treg cells (Ariizumi et al. 2000).

3.5 SR-A, CD36, CD14 Receptors: Macrophages play a role in tissue maintenance and homeostasis, for example, the involvement of several macrophage receptors (e.g., **SR-A, CD36, CD14**) in the clearance of apoptotic cells, such as apoptotic thymocytes.

3.6 TAM Receptor Tyrosine Kinases: Tyro3, Axl, and Mer and their ligands (Gas6 and Protein S) are essential for the phagocytosis of apoptotic cells and membranes. TAM signaling is normally activated by TLR and type I interferon signaling, as part of the innate inflammatory response in macrophages and DCs (Jain et al. 2013). The recognition of apoptotic cells by CD36 has been shown to contribute to peripheral tolerance and prevention of autoimmunity by impairing DC maturation (e.g., ligation of cell surface CD36 negatively regulates DC maturation by enhancing IL-10 secretion and reducing the secretion of TNF-α and IL-1β) (Gorantla et al. 2020).

3.7 Folate Receptor: Macrophages express key potential targets for folate-based therapeutics as they express folic acid binding receptor, that is, folate receptor beta (FR-β), having very high affinity for folic acid overexpressed only by activated macrophages (Nogueira et al. 2015). Therefore, a folate-based nanoformulation will possibly deliver the therapeutic agents to activated macrophages without affecting normal cells.

3.8 CD44 (Glycoprotein): These are overexpressed on the surface of activated macrophages at the pannus and synovial region in RA condition. Hence, folic acid and hyaluronic acid molecules that can easily bind to folate receptor and CD44 receptor, respectively, can be used as potential ligands for drug targeting synovial macrophages (Kottarath et al. 2020). Nogueira et al. explored folic acid receptors linked with methotrexated-loaded liposome by using SP-DS3 peptide-enhanced discrete internalization in activated macrophages for the treatment of RA, compared to macrophages that lack of receptors (Li et al. 2020).

3.9 E-Selectin (Glycoprotein): These receptors are expressed on vascular endothelium upregulated in RA condition likewise overexpression of vasoactive intestinal peptides on the macrophages and synoviocytes during the proliferating stage can be used as targeting approach. Upregulation of other biomarkers like alpha V and beta 3 integrins at the time of hypoxia condition at the inflamed synovium targeted can be a good approach to avert angiogenesis and bone resorption (Kottarath et al. 2020). Furthermore, another study done by Kottarath et al. fabricated methotrexate (MTX) encapsulated cholesterol grafted chitosan nanocarrier and tagged with antifolate receptor-β antibody to facilitate the active targeting in RA management (Chalmers et al. 2015). SLE progresses due to vCD4 and CD8 negative T-cells and loss of function in splenic macrophages, which is the leading cause of morbidity and mortality for SLE patients (Jahagirdar et al. 2019). Congenital deletion of CSF-1 (colony stimulating factor 1) or CSF-1R (colony stimulating factor 1 receptor) in lupus macrophages ameliorated kidney disease. Therefore, macrophages can be targeted by targeting CSF-1/CSF-1R signaling, CX3CR1, MCP-1 (macrophage chemotaxic protein-1)/CCR2 (receptor of MCP-1) axis, and Bruton's Tyrosine Kinase (BTK) for effective SLE therapy (Suzuki et al. 2018).

3.10 Mannose Receptor: It is a C-type lectin, composed of carbohydrate recognition domain, the carboxyl-terminal domain, cysteine-rich region, and the fibronectin region. Because of these domains, it can recognize and attach to a variety of ligands (Mukhtar et al. 2020). Mannose receptors/mannose-fucose receptor has affinity for mannose and fucose moieties of microorganisms. CD206, a mannose receptor, plays a vital role in the recognition of pathogens via its pattern recognition domains (He et al. 2011). Therefore, it is used as a tool for studying endocytosis in macrophages and targeting site for drug and vaccine delivery.

3.11 Toll-like Receptors (TLRs): These include leucine-rich domain and intracellular IL-1 receptor region. TLR4 is present on macrophages and considered to be the first established mammalian toll-like receptors (Yoshitomi et al. 2005). Moreover, it is believed that TLRs enhance the innate immune system by initiating macrophage necrosis signal by receptor-interacting kinase-3-dependent pathway response to viruses and pathogen (Murray and Wynn 2011). TLRs play a dominant role in atherosclerosis by direct stimulation of macrophages with ligands TLR-2, TLR-4, TLR-9, and TLR-7 for SLE that promote the uptake of lipid. Another receptor expressed by macrophage is Glucan receptor (Dectin-1) and its affinity toward fungal β-glucan causes autoimmune arthritis (Nahar and Jain 2009). This glucan is present in fungal cell walls, hence allowing Dectin-1 to identify the fungus. This receptor is also involved in endocytosis of the attached moiety and this endocytosis may serve as a potential uptake mechanism for targeted delivery vehicles (Yoshitomi et al. 2005).

4 Nanoparticles-Mediated Macrophage Targeting

Nanoparticles are the drug delivery systems that carry the drug in the surrounding areas of preferred target site and enhance the therapeutic efficiency of drugs. Macrophages can be exploited as Trojan horses for targeted drug delivery. Nanocarriers can migrate across the different membrane barriers and release their drug cargo at sites of infection. Dysregulated macrophage functions can be associated with a varied range of autoimmune disorders such as RA, SLE, AIDS, MS, and SS (Prakash et al. 2010). Some receptors are reported to be overexpressed on infected macrophages that can effectively be targeted with suitable nanocarriers (Pruthi et al. 2012; Jain et al. 2013). In vaccination, where macrophage functions as the therapeutic target, higher uptake of surface modified nanocarriers should be optimized (Malam et al. 2009). Table 2 represents various nanoparticulate delivery systems targeting macrophages in autoimmune disease.

4.1 Liposomes-Mediated Macrophage Targeting

Liposomes are microscopic, self-assembled vesicles with an inner aqueous core surrounded with an outer lipid bilayer that can be natural or synthetic phospholipids. The different chemical and physical characteristics of liposomes are stearic hindrance, permeability, and charge density. These properties of liposomes allow them to be used as efficient drug delivery systems (DDS) and hydrophobic drugs can be loaded in the outer lipid bilayer while hydrophilic drugs can be incorporated in the central aqueous compartment (Mamo et al. 2010; Chen et al. 2019). Human body recognizes liposomes as the foreign bodies and thus can be easily taken up by mononuclear phagocytic cells after entry (Nogueira et al. 2015). Liposomes coated with oligo mannose showed better cellular uptake efficiency by the peritoneal macrophages (M1) and induced systemic immune response, and the enclosed antigen was efficiently displayed by both MHC-I and MHC-II (Tsujimura et al. 2009). Macrophage-liposome binding equilibrium is the main reason behind the established in vivo events where the nature of target cells determines the particle size preferences. Moreover, modifying surface properties of liposomes makes it feasible to match the carrier to the site of drug action. A study taking advantage of the above interaction shows that hyaluronan-modified liposomes are effective drug delivery systems to deliver the anti-inflammatory drugs to the mouse macrophages cells (RAW264.7) surface or its inside (Glucksam-Galnoy et al. 2012). Williams et al. explored the effect of methotrexate-loaded liposome in the inflamed joints of induced arthritis rats to conclude whether this formulation would be able to reduce the severity of arthritis. This formulation had been revealed to prevent the release of two pro-inflammatory mediators, that is, prostaglandin (PGE2) and TNF-activated rat peritoneal macrophages in vitro (Williams et al. 1994).

4.2 Polymeric Nanoparticles-Mediated Macrophage Targeting

The versatile nature of polymers offers a distinctive opportunity for the development of drug delivery systems. Nanoparticles are colloidal particles containing either biodegradable or nondegradable polymeric components. A large variety of natural and synthetic polymeric materials like albumin, gelatin, chitosan, poly(ε-caprolactone), and poly(D, L-lactide-co-glycolide) are used for the preparation of nanocarriers. The drug is either adsorbed or encapsulated in the nanocarriers. Polymeric nanocarriers can be modified with their size and surface charge. The problem of aggregation of nanoparticles can be resolved by surface modification with surfactants like poloxamers or poloxamines (Parveen et al. 2012).

Zhao et al. developed dual-functionalized lipid-polymeric hybrid pH-responsive nanoparticles encapsulated with MTX that shows faster release of MTX in phosphate buffered solution (PBS) of pH 5.0 than at pH 7.4. The cellular uptake study revealed that these nanoparticles

Table 2 Nanoparticulate delivery targeting macrophages

Nanocarrier	Delivery system	Cargo	Targeting agent	Target	Effect	Disease	References
Liposomes	PEGylated liposome	Methylprednisolone	PEG + glutathione	–	Bigger reduction in disease score with the targeted vs non targeted liposome	MS	Gaillard et al. (2012)
	Liposomes	Aceclofenac	–	–	Enhanced release of drug in presence of nanocarrier	RA	Sharma et al. (2017)
	Phosphatidylserine (PS) modified liposome-coated AuNCs	–	–	Macrophages	Anti-dsDNA autoantibodies, reduced inflammatory response	SLE	Xu et al. (2019)
Polymeric	PLGA	(tNP) PLP + rapamycin	–	Macrophage	In vivo trafficking—IV – accumulation in liver and spleen most localisation to macrophages and DCs in the spleen, but SC goes to the draining lymphnodes	MS	Maldonado et al. (2015)
	Glycol chitosan	Cyanine 5.5	Cyanine 5.5	Macrophage expressing Mac-1 molecules	Selective accumulation of HGC-Cy5.5 within synovial tissues included increased phagocytosis and permeability through leaky vessels	RA	Park et al. (2012)
	mPEG-PLGA-PLL	miR-125a	PEAL cy5-miNC	Splenic B cells macrophages DCs neutrophil	Enhanced biocompatibility and protect miR-125a from degradation retention time of miRNA	SLE	Zhang et al. (2020)
	Chitosan or folate-chitosan nanoparticles	siRNA	Folic acid	Folate receptors	Folate-chitosan illustrate likely safe and efficient siRNA delivery	SS	Fernandes et al. (2012)
Dendrimers	GO/pPG/anti-MBP/anti-tau	–	Anti-MBP/anti-tau	MS specific protein levels: Myelin basic protein (MBP) and tau proteins	Quantification of MS specific protein levels: Myelin basic protein (MBP) and tau proteins in cerebrospinal fluid and serum	MS	Derkus et al. (2017)
	Folate-PEG-PAMAM	Indomethacin	Folic acid	Folate receptors	Drug targeting efficiency enhanced with folate-PEG conjugate comparison to native dendrimer	RA	Chandrasekar et al. (2007)

Nanocarrier	Delivery system	Cargo	Targeting agent	Target	Effect	Disease	References
Metallic	Au (gold)	(tNP) MOG + small molecule (ITE)	PEG(to stabilize)	Dendritic cells	ITE ligand activates the aryl hydrocarbon receptor (Ahr), which can induce tolerogenic DCs Observed Ahr activation in macrophages in vivo	MS	Yeste et al. (2012)
	Ag (silver)	–	Folic acid	M1 macrophage	M1 macrophages apoptosis and reactive oxygen species (ROS) scavenging to facilitate M2 macrophages polarization	RA	Yang et al. (2021)
Lipid nanoparticles	PEGylated-triglycerol monostearate (TGMS)	Dexamethasone	–	Macrophage	Reduce the degree of joint swelling and inhibit the production of TNF-α and IL-1β in joint tissues	RA	He et al. (2020)
	Solid lipid nanoparticles	Dimethyl fumarate	–	–	DMF loaded controlled release SLNs could be a promising candidate for the better management of the MS	MS	Ojha and Kumar (2018)

were internalized into activated macrophages through folate receptor-mediated endocytosis (Zhao et al. 2018). A different study synthesized folate-conjugated PLGA nanoparticles loaded with dexamethasone phosphate (FA-DPNP) to selectively target the activated macrophages for improvement of inflammatory responses. The results show that FA-DPNP inhibit the production of nitic oxide and pro-inflammatory cytokines (TNF-α and IL-6) from activated macrophages more significantly than by DPNP and free DP. Together, the FA-DPNP were selectively internalized by activated macrophages and efficiently inhibited inflammatory responses (Cao et al. 2015). A different study indicated spatial delivery, targeting prospective, enhanced retention potential, and elevated bio-availability of the folic acid-conjugated nanoparticles loaded with etoricoxib to the activated macrophages (Bilthariya et al. 2015). Another group investigated the role of macrophage selective gene delivery and transfection of reporter (GFP expressing) and plasmid DNA (mIL-10 expressing) with the help of tuftsin modified alginate nanocarriers. These nanoparticles show swift internalization in murine macrophages (J774A.1) resulting into higher transfection potential without any toxic effects. The qualitative and quantitative analysis of mIL-10 and GFP transgene expression proved efficient delivery of DNA with tuftsin-alginate nanocarriers. Moreover, the macrophage cells were activated with LPS for pro-inflammatory cytokine (TNF-α) production to confirm the anti-inflammatory ability of mIL-10 transfection. The macrophage cells previously transfected with mIL-10 expressing plasmid DNA have reduced level of pro-inflammatory cytokines (Jain and Amiji 2012). Yang et al. synthesized folate-conjugated dextran-methotrexate nanoparticles (Dex-*g*-MTX/FA) and dextran-methotrexate conjugate (Dex-*g*-MTX). Their study revealed the enhanced cellular uptake and more cytotoxicity of Dex-*g*-MTX/FA compared to Dex-*g*-MTX toward the lipopolysaccharide (LPS) activated macrophages. In addition, significant suppression of inflammation was observed with Dex-g-MTX/FA due to target specificity of folate, reduced production of pro-inflammatory cytokine, and selective accumulation. The results suggest that Dex-g-MTX/FA that is macrophage-targeted prodrug have promising potential for targeted therapy of RA (Yang et al. 2016).

Howard et al. studied reduced TNF-α production in macrophages with chitosan-siRNA nanoparticle for anti-inflammatory treatment in CIA mice. The histological assessment of joints in anti-TNF-α-treated mice revealed inflammatory cell infiltration and slight destruction in cartilage. This study demonstrated that TNF-α knockdown facilitated by chitosan-siRNA nanoparticles decreases both systemic and local inflammation (Howard et al. 2009). Schmitt et al. studied the targeting of macrophages with the hydrophilic nanogels based on chitosan with hyaluronate surface, encapsulated with photosensitizers. This investigation showed the selective uptake of nanoparticles by macrophages and enhanced retention of drugs in the inflamed tissues. The nanogel loaded with photosensitizers stayed in the inflamed joints for a long time while injection free photosensitizers gets liberated from the inflamed joints rapidly (Schmitt et al. 2010).

4.3 Dendrimers-Mediated Macrophage Targeting

Dendrimers are one of the most recent and interesting nanocarrier systems used for the delivery of various therapeutic agents. These are synthetic, low dispersed, highly branched, and 3-dimensional macromolecules approaching the architecture of a tree (Oliveira et al. 2018). A usual dendrimer is mainly composed of three components (that presents several benefits) including multifunctional central core, surrounding branches, and functional surface groups (Yu et al. 2015). Dendrimers have several unique properties such as ability to bind to a variety of targeting agents with the help of surface groups, uniform particle size, and multivalences of end groups that helps to bind with different receptors, biodegradability, and solubility in various media (Wu et al. 2015).

A study investigated the folate-conjugated poly (amidoamine) dendrimer G5 (generation 5) nanoparticle loaded with MTX (G5-FA-MTX) as a therapeutic drug delivery system for the inflammatory arthritis. The folate-conjugated dendrimers were internalized by both primary mouse macrophage and folate receptor β-expressing macrophage cell lines in a receptor-specific manner. The G5-FA-MTX conjugate performed as an effective anti-inflammatory agent and reduced arthritis-induced parameters of inflammation like bone resorption, cartilage damage, paw volume, and body weight decrease (Thomas et al. 2011). Qi et al. developed generation 5 (G5) folate-conjugated PAMAM dendrimers encapsulated with methotrexate (G5-FA-MTX) and investigated their therapeutic efficacy to target inflammation stimulated folate receptors overexpressing macrophage cell lines. These macrophages have very crucial role in RA development. The binding affinities of G5-FA-MTX, G5-MTX, and control G5 to activated macrophages were investigated in primary peritoneal macrophages, NR8383, and RAW264.7. The in vitro results showed that the binding of nanocarriers to macrophages was either temperature or concentration dependent and the presence of 6.25 mM free folic acid ($p < 0005$) can block this interaction. The inhibitory effects of G5-FA-MTX and G5-MTX formulations on the arthritis development were investigated on an inflammatory arthritis model (adjuvant-induced) and had similar inhibitory results in inflammatory arthritis model at free MTX dose of 4.95 mol/kg. These results showed that dendritic polymers conjugated with multiples of MTX specifically bind to folate receptor overexpressing macrophages and also have similar anti-inflammatory effect to folate-conjugated MTX-loaded polymers (Qi et al. 2015). Hayder et al. investigated the therapeutic potential of azabisphosphonate (ABP) dendrimer to target monocytes in the treatment of rheumatoid arthritis (Hayder et al. 2011).

4.4 Metallic Nanoparticles-Mediated Macrophage Targeting

Metallic nanoparticles are the class of materials with very valuable properties. These nanoparticles can be prepared and modified with variety of functional groups that allow them to conjugate with drugs, ligands and antibodies. These nanoparticles have various therapeutic applications, such as vehicle for drug and gene delivery. Due to unique properties of metallic nanoparticles (for instance, optical and surface plasmon resonance properties), they has gained a distinct status in the field of nanotechnology (Kumar et al. 2018). There have been some studies regarding the use of metallic nanoparticles such as gold nanoparticles (AuNPs), iron oxide nanoparticles, and silver nanoparticles for the treatment of diseases (Howard et al. 2009).

A study explored the effect of AuNP and hyaluronic acid (HA) nanoparticles on the arthritic animal model, where the local inflammation was visibly labeled upon systematic administration of AuNP. The findings of this study propose that the above nanoparticles can be used as a potential imaging agent for the detection of RA (Lee et al. 2008). In another study by Huang et al., CIA rats were intra-articularly administrated with Galectin-1-nanogold (Au-GAL1) nanoparticles. This formulation endorsed reduction in pro-inflammatory cytokines concentration in ankle joints, killing of CD4+ T cells, and improvement of clinical symptoms of RA (Huang et al. 2012). Kim et al. studied multimodal nanoparticles (MNP) that act as contrast agents of inflamed synovium joints in arthritic model. The MNP comprises of Ru(bpy)3Cl2 central core surrounded by a coating of paramagnetic gadolinium chelates. After intravenous injection of these nanoparticles, magnetic resonance and optical images were acquired and the disease was detected in MNP-labeled cells (macrophages/monocytes) within the inflamed synovium joints. During the course of RA, magnetic resonance confirmed the lessening in inflamed joints that was obvious earlier. The MNP thus demonstrated as a possible contrast agent in arthritis and restrict to macrophages/monocytes within the inflamed joints (Kim et al. 2009). A different study investigated azathioprine-encapsulated silver nanoparticles for RA management by a green approach. The 3 T3 NIH fibroblast cell line was used to study the in vitro toxicity of above formulation. This formulation releases the encapsulated drug at the target site in a controlled manner and delivers collective effect on the inflammatory target sites (Prasad et al. 2013).

4.5 Lipid Nanoparticles–Mediated Macrophage Targeting

Lipid nanoparticles are the most clinically advanced nonviral gene delivery system. Lipid nanoparticles can be optimized for oral, parenteral, topical, and pulmonary delivery as they offer various advantages such as versatility and high biocompatibility. First generation lipid nanoparticles were solid-lipid nanoparticles (SLN) and nanostructured lipid carriers (NLC) are being investigated as second-generation lipids due to their potential applications in clinical medicine and drug delivery. Ease of surface modification, nonimmunogenicity, high encapsulation efficiency, reduced toxicity, and enhanced circulation times with altered biodistribution of drug are added advantages of using lipid nanoparticles (Attama et al. 2012). Zhou et al. synthesized hyaluronic acid (HA)-coated solid lipid nanoparticles loaded with glucocorticoid prednisolone (HA-SLNs/PD). The nano-formulation binds to hyaluronic receptor CD44 through HA, which is overexpressed on the surface of fibroblasts, macrophages, and synovial lymphocytes in RA. The intravenous administration of HA-SLNs/PD nano-conjugates in collagen-induced arthritis (CIA) mice shows accumulation in the inflamed joint tissue. HA-SLNs/PD nano-conjugates remains longer in circulation than drug loaded in SLN without HA or free drug, thereby preserving both the cartilage and bone architecture. HA-SLNs/PD results in reduced levels of inflammatory cytokines, bone erosion, and joint swelling. These findings suggest that loading PD in SLN coated with HA may make them effective and safe for managing inflammatory disorders (Zhou et al. 2018). Another study by Mahajan et al. demonstrated that cationic nanostructured lipid carrier loaded with efavirenz (EFV-NLC) had higher and sustained inhibition of p24 antigen for 6 days implying successful anticipation of viral replication in HIV-1-infected macrophages and therefore can be used as promising strategy for anti-HIV therapy. This study concluded that targeting infected macrophages with these nanoparticles resulted in a 2.29-fold increase in the anti-HIV effectiveness of EFV (Mahajan et al. 2020).

5 Conclusion

Macrophages have twin-roles in progression and regulation of inflammation. The differentiation, diversity, and circulation of macrophages critically influence the onset and progression of systemic as well as organ-specific autoimmune diseases. Macrophages act as a link between innate and adoptive immunity to sustain the immunological homeostasis. Macrophage impairment/dysfunction causes triggering of severe immune disorders. Clinical interventions that target monocytes or macrophages may well lead to novel therapeutics targeting autoimmune disorders. Improved understanding of the primary mechanisms underlying the macrophage plasticity and the identification of requisite pathological and physiological macrophage subsets in health and diseases will help develop new molecular tools to achieve the aforementioned challenges.

Nano-delivery systems enable targeting of drug(s) facilitating sustained drug delivery to the target tissues, extend duration of action, reduce the therapeutic dose, improve patient compliance, and reduce the adverse effects of highly toxic drugs. However, there is a need to identify the receptors that are present exclusively on macrophages. The identification of such receptors may further facilitate drug targeting to various parts/organs/cells of body possessing different type of macrophages. Although targeted drug delivery approach pledges to enhance the therapeutic window of drugs, it may facilitate calculation of the target–nontarget tissue ratio resulting in reduction in the minimum effective dose of the drug and the associated toxicity of drug. Owing to the presence of a limited number of receptor sites on any given tissue, targeted delivery emerges as an attractive approach for optimizing the therapeutic window by using very low concentration of drugs. Further research efforts are desired to ensure the safety of long-term in vivo applications. There is an imperative requirement for cautiously designed comparative toxicology as well as toxicokinetic studies for all types of nanocarriers for their likely clinical use. Any synthetic, biodegradable polymeric nanocarrier, which shall succeed in attaining the "Generally Regarded As Safe" (GRAS) status, will receive an overwhelming therapeutic acceptance in terms of safety and efficacy. Despite the encouraging results from most studies, some adjustments in these approaches are still required, especially with respect to efficacy and safety of the developed nano-delivery systems, before being applied in humans. Nonetheless, it can be stated that advanced nanomaterials will certainly play a crucial role in the future, to develop nanopharmaceutics, and in particular, in patient-specific approaches for the treatment of human autoimmune disorders.

References

Arango Duque G, Descoteaux A. Macrophage cytokines: involvement in immunity and infectious diseases. Front Immunol. 2014;5:491.

Ariizumi K, Shen GL, Shikano S, Xu S, Ritter R III, Kumamoto T, et al. Identification of a novel, dendritic cell-associated molecule, dectin-1, by subtractive cDNA cloning. J Biol Chem. 2000;275(26):20157–67.

Attama AA, Momoh MA, Builders PF. Lipid nanoparticulate drug delivery systems: a revolution in dosage form design and development. Recent Adv Novel Drug Carrier Syst. 2012;5:107–40.

Banwell B, Bar-Or A, Arnold DL, Sadovnick D, Narayanan S, McGowan M, et al. Clinical, environmental, and genetic determinants of multiple sclerosis in children with acute demyelination: a prospective national cohort study. Lancet Neurol. 2011;10(5):436–45.

Berwin B, Delneste Y, Lovingood RV, Post SR, Pizzo SV. SREC-I, a type F scavenger receptor, is an endocytic receptor for calreticulin. J Biol Chem. 2004;279(49):51250–7.

Bilthariya U, Jain N, Rajoriya V, Jain AK. Folate-conjugated albumin nanoparticles for rheumatoid arthritis-targeted delivery of etoricoxib. Drug Dev Ind Pharm. 2015;41(1):95–104.

Biswas SK, Mantovani A. Macrophage plasticity and interaction with lymphocyte subsets: cancer as a paradigm. Nat Immunol. 2010;11(10):889–96.

Bramow S, Frischer JM, Lassmann H, Koch-Henriksen N, Lucchinetti CF, Sørensen PS, Laursen H. Demyelination versus remyelination in progressive multiple sclerosis. Brain. 2010;133(10):2983–98.

Cao J, Naeem M, Noh JK, Lee EH, Yoo JW. Dexamethasone phosphate-loaded folate-conjugated polymeric nanoparticles for selective delivery to activated macrophages and suppression of inflammatory responses. Macromol Res. 2015;23(5):485–92.

Carroll MC. The complement system in regulation of adaptive immunity. Nat Immunol. 2004;5(10):981–6.

Chalmers SA, Chitu V, Ramanujam M, Putterman C. Therapeutic targeting of macrophages in lupus nephritis. Discov Med. 2015;20(108):43–9.

Chandrasekar D, Sistla R, Ahmad FJ, Khar RK, Diwan PV. Folate coupled poly (ethyleneglycol) conjugates of anionic poly (amidoamine) dendrimer for inflammatory tissue specific drug delivery. J Biomed Mater Res A. 2007;82(1):92–103.

Chen S, Chen B, Wen Z, Huang Z, Ye L. IL-33/ST2-mediated inflammation in macrophages is directly abrogated by IL-10 during rheumatoid arthritis. Oncotarget. 2017;8(20):32407.

Chen M, Daddy JCKA, Su Z, Guissi NEI, Xiao Y, Zong L, Ping Q. Folate receptor-targeting and reactive oxygen species-responsive liposomal formulation of methotrexate for treatment of rheumatoid arthritis. Pharmaceutics. 2019;11(11):582.

Cho JH, Feldman M. Heterogeneity of autoimmune diseases: pathophysiologic insights from genetics and implications for new therapies. Nat Med. 2015;21(7):730–8.

Denney L, Kok WL, Cole SL, Sanderson S, McMichael AJ, Ho LP. Activation of invariant NKT cells in early phase of experimental autoimmune encephalomyelitis results in differentiation of Ly6Chi inflammatory monocyte to M2 macrophages and improved outcome. J Immunol. 2012;189(2):551–7.

Derkus B, Bozkurt PA, Tulu M, Emregul KC, Yucesan C, Emregul E. Simultaneous quantification of myelin basic protein and tau proteins in cerebrospinal fluid and serum of multiple sclerosis patients using nano-immunosensor. Biosens Bioelectron. 2017;89:781–8.

Dongsheng Z, Zhiguang F, Junfeng J, Zifan L, Li W. Cyclophilin A aggravates collagen-induced arthritis via promoting classically activated macrophages. Inflammation. 2017;40(5):1761–72.

Drummond RA, Saijo S, Iwakura Y, Brown GD. The role of Syk/CARD9 coupled C-type lectins in antifungal immunity. Eur J Immunol. 2011;41(2):276–81.

Epelman S, Lavine KJ, Randolph GJ. Origin and functions of tissue macrophages. Immunity. 2014;41(1):21–35.

Fernandes JC, Qiu X, Winnik FM, Benderdour M, Zhang X, Dai K, Shi Q. Low molecular weight chitosan conjugated with folate for siRNA delivery in vitro: optimization studies. Int J Nanomedicine. 2012;7:5833.

Fiete DJ, Beranek MC, Baenziger JU. A cysteine-rich domain of the "mannose" receptor mediates GalNAc-4-SO4 binding. Proc Natl Acad Sci. 1998;95(5):2089–93.

Fox RI, Howell FV, Bone RC, Michelson P. Primary Sjögren's syndrome: proposed criteria for classification. Arthritis Rheum. 1984;14:77–105.
Fu W, Wojtkiewicz G, Weissleder R, Benoist C, Mathis D. Early window of diabetes determinism in NOD mice, dependent on the complement receptor CRIg, identified by noninvasive imaging. Nat Immunol. 2012;13(4):361.
Funes SC, Rios M, Escobar-Vera J, Kalergis AM. Implications of macrophage polarization in autoimmunity. Immunology. 2018;154(2):186–95.
Gaillard PJ, Appeldoorn CC, Rip J, Dorland R, van der Pol SM, Kooij G, et al. Enhanced brain delivery of liposomal methylprednisolone improved therapeutic efficacy in a model of neuroinflammation. J Control Release. 2012;164(3):364–9.
Gazi U, Martinez-Pomares L. Influence of the mannose receptor in host immune responses. Immunobiology. 2009;214(7):554–61.
Glucksam-Galnoy Y, Zor T, Margalit R. Hyaluronan-modified and regular multilamellar liposomes provide sub-cellular targeting to macrophages, without eliciting a pro-inflammatory response. J Control Release. 2012;160(2):388–93.
Godsell J, Rudloff I, Kandane-Rathnayake R, Hoi A, Nold MF, Morand EF, Harris J. Clinical associations of IL-10 and IL-37 in systemic lupus erythematosus. Sci Rep. 2016;6(1):1–10.
Gorantla S, Singhvi G, Rapalli VK, Waghule T, Dubey SK, Saha RN. Targeted drug-delivery systems in the treatment of rheumatoid arthritis: recent advancement and clinical status. Ther Deliv. 2020;11(4):269–84.
Greenwell-Wild T, Moutsopoulos NM, Gliozzi M, Kapsogeorgou E, Rangel Z, Munson PJ, et al. Chitinases in the salivary glands and circulation of patients with Sjögren's syndrome: macrophage harbingers of disease severity. Arthritis Rheum. 2011;63(10):3103–15.
Hayder M, Poupot M, Baron M, Nigon D, Turrin CO, Caminade AM, et al. A phosphorus-based dendrimer targets inflammation and osteoclastogenesis in experimental arthritis. Science translational medicine. 2011;3(81):81ra35.
He S, Liang Y, Shao F, Wang X. Toll-like receptors activate programmed necrosis in macrophages through a receptor-interacting kinase-3–mediated pathway. Proc Natl Acad Sci. 2011;108(50):20054–9.
He L, Fan D, Liang W, Wang Q, Fang J. Matrix metalloproteinase-responsive PEGylated lipid nanoparticles for controlled drug delivery in the treatment of rheumatoid arthritis. ACS Appl Bio Mater. 2020;3(5):3276–84.
Hogquist KA, Jameson SC. The self-obsession of T cells: how TCR signaling thresholds affect fate 'decisions' and effector function. Nat Immunol. 2014;15(9):815.
Howard KA, Paludan SR, Behlke MA, Besenbacher F, Deleuran B, Kjems J. Chitosan/siRNA nanoparticle–mediated TNF-α knockdown in peritoneal macrophages for anti-inflammatory treatment in a murine arthritis model. Mol Ther. 2009;17(1):162–8.
Huang YJ, Shiau AL, Chen SY, Chen YL, Wang CR, Tsai CY, et al. Multivalent structure of galectin-1-nanogold complex serves as potential therapeutics for rheumatoid arthritis by enhancing receptor clustering. Eur Cell Mater. 2012;23:170–81.
Jahagirdar P, Lokhande AS, Dandekar P, Devarajan PV. Mannose receptor and targeting strategies. In: Targeted intracellular drug delivery by receptor mediated endocytosis. Cham: Springer; 2019. p. 433–56.
Jain S, Amiji M. Tuftsin-modified alginate nanoparticles as a noncondensing macrophage-targeted DNA delivery system. Biomacromolecules. 2012;13(4):1074–85.
Jain NK, Mishra V, Mehra NK. Targeted drug delivery to macrophages. Expert Opin Drug Deliv. 2013;10(3):353–67.
Janossy G, Duke O, Poulter LW, Panayi G, Bofill M, Goldstein G. Rheumatoid arthritis: a disease of T-lymphocyte/macrophage immunoregulation. Lancet. 1981;318(8251):839–42.
Jiang HR, Milovanović M, Allan D, Niedbala W, Besnard AG, Fukada SY, et al. IL-33 attenuates EAE by suppressing IL-17 and IFN-γ production and inducing alternatively activated macrophages. Eur J Immunol. 2012;42(7):1804–14.
Katsifis GE, Moutsopoulos NM, Wahl SM. T lymphocytes in Sjögren's syndrome: contributors to and regulators of pathophysiology. Clin Rev Allergy Immunol. 2007;32(3):252–64.
Kawane K, Fukuyama H, Kondoh G, Takeda J, Ohsawa Y, Uchiyama Y, Nagata S. Requirement of DNase II for definitive erythropoiesis in the mouse fetal liver. Science. 2001;292(5521):1546–9.
Kawane K, Ohtani M, Miwa K, Kizawa T, Kanbara Y, Yoshioka Y, et al. Chronic polyarthritis caused by mammalian DNA that escapes from degradation in macrophages. Nature. 2006;443(7114):998–1002.
Kerrigan AM, Brown GD. Syk-coupled C-type lectins in immunity. Trends Immunol. 2011;32(4):151–6.
Kim JS, An H, Rieter WJ, Esserman D, Taylor-Pashow KM, Sartor RB, et al. Multimodal optical and Gd-based nanoparticles for imaging in inflammatory arthritis. Clin Exp Rheumatol. 2009;27(4):580–6.
Kottarath SK, Bhat M, Verma C, Bhattacharya S, Kaul A, Kumar U, Dinda AK. Folate receptor-β targeted cholesterol-chitosan nanocarrier for treatment of rheumatoid arthritis: An animal study. J Drug Deliv Sci Technol. 2020;60:101946.
Kumar H, Venkatesh N, Bhowmik H, Kuila A. Metallic nanoparticle: a review. Biomed J Sci Tech Res. 2018;4(2):3765–75.
Laria A, Lurati A, Marrazza M, Mazzocchi D, Re KA, Scarpellini M. The macrophages in rheumatic diseases. J Inflamm Res. 2016;9:1.
Lee H, Lee K, Kim IK, Park TG. Synthesis, characterization, and in vivo diagnostic applications of hyaluronic acid immobilized gold nanoprobes. Biomaterials. 2008;29(35):4709–18.
Li N, Hua J. Interactions between mesenchymal stem cells and the immune system. Cell Mol Life Sci. 2017;74(13):2345–60.

Li H, Adamopoulos IE, Moulton VR, Stillman IE, Herbert Z, Moon JJ, et al. Systemic lupus erythematosus favors the generation of IL-17 producing double negative T cells. Nat Commun. 2020;11(1):1–12.
Lin NY. OP0087 autophagy contributes to TNFα-mediated joint destruction. Ann Rheum Dis. 2013;71(Suppl 3):82–3.
Liston A, Gray DH. Homeostatic control of regulatory T cell diversity. Nat Rev Immunol. 2014;14(3):154–65.
Littlewood-Evans A, Sarret S, Apfel V, Loesle P, Dawson J, Zhang J, et al. GPR91 senses extracellular succinate released from inflammatory macrophages and exacerbates rheumatoid arthritis. J Exp Med. 2016;213(9):1655–62.
Liu C, Li Y, Yu J, Feng L, Hou S, Liu Y, et al. Targeting the shift from M1 to M2 macrophages in experimental autoimmune encephalomyelitis mice treated with fasudil. PLoS One. 2013;8(2):e54841.
Ma WT, Chang C, Gershwin ME, Lian ZX. Development of autoantibodies precedes clinical manifestations of autoimmune diseases: a comprehensive review. J Autoimmun. 2017;83:95–112.
Ma WT, Gao F, Gu K, Chen DK. The role of monocytes and macrophages in autoimmune diseases: a comprehensive review. Front Immunol. 2019;10:1140.
Mahajan K, Rojekar S, Desai D, Kulkarni S, Vavia P. Efavirenz loaded nanostructured lipid carriers for efficient and prolonged viral inhibition in HIV-infected macrophages. Pharma Sci. 2020; https://doi.org/10.34172/PS.2020.96.
Malam Y, Loizidou M, Seifalian AM. Liposomes and nanoparticles: nanosized vehicles for drug delivery in cancer. Trends Pharmacol Sci. 2009;30(11):592–9.
Maldonado RA, LaMothe RA, Ferrari JD, Zhang AH, Rossi RJ, Kolte PN, et al. Polymeric synthetic nanoparticles for the induction of antigen-specific immunological tolerance. Proc Natl Acad Sci. 2015;112(2):E156–65.
Malkiel S, Barlev AN, Atisha-Fregoso Y, Suurmond J, Diamond B. Plasma cell differentiation pathways in systemic lupus erythematosus. Front Immunol. 2018;9:427.
Mamo T, Moseman EA, Kolishetti N, Salvador-Morales C, Shi J, Kuritzkes DR, et al. Emerging nanotechnology approaches for HIV/AIDS treatment and prevention. Nanomedicine. 2010;5(2):269–85.
McInnes IB, Schett G. Cytokines in the pathogenesis of rheumatoid arthritis. Nat Rev Immunol. 2007;7(6):429–42.
Mellor-Pita S, Citores MJ, Castejon R, Yebra-Bango M, Tutor-Ureta P, Rosado S, et al. Monocytes and T lymphocytes contribute to a predominance of interleukin 6 and interleukin 10 in systemic lupus erythematosus. Cytometry B Clin Cytom. 2009;76(4):261–70.
Misharin AV, Cuda CM, Saber R, Turner JD, Gierut AK, Haines GK III, et al. Nonclassical Ly6C− monocytes drive the development of inflammatory arthritis in mice. Cell Rep. 2014;9(2):591–604.
Mueller DL. Mechanisms maintaining peripheral tolerance. Nat Immunol. 2010;11(1):21–7.
Mukhtar M, Ali H, Ahmed N, Munir R, Talib S, Khan AS, Ambrus R. Drug delivery to macrophages: a review of nano-therapeutics targeted approach for inflammatory disorders and cancer. Expert Opin Drug Deliv. 2020;17(9):1239–57.
Mulherin D, Fitzgerald O, Bresnihan B. Synovial tissue macrophage populations and articular damage in rheumatoid arthritis. Arthritis Rheum. 1996;39(1):115–24.
Murray PJ, Wynn TA. Protective and pathogenic functions of macrophage subsets. Nat Rev Immunol. 2011;11(11):723–37.
Nahar M, Jain NK. Preparation, characterization and evaluation of targeting potential of amphotericin B-loaded engineered PLGA nanoparticles. Pharm Res. 2009;26(12):2588–98.
Navegantes KC, de Souza Gomes R, Pereira PAT, Czaikoski PG, Azevedo CHM, Monteiro MC. Immune modulation of some autoimmune diseases: the critical role of macrophages and neutrophils in the innate and adaptive immunity. J Transl Med. 2017;15(1):1–21.
Nogueira E, Lager F, Le Roux D, Nogueira P, Freitas J, Charvet C, et al. Enhancing methotrexate tolerance with folate tagged liposomes in arthritic mice. J Biomed Nanotechnol. 2015;11(12):2243–52.
Ojha S, Kumar B. Preparation and statistical modeling of solid lipid nanoparticles of dimethyl fumarate for better management of multiple sclerosis. Adv Pharm Bull. 2018;8(2):225.
Oliveira IM, Gonçalves C, Reis RL, Oliveira JM. Engineering nanoparticles for targeting rheumatoid arthritis: past, present, and future trends. Nano Res. 2018;11(9):4489–506.
Orme J, Mohan C. Macrophages and neutrophils in SLE—an online molecular catalog. Autoimmun Rev. 2012;11(5):365–72.
Ousman SS, Kubes P. Immune surveillance in the central nervous system. Nat Neurosci. 2012;15(8):1096–101.
Park KS, Kang JH, Sa KH, Koo HB, Cho HJ, Nam EJ, Youn IC, Kim KM, Kim IS, Kwon IC, et al. In vivo quantitative measurement of arthritis activity based on hydrophobically modified glycol chitosan in inflammatory arthritis: more active than passive accumulation. Mol Imaging. 2012;11:389–400.
Parveen S, Misra R, Sahoo SK. Nanoparticles: a boon to drug delivery, therapeutics, diagnostics and imaging. Nanomedicine. 2012;8(2):147–66.
Peter JM, Thomas AW. Protective and pathogenic function of macrophage subsets. Nat Rev Immunol. 2011;11(11):723–37.
Piccolo V, Curina A, Genua M, Ghisletti S, Simonatto M, Sabò A, et al. Opposing macrophage polarization programs show extensive epigenomic and transcriptional cross-talk. Nat Immunol. 2017;18(5):530–40.
Plüddemann A, Neyen C, Gordon S. Macrophage scavenger receptors and host-derived ligands. Methods. 2007;43(3):207–17.
Prakash J, Beljaars L, Harapanahalli AK, Zeinstra-Smith M, de Jager-Krikken A, Hessing M, et al. Tumor-targeted intracellular delivery of anticancer drugs

through the mannose-6-phosphate/insulin-like growth factor II receptor. Int J Cancer. 2010;126(8):1966–81.

Prasad SR, Elango K, Damayanthi D, Saranya JS. Formulation and evaluation of azathioprine loaded silver nanopartilces for the treatment of rheumatoid arthritis. Asian J Biomed Pharm Sci. 2013;3(23):28–32.

Pruthi J, Mehra NK, Jain NK. Macrophages targeting of amphotericin B through mannosylated multiwalled carbon nanotubes. J Drug Target. 2012;20(7):593–604.

Pyonteck SM, Akkari L, Schuhmacher AJ, Bowman RL, Sevenich L, Quail DF, et al. CSF-1R inhibition alters macrophage polarization and blocks glioma progression. Nat Med. 2013;19(10):1264.

Qi R, Majoros I, Misra AC, Koch AE, Campbell P, Marotte H, et al. Folate receptor-targeted dendrimer-methotrexate conjugate for inflammatory arthritis. J Biomed Nanotechnol. 2015;11(8):1431–41.

Ravetch JV. Fcgamma receptors as regulators of immune responses. Nat Rev Immunol. 2008;8:34–47.

Roszer T. Understanding the biology of self-renewing macrophages. Cell. 2018;7(8):103.

Rothlin CV, Carrera-Silva EA, Bosurgi L, Ghosh S. TAM receptor signaling in immune homeostasis. Annu Rev Immunol. 2015;33:355–91.

Sack U, Stiehl P, Geiler G. Distribution of macrophages in rheumatoid synovial membrane and its association with basic activity. Rheumatol Int. 1994;13(5):181–6.

Schmitt F, Lagopoulos L, Käuper P, Rossi N, Busso N, Barge J, et al. Chitosan-based nanogels for selective delivery of photosensitizers to macrophages and improved retention in and therapy of articular joints. J Control Release. 2010;144(2):242–50.

Sharma G, Saini MK, Thakur K, Kapil N, Garg NK, Raza K, et al. Aceclofenac cocrystal nanoliposomes for rheumatoid arthritis with better dermatokinetic attributes: a preclinical study. Nanomedicine. 2017;12(6):615–38.

Shi C, Pamer EG. Monocyte recruitment during infection and inflammation. Nat Rev Immunol. 2011;11(11):762–74.

Shimaoka T, Nakayama T, Fukumoto N, Kume N, Takahashi S, Yamaguchi J, et al. Cell surface-anchored SR-PSOX/CXC chemokine ligand 16 mediates firm adhesion of CXC chemokine receptor 6-expressing cells. J Leukoc Biol. 2004;75(2):267–74.

Shin TH, Kim HS, Kang TW, Lee BC, Lee HY, Kim YJ, et al. Human umbilical cord blood-stem cells direct macrophage polarization and block inflammasome activation to alleviate rheumatoid arthritis. Cell Death Dis. 2016;7(12):–e2524.

Soldano S, Trombetta AC, Contini P, Tomatis V, Ruaro B, Brizzolara R, et al. Increase in circulating cells coexpressing M1 and M2 macrophage surface markers in patients with systemic sclerosis. Ann Rheum Dis. 2018;77(12):1842–5.

Sun W, Zhang H, Wang H, Chiu YG, Wang M, Ritchlin CT, et al. Targeting notch-activated M1 macrophages attenuates joint tissue damage in a mouse model of inflammatory arthritis. J Bone Miner Res. 2017;32(7):1469–80.

Suzuki Y, Shirai M, Asada K, Yasui H, Karayama M, Hozumi H, et al. Macrophage mannose receptor, CD206, predict prognosis in patients with pulmonary tuberculosis. Sci Rep. 2018;8(1):1–9.

Ta W, Chawla A, Pollard JW. Origins and hallmarks of macrophages: development, homeostasis, and disease. Nature. 2013;496:445–55.

Thomas TP, Goonewardena SN, Majoros IJ, Kotlyar A, Cao Z, Leroueil PR, Baker JR Jr. Folate-targeted nanoparticles show efficacy in the treatment of inflammatory arthritis. Arthritis Rheum. 2011;63(9):2671–80.

Tsujimura K, Ikehara Y, Nagata T, Koide Y, Kojima N. Induction of anti-tumor immune responses with oligomannose-coated liposomes targeting to peritoneal macrophages. Procedia Vaccinol. 2009;1(1):127–34.

Upchurch K, Oh S, Joo H. Dectin-1 in the control of Th2-type T cell responses. Receptors Clin Investig. 2016;3(1):e1094.

Ushio A, Arakaki R, Yamada A, Saito M, Tsunematsu T, Kudo Y, Ishimaru N. Crucial roles of macrophages in the pathogenesis of autoimmune disease. World J Immunol. 2017;7(1):1.

Vogel DY, Vereyken EJ, Glim JE, Heijnen PD, Moeton M, van der Valk P, et al. Macrophages in inflammatory multiple sclerosis lesions have an intermediate activation status. J Neuroinflammation. 2013;10(1):1–12.

Waldmann H. Tolerance: an overview and perspectives. Nat Rev Nephrol. 2010;6(10):569–76.

Wildenberg ME, Welzen-Coppens JM, van Helden-Meeuwsen CG, Bootsma H, Vissink A, van Rooijen N, et al. Increased frequency of CD16+ monocytes and the presence of activated dendritic cells in salivary glands in primary Sjögren syndrome. Ann Rheum Dis. 2009;68(3):420–6.

Williams AS, Camilleri JP, Williams BD. Suppression of adjuvant-induced arthritis by liposomally conjugated methotrexate in the rat. Rheumatology. 1994;33(6):530–3.

Wu LP, Ficker M, Christensen JB, Trohopoulos PN, Moghimi SM. Dendrimers in medicine: therapeutic concepts and pharmaceutical challenges. Bioconjug Chem. 2015;26(7):1198–211.

Xu N, Li J, Gao Y, Zhou N, Ma Q, Wu M, et al. Apoptotic cell-mimicking gold nanocages loaded with LXR agonist for attenuating the progression of murine systemic lupus erythematosus. Biomaterials. 2019;197:380–92.

Yang M, Ding J, Zhang Y, Chang F, Wang J, Gao Z, et al. Activated macrophage-targeted dextran–methotrexate/folate conjugate prevents deterioration of collagen-induced arthritis in mice. J Mater Chem B. 2016;4(12):2102–13.

Yang Y, Guo L, Wang Z, Liu P, Liu X, Ding J, Zhou W. Targeted silver nanoparticles for rheumatoid arthritis therapy via macrophage apoptosis and Re-polarization. Biomaterials. 2021;264:120390.

Yarilina A, Xu K, Chan C, Ivashkiv LB. Regulation of inflammatory responses in tumor necrosis factor–activated and rheumatoid arthritis synovial macrophages by JAK inhibitors. Arthritis Rheum. 2012;64(12):3856–66.

Ye L, Wen Z, Li Y, Chen D, Yu T, Liu L, et al. Interleukin-10 attenuation of collagen-induced arthritis is associated with suppression of interleukin-17 and retinoid-related orphan receptor γt production in macrophages and repression of classically activated macrophages. Arthritis Res Ther. 2014;16(2):1–14.

Yeste A, Nadeau M, Burns EJ, Weiner HL, Quintana FJ. Nanoparticle-mediated codelivery of myelin antigen and a tolerogenic small molecule suppresses experimental autoimmune encephalomyelitis. Proc Natl Acad Sci U S A. 2012;109:11270–5. https://doi.org/10.1073/pnas.1120611109.

Yilmaz B, Spalinger MR, Biedermann L, Franc Y, Fournier N, Rossel JB, et al. The presence of genetic risk variants within PTPN2 and PTPN22 is associated with intestinal microbiota alterations in Swiss IBD cohort patients. PLoS One. 2018;13(7):e0199664.

Yoshitomi H, Sakaguchi N, Kobayashi K, Brown GD, Tagami T, Sakihama T, et al. A role for fungal β-glucans and their receptor Dectin-1 in the induction of autoimmune arthritis in genetically susceptible mice. J Exp Med. 2005;201(6):949–60.

Yu M, Jie X, Xu L, Chen C, Shen W, Cao Y, et al. Recent advances in dendrimer research for cardiovascular diseases. Biomacromolecules. 2015;16(9):2588–98.

Zani IA, Stephen SL, Mughal NA, Russell D, Homer-Vanniasinkam S, Wheatcroft SB, Ponnambalam S. Scavenger receptor structure and function in health and disease. Cell. 2015;4(2):178–201.

Zhang J, Chen C, Fu H, Yu J, Sun Y, Huang H, et al. MicroRNA-125a-loaded polymeric nanoparticles alleviate systemic lupus erythematosus by restoring effector/regulatory T cells balance. ACS Nano. 2020;14(4):4414–29.

Zhao J, Zhang X, Sun X, Zhao M, Yu C, Lee RJ, et al. Dual-functional lipid polymeric hybrid pH-responsive nanoparticles decorated with cell penetrating peptide and folate for therapy against rheumatoid arthritis. Eur J Pharm Biopharm. 2018;130:39–47.

Zhou M, Hou J, Zhong Z, Hao N, Lin Y, Li C. Targeted delivery of hyaluronic acid-coated solid lipid nanoparticles for rheumatoid arthritis therapy. Drug Deliv. 2018;25(1):716–22.

Macrophage Targeting for Therapy of Cardiovascular Diseases (CVD)

Zhuqiu Jin

Abstract

Cardiac macrophages can be classified into primitive resident macrophages and blood-borne infiltrating macrophages. Primitive resident macrophages (CCR2−) originate from yolk sac-derived erythromyeloid and perform homeostatic functions and depend on local proliferation in the heart. They play an indispensable role in coronary artery development and maturation, valvular development and remodeling, facilitation of electrical conduction, and cardiac regeneration in neonatal heart. Blood-borne macrophages (CCR2+) originate from bone marrow and spleen and maintain through a combination of monocyte recruitment and proliferation. There are three subtypes of macrophages (CCR2-MHC-II^{low}, CCR2-MHC-II^{high}, and CCR2 + MHC-II^{high}) and one subtype of monocyte (CCR2 + MHC-II^{low}) in the adult heart. Heart-resident macrophages are critical in cardiovascular diseases, such as chronic heart failure, myocardial infarction, diabetic cardiomyopathy, hypertension, stroke, and atherosclerosis. In general, M1 macrophages are pro-inflammatory whereas M2 macrophages are anti-inflammatory. Selective antagonism of CCR2, competitive inhibition of CCL2, and M2 macrophages plasticity modulation may shed light for future drug discovery and development to prevent and/or treat cardiovascular disease.

Keywords

Macrophages · Cardiovascular disease · CCR2 · CCL2 · M2 plasticity

Abbreviations and Acronyms

AZM	Azithromycin
CCL2	C-C motif chemokine ligand-2
CCR2	C-C chemokine receptor type 2
CVD	Cardiovascular disease
CXCL-1	Chemokine (C-X-C motif) ligand 1
CX3CL-3	Chemokine (C-X3-C motif) ligand-3
CX3CR1	Chemokine (C-X3-C motif) receptor type 1
DCM	Dilated cardiomyopathy
DT	Diphtheria toxin
HFpEF	Heart failure with preserved ejection fraction
HFrEF	Heart failure with reduced ejection fraction
HLA-DR	Human leukocyte antigen-DR isotype
ICH	Intracerebral hemorrhage
ICM	Ischemic cardiomyopathy

Z. Jin (✉)
Department of Pharmaceutical & Biomedical Sciences, College of Pharmacy, California Northstate University, Elk Grove, CA, USA
e-mail: zhuqiu.jin@cnsu.edu

S. Gupta, Y. V. Pathak (eds.), *Macrophage Targeted Delivery Systems*,
https://doi.org/10.1007/978-3-030-84164-5_15

IR	Ischemia/reperfusion
KLF-4	Kruppel-like factor 4
IL-1R	IL-1 receptor
IL-RAcP	IL-1 receptor accessory protein
IRAK	IL-1R-associated kinase
LAD	Left anterior descending artery
LVEDP	Left ventricular end-diastolic pressure
M1	Classically activated macrophages
M2	Alternatively activated microphages
MCP	Monocyte chemoattractant protein-1
MI	Myocardial infarction
MYD88	Myeloid differentiation primary response 88
NICM	Nonischemic cardiomyopathy
P1	Postnatal day 1
PAR-1	Protease activated protein-1
RIP	Remote ischemic postconditioning
TAC	Transverse aortic constriction
TLR	Toll-like receptor
TRM	Tissue-resident macrophages

According to American Heart Association's 2020 heart disease and stroke statistical update, cardiovascular disease (CVD) accounted for 859,125 deaths in the United States in 2017. Coronary heart disease is the primary cause of mortality and morbidity attributable to CVD in the United States, followed by stroke, hypertension, chronic heart failure, diseases of the arteries, and others (Virani et al. 2020). CVD is also the leading cause of mortality and morbidity worldwide. It accounted for approximately 17.8 million deaths globally in 2017, which indicated an increase of 21% from 2007. Targeting CVD is a global emerging task and challenge. Seeking effective strategies and novel targets to tackle CVD are becoming prominent and obligatory.

Macrophages are essential elements of the body's innate immune system to sense and respond to invasion from infectious microorganisms and tissue injury through various opsonin receptors, scavenger receptors, pattern recognition receptors, or other phagocytic receptors. Macrophages are critical for responses to homeostatic or tissue-damaging signals. Macrophages also have important functions to keep homeostasis of the living body by acting as sentinels and responding to changes in physiological condition as well as stress condition (Hussell and Bell 2014). Nevertheless, under unfavorable circumstance, these homeostatic and reparative functions can be subverted to result in causal association of macrophages with pathophysiological changes in a variety of diseases, such as fibrosis, inflammation, and insulin resistance by continuous insult (Wynn et al. 2013). The role of macrophages from physiological engagement to pathophysiologic adaptation depends on subtypes of macrophages and the cytokines or other bioactive molecules released from these cells. For example, macrophages are the major type of immune cells infiltrated into adipose tissue. In healthy or lean mice with normal diet, alternatively activated macrophages (M2) of adipose tissue secrete bioactive cytokines and growth factors, such as interleukin (IL)-10 and TGF-β to protect tissue from inflammation. These cells are involved in tissue homeostasis, tissue remodeling, and anti-inflammation. Under high-fat diet-induced obesity, interferon-γ or lipopolysaccharide stimulates accumulation of classically activated macrophages (M1). These cells induce secretion of bioactive cytokines, such as IL-1β, IL-6, IL-12, iNOS, and TNF-α to initiate inflammation (Castoldi et al. 2016).

1 Origins and Functions of Cardiac Tissue-Resident Macrophages in the Normal Heart Under Healthy and Physiological Condition

1.1 Origin of Cardiac-Resident Macrophages in the Normal Heart (Healthy and Physiological Condition)

The heart is a vital organ composed of multiple types of cells: cardiac myocytes, fibroblasts, endothelial cells, microvascular smooth muscle cells, etc. With immunohistochemistry and typical cellular markers, the components and cellular populations of the heart have been determined. Cardiomyocytes account for about 20–35% of cells in the mouse and human hearts. The largest proportion (60%) of non-myocytes in the adult heart belongs to endothelial cells located in endocardium and blood vessels. Less than 20% of cells belong to fibroblasts. Vascular smooth muscle cells make up 6% of total cells in the heart. Around 8% of cells are identified as leukocytes (Pinto et al. 2016; Gray et al. 2018). Tissue-resident macrophages (TRM) are present in the myocardium and the distal atrioventricular (AV) node as well as areas adjacent to blood vessels of the murine and human heart during homeostasis, comprising 7–8% of non-cardiomyocytes in the steady condition. Murine monocytes are identified as lineage (CD90/CD19/NK1.1/Ly-6G/Ter119) ^{low}, CD11b^{+}, F4/80 ^{low}, CD115^{high}, and Ly-6C $^{low/high}$. Macrophages are identified as lineage (CD90/CD19/NK1.1/Ly-6G/Ter119) ^{int}, CD11b^{+}, F4/80^{high}, and Ly-6C ^{low}. Among these cells, CD45 $^{+}$CD11b^{+}F4/80^{+}Ly6C^{low} macrophages appear to be the largest population among leukocytes in the heart by using flow cytometry (Heidt et al. 2014).

C-C chemokine receptor type 2 (CCR2) is a protein encoded by the CCR2 gene. CCR2 is a receptor for C-C motif chemokine ligand-2 (CCL2) or monocyte chemoattractant protein-1 (MCP-1), a ligand that is involved in monocyte infiltration in inflammatory diseases. Based on the origins of macrophages, heart-resident macrophages include two classes: primitive resident macrophages and blood-borne infiltrating macrophages. Primitive resident macrophages originate from yolk sac-derived erythromyeloid progenitors or others. These cells perform homeostatic functions and self-maintain locally. Infiltrating macrophages originate from circulating classic Ly6C^{hi} (inflammatory) monocytes and are recruited under pathophysiological condition. Ly6C is a monocyte/macrophage differentiation antigen regulated by interferon gamma. CCL2-CCR2 chemotaxis pathway is involved in the recruitment of monocytes/macrophages to the heart.

Based on presence of CCR2, tissue-resident macrophages in the heart can be classified into two subtypes: CCR2− resident macrophages and CCR2+ resident macrophages. According to expressing levels of major histocompatibility complex (MHC) class II, macrophages can be further divided into MHC-II^{high}CCR2−; MHC-II^{low}CCR2−; and MHC-II^{high}CCR2+ in the mouse heart (Li et al. 2016). Subsets of CCR2− and CCR2+ macrophages are also detected in the human myocardium. CCR2− and CCR2+ macrophages have unique molecular profiles and functional characteristics, analogous to reparative M2 and inflammatory M1 macrophages in the mouse heart (Bajpai et al. 2018). CCR2− resident macrophages are embryonic-derived ones and originate from primitive yolk sac progenitors and fetal liver progenitors as well as hemogenic endocardium of the heart (Gomez Perdiguero et al. 2015). They seed the heart during embryonic and early postnatal development. CCR2− macrophages are maintained through local proliferation independent of peripheral monocyte recruitment. CCR2+ resident macrophages in the heart originate from bone marrow and spleen or others containing hematopoietic progenitors. They enter the heart from the initial few weeks of life after birth and are maintained under steady state and inflammatory conditions through monocyte recruitment in a mechanism depended on CCR2. Accordingly, embryonic-derived macrophage lineages exist within the adult heart in concert with monocyte-derived subsets. (Gomez Perdiguero et al. 2015; van de Laar et al. 2016).

CCR2− macrophages are mostly located around coronary endothelial cells. In contrast, CCR2+ macrophages are distributed areas and fields containing scar and fibrotic tissue. Heart macrophages are spindle-like shape that resemble fibroblasts and are different from typical irregular round shape of macrophages (Heidt et al. 2014).

The turnover kinetics of cardiac-resident macrophages can be estimated based on a pulse-chase experiment with BrdU labeling (Heidt et al. 2014). The population of BrdU-positive macrophages decreases to 2.6-fold within 3 weeks indicates that TRM turn over slowly in steady state. It would take two additional weeks until the disappearance of all BrdU-positive cells. Heart resident CCR2− macrophages are replenished through a mechanism of local proliferation. Heart resident CCR2+ macrophages are maintained through a combination of monocyte recruitment and proliferation.

1.2 Functions of Cardiac-Resident Macrophages in Physiological Condition

1.2.1 Involvement of Coronary Artery Development and Maturation

It has been reported that yolk sac-derived CCR2− macrophages play a crucial role in coronary development of the heart. These cells are recruited to coronary blood vessels and mediate coronary plexus remodeling through expansion of perfused vasculature and pro-angiogenetic properties mediated by insulin-like growth factor (Leid et al. 2016).

1.2.2 Obligatory Role in Valvular Development and Remodeling

The valves of the heart are critical for maintenance of unidirectional blood flow in the heart and circulation. Hemogenic endocardium is an important source of cardiac residential macrophages. Endocardially-derived macrophages play a phagocytic and antigen presenting role in normal valvular tissue remodeling in the heart. Genetic ablation of endocardially derived macrophages causes severe valve malformation (Shigeta et al. 2019).

1.2.3 Facilitation of Electrical Conduction in the Heart

Cardiac macrophages exist in distal of atrioventricular (AV) node. Both AV node and left ventricular (LV) macrophages express ion channels and exchangers that are involved in electrical conduction. The expression profile of AV node macrophages is similar to LV resident macrophages. Connexin 43, one of the key proteins in gap junctions, is mainly expressed in AV node macrophages and specifically in the interacting sites between macrophages and cardiomyocytes to synchronize the action potentials and cardiac conduction. Macrophages render resting membrane potentials of cardiomyocytes more positive and more depolarized. Depletion of connexin 43 or congenital deficiency of macrophages cause impaired AV node conduction (Hulsmans et al. 2017).

1.2.4 Promotion of Cardiac Regeneration

The mammalian adult heart does not have capability to regenerate itself following injury of the heart. Instead, fibrotic scar is usually formed after loss of cardiomyocytes because of the injury. In contrast, neonatal mammalian heart retains the capability to regenerate itself following injury for a short period after birth. One-day old neonatal mice with myocardial apical resection or myocardial infarction (MI) regenerate the injured myocardium through inflammatory cell infiltration, clot formation, and cardiomyocyte dedifferentiation with activation of the cardiomyocyte fetal gene program (Porrello et al. 2011; Porrello et al. 2013). This capability of regeneration of the injured heart is rapidly lost if neonatal mice are more than 7-day old. Neonatal porcine hearts with age from 1 to 3-day old also have the capability to regenerate injured myocardium (Zhu et al. 2018). Interestingly, cardiac regenerative capability is abolished when macrophages are selectively depleted with clodronate liposomes during early phase of repair in neonatal mice (Aurora et al. 2014).

2 Origins and Functions of Cardiac Tissue-Resident Macrophages in the Heart Subjected to Pathophysiological Condition

2.1 Origin of Cardiac-Resident Macrophages in Pathophysiological Condition

Macrophages also exist in the human heart and play an indispensable role in heart injury or cardiac protection under pathophysiological condition. Based on expression of human leukocyte antigen-DR isotype (HLA-DR, an MHC class II cell surface receptor encoded by the human leukocyte antigen complex) and expression of CCR2, heart-resident macrophages isolated from patients with dilated cardiomyopathy (DCM) and ischemic cardiomyopathy (ICM) can be divided into three subgroups: CCR2 + HLA-DR^{low}, CCR2 + HLA-DR^{high}, and CCR2-HLA-DR^{high}. CCR2− macrophages are mostly detected around coronary endothelial cells whereas CCR2+ macrophages exist in some fibrotic areas or scar tissue. CCR2− resident macrophages are embryonic-derived ones and originate from primitive yolk sac progenitors or other fetal tissues. They are maintained through cell proliferation independent of peripheral monocyte recruitment. CCR2+ resident macrophages are maintained through a mechanism of monocyte recruitment and cell proliferation. The morphology of CCR2 + HLA-DR^{high} macrophages is different from CCR2 + HLA-DR^{low} and CCR2-HLA-DR^{high}. CCR2 + HLA-DR^{high} macrophages exhibit larger size than CCR2-HLA-DR^{high} in the failing heart but both types of cells contain increased granularity (Bajpai et al. 2018).

CCR2− and CCR2+ macrophages isolated from the specimens of patients of DCM and ICM present distinct gene expression profiles. Upregulated genes in CCR2+ macrophages are associated with inflammatory pathways, such as TNFα/NF-κB signaling, IL2/STAT5, IL6/STAT3, interferon γ, K-RAS, adhesion molecule PLEKHA7, or other inflammatory reaction pathways. CCR2+ macrophages also express some growth factors related to fibrosis and hypertrophy. Upregulated genes in CCR2− macrophages are associated with coagulation, epithelial mesenchymal transition, myogenesis, signaling pathways of P53 and IL2/STAT5, etc. CCR2− cells also express some genes, such as growth factors (IGF1, PDGFC, EGFL7, GDF15, EGF13), extracellular matrix genes, and conduction genes. From the gene pattern, the populations of CCR2+ macrophages are involved in pro-inflammatory process and tissue damage whereas the population of CCR2− macrophages is associated with anti-inflammatory or tissue repair process.

Ang II is the major bioactive molecule to trigger cardiac hypertrophy and mediates hypertension as well as chronic heart failure. Infusion of Ang II induces substantial expansion and replenishment of cardiac macrophages through monocyte recruitment and local cell proliferation (Epelman et al. 2014).

CCL2 (MCP-1) is the major molecule for recruitment of monocytes. Clinical observation indicates that elevated levels of MCP-1 are associated with atherosclerosis, chronic heart failure, coronary artery disease, diabetes, hypertension, and others (de Lemos et al. 2003).

The ontogenesis of macrophages in the injured heart has been determined in three murine models with cardiomyocyte death: permanent myocardial infarction [left anterior descending artery ligation (LAD)], ischemia/reperfusion (IR) heart injury, diphtheria toxin (DT), and cardiomyocyte ablation. Populations of $Ly6C^{high}$CCR2 + MHC-II^{low} monocytes and CCR2 + MHC − II^{high} macrophages were increased across each model in the murine heart (Bajpai et al. 2019). To the response of heart injury, recruitment of monocyte and accumulation of CCR2+ macrophages in the heart are accelerated 4 days after ischemic heart. New recruitment of monocytes from circulation contributes to this enhancement of cell populations. $Ly6C^{high}$CCR2 + MHC-II^{low} monocytes and monocyte-derived CCR2 + MHC − II^{high} macrophages are the two major subsets of phagocytes in the heart to replace cardiac resident CCR2− macrophages to cell death. These cells mostly locate on infarct area and border zone. The increased CCR2 + $Ly6C^{high}$ monocyte recruitment and CCR2 + MHC − II^{high} macrophages represent common response to cardiomyocyte death in heart injury. Recruited CCR2+ macrophages express higher levels of inflammatory chemokines, such as chemokine C-X-C motif ligand 1 (CXCL1), CXCL2, CCL2, CCL7, and CCL9; cytokines, such as IL1β and IL10; and genes involved in pathophysiological cardiac remodeling, such as amphiregulin (Areg)], epiregulin (Ereg), arginase1 (Arg1), and growth and differentiation factor (GDF). Recruited CCR2+ macrophages are involved in allograft rejection, inflammatory response, TNF and NF-κβ signaling, mTORC1 signaling, IL2/STAT5 signaling, and K-Ras signaling.

Activation of resident CCR2+ macrophages is associated with allograft rejection, apoptosis, inflammatory response, interferon signaling, and

IL6/STAT3 signaling (Bajpai et al. 2019). Both recruited CCR2+ macrophages and resident CCR2+ macrophages are involved in inflammatory tissue injury. Both share the similar pathways. Nevertheless, recruited CCR2+ macrophages are more potent in secretion of inflammatory cytokines and chemokines than resident CCR2− macrophages. They are involved in pro-inflammatory processes. In contrast to CCR2+ macrophages, resident CCR2− macrophages express higher levels of growth factors, such as IGF1, PDGF-C, HB-EGF, and collagens and extracellular matrix-associated proteins, such as Coll3/4/6 and cysteine-rich angiogenic inducer (Cyr61), to exert anti-inflammatory capabilities. Based on the Gene Set Enrichment Analysis, resident CCR2− macrophages are involved in myogenesis, mitosis, epithelial mesenchymal transition, K-Ras signaling, or other tissue repair process.

2.2 Functions of Macrophages in Pathophysiological Condition

Monocytes and monocyte-derived macrophages play critical role in pathogenesis of cardiovascular diseases through release of cytokines or other bioactive molecules.

2.2.1 Recruitment of Monocytes to the Heart

Tissue CCR2+ macrophages promote recruitment of monocytes into the heart. Deletion of tissue-resident CCR2+ macrophages with diphtheria toxin (DT) for 4 days prior to heart transplantation markedly reduced accumulation of monocytes and macrophages in the heart after transplantation. In contrast, donor hearts lacking tissue-resident CCR2− macrophages exhibited substantial increase in recruitment of monocytes/macrophages in the heart (Bajpai et al. 2019).

Conditional deletion of cardiac myeloid differentiation primary response 88 (MYD88) in macrophages diminished the accumulation of monocytes and macrophages in the heart (Bajpai et al. 2019). It provides direct evidence that MYD88 is crucial for recruitment of monocytes and macrophages into the heart. MYD88 is a protein as the canonical adaptor for inflammatory signaling pathway. It connects Toll-like receptor (TLR) or IL-1 receptor (IL-1R) to IL-1R-associated kinase (IRAK) 1, 2, or 4. IL-1R forms a heterodimer with the IL-1 receptor accessory protein (IL-RAcP) upon binding of IL-1α and IL-1β leading to intracellular recruitment of MyD88, which activates NF-κB and P38 microtubule-associated protein (MAP) and increases production of monocyte/macrophage chemokines (CCL2/MCP1 and CCL7/MCP3) (Fig. 1).

2.2.2 Cardiac Diastolic Dysfunction in Hypertension-Induced Chronic Heart Failure

Hypertensive heart disease can lead to either systolic heart failure or diastolic heart failure or a combination of the two cardiac dysfunctions. Chronic heart failure is divided into heart failure with reduced ejection fraction (HFrEF) or systolic heart failure and heart failure with preserved ejection fraction (HFpEF) or diastolic heart failure. Nearly 50% of heart failure belong to HFpEF. The etiology of HFpEF is associated with stiffer cardiomyocytes and interstitial fibrosis. So far, there is no specific intervention to HFpEF. Diastolic heart failure is often associated with high mortality and morbidity comparable with systolic heart failure. Patients with HFpEF and hypertension contain higher density of macrophages in myocardium than age-matched subjects. This expansion of cardiac macrophages relies on CCR2-dependent monocyte migration.

A recent study unravels that the population of $Ly6^{high}$ monocyte and CCR2+ macrophages in the left ventricle myocardium increases significantly in mice injected with salty drinking water, unilateral nephrectomy, and chronic exposure to aldosterone (SAUNA) or aging mouse model (30 months old) with high blood pressure, cardiac fibrosis, increased left ventricular mass, and reduced cardiac diastolic function (increased LVEDP and reduced -dP/dt] were manifested in these mice. Deletion of IL-10 in cardiac-resident macrophages, the dominant source of IL-10 in the heart, reduced infiltration of macrophages in the heart and improved diastolic dysfunction with enhanced -dP/dt and reduced LVEDP

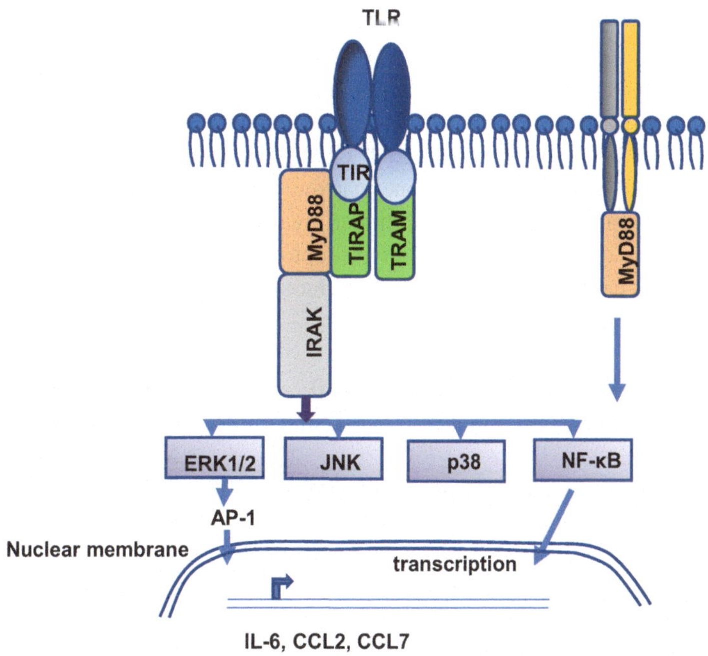

Fig. 1 Signal pathway of MyD88-mediated recruitment of monocytes/macrophages

when these conditional knockout mice were exposed to SAUNA. This protection is associated with less activation of cardiac fibroblasts (Hulsmans et al. 2018).

2.2.3 Nonischemic Cardiomyopathy

Cardiomyopathy is a type of progressive heart disease that reduces myocardial contractile force and insufficiently eject the blood into circulation to maintain blood flow to vital organs. Based on the etiology, cardiomyopathy is divided into ischemic cardiomyopathy (ICM) or nonischemic cardiomyopathy (NICM). NICM refers to all abnormal conditions that cause muscle dysfunction and structural changes such as intrinsic gene mutation (familial dilated cardiomyopathy, etc.) or extrinsic stress, such as pressure overload secondary to chronic hypertension or cardiac valve diseases.

Transverse aortic constriction (TAC) is a frequently used animal model for studies of pressure overload-induced cardiac hypertrophy and chronic heart failure. It can be divided into two phases: compensatory hypertrophy with normal function or decompensatory hypertrophy with decreased cardiac contractile force. The compensatory cardiac hypertrophy with preserved cardiac contractility usually appears at the first 7–10 days following TAC, followed by decompensation and development of chronic heart failure between 2 and 4 weeks after TAC (Liao et al. 2018). There are two phases with increase of cardiac macrophages in this model. The early phase appears 3 days post-TAC and reaches the peak at 7 days. The late phase occurs 4 weeks after TAC. Local cell proliferation of cardiac resident macrophages contributes to the increase in mice with pressure overload at early phase.

The majority macrophages resided in the heart at this stage belong to CCR2− macrophages since CCR2 antagonist, RS-504393, has no effects on the increased population of macrophages. Kruppel-like factor 4 (KLF-4), a member of KLF family of zinc finger transcription factors, regulates proliferation of cardiac resident macrophages. The macrophage-depleted KLF-4-deficient macrophages exhibit increased inflammation pathway and cardiac fibrosis as well as dilated cardiomyopathy. These macrophages are essential for myocardial adaptation to overload pressure. Depletion of macrophages at this phase with clodronate liposomes developed acute heart failure following TAC and around 80% of mice dead within 6 days. Cardiac resident CCR2− macrophages are also critical to maintain cardiac protection at late stage. In contrast to functions of cardiac resident macrophages, infiltrating CCR2+ macrophages play detrimental role in cardiac function and survival. CCR2-KO mice, which demonstrate lower overall macrophage numbers in the heart due to impaired CCL2/CCR2-mediated trafficking from bone marrow, exhibit normal cardiac functions and higher myocardial capillary density whereas wild-type control mice present cardiac dysfunction with reduced ejection fraction after 8 weeks of TAC.

2.2.4 Myocardial Infarction (MI)

(i) Ischemia/Reperfusion Mouse MI Model.

Effects of tissue-resident macrophages on myocardial infarction have been investigated substantially in a variety of animal models. In a myocardial ischemia/reperfusion (IR) murine model (90 minutes of ischemia followed by 28 days of reperfusion), depletion of tissue-specific CCR2+ macrophages by using single injection of diphtheria toxin (DT) to CCR2-DTR mice for 4 days resulted in small infract size and reduced cardiac hypertrophy. Depletion of tissue-specific CCR2− macrophages by using DT to CD169-DTR mice for 1 week before IR injury displayed larger infarct size and increased cardiac hypertrophy. These opposing effects may relate to subtypes of macrophage and different pathways. Manipulation of CCR2+ or CCR2− macrophages affects the fate of monocytes in the heart. Depletion of tissue-resident CCR2+ macrophages reduces the numbers of $Ifit3^{+}$macrophages and increases numbers of $Tnip3/Itgb7^{+}$ macrophages before IR. Depletion of tissue-resident CCR2− macrophages increases the numbers of $Arg1^{+}$, $Cxcl1/Ccrl2^{+}$, and proliferating macrophages and reduces the numbers of $Lyve1^{+}$ macrophages (Bajpai et al. 2019).

(ii) Permanently-Ligated Myocardial Ischemia Mouse Model in Adult Mice.

M1 macrophages and M2 macrophages play distinct role in ischemic heart injury. Mice with genetic removal of *Trib1* [a member of the Ca^{2+}/calmodulin-dependent protein kinase (CAMK)] selectively displayed elimination of the accumulation of M2 macrophages without alternation of M1 macrophage in the heart after MI. Increased mortality and myocardial infarct area are observed in these genetic mice. As expected, cardiac pathophysiological remodeling and tissue rupture are escalated whereas cardiac functions are reduced in these mice. Supplement of M2-like macrophages into *Trib1* knockout mice markedly reduced infarct size and promoted cardiac functions (Shiraishi et al. 2016).

IL-4 is a Th2 cytokine that promotes the differentiation of monocytes and macrophages to M2 subtypes and local proliferation of these macrophages in the heart or other organs after MI. Administration of IL-4 (i.p.) before induction of MI substantially improved the survival of mice after MI and reduced cardiac rupture and remodeling. Both cardiac systolic and diastolic functions of the heart were increased in mice treated with IL-4 at day 28 of MI.

The population of $CD11b^{+}F4/80^{+}CD206^{+}$ macrophages or named alternatively activated macrophages (M2) decreases from 90% of $F4/80^{+}CD11^{+}$ to 78% at the day 7 after myocardial infarction induced by ligating left coronary artery in mice. $CD11b^{+}F4/80^{+}CD206^{+}$ are primarily (>90%) negative for CCR2 and Ly6C (Shiraishi et al. 2016). $CD11b^{+}F4/80^{+}CD206^{-}$ macrophages or named classically activated macrophages (M1) with positive to CCR2 and Ly6C expanded and occupied in the heart after myocardial infarction. Some cytokines and growth factors that are associated with anti-inflammatory and tissue-repair, such as IL10, IL1rn, IL1α, VEGFα, and osteopontin, were up-regulated in M2-like macrophages after myocardial infarction. M2-like macrophages are cardioprotective and critical for cardiac repair function (Fig. 2).

(iii) Myocardial Infarction Induced by Permanent Ligation of Left Anterior Descending Coronary Artery in Neonatal Mice.

To explore the role of macrophages in heart regeneration following myocardial infarction (MI), postnatal day 1 (P1) regenerative mice and postnatal day 14 (P14) nonregenerative mice were used in this study (Aurora et al. 2014). At baseline and 7 days following MI, robust population of

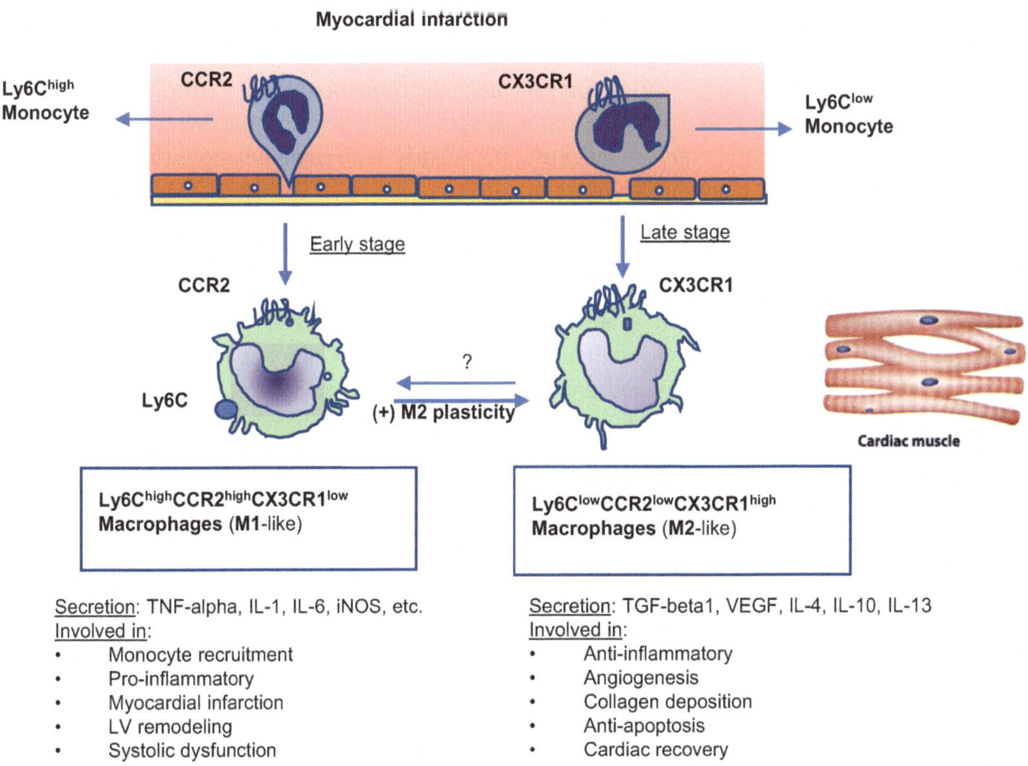

Fig.2 Comparison of M1 macrophages with M2 macrophages in myocardial infarction

$Ly6C^{lo}(F4/80/I\text{-}A^{b}/CD11c)^{hi}$ macrophages/dendritic cells (DCs) occurred and distributed throughout the myocardium. P14 mice had few macrophages and majority of macrophages concentrated on infarct area.

P1 control mice had substantial regeneration of the heart at 7 days and completely regenerated the heart at 21 days. Nevertheless, P1 mice treated with clodronate liposome, which had a specific depletion of cardiac mononuclear phagocytes ($CD11b^{+}Ly\text{-}6C^{-}$) and macrophage population, had visible transmural infarct area at day 7 and fibrotic scar formation and cardiac function. Results of transcriptional profiles indicated that 42 genes were markedly upregulated and 41 genes were downregulated in monocytes/macrophages of P1 mice. These upregulated genes are associated with angiogenesis, inflammation or immune function, and oxidative stress response.

These results clearly indicate that P1 monocytes/macrophages directly or indirectly accelerate neovascularization and recovery of cardiac functions during cardiac regeneration during MI.

(iv) Comparison Between Neonatal Mice and Adult Mice in Heart Injury – In Vivo Cardiomyocyte Cell Ablation Induced by Injection of Diphtheria Toxin.

The lineage of cardiac-resident macrophages in the neonatal mice is different from lineage of macrophages in the adult mice. The neonatal heart contains one macrophage (CCR2 − MHC − II^{low}) and monocyte (CCR2 + MHC − II^{low}). There are three subtypes of macrophages (CCR2 − MHC − II^{low}, CCR2 − MHC − II^{high}, and CCR2 + MHC − II^{high}) and one subtype of monocyte (CCR2 + MHC − II^{low}) in the adult heart. Some studies suggest that there is a single CCR2 − MHC − I^{low} macrophage subset in neonatal P14 mice (Dick et al. 2019). Postnatal day 1 (P1) neonatal mice were resistant to heart injury with minimal mortality in an in vivo cardiomyocyte cell ablation model induced by injection of diphtheria toxin (20 ng/kg, i.p.) to Rosa26-$DTR^{Mlc2v\text{-}Cre}$ mice. The population of cardiac-resident macrophages increased substantially after heart injury. Local proliferation of CCR2− macrophages in neonatal mice was observed within 5 days after injury. For adult mice, recruitment of CCR2+ monocytes and monocyte-derived macrophages is the major mechanism for replenishment of macrophages. The mortality is significantly higher (around 75%) in P14 and adult mice. Neonatal mice display normal cardiac tissue architecture and minimal left ventricular remodeling and fibrosis in this model. Adult mice present marked cardiac hypertrophy and pathological remodeling as well as interstitial fibrosis. Depletion of macrophages with clodronate liposomes in neonatal mice results in increased mortality and hypertrophy. Inhibition of recruitment of monocytes into the heart with RS504393, a selective CCR2 antagonist, reduces inflammation and improves recovery of the heart. The DT animal model appears differently between neonatal and adult mice (Lavine et al. 2014).

2.2.5 Microglia/Macrophages and Ischemic Brain Injury (Focal Cerebral Ischemia) – Stroke

Stroke is the leading cause of physical disability globally. According to Center for Disease Control and Prevention updated report, one in every six deaths from cardiovascular disease was due to stroke in 2018. Every year, more than 795,000 people in the United States have a stroke. Currently, there is no specific molecular target to restore brain function and enhancement of neuron survival. Microglia cells are the primary innate immune cells within the brain and spinal cord. Around 10 – 15% of all cells found within the brain belong to microglia. Microglia originate from yolk sac erythromyeloid precursors under normal condition (Li and Barres 2018). These precursors migrate to the brain parenchyma and renew through local proliferation. High levels of MHC-II for antigen presentation are expressed in perivascular microglia cells derived from bone marrow. As the resident macrophage cells, they play a major role in active immune defense in the central nervous system.

There are two main subtypes of microglia/macrophages: classically activated microglia/macrophages (M1) and alternatively activated microglia/macrophages (M2) (Frieler et al. 2011). Lipopolysaccharide and interferon-γ stimulate the formation of M1 cells. These cells synthesize and release some molecules that are associated with inflammation, such as iNOS, IL-12, CD11b, CD 16, CD32, and CD86. Interleukin (IL)-4 and IL-10 promote the formation of some chemokines or cytokines that are associated with tissue repair and antiinflammation, such as Arg1, TGF-β, IL-10, CD 206, CCL-32, and Ym1/2 (Cherry et al. 2014).

M1/M2 polarization is determined by measurement of M1 phenotypic pattern (iNOS and CD16) and M2 phenotypic pattern (CD206, YM-1).

By characterizing signature genes for M1 microglia/macrophages and M2 microglia/macrophages, the population of microglia/macrophages is able to be identified with RT-PCR. M1 cells emerged from day 3 and remained elevated through the whole time (14 days) window after middle cerebral artery occlusion to cause ischemic brain. M2 microglia/macrophages emerged from day 1 to day 3 and peaked by 3 to 5 days after ischemic injury. These genes returned to preischemic levels at the end of ischemic brain injury (Hu et al. 2012). Similar to the heart injury model, the primary type of microglia/macrophages recruited to the site of brain injury belongs to M2 cells at early stage of tissue injury. As the consequence of pathophysiological aggravated, M1 replaced M2 in the infarct or border zone at late stage.

M1 Microglia/macrophages are critical in cell death and injury. By using lipopolysaccharide plus IFNγ 48 hours, M1-polarized cells increased release of lactate dehydrogenase and reduced expression of microtubule-associated protein 2 (MAP2) from oxygen-glucose-deprivation model. Diminished expression of MAP2 is an early and sensitive marker of ischemic stroke. In contrast to M1 cells, IL-4-stimulated M2-polarized cells decrease the release of LDH and enhance expression of MAP2.

CCR2 is predominantly expressed on $Ly6C^{hi}$ monocytes. CCR2 is required for infiltration of monocytes into organs, including the brain. INCB3344, a selective CCR2 antagonist, markedly reduces circulating and infiltrating $Ly6C^{hi}$ monocytes into the brain and decreases gene expression of M2 markers (*Ym1*, *Arg1*, and *Il10*) while no changes remain on M1 markers in an experimental ischemic stroke model induced by occlusion of middle cerebral artery. Treatment with INCB3344 exacerbated brain function impairment and neurological deficit. The brain infarct size was substantially increased in ischemic stroke mice treated with INCB3344 (Chu et al. 2015a). These results suggest that M2 macrophages may have neuroprotection against ischemic stroke.

Interleukin-4 (IL-4) is a multifunctional cytokine secreted by some immune cells. IL-4 is one of the best characterized promoters for M2 polarization in microglia and macrophages. Deficiency of IL-4, exacerbated brain injury and impaired neurological functional outcomes at 24 hours in ischemic stroke mouse model (Xiong et al. 2011). Reduced gene expression of IL-4 and enhanced IL-4 receptor expression appeared at day 14 after ischemic stroke. Loss of IL-4 enhanced expression of M1 microglia/macrophage markers (iNOS, TNF-α, and CD16) and decreased expression of M2 microglia/macrophage markers (IL-10 and CD206) in mice with ischemic stroke. Loss of IL-4 impaired cognitive functions and exacerbated neurological deficits at 21 days post ischemic injury. Infusion of recombinant IL-4 improved neurological performance in the Rotarod and foot fault tests exhibited improvement in cognitive functions (Liu et al. 2016).

The neuroprotection of M2 macrophages has been confirmed in a clinical study. Patients who received one injection of autologous M2 macrophages (22.0 x 10^6) resulted in a significant improvement of the NIH Stroke Scale score from 11 before treatment to 6 at six months after treatment (Chernykh et al. 2016). Azithromycin (AZM), a macrolide antibiotic, concentrates in macrophages and neutrophils because of low pH values in lysosomes. Treatment with azithromycin increases F4/80^+/Ym1^+ population of M2 macrophages in the ischemic area but reduces F4/80^+/Gr1^+ infiltrating inflammatory macrophages in the ischemic hemisphere. Azithromycin reduces infarction volume and neurological deficit in an acute ischemic stroke model (Amantea et al. 2016a).

Chemokine (C-X3-C motif) receptor (CX3CR1) is a chemokine receptor for binding single ligand – CX3CL1. CX3CR1 is expressed in immune cells, including monocytes. CX3CL1/CX3CR1 pathway is also involved in recruitment of monocytes to local tissues. Deletion of CX3CR1 decreases Iba-1 positive cells, mostly monocytes and macrophages in hippocampus, striatum, cortex, and peri-infarct zone. To characterize M1/M2 polarization pattern, representative gene pattern is detected by RT-PCR. M2-type genes (Ym1 and Mcr1) were increased whereas M1-type gene (iNOS) was reduced within CX3CR$1^{-/-}$ microglia/macrophages. Deletion of CXC3CR1 may promote the formation of M2-like macrophages in the brain after stroke. Small infarct volume and improved neurological deficits were observed in CXC3CR$1^{-/-}$ mice (Tang et al. 2014).

The modulation of M2 macrophages in ischemic stroke has also been determined in several types of endogenous intervention. Ischemic postconditioning is referred as a series of brief period of ischemia and reperfusion cycles applied immediately at the site of ischemic tissue/organ after reperfusion to protect organs, such as brain and heart, from ischemic injury. This potent endogenous self-adaptation has been confirmed in several animal models and clinical studies (Hausenloy and Yellon 2006). Postconditioning with 10 min of reperfusion and 10 min of ischemia after 100 min of middle cerebral artery occlusion promotes activation and polarization of brain-resident microglia and increases expression of VEGF and population of M2-like macrophages at day 3 of reperfusion. Postconditioning reduces infarct volume and improves neurological scores. Selective antagonists of Flk1, one of the receptors for VEGF, abolish neuroprotection induced by postconditioning.

2.2.6 Hemorrhagic Stroke – Spontaneous Intracerebral Hemorrhage

Hemorrhagic stroke is a sudden bleeding caused by rupture in a weakened blood vessel in the brain. This type of stroke is less common and makes up about 13% of total strokes. Spontaneous intracerebral hemorrhage (ICH) produces the highest acute mortality and worst outcomes of all stroke patients and needs emergent intervention. There is no special treatment for hemorrhagic stroke.

M1/M2 polarization is associated with hemorrhagic stroke. Intracerebral hemorrhage (ICH) activates resident microglia and accelerates the formation of M1 macrophages in the injured tissue. Increased M1 phenotypic markers (iNOS and CD16) reached the peak at as early as 4 hours and maintained for 3 days. M2 macrophage markers (CD 206, and YM1) reached the peak at day 1 and declined at day 7. Deletion of protease-activated receptor-1 (PAR-1) reduced pro-inflammatory cytokines (TNF α and IL-1β) release and increased IL-10 and TGF-β expression. PAR-1 knockout mice exhibited neuroprotection with less brain swelling and neuronal death although there was no difference for lesion volume (Wan et al. 2016).

Remote ischemic postconditioning (RIP) provides neuroprotection and neuronal survival. It is feasible for translational research. The protocol and procedure of RIP is different from traditional ischemic postconditioning. RIP is applied in remote site distant from the injury tissue. For instance, to explore the protective effects of RIP on hemorrhagic stroke, four cycles of brief period of ischemia (5 min/each) and reperfusion (5 min/each) were applied in a mouse intracerebral hemorrhage model induced by loading of collagenase. RIP-promoted injury-independent polarization of microglia to M2 macrophages is characterized by anti-inflammatory phenotype (e.g., CD11b, CD206, F4/80, and IL-10). Proinflammatory phenotype was reduced or no changes were observed. RIP decreased hematoma volume and improved neurobehavioral outcomes. RIP-induced neuroprotection was abolished in CCR2 knockout mice or CD36 (scavenger receptor) knockout mice or myeloid-specific AMPKα knockout mice. These results suggest that CCR2-CD36-AMPKα pathway is essential for neuroprotection induced by remote ischemic postconditioning (Vaibhav et al. 2018).

2.2.7 Macrophages and Atherosclerosis

Atherosclerosis is a chronic progressive inflammatory disorder of the blood vessel wall. Accumulation of foam cells and accelerated formation of plaques of blood vessel walls often cause reduction of blood flow to vital organs, such as the heart and the brain, etc., and triggers ischemic injury, such as coronary artery disease and ischemic stroke, the major causes of mortality and morbidity. Multiple factors, such as lipids, lipoproteins, monocytes/macrophages, and endothelial cells are involved in initiation and progression of atherosclerosis. Monocytes are initially recruited to subendothelial area of the blood vessel wall by entrapped oxidized low-density lipoprotein. Monocytes are differentiated to macrophages by factors such as some cytokines or chemokines. Macrophages become foam cells, which secrete proinflammatory molecules to accelerate the formation of plaque in the blood vessel wall.

The progression of atherosclerosis can be described as early phase (initial xanthoma), advanced stage (fibrous cap atheroma), or hemorrhaged (fibrous cap atheroma with intraplaque hemorrhage) (Stöger et al. 2012). Relatively, both early phase and advanced stages belong to stable plaque phenotypes. Intraplaque hemorrhage represents unstable phenotype.

Macrophages are critical in pathogenesis and progression of atherosclerosis. Macrophages are involved in almost all phases. M1 and M2 macrophages are present throughout atherogenesis (Wan et al. 2016). Macrophages are precursors of foam cells. M2 macrophages are more likely to form foam cells than M1 macrophages containing low density lipoproteins (LDL) and uptake oxLDL cholesterol because of higher expression of scavenger receptors (CD36 and SR-A1) (van Tits et al. 2011).

Circulating monocytes can infiltrate into the blood vessel walls and be converted to mature macrophages. Macrophages are able to internalize oxidized low-density lipoprotein (OxLDL) or other ApoB lipoproteins through membrane scavenger receptors such as CD36. Monocytes are able to differentiate into both types of macrophages under different microenvironment to become M1 macrophages through granulocyte-macrophage colony-stimulating factor (GM-CSF) and become M2 macrophages through M-CSF (macrophage colony stimulating factor). M2 macrophages uptake more OxLDL and display higher intracellular cholesterol ester formation than M1 macrophages. Upon stimulation with lipopolysaccharide (LPS), M1 macrophages secrete more IL-6 and IL-8 whereas M2 macrophages secrete high levels of IL-10 (van Tits et al. 2011). Kruppel-like factors (KLFs) are involved in regulation of monocyte activation and induction of macrophages. KLF2 expresses in monocytes and inhibits activation of monocytes (Das et al. 2006). KLF4 is critical in inflammatory monocyte differentiation and induction of M1/M2 macrophage polarization through IL-4/Stat6 pathway, a transcriptional control of macrophage polarization (Alder et al. 2008; Liao et al. 2011). Exposure of oxidized LDL reduces expression of KFL2 in M2 macrophages and enhance the inflammatory capacity of M2 macrophages (van Tits et al. 2011). M2 macrophages play a critical role in the formation of foam cells and atherosclerosis. In general, M1 macrophages promote formation of an unstable plaque and are involved in progression lesions and pro-atherogenic. M2 macrophages are enriched in regression plaques and promote tissue repair process (Barrett 2020).

Since M1 and M2 have opposing effects on inflammation, the polarization of M1 and M2 depends on the microenvironment and plays a critical role in progression of atherosclerosis. Foam cells can either be degraded to release cholesterol through de-esterification or further develop to become plaque in the blood vessel wall. Both mRNA and protein expressions of M1 and M2 macrophage markers are upregulated in ruptured plaque segment in human atherosclerotic arteries and equally distributed at the fibrous cap region. Dominant expression of M1 macrophage population is usually associated with inflammatory plaque shoulder. The majority of macrophages located in the perivascular adventitial tissue belong to M2 macrophages while Hb-associated macrophages expressing CD163 are predominant in areas of plaque hemorrhage (Leitinger and Schulman 2013).

Endoplasmic reticulum (ER) is an important organelle involved in protein folding and transport. ER stress is critical in M2 differentiation via JNK/PPARγ pathway in human diabetic macrophages or apoE knockout mice fed with high-fat diet (Oh et al. 2012). Induction of ER stress with thapsigargin stimulates M2 express pattern (increased CD163 and MR). Inhibition of ER stress with 4-phenylbutyrate (4-PBA) or c-Jun N-terminal kinase (JNK) inhibitor SP600125 promotes the conversion of M2 to M1 macrophages and reduces foam cell formation.

Intraplaque hemorrhage stimulates monocyte differentiation to macrophages for hemoglobin disposal. This kind of hemorrhage contributes atherosclerotic lesion development and destabilization. A new type of monocyte-driven CD163-positive macrophages are located in atherosclerotic lesions, especially within intraplaque hemorrhage. CD163 is a scavenger receptor to bind with the hemoglobin-haptoglobin complex. A novel hemorrhage-associated macrophage (HA mac) population is characterized with high levels of CD163 and low levels of human leukocyte antigen (HLA)-DR. These macrophages are designated as Mhem macrophages and featured with a high iron load and HO-1 activity in contrast to the low content of iron and HO-1 in M1, M2, and Mox (Boyle et al. 2009; Etzerodt and Moestrup 2013). IL-6, IL-10, and Hb promote upregulated expression of CD163 whereas IFN-γ, TNF-α, GM-CSF, and CXCL4 downregulated expression of CD163.

Hemoglobin-related macrophages (Mhem) are positively associated with atherosclerotic disease progression. Activation of $CD163^{+}$Mhem macrophages increases plaque angiogenesis and enhances microvascular permeability through increased HIFα and VEGF-A expression as well as reduced expression of VE-cadherin. Selective inhibition of VEGF-A/VEGFR2 pathway abolished the increased endothelial cell permeability. Genetic deletion of CD163 in mice decreases intraplaque neovascularization and plaque progression. Hemoglobin and CD163 are critical for M(Hb) macrophage-mediated intraplaque hemorrhage and atherosclerotic lesion (Guo et al. 2018).

3 Pharmacological Interventions for Targeting Macrophages for CVD

Mounting evidence indicates that tissue resident CCR2− macrophages in the heart promote heart regeneration and recovery whereas monocyte-derived CCR2+ macrophages from circulation in the heart mediate heart injury and cardiac dysfunction. Since CCR2 is critical for regulation of mobilization and recruitment of monocyte-derived macrophages in response to CCL2/MCP-1, agents that antagonize CCR2 or reduce CCR2 expression or CCL2 formation have the potential to be developed to new medication for cardiovascular diseases. M1 macrophages are pro-inflammatory whereas M2 macrophages are anti-inflammatory. Agents that favor M2 plasticity have been shown to be anti-inflammatory and cardioprotective.

3.1 CCR2 Antagonists

Chemokines, the key small proteins for attraction of leukocytes to inflammatory sites, are divided into four subfamilies based on the number of residues between the first and second cysteine: CXC, CC, CX3C, and XC. Some chemokines are associated with homeostasis (CCL14, CCL19, CCL20, etc.). Some chemokines are relevant to inflammation (CCL2, CCL3, CCL4, etc.). Among these chemokines, monocyte chemoattractant protein-1 (MCP-1 or CCL2) is one of the most important chemokines because of its involvement in several diseases.

Increased levels of MCP-1 are observed in atherosclerosis, myocarditis, myocardial ischemia/reperfusion injury, peripheral artery disease, etc. CCR2, a cognate receptor for MCP-1, is a chemokine membrane receptor on cellular surface of monocyte with seven transmembrane domains. CCR-2 is predominantly expressed on Ly6C^{hgh} monocytes. Low level of expression of CCR-2 is also detected in other immune cells, such as Ly6C^{low} monocytes, activated T cells, dendritic cells, natural killer cells, and neutrophils. Mutations at positions 284 (Asp^{284}) and 291 (Glu^{291}) significantly reduce MCP-1 binding affinity to CCR2. Activation of MCP-1/CCR2 axis attracts monocyte/macrophages to the site of inflammation and triggers tissue injury. Selective antagonism of CCR2 or intervention of MCP-1 binding to CCR2 provides novel strategies to prevent or treat inflammatory diseases.

3.1.1 RS-504393

RS-504393 is a selective and potent CCR2 receptor antagonist with IC50 (89 nM) to CCR2 binding (Mirzadegan et al. 2000). It is a spiropiperidine small molecule with molecular weight 417.51. This antagonist selectively antagonizes this receptor by occupation of a binding site that includes acidic residue Glu^{291}. Pharmacological effects of RS-504393 on inflammatory diseases have been tested in kidney, liver, lung, and the brain. Recently, one animal study unravels cardiac effects of RS-505393 on morphology and functional changes of chronic heart failure. By using chronic heart failure mouse model induced by pressure overload (transverse aortic constriction, TAC), treatment with RS-504393 (2 mg/kg, i.p.) significantly abolished infiltration and expansion of CCR2+ macrophages, CD4$^+$, and CD8$^+$ T cells in the heart at end of one-week of post-TAC. RS-504393 also reduced cardiac interstitial fibrosis and left ventricular hypertrophy as well as improved cardiac dysfunction (ejection fraction, left ventricular end-systolic, and diastolic volume) at the end of 4-week in chronic heart failure (Patel et al. 2018). These results suggest that CCR2-dependent monocyte/macrophage signaling is critical for the early infiltration and expansion of macrophages in the heart, activated adaptive immunity, and compensatory development of left ventricular hypertrophy after pressure overload.

3.1.2 INCB3344

INCB3344 is a novel and specific CCR-2 antagonist, which blocks the binding of CCL2 to CCR-2 on the membrane surface of monocytes and other leukocytes.

(i) Ischemic Stroke.

In a murine model of brain ischemia/reperfusion model, INCB3344 (30 mg/kg, IP, daily) reduced numbers of CCR2+ monocytes and leukocytes in the blood and in the brain. INCB3344 aggravated the brain functional deficits and increased the brain infarct size after stroke. This detrimental effect induced by INCB3344 may associate with reduced numbers of M2-polarized macrophages. As monocytes, especially LyC6^{hi} monocytes, are a critical type of cells on the polarization state of macrophages, the levels of genes associated with M1 or M2 were determined. Remarkable increase of both M1-associated genes (e.g., *Tnf, IL-6,* and *IL-1β*) and M2-associated genes (e.g., *Ym1, Arg1,* and *IL-10*) was detected in the ischemic brain. INCB3344 treatment prevented the increase of expression of the M2 markers but had no effect on M1 markers after cerebral ischemia (Chu et al. 2015b). It suggests that Ly6Chi monocytes provide protection in acute ischemic stroke by M2 macrophage polarization-dependent mechanism.

(ii) Hypertension.

In an uninephrectomy and DOCA-salt hypertension model, both CCL2 and CCR-2 mRNA levels were increased in the mouse aorta. Treatment with INCB3344 reduced accumulation of macrophages in DOCA/salt-treated mouse aorta without effects on total leukocytes. After 14-day intervention period, INCB3344 reduced systolic blood pressure significantly. It suggests that targeting CCR2 may provide a novel strategy for development of antihypertensive agents (Chan et al. 2012).

(iii) Diabetic Cardiomyopathy.

CCR2+ monocytes/macrophages are recruited into diabetic heart. INCB3344 (30 mg/kg/day for 5 days, i.p.) significantly reduced levels of serum triglyceride in db/db mice but without effects on levels of glucose. INCB3344 increased cardiac output and ejection fraction as well as fractional shortening (Tan et al. 2019). Selective antagonism of CCR-2 receptor has potential effects for development of new drugs on diabetic cardiomyopathy.

3.2 MCP-1/CCL2 Competitive Inhibitor – PA508

MCP-1/CCL2 is highly induced in some common cardiovascular diseases, such as atherosclerosis, chronic heart failure, myocardial ischemia/reperfusion injury, etc. CCL2 specifically recruits and activates CCR-2 positive cells. Glycosaminoglycan (GAG) binding to chemokines is essential for enhancement of monocyte/macrophage or other leukocytes migration into tissues. Prohibition of this binding reduces chemotaxis of leukocytes and may exert anti-inflammatory effects.

CCL2 is a monomeric polypeptide containing 76 amino acids in mature form with molecular weight of 13 kDa. PA508 is a CCL2-based decoy protein, a recombinant CCL2 inhibitor (competitor) with amino acid mutations in human CCL2 complement sequence (Fig. 3).

PA508 can compete endogenous CCL2 for binding of CCR2 and display a four-fold higher affinity toward natural CCL2 glycosaminoglycans (GAGs) ligand, heparan sulfate, and thereby displays natural CCL2 from heparan sulfate. PA 508 lacks activation of CCR2 (Piccinini et al. 2010).

PA508 exhibits about 20-fold increase in affinity for heparin with calculated dissociation constants (kilodaltons) of 5.2×10^{-8} M for PA508 and 9.4×10^{-7} M for wild-type CCL2.

Safety: PA508 (i.p.) is rapidly absorbed from injection site into circulation. It has no effects on

CCL2 1 QPDAINAPVTCC**Y**NFTNRKI**S**V**Q**RLASYRRITSSKCPK 38
PA508 **M**QPDAINAPVTCC**A**NFTNRKI**K**V**R**RLASYRRITSSKCPK

CCL2 39 EAVIFKTIVAKEICADPKQKWVQDSMDHLDKQTQTPKT 76
PA508 EAVIFKTIVAKEICADPKQKWVQDSMDHLDKQTQTPKT

Fig. 3 Sequence comparison of human CCL2 and PA508 with modified amino acid residues in bold type and underline

circulating levels of endogenous CCL2. Liver and kidney functions and histology remain unchanged after administration of PA508.

By using site-directed mutagenesis, PA508 is a well-designed competitive antagonist to CCL2/CCR-2 axis. It has more potent affinity with GAG and competes with CCL2 thus inactivates CCR-2 binding with CCL2 (53). PA508 inhibits leukocyte recruitment and trans-endothelial migration toward CCL2. In a mouse myocardial ischemia/reperfusion model, PA508 administered after ischemia (10 μg, intravenous route) decreased numbers of inflammatory monocytes in the heart and reduced infarct size. Both cardiac systolic function and diastolic function were improved significantly. In a mouse wire-induced neointima formation model, PA508 reduces accumulation of macrophages whereas increases the numbers of smooth muscle cells in neointima (Liehn et al. 2010).

3.3 M2 Macrophage Plasticity Modulator – Azithromycin

It has been known that macrophages exert significant plasticity in response to different environment exposures. In general, macrophages are divided into the classically activated macrophages (M1 macrophages) and the alternatively activated macrophages (M2 macrophages). M1 macrophages are associated with pro-inflammation and cell injury whereas M2 macrophages are relevant to cell survival and antiinflammation. M2 macrophages display high levels of scavenger, mannose, and galactose type receptors on the surface to accelerate debris clearance. M2 macrophages also constitutively express TGF-beta and Arginase 1 (Arg1) to hydrolyze L-arginine to L-ornithine, the main precursor for polyamines for cell survival (Ruytinx et al. 2018). To enhance M2 population or redirect plasticity from M1 to M2 in the heart may provide a novel strategy for cardiac protection.

Usually, Th1 cytokines, such as TNF-alpha, interferon (IFN)-gamma, and bacterial components such as LPS-induced conversion of M0 to M1. M2 macrophages are induced by stimulation with Th2 cytokines, such as IL-4, IL-10, and IL-13.

Azithromycin (AZM), the antimicrobial azalide, promotes induction and polarization of macrophages to M2 phenotype both in vitro and in vivo. AZM decreases NF-κB/STAT-1 activation and prevents P65 translocation from cytoplasm to nuclei via an IKKβ-dependent mechanism. AZM also increases Arg1 gene expression and enzyme activity (Haydar et al. 2019). Azithromycin promotes transition to M2 (Amantea et al. 2016b).

Recently, it was reported that oral treatment with AZM (160 mg/kg/day) 3 days prior to myocardial infarction and continued to 7 days post-myocardial infarction induced by permanent ligation of left anterior descending coronary artery in anesthetic mouse model significantly reduced pro-inflammatory macrophages (CD45+/Ly6G+/F4-80+/CD86+) at day 1 of MI and increased anti-inflammatory macrophages (CD45+/Ly6G+/F4-80+/CD206+) at day 3. AZM therapy shifts macrophages from pro-inflammatory to anti-inflammatory after MI. AZM also reduces inflammatory monocytes and inflammatory cytokines with no changes for IL-10 in the heart after MI. Remarked reduction of myocardial infarct size and scar size as well as mortality was observed in the mice treated with AZM. Cardiac systolic function and diastolic function were preserved in the mice treated with AZM (Al-Darraji et al. 2018).

In a separate study for experimental translational research, liposomal formulations of AZM (10 mg/kg/day and 40 mg/kg/day) were delivered (i.v.) immediately post-myocardial infarction in a permanent murine myocardial infarction model for 7 days. AZM increased survival rate and enhanced cardiac functions. The infarct size and apoptotic cells were reduced by AZM. As expected, AZM enhanced the population of anti-inflammatory macrophages and reduced circulating monocytes and pro-inflammatory macrophages (Al-Darraji et al. 2020).

References

Al-Darraji A, Haydar D, Chelvarajan L, et al. Azithromycin therapy reduces cardiac inflammation and mitigates adverse cardiac remodeling after myocardial infarction: potential therapeutic targets in ischemic heart disease. PLoS One. 2018;13(7):e0200474.

Al-Darraji A, Donahue RR, Tripathi H, et al. Liposomal delivery of azithromycin enhances its immunotherapeutic efficacy and reduces toxicity in myocardial infarction. Sci Rep. 2020;10(1):16596.

Alder JK, Georgantas RW 3rd, Hildreth RL, Kaplan IM, Morisot S, Yu X, McDevitt M, Civin CI. Kruppel-like factor 4 is essential for inflammatory monocyte differentiation in vivo. J Immunol. 2008;180(8):5645–52.

Amantea D, Certo M, Petrelli F, et al. Azithromycin protects mice against ischemic stroke injury by promoting macrophage transition towards M2 phenotype. Exp Neurol. 2016a;275(Pt 1):116–25.

Amantea D, Certo M, Petrelli F, et al. Azithromycin protects mice against ischemic stroke injury by promoting macrophage transition towards M2 phenotype. Exp Neurol. 2016b;275(Pt 1):116–25.

Aurora AB, Porrello ER, Tan W, et al. Macrophages are required for neonatal heart regeneration. J Clin Invest. 2014;124(3):1382–92.

Bajpai G, Schneider C, Wong N, et al. The human heart contains distinct macrophage subsets with divergent origins and functions. Nat Med. 2018;24(8):1234–45.

Bajpai G, Bredemeyer A, Li W, et al. Tissue resident CCR2- and CCR2+ cardiac macrophages differentially orchestrate monocyte recruitment and fate specification following myocardial injury. Circ Res. 2019;124(2):263–78.

Barrett TJ. Macrophages in atherosclerosis regression. Arterioscler Thromb Vasc Biol. 2020;40(1):20–33.

Boyle JJ, Harrington HA, Piper E, et al. Coronary intraplaque hemorrhage evokes a novel atheroprotective macrophage phenotype. Am J Pathol. 2009;174(3):1097–10108.

Castoldi A, Naffah de Souza C, Câmara NO, Moraes-Vieira PM. The macrophage switch in obesity development. Front Immunol. 2016;6:637.

Chan CT, Moore JP, Budzyn K, et al. Reversal of vascular macrophage accumulation and hypertension by a CCR2 antagonist in deoxycorticosterone/salt-treated mice. Hypertension. 2012;60(5):1207–12.

Chernykh ER, Shevela EY, Starostina NM, et al. Safety and therapeutic potential of M2 macrophages in stroke treatment. Cell Transplant. 2016;25(8):1461–71.

Cherry JD, Olschowka JA, O'Banion MK. Neuroinflammation and M2 microglia: the good, the bad, and the inflamed. J Neuroinflammation. 2014;11:98.

Chu HX, Broughton BR, Kim HA, Lee S, Drummond GR, Sobey CG. Evidence that Ly6C(hi) monocytes are protective in acute ischemic stroke by promoting M2 macrophage polarization. Stroke. 2015a;46(7):1929–37.

Chu HX, Broughton BR, Kim HA, Lee S, Drummond GR, Sobey CG. Evidence that Ly6C(hi) monocytes are protective in acute ischemic stroke by promoting M2 macrophage polarization. Stroke. 2015b;46(7):1929–37.

Das H, Kumar A, Lin Z, Patino WD, et al. Kruppel-like factor 2 (KLF2) regulates proinflammatory activation of monocytes. Proc Natl Acad Sci U S A. 2006;103(17):6653–8.

de Lemos JA, Morrow DA, Sabatine MS, et al. Association between plasma levels of monocyte chemoattractant protein-1 and long-term clinical outcomes in patients with acute coronary syndromes. Circulation. 2003;107(5):690–5.

Dick SA, Macklin JA, Nejat S, et al. Self-renewing resident cardiac macrophages limit adverse remodeling following myocardial infarction. Nat Immunol. 2019;20(1):29–39.

Epelman S, Lavine KJ, Beaudin AE, Sojka DK, Carrero JA, Calderon B, et al. Embryonic and adult-derived resident cardiac macrophages are maintained through distinct mechanisms at steady state and during inflammation. Immunity. 2014;40(1):91–104.

Etzerodt A, Moestrup SK. CD163 and inflammation: biological, diagnostic, and therapeutic aspects. Antioxid Redox Signal. 2013;18(17):2352–63.

Frieler RA, Meng H, Duan SZ, et al. Myeloid-specific deletion of the mineralocorticoid receptor reduces infarct volume and alters inflammation during cerebral ischemia. Stroke. 2011;42(1):179–85.

Gomez Perdiguero E, Klapproth K, Schulz C, et al. Tissue-resident macrophages originate from yolk-sac-derived erythro-myeloid progenitors. Nature. 2015;518(7540):547–51.

Gray GA, Toor IS, Castellan R, Crisan M, Meloni M. Resident cells of the myocardium: more than spectators in cardiac injury, repair and regeneration. Curr Opin Physiol. 2018;1:46–51.

Guo L, Akahori H, Harari E, et al. CD163+ macrophages promote angiogenesis and vascular permeability accompanied by inflammation in atherosclerosis. J Clin Invest. 2018;128(3):1106–24.

Hausenloy DJ, Yellon DM. Survival kinases in ischemic preconditioning and postconditioning. Cardiovasc Res. 2006;70(2):240–53.

Haydar D, Cory TJ, Birket SE, et al. Azithromycin polarizes macrophages to an M2 phenotype via inhibition of the STAT1 and NF-κB signaling pathways. J Immunol. 2019;203(4):1021–30.

Heidt T, Courties G, Dutta P, et al. Differential contribution of monocytes to heart macrophages in steady-state and after myocardial infarction. Circ Res. 2014;115(2):284–95.

Hu X, Li P, Guo Y, et al. Microglia/macrophage polarization dynamics reveal novel mechanism of injury expansion after focal cerebral ischemia. Stroke. 2012;43(11):3063–70.

Hulsmans M, Clauss S, Xiao L, et al. Macrophages facilitate electrical conduction in the heart. Cell. 2017;169(3):510–22.

Hulsmans M, Sager HB, Roh JD, et al. Cardiac macrophages promote diastolic dysfunction. J Exp Med. 2018;215(2):423–40.
Hussell T, Bell TJ. Alveolar macrophages: plasticity in a tissue-specific context. Nat Rev Immunol. 2014;14(2):81–93.
Lavine KJ, Epelman S, Uchida K, et al. Distinct macrophage lineages contribute to disparate patterns of cardiac recovery and remodeling in the neonatal and adult heart. Proc Natl Acad Sci U S A. 2014;111(45):16029–34.
Leid J, Carrelha J, Boukarabila H, Epelman S, Jacobsen SE, Lavine KJ. Primitive embryonic macrophages are required for coronary development and maturation. Circ Res. 2016;118(10):1498–511.
Leitinger N, Schulman IG. Phenotypic polarization of macrophages in atherosclerosis. Arterioscler Thromb Vasc Biol. 2013;33(6):1120–6.
Li Q, Barres BA. Microglia and macrophages in brain homeostasis and disease. Nat Rev Immunol. 2018;18(4):225–42.
Li W, Hsiao HM, Higashikubo R, et al. Heart-resident CCR2+ macrophages promote neutrophil extravasation through TLR9/MyD88/CXCL5 signaling. JCI Insight. 2016;1(12):e87315.
Liao X, Sharma N, Kapadia F, et al. Krüppel-like factor 4 regulates macrophage polarization. J Clin Invest. 2011;121(7):2736–49.
Liao X, Shen Y, Zhang R, et al. Distinct roles of resident and nonresident macrophages in nonischemic cardiomyopathy. Proc Natl Acad Sci U S A. 2018;115(20):E4661–9.
Liehn EA, Piccinini AM, Koenen RR, et al. A new monocyte chemotactic protein-1/chemokine CC motif ligand-2 competitor limiting neointima formation and myocardial ischemia/reperfusion injury in mice. J Am Coll Cardiol. 2010;56(22):1847–57.
Liu X, Liu J, Zhao S, et al. Interleukin-4 is essential for microglia/macrophage M2 polarization and long-term recovery after cerebral ischemia. Stroke. 2016;47(2):498–504.
Mirzadegan T, Diehl F, Ebi B, et al. Identification of the binding site for a novel class of CCR2b chemokine receptor antagonists: binding to a common chemokine receptor motif within the helical bundle. J Biol Chem. 2000;275(33):25562–71.
Oh J, Riek AE, Weng S, Petty M, et al. Endoplasmic reticulum stress controls M2 macrophage differentiation and foam cell formation. J Biol Chem. 2012;287(15):11629–41.
Patel B, Bansal SS, Ismahil MA, et al. CCR2+ monocyte-derived infiltrating macrophages are required for adverse cardiac remodeling during pressure overload. JACC Basic Transl Sci. 2018;3(2):230–44.
Piccinini AM, Knebl K, Rek A, Wildner G, Diedrichs-Möhring M, Kungl AJ. Rationally evolving MCP-1/CCL2 into a decoy protein with potent anti-inflammatory activity in vivo. J Biol Chem. 2010;285(12):8782–92.
Pinto AR, Ilinykh A, Ivey MJ, et al. Revisiting cardiac cellular composition. Circ Res. 2016;118(3):400–9.
Porrello ER, Mahmoud AI, Simpson E, et al. Transient regenerative potential of the neonatal mouse heart. Science. 2011;331(6020):1078–80.
Porrello ER, Mahmoud AI, Simpson E, et al. Regulation of neonatal and adult mammalian heart regeneration by the miR-15 family. Proc Natl Acad Sci U S A. 2013;110(1):187–92.
Ruytinx P, Proost P, Van Damme J, Struyf S. Chemokine-induced macrophage polarization in inflammatory conditions. Front Immunol. 2018;9:1930.
Shigeta A, Huang V, Zuo J, et al. Endocardially derived macrophages are essential for Valvular remodeling. Dev Cell. 2019;48(5):617–30.
Shiraishi M, Shintani Y, Shintani Y, et al. Alternatively activated macrophages determine repair of the infarcted adult murine heart. J Clin Invest. 2016;126(6):2151–66.
Stöger JL, Gijbels MJ, van der Velden S, et al. Distribution of macrophage polarization markers in human atherosclerosis. Atherosclerosis. 2012;225(2):461–8.
Tan X, Hu L, Shu Z, Chen L, et al. Role of CCR2 in the development of Streptozotocin-treated diabetic cardiomyopathy. Diabetes. 2019;68(11):2063–73.
Tang Z, Gan Y, Liu Q, et al. CX3CR1 deficiency suppresses activation and neurotoxicity of microglia/macrophage in experimental ischemic stroke. J Neuroinflammation. 2014;11:26.
Vaibhav K, Braun M, Khan MB, et al. Remote ischemic post-conditioning promotes hematoma resolution via AMPK-dependent immune regulation. J Exp Med. 2018;215(10):2636–54.
van de Laar L, Saelens W, De Prijck S, et al. Yolk sac macrophages, fetal liver, and adult monocytes can colonize an empty niche and develop into functional tissue-resident macrophages. Immunity. 2016;44(4):755–68.
van Tits LJ, Stienstra R, van Lent PL, Netea MG, Joosten LA, Stalenhoef AF. Oxidized LDL enhances pro-inflammatory responses of alternatively activated M2 macrophages: a crucial role for Krüppel-like factor 2. Atherosclerosis. 2011;214(2):345–9.
Virani SS, Alonso A, Benjamin EJ, et al. American Heart Association Council on epidemiology and prevention statistics committee and stroke statistics subcommittee. Heart disease and stroke Statistics-2020 update: a report from the American Heart Association. Circulation. 2020;141(9):e139–596.
Wan S, Cheng Y, Jin H, et al. Microglia activation and polarization after intracerebral hemorrhage in mice: the role of protease-activated Receptor-1. Transl Stroke Res. 2016;7(6):478–87.
Wynn TA, Chawla A, Pollard JW. Macrophage biology in development, homeostasis and disease. Nature. 2013;496(7446):445–55.
Xiong X, Barreto GE, Xu L, Ouyang YB, Xie X, Giffard RG. Increased brain injury and worsened neurological outcome in interleukin-4 knockout mice after transient focal cerebral ischemia. Stroke. 2011;42(7):2026–32.
Zhu W, Zhang E, Zhao M, et al. Regenerative potential of neonatal porcine hearts. Circulation. 2018;138(24):2809–16.

Delivery of Oligonucleotide Therapeutics for Macrophage Reprogramming in Inflammatory Diseases

Dhaval Oza and Mansoor M. Amiji

Abstract

Macrophages are one of the most plastic and fascinating cells of the innate immune system. From phagocytosis of invading pathogens and dead cells, maintaining tissue homeostasis, to tissue development and repair, they carry out a range of functions. In this chapter, we cover the recent understanding of macrophage biology based on their ontogeny and location that has helped to decipher their various phenotypic states and exposed the enormous complexity within the cells as they adapt to their environmental stimuli and pathological context. Abnormal macrophage polarization can lead to varying pathophysiologies as it deceives macrophage functional and secretory pathways and subsequently their regulation of the surrounding microenvironment. In the context of aberrant phenotypic state, a central paradigm of genetic reprogramming macrophages to restore homeostasis and alleviate the disease is an area of intense research. We discuss novel approaches to reprogram tissue-resident and infiltrating macrophages by utilizing oligonucleotides modalities. Lastly, we exemplify the differing yet intertwined roles of both tissue-resident and infiltrating macrophage populations in context of acute and chronic inflammation in the liver and highlight some contemporary findings in the areas of macrophage-specific delivery utilizing nanoparticle-based and conjugate-based oligonucleotide delivery.

Keywords

Resident and infiltrating macrophages · Acute and chronic liver injury · Oligonucleotide therapeutics · RNA interference · Nanoparticle formulation · Conjugates · Receptor-mediated endocytosis

D. Oza
Department of Pharmaceutical Sciences, School of Pharmacy, Northeastern University,
Boston, MA, USA

Alnylam Pharmaceuticals, Inc.,
Cambridge, MA, USA

M. M. Amiji (✉)
Department of Pharmaceutical Sciences, School of Pharmacy, Northeastern University,
Boston, MA, USA
e-mail: m.amiji@northeastern.edu

1 Macrophage Biology and Phenotypes in Health and Disease

Since being discovered by Elie Metchnikoff more than a century ago, which led to him winning the Nobel prize in 1908, macrophages have always been at the forefront of immunology research and have been known as the prodigious innate

S. Gupta, Y. V. Pathak (eds.), *Macrophage Targeted Delivery Systems*,
https://doi.org/10.1007/978-3-030-84164-5_16

immune cells involved in phagocytosis (Barreda et al. 2017). Immunologists for a long time have focused on the role of macrophages in context of host defense (Yona and Gordon 2015). It has been well characterized that macrophages play a "janitorial" role of phagocytosing not just invading pathogens but also clearing up dead cells and cellular debris. However, along the years as more light has been shed on these cells of the innate immune system, their more versatile roles in maintaining tissue homeostasis in health and disease and in propagation of inflammatory diseases have been identified (Teti et al. 2017; Nathan 2008).

Over the years, macrophages have been classified based on a number of defining characteristics like their ontogeny, location, distribution, plasticity, and phenotypic properties (Gordon and Plüddemann 2017). They are one of the most versatile cell types in eukaryotic organisms and have been classified into subsets of tissue-resident macrophages (TRM) and monocyte-differentiated macrophages based on their origin and differential localization in specific tissues in the body (Okabe and Medzhitov 2016; Gordon et al. 2014). Lineage tracing by gene fate-mapping techniques have given vital insights into precisely tracking the various embryonic macrophage populations and accurately bifurcating the distinct tissue-resident macrophages that have an embryonic origin versus monocyte-derived macrophage that have their origin in hematopoietic stem cells (Gordon et al. 2014; Cochain et al. 2018). Additionally, recent advances in single cell RNA sequencing have further helped elucidate specific transcriptomic profiles of different TRMs (Cochain et al. 2018). Different TRMs have different transcriptional profiles, leading to slightly different phenotypes depending on their tissue localization (Gordon et al. 2014). For example, liver tissue-resident macrophages, also known as Kupffer cells, have completely different transcriptional profile and phenotypic characteristics versus brain-resident macrophages or microglia (Gordon and Plüddemann 2017; Okabe and Medzhitov 2016; Gordon et al. 2014; Epelman et al. 2014; Liao et al. 2018).

It is now well established that in healthy tissue, TRMs maintain tissue homeostasis by regulating the finely tuned balance between clearing senescent and necrotic cells and cell metabolism (Gordon and Plüddemann 2017; Okabe and Medzhitov 2016; Gordon et al. 2014; Epelman et al. 2014; Liao et al. 2018). Besides classifying macrophages based on their lineage, one of the most well-characterized classification is based on macrophage activation and polarization states. Further in this section, we will look into how this concept of macrophage activation has changed as we have understood more about this complex and fascinating cell type (Martinez and Gordon 2014).

For many years, it was generally accepted that macrophages exist as two extreme opposite phenotypes, one being responsible for propagation of inflammation, and the other, its resolution (Martinez and Gordon 2014; Edwards et al. 2006). This classification based on their activation states was initially done to imitate T-cell literature (Martinez and Gordon 2014; Edwards et al. 2006). With the evolution of acquired immunity mediated by activated T and B lymphocytes, reciprocal interaction with cell-mediated immunity chiefly driven by circulating monocyte-derived macrophages provided novel understanding of their roles as mediators of both host defense against intracellular pathogenic invasion, where they would need to propagate inflammation, and resolution of host response to infection by clearing extracellular cell debris, where they would mostly assume the role of resolving inflammation (Mosser and Edwards 2008). With this concept in mind, macrophages, historically, were branched into two polarization states, as being either classically activated pro-inflammatory macrophages that emulated Th-1-derived cell-mediated immunity and alternatively activated anti-inflammatory macrophages that emulated Th-2-like immunity (Martinez and Gordon 2014; Edwards et al. 2006; Mosser and Edwards 2008; Mills and Ley 2014).

However, recent advances in macrophage biology in context of various autoimmune, inflammatory, and oncogenic diseases have led to the discovery that these two phenotypic activation states of macrophages are two extreme ends of a spectrum of activation states of macrophages, that in fact are a continuum of various interlaced phenotypic states rather than two distinct extreme phenotypes (Mosser and Edwards 2008). To characterize this extremely complex cell type by just two phenotypic states would not just be an oversimplification of a multitude of existing phenotypes but also a seemingly inaccurate and outdated way of representing this extremely versatile and plastic cell type (Martinez and Gordon 2014; Mosser and Edwards 2008). Additional contextualization of their role in health and pathophysiology has also made us appreciate the various hues of macrophages and has made us understand that pro- and anti-inflammatory states are merely the ends of an entire spectrum of macrophage phenotypes (Mosser and Edwards 2008; Mills and Ley 2014; Murray 2017; Italiani and Boraschi 2014; Mantovani et al. 2004). Figure 1 summarizes the different macrophage phenotypic states (Martinez and Gordon 2014; He et al. 2020).

Depending on the preceding stimuli, like stress, pathogenic infection, tissue damage, or other homeostatic processes, macrophages can adapt accordingly (Italiani and Boraschi 2014). Another important consideration while phenotyping macrophages is adding immunological as well as pathological contextualization (He et al. 2020; Bashir et al. 2016). Historically, in the pregenomic era, the concept of macrophage polarization was first established just by a few markers that were used to establish the differences and similarities in macrophage responses to various stimuli (Mantovani et al. 2004). However, updated knowledge about the intricate roles and interplay of various cytokines and their signaling pathways in development and disease has revealed a much more complicated and disparate universe of stimuli that can alter and affect macrophage polarization states. Transcriptomic and proteomic analyses along with genetically modified models have revealed a far more elaborate picture with significant challenges in identifying markers of macrophage modulation (Martinez and Gordon 2014; Edwards et al. 2006; Mosser and Edwards 2008; Mills and Ley 2014; Murray 2017; Italiani and Boraschi 2014).

The pro-inflammatory activation of macrophages is a representation of the putative activation of macrophages and this term is used to identify the effector macrophage population that gets activated as the host cell-mediated immune response (Italiani and Boraschi 2014). Originally when this subset of macrophages was characterized, chiefly two markers were identified that differentiated macrophages to a phenotype that had enhanced microbicidal, pathogen-clearing, and tumoricidal traits. These markers were identified as a combination of interferon-γ (IFN-γ) and tumor necrosis factor (TNF) that could be secreted by both innate and adaptive immune cells (Mantovani et al. 2004; Rath et al. 2014; Dransfield et al. 2015). For example, large amounts of IFN-γ and TNF can be secreted by natural killer (NK) innate immune cells as a response to either infection or tumor. This in turn activates macrophages to a pro-inflammatory phenotype and further leads to secretions of inflammation-inducing cytokines, superoxide anions, and oxygen- and nitrogen-free radicals, priming them for enhanced killing and phagocytosing ability (Mantovani et al. 2004; Rath et al. 2014; Dransfield et al. 2015). The secreted IFN-γ can bind to toll-like receptors (TLRs) expressed on macrophage surface, and through various pathways, including MyD88-dependent and TIR-domain-containing adaptor protein inducing IFN-β (TRIF)-dependent pathways, can lead to increased transcription of TNF, which in turn can turn on transcription of a lot of downstream genes for pro-inflammatory cytokines. Some of the well-characterized pro-inflammatory cytokines that are released upon macrophage activation are IL-6, IL-1, and IL-23 (Mantovani et al. 2004; Rath et al. 2014; Wang et al. 2014). It has been well established by now that classically activated macrophages confer host defense against intracellular pathogens. It has also been established by various genetic mouse models that the lack of IFN-γ makes them susceptible to various pathogenic infection (Mantovani et al. 2004; Rath et al. 2014; Wang et al. 2014). More importantly, even in humans, various mutations in these pathways have been associated with more susceptibility to bacterial, viral, and protozoal infections. Hence,

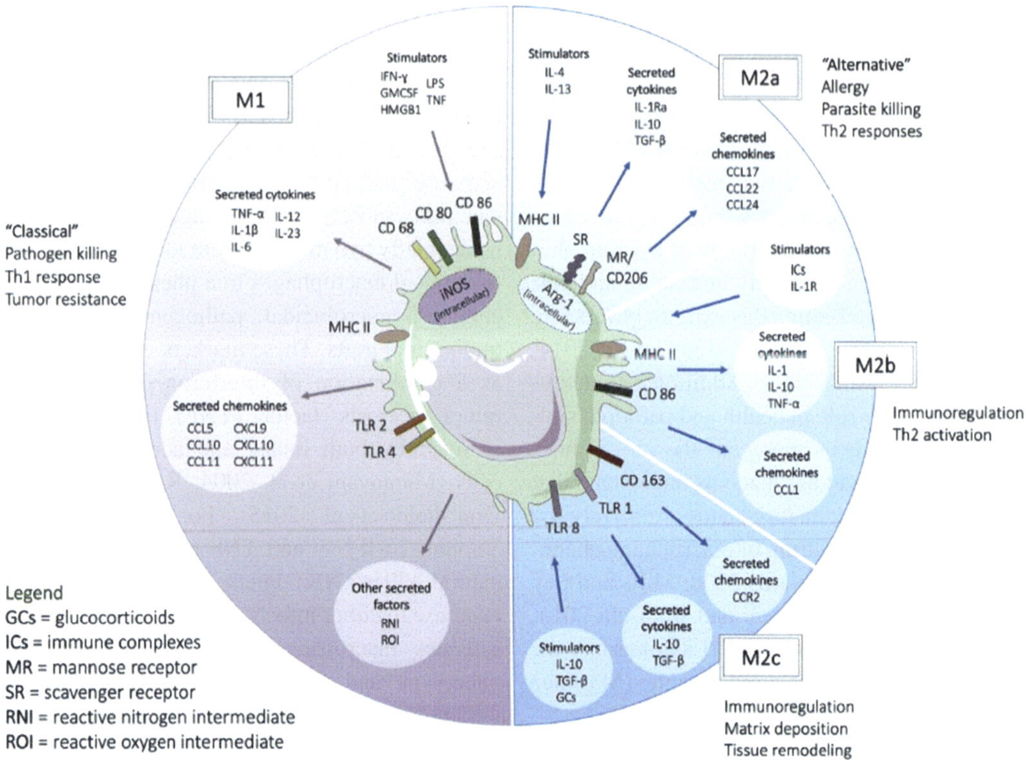

Fig. 1 Spectrum of macrophage activation and corresponding stimulators, markers, and secretory outputs. Macrophage polarization is best characterized by a multidimensional spectrum, but it is often simplified into inflammatory (M1) and anti-inflammatory (M2) phenotypes. The inflammatory responses of M1 macrophages (classically activated) encompass pathogen killing, Th1 activation, and tumor resistance. The anti-inflammatory response of M2 macrophages (alternatively activated) can be further categorized as M2a (allergy, parasite killing, Th2 responses), M2b (immunoregulation, Th2 activation), and M2c (immunoregulation, matrix deposition, tissue remodeling). Although the corresponding stimulators, cytokines, and chemokines serve as general hallmarks for each activation state, macrophages may express a mix of these markers, regardless of function
Adapted from *Wei et al* (He et al. 2020); https://doi.org/10.1016/j.addr.2019.12.001., Copyright permission granted from Elsevier; Journal: Adv drug delivery reviews; Original figure adapted from Martinez FO, Gordon S. doi:10.12703/P6-13 Copyright © 2014 Faculty of 1000 Ltd; Open Access article distributed under Creative Commons Attribution license

the classical pro-inflammatory activated state of macrophages is indeed one of hosts' first line of defense against invading pathogens (Mantovani et al. 2004; Rath et al. 2014; Wang et al. 2014).

However, this kind of innate immune-mediated activation of macrophages is short term as production of IFN-γ by NK cells by itself cannot support long-term macrophage activation (Wang et al. 2014). Hence, in order to maintain this inflammatory phenotype, after that initial bout of innate immune-mediated activation, an adaptive response takes over and is generally required to maintain an inflammation inducing state. This in turn leads to a sustained host-immune response against invading pathogens. The major cell types involved here which produce prolonged levels of IFN-γ are Th-1 cells that help in maintaining a long-term inflammatory response (Wang et al. 2014).

One of the major drawbacks of sustained pro-inflammatory-like activation of macrophages is that despite Th-1 cells themselves being antigen specific in their response, their downstream long-term activation of macrophages can lead to an indiscriminate non-cell-specific killing response, which can ultimately cause damage to healthy tissue by this haphazard killing response of activated macrophages (Rath et al. 2014; Wang et al. 2014). Hence, along with being important for host defense, an unchecked long-term activation of macrophages in a pro-inflammatory state can be a potential liability. One downstream mechanism by which this pro-inflammatory cytokine release can contribute to pathophysiological consequences is by activation of Th-17 helper T cells, which in turn leads to further recruitment of polymorphonuclear leukocytes (PMN) to healthy tissues (Rath et al. 2014; Wang et al. 2014). These recruited PMNs can then attack normal healthy tissues in their vicinity. Prolonged activated macrophages are important contributors to not just autoimmune-related disorders but also acute and chronic inflammatory indications (He et al. 2020; Bashir et al. 2016). Later in this chapter, we will look into a specific example of how a liver-resident macrophage population, also known as Kupffer cells, can chronically get activated into a pro-inflammatory macrophage state and helps drive the pathophysiology of a chronic liver disease called non-alcoholic steatohepatitis (NASH) (Krenkel and Tacke 2017). This phenomenon can contribute to random indiscriminate damage to the liver tissue (Krenkel and Tacke 2017).

At the other end of this macrophage activation spectrum lies the alternately activated anti-inflammatory macrophage differentiation that mimics a Th-2-mediated response and carries out wide-ranging functions from wound healing to maintaining tissue homeostasis (Mosser and Edwards 2008; Mills and Ley 2014; Murray 2017; Italiani and Boraschi 2014). These macrophages were initially coined as "M2" or alternately activated macrophages. However, in recent time, a more nuanced and functional based characterization of these macrophages has been established (Italiani and Boraschi 2014). Depending on their underlying roles, the classically known M2 macrophages have now been further subcharacterized into M2a or wound-healing macrophages, M2b or regulatory macrophages, and M2c macrophages that have more of an intermediary phenotype as they are anti-inflammatory, but also pro-fibrotic and phagocytic (Fig. 1). Other phenotypes like M2d have also been identified, that are also known as tumorassociated macrophages (Allavena et al. 2008; Wang et al. 2010; Hou et al. 2018).

Just like pro-inflammatory macrophages, alternatively activated macrophages can also be activated by both innate and adaptive immune cells. The signature marker of alternative macrophage activation is IL-4 (Mantovani et al. 2004). Among the innate immune cells, both mast cells and basophils can lead to secretion of IL-4 as a wound-healing response of the immune system. IL-4 can activate macrophages to a wound-healing M2a phenotype, which promotes arginase metabolism in macrophages to ornithine (Rath et al. 2014). Ornithine is a precursor to collagen, the chief structural extracellular matrix protein. Hence, as a response to IL-4 stimulation, macrophages can exert the secretion of extracellular matrix and impart a wound-healing response. Along with IL-4, other anti-inflammatory cytokines like IL-10 and IL-13 along with glucocorticoids and LPS can also contribute to activating macrophages to an anti-inflammatory phenotype (Mantovani et al. 2004; Rath et al. 2014). However, Th-2 helper T-cell-mediated adaptive response is chiefly responsible for induction and maintenance of the anti-inflammatory phenotypic state of macrophages. Upon IL-4 and IL-13 secretion from Th-2 cells as a response to disturbances or injury to mucosal layer, macrophages get differentiated to the wound-healing state and eventually aid in secreting extracellular matrix (Martinez and Gordon 2014; Mosser and Edwards 2008; Mills and Ley 2014; Murray 2017; Mantovani et al. 2004; He et al. 2020; Rath et al. 2014; Wang et al. 2014).

Long-term macrophage activation to an anti-inflammatory phenotypic state has a well-characterized role of playing a pro-fibrotic role, hence can also lead to detrimental effects on the host if they are overactivated. To exemplify this in context of a pathophysiological condition, it has been observed that overactivation of wound-healing macrophages has an association with idiopathic pulmonary fibrosis (IPF) in lungs (Hou et al. 2018). Tissue biopsy samples of IPF patients have had a confirmed presence of infiltrating macrophages with an alternate wound healing like phenotype (Hou et al. 2018).

Additionally, some evidence from in vitro systems points out that upon IL-4 or IL-13 treatment, macrophages become more susceptible to invading pathogens (Mantovani et al. 2004). There are indeed some caveats to these studies, one being that the conditions in which IL-4 was generated were the same very conditions where there was low IFN-γ, a powerful pathogen clearing cytokine; hence, it is difficult to clearly determine if the macrophages become more susceptible to infection due to high IL-4 or low IFN-γ (Mantovani et al. 2004). However, just the fact that the same cell type adopts opposing phenotypes leading to both immune defense and susceptibility to infection is ironical. This indeed makes these cells fascinating and complicated to study and further shows how plastic macrophages are in reality (Martinez and Gordon 2014; Mosser and Edwards 2008; Mills and Ley 2014; Murray 2017; Mantovani et al. 2004; He et al. 2020; Rath et al. 2014; Wang et al. 2014).

Moreover, a plethora of other macrophage phenotypic states have also been identified and coined, depending upon their associating tissue and their functionalities (Skuratovskaia et al. 2020). For example, macrophages associated with atherosclerotic plaque can be defined into several distinct but intertwined populations of macrophages like M1 pro-inflammatory and pro-atherogenic phenotype; M4 pro-inflammatory and low phagocytic macrophages; Mhem and M(Hb) or hemoglobin-related anti-inflammatory macrophages that stop formation of foam cells and are atheroprotective; M2 anti-inflammatory and atheroprotective macrophages; and Mox or oxidized macrophages with antioxidant and low phagocytic capacity (Chinetti-Gbaguidi et al. 2015). Likewise, several other phenotypic states exist depending on associations with respective tissues and functions (Skuratovskaia et al. 2020).

This reiterates the fact that macrophages can don a multitude of phenotypes, ranging from pro-inflammatory, to anti-inflammatory, to immune regulatory phenotypes (Gordon et al. 2014; Martinez and Gordon 2014). This depends on the physiological or pathophysiological context and their local environmental stimuli (Gordon et al. 2014; Martinez and Gordon 2014). Due to this complicated nature of macrophages, it has been notoriously difficult to develop macrophage-targeting therapies, especially as there is lack of clearly defined markers to separate pathological macrophages from their normal physiological lookalikes. Out of the many strategies to target macrophages therapeutically, the most frequently adopted and important strategy to manipulate macrophages in order to mitigate inflammation is by modulating macrophage polarization (Ponzoni et al. 2018). To attain balance between these phenotypes is often the determinant to the outcome of macrophage modulation strategies. Accordingly, various macrophage-targeting therapies either aim to achieve modulation by targeting proteins on macrophage itself or by targeting stimuli secreted from the macrophages or acting on the macrophages (He et al. 2020; Ponzoni et al. 2018). One of the biggest challenges, however, is macrophage-specific delivery, which is almost impossible to achieve by systemically administered small molecule drugs or antibodies.

2 Targeting Tissue-Resident Versus Infiltrating Macrophages

2.1 Tissue-Resident Macrophages (TRMs) and Infiltrating Macrophages (IMs) in Disease Progression and Therapy

In this section, we will look at the specific roles of both a liver-resident macrophage population and infiltrating macrophage population in driving both acute and chronic liver inflammatory diseases. Targeting macrophages therapeutically in context of cancer, autoimmune diseases, and infectious diseases has been illustrated in depth in the previous chapters. We want to highlight how macrophage activation and polarization can also drive the progression of acute and chronic inflammation in the liver. Here, we focus on liver to study the roles of both TRMs and IMs in promoting inflammation (van der Heide et al. 2019). The liver is where around 90% of all the macrophages in the human body reside, making them a nice organ to capture various macrophage roles and phenotypes (Tacke 2017; Tacke and Zimmermann 2014).

2.1.1 IMs in Propagating Acute Liver Injury

Apart from its own TRM population of Kupffer cells (KCs), liver is a node of infiltrating macrophages, out of which the most well-defined ones are the circulating monocyte-derived macrophages (MoMFs). Owing to its rich vasculature, MoMFs can be quickly recruited to the liver upon liver injury, metabolic damage, or toxic damage. Primarily, recruitment of MoMF is initiated via secretion of chemokine ligand 2 (CCL2) or monocyte chemoattractant protein 1 (MCP-1) from activated KCs or hepatic stellate cells (HSCs). These MoMFs are derived from BM progenitor cells and can be identified by expression of chemokine receptor-2 (CCR-2), lymphocyte antigen 6 complex locus C1 (Ly-6C), chemokine CX3-C motif receptor 1 (CXCR1), and CD11b (Fogg et al. 2006; Baeck et al. 2012; Karlmark et al. 2009). Depending on the environmental cues, MoMFs further differentiate in macrophages with distinct phenotypes. Various preclinical acute injury models have been studied to characterize the role of MoMFs in propagation of inflammation. It is well known that in acute liver failure, liver-infiltrating MoMFs secrete large amounts of pro-inflammatory cytokines, thus promoting inflammation (Antoniades et al. 2008; Wu et al. 2010). Ly-6C^{hi} MoMFs express a high level of CCR2 proteins. Upon release of CCL2 from KCs and HSCs, MoMFs are recruited to the liver predominantly via CCL2/CCR2 pathway (Karlmark et al. 2009; Galastri et al. 2012). Hence, this pathway has been known to critically regulate infiltration of MoMF into the liver and highlights the therapeutic potential of inhibiting CCR2 in inhibiting infiltration of monocytes into the liver. However, there is also some contradictory evidence which points out to a protective role of Ly-6C^{hi} monocytes in preventing acute liver injury. Ramachandran et al. in a preclinical model of acute carbon tetrachloride (CCl4)-induced liver fibrosis characterized that these infiltrated monocytes could undergo a phenotypic switch to a Ly-6C^{lo} phenotype, which can help restore liver damage (Ramachandran et al. 2012). Moreover, it has also been observed in an acetaminophen-induced liver injury mouse model that IMs in fact help resolve acute injury by helping with hepatic blood vessel repair (You et al. 2013). This also points to the fact that even within liver-infiltrating MoMFs, there is substantial heterogeneity, and based on the environmental stimuli, these macrophages can adopt a complex spectrum of polarization states.

In recent times, an intriguing finding by Wang and Kubes identified a distinct infiltrating macrophage population in the liver. After a series of elegant experiments, they identified this population as mature large peritoneal macrophages (LPMs) (Wang and Kubes 2016). These macrophages were identified as distinct from liver-resident macrophages, or MoMFs, and helped restore acute liver injury. These cells had the characteristic LPM signature of F4/80^{hi} CD11b^{hi} MHCII^{lo} and also expressed LPM-specific transcription factor GATA6 (Wang and Kubes 2016). Within an hour of eliciting liver injury, they were found to rapidly infiltrate to the site of thermal injury in the liver. This was a fascinating finding, which subsequently has also been observed in other acute injury models. Parayath et al. have also shown this unique migration pattern of LPMs to lung tissue in mice upon depleting alveolar macrophages in the lungs by intranasal chlodronate administration (Parayath et al. 2018). The details of LPM migration and infiltration to both the liver and distal organs are beyond the scope of this chapter; however, this surely is an intriguing finding that needs further exploration. Modulation of LPM polarization can potentially be another way that can be therapeutically explored to mitigate acute injury/inflammation.

2.1.2 Illustrative Example of TRM/Kupfer Cell-Targeted Therapy for NAFLD

Nonalcoholic fatty liver diseases (NAFLD) are essentially an array of diseases ranging from simple liver steatosis to nonalcoholic steatohepatitis (NASH). NASH has a tremendous burden on health and economics in liver disease (Sumida and Yoneda 2018; Drescher et al. 2019; Younossi et al. 2019; Shetty and Syn 2019). American Liver Foundation estimates that by 2030, it will become the leading cause of liver transplantation in the United States and as of now, it roughly affects between 2 and 5% of Americans, which roughly estimates a staggering 6.5–16.3 million people (*Source:* liverfoundation.org). It is chiefly characterized by excessive deposition of triglycerides in hepatocytes over a period of time, which leads to hepatocyte injury and death by various mechanisms leading to downstream inflammation and subsequently, hepatic fibrosis and cirrhosis and/or hepatocellular carcinoma (HCC). NASH is closely associated with metabolic syndrome and increases the risk of other non-liver-related co-morbidities. While fatty liver itself is a relatively benign indication, what drives this series of pathophysiological events is the ensuing inflammation (Sumida and Yoneda 2018; Drescher et al. 2019; Younossi et al. 2019; Shetty and Syn 2019; Tanaka et al. 2019; Brunt et al. 2020; Connelly et al. 2020; Muthiah and Sanyal 2020; Cariou et al. 2021; Anstee et al. 2019). A detailed discussion of pathophysiology of NAFLD and NASH is not the central focus of this chapter; however, we will briefly elaborate the role of the TRMs residing in the liver called Kupffer cells in driving NASH and how modulating them can be beneficial to mitigate inflammation.

While KCs are merely a piece in the complex pathogenesis of chronic inflammation driving progression of NASH, they most certainly are the most important cells in driving liver inflammation in not just NASH but also other liver-related inflammatory diseases like acute liver failure, alcoholic liver disease, and viral hepatitis (Krenkel and Tacke 2017; van der Heide et al. 2019). They are the TRM population residing in the liver sinusoidal lining. In the physiological state, KCs play many important roles as accessory cells to hepatocytes. To name a few, they help maintain iron load in the liver, scavenge pathogens arriving from gut via portal circulation, help with cholesterol storage and metabolism, and are also involved in active immune surveillance and maintaining homeostasis (Peterson et al. 2018).

Massive lipid buildup in hepatocytes beyond its storage capacity triggers hepatocyte injury and death through lipid peroxidation, ER stress, mitochondrial and DNA damage, autophagy, and various other mechanisms of cell death (Noureddin and Sanyal 2018; Kim and Lee 2018; Parthasarathy et al. 2020; Lefere and Tacke 2019; Kazankov et al. 2019). These injured/dead hepatocytes trigger a "sterile inflammation" by releasing damage-associated molecular patterns (DAMPs) like nuclear and mitochondrial DNA, high mobility group box-1 (HMGB1) and others, that are taken up by KCs through various toll-like receptors (TLRs) and receptor for advanced glycation end products (RAGE). Through NRLP3-inflammasome activation and other M1-like macrophage-activating cellular pathways that we discussed in more detail in the previous section, resident KCs get differentiated to a pro-inflammatory macrophage phenotype, which leads to activation and maturation of pro-IL-1β and pro-IL-18 in a caspase-dependent manner and a further release of pro-inflammatory cytokines like IL-1β and IL-18, triggering further liver damage (Noureddin and Sanyal 2018; Kim and Lee 2018; Parthasarathy et al. 2020; Lefere and Tacke 2019; Kazankov et al. 2019). Further, other cytokines like TNF-α and IL-6 are also upregulated as a response to an overall pro-inflammatory stimulation. These also bind to various pattern recognition receptors (PRRs) and toll-like receptors (TLRs) expressed on KCs, hepatocytes, hepatic stellate cells (HSCs), biliary epithelial cells, and sinusoidal endothelial cells, further driving more leukocyte infiltration and an uncontrolled amplification of inflammation, leading to downstream transdifferentiation of HSCs and portal fibroblasts into active myofibroblasts (Noureddin and Sanyal 2018; Kim and Lee 2018;

Parthasarathy et al. 2020, Lefere and Tacke 2019; Kazankov et al. 2019). This drives fibrogenesis and ultimately leads to hepatic fibrosis and liver cirrhosis (Fig. 2). The above description is an oversimplified explanation of the role of macrophages in driving NASH, as in reality, the picture is far more complex, as there are different phenotypes of KCs present at various stages of NASH, with evidence also pointing out to the role of a wound-healing phenotype in driving late-stage liver fibrosis and cirrhosis as the diseases progresses (Kazankov et al. 2019; Chen et al. 2020; Duarte et al. 2015). This reiterates the fact that macrophages are extremely adaptable and understanding the complexity of various phenotypes and their role in health and disease is context dependent and diverse.

NAFLD is closely associated with other non-liver manifestations of metabolic syndrome, inciting an active debate in this space over the causality and direction of progression of metabolic syndrome: The chicken and egg conundrum of whether NAFLD drives type II diabetes and metabolic syndrome or whether the reverse is more prominently seen (Sumida and Yoneda 2018; Drescher et al. 2019; Younossi et al. 2019; Shetty and Syn 2019). While this is a bigger discussion not warranted to be discussed here, one thing is certain: Chronic inflammation and macrophages have been involved in driving a lot of adverse pathophysiology in metabolic diseases.

Due to the important contribution of KCs in the amplification of inflammation, modulating their polarization to an anti-inflammatory phenotype has been an attractive strategy to mitigate NASH-associated inflammation. This would, in theory, also reprogram the KCs from a pathogenic to a restorative phenotype. Various immunomodulatory cytokines like IL-4, IL-10, prostaglandin E2, steroids, and colony-stimulating factor I receptor (CSF-1R) have shown some success in modulating macrophages to treat various liver diseases, including acute liver failure and chronic metabolic indications like NASH (van der Heide et al. 2019; Triantafyllou et al. 2018a). Utilizing some of the novel ligands for targeted delivery of oligonucleotides to KCs to modulate their polarization holds a lot of promise. Table 1 highlights a list of some current or past assets explored to target both TRM and MoMF for mitigating acute as well as chronic inflammation in the liver (non-exhaustive).

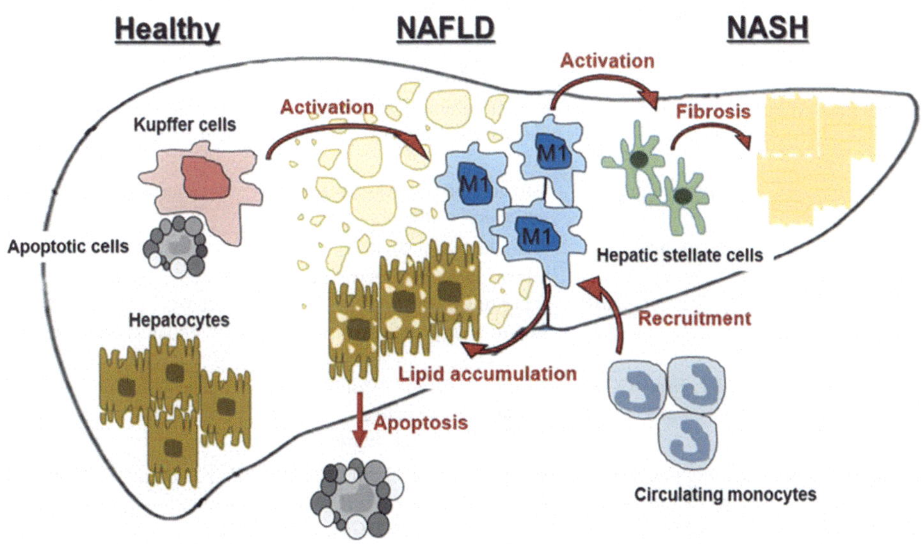

Fig. 2 Involvement of hepatic macrophages in progression of nonalcoholic fatty liver disease (NAFLD) and nonalcoholic steatohepatitis (NASH). Adapted from *Ji-Young Cha, Da-Hyun Kim and Kyung-Hee Chun* (2018); https://doi.org/10.5625/lar.2018.34.4.133;

3 Oligonucleotide Therapeutics

Antisense oligonucleotides (ASOs) were discovered back in 1978, when a 13-nucleotide DNA oligonucleotide was found to inhibit Rous sarcoma virus translation in a sequence-specific manner. Ever since their discovery, ASOs have paved the way for a ground-breaking class of medicine (Setten et al. 2019; Wilson and Doudna 2013). This story took a new exciting turn upon discovery of RNA interference approximately 20 years ago, which has shed light upon utilizing ASOs therapeutically for gene silencing (Fire et al. 1998; Elbashir et al. 2001). In this section, we will briefly touch base on the broad classification of oligonucleotide-based therapies and their mechanisms with an emphasis on small interfering RNAs. We will then evaluate some exciting avenues as to how siRNAs can be utilized therapeutically to target macrophages, with a specific dive into targeting the liver-resident macrophages—Kupffer cells, as a potential therapeutic strategy to mitigate non-alcoholic steatohepatitis (NASH).

Table 1 Some highlighted MoMF and KC-targeted therapies for mitigating inflammation

Strategy	Modality	Mechanism of action	References
Reprogramming KCs to an anti-inflammatory phenotype	Glucocorticoids	Broad anti-inflammatory	Triantafyllou et al. (2018b, LiverTox (2012)
	CSF-1R agonists	Inhibition of resident macrophage proliferation and recruitment of monocytes	Stutchfield et al. (2015)
	MTC-TNF-α siRNA NPs	Inhibition of TNF-α	He et al. (2013)
	Dexamethasone liposomes	Anti-inflammatory, induction of T-cell apoptosis	Bartneck et al. (2015)
	Galectin-3	Inhibition of pro-inflammatory macrophage functions	Traber and Zomer (2013)
	SYK pathway inhibitor R406	Anti-inflammatory	Bukong et al. (2016)
Blocking KC activation	Selonsertib	ASK-1 inhibitor, anti-inflammatory	Loomba et al. (2018)
	HMGB1 antagonists	DAMP inhibitor, attenuation of DAMP-mediated activation of KCs	Triantafyllou et al. (2018a, Liu et al. (2015)
	PRR antagonists	Attenuation of DAMP-mediated activation of KCs	Brenner et al. (2013)
	Curcumin	Immunomodulatory	Lefebvre et al. (2016)
Inhibiting monocyte recruitment and infiltration	Cenicriviroc, CCR2 antagonist	Blocks monocyte recruitment via inhibiting CCR2, anti-inflammatory, anti-fibrotic	Lefebvre et al. (2016, Krenkel et al. (2018), Friedman et al. (2018), Mulder et al. (2017)
	Maraviroc, CCR5 antagonist	Blocks monocyte recruitment via inhibiting CCR5, anti-inflammatory, anti-fibrotic	Pérez-Martínez et al. (2014)

ASK-1 apoptosis signal-regulating kinase 1, *CCL* chemokine (C–C motif) ligand, *CCR* chemokine (C–C motif) receptor, *CSF-1R* colony-stimulating factor 1 receptor, *KCs* Kupffer cells, *PRR* pattern recognition receptors, *TNF-α* tumor necrosis factor-α, *DAMP* damage-associated molecular patterns, *HMGB1* high mobility group box 1 protein, *SYK* spleen tyrosine kinase, *MTC* mannose-modified trimethyl chitosan-cysteine, *NPs* nanoparticles, *siRNA* small interfering RNA

Adapted from van der Heide D, Weiskirchen R and Bansal R (van der Heide et al. 2019); https://doi.org/10.3389/fimmu.2019.02852;

3.1 Classification of Oligonucleotide-Based Therapies

Oligonucleotides, as the name suggest, are short length DNA or RNA molecules. They also go by the name "oligomers," with the term "mer" used to describe the length of an oligonucleotide sequence. To exemplify, an oligonucleotide with a length of 21 nucleotides would usually be referred to as a "21-mer" (Roberts et al. 2020). The central dogma of any oligonucleotide would be specific complementary binding to their respective DNA or RNA molecules. Because of this mechanism, oligos have wide ranging applications from genetic testing to polymerase chain reaction (PCR), and from DNA and RNA sequencing to even farther reaching applications like data storing, owing to the capacity of DNA to store huge amounts of data (Roberts et al. 2020). Many of these applications are beyond the scope of this chapter; however, we will focus on their utility as a gene-silencing modality that has completely revolutionized how we make medicines. Broadly, oligonucleotide-based therapies can be classified into four main categories.

3.1.1 RNase-H-Dependent Antisense Oligonucleotides (ASOs)

ASOs are single-stranded, chemically synthesized oligonucleotides that bind to complementary sequences of their target mRNAs and reduce expression in an RNase-H-dependent manner. RNase-H is a ubiquitously expressed ribonuclease which recognizes a DNA-RNA complementary pair and degrades the mRNA strand of this complex. Along with this, inhibition of target mRNAs also happens partly by inhibition of translation by steric blockage of ribosomes. (Fig. 3) Nuclear retained RNA degradation mediated by RNase-H-dependent ASOs has been well established, albeit there is recent evidence also pointing to activity of ASOs in silencing cytoplasmic mRNAs (Wittrup and Lieberman 2015).

3.1.2 Exon Skipping Antisense Oligonucleotides

Exon skipping ASOs target specific intron-exon splice sites and by deleting specific intron-exon junctions, they essentially force the choice of an alternative splice site. This would lead to translation of an alternative protein isoform that can restore function or stability to the translated protein (Fig. 3). Successful implementation has led to groundbreaking medicines that can rectify or bypass exon deletions caused by genetic mutations. A well-known example is utilization of exon skipping ASOs to mitigate Duchenne Muscular Dystrophy (DMD) (Relizani and Goyenvalle 2018). DMD is the downstream disease manifestation due to mutation in the dystrophin gene, which leads to one or more exons in the gene missing, leading to a disrupted reading frame and lack of translation of sufficient amount of dystrophin protein. Exon skipping ASOs have been implemented to bypass this by essentially skipping more exons and leading to generation of internally deleted and a shorter, albeit a largely functional dystrophin protein. This would convert a severe DMD into a milder Becker muscular dystrophy phenotype (Setten et al. 2019; Wilson and Doudna 2013; Wittrup and Lieberman 2015; Relizani and Goyenvalle 2018).

3.1.3 Small Interfering RNAs (siRNAs)

siRNAs exist as endogenously expressed noncoding RNAs enabling a biological process called RNA interference (RNAi). Hence, drug siRNA molecules essentially harness a natural cellular pathway in which they inhibit target mRNA molecules of interest by exhibiting perfect sequence complementarity to the target sequence. Naturally occurring siRNAs exist as long double-stranded RNA (dsRNA) that are cleaved off by the enzyme Dicer into smaller 21–23 nucleotide duplex, now known as small interfering RNA (siRNA) (Setten et al. 2019). Synthetic siRNA drugs usually are 21–23-mer double-stranded molecules, with a passenger strand (sense strand) and the guide strand (antisense strand), annealed together as a double-

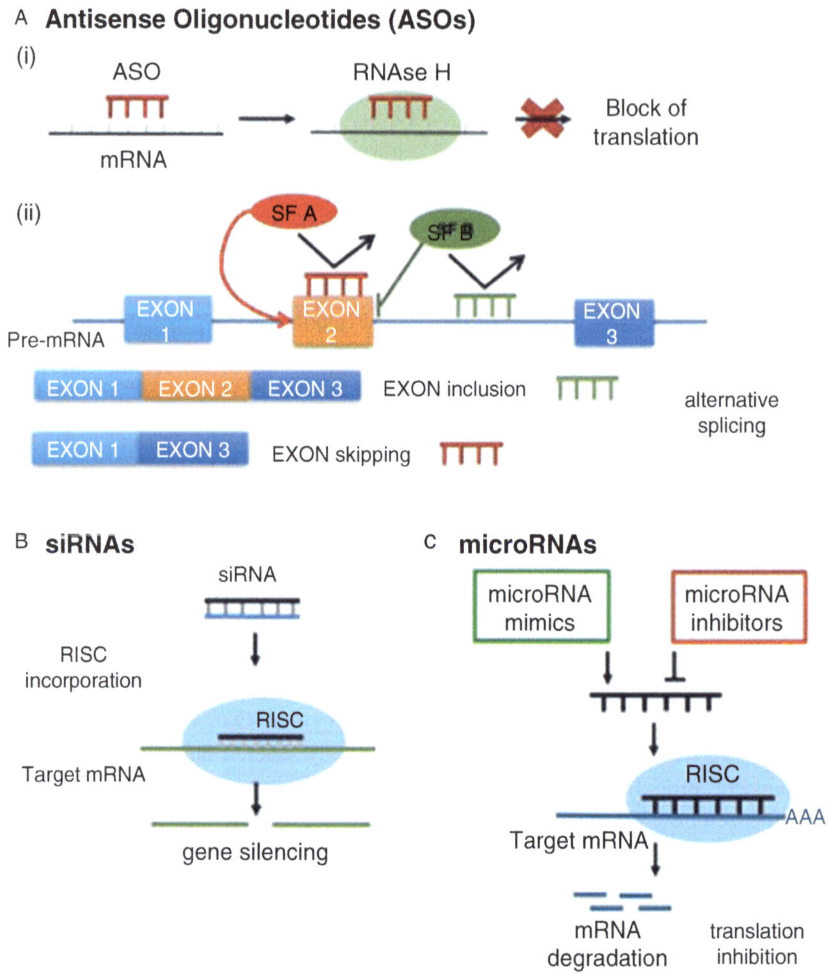

Fig. 3 RNA therapeutics in action. (A) Antisense oligonucleotides (ASOs); short, synthetic, single-stranded oligodeoxynucleotides that modify protein expression through the following mechanisms; (Ai) inhibition of protein production by antisense gapmers through activation of the ribonuclease RNAase H resulting in target mRNA degradation; (Aii) control of splicing by ASOs in alternative splicing. ASOs can modulate alternative splicing by preventing the binding of splicing factors (SF) resulting in translational arrest through ribosome attachment blocking; (B) siRNAs. Double-stranded (ds) RNA is processed by Dicer, a dsRNA-specific ribonuclease, into 21–25 nucleotide-long ds siRNAs with two nucleotides in their 3′ overhang and 5′ phosphate groups. siRNAs are then recognized by the Argonaute 2 (AGO2) and loaded into the RNA-induced silencing complex (RISC) and unwind into their single-stranded components. AGO2, which is a component of RISC, cleaves the sense strand of siRNA and the antisense strand binds with perfect complementarity to the target mRNA resulting in target mRNA cleavage; (C) microRNAs. Induction or inhibition of gene expression by microRNA mimics or inhibitors. Adapted from *Laina et al* (2018); l doi: 10.3389/fphys.2018.00953; Open Access article distributed under Creative Commons Attribution license

stranded duplex. Upon being released into the cytosol after being taken up by cells, siRNAs activate the RNA-induced silencing complex (RISC), during which the passenger strand is degraded and the guided strand then locates and binds with a complementary sequence in an mRNA molecule and induces cleavage (Setten et al. 2019; Wilson and Doudna 2013) (Fig. 3). The cleavage of target mRNA is carried out by the endonuclease protein Argonaute 2, which catalyzes the target mRNA degradation. Unlike single-stranded ASOs, the site for target mRNA cleavage for siRNAs is almost entirely cytoplasmic, although recent evidence points out to some nuclear siRNA activity too (Setten et al. 2019; Wilson and Doudna 2013; Wittrup and Lieberman 2015; Meister 2013).

3.1.4 MicroRNA Mimics and Anti-MicroRNAs

Lastly, microRNAs (miRNA) are another type of non-coding RNAs that also exist naturally and follow a similar intracellular pathway as siRNA and also exert target mRNA silencing via RNAi. However, unlike siRNAs, microRNAs are single-stranded RNAs that do not have a perfect sequence complementarity to the target mRNA. Synthetic oligonucleotides can be made to modulate miRNA expression, either by antagonizing or mimicking their function. Just like siRNA, miRNAs can inhibit mRNA expression through the RISC machinery. However, they suppress RNA expression by partial complementary sequence-match, accelerating their degradation. This makes them more promiscuous but less efficient than siRNAs (Setten et al. 2019; Wilson and Doudna 2013; Wittrup and Lieberman 2015; Meister 2013).

3.2 Challenges to Oligonucleotide-Based Delivery Approach

Oligonucleotides have the potential to reach targets that are "undruggable" by more conventional small molecule, biologics, and monoclonal antibody-based therapeutic approaches. As every cell has RNAi machinery, this essentially opens up the potential for any disease-causing gene to be silenced in any cell type, widely expanding the drug development net (Roberts et al. 2020). However, despite being a powerful tool for knocking down messages that translate for potentially disease-causing proteins, this also increases chances of drug-related direct on-target cytotoxicity. Along with this, there are still huge challenges to selectively deliver these large anionic molecules to cell types of interest along with huge roadblocks of getting across cell membranes into endosomal compartments, and from endosome to cytosol (for RNAi) and nucleus (for ASOs) (Setten et al. 2019; Roberts et al. 2020; Wittrup and Lieberman 2015; Aigner 2019). This makes for some pretty significant rate limiting steps to oligonucleotide-based drug delivery that need to be surmounted to fully realize the huge potential they have. Table 2 highlights some of the common challenges faced by oligonucleotide therapies and some possible strategies to overcome them.

4 Macrophage-Specific Delivery Strategies

Ever since the groundbreaking discovery of ASOs and RNAi, there has been a plethora of molecules in the clinic and already some drug approvals utilizing oligonucleotides. Most notably, siRNA drugs have had multiple approvals just within the past 3 years, with a burgeoning pipeline of more assets in clinical development (No Authors 2020; Kim 2020; Fitzgerald et al. 2016; Balwani et al. 2020; Garber 2018).

Despite impressive progress of oligonucleotide delivery and functionality in hepatocytes, it still remains extremely challenging to efficiently deliver these huge anionic molecules to other extrahepatic cells. With their extraordinary capabilities as a cell type to don a wide variety of roles ranging from host defense to maintaining tissue homeostasis, macrophages are excellent candidates for cell-type-specific delivery and activity of oligonucleotide therapies in order to not only treat cancer, infectious diseases, and autoimmune diseases but also inflammatory diseases in general (Novobrantseva et al. 2012).

Targeting macrophages by utilizing gene silencing approach definitely offers a unique opportunity to modulate polarization of macrophages and also attain a controlled regulation of macrophage proliferation by selective silencing of specific genes rather than complete ablation of general macrophage function. Selective pharmacological targeting of macrophages by oligonu-

Table 2 Challenges to oligonucleotide therapies and strategies to overcome them

Challenge	Strategies to overcome
1. How to achieve better access to target tissues?	Utilization of intravenous or subcutaneous dosing for entry into systemic circulation
	Local administration for direct uptake in skin, eye, and mucosal tissues
2. How to get a longer circulation time to reach target tissues and avoid excretion?	Increasing PEGylation improves circulation time due to increased molecular weight
	Increasing cholesterol content improves circulation time due to more binding with circulating lipoproteins
3. How to avoid nuclease digestion once in circulation?	Encapsulation of non-modified oligonucleotides in various LNP-based formulations
	Chemically modifying nucleic acid backbones. Some well-known backbone modifications are: deoxynucleotide overhangs, phosphorothioate linkages
4. How to avoid immunogenicity?	Encapsulation in LNP-based formulations utilizing relatively non-immunogenic excipients
	Nucleic acid backbone modifications like 2′-Fluoro and 2′-Omethyl are well-known approaches to minimize innate immune stimulation
5. How to achieve extravasation into tissues of interest?	Targeting tissues with rich vascularization usually leads to more extravasation into these tissues. Liver has been a prime example of a highly perfused organ with fenestrated endothelium. Other potential organs: kidneys, spleen
	Targeting blood cells or endothelial cells, which would not need extravasation
6. How to maximize cellular uptake?	LNPs: Well-known approach to encapsulate oligos in LNP-based formulations. One potential mechanism of cellular uptake: Binding of LNPs to serum apolipoproteins leading to LDLR-mediated cellular uptake
	Bio-conjugation of oligos to peptides, antibody, sugar, and lipid ligands specific to a receptor to enhance cell-specific delivery and uptake (GalNAc ligands are well-known example covered later in this section)
7. How to achieve endosomal release into cytosol?	Utilizing LNPs with membrane destabilizing lipids that can be destabilized once inside relatively acidic endosomal compartment
	Utilizing destabilizing endosomal cleavable peptide or polymer linkers that become charged once inside acidic endosomal compartment and facilitate release of oligomers
	Other approach could be just more endosomal uptake which facilitates slow albeit sustained release of oligomer over a period of time

LNP Lipid nanoparticles, *PEG* Polyethylene glycol, *GalNAc* N-Acetyl galactosamine
Adapted from *Wittrup, A., Lieberman, J.* Journal: NATURE REVIEWS | **GENETICS** VOLUME 16 | SEPTEMBER 2015 (Wittrup and Lieberman 2015)

cleotide therapies certainly is not merely an innovative approach, but it truly has the potential to mitigate systemic off-target effects posed by more traditional immunosuppressive therapies like glucocorticoids. Glucocorticoids, despite being a powerful mediator of repolarizing macrophages to an anti-inflammatory phenotype, have huge off-target liabilities of non-specifically suppressing non-macrophage cells (He et al. 2020; Ponzoni et al. 2018).

4.1 Nanoparticle-Based Macrophage Delivery

Within the past two decades, CD44 receptor targeting hyaluronic acid (HA)-based nanoparticles (NP) have been extensively utilized as novel modalities to deliver oligonucleotide, as well as other nucleic acid payloads to macrophages (Lee et al. 2015; Rangasami et al. 2021). Amiji and colleagues have been instrumental in demonstrating in vivo proof of concept (POC) to deliver siRNAs, miRNA, and even gene editing modalities to different macrophage types by utilizing HA-based NPs over the past decade (Parayath et al. 2018; Aldawsari et al. 2019; Tran et al. 2015; Parayath and Amiji 2020; Su et al. 2016; Tran et al. 2016; Parayath et al. 2019; Mattheolabakis et al. 2015; Kosovrasti et al. 2016). To exemplify some pivotal work done in this space, Tran et al. showed successful modulation of macrophage functional polarity toward anti-inflammatory phenotype by delivering plasmid DNA expressing IL-4 encapsulated in hyaluronic acid-poly(ethyleneimine) (HA-PEI) NPs in a rodent model of stimulated peritoneal macrophages (Tran et al. 2015). Subsequently, Parayath et al. also successful displayed in vivo POC by modulation of macrophage polarization to a proinflammatory phenotype by delivering miR-125 encapsulated in HA-PEI to tumor-associated macrophages (TAMs) in a genetically engineered non-small cell lung cancer model (Parayath et al. 2018). Some novel cationic lipid-based formulations have also successfully demonstrated macrophage-specific RNAi uptake and gene silencing. Novobrantseva et al. have shown a robust uptake and functionality of siRNAs in cells of myeloid origin in both rodent and nonhuman primates by utilizing novel ionizable lipids like DLin-KC2-DMA and cationic lipids like C12-200 (Novobrantseva et al. 2012; Love et al. 2010; Semple et al. 2010).

Several other macrophage markers have also been utilized to deliver macrophage-specific therapeutic modalities. Right from targeting hemoglobin scavenger receptor CD163 to internal membrane glycoprotein receptor CD64 that gets upregulated in pro-inflammatory macrophages, to targeting the folate receptor, various macrophage surface receptors have been explored to deliver different modalities into macrophages (Ponzoni et al. 2018; Lee et al. 2015; Moestrup and Møller 2004; Zhao et al. 2018; He et al. 2018; Kim et al. 2018; Poh et al. 2017).

The earlier chapters already cover NP-based formulations for macrophage delivery, however, owing to non-specificity and less biodegradability of lipid NP-based formulations, there is huge room for improvement in having better delivery modalities to target macrophages more efficiently and safely (Roberts et al. 2020; Barba et al. 2019; Wang et al. 2015).

4.2 Receptor-Mediated Oligonucleotide Delivery Using Ligand Conjugates

Bio-conjugation of oligonucleotides to unlock their therapeutic potential and enhance uptake to various cell types has been both a huge opportunity as well challenge for specific cell type delivery. Craig et al. have recently published a review of a variety of bio-conjugates like lipids, cell-penetrating peptides, polymers, antibodies, and carbohydrates that have been explored for oligonucleotide delivery (Craig et al. 2018). Perhaps the most successful cell types for oligonucleotide-conjugate mediated delivery are the hepatocytes, the primary parenchymal cells of the liver, where N-acetyl galactosamine (GalNAc) ligand has been widely used as the ligand of choice with huge preclinical and clinical success (Springer and Dowdy 2018; Debacker et al. 2020; Zimmermann et al. 2017). A single molecule, chemically modified oligomer is usually conjugated to a trivalent GalNAc ligand, targeting the asialoglycoprotein (ASGPR) receptor, specifically expressed by hepatocytes (Rajeev et al. 2015). Along with cellular uptake, receptor internalization and intracellular trafficking efficiency have made GalNAc ligands efficient at delivering oligonucleotides to hepatocytes (Rajeev et al. 2015). While hepatocyte delivery is not the focus of this chapter, GalNAc ligands surely make for an ideal example of what a ligand-receptor-mediated cellular uptake and cytosolic release of oligonucleotides should look like. Over the years, some very promising candidates for macrophage-specific delivery have also been identified and characterized over the years (He et al. 2020; Setten et al. 2019; Wilson and Doudna 2013;

Roberts et al. 2020; Wittrup and Lieberman 2015; Meister 2013). Here, we dive into one such example of an exciting and increasingly popular ligand-receptor pair for macrophage-specific delivery: macrophage mannose CD206 receptor that selectively binds to mannosylated ligands.

Over the last decade, the mannose receptor, also known as CD206, belonging to the C-type lectin superfamily, has been increasingly used as a macrophage delivery vehicle for oligonucleotide-based therapy (Jaynes et al. 2020; Movahedi et al. 2012; Varasteh et al. 2019; Costa et al. 2018; Fiani et al. 2020; Kawakami et al. 2000). Mannose receptor was initially identified on macrophage cells; hence, they are also called macrophage mannose receptors. However, it has now been established that it is also expressed on dendritic cells and endothelial cells along with macrophages (Dalle Vedove et al. 2018; Gazi and Martinez-Pomares 2009). It has 8–10 carbohydrate recognition domains (CRDs) along with a short cytoplasmic tail that has two endocytosis motifs. CRD 4 and 5 help it to identify and bind to a number of sugar ligands containing mannose, fucose, and N-acetylglucosamine (Dalle Vedove et al. 2018). It binds with particularly high affinity to mannosylated ligands usually expressed by various pathogens on their surface. Hence, it can mediate efficient phagocytosis and endocytosis of mannosylated ligands. One of the other nice features about it which can be exploited to deliver oligonucleotide payloads is that it recycles back between the plasma membrane and endosomal compartments, facilitating release of extracellular cargo into the cell (Dalle Vedove et al. 2018; Gazi and Martinez-Pomares 2009). A lot of preclinical data have been published in context of delivering oligonucleotide therapies via both mannosylated ligand conjugates, and also via decorating various nanoparticle-based formulations with mannose ligands. Tumor-associated macrophages (TAMs) that exert tumor propagation strongly express CD206 (Jaynes et al. 2020). For this reason, various groups have tried targeting them utilizing mannosylated ligands. Mohadevi et al. utilized the antibody approach by making nanobodies derived from *Camelidae* heavy chain antibodies, specifically recognizing the CD206 domains and showed both in vitro POC by delivering CD206-targeting nanobodies in lung and breast cancer single-cell suspensions and in vivo POC by successfully getting into TAMs after injecting ^{99m}Tc-labeled anti-CD206 antibodies (Movahedi et al. 2012). Subsequently, Varasteh et al. published recently that anti-CD206 nanobodies labeled with Gallium-68 can also be successfully delivered into atherosclerotic plaque-associated macrophages in the apolipoprotein E (Apo-E) knockout mouse model of atherosclerosis (Varasteh et al. 2019). Besides the nanobodies-based approach, solid lipid nanoparticles (SLN) functionalized with mannose on their surface have also been characterized to deliver into CD206+ macrophages. Costa et al. utilized this approach to deliver isoniazid-loaded SLN functionalized with mannose to effectively target alveolar macrophages (Costa et al. 2018). Very recently, some more intriguing work has been published on targeting TAMs. Fiani et al., by some elegant work, have demonstrated that small extracellular vesicles (sEVs) derived from TAMs highly express mannose glycans that can efficiently bind to CD206 receptors (Fiani et al. 2020). This makes sEVs attractive delivery vectors that could potentially be used to encapsulate oligonucleotide payloads. This could be a novel and effective way to also target immunosuppressive TAMs in the tumor microenvironment.

Table 3 summarizes some other unique modalities that can also be harnessed to deliver oligonucleotides to macrophages. By no means is this an exhaustive list, but it exemplifies some key recent delivery and macrophage modulation strategies.

Although there is a plethora of preclinical data supporting the case for macrophage mannose receptors as a suitable target to deliver oligonucleotide payloads into macrophages in the past decade, the paucity of active clinical trials utilizing mannose receptor-targeting modalities is also telling (Jaynes et al. 2020; Movahedi et al. 2012; Varasteh et al. 2019; Costa et al. 2018; Fiani et al. 2020; Kawakami et al. 2000; Dalle Vedove et al. 2018; Gazi and Martinez-Pomares 2009). Having said that, not a lot was also known about oligo-

Table 3 Some preclinical examples of novel delivery strategies to target macrophages (non-exhaustive)

Therapy	Strategy	Preclinical model	References
C-type lectin family	Altering activation state	Review on preclinical models for IRD	Frenz et al. (2015)
SPIONs	Reprogramming TAMs from M2 to M1	Muse primary IL-4 activated BMDM model; human M2-like differentiated THP-1 cells)	Rojas et al. (2016)
Man-HA-MnO2 NPs	M2 targeting	Review on preclinical models for AT	Song et al. (2016)
PEG- and mannose-NP	M2 targeting	B16-F10 mouse melanoma model in C57Bl/6 mice	Zhu et al. (2013)
Mannosylated siRNA	M2 targeting	Preclinical model of RA	Yu et al. (2013)
siRNA-NPs	Targeting Notch-1 signaling	CIA mouse model of cancer	Kim et al. (2015)
AM NPs	Suppress uptake of OxLDL by macrophage	ApoE −/− mice	Lewis et al. (2015)
DNP	LXR activation	LDLR −/− mice	He et al. (2017)
Statin-loaded HDL	Reduced macrophage accumulation in plaques	ApoE −/− mice	Tang et al. (2015)
Lyp-1 (CGNKRTRGC)	Reduced macrophage accumulation in plaques	Mouse model of macrophage-rich lesions in left common carotid arteries	Song et al. (2019)
rHDL Fluo	Reduced macrophage accumulation in plaques	ApoE −/− mice	Duivenvoorden et al. (2014)
LT rHDL	Macrophage infiltration and expression of matrix metalloproteinase	Preclinical model for AT	Liu et al. (2014)
TLR7/8 agonist-loaded cyclodextrin NPs	Reprogramming TAMs from M2 to M1	Orthotopic breast cancer mouse model	Rodell et al. (2018)

AM NPs sugar-based amphiphilic core shell-layered nanoparticles, *ApoE−/−* apolipoprotein E-deficient mice, *AT* atherosclerosis, *CIA* collagen-induced arthritis, *DNP* mannose-functionalized dendrimer nanoparticles, *IRD* inflammatory-related diseases, *LDLR* low-density lipoprotein, LDL receptor, *LTrHDL* t lovastatin (LT) delivered by HA-modified rHDL, *LXR* liver X receptor; Lyp-1, cyclic peptide, Lup-1 (CGNKRTRGC), *M2pep* peptide designed to recognize specifically M2-like macrophages, *Man-HA-MnO2 NPs* mannan-conjugated MnO2 particles with hyaluronic acid (HA) modification, *OxLDL* oxidized low-density lipoprotein, *PEG-and mannose-NP* polyethylene glycol (PEG)-sheddable and mannose-modified nanoparticle delivery system, *RA* rheumatoid arthritis, *rHDL Fluo* reconstituted HDL (rHDL) nanoparticles to deliver statins to atherosclerotic plaques. rHDL labeled with Cy5.5 (lipid monolayer) and DiR (hydrophobic core), *siRNA-NPs* siRNA against Notch1 (siRNA-NPs) through self-assembled poly-siRNA and thiolated-glycol chitosan nanoparticle, *SPIONs* superparamagnetic iron oxide nanoparticles, *TLR7/8* toll-like receptor type 7/8
Adapted from *Castegna et al.* (2020); https://doi.org/10.1016/j.pharmthera.2020.107521 Copyright permission granted from Elsevier; Journal: Pharmacology & Therapeutics

nucleotide therapies until the last decade. Advancements in making better oligonucleotide medicines paired with a renewed interest in drug delivery to immune cells, especially macrophages, make this space exciting to follow and will be intriguing to see if any of these CD206 targeting modalities translate into clinical development (Roberts et al. 2020; Wittrup and Lieberman 2015; Meister 2013; Aigner 2019; No Authors 2020; Kim 2020; Fitzgerald et al. 2016; Balwani et al. 2020; Garber 2018; Mathew and Wang 2019).

5 Inference

Macrophages are indeed one of the most versatile, plastic, and heterogenous cells. Evolution of our understanding of their diverse roles in both homeostasis and disease has led to a renaissance of this field. The classical concepts of both macrophage ontogeny and polarization have been revisited and revised as we are truly beginning to understand the complex phenotypic states of macrophages which demonstrate mosaicism rather than two extremes of M1 and M2 macrophages, based on the physiological and pathophysiological context and their environmental stimuli (Mosser and Edwards 2008; Mills and Ley 2014; Murray 2017; Italiani and Boraschi 2014; Mantovani et al. 2004). Numerous macrophage-targeting and macrophage-modulation strategies have evolved, right from small molecules and biologics, including cells, recombinant proteins, antibodies, to newer groundbreaking modalities like oligonucleotides (He et al. 2020; Ponzoni et al. 2018; Castegna et al. 2020).

With numerous high-profile drug approvals and an endless potential to silence any gene, harnessing oligonucleotides as macrophage-targeting modalities truly has tremendous value (Setten et al. 2019). Despite substantial bottlenecks like macrophage-specific delivery, avoiding systemic immunogenicity and uptake and endosomal release to name a few, an explosion of research is currently undertaken in this space (Roberts et al. 2020; Wittrup and Lieberman 2015). Advanced drug delivery systems right from LNP-based formulations, to novel liposomes, to bio-conjugates with ligands for targeted receptor-mediated delivery are the perfect match to advancements in oligonucleotide chemistry and better target engagement strategies (Craig et al. 2018). This will enable therapies to reach previously hard-to-reach macrophages and will herald a new revolution in ultimately providing therapies for not just cancer, autoimmune diseases, and infectious diseases but also acute and chronic inflammatory diseases.

References

Aigner A. Perspectives, issues and solutions in RNAi therapy: the expected and the less expected. Nanomedicine (Lond). 2019;14(21):2777–82.

Aldawsari HM, Dhaliwal HK, Aljaeid BM, Alhakamy NA, Banjar ZM, Amiji MM. Optimization of the conditions for plasmid DNA delivery and transfection with self-assembled hyaluronic acid-based nanoparticles. Mol Pharm. 2019;16(1):128–40.

Allavena P, Sica A, Garlanda C, Mantovani A. The Yin-Yang of tumor-associated macrophages in neoplastic progression and immune surveillance. Immunol Rev. 2008;222:155–61.

Anstee QM, Reeves HL, Kotsiliti E, Govaere O, Heikenwalder M. From NASH to HCC: current concepts and future challenges. Nat Rev Gastroenterol Hepatol. 2019;16(7):411–28.

Antoniades CG, Berry PA, Wendon JA, Vergani D. The importance of immune dysfunction in determining outcome in acute liver failure. J Hepatol. 2008;49(5):845–61.

Baeck C, Wehr A, Karlmark KR, Heymann F, Vucur M, Gassler N, et al. Pharmacological inhibition of the chemokine CCL2 (MCP-1) diminishes liver macrophage infiltration and steatohepatitis in chronic hepatic injury. Gut. 2012;61(3):416–26.

Balwani M, Sardh E, Ventura P, Peiró PA, Rees DC, Stölzel U, et al. Phase 3 trial of RNAi therapeutic givosiran for acute intermittent porphyria. N Engl J Med. 2020;382(24):2289–301.

Barba AA, Bochicchio S, Dalmoro A, Lamberti G. Lipid delivery systems for nucleic-acid-based-drugs: from production to clinical applications. Pharmaceutics. 2019;11(8):360.

Barreda DR, Neely HR, Flajnik MF. Evolution of myeloid cells. Myeloid cells in health and disease: American Society of Microbiology. 2017.

Bartneck M, Scheyda KM, Warzecha KT, Rizzo LY, Hittatiya K, Luedde T, et al. Fluorescent cell-traceable dexamethasone-loaded liposomes for the treatment of inflammatory liver diseases. Biomaterials. 2015;37:367–82.

Bashir S, Sharma Y, Elahi A, Khan F. Macrophage polarization: the link between inflammation and related diseases. Inflamm Res. 2016;65(1):1–11.

Brenner C, Galluzzi L, Kepp O, Kroemer G. Decoding cell death signals in liver inflammation. J Hepatol. 2013;59(3):583–94.

Brunt EM, Kleiner DE, Carpenter DH, Rinella M, Harrison SA, Loomba R, et al. Nonalcoholic fatty liver disease: reporting histologic findings in clinical practice. Hepatology 2020.

Bukong TN, Iracheta-Vellve A, Saha B, Ambade A, Satishchandran A, Gyongyosi B, et al. Inhibition of spleen tyrosine kinase activation ameliorates inflammation, cell death, and steatosis in alcoholic liver disease. Hepatology. 2016;64(4):1057–71.

Cariou B, Byrne CD, Loomba R, Sanyal AJ. Nonalcoholic fatty liver disease as a metabolic disease in humans: a literature review. Diabetes Obes Metab. 2021;23(5):1069–83.

Castegna A, Gissi R, Menga A, Montopoli M, Favia M, Viola A, et al. Pharmacological targets of metabolism in disease: opportunities from macrophages. Pharmacol Ther. 2020;210:107521.

Cha J-Y, Kim D-H, Chun K-H. The role of hepatic macrophages in nonalcoholic fatty liver disease and nonalcoholic steatohepatitis. Lab Anim Res. 2018;34(4):133–9.

Chen J, Deng X, Liu Y, Tan Q, Huang G, Che Q, et al. Kupffer cells in non-alcoholic fatty liver disease: friend or foe? Int J Biol Sci. 2020;16(13):2367–78.

Chinetti Gbaguidi G, Colin S, Staels B. Macrophage subsets in atherosclerosis. Nat Rev Cardiol. 2015;12(1):10–7.
Cochain C, Vafadarnejad E, Arampatzi P, Pelisek J, Winkels H, Ley K, et al. Single-cell RNA-Seq reveals the transcriptional landscape and heterogeneity of aortic macrophages in murine atherosclerosis. Circ Res. 2018;122(12):1661–74.
Connelly MA, Velez Rivera J, Guyton JR, Siddiqui MS, Sanyal AJ. Review article: the impact of liver-directed therapies on the atherogenic risk profile in non-alcoholic steatohepatitis. Aliment Pharmacol Ther. 2020;52(4):619–36.
Costa A, Sarmento B, Seabra V. Mannose-functionalized solid lipid nanoparticles are effective in targeting alveolar macrophages. Eur J Pharm Sci. 2018;114:103–13.
Craig K, Abrams M, Amiji M. Recent preclinical and clinical advances in oligonucleotide conjugates. Expert Opin Drug Deliv. 2018;15(6):629–40.
Dalle Vedove E, Costabile G, Merkel OM. Mannose and mannose-6-phosphate receptor–targeted drug delivery systems and their application in cancer therapy. Adv Healthc Mater. 2018;7(14):1701398.
Debacker AJ, Voutila J, Catley M, Blakey D, Habib N. Delivery of oligonucleotides to the liver with GalNAc: from research to registered therapeutic drug. Mol Ther. 2020;28(8):1759–71.
Dransfield I, Zagórska A, Lew ED, Michail K, Lemke G. Mer receptor tyrosine kinase mediates both tethering and phagocytosis of apoptotic cells. Cell Death Dis. 2015;6(2):e1646.
Drescher HK, Weiskirchen S, Weiskirchen R. Current status in testing for Nonalcoholic Fatty Liver Disease (NAFLD) and nonalcoholic steatohepatitis (NASH). Cells. 2019;8(8):845.
Duarte N, Coelho IC, Patarrão RS, Almeida JI, Penha-Gonçalves C, Macedo MP. How inflammation impinges on NAFLD: a role for Kupffer cells. BioMed Res Int. 2015;2015:984578.
Duivenvoorden R, Tang J, Cormode DP, Mieszawska AJ, Izquierdo-Garcia D, Ozcan C, et al. A statin-loaded reconstituted high-density lipoprotein nanoparticle inhibits atherosclerotic plaque inflammation. Nat Commun. 2014;5(1):3065.
Edwards JP, Zhang X, Frauwirth KA, Mosser DM. Biochemical and functional characterization of three activated macrophage populations. J Leukoc Biol. 2006;80(6):1298–307.
Elbashir SM, Harborth J, Lendeckel W, Yalcin A, Weber K, Tuschl T. Duplexes of 21-nucleotide RNAs mediate RNA interference in cultured mammalian cells. Nature. 2001;411(6836):494–8.
Epelman S, Lavine Kory J, Randolph Gwendalyn J. Origin and functions of tissue macrophages. Immunity. 2014;41(1):21–35.
Fiani ML, Barreca V, Sargiacomo M, Ferrantelli F, Manfredi F, Federico M. Exploiting manipulated small extracellular vesicles to subvert immunosuppression at the tumor microenvironment through mannose receptor/CD206 targeting. Int J Mol Sci. 2020;21(17):6318.
Fire A, Xu S, Montgomery MK, Kostas SA, Driver SE, Mello CC. Potent and specific genetic interference by double-stranded RNA in caenorhabditis elegans. Nature. 1998;391(6669):806–11.
Fitzgerald K, White S, Borodovsky A, Bettencourt BR, Strahs A, Clausen V, et al. A highly durable RNAi therapeutic inhibitor of PCSK9. N Engl J Med. 2016;376(1):41–51.
Fogg DK, Sibon C, Miled C, Jung S, Aucouturier P, Littman DR, et al. A clonogenic bone marrow progenitor specific for macrophages and dendritic cells. Science. 2006;311(5757):83–7.
Frenz T, Grabski E, Durán V, Hozsa C, Stępczyńska A, Furch M, et al. Antigen presenting cell-selective drug delivery by glycan-decorated nanocarriers. Eur J Pharm Biopharm. 2015;95:13–7.
Friedman SL, Ratziu V, Harrison SA, Abdelmalek MF, Aithal GP, Caballeria J, et al. A randomized, placebo-controlled trial of cenicriviroc for treatment of non-alcoholic steatohepatitis with fibrosis. Hepatology. 2018;67(5):1754–67.
Galastri S, Zamara E, Milani S, Novo E, Provenzano A, Delogu W, et al. Lack of CC chemokine ligand 2 differentially affects inflammation and fibrosis according to the genetic background in a murine model of steatohepatitis. Clin Sci (Lond). 2012;123(7):459–71.
Garber K. Alnylam launches era of RNAi drugs. Nat Biotechnol. 2018;36(9):777–8.
Gazi U, Martinez-Pomares L. Influence of the mannose receptor in host immune responses. Immunobiology. 2009;214(7):554–61.
Gordon S, Plüddemann A. Tissue macrophages: heterogeneity and functions. BMC Biol. 2017;15(1):53.
Gordon S, Plüddemann A, Martinez Estrada F. Macrophage heterogeneity in tissues: phenotypic diversity and functions. Immunol Rev. 2014;262(1):36–55.
He C, Yin L, Tang C, Yin C. Multifunctional polymeric nanoparticles for oral delivery of TNF-α siRNA to macrophages. Biomaterials. 2013;34(11):2843–54.
He H, Ghosh S, Yang H. Nanomedicines for dysfunctional macrophage-associated diseases. J Control Release. 2017;247:106–26.
He H, Yuan Q, Bie J, Wallace RL, Yannie PJ, Wang J, et al. Development of mannose functionalized dendrimeric nanoparticles for targeted delivery to macrophages: use of this platform to modulate atherosclerosis. Transl Res. 2018;193:13–30.
He W, Kapate N, Shields CW, Mitragotri S. Drug delivery to macrophages: a review of targeting drugs and drug carriers to macrophages for inflammatory diseases. Adv Drug Deliv Rev. 2020;165–166:15–40.
Hou J, Shi J, Chen L, Lv Z, Chen X, Cao H, et al. M2 macrophages promote myofibroblast differentiation of LR-MSCs and are associated with pulmonary fibrogenesis. Cell Commun Signal. 2018;16(1):89.
Italiani P, Boraschi D. From monocytes to M1/M2 macrophages: phenotypical vs. functional differentiation. Front Immunol. 2014;5:514.
Jaynes JM, Sable R, Ronzetti M, Bautista W, Knotts Z, Abisoye-Ogunniyan A, et al. Mannose receptor

(CD206) activation in tumor-associated macrophages enhances adaptive and innate antitumor immune responses. Sci Transl Med. 2020;12(530):eaax6337.
Karlmark KR, Weiskirchen R, Zimmermann HW, Gassler N, Ginhoux F, Weber C, et al. Hepatic recruitment of the inflammatory Gr1+ monocyte subset upon liver injury promotes hepatic fibrosis. Hepatology. 2009;50(1):261–74.
Kawakami S, Sato A, Nishikawa M, Yamashita F, Hashida M. Mannose receptor-mediated gene transfer into macrophages using novel mannosylated cationic liposomes. Gene Ther. 2000;7(4):292–9.
Kazankov K, Jørgensen SMD, Thomsen KL, Møller HJ, Vilstrup H, George J, et al. The role of macrophages in nonalcoholic fatty liver disease and nonalcoholic steatohepatitis. Nat Rev Gastroenterol Hepatol. 2019;16(3):145–59.
Kim Y-K. RNA therapy: current status and future potential. Chonnam Med J. 2020;56(2):87–93.
Kim KH, Lee M-S. Pathogenesis of nonalcoholic steatohepatitis and hormone-based therapeutic approaches. Front Endocrinol (Lausanne). 2018;9:485.
Kim MJ, Park J-S, Lee SJ, Jang J, Park JS, Back SH, et al. Notch1 targeting siRNA delivery nanoparticles for rheumatoid arthritis therapy. J Control Release. 2015;216:140–8.
Kim M, Sahu A, Kim GB, Nam GH, Um W, Shin SJ, et al. Comparison of in vivo targeting ability between cRGD and collagen-targeting peptide conjugated nano-carriers for atherosclerosis. J Control Release. 2018;269:337–46.
Kosovrasti VY, Nechev LV, Amiji MM. Peritoneal macrophage-specific TNF-α gene silencing in LPS-induced acute inflammation model using CD44 targeting hyaluronic acid nanoparticles. Mol Pharm. 2016;13(10):3404–16.
Krenkel O, Tacke F. Liver macrophages in tissue homeostasis and disease. Nat Rev Immunol. 2017;17(5):306–21.
Krenkel O, Puengel T, Govaere O, Abdallah AT, Mossanen JC, Kohlhepp M, et al. Therapeutic inhibition of inflammatory monocyte recruitment reduces steatohepatitis and liver fibrosis. Hepatology. 2018;67(4):1270–83.
Laina A, Gatsiou A, Georgiopoulos G, Stamatelopoulos K, Stellos K. RNA therapeutics in cardiovascular precision medicine. Front Physiol. 2018;9:953.
Lee GY, Kim J-H, Choi KY, Yoon HY, Kim K, Kwon IC, et al. Hyaluronic acid nanoparticles for active targeting atherosclerosis. Biomaterials. 2015;53:341–8.
Lefebvre E, Moyle G, Reshef R, Richman LP, Thompson M, Hong F, et al. Antifibrotic effects of the dual CCR2/CCR5 antagonist cenicriviroc in animal models of liver and kidney fibrosis. PLOS ONE. 2016;11(6):e0158156.
Lefere S, Tacke F. Macrophages in obesity and non-alcoholic fatty liver disease: crosstalk with metabolism. JHEP Rep. 2019;1(1):30–43.
Lewis DR, Petersen LK, York AW, Zablocki KR, Joseph LB, Kholodovych V, et al. Sugar-based amphiphilic nanoparticles arrest atherosclerosis in vivo. Proc Natl Acad Sci USA. 2015;112(9):2693–8.
Liao X, Shen Y, Zhang R, Sugi K, Vasudevan NT, Alaiti MA, et al. Distinct roles of resident and nonresident macrophages in nonischemic cardiomyopathy. Proc Natl Acad Sci USA. 2018;115(20):E4661–E9.
Liu L, He H, Zhang M, Zhang S, Zhang W, Liu J. Hyaluronic acid-decorated reconstituted high density lipoprotein targeting atherosclerotic lesions. Biomaterials. 2014;35(27):8002–14.
Liu Y, Yan W, Tohme S, Chen M, Fu Y, Tian D, et al. Hypoxia induced HMGB1 and mitochondrial DNA interactions mediate tumor growth in hepatocellular carcinoma through Toll-like receptor 9. J Hepatol. 2015;63(1):114–21.
LiverTox: Clinical and research information on drug-induced liver injury. Bethesda: National Institute of Diabetes and Digestive and Kidney Diseases; 2012.
Loomba R, Lawitz E, Mantry PS, Jayakumar S, Caldwell SH, Arnold H, et al. The ASK1 inhibitor selonsertib in patients with nonalcoholic steatohepatitis: a randomized, phase 2 trial. Hepatology. 2018;67(2):549–59.
Love KT, Mahon KP, Levins CG, Whitehead KA, Querbes W, Dorkin JR, et al. Lipid-like materials for low-dose, in vivo gene silencing. Proc Natl Acad Sci USA. 2010;107(5):1864–9.
Mantovani A, Sica A, Sozzani S, Allavena P, Vecchi A, Locati M. The chemokine system in diverse forms of macrophage activation and polarization. Trends Immunol. 2004;25(12):677–86.
Martinez FO, Gordon S. The M1 and M2 paradigm of macrophage activation: time for reassessment. F1000Prime Rep. 2014;6:13.
Mathew V, Wang AK. Inotersen: new promise for the treatment of hereditary transthyretin amyloidosis. Drug Des Dev Ther. 2019;13:1515–25.
Mattheolabakis G, Milane L, Singh A, Amiji MM. Hyaluronic acid targeting of CD44 for cancer therapy: from receptor biology to nanomedicine. J Drug Target. 2015;23(7-8):605–18.
Meister G. Argonaute proteins: functional insights and emerging roles. Nat Rev Genet. 2013;14(7):447–59.
Mills CD, Ley K. M1 and M2 macrophages: the chicken and the egg of immunity. J Innate Immun. 2014;6(6):716–26.
Moestrup S, Møller H. CD163: a regulated hemoglobin scavenger receptor with a role in the anti-inflammatory response. Ann Med. 2004;36(5):347–54.
Mosser DM, Edwards JP. Exploring the full spectrum of macrophage activation. Nat Rev Immunol. 2008;8(12):958–69.
Movahedi K, Schoonooghe S, Laoui D, Houbracken I, Waelput W, Breckpot K, et al. Nanobody-based targeting of the macrophage mannose receptor for effective in vivo imaging of tumor-associated macrophages. Cancer Res. 2012;72(16):4165–77.
Mulder P, van den Hoek AM, Kleemann R. The CCR2 inhibitor propagermanium attenuates diet-induced insulin resistance, adipose tissue inflammation

and non alcoholic steatohepatitis. PLOS ONE. 2017;12(1):e0169740.
Murray PJ. Macrophage polarization. 2017;79(1):541–66.
Muthiah MD, Sanyal AJ. Burden of disease due to non-alcoholic fatty liver disease. Gastroenterol Clin North Am. 2020;49(1):1–23.
Nathan C. Metchnikoff's Legacy in 2008. Nat Immunol. 2008;9(7):695–8.
No Authors. Second RNAi drug approved. Nat Biotechnol. 2020;38(4):385.
Noureddin M, Sanyal AJ. Pathogenesis of NASH: the impact of multiple pathways. Curr Hepatol Rep. 2018;17(4):350–60.
Novobrantseva TI, Borodovsky A, Wong J, Klebanov B, Zafari M, Yucius K, et al. Systemic RNAi-mediated gene silencing in nonhuman primate and rodent myeloid cells. Mol Ther Nucleic Acids. 2012;1(1):e4-e.
Okabe Y, Medzhitov R. Tissue biology perspective on macrophages. Nat Immunol. 2016;17(1):9–17.
Parayath NN, Amiji MM. Preparation of hyaluronic acid-based nanoparticles for macrophage-targeted MICRORNA delivery and transfection. In: Ferrari E, Soloviev M, editors. Nanoparticles in biology and medicine: methods and protocols. New York: Springer; 2020. p. 99–110.
Parayath NN, Parikh A, Amiji MM. Repolarization of tumor-associated macrophages in a genetically engineered nonsmall cell lung cancer model by intraperitoneal administration of hyaluronic acid-based nanoparticles encapsulating microrna-125b. Nano Lett. 2018;18(6):3571–9.
Parayath NN, Gandham SK, Leslie F, Amiji MM. Improved anti-tumor efficacy of paclitaxel in combination with MicroRNA-125b-based tumor-associated macrophage repolarization in epithelial ovarian cancer. Cancer Lett. 2019;461:1–9.
Parthasarathy G, Revelo X, Malhi H. Pathogenesis of nonalcoholic steatohepatitis: an overview. Hepatol Commun. 2020;4(4):478–92.
Pérez-Martínez L, Pérez-Matute P, Aguilera-Lizarraga J, Rubio-Mediavilla S, Narro J, Recio E, et al. Maraviroc, a CCR5 antagonist, ameliorates the development of hepatic steatosis in a mouse model of non-alcoholic fatty liver disease (NAFLD). J Antimicrob Chemother. 2014;69(7):1903–10.
Peterson KR, Cottam MA, Kennedy AJ, Hasty AH. Macrophage-targeted therapeutics for metabolic disease. Trends Pharmacol Sci. 2018;39(6):536–46.
Poh S, Putt KS, Low PS. Folate-targeted dendrimers selectively accumulate at sites of inflammation in mouse models of ulcerative colitis and atherosclerosis. Biomacromolecules. 2017;18(10):3082–8.
Ponzoni M, Pastorino F, Di Paolo D, Perri P, Brignole C. Targeting macrophages as a potential therapeutic intervention: impact on inflammatory diseases and cancer. Int J Mol Sci. 2018;19(7):1953.
Rajeev KG, Nair JK, Jayaraman M, Charisse K, Taneja N, O'Shea J, et al. Hepatocyte-specific delivery of siRNAs conjugated to novel non-nucleosidic trivalent N-acetylgalactosamine elicits robust gene silencing in vivo. Chembiochem. 2015;16(6):903–8.
Ramachandran P, Pellicoro A, Vernon MA, Boulter L, Aucott RL, Ali A, et al. Differential Ly-6C expression identifies the recruited macrophage phenotype, which orchestrates the regression of murine liver fibrosis. Proc Natl Acad Sci USA. 2012;109(46):E3186–E95.
Rangasami VK, Samanta S, Parihar VS, Asawa K, Zhu K, Varghese OP, et al. Harnessing hyaluronic acid-based nanoparticles for combination therapy: a novel approach for suppressing systemic inflammation and to promote antitumor macrophage polarization. Carbohydr Polym. 2021;254:117291.
Rath M, Müller I, Kropf P, Closs EI, Munder M. Metabolism via arginase or nitric oxide synthase: two competing arginine pathways in macrophages. Front Immunol. 2014;5:532.
Relizani K, Goyenvalle A. The use of antisense oligonucleotides for the treatment of duchenne muscular dystrophy. Methods Mol Biol. 2018;1687:171–83.
Roberts TC, Langer R, Wood MJA. Advances in oligonucleotide drug delivery. Nat Rev Drug Discov. 2020;19(10):673–94.
Rodell CB, Arlauckas SP, Cuccarese MF, Garris CS, Li R, Ahmed MS, et al. TLR7/8-agonist-loaded nanoparticles promote the polarization of tumour-associated macrophages to enhance cancer immunotherapy. Nat Biomed Eng. 2018;2(8):578–88.
Rojas JM, Sanz-Ortega L, Mulens-Arias V, Gutiérrez L, Pérez-Yagüe S, Barber DF. Superparamagnetic iron oxide nanoparticle uptake alters M2 macrophage phenotype, iron metabolism, migration and invasion. Nanomedicine. 2016;12(4):1127–38.
Semple SC, Akinc A, Chen J, Sandhu AP, Mui BL, Cho CK, et al. Rational design of cationic lipids for siRNA delivery. Nat Biotechnol. 2010;28(2):172–6.
Setten RL, Rossi JJ, Han S-p. The current state and future directions of RNAi-based therapeutics. Nat Rev Drug Discov. 2019;18(6):421–46.
Shetty A, Syn W-K. Health and economic burden of non-alcoholic fatty liver disease in the United States and its impact on veterans. Fed Pract. 2019;36(1):14–9.
Skuratovskaia D, Vulf M, Khaziakhmatova O, Malashchenko V, Komar A, Shunkin E, et al. Tissue-specific role of macrophages in noninfectious inflammatory disorders. Biomedicines. 2020;8(10):400.
Song M, Liu T, Shi C, Zhang X, Chen X. Bioconjugated manganese dioxide nanoparticles enhance chemotherapy response by priming tumor-associated macrophages toward M1-like phenotype and attenuating tumor hypoxia. ACS Nano. 2016;10(1):633–47.
Song N, Zhao L, Zhu M, Zhao J. Recent progress in LyP-1-based strategies for targeted imaging and therapy. Drug Deliv. 2019;26(1):363–75.
Springer AD, Dowdy SF. GalNAc-siRNA conjugates: leading the way for delivery of RNAi therapeutics. Nucleic Acid Ther. 2018;28(3):109–18.
Stutchfield BM, Antoine DJ, Mackinnon AC, Gow DJ, Bain CC, Hawley CA, et al. CSF1 restores innate immunity after liver injury in mice and serum levels

indicate outcomes of patients with acute liver failure. Gastroenterology. 2015;149(7):1896–909.e14.

Su M-J, Aldawsari H, Amiji M. Pancreatic Cancer Cell Exosome-Mediated Macrophage Reprogramming and the Role of MicroRNAs 155 and 125b2 Transfection using Nanoparticle Delivery Systems. Scientific Reports. 2016;6(1):30110.

Sumida Y, Yoneda M. Current and future pharmacological therapies for NAFLD/NASH. J Gastroenterol. 2018;53(3):362–76.

Tacke F. Targeting hepatic macrophages to treat liver diseases. J Hepatol. 2017;66(6):1300–12.

Tacke F, Zimmermann HW. Macrophage heterogeneity in liver injury and fibrosis. J Hepatol. 2014;60(5):1090–6.

Tanaka N, Kimura T, Fujimori N, Nagaya T, Komatsu M, Tanaka E. Current status, problems, and perspectives of non-alcoholic fatty liver disease research. World J Gastroenterol. 2019;25(2):163–77.

Tang J, Lobatto ME, Hassing L, van der Staay S, van Rijs SM, Calcagno C, et al. Inhibiting macrophage proliferation suppresses atherosclerotic plaque inflammation. Sci Adv. 2015;1(3):e1400223.

Teti G, Biondo C, Beninati C. The phagocyte, metchnikoff, and the foundation of immunology. Myeloid cells in health and disease: American Society of Microbiology. 2017.

Traber PG, Zomer E. Therapy of experimental NASH and fibrosis with galectin inhibitors. PLOS ONE. 2013;8(12):e83481.

Tran T-H, Rastogi R, Shelke J, Amiji MM. Modulation of macrophage functional polarity towards anti-inflammatory phenotype with plasmid DNA delivery in CD44 targeting hyaluronic acid nanoparticles. Sci Rep. 2015;5(1):16632.

Tran T-H, Krishnan S, Amiji MM. MicroRNA-223 induced repolarization of peritoneal macrophages using CD44 targeting hyaluronic acid nanoparticles for anti-inflammatory effects. PLOS ONE. 2016;11(5):e0152024.

Triantafyllou E, Woollard KJ, McPhail MJW, Antoniades CG, Possamai LA. The role of monocytes and macrophages in acute and acute-on-chronic liver failure. Front Immunol. 2018a;9:2948.

Triantafyllou E, Woollard KJ, McPhail MJW, Antoniades CG, Possamai LA. The role of monocytes and macrophages in acute and acute-on-chronic liver failure. Front Immunol. 2018b;9:2948.

van der Heide D, Weiskirchen R, Bansal R. Therapeutic targeting of hepatic macrophages for the treatment of liver diseases. Front Immunol. 2019;10:2852.

Varasteh Z, Mohanta S, Li Y, López Armbruster N, Braeuer M, Nekolla SG, et al. Targeting mannose receptor expression on macrophages in atherosclerotic plaques of apolipoprotein E-knockout mice using 68Ga-NOTA-anti-MMR nanobody: non-invasive imaging of atherosclerotic plaques. EJNMMI Res. 2019;9(1):5.

Wang J, Kubes P. A reservoir of mature cavity macrophages that can rapidly invade visceral organs to affect tissue repair. Cell. 2016;165(3):668–78.

Wang Q, Ni H, Lan L, Wei X, Xiang R, Wang Y. Fra-1 protooncogene regulates IL-6 expression in macrophages and promotes the generation of M2d macrophages. Cell Res. 2010;20(6):701–12.

Wang N, Liang H, Zen K. Molecular mechanisms that influence the macrophage M1–M2 polarization balance. Front Immunol. 2014;5:614.

Wang Y, Miao L, Satterlee A, Huang L. Delivery of oligonucleotides with lipid nanoparticles. Adv Drug Deliv Rev. 2015;87:68–80.

Wilson RC, Doudna JA. Molecular mechanisms of RNA interference. Annu Rev Biophys. 2013;42(1):217–39.

Wittrup A, Lieberman J. Knocking down disease: a progress report on siRNA therapeutics. Nat Rev Genet. 2015;16(9):543–52.

Wu Z, Han M, Chen T, Yan W, Ning Q. Acute liver failure: mechanisms of immune-mediated liver injury. Liver Int. 2010;30(6):782–94.

Yona S, Gordon S. From the reticuloendothelial to mononuclear phagocyte system – the unaccounted years. Front Immunol. 2015;6:328.

You Q, Holt M, Yin H, Li G, Hu C-J, Ju C. Role of hepatic resident and infiltrating macrophages in liver repair after acute injury. Biochem Pharmacol. 2013;86(6):836–43.

Younossi ZM, Tampi R, Priyadarshini M, Nader F, Younossi IM, Racila A. Burden of illness and economic model for patients with nonalcoholic steatohepatitis in the United States. Hepatology. 2019;69(2):564–72.

Yu SS, Lau CM, Barham WJ, Onishko HM, Nelson CE, Li H, et al. Macrophage-specific RNA interference targeting via "click", mannosylated polymeric micelles. Mol Pharm. 2013;10(3):975–87.

Zhao Y, He Z, Gao H, Tang H, He J, Guo Q, et al. Fine tuning of core–shell structure of hyaluronic acid/cell-penetrating peptides/siRNA nanoparticles for enhanced gene delivery to macrophages in antiatherosclerotic therapy. Biomacromolecules. 2018;19(7):2944–56.

Zhu S, Niu M, O'Mary H, Cui Z. Targeting of tumor-associated macrophages made possible by PEG-sheddable, mannose-modified nanoparticles. Mol Pharm. 2013;10(9):3525–30.

Zimmermann TS, Karsten V, Chan A, Chiesa J, Boyce M, Bettencourt BR, et al. Clinical proof of concept for a novel hepatocyte-targeting GalNAc-siRNA conjugate. Mol Ther. 2017;25(1):71–8.

Macrophage-Targeted Chemotherapy for Tuberculosis

Priya Shrivastava, Laxmikant Gautam, Sonal Vyas, and Suresh P. Vyas

Abstract

The emergence and frequent occurrence of multidrug-resistant strains of *Mycobacterium tuberculosis* (the causative pathogen for the deadly infectious disease, i.e., tuberculosis) present a severe health threat to human beings. It raises the demand for short-term and less toxic drug regimens for improvement of the existing chemotherapies. It is difficult to develop new bioactive(s) because the process is time-consuming and more often the developed drugs are associated with side effects. To surmount these challenges, nanobiotechnological modules/nanoarchitectures have proven to be in seminal (pivotal) as a molecular target tracker with an ability to improve the existing treatment regimens. Furthermore, the approach could improve pharmacodynamics of bioactive(s) that suffer poor aqueous solubility and higher toxicity. Bioactive(s)/antimicrobials loaded delivery modules are designed to be avidly uptaken by the host cell, i.e. macrophages (tropics of *Mycobacterium tuberculosis*). The macrophages are recruited to the containment zone (infected area). Additionally, they are able to carry and transport bioactive(s)/antimicrobials accumulated within them, enabling passive targeting, a successful strategy for the treatment of tuberculosis. Another strategy, that is, active targeting, involves the use of engineered or surface ameliorated nanoconstructs with the ligands specific to the receptors exclusively or differently overexpressed on the membrane of macrophages. Thus, receptor-mediated endocytosis increases the concentration of the bioactive(s) within the target site resulting in improved treatment efficacy. This chapter highlights different targeting strategies explored for intracellular delivery of bioactive(s) to address the challenges and problems associated drug-resistant tuberculosis.

Keywords

Tuberculosis · Infectious disease · *Mycobacterium tuberculosis* · Macrophages · Chemotherapy · Passive targeting · Active targeting

P. Shrivastava · L. Gautam · S. P. Vyas (✉)
Drug Delivery Research Laboratory, Department of Pharmaceutical Sciences, Dr. Harisingh Gour Vishwavidyalaya, Sagar, MP, India

S. Vyas
Director, Sampoorna Path Care Labs, Sagar, MP, India

1 Tuberculosis (TB): An Insight

Globally, TB is now considered a dreadful health threat and is one of the leading causes of mortality and morbidity from infectious diseases. It is a

S. Gupta, Y. V. Pathak (eds.), *Macrophage Targeted Delivery Systems*,
https://doi.org/10.1007/978-3-030-84164-5_17

communicable airborne disease caused by an opportunistic infectious intracellular pathogen, *Mycobacterium tuberculosis* (Shrivastava et al. 2020a). It is primarily a disease of the respiratory system and spreads by inhaling the droplets released by the infected person who expels bacteria into the air, for example, by coughing. *Mycobacterium tuberculosis* not only attacks the lungs, but other organs are also affected (Smith 2003). The TB which commonly affects the body's pulmonary system (lungs) is referred to as pulmonary TB whereas the TB which affects the organs of the body other than the lungs is known as extrapulmonary TB (Hasnain et al. 2019a). The organs other than the lungs which get affected by the extrapulmonary TB include the central nervous system (TB meningitis), the lining covering the lungs (pleural TB), lymph nodes and abdomen (abdominal TB), bone and joints (mucoskeletal TB), and kidney and bladder (urogenital TB). In 2019, approximately 1.2 million deaths were estimated due to TB among HIV-negative people and an additional 251,000 deaths among HIV-positive people were reported in 2018. Furthermore, it is worth noting that one-third of the world population harbors *Mycobacterium tuberculosis* in the latent state with a probability of developing the active disease (World Health Organization 2019).

The development of the pathological basis for tuberculosis in a progressive manner has enabled the essential chemotherapeutic strategies broadly divided as one of those that were used in the pre-antibiotic and others that are being used in post-antibiotic era (Armocida and Martini 2020). Specifically, "antimicrobial chemotherapy" could circumvent the problem as a clinical solution to the challenge from *Mycobacterium tuberculosis*. DOTS, that is, directly observed treatment short course, is currently recommended as a standard TB chemotherapy for bioactive susceptible TB (Falzon et al. 2017). It consists of two phases: the initial phase is bactericidal or intensive, which comprises the treatment regimen with the four first-line antitubercular bioactives (isoniazid, rifampicin, pyrazinamide, and ethambutol) for 4 months, followed by the sterilizing phase for an additional 2 months of therapy with isoniazid and rifampicin. During the intensive phase, the majority of the bacilli are killed, the clinical symptoms are subsided, while transmission risk and development of bioactive(s) resistance are minimized. During the sterilization phase, the remaining bioactive-tolerant persisters are sterilized, enabling a reduction in the relapse episodes (Poce et al. 2014; Sotgiu et al. 2015). Although DOTS can treat TB, it often results in patient nonadherence due to long-term periods and complex treatment regimens. As a consequence of prolonged therapy, most of the patients either fail or are unable to follow the chemotherapy based on multiple drugs. Undoubtedly, it could be one of the key factors that the majority of the patients do not receive appropriate antitubercular therapy among the prevalent active cases. The discontinuance to therapy occurs due to (1) incomplete or unmonitored therapy in terms of bioactive(s), durations, and dosing intervals and (2) incorrect or improper use of antitubercular bioactive(s). This patient's nonadherence to antitubercular bioactive(s) could be one of the contributing factors for the development of bioactive(s) resistance toward tuberculosis and turns to be a public health menace in both the developing and developed world. It is worth noting that the worldwide misuse or mismanagement of antimicrobials gives rise to the development of multidrug-resistant tuberculosis (MDR-TB), extensively drug-resistant tuberculosis (XDR-TB), and total drug-resistant tuberculosis (TDR-TB) (Kesharwani 2020).

Thus, this threat poses a demand for new bioactive regimens of shorter duration, especially and broadly applicable for MDR and XDR tuberculosis, simple dosing schedule, minimal interaction with other antitubercular bioactive(s) without compromising with the therapeutic outcomes. This would make the therapy more appropriate and widely available. In this context, it is difficult to figure out how existing antitubercular bioactive(s) could be used more effectively. Moreover, developing a new bioactive molecule is challenging owing to the side effects which are often associated with the new bioactive molecule (Baranyai et al. 2020). To conquer this dilemma, receptor-mediated macrophage targeting by using drug delivery modules and combination chemotherapy could be the promising approaches for tuberculosis chemotherapy. Receptor mediated internalization of the carrier-drug(s) composite results in accumulation of the bioactive(s) in the target cells with the simultaneous exclusion of nontarget cells as they lack the receptor involved in drug targeting. The use of two or more chemotherapeutic drugs thus reduces the risk of resistance to the treatment by being able to address several targets involved in pulmonary tuberculosis. It also reduces the active dose of the drugs; hence, resultantly reduces the toxic effects.

These approaches bridge the gap between physical and biological sciences by applying microstructures, nano-architectures, and nanophases in multiple arenas of science, particularly in nanomedicine-based bioactive delivery modules, where these nanocarriers are of profound interest (Liu et al. 2009). The objective of the present chapter is to discuss the multiple strategies for macrophage targeting, have been developed to conquer tuberculosis.

2 Macrophage: The Cellular Tropics of *Mycobacterium tuberculosis*

Macrophages are the cells of the host immune system developed with the objective to clear the pathogenic burden. These are the target cells or the tropics of *Mycobacterium tuberculosis*. The host foe (*Mycobacterium tuberculosis*) attacks and replicates within the host macrophages. The ability of the tubercle bacilli to adapt the hostile surroundings of the macrophage has been instrumental in its survival as a pathogenic microorganism (Upadhyay et al. 2018). The tubercle bacilli interfere with the host signaling/trafficking pathways by altering the sequential events of the phagosomal or endosomal maturation pathways to build a protected niche within phagosomes. The phagosome, which contains live pathogen (tubercle bacilli), while associated with the phagocytic pathway, does not fuse with the lysosomes, due to the deposition of a protein named tryptophan aspartate containing coat protein (TACO) on the phagosomal membrane. The TACO also recognized as coronin 1 is reported to be recruited and retained on the phagosomal surface of the infected macrophages where it activates calcium-calcineurin signaling to prevent the maturation of phagosomes and subsequent fusion with lysosomes. The TACO protein is normally released from the phagosomes enclosing bacilli (Pieters 2008). The non-fusogenicity of the phagosome is considered to be a pivotal factor for the survival of the tubercle bacilli within the hostile environment of the host cell (macrophages).

Being *Mycobacterium tuberculosis* an obligate or opportunistic intracellular pathogenic species, it can infect a range of multiple cell types that are highly diverse, involving both myeloid and lymphoid origin. The dominant cell types that are generally infected by the tubercle bacilli are alveolar macrophages. These cells engulf the *Mycobacterium tuberculosis* by the process referred to as phagocytosis, upon its entry into the lungs. Furthermore, it can also infect other cell types including dendritic cells, foamy macrophages, neutrophils, lymphocytes, epithelioid macrophages, myeloid-derived suppressor cells, and interstitial macrophages. Apart from these cells, other multipotent cell types such as human and mice mesenchymal stem cells (MSCs) are also infected by the *Mycobacterium tuberculosis*. These multipotent cells can further differentiate into multiple cell phenotypes. MSCs phagocytose *Mycobacterium tuberculosis* via scavenger receptor-A-mediated phagocytosis. The process of nitric oxide production and autophagy is triggered by these cells to kill the replicating bacilli. The dormant tubercle bacilli, however, cannot be easily removed. *Mycobacterium tuberculosis* attains a new habitat to persist, when the migration of MSCs to bone marrow occurs. As a matter of fact, tubercle bacilli still could be observed in CD27+ bone marrow MSCs in patients that have successfully completed the regimen of antitubercular chemotherapy (Pai et al. 2016a; Cheung et al. 2019).

In each of these instances, the expression of pattern recognition receptors (PRRs) on the host cell enables them to recognize the specific antigens, thus endowing a degree of specificity to the innate immune system. Various PRRs/biosignatures including toll-like receptors, mannose receptors, scavenger receptors, dectin I and II receptors, mincle receptors, NOD proteins (nucleotide-binding oligomerization domain-containing protein-like receptors), complement receptors, dendritic cell-specific intercellular adhesion molecule (ICAM)-3-grabbing nonintegrin (DC-SIGN), and F_c receptors are involved in the ligand recognition and response to the tubercle bacilli. These biosignatures/PRRs recognize molecular patterns that are associated with the pathogen and trigger signaling events or inflammatory responses (Mishra et al. 2017; Rajaram et al. 2014). The initiation of the signaling cascade occurs in response to the recognition of the ligand and engulfment/destruction of the pathogen via receptor-mediated endocytosis or phagocytosis. The molecular mechanism of the receptor-mediated endocytosis or phagocytosis are described below.

2.1 Receptor-Mediated Endocytosis

Receptor-mediated endocytosis is a cellular mechanism that is constitutive by character. It involves cellular uptake of a large range of macromolecules from the extracellular domain (environment). It is triggered or initiated by the clustering of the receptors in the peculiar regions on the cell surface referred to as coated pits. These coated pits contain clathrin and other structural proteins. This is now largely accepted that the cytoplasmic tails of the receptors comprise of structural motifs and are destined for internalization owing to their cross talk with certain clusters of intracellular proteins (adaptin) in the coated pits. It is not clearly understood whether or not the coated pits which contain clathrin proteins start up the vesiculation process. However, clathrin appears to limit the size of the endosomal vesicles. The coated vesicles are temporary structures that shed their coat proteins immediately after their constitution. The receptors or ligands are present on these constituted smooth vesicles. After their exposure to endosomes, these vesicles become acidic due to the proton influx by proton pumps on the endosomes membrane. Subsequently, endosomes-lysosomes fusion occurs, and consequently, larger compartment is formed where membrane and protein sorting does take place followed by the process of antigen processing and presentation (Fig. 1). The process of endosomal sorting in macrophages and other nonpolarized cells seems to entail at least three pivotal mechanisms: (a) the constitution of a subset of vesicles that mediate the recycling of unoccupied receptors and other membrane molecules back to the cell surface; (b) the second component is a vesicular portion/component that enables the return of a selected group of molecules to the trans Golgi network/apparatus; (c) the third component is that which enables the transfer of membrane or solutes to the lysosomal compartment (Pieters 2008; Mayorga et al. 1992; Gupta et al. 2019). It is not clear yet whether these sorting processes occur consecutively or concomitantly.

2.2 Receptor-Mediated Phagocytosis

Receptor-mediated phagocytosis is another cellular mechanism that also depends on receptor-ligand engagement. It is triggered when the ligands bind or interact with the receptor molecules which exist on the surface of the phagocytes. Several receptors including Fc receptors, complement receptors, mannose receptors, and others are involved in the mediation of this engagement. The binding of the ligand decorated particles with the receptors (receptor-ligand engagement) results in the internalization of the particle. Subsequently, after internalization, a sequence of membrane fusion and budding events occur. It results in the phagosome maturation and the delivery of the particles to the lysosomal compartment where extensive degradation of the cargo can occur. The process of endocytosis and phagocytosis shares several parallels/analogies in terms of selective internalization, membrane fusion, and budding events, membrane exchange with elements of the Trans-Golgi apparatus, and selective fusion or interaction with the lysosomal compartment (Fig. 2) (Chaurasiya 2018; Uribe-Querol and Rosales 2017).

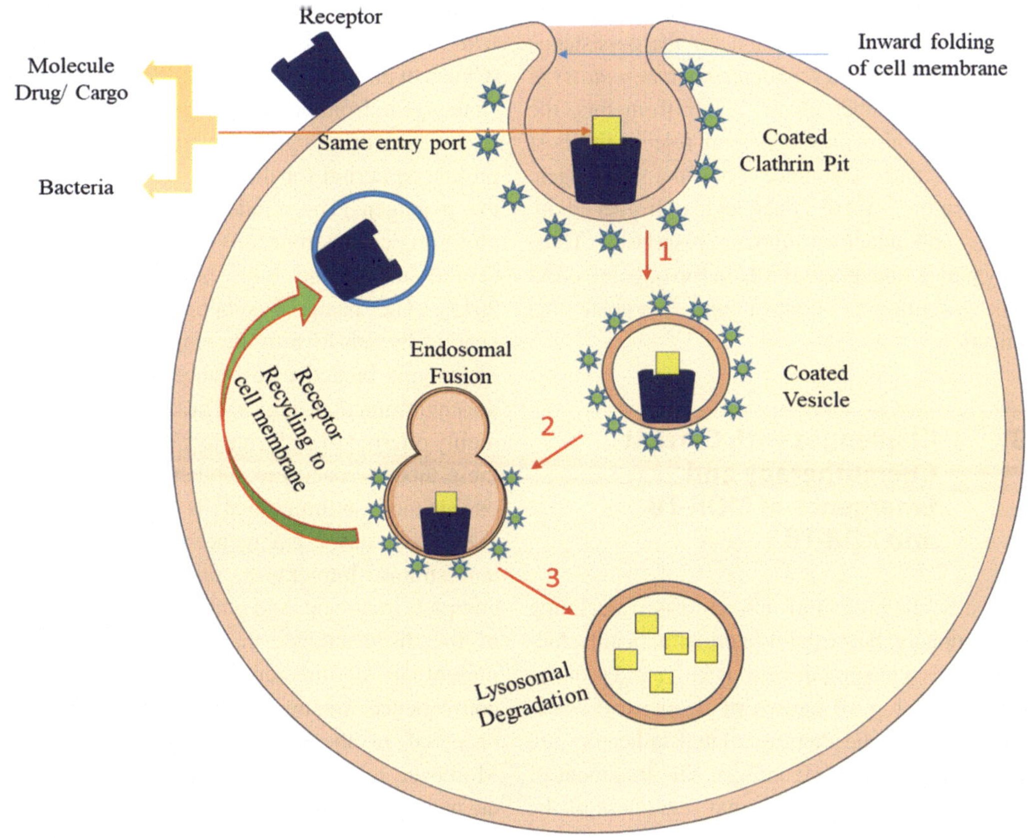

Fig. 1 Schematic representation of receptor-mediated endocytosis

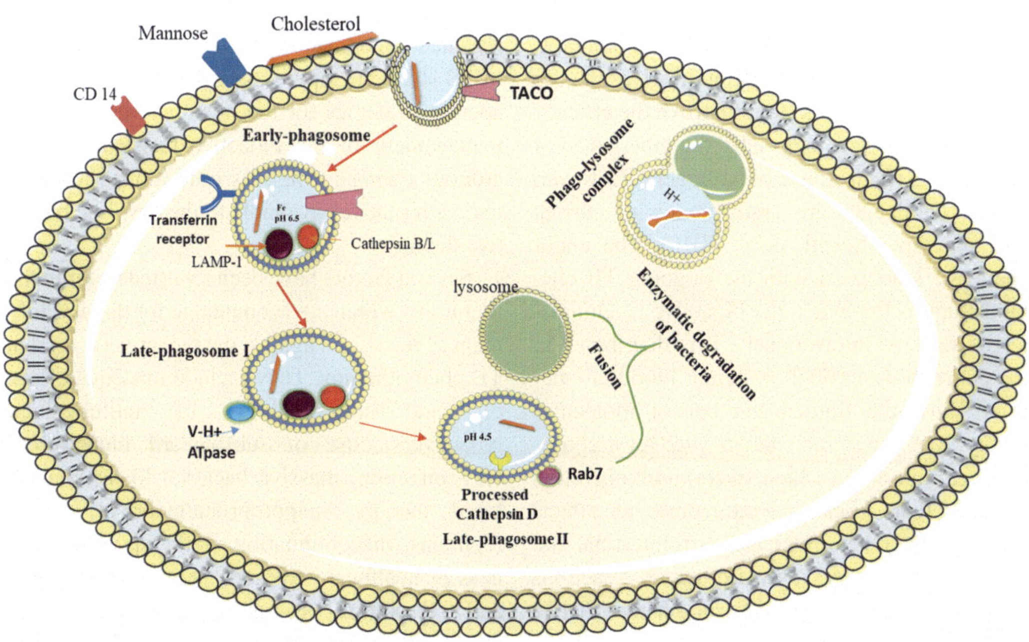

Fig. 2 Schematic representation of receptor-mediated phagocytosis

Thus, phagocytes (macrophages) present as the primary host cell habitat for the persistence and growth of the *Mycobacterium tuberculosis* in all stages of tuberculosis. Apart from this, the phagocytes also play a pivotal role in the host defense and clearance of the pathogen. Furthermore, these phagocytes trigger both innate and adaptive immune responses. Thus, they play a fundamental role in the ongoing cross talk essential to control and eradicate the infection.

3 Challenges with Current Chemotherapy and Emergence of MDR-TB and XDR-TB

Globally, the burden of tuberculosis is declining continuously; but still, it is a dreadful health challenge. It is a leading infectious killer and remains one of the top 10 causes of death worldwide. Moreover, all the therapy-related indicators are confronted with shortcomings. The treatment of tuberculosis is still suboptimal regardless of the advancements in the pharmacological field and the noticeable advantages associated with the antitubercular bioactive(s). The ongoing/current therapy for tuberculosis is DOTS therapy and is primarily based on combination chemotherapy. This combination chemotherapy involves the use of multiple bioactive(s) that improve the efficacy and also counteract especially the emergence of bioactive resistant strains of pathogens. However, several challenges are associated with current chemotherapy (Sosnik et al. 2010). The major challenge associated with the ongoing TB chemotherapy is that when the bioactive is administered orally or intravenously, it is distributed in the entire body through systemic blood circulation. Thus, the limited number of molecules finally does reach their target site. As a consequence, high dose of bioactive(s) administration is needed that causes some intolerable side effects including liver, neurological, gastrointestinal disorders, and other dermatological reactions (Shrivastava et al. 2020a; Nasiruddin et al. 2017). Apart from this, other factors such as short plasma half-life and rapid clearance associated with some bioactive(s) reduce their efficacy (Chuan et al. 2013; Singh et al. 2019). The major challenge in front of optimal TB chemotherapy is the existing drug regimen that is required over a prolonged period for the effective eradication of the pathogen. This could be attributed to the physiological heterogeneity of the tubercle bacilli (Traini and Young n.d.; Jawahar and Reddy 2012). The bacterial subpopulations that are genetically sensitive to the bioactive(s) manifest phenotypic bioactive resistance in retaliation to the environmental signals. These bacilli are commonly referred to as "persisters." The exposure of these tubercle bacilli to subtherapeutic levels of one or more antimicrobials is the major or primary cause of the emergence of bioactive resistant strains. Intolerance, nonadherence to the complete treatment, and other toxicities are some of the disadvantages that are associated with present-day antitubercular chemotherapy. As a consequence, treatment failure and relapse are observed in the vast majority of the cases. Moreover, unmonitored therapy and inappropriate management of the antitubercular bioactive(s) results in the development of the bioactive resistance that further demands the use of second- and third-line antitubercular bioactive(s). To achieve acceptable levels of bioactive(s) in the blood, the patient's adherence to the duration and regimen of TB chemotherapy is essential. The chemotherapeutic strategies for the treatment, control, and management of the bioactive resistant TB are arduous owing to the less efficacious, however, toxic alternatives available to the first-line bioactive therapy.

Several factors have been reported so far in the literatures which are accountable for the development of resistance against the bioactive(s) used in TB chemotherapy. They include inadequate supply and nonavailability of antitubercular bioactive(s), use of substandard bioactive(s), HIV infection, massive bacterial load, unmonitored therapy, inappropriate/incorrect dosage regimens, discontinuation of therapy, unawareness of healthcare workers, lack of supervision, and TB control programs (Rawal and Butani 2016). In general terms, bioactive resistance is a human-made dilemma that emerges from the misuse or abuse of existing antitubercular dosage regimens, and mismanagement of disease course or unmonitored therapy. In nutshell, the poor management or mismanagement of the first-line antitubercular bioactive(s) (Isoniazid and Rifampicin) leads to the development of MDR-TB whereas when the second-line antitubercular bioactives (fluoroquinolones and one or more second-line injectable bioactive(s) including amikacin, capreomycin, and kanamycin) are misused or abused, XDR-TB develops, specifically in HIV-positive individuals. This renders the patient condition cursed that further prolongs the therapy needed to cure the disease (Singh et al. 2016; Kaur and Singh 2014; Thakare et al. 2012; Yashodhara et al. 2010).

Conventionally, the main focus of the TB control efforts has largely been on the progress of the cure rates for the bioactive-susceptible disease to decrease the number of bioactive resistant cases. Currently, it has been aimed to address these problems and to avert and avoid MDR in TB chemotherapy, and also to improve clinical outcome through new antitubercular bioactive(s) discovery along with their delivery mechanisms/devices that specifically target infected or diseased cells with safe and limited exposure of nontarget cells.

4 Need for Drug Designing, Engineered Nanoconstructs, and Macrophage-Targeted Chemotherapy

In the present scientific scenario, drug designing is an emerging paradigm rewarding as a science. This could be attributed to the contemporary advancement in the understanding of disease pathogenesis specifically at the molecular level. In general, bioactive(s) are taken in a formulated state. A bioactive(s) dosage preparation encompasses one or multiple active medicaments and the other excipients. Now, the excipients are no longer considered inert ingredients and are equally important as the bioactive itself. Therefore, they can play a pivotal role in the formulation development. Apart from this, they can affect a controlled and constant release of the bioactive(s) and can circumvent its premature biodegradation process. These are the key parameters that influence the rate and degree of the absorption of the bioactive(s). It is therefore evident that they can directly affect the bioavailability of the bioactive(s). In designing and developing a pharmaceutical dosage formulation, the bioavailability of the bioactive(s) is of paramount importance. The optimum systemic concentration of the bioactive(s) is vital in order to ensure its desirable level at the site of action for significant therapeutic benefits (Pandey and Khuller 2004). A sustained or controlled release system for a protracted duration of action should be developed. This could be accomplished by the selection of appropriate material for carrier construction and also the process and carrier for encapsulation or loading of bioactive(s).

Apart from this, novel bioactive(s) and drug delivery system(s) that are safe, effective, or better tolerated are needed to be developed for optimal and short-term TB chemotherapy. We could rely on the concept of "Old drugs in new clothing" (Gregoriadis 1984). The biodegradable polymer-based drug delivery modules have been advocated to be advantageous which may reduce the dose, frequency of administration, and toxicity issues associated with the bioactive(s) with manifold effective therapeutic outcome. As of now, nanotechnology offers multiple options and benefits to square up the challenges associated with the current chemotherapy. Scientists are taking interest in placing reliance on the development of nanotechnology-based drug delivery platforms to reduce the toxicities and enhance the therapeutic efficacy (Joseph and Venkatraman 2017). The ingenious designing and judicious use of nanotech-based versatile approaches could be employed for controlled and targeted antitubercular chemotherapy (Niu et al. 2015). There appears a great therapeutic potential in nanomedicine-based promising approaches which could be advantageous or efficacious particularly in the pulmonary administration of the bioactive(s) (Fig. 3).

With the advancement and progress in nanotechnology, substantial developments have been

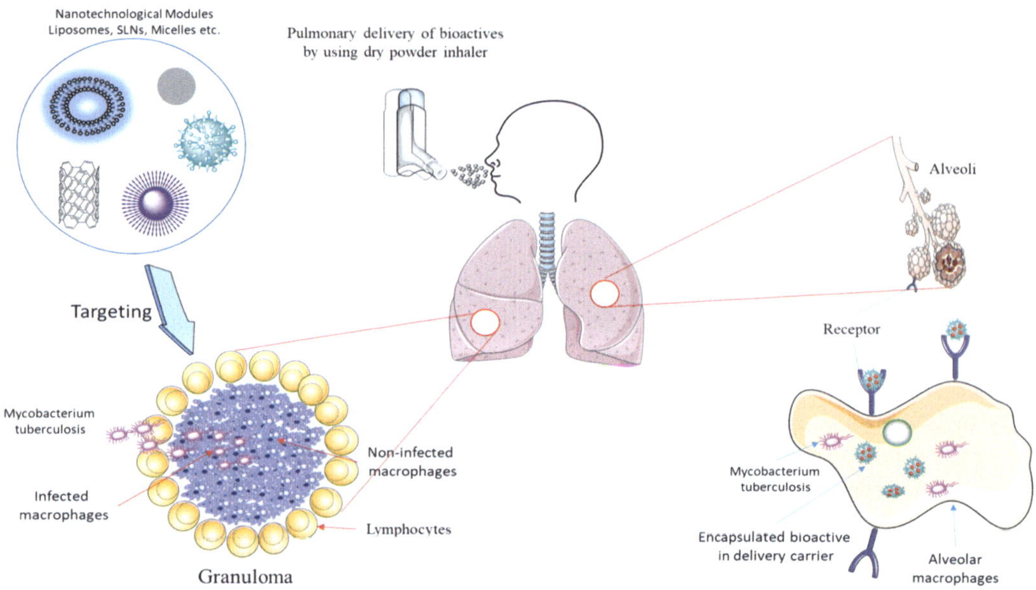

Fig. 3 Pulmonary delivery of the bioactive(s) by using nanotechnological bioactive delivery modules

made in the designing and development of nanoplatforms for their safe and effective use in clinical therapy. In regard to tuberculosis, the realization of the central role of the macrophages has revitalized our interest in drug targeting to macrophages. These phagocytic cells (macrophages) play a pivotal role in the stimulation of the antigen/pathogen-specific immune system and also in innate immunological defense mechanisms. Therefore, by tailoring these drug delivery modules, it could be possible to specifically target the macrophages to treat phagosomotropic diseases.

In addition, host-directed therapies and other supportive therapies based on immunological mechanisms are needed as adjunct therapy to improve the host potentiality in clearing *Mycobacterium tuberculosis* infection, reducing the period of therapy, and preventing the development of bioactive(s) resistance (Wallis and Hafner 2015). Repurposed bioactive(s) are desired to surmount the costly process of designing and developing new bioactive(s) and nanoplatforms to further improve the therapy (Wong et al. 2013). There is a compelling need of preventive therapy or more specifically translational therapy using safer bioactive(s) to face the challenge of latent TB (Kesharwani 2020). In nutshell, the alarming threat of MDR-TB and XDR-TB and the failure to provide successful chemotherapy necessitate the patient-oriented research for the complete eradication of tuberculosis.

4.1 Passive Drug Targeting

The preferential or favored natural distribution of the bioactive(s) loaded drug carrier to the domain of interest or target site without the functionalization or surface modification of carrier(s) with specific ligands for receptor-mediated endocytosis is referred to as passive targeting. It relies on the natural course of biodistribution of the nanotherapeutics. Concerning TB, two key approaches could be employed to achieve passive targeting. The first approach that could be advantageous is the avid uptake of the drug delivery modules by the mononuclear phagocytic system (macrophages, monocytes, and dendritic cells) via phagocytosis or endocytosis. The approach might be successful when infectious agents dwell within macrophagic tropics, which not only harbor the infection but also cluster together to constitute granulomas. Furthermore, drug delivery modules could be tailored to increase their interception and uptake by the mononuclear phagocytic system (MPS). The augmented capture could be achieved by tuning or altering their properties such as size, shape, porosity, elasticity, charge, lipophilicity, rigidness, etc. The mechanism of internalization (phagocytosis or endocytosis) can vary because it solely depends on the properties of the drug delivery module. The drug delivery modules that are taken up by the macrophages subsequently transport the drug to cytoplasmic domain and other organelles of the cells. The factors which reportedly facilitate the process of phagocytosis by the MPS are opsonization and protein corona accumulation (Monopoli et al. 2012; Gustafson et al. 2015; Clemens et al. 2012).

The second approach corresponds to the enhanced permeability and retention effect (EPR) (Nakamura et al. 2016). It involves the favored accumulation of nano drug delivery modules due to leaky vasculature in the tumor microenvironment compared to other sites on account of their size. Previously it was reported that the enhanced permeability and retention phenomenon may also operate in the infected tissues (Azzopardi et al. 2013). This could be attributed to the enhanced microvascular permeability of the capillaries as a consequence of the inflammatory reaction. Thus, as a general rule, this phenomenon could be advantageous for improvising nanocarrier accumulation at the site of infection or infected area, particularly in the granuloma. Therefore, nano drug delivery modules with extended systemic circulation are preferred in order to achieve their maximum accumulation in an infected area.

Another well-known approach that involves passive accumulation is the prodrug design. It is a molecular modification aimed at optimizing and improving the physicochemical and pharmacological characteristics of the bioactive(s) to enhance their pharmacokinetic and solubility characters and reduce their toxicities issues (Jornada et al. 2016). One such approach is lipid drug conjugates. Lipid drug conjugates (LDCs) are chemical entities that are often addressed as a lipoidal prodrug(s). They encompass bioactive moieties, covalently or non-covalently linked with lipids like fatty acids, glycerides, or phospholipids. Lipid drug conjugates are fabricated/tailored to increase bioactive(s) payload. Leakage of a highly polar bioactive(s) from the lipophilic matrix could be prevented by LDCs. The hydrophobicity of the bioactive(s) is enhanced by linking or conjugating the lipidic moieties to bioactive molecules. These conjugates possess numerous merits including enhanced tumor targeting, lymphatic system targeting, systemic bioavailability, and decreased toxicity (Shrivastava et al. 2020b; Irby et al. 2017). One similar study is reported by Pandit et al. (2020). The author worked on isoniazid (hydrophilic drug) loaded lipid drug conjugate nanoparticles to enhance its intracellular delivery to human macrophages. The drug isoniazid is associated with several limitations owing to its hydrophilic nature, including poor bioavailability, failure to cross lipophilic blood-brain barrier, poor gut permeability, etc., and these drawbacks, in turn, limit its clinical efficacy. To conquer these limitations and to enhance its bioavailability in the blood stream, the author linked hydrophobic moiety to this molecule (Isoniazid). Isoniazid was conjugated with a short lipid chain of stearoyl chloride to form a stable covalently linked lipid drug conjugate. The lipid drug conjugate-based nanoparticles were fabricated by using cold-high pressure homogenization technique and extensively characterized. The size of the nanoparticles was in the range of 124.60 ± 5.56 nm. The entrapment efficiency of the isoniazid conjugate was $92.73 \pm 6.31\%$. The in vitro release study indicates sustained release behavior of the isoniazid for up to 72 h. Furthermore, confocal microscopy studies revealed enhanced uptake of developed nanoparticles by THP-1 macrophages. This could be attributed to the increased lipophilicity and anionic surface charge. Consequently, intracellular trafficking into endosomal and lysosomal vesicles occurs progressively (Pandit et al. 2020). Thus, the lipid drug conjugate-loaded nanoparticles demonstrated considerable potential as an ability to affect enhanced/preferential intracellular delivery of isoniazid (a hydrophilic bioactive).

Multiple studies have reported the in vitro and in vivo passive targeting of nanoplatforms for TB chemotherapy as discussed here in Table 1.

Table 1 Some recent contributions of bioactive(s) delivery modules via passive targeting-based approaches in the treatment of TB

Bioactive delivery modules	Loaded bioactive	Major outcomes/findings	References
Chitosan nanoparticles (dry powder inhalers)	Bedaquiline	Higher % cell viability and higher accumulation of bedaquiline in lungs.	Rawal et al. (2018)
Respirable graft copolymer-based polymeric micelles	Rifampicin	The in vitro drug antitubercular activity in *Mycobacterium tuberculosis*-infected THP-1 macrophages was found to be increased (up to 2.5-fold) in the case of micellar formulation as compared to free drug solution.	Grotz et al. (2019)
Poly(ε-caprolactone)-*b*-PEG-*b*-poly(ε-caprolactone) "flower-like" polymeric micelles	Rifampicin and isoniazid	Nanoencapsulation reduced the rate of degradation of rifampicin. The oral bioavailability of rifampicin was increased (up to 3.3 times) as compared to the free drug in the presence of isoniazid.	Moretton et al. (2014)
HPMA-PLA Nano-polymeric micelles	Rifampicin and isoniazid	Hemolytic toxicity study showed that toxicity of bioactives-loaded polymeric micelles was decreased as compared to free bioactives.	Upadhyay et al. (2017)
Chitosan nanoparticles (Dry powder inhalers)	Rifampicin	Prolonged residence time and slow clearance of rifampicin from the lungs.	Rawal et al. (2017)
Chitosan microparticles (Dry powder inhalers)	Rifampicin and rifabutin	Good aerodynamic characteristics. Microparticles were nontoxic to U937 human macrophage cells.	Pai et al. (2016b)
Niosomes	Ethambutol	Niosomal encapsulation leads to the controlled release of the bioactive(s), enhanced efficiency in decreasing bacterial counts in the lungs of guinea pigs infected by *Mtb* H37Rv.	El-Ridy et al. (2015)
Inhalable PLGA nanoparticles	Linezolid	Sustained release of the bioactive up to 120 h in simulated lung fluid.	Shah et al. (2020)
PLGA nanoparticles	TB515	Enhanced cellular uptake and increased antitubercular effect observed in the case of TB515-loaded PLGA nanoparticles.	Kiss et al. (2014)

4.2 Active Drug Targeting

The most studied approach that has extensively been explored to effectively target the *Mycobacterium tuberculosis* infected cellular niche, that is, macrophage targeting using functionalized nanocarriers is referred to as active targeting. It depends on the cross talks between the receptors on the target cell and the specific ligands on the surface of the nanoplatforms, resulting in receptor-mediated endocytosis or phagocytosis. Consequently, internalization of the nanocarrier occurs that further increases the intracellular bioactive(s) accumulation in the target cells/tissues. Concerning tuberculosis, the characteristics receptors on MPS cells (major host cell) could be advantageous in macrophagic targeting of antitubercular bioactive(s). Thus, the PRRs present on the macrophages are the prime receptors for receptor mediated drug targeting. These receptors play a significant role in recognizing pathogen-associated molecular patterns (PAMPs), and as a result, their cross talk(s) interaction and interception follow the internalization of the microbe. The PRRs are classified on the basis of specificity to the ligand, localization, function, and evolutionary relationships. On the basis of their localization in the cell, they are further categorized into two main categories:

(a) Membrane-bound PRRs: Toll-like receptors and C-type lectin receptors.
(b) Cytoplasmic PRRs: Nucleotide-binding oligomerization domain (NOD)-like receptors and retinoic acid-inducible gene I (RIG-I)-like receptors.

The membrane-bound PRRs are of paramount significance for receptor recognition and targeting. The toll-like receptors are primarily capable of recognizing heat shock proteins, lipoproteins, lipopolysaccharides, and flagellar proteins whereas the C-type lectin receptors can recognize carbohydrate derivatives including fucose, mannose, β-glucans, and glycolipids (Areschoug and Gordon 2008; Takeuchi and Akira 2010; Kumar 2020). These carbohydrate derivatives are expressed on the microbe's surface. Apart from this, another receptor that also constitutes a subset of membrane-bound PRRs are referred to as scavenger receptors (Zani et al. 2015). They could be advantageous candidates for site-specific delivery of the bioactive(s) due to their overexpression on the membrane of phagocytes. These receptors possess specificity for a wide range of ligands including proteoglycans, lipoproteins, polysaccharides, phospholipids, and bacterial components (Postlethwait 2007). Apart from the PRRs, other types of the receptor such as Fc receptors and complement receptors are also expressed on the surface of the phagocytes. They are involved in the opsonin-dependent phagocytosis and several other immunological responses as a result of which clearance of the opsonized microbe does occur. The complement receptors are capable of recognizing serum complement proteins that are critically involved in bacterial opsonization whereas the Fc receptors are specific for Fc fragment of the antibodies (Collins and Kaufmann 2001; Costa et al. 2015). The schematic representation of the principle of active and passive targeting along with the phagocytic receptors exploitable for active targeting is presented in Fig. 4.

A diverse range of ligands has been investigated and reported in the literature that could be used to facilitate the active targeting of drugs to the macrophages. Examples of such ligands are peptides, antibodies, proteins, nucleic acid aptamers, carbohydrates, and others. Recent studies on active targeting of antitubercular bioactive(s)-loaded nanoplatforms for TB chemotherapy are discussed in Table 2. Regardless of the fact that a large number of studies on the targeted drug delivery modules are focused on cancer treatment; however, this approach could effectively be implemented for host cell-targeted delivery of the bioactive(s) specifically against intracellular microbe such as *Mycobacterium tuberculosis*. In particular, phagocytes are extensively studied as therapeutic target cells due to their involvement in diseases like cancer, tuberculosis, and other infectious or inflammatory infections (Jin et al. 2019; Ponzoni et al. 2018). The progress in the domain of macrophagic active targeting-based strategies for the treatment of TB is discussed here.

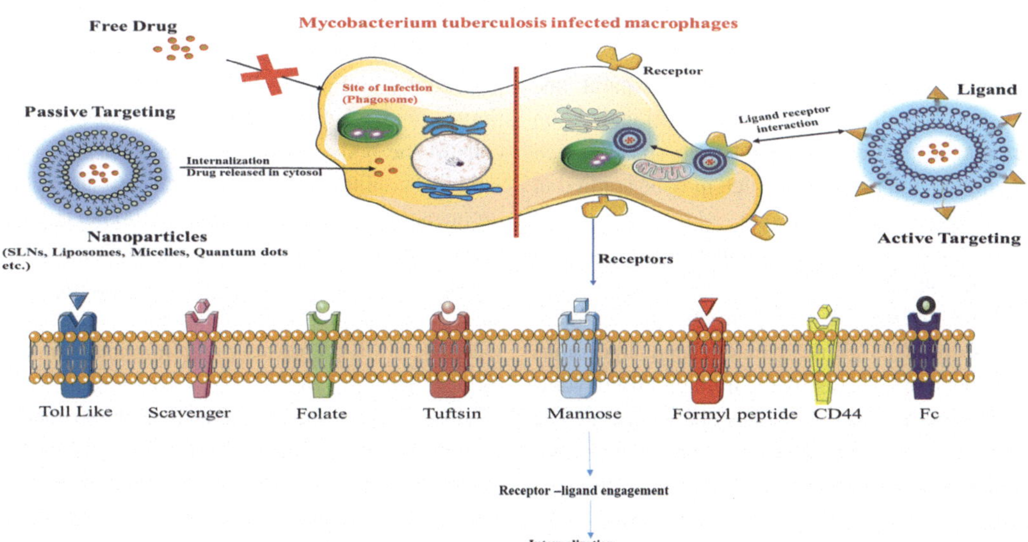

Fig. 4 Schematic representation of the principle of active and passive targeting along with the phagocytic receptors exploitable for active targeting

4.2.1 Mannose Receptor-Mediated Macrophage Targeting

The mannose receptor is also known as CD206 that belongs to a C-type lectin receptor family. It is highly expressed on the surface of alveolar macrophages and dendritic cells. These receptors are capable of recognizing targeting moieties (ligand) with a terminal mannose, N-acetylglucosamine, or fucose moiety (Azad et al. 2014). The processes that are mediated by mannose receptors are endocytosis, phagocytosis, and other inflammatory consequences as well as intracellular trafficking pathways. In addition to this, it is also involved in the uptake of *Mycobacterium tuberculosis* by the process termed as phagocytosis, in the phagosome-lysosome fusion inhibition, consequently facilitates the bacterial survival within the host cell. Apart from this, it also facilitates granuloma formation (Kang et al. 2005; Rajaram et al. 2017). Based on these studies, it may be realized that mannose receptor-mediated targeting holds the potential for effective TB therapy. It may also facilitate the intracellular colocalization of the tubercle bacilli and bioactive(s)-loaded delivery modules since they could share a common entry passage with the phagosomes of macrophages. Several studies on mannose receptor-mediated targeting to macrophages are reported in the literature which exhibited an increased uptake of bioactive(s)-loaded mannosylated delivery modules as compared to nontargeted preparations. Several modifications or functionalization approaches for mannosylation have been attempted and proposed which may be applied in drug delivery module design.

Shrivastava et al. (2020a) developed isoniazid and rifampicin co-loaded mannosylated liposomes for macrophage targeting for the treatment of tuberculosis. The mannosylated liposomes were developed by using the lipid thin film hydration technique and extensively characterized. The average vesicle diameter of mannosylated liposomes was in the range of 1.29 ± 0.24 μm. The drug entrapment efficiency was recorded to be 31.8 ± 0.12% for isoniazid and 84.7 ± 1.25% for rifampicin. In vitro bioactive(s) release study suggested that the hydrophilic drug (Isoniazid) releases at a faster rate than lipophilic drug (Rifampicin) and the release pattern of either of bioactive(s) follows the Higuchi diffusion model. The ex vivo studies were performed on J774A.1 macrophage cells. The antitubercular activity data revealed that the colony-forming units in the case of mannosylated liposomes were less as compared to non-mannosylated liposomes and free drug(s) combination. The FACS analysis and confocal microscopy study showed that the cellular localization was significantly high in the case of mannosylated liposomes. The biodistribution study also demonstrated that a higher accumulation or concentration of the bioactive(s) was maintained for a prolonged duration (Shrivastava et al. 2020a). The above results suggested that the mannosylated liposomes are a promising carrier for macrophage-targeted chemotherapy and for dual delivery of the bioactive(s). Recent studies on mannose receptor-mediated targeting of antitubercular bioactive(s)-loaded delivery modules for TB chemotherapy are discussed in Table 2.

4.2.2 Hyaluronic Acid Receptor-Mediated Macrophage Targeting

Hyaluronic acid or hyaluronan belongs to the glycosaminoglycan category and is composed of repeating disaccharide units of D-glucuronic acid and N-acetyl-D-glucosamine. It is present in the pericellular and extracellular matrix, in all body tissues, and when it interacts with the immune cells of the body, it exhibits normal (healthy) or inflamed conditions (Liang et al. 2016). The hyaluronic acid of molecular weight > 1000KDa possesses immunosuppressive and anti-inflammatory activity and is largely present in the healthy tissues, whereas the hyaluronic acid of molecular weight < 500KDa possesses the immunostimulatory activity and forms as a consequence of the fragmentation of the components of the extracellular matrix, tissue damage, or infection. The cells of the immune system are activated upon an inflammatory response and overexpressed hyaluronic acid receptor which is also known as CD44. Apart from this, in a homeostatic state (noninflammatory), the alveolar macrophages can

Table 2 Some recent contributions of bioactive(s) delivery modules via active targeting-based approaches in the treatment of TB

Bioactive delivery modules	Loaded bioactive	Targeting moiety/ligand	Major outcomes/findings	References
Mannosylated nanoliposomes	Moxifloxacin	4-aminophenyl-alpha-D–manno-pyranoside (PAM)	The higher antitubercular activity enhanced alveolar macrophage uptake and deep lung deposition was observed.	Hamed et al. (2019)
Mannosylated solid lipid nanoparticles (SLNs)	Rifampicin	Mannose	Internalization of the drug-carrier complex in macrophages was improved with mannosylation.	Pi et al. (2019)
Mannosylated graphene oxide	Rifampicin	Mannose	Intracellular delivery of rifampicin was improved. The killing efficiency of rifampicin was enhanced against intracellular *Mycobacterium tuberculosis*.	Pi et al. (2019)
Mannosylated gelatin nanoparticles	Linezolid	Mannose	Sustained release of the bioactive with a significant increase in the mean residence time and the half-life.	Patil et al. (2020)
Mannose-functionalized solid lipid nanoparticles (SLNs)	Isoniazid	Mannose	Intrinsic cytotoxicity of isoniazid was reduced. Isoniazid-loaded mannosylated SLNs were internalized more efficiently than plain SLNs.	Costa et al. (2018)
Chitosan-Hyaluronic acid-conjugated nanoparticles	Isoniazid	Chitosan-Hyaluronic acid	Functionalized nanoparticles were compatible with the A549 cell line and were found nontoxic.	Mukhtar et al. (2020)
Liquid-crystalline folate nanoparticles	Rifampicin	Folic acid	Sustained release of the bioactive, low cytotoxicity was observed on NR8383 cells.	Parmar et al. (2015)
Functionalized nanostructured lipid carriers	Rifampicin	Tuftsin	The developed formulation was twofold more effective against *Mycobacterium tuberculosis* as compared to the free drug.	Carneiro et al. (2019)
Cyclodextrin-conjugated curdlan nanoparticles	Rifampicin and levofloxacin	Curdlan (Dectin-1 receptor ligand)	Sustained release of both the bioactive(s) for a protracted period, nontoxic to RAW 264.7 cells and L929 cells.	Basha et al. (2019)
Transferrin-conjugated silver quantum-dots	Rifampicin	Transferrin	The drug was released in a sustained manner. The ten-fold higher antitubercular activity was observed against *Mycobacterium smegmatis* and *Mycobacterium bovis-BCG* as compared to the free drug.	Pati et al. (2016)

also bind to hyaluronic acid. The uptake of hyaluronic acid by the macrophages occurs in a CD44-dependent manner, followed by its transportation to the lysosomes. Moreover, hyaluronic acid receptors or CD44 that are present on the macrophages are also a binding site for *Mycobacterium tuberculosis*. The tubercle bacilli require hyaluronic acid as a carbon source for multiplication. Characteristics such as biocompatibility, biodegradability, and presence of multiple modification sites make hyaluronic acid a promising and potential candidate that could be used as a targeting ligand in bioactive(s) delivery (Lee-Sayer et al. 2015; Lallana et al. 2017; Leemans et al. 2003). Rossi et al. (2019) worked on rifampicin-, isoniazid-, and verapamil-loaded sodium hyaluronate inhalable microparticulate system to treat pulmonary tuberculosis. Verapamil was used here as an efflux pump inhibitor. The aerodynamic diameter of particles was in the range of 0.94 ± 0.15 which was found suitable for inhalation and deposition in the alveoli. The respirable sodium hyaluronate microparticles demonstrated that the developed formulation is safe for macrophages. The enhanced antitubercular activity was recorded in the case of bioactive(s) encapsulated respirable sodium hyaluronate microparticles as compared to non-encapsulated formulations against both bioactive-susceptible and bioactive-resistant *Mycobacterium tuberculosis* (Rossi et al. 2019). The carrier seems promising and holds the potential for alveolar macrophage targeting of antitubercular bioactive(s). Recent studies on hyaluronic acid receptor-mediated targeting of antitubercular bioactive(s)-loaded delivery modules for TB chemotherapy are discussed in Table 2.

4.2.3 Folate Receptor-Mediated Macrophage Targeting

As an essential vitamin, folic acid is required in the biosynthesis of precursors of nucleotides in the mammalian cells and is also pivotal for single carbon transfer reactions. The folic acid derivatized molecule or nanocarriers are taken up by the cells via folate receptor-mediated endocytosis. The two major isoforms of folate receptors that are commonly expressed on the surface of the macrophages are folate receptor α and folate receptor β. The activated macrophages that are involved in autoimmune and inflammatory diseases overexpress folate receptor β (Gaspar et al. 2019). The overexpression of folate receptor α is commonly seen in cancer cells. Thus, folic acid could be used as a ligand in the targeted delivery of therapeutics to activated macrophages (in case of inflammatory diseases) and tumor cells (Yi 2016). However, only a few studies were reported in the previous literature on folate receptor-mediated macrophage targeting in the domain of intracellular infections. Folate receptor has not been largely exploited concerning TB. In one of the studies, Shah et al. (2017) reported rifampicin-loaded nanoemulsions for pulmonary delivery for the treatment of tuberculosis. The author developed chitosan-folate conjugate and utilized it for the preparation of nanoemulsion. The nanoemulsion was formulated by using the emulsification technique and extensively characterized. The average droplet size of the nanoemulsion was in the range of 40–60 nm. The nanoemulsion was delivered via nebulization technique and the obtained aerosol offered optimal characteristics for profound lung accumulation. The cytotoxicity assay demonstrated that no cytotoxicity is observed on alveolar macrophages in the case of developed nanoemulsion. The cellular uptake was better in the case of chitosan-folate conjugated nanoemulsion as compared to nonconjugated nanoemulsion. This could be attributed to the involvement of the folate and mannose receptors that facilitate the internalization of the nanoemulsion into the cell. Moreover, the *N*-acetylglucosamine residues that are present in the chitosan molecule as structural motifs could be recognized by the mannose receptors. In vivo study revealed that in the case of chitosan-folate conjugated nanoparticles, higher bioactive(s) accumulation in the lungs and decreased bioactive concentration in plasma were obtained as compared to nonconjugated formulations (Shah et al. 2017). The folate receptor could further be explored in the domain of macrophage targeting concerning tuberculosis.

4.2.4 Tuftsin Receptor-Mediated Macrophage Targeting

Tuftsin, that is, Thr-Lys-Pro-Arg, is a basic and naturally occurring tetrapeptide. It is being widely exploited as a targeting ligand in the field of drug delivery to improve the targeting of bioactive(s) because it selectively binds to the cells of the MPS. It is formed as a result of the enzymatic cleavage of the Fc domain of the immunoglobulin G. Two enzymes facilitate the enzymatic cleavage including leukokininase (neutrophil-derived enzyme) and spleen tuftsin endocarboxypeptidase. Tuftsin and its analogs were found to reveal a wide array of biological activities including immunostimulatory activity. It is worth noting that tuftsin improves the phagocytic activity of the cells of the MPS (monocytes and macrophages). Moreover, it also functions as an immunomodulator by stimulating or activating the members of the MPS nonspecifically against intracellular infections (Fridkin and Najjar 1989; Siebert et al. 2017). Several tuftsin-decorated nanoplatforms were reported in the previous literatures that are used for macrophage targeting in TB as drug-peptide conjugates.

Horváti et al. (2014) designed and developed lipopeptide conjugates of isoniazid. The drug, that is, isoniazid was covalently conjugated to a fatty acid-derivatized tuftsin (palmitoylated tuftsin). Cell line studies were performed on MonoMac6 monocytes. The developed conjugate demonstrated improved antitubercular activity on Mtb H37Rv culture. Moreover, it showed reduced cytotoxicity and hemolytic activity on MonoMac6 monocytes cells. The conjugate was found efficacious in killing intracellular tubercle bacilli as compared to free isoniazid. The author also encapsulated the developed conjugate into poly(lactide-co-glycolide) nanoparticles. In vivo studies were performed on the infected guinea pigs model owing to their susceptibility to infection with Mycobacterium tuberculosis and develop the disease in a similar way as in human beings. The conjugate-loaded nanoparticles were administered orally. The bacterial level was found to be reduced and no cytotoxicity was observed. However, the development of the disease (severe lesions, parenchymal involvement, necrosis, intralesional mineralization) was observed in untreated control guinea pigs (Horváti et al. 2014).

4.2.5 Formyl Peptide Receptors-Mediated Macrophage Targeting

Formyl peptide receptors are the pivotal players that regulate innate immune responses. They are PRRs involved in stimulating inflammatory responses and host defense mechanisms. The three subtypes of formyl peptide receptors that are identified in humans are FPR1, FPR2, and FPR3. The cells of the MPS (macrophages, neutrophils, and monocytes) express these chemoattractant receptors. Although, these receptors are also expressed by the other cell types. N-formyl peptides such as *N*-formylated methionine containing peptides (e.g., *N*-formylmethionyl-leucyl-phenylalanine (fMLF)) were the first ligands identified for the formyl peptide receptors. These products are formed as a result of the cleavage of the bacterial or mitochondrial proteins. Several other ligands have also been reported for these receptors in the previous literatures. The availability of the wide array of ligands for the formyl peptide receptors provides the possibility to develop novel strategies for formyl peptide receptor-mediated macrophage targeting via nanoplatforms (Lee et al. 2017).

Mir and Sharma (2019) reported the immunotherapeutic effect of N-formylated peptide in the case of experimental TB. The author worked on a synthetic peptide that comprises of N-terminally formylated listerial peptide (LemA) with the amino acid sequence "f-MIGWII." The developed peptide was tested along with the combination of antitubercular bioactive(s) in the mouse model of tuberculosis. The formulation was administered via the subcutaneous route in the *Mycobacterium tuberculosis*-infected mice. In the mouse neutrophils, a significant rise in the intracellular reactive oxygen species level was observed. Moreover, a marked reduction in the bacterial load (colony forming units) in the lungs and spleen of the *Mycobacterium tuberculosis*-infected mice was obtained. The therapeutic efficacy was enhanced when antitubercular bioactive(s) was given in combination with peptide as compared to antitubercular drugs alone (Mir and Sharma 2019). This therapy might be advantageous and could be used as an adjunct therapy to conventional therapy for the treatment of tuberculosis. The above results demonstrated the potential of formyl receptor-mediated macrophage targeting. This domain, however, is still far from being exploited.

4.2.6 Other Receptors-Mediated Macrophage Targeting

Apart from the above receptors that are presented here for macrophage targeting, several other receptors could also be used for macrophage targeting. Based on the developmental origin, native environment, and disease progression in macrophages, they possess distinctive characteristics like phenotypic heterogeneity and functional plasticity. Thus, a diverse range of receptors could be significant, apart from the ones that are presented above based on the characteristic macrophage populations that are intended to be targeted. Multiple receptors are present on the alveolar or pulmonary macrophages that could be used for macrophage targeting in the case of TB including CD71 (transferrin receptor), CD200R, CD115 (colony-stimulating factor 1 receptor), CD14 (lipopolysaccharide (LPS) receptor), scavenger receptors (SR-A, CD36 (SR-B), toll-like receptors (TLR4, TLR9), Dectin-1/2 (β-glucan receptor), Fc receptors (CD64, CD32, CD16), and macrophage receptor with collagenous structure (MARCO), CD163, CD204, CD68) (Hasnain et al. 2019b; Basha et al. 2019; Pati et al. 2016). These receptors hold the potential and could pave the way for novel strategies for macrophage targeting. Besides alveolar or pulmonary macrophages, certain receptors are also present on the other host cells that play a pivotal role in the recognition of *Mycobacterium tuberculosis* and they could also be targeted to overcome the disease. For example, dendritic cell-specific intercellular adhesion molecule-3 grabbing non-integrin (DC-SIGN) is the main receptor that is present on the dendritic cells. It belongs to the C-type lectin receptors family. It is involved in the recognition of *Mycobacterium tuberculosis* and could be used for targeting (Maeda et al. 2003). Besides the targeting moieties presented above, a wide variety of other targeting moieties (ligands) were also exploited to target macrophages in TB by using bioactive(s) delivery modules, including mycolic acids, lactose, *O* stearoyl Amylopectin, fucoidan, maleylated bovine serum albumin (MBSA), immunoglobulin G, etc. These targeting moieties or ligands have been observed to be promising candidates to target macrophages to treat TB. In a nutshell, the information/knowledge acquired so far in regard to the macrophagic receptors and targeting moieties (ligands) offers an opportunity to develop new strategies for the effective delivery of the bioactive(s) in TB chemotherapy.

5 Concluding Remarks and Future Prognosis

Over the past few years, nanotherapeutics has emerged as a powerful tool to strengthen the delivery of the bioactive(s) and efficiency of disease therapy. Nevertheless, the majority of them are directed for cancer therapy and only a few nanoformulations are focused on infectious diseases. However, nanotechnology-based nanoplatforms could be advantageous in the fight against TB. Furthermore, integrating the nanotherapeutics with the pulmonary delivery route bestows the most promising strategy in TB chemotherapy. The application of pulmonary bioactive(s) delivery devices provides a noninvasive route of administration of the bioactive(s). This approach could be promising even for the delivery of bioactive(s) for systemic action. It is worth noting that, in developing countries, where a large number of the population does not have access to needed health treatment, so in that case, the development of these bioactive(s) delivery modules or nanotherapeutics assisted with pulmonary administration could be the fundamental factor to control the disease like TB. The administration of the nanotherapeutics by using aerosol or inhaler devices could offer an opportunity to make the TB treatment simpler, accessible, and more affordable. Furthermore, it also simplifies the complex dosage regimens of TB chemotherapy, reduces the medical personnel and also the cost of the treatment.

It is necessary to combine different key enabling technologies to build a quantum leap to control the pandemics. In this chapter, we have highlighted several macrophage targeting strategies along with the application of nano drug delivery modules for TB chemotherapy. A diverse range of advantages is associated with these carrier systems including enhanced stability of the bioactive(s), the possibility of drug targeting, increased bioavailability, and biodistribution of the bioactive(s), reduced the active dose, and toxicity associated with the bioactive(s). Consequently, bioactive(s) resistance also gets reduced. Nanotechnology appears to be a promising approach; however, no nanotechnology-based commercial product is available in the market for TB disease so far. Besides this, the number of patents in this domain is also very few compared with nanomedicines for cancer. A multidisciplinary approach that involves the integration of nanotechnology, medicine, and engineering is needed to accomplish revolutionary progress in

TB chemotherapy and to make pulmonary delivery of the nanotherapeutics executable for noninvasive clinical trials.

Conflict of Interest The authors declare no competing financial/personal interest whatsoever.

References

Areschoug T, Gordon S. Pattern recognition receptors and their role in innate immunity: focus on microbial protein ligands. In: Trends in innate immunity, vol. 15. Karger Publishers; 2008. p. 45–60.

Armocida E, Martini M. Tuberculosis: a timeless challenge for medicine. J Prev Med Hyg. 2020 Jun;61(2):E143.

Azad AK, Rajaram MV, Schlesinger LS. Exploitation of the macrophage mannose receptor (CD206) in infectious disease diagnostics and therapeutics. J Cytol Mol Biol. 2014 Jan;10:1(1).

Azzopardi EA, Ferguson EL, Thomas DW. The enhanced permeability retention effect: a new paradigm for drug targeting in infection. J Antimicrob Chemother. 2013 Feb 1;68(2):257–74.

Baranyai Z, Soria-Carrera H, Alleva M, Millán-Placer AC, Lucía A, Martín-Rapún R, Aínsa JA, de la Fuente JM. Nanotechnology-based targeted drug delivery: an emerging tool to overcome tuberculosis. Adv Therap. 2020;2000113

Basha RY, Sampath Kumar TS, Doble M. Dual delivery of tuberculosis drugs via cyclodextrin conjugated curdlan nanoparticles to infected macrophages. Carbohydr Polym. 2019 Aug 15;218:53–62.

Carneiro SP, Carvalho KV, Soares RD, Carneiro CM, de Andrade MH, Duarte RS, Dos Santos OD. Functionalized rifampicin-loaded nanostructured lipid carriers enhance macrophages uptake and antimycobacterial activity. Colloids Surf B: Biointerfaces. 2019 Mar 1;175:306–13.

Chaurasiya SK. Tuberculosis: smart manipulation of a lethal host. Microbiol Immunol. 2018 Jun;62(6):361–79.

Cheung LS, Srikrishna G, Bishai WR. Role of myeloid-derived suppressor cells and regulatory T-cells in the tuberculous granuloma. In: Tuberculosis host-pathogen interactions. Cham: Springer; 2019. p. 63–93.

Chuan J, Li Y, Yang L, Sun X, Zhang Q, Gong T, Zhang Z. Enhanced rifampicin delivery to alveolar macrophages by solid lipid nanoparticles. J Nanopart Res. 2013 May 1;15(5):1634.

Clemens DL, Lee BY, Xue M, Thomas CR, Meng H, Ferris D, Nel AE, Zink JI, Horwitz MA. Targeted intracellular delivery of antituberculosis drugs to Mycobacterium tuberculosis-infected macrophages via functionalized mesoporous silica nanoparticles. Antimicrob Agents Chemother. 2012 May 1;56(5):2535–45.

Collins HL, Kaufmann SH. The many faces of host responses to tuberculosis. Immunology. 2001 May;103(1):1.

Costa A, Sarmento B, Seabra V. Targeted drug delivery systems for lung macrophages. Curr Drug Targets. 2015 Dec 1;16(14):1565–81.

Costa A, Sarmento B, Seabra V. Mannose-functionalized solid lipid nanoparticles are effective in targeting alveolar macrophages. Eur J Pharm Sci. 2018 Mar 1;114:103–13.

El-Ridy MS, Yehia SA, Kassem MA, Mostafa DM, Nasr EA, Asfour MH. Niosomal encapsulation of ethambutol hydrochloride for increasing its efficacy and safety. Drug Deliv. 2015 Jan 2;22(1):21–36.

Falzon D, Schünemann HJ, Harausz E, González-Angulo L, Lienhardt C, Jaramillo E, Weyer K. World Health Organization treatment guidelines for drug-resistant tuberculosis, 2016 update. Eur Respir J. 2017 Mar;1:49(3).

Fridkin M, Najjar VA. Tuftsin: its chemistry, biology, and physiological function. Crit Rev Biochem Mol Biol. 1989;24:1–40.

Gaspar N, Zambito G, Löwik CM, Mezzanotte L. Active Nano-targeting of Macrophages. Curr Pharm Des. 2019 May 1;25(17):1951–61.

Gregoriadis G. Drug delivery: old drugs in new clothing. Nature. 1984 Jul;310(5974):186–7.

Grotz E, Tateosian NL, Salgueiro J, Bernabeu E, Gonzalez L, Manca ML, Amiano N, Valenti D, Manconi M, García V, Moretton MA. Pulmonary delivery of rifampicin-loaded soluplus micelles against Mycobacterium tuberculosis. J Drug Del Sci Technol. 2019 Oct 1;53:101170.

Gupta K, Saldanha M, Parasnis M, Devarajan PV, Jain R, Dandekar P. Toll-like receptor-mediated endocytosis in infectious disease. In: Targeted intracellular drug delivery by receptor mediated endocytosis. Cham: Springer; 2019. p. 323–49.

Gustafson HH, Holt-Casper D, Grainger DW, Ghandehari H. Nanoparticle uptake: the phagocyte problem. Nano Today. 2015 Aug 1;10(4):487–510.

Hamed A, Osvman R, Al-Jamal KT, Holayel SM, Geneidi AS. Enhanced antitubercular activity, alveolar deposition and macrophages uptake of mannosylated stable nanoliposomes. J Drug Del Sci Technol. 2019 Jun 1;51:513–23.

Hasnain SE, Ehtesham NZ, Grover S, editors. Mycobacterium tuberculosis: molecular infection biology, pathogenesis, diagnostics and new interventions. Springer; 2019a Nov 30.

Hasnain SE, Ehtesham NZ, Grover S. Mycobacterium tuberculosis: molecular infection biology, pathogenesis, diagnostics and new interventions. Springer; 2019b Nov 30.

Horváti K, Bacsa B, Kiss E, Gyulai G, Fodor K, Balka G, Rusvai M, Szabó E, Hudecz F, Bősze S. Nanoparticle encapsulated lipopeptide conjugate of antitubercular drug isoniazid: in vitro intracellular activity and in vivo efficacy in a Guinea pig model of tuberculosis. Bioconjug Chem. 2014 Dec 17;25(12):2260–8.

Irby D, Du C, Li F. Lipid–drug conjugate for enhancing drug delivery. Mol Pharm. 2017 May 1;14(5):1325–38.
Jawahar N, Reddy G. Nanoparticles: a novel pulmonary drug delivery system for tuberculosis. J Pharm Sci Res. 2012 Aug 1;4(8):1901.
Jin Y, Hu G, Guo M, Xu J, Wu F, Fan J, Huang Q, Yang G, Lv Z, Wang X. Nanoparticles targeting macrophages as potential clinical therapeutic agents against cancer and inflammation. Front Immunol. 2019;10:1998.
Jornada DH, dos Santos Fernandes GF, Chiba DE, De Melo TR, Dos Santos JL, Chung MC. The prodrug approach: a successful tool for improving drug solubility. Molecules. 2016 Jan;21(1):42.
Joseph RR, Venkatraman SS. Drug delivery to the eye: what benefits do nanocarriers offer? Nanomedicine. 2017 Mar;12(6):683–702.
Kang PB, Azad AK, Torrelles JB, Kaufman TM, Beharka A, Tibesar E, DesJardin LE, Schlesinger LS. The human macrophage mannose receptor directs Mycobacterium tuberculosis lipoarabinomannan-mediated phagosome biogenesis. J Exp Med. 2005 Oct 3;202(7):987–99.
Kaur IP, Singh H. Nanostructured drug delivery for better management of tuberculosis. J Control Release. 2014 Jun 28;184:36–50.
Kesharwani P, editor. Nanotechnology based approaches for tuberculosis treatment. Academic Press; 2020 Jun 10.
Kiss É, Gyulai G, Pénzes CB, Idei M, Horváti K, Bacsa B, Bősze S. Tuneable surface modification of PLGA nanoparticles carrying new antitubercular drug candidate. Colloids Surf A Physicochem Eng Asp. 2014 Sep 20;458:178–86.
Kumar H. Metabolic pathways and metabolites shaping innate immunity. Int Rev Immunol. 2020 May 3;39(3):81–2.
Lallana E, de la Rosa JM R, Tirella A, Pelliccia M, Gennari A, Stratford IJ, Puri S, Ashford M, Tirelli N. Chitosan/hyaluronic acid nanoparticles: rational design revisited for RNA delivery. Mol Pharm. 2017 Jul 3;14(7):2422–36.
Lee HY, Lee M, Bae YS. Formyl peptide receptors in cellular differentiation and inflammatory diseases. J Cell Biochem. 2017 Jun;118(6):1300–7.
Leemans JC, Florquin S, Heikens M, Pals ST, van der Neut R, van der Poll T. CD44 is a macrophage binding site for Mycobacterium tuberculosis that mediates macrophage recruitment and protective immunity against tuberculosis. J Clin Invest. 2003 Mar 1;111(5):681–9.
Lee-Sayer SS, Dong Y, Arif AA, Olsson M, Brown KL, Johnson P. The where, when, how, and why of hyaluronan binding by immune cells. Front Immunol. 2015 Apr 14;6:150.
Liang J, Jiang D, Noble PW. Hyaluronan as a therapeutic target in human diseases. Adv Drug Deliv Rev. 2016 Feb 1;97:186–203.
Liu Z, Tabakman S, Welsher K, Dai H. Carbon nanotubes in biology and medicine: in vitro and in vivo detection, imaging and drug delivery. Nano Res. 2009 Feb 1;2(2):85–120.
Maeda N, Nigou J, Herrmann JL, Jackson M, Amara A, Lagrange PH, Puzo G, Gicquel B, Neyrolles O. The cell surface receptor DC-SIGN discriminates between mycobacterium species through selective recognition of the mannose caps on Lipoarabinomannan. J Biol Chem. 2003 Feb 21;278(8):5513–6.
Mayorga LS, Pitt A, Stahl PD. Receptor-mediated endocytosis and phagocytosis share common determinants for intracellular traficking. In: Mononuclear phagocytes. Dordrecht: Springer; 1992. p. 214–23.
Mir SA, Sharma S. Immunotherapeutic potential of an N-formylated peptide of listeria monocytogenes in experimental tuberculosis. Immunopharmacol Immunotoxicol. 2019 Mar 4;41(2).292–8.
Mishra A, Akhtar S, Jagannath C, Khan A. Pattern recognition receptors and coordinated cellular pathways involved in tuberculosis immunopathogenesis: emerging concepts and perspectives. Mol Immunol. 2017 Jul 1;87:240–8.
Monopoli MP, Åberg C, Salvati A, Dawson KA. Biomolecular coronas provide the biological identity of nanosized materials. Nat Nanotechnol. 2012 Dec;7(12):779–86.
Moretton MA, Hocht C, Taira C, Sosnik A. Rifampicin-loaded 'flower-like'polymeric micelles for enhanced oral bioavailability in an extemporaneous liquid fixed-dose combination with isoniazid. Nanomedicine. 2014;9(11):1635–50.
Mukhtar M, Pallagi E, Csóka I, Benke E, Farkas Á, Zeeshan M, Burián K, Kókai D, Ambrus R. Aerodynamic properties and in silico deposition of isoniazid loaded chitosan/thiolated chitosan and hyaluronic acid hybrid nanoplex DPIs as a potential TB treatment. Int J Biol Macromol. 2020 Dec 15;165:3007–19.
Nakamura Y, Mochida A, Choyke PL, Kobayashi H. Nanodrug delivery: is the enhanced permeability and retention effect sufficient for curing cancer? Bioconjug Chem. 2016 Oct 19;27(10):2225–38.
Nasiruddin M, Neyaz M, Das S. Nanotechnology-based approach in tuberculosis treatment. Tuberc Res Treat. 2017 Jan;22:2017.
Niu NK, Yin JJ, Yang YX, Wang ZL, Zhou ZW, He ZX, Chen XW, Zhang X, Duan W, Yang T, Zhou SF. Novel targeting of PEGylated liposomes for codelivery of TGF-β1 siRNA and four antitubercular drugs to human macrophages for the treatment of mycobacterial infection: a quantitative proteomic study. Drug Desig Dev Ther. 2015;9:4441.
Pai M, Behr MA, Dowdy D, Dheda K, Divangahi M, Boehme CC, Ginsberg A, Swaminathan S, Spigelman M, Getahun H, Menzies D. Tuberculosis. Nat Rev Dis Primers. 2016a Oct 27;2:16076.
Pai RV, Jain RR, Bannalikar AS, Menon MD. Development and evaluation of chitosan microparticles based dry powder inhalation formulations of rifampicin and rifabutin. J Aerosol Med Pulm Drug Deliv. 2016b Apr 1;29(2):179–95.
Pandey R, Khuller GK. Polymer based drug delivery systems for mycobacterial infections. Curr Drug Deliv. 2004 Jul 1;1(3):195–201.

Pandit S, Roy S, Pillai J, Banerjee S. Formulation and intracellular trafficking of lipid–drug conjugate nanoparticles containing a hydrophilic Antitubercular drug for improved intracellular delivery to human macrophages. ACS Omega. 2020 Feb 26;5(9):4433–48.

Parmar R, Misra R, Mohanty S. In vitro controlled release of rifampicin through liquid-crystalline folate nanoparticles. Colloids Surf B: Biointerfaces. 2015 May 1;129:198–205.

Pati R, Sahu R, Panda J, Sonawane A. Encapsulation of zinc-rifampicin complex into transferrin-conjugated silver quantum-dots improves its antimycobacterial activity and stability and facilitates drug delivery into macrophages. Sci Rep. 2016 Apr 26;6:24184.

Patil KD, Bagade SB, Bonde SC. Biodistribution, pharmacokinetics and toxicity evaluation of mannosylated gelatin nanoparticles of linezolid for anti-tubercular therapy. Mater Technol. 2020 Sep;5:1–9.

Pi J, Shen L, Shen H, Yang E, Wang W, Wang R, Huang D, Lee BS, Hu C, Chen C, Jin H. Mannosylated graphene oxide as macrophage-targeted delivery system for enhanced intracellular M. tuberculosis killing efficiency. Mat Sci Engin C. 2019 Oct 1;103:109777.

Pieters J. Mycobacterium tuberculosis and the macrophage: maintaining a balance. Cell Host Microbe. 2008 Jun 12;3(6):399–407.

Poce G, Cocozza M, Consalvi S, Biava M. SAR analysis of new anti-TB drugs currently in pre-clinical and clinical development. Eur J Med Chem. 2014 Oct 30;86:335–51.

Ponzoni M, Pastorino F, Di Paolo D, Perri P, Brignole C. Targeting macrophages as a potential therapeutic intervention: impact on inflammatory diseases and cancer. Int J Mol Sci. 2018 Jul;19(7):1953.

Postlethwait EM. Scavenger receptors clear the air. J Clin Invest. 2007 Mar 1;117(3):601–4.

Rajaram MV, Ni B, Dodd CE, Schlesinger LS. Macrophage immunoregulatory pathways in tuberculosis. Semin Immunol. 2014 Dec 1;26(6):471–85.

Rajaram MV, Arnett E, Azad AK, Guirado E, Ni B, Gerberick AD, He LZ, Keler T, Thomas LJ, Lafuse WP, Schlesinger LS. M. tuberculosis-initiated human mannose receptor signaling regulates macrophage recognition and vesicle trafficking by FcRγ-Chain, Grb2, and SHP-1. Cell Rep. 2017 Oct 3;21(1):126–40.

Rawal T, Butani S. Combating tuberculosis infection: a forbidding challenge. Indian J Pharm Sci. 2016 Jan;78(1):8.

Rawal T, Parmar R, Tyagi RK, Butani S. Rifampicin loaded chitosan nanoparticle dry powder presents an improved therapeutic approach for alveolar tuberculosis. Colloids Surf B: Biointerfaces. 2017 Jun 1;154:321–30.

Rawal T, Patel S, Butani S. Chitosan nanoparticles as a promising approach for pulmonary delivery of bedaquiline. Eur J Pharm Sci. 2018 Nov 1;124:273–87.

Rossi I, Buttini F, Sonvico F, Affaticati F, Martinelli F, Annunziato G, Machado D, Viveiros M, Pieroni M, Bettini R. Sodium hyaluronate nanocomposite respirable microparticles to tackle antibiotic resistance with potential application in treatment of mycobacterial pulmonary infections. Pharmaceutics. 2019 May;11(5):203.

Shah K, Chan LW, Wong TW. Critical physicochemical and biological attributes of nanoemulsions for pulmonary delivery of rifampicin by nebulization technique in tuberculosis treatment. Drug Deliv. 2017 Jan 1;24(1):1631–47.

Shah S, Cristopher D, Sharma S, Soniwala M, Chavda J. Inhalable linezolid loaded PLGA nanoparticles for treatment of tuberculosis: design, development and in vitro evaluation. J Drug Del Sci Technol. 2020 Dec 1;60:102013.

Shrivastava P, Gautam L, Sharma R, Dube D, Vyas S, Vyas SP. Dual antitubercular drug loaded liposomes for macrophage targeting: development, characterization, ex vivo and in vivo assessment. J Microencapsul. 2020a Nov;28:1–40.

Shrivastava P, Gautam L, Jain A, Vishwakarma N, Vyas S, Vyas SP. Lipid drug conjugates for improved therapeutic benefits. Current pharmaceutical design. 2020b.

Siebert A, Gensicka-Kowalewska M, Cholewinski G, Dzierzbicka K. Tuftsin-properties and analogs. Curr Med Chem. 2017 Oct 1;24(34):3711–27.

Singh J, Garg T, Rath G, Goyal AK. Advances in nanotechnology-based carrier systems for targeted delivery of bioactive drug molecules with special emphasis on immunotherapy in drug resistant tuberculosis–a critical review. Drug Deliv. 2016 Jun 12;23(5):1676–98.

Singh M, Guzman-Aranguez A, Hussain A, Srinivas CS, Kaur IP. Solid lipid nanoparticles for ocular delivery of isoniazid: evaluation, proof of concept and in vivo safety & kinetics. Nanomedicine. 2019 Feb;14(4):465–91.

Smith I. Mycobacterium tuberculosis pathogenesis and molecular determinants of virulence. Clin Microbiol Rev. 2003 Jul 1;16(3):463–96.

Sosnik A, Carcaboso ÁM, Glisoni RJ, Moretton MA, Chiappetta DA. New old challenges in tuberculosis: potentially effective nanotechnologies in drug delivery. Adv Drug Deliv Rev. 2010 Mar 18;62(4–5):547–59.

Sotgiu G, Centis R, D'ambrosio L, Migliori GB. Tuberculosis treatment and drug regimens. Cold Spring Harb Perspect Med. 2015 May 1;5(5):a017822.

Takeuchi O, Akira S. Pattern recognition receptors and inflammation. Cell. 2010 Mar 19;140(6):805–20.

Thakare V, Nayak U, Lacchiramka P. Newer drugs for multi drug resistant tuberculosis and special emphasis on linezolid. Biol Sci. 2012;2:37–46.

Traini D, Young PM. Drug delivery for tuberculosis: is inhaled therapy the key to success?

Upadhyay S, Khan I, Gothwal A, Pachouri PK, Bhaskar N, Gupta UD, Chauhan DS, Gupta U. Conjugated and entrapped HPMA-PLA Nano-polymeric micelles based dual delivery of first line anti TB drugs: improved and safe drug delivery against sensitive and resistant Mycobacterium tuberculosis. Pharm Res. 2017 Sep 1;34(9):1944–55.

Upadhyay S, Mittal E, Philips JA. Tuberculosis and the art of macrophage manipulation. Pathog Dis. 2018 Jun;76(4):fty037.

Uribe-Querol E, Rosales C. Control of phagocytosis by microbial pathogens. Front Immunol. 2017 Oct 24;8:1368.

Wallis RS, Hafner R. Advancing host-directed therapy for tuberculosis. Nat Rev Immunol. 2015 Apr;15(4):255–63.

Wong EB, Cohen KA, Bishai WR. Rising to the challenge: new therapies for tuberculosis. Trends Microbiol. 2013 Sep 1;21(9):493–501.

World Health Organization. The WHO global task force on TB impact measurement. World Health Organization; 2019.

Yashodhara BM, Huat CB, Naik LN, Umakanth S, Hande M, Pappachan JM. Multidrug and extensively drug-resistant tuberculosis from a general practice perspective. Inf Drug Res. 2010;3:115.

Yi YS. Folate receptor-targeted diagnostics and therapeutics for inflammatory diseases. Immun Net. 2016 Dec 1;16(6):337–43.

Zani IA, Stephen SL, Mughal NA, Russell D, Homer-Vanniasinkam S, Wheatcroft SB, Ponnambalam S. Scavenger receptor structure and function in health and disease. Cell. 2015 Jun;4(2):178–201.

Macrophage-Targeted Chemotherapy for Respiratory Diseases Other Than Tuberculosis

Seema Kohli

Abstract

Respiratory system is unremittingly exposed to a number of microorganisms and air pollutants that may influence the normal functioning of the system. Every day the respiratory tract and the alveoli are exposed to enormous volumes of air containing countless dust bits, chemical substances, and infectious material. In spite of this, the lungs are sterile and are managed by the filtering and cleaning mechanisms – nasal clearance and tracheobronchial clearance. This is attributed to respiratory macrophage that presents a vital role in defense. Alveolar macrophages are considered the most important means of eliminating pathogenic organisms that macrophages have gained immense interest of researchers as potential therapeutic targets. The extreme efficiency of macrophages in terms of particle uptake is because of occurrence of wide range of receptors on the membrane that facilitate their reuptake. The tendency of macrophages for the phagocytic/pinocytic clearance of foreign particles offers a sensible methodology to macrophage-specific targeting through suitable particulate carrier. The particulate matter targeting of macrophages is noteworthy as multiple novel delivery system are available viz. liposomes, niosomes, polymeric nanospheres, and oil-in-water microemulsions. The chapter entails different strategies for targeting the macrophages in various respiratory disease/disorders/infections employing nanotechnology techniques.

Keywords

Macrophages · Alveolar macrophages · Respiratory macrophages · Liposomes · Solid lipid nanoparticles · Polymeric nanoparticles · Macrophage targeting

1 Introduction

Macrophages have been first described by Metchnikoff in 1892 and since then macrophages have gained prime importance in immunity system. Macrophages are present throughout the body and they perform a variety/multiple functions – homeostatic, physiological, and immunological. Out of these, phagocytosis is the prime function of macrophages attributed to the huge count of specialized receptors that acknowledge the plasma membrane. These receptors make them suitable to arrest and eliminate senescent and damaged cells, dust particles, debris, and the microorganisms (Byrne et al. 2015)

S. Kohli (✉)
Department of Pharmaceutical Sciences, Kalaniketan Polytechnic College,
Jabalpur, Madhya Pradesh, India

S. Gupta, Y. V. Pathak (eds.), *Macrophage Targeted Delivery Systems*,
https://doi.org/10.1007/978-3-030-84164-5_18

Macrophages possess a huge army of hydrolytic enzymes which acknowledge their ability to degrade rapidly various materials.

Besides macrophages delivers antigens to T cells; they also induce the process of inflammation by triggering other cells. Macrophages are formed from circulating macrophages and then move off from blood circulation for further differentiation. There is substantial diversification among each macrophage. Macrophage migrate to and circulate in almost every cells/tissue, defending from infecting micro-organisms and also removes dead cells. The location and functions of macrophages in the body are summarized in Fig. 1.

Macrophage recognizing receptors like Toll-like receptors (TLRs) have the potential to bind specially to the sugar component of pathogen and thus able to eliminate them.

The macrophages have interference in almost all aspects of inflammatory response.

(i) Phagocytosis and digestion of invading organisms or foreign particles.
(ii) Release of potent enzymes that may degrade connective tissue. These first two effects may be helpful, since they tend to clear the inflammatory sites, a process akin to the "debridement" that surgeons perform when they clean dirty wounds. It is obvious, however, that these enzymes might also increase tissue damage.
(iii) Release of chemotactic and permeability factors that may prolong inflammation.
(iv) Release of substances responsible for leukocytosis and fever (prostaglandins and endogenous pyrogen) of inflammation.
(v) Release of factors that aid in healing.
(vi) Secretion of proteins that are important in defense mechanisms, such as antibacterial (lysozyme) and antiviral (interferon) substances (Robbins and Cortran 1979).

Significantly, few of these effects may seem to be harmful to the person; however, the overall effect is good. Therefore, macrophages demand due responsiveness and respect. Hence the macrophages act as a friend and foe (Lei et al. 2018).

With this brief background of macrophages, let us discuss about their functioning as a pharmaceutical potential in the treatment of respiratory diseases.

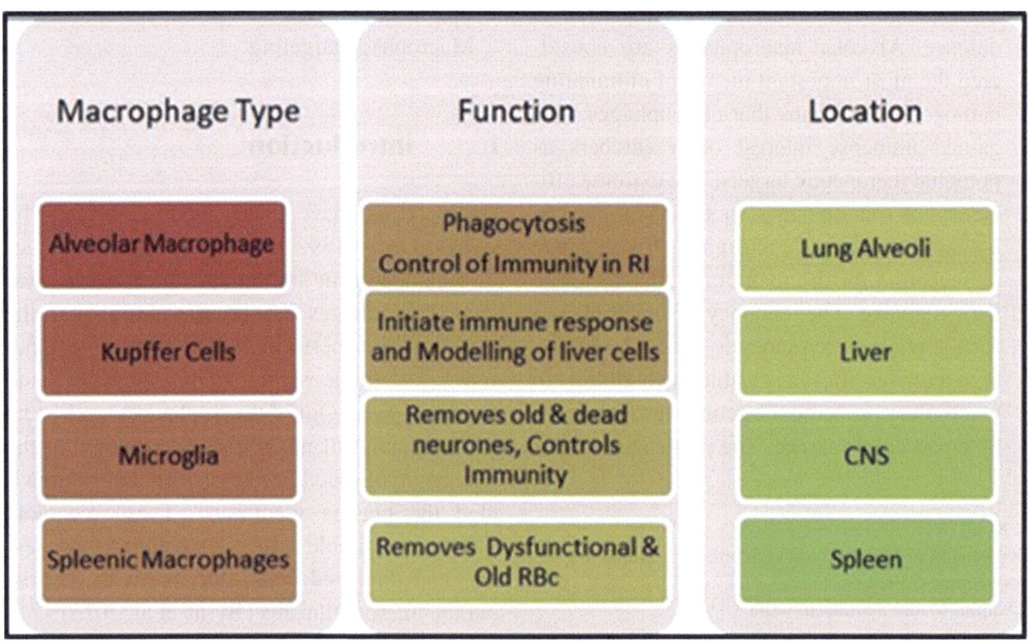

Fig. 1 Location and function of macrophages

1.1 Respiratory System

Respiratory system is incessantly opened to a number of microorganisms and air pollutants that may influence the normal functioning of the system. Each day the respiratory airways and alveoli are exposed to large volumes of air containing countless dust particles, chemicals plus infection-creating material. In spite of this, the lungs are sterile and are maintained by filtering and cleaning mechanisms – nasal clearance and tracheobronchial clearance. This organ system is strengthened with a strong immune retort – innate and acquired immunity – that presents a vital role in defense.

Innate Immunity: It is an important component of host defense mechanism. Innate immunity is initiated within hours and provides a rapid array of defenses.

Adaptive Immunity: Adaptive immunity is initiated on an antigen. Specific response is induced after a few weeks of infection.

The short-lived innate immunity cells, for example, neutrophils in lung infections or acute respiratory problems, provide protection from any kind of damage. Even epithelial cells present in the respiratory system also constitute an important part of the immune system. Respiratory system possesses a diversified epithelium cells.

- Cubic and non-ciliated cell at lower airways.
- Pseudo-stratified columnar and ciliated cells at upper airways.
- Tracheobronchial tree: non-ciliated secretory cell for mucous production, ciliated cells intercalated mucous secretory cell, and non-mucus secretory cells.

The epithelial cell forms the foremost shield against wide environmental factors and microorganisms and thus acts as a link between innate immune response besides adaptive immune reactions (Gordon and Read 2002).

2 Population of Respiratory Macrophage

There are at least two populations of respiratory macrophages that exist in lungs. These are characterized by their location, function, and properties as:

2.1 Alveolar Macrophages (AMs)

In fact, alveolar macrophages (AMs) are a kind of white blood cells. They are also known as dust cells. AMs constitute the baseline of defense against invading respiratory pathogens. AM forms 95% of the cell burden in broncho-alveolar lavage, the remainder being lymphocytes. All cell type of mononuclear phagocyte system originates from the hematopoietic stem cells present in bone marrow and ends up in monocyte. The monocytes have a half-life of a day, whereas the macrophages can live up to months or year in a tissue. The size and shape of AM may differ in accordance with the conditions. Generally, the alveolar macrophages phenotypes have been acknowledged.

- Activated macrophage (M1 macrophage).
- Alternatively activated macrophage (M2 macrophage).

The M1 macrophages react to microbial factors and Th1 proinflammatory cytokines that are allied with inflammatory cytokine release and boosted bacterial killing. They are also related to employment of immunity cells in the parenchymatous layer of the lung and also in alveolus. Compared to M1 macrophages, the M2 macrophages are tempted by acquaintance to the Th2 cytokines to undergo oxidative metabolism. The M2 macrophages are linked with anti-inflammatory cytokine release, phagocytosis of apoptotic cells (efferocytosis), and collagen deposition, responsible to resolve the inflammation and healing of injured tissues.

AMs have an indispensable position in scavenging various infecting organisms – bacteria, virus, fungi, and other environmental pollutants, dust particles, tissue debris, etc.

AMs perform their defense action by receptor-mediated phagocytosis using toll-like receptors (TLR) that interact with pathogen-associated molecular pattern receptor (PAMP) present on the surface of micro-organisms. AMs have been shown to mediate innate response to *Mycobacterium tuberculosis* and *Streptococcus pneumoniae*. AMs are active producers of cytokines and leukotriene, and possess paramount action as pro and anti-inflammatory activity.

2.2 Interstitial Macrophages (IMs)

The interstitial macrophage (IM) comprise 30–40% of lung macrophages. IMs are situated in parenchyma layer of lungs. It is the fine wall of alveolus that separates alveolar macrophages from interstitial macrophages, which occupy the same narrow interalveolar space as alveolar capillaries, fibroblasts, and other mesenchyme cells. Interstitial macrophages are believed to possess key functioning in the remodeling of tissue and maintenance along with antigen presentation and influencing dendritic cell functions to prevent airway allergy. Contemporarily, the studies endorse that interstitial macrophages accomplish vital immune functions, comprising the upkeep of lung homeostasis and preclusion of immunity intermediated allergic inflammation in the airways.

2.3 Dendritic Cells

Dendritic cells (DCs) are the most professional antigen-presenting cells (APCs). They are immune cells that sustains the innate and acquired immunity. Dendritic cells are considered as professional APCs because of their ability to induce the activation and differentiation of naïve T lymphocytes. DCs share common substrate with AMs. DCs are present in the cells which exist in contact with external aura, such as skin, lungs, nose, stomach, and intestines. They develop finger like projection called dendritic during their development, hence received the term DC. A dendritic cell holds the ability to induce a primary immune reaction in the inactive naïve T lymphocytes. Apart from this, DCs also endow the function of the B cells and preserve their immune memory. However, DCs are poor phagocytes (Simons et al. 2010).

2.4 Monocytes

Monocytes and macrophages are well related to one another. In very simple words, monocytes are macrophages in blood while macrophages are monocytes in tissues.

Monocytes are present in the circulating blood, bone-marrow, and spleen. Monocytes are equipped with chemokine and pathogen-recognizing receptors that moves from blood to tissue during the condition of infection. Monocytes are betrothed in the cytokines (inflammatory) production and furthermore they take up cells and toxic molecules (Stephen et al. 2016). Monocytes in lungs act as reserve component of immune system that prevent infection. In lungs, monocytes are able to differentiate into DCs and macrophages.

3 Respiratory Tract and Infection

The respiratory tract is the most easily accessible site prone to infections by pathogens. Annually, children acquire two to five respiratory infections, whereas the adults are attacked by just one or two. This is because of the fact that respiratory tract or the airways is exposed to the outside environmental conditions like dust, moisture, and also open to various pathogenic organisms (Blasi et al. 2004). An extensive range of micro-organisms can infect the respiratory tract, including viruses, bacteria, fungi, and parasites are presented in Table 1.

Table 1 Microbial infection affecting respiratory system

S No.	Disease	Pathogen	Sign and symptoms	Transmission
1	Acute otitis media (AOM)	*Haemophilus influenzae*, *Streptococcus pneumoniae*, *Moraxella catarrhalis*, others	Earache, possible effusion; may cause fever, nausea, vomiting, and diarrhea	Often a secondary infection; bacteria from respiratory tract become trapped in eustachian tube, cause infection
2	Diphtheria	*Corynebacterium diphtheria*	Pseudomembrane on throat, possibly leading to suffocation and death	Inhalation of respiratory droplets or aerosols from infected person
3	Legionnaires disease	*Legionella pneumophila*	Cough, fever, muscle aches, headaches, nausea, vomiting, and confusion; sometimes fatal	Inhalation of aerosols from contaminated water reservoirs
4	Pertussis (whooping cough)	*Bordetella pertussis*	Severe coughing with "whoop" sound; chronic cough lasting several months; can be fatal in infants	Inhalation of respiratory droplets from infected person
5	Q fever	*Coxiella burnetii*	High fever, coughing, pneumonia, and malaise; in chronic cases, potentially fatal endocarditis	Inhalation of aerosols of urine, feces, milk, or amniotic fluid of infected cattle, sheep, and goats
6	Streptococcal pharyngitis, scarlet fever	*Streptococcus pyogenes*	Fever, sore throat, inflammation of pharynx and tonsils, petechiae, swollen lymph nodes; skin rash (scarlet fever), and strawberry tongue	Direct contact, inhalation of respiratory droplets or aerosols from infected person
7	Tuberculosis	*Mycobacterium tuberculosis*	Formation of tubercles in lungs; rupture of tubercles leading to chronic, bloody cough; healed tubercles (Ghon complexes) visible in radiographs; can be fatal	Inhalation of respiratory droplets or aerosols from infected person
8	Chlamydial pneumonia	*Chlamydophila pneumoniae*, *C. psittaci*, *chlamydia trachomatis*	Bronchitis; mild-to-severe respiratory distress	Inhalation of respiratory droplets or aerosols from infected person (*C. pneumoniae*); exposure to infected bird (*C. psittaci*); exposure in the birth canal (*chlamydia trachomatis*)

The internal structure of the URT (upper respiratory tract) assists to purge the system of physical particulate matter and disease-causing microbes. The nasal cavity has a mucus and ciliary lining similar to that of the LRT (lower respiratory tract). The inner part of the nose is lined with hairs, which perform to filter larger particles that are breathed in. The turbinate bones ("baffle plates") are shielded with mucus that gathers particles which are not filtered by nasal hairs. As the inhaled air moves from nasal passage, it changes direction and results in larger airborne particles to meddle on the rear of the throat. The adenoids and tonsils constitute an imperative portion of URT immune system.

The lower respiratory tract has sheath of mucus and is provided with cilia cells. In case the infectious agents approach the LRT then by virtue of the ciliary action they are wafted upwards to the throat. Furthermore, the sneeze and cough reflexes constitute an important defense mechanisms for respiratory tract in clearing the dust and other pathogens (courses.lumenlearning.com).

3.1 Microbiota

Virtually, all of the parts of the respiratory tract, explicitly nasal and oral passages, nasopharynx, oropharynx, trachea, bronchi, bronchioles, and alveolar sacs are populated by the host microbiota. These organisms are occupants of the respiratory tract and seldom cause disease. The microbiota of the respiratory passage performs the following functions that assist in boosting the healthy status of the host:

1. Firstly, competition of microbiota with pathogens for prospective attachment sites.
2. Secondly, microbiota has property to produce bactericidal substances and hence protect from infection by pathogens.

Earlier it was presumed that microbiota are absent in the lower respiratory tract. However, from the present research work, this has been learnt that there are about 10–100 bacterial cells per 1000 lung cells.

The microbiota in LRT (lower respiratory tract) grows on different media as used by commonly growing microbes. Therefore, it is apparent that predominantly lung microbiota emanates from the oropharynx region only.

In the LRT (lower respiratory tract) alveolar macrophages usually eliminate the aspirated pathogenic organisms. Alveolar macrophages are considered the most important means of eliminating pathogenic organisms that enter the lungs (Man et al. 2017; Dickson et al. 2016).

To protect themselves from the alveolar macrophages, most pathogenic bacteria (e.g., *Streptococcus pneumoniae, Klebsiella pneumonia,* and *Haemophilus influenzae*) produce a casing that constrains phagocytosis. Other pathogenic microorganisms also escape alveolar macrophage killing either by reproducing/surviving in the cells lining of the respiratory tree (e.g., Influenza virus) or by reproducing/surviving in the alveolar macrophages (e.g., *Mycobacterium tuberculosis*).

4 Respiratory Disorders/Diseases

Respiratory conditions impose an enormous burden on society. According to the WHO World Health Report (2000), there are five topmost respiratory diseases that account for 17.4% of all deaths and 13.3% of all disability-adjusted life years (DALYs). Lower respiratory tract infections, tuberculosis, chronic obstructive pulmonary disease (COPD), and lung cancer are among the leading 10 causes of death worldwide. Various disorders and diseases can undesirably infect the respiratory tract and hence its functioning. These disorders may be congenital or acquired. A congenital respiratory disorder impacts the infant in the uterus before their birth. Hyaline membrane disease or cystic fibrosis exemplifies this problem.

Acquired respiratory diseases and disorders may happen at any stage of life (Fig. 2).

4.1 Understanding Respiratory Disorders

The chronic diseases that affect the respiratory system are the diseases that are long affecting the air passage and associated structures of the lung. Commonly occurring disorder includes asthma, COPD, pulmonary hypertension, and other occupational lung diseases.

4.2 Asthma

Asthma is one of the major noncommunicable diseases. It is a chronic disease of the air passages of the lungs which inflames and narrows them. WHO estimates that about 235 million people suffer from asthma. The disease is characterized by breathlessness, wheezing, and cough. Asthma impacts the people even in childhood or adult younghood. Asthma may possibly occur due to some allergenic reactions as well (Fricker et al. 2017). The pathogenesis of asthma is depicted in Fig. 3.

Fig. 2 Respiratory diseases and disorders

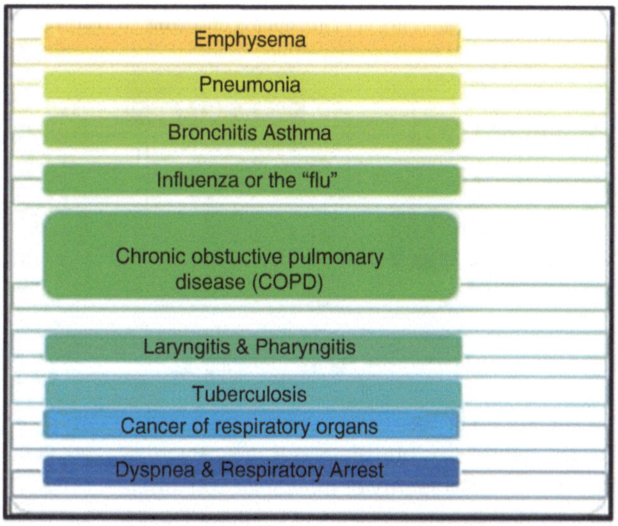

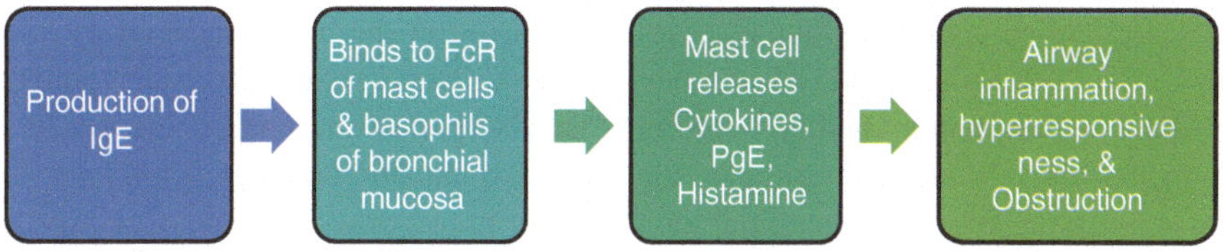

Fig. 3 Pathogenesis of asthma

4.3 COPD

COPD is considered not to be a solitary disease rather embraces a collection. The term chronic obstructive pulmonary disease (COPD) speak of a group of conditions – emphysema, chronic bronchitis, and bronchiectasis that are associated with repeated obstruction to air flow within the lungs (WHO fact sheet). The increase in environmental pollution, cigarette smoking, and exposure to other noxious material has significantly raised the incidences of COPD. The more a person smokes, the chances of COPD increases. But in case of non-smokers, due to the absence of alpha-1 trypsin, emphysema develops. COPD makes breathing difficult. There are predominantly two forms of COPD:

- Chronic bronchitis, which implicates a long-term cough along with mucus.
- Emphysema, in which there is damage to the lungs over a period of time.

The most common symptoms of COPD are breathlessness (or a "need for air"), chronic cough, and sputum (mucous) production (Barnes et al. 2015).

Several mechanisms for COPD pathogenesis have been brought forward; the most relevant one involves protease–anti protease imbalance, in particular, alpha 1 antitrypsin deficiency. The COPD pathophysiology embraces increased number of AMs and increased secretion of chemokines, cytokines, and elastolytic enzyme (Abboud et al. 2008).

4.4 Pulmonary Fibrosis

Pulmonary fibrosis (PF) in a very simple manner is scarring of the lungs. And this, over a prolonged period of time, damages the lungs, eventually posing a problem for oxygen to mix into the blood, and the patient feels difficulty in breathing, mainly while walking and exercising.

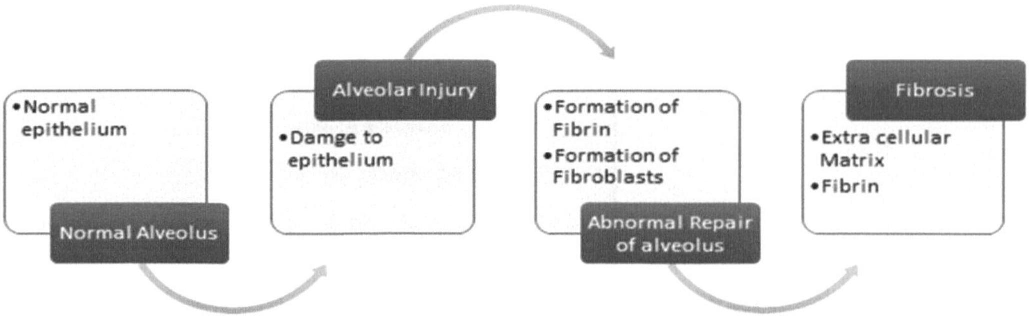

Fig. 4 Process of pulmonary fibrosis formation

In fact, pulmonary fibrosis is not mere one disease, rather it is a family of more than 200 different lung diseases that appears to be alike. The PF family of lung diseases falls into an even larger group of diseases called interstitial lung diseases (also known as ILD), encompassing all the diseases that have inflammation and/or scarring symptoms in the lung. Some ILD excludes the presence of scar tissue. When an interstitial lung disease does include scar tissue in the lung, it is named as pulmonary fibrosis. The scars in lungs are caused by exposure to different environmental agents like silica, asbestos, hard metals dust, etc. (Putter et al. 2019). The process of PF formation is given in Fig. 4.

4.5 Lung Cancer

Lung cancer is a type of malignancy that affects the lungs. Lung cancers typically commences by making alterations in the bronchial cell lining and then progressing to lungs. The lung cancer is specified as – small cell lung cancer (SCLC) and non-small cell lung cancer (NSCLC). Causes of lung cancer include smoking, second-hand smoke, exposure to certain toxins, and family history. Symptoms include a cough (often with blood), chest pain, wheezing, and weight loss. These symptoms often don't appear until the cancer is advanced.

The treatments may differ but comprise surgery, chemotherapy, radiation therapy, targeted drug therapy, and immunotherapy.

Macrophages are the determining portion of infiltrating leukocytes in all tumors, where they are demarcated as tumor-associated macrophages (TAMs) and are predominately characterized by an M2-like phenotype (Lewis et al. 2006). They are mainly obtained from circulating monocytes, employed at the tumor location by chemotactic factors. Tumor-associated macrophages (TAM) are also present in lung cancer, possessing both pro and anti-tumor effect. TAMs are engaged in tumor in response to chemotactic aspects and are present in oxygen-deficient area of the tumor. The study performed by Pouniotis et al. (2006) evaluated the functional activity of AMs and phenotype in different primary lung cancers (e.g., squamous cell carcinoma, adenocarcinoma, and small-cell lung carcinoma plus large-cell undifferentiated lung carcinoma) with a purpose to assess if immune response of AMs changes according to the type of tumor (Fig. 5).

5 Way to Reach Macrophages

Macrophages are immensely adaptable in nature and has developed as professional phagocytes. Their phagocytosis capacity is pivotal for the uptake and degradation of infectious agents and senescent or damaged cells of the body. This makes them essential cells in tissue remodeling and repairs, as well as key players in immune and inflammatory reactions. The mode of activation of the macrophage that can occur after engagement of the multiple types of receptors is dependent on:

- Composition of pathogen.
- Properties of the particle to be internalized.

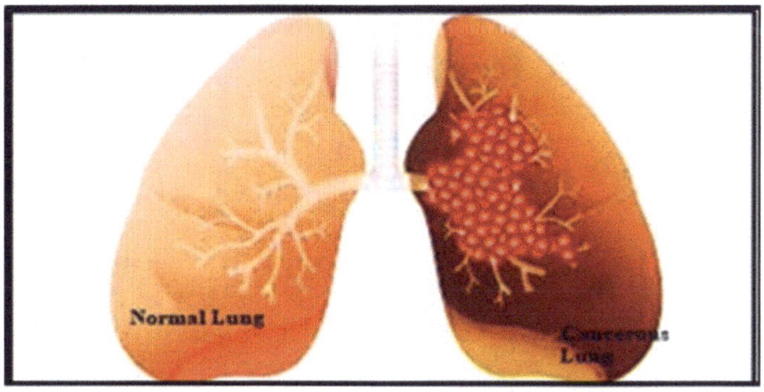

Fig. 5 Normal lung and cancerous lung

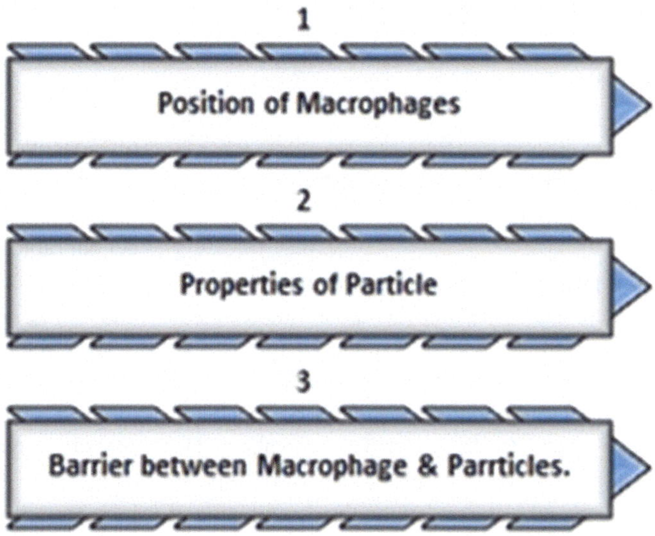

Fig. 6 Factors affecting accessibility of macrophages

The accessibility of different macrophages to the molecules/molecular complex/particulate carrier is reliant on certain factors (Fig. 6):

All macrophages are accessible to small molecules when these are able to pass the capillary networks in order to penetrate into the tissues. But for larger molecules, molecular complexes, or particles, the approach to macrophage can be made easily if there is no physical obstruction between the injection location and the macrophage. For example, the barrier could be formed by endothelial cells in the wall of blood vessels, by alveolar epithelial cells in the lung, and by reticular fibers or collagen fibers in the spleen. Also the closely packed cells like lymphocytes in the white pulp of the spleen or in the paracortical fields of lymph nodes constitute the physical barrier. This physical barrier can be kept at a minimum by selection of the suitable injectable route for the administration of the materials. For appropriate reach to macrophages, liposomes have been studied to be suitable for targeting macrophages via adjustable routes (Fig. 7). Liposomes, the artificially prepared phospholipid vesicles, encapsulating the bisphosphonate clodronate can be used to deplete macrophages in various organs and tissues (Rosen et al. 2012).

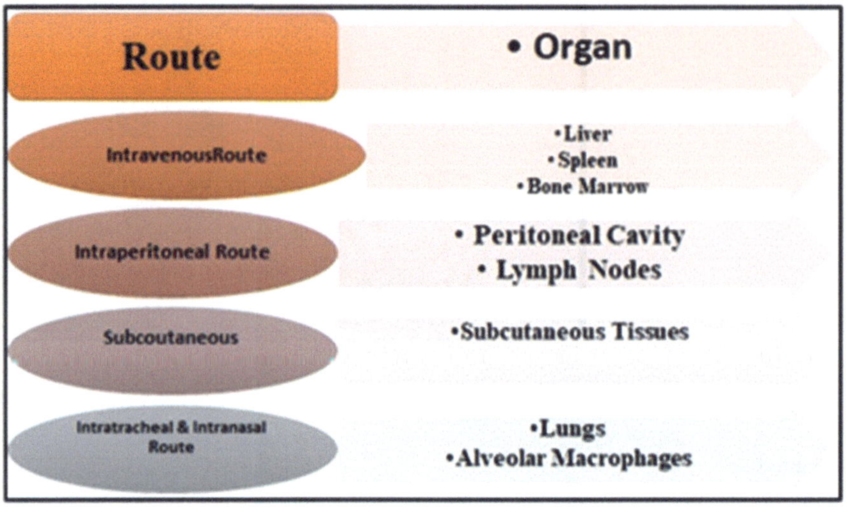

Fig. 7 Routes of administration to reach macrophages

6 Macrophage as Therapeutic Target

Macrophages have gained immense interest to researchers as potential therapeutic targets. The extreme efficiency of macrophages in terms of particle uptake may be attributed to the occurrence of large count of receptors on the membrane that facilitate their reuptake. The macrophages are activated by multiple type of receptors and is dependent on the composition of the pathogen or particle to be internalized. The various macrophage receptors involved in targeting are:

- Sialoadhesin receptors.
- Folate receptors.
- Galactose receptors.
- Mannose receptors (MRs).
- β-Glucan receptors,
- Scavenger receptors.
- Tuftsin receptors.
- Carboxylesterase-1, macrophage enzyme (Veaceslav et al. 2013).

Macrophage's reactions to stimuli like tissue dependent specialization adds to the complexity of the cell and this constitute an imperative factor in targeting. Besides the phagocytic capacity of macrophages also causes unwanted degradation of bioactive products prior to their destination, the target site cells. The proficient uptake of particles by macrophages makes them a perfect target but rapid degradation may lead to inept effect of bioactive material or undesirable effect on macrophages.

The aptness of macrophages for the phagocytic/pinocytic elimination of external particles offers a sensible methodology to macrophage-specific targeting through suitable particulate carrier. The particulate targeting of macrophages is noteworthy as multiple delivery systems are available viz. liposomes, niosomes, polymeric nanospheres, and oil-in-water microemulsions.

The particle with variable physical and chemical characters and loading efficiency can be designed and tailored in these systems (Kraal and Rooijen 2012).

The particulate entities in use include liposomes, niosomes, polymeric nanospheres, oil-in-water microemulsions, and even natural constructs such as lipoproteins and erythrocytes.

Macrophages attain several phenotypes in accordance with the subjected environmental stimuli. Because of this reason, macrophages present challenge in design as therapeutic target. Therefore, copious tactics have been implemented with the objective of manipulating or redesign the macrophages and their re-education have been considered as the most significant stride in this direction. Here, nanotechnology-based approaches have been implicated with prominent result in macrophage targeting.

6.1 Nanocarriers for Macrophage Targeting

Nanotechnology is an innovative technology that is bashing at the door of therapeutic drug delivery system. It has wider applications and is the central focus for many technologies to converge and open a large number of applications. It deals with things smaller than 100 nanometers size. Nanos means dwarf. This technology is concerned with material science and its usages at the nano meter scale (one billionth of 1 m). Nanoparticle (NP) is a microscopic particle whose size is measured in nanometers. It is defined as a particle with at least one dimension less than 100 nm. Nanoparticles are often referred to as clusters. Nanospheres, nanorods, and nanocups are few shapes that have developed. The nanoparticles have a greater surface area per weight than larger particles which makes them to be more reactive to some other molecules.

Nanoparticles are presented in the form of suspensions, colloids, or dispersed aerosols depending on their chemical and electromagnetic properties. Nanoparticles can be tailor-made for their dimensions and surficial features for entrapment of drug. The nanostructures have the ability to enter cells that typically internalize materials below 100 nm. When incorporated materials are produced from nanoparticles in the 1–100 nm size range instead of bigger microparticles, they have an enormous surface area for the same volume, smaller pore size, improved solubility, and altered structural properties. These features can ameliorate the diffusion and degradation properties of loaded drug.

6.1.1 Criteria for Macrophage Targeting

For targeting, three criteria must be considered.

- The primary dimension to be considered is the distribution of accessible macrophages in tissues to various physiological entrances.
- The second involves elements of phagocytic recognition and ingestion that includes macrophage phagocytic/endocytic receptors regarding their nature, density, and their state of activation and also the influence of environmental aspects on phagocytic utilities of macrophages.
- The third criterion is the physicochemical characteristics of the particles to be ingested and includes particle morphology, hydrodynamic size and surface characteristics exemplified by ligand expression, and bound opsonins.

6.1.2 Lipid-Based Nanoparticles

Liposomes

Liposomes are the most compatible nanocarrier system for the delivery of drugs and cosmetics. They have the potential to enhance the drug delivery, improving bioavailability, and efficacy of the drug. Besides, liposomes also reduce the adverse reactions and promote drug stability and drug release properties. Liposomes are colloidal carriers, having a size of 0.01–5.0 μm in diameter. Actually, liposomes are bilayered vesicles that are formed when phospholipids are hydrated in excess of aqueous medium. They have the capacity to encapsulate hydrophilic along with hydrophobic drug and ability to target the drug moiety at the desired site of action. Liposomes have varied physical and chemical properties depending on their composition and formulation parameters. The major therapeutic applications of liposomes in drug delivery are summarized in the Fig. 8.

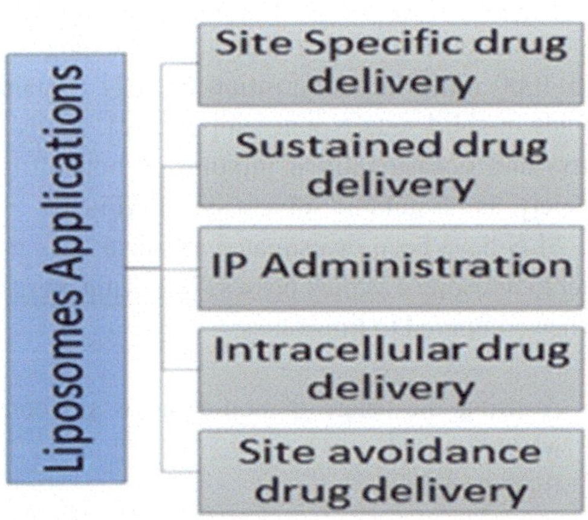

Fig. 8 Applications of liposomes in drug delivery

Liposomes have been comprehensively envisioned for the targeting of alveolar macrophages (AM) by coating them with specific ligands having affinity for the mannose receptors (Smola et al. 2008) In a substantial study made by Vyas et al. (2000), mannosylated liposomes showed a good uptake by AMs when compared with uncoated liposomes or simple drug solution. In a similar attempt made by Vyas et al. (2005), it was observed that amphotericin B loaded liposomes coated with O-palmitoyl mannon and O-palmitoylated pullulan as ligands when administered through aerosol showed high accumulation in lungs as compared to plain liposomes and free drug.

AMs targeting employing two different approaches were conducted by Vyas et al. (2004). This was done by modifying lipids of formulation-making negative charge to liposomes by incorporating dicetylphosphate and coating phospholipids with specific ligands for AM targeting. Conjugation with phospholipid tails was made employing O-steroyl amylopectin (O-SAP) and maleylated bovine serum albumin (MBSA). The ligand-mediated liposomes exhibited high accumulation in lungs. Shao and Ma (2008) investigated the mannose-conjugated liposomes for intracellular drug delivery to AMs. A glycosylphospholipid with a terminal mannose moiety was synthesized and used in the preparation of liposomes. The effect of the conjugated mannose on the uptake of liposome by AMs was investigated.

Solid Lipid Nanoparticles (SLN)

Solid lipid nanoparticles (SLNs) are commonly spherical in shape with a diameter in the range of 50–1000 nm. The composition of SLN contain lipids in solid state (room temperature), emulsifiers and occasionally a mixture of both drug moiety, and a suitable solvent system (Fig. 9).

SLN have been investigated as substitute carrier to liposomes as they possess following merits when compared to liposomes:

- Averting the usage of non-aqueous solvents when required.
- Biodegradable.
- Large-scale production is achievable.

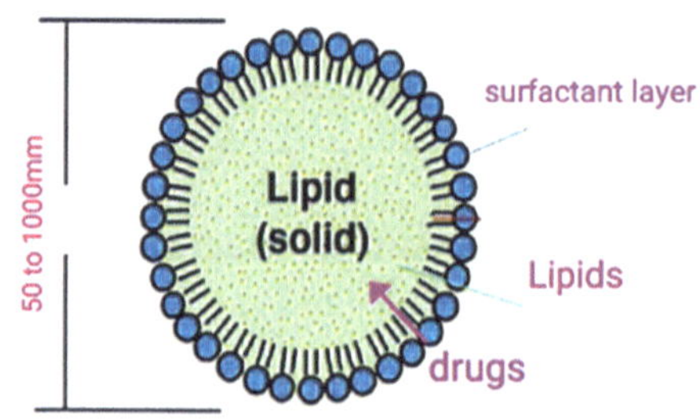

Fig. 9 Structure of Solid Lipid Nanoparticle

- Stability of active ingredient.
- Greater entrapment efficiency of hydrophobic drugs.
- Controlled and targeted release of drugs is feasible using binding with ligands.

SLN can be functionalized with mannose residues through the opening of mannose ring, followed by the reaction with aldehyde group of mannose with free amines of lipids, presented on the surface of the particles (Jain et al. 2010). In vivo studies completed by Sahu et al. (2015) revealed that mannosylated SLN had higher amassing in lung tissues as compared to control (Rhodamine-B-loaded placebo SLN). Furthermore, the percentage of paclitaxel recovery in lungs was greater with mannosylated SLN (26%) than with plain SLN (18%) and plain drugs (4%) after 24-h of intravenous administration, signifying that mannosylated SLN is a worthy tactic for targeting lung diseases. SLN functionalized with mannose or other related compounds, such as mannan also showed ability to target AMs with a favorable in vitro cell uptake than with bare SLN, however, using functionalized SLN for alveolar targeting remains to be explored (Yu et al. 2010).

Polymeric Nanoparticles for AMs Targeting

Polymeric nanoparticle (NPs) ranges from 1 to 1000 nm in size. They can be loaded with active pharmaceutical ingredient either entrapped inside or surface adsorbed in the polymeric core. These polymeric nanoparticles have significant role in targeting AMs. The alveolar macrophages (AMs) have mannose receptors (MRs) that have the capability to recognize the glycoprotein residues, where terminal ending constitutes of glucose, L-fucose residues, and N-acetyl-D-glucosamine (Aleksandra et al. 2020). In this context, Song et al. (2012), synthesized a glycopolymer using functionalized carbohydrates as precursors. Ethyl methacrylate (EMA) was used to functionalize each sugar monomer's subsequent polymerization and conjugation with maleimide-containing fluorophore. Glycopolymers containing mannose-EMA and N-acetylglucosamine-EMA were internalized by AMs after administration in mice, but glycopolymers containing galactose-EMA had less internalization.

In a study conducted by Park et al. (2013), it is established that ofloxacin-loaded-glutaraldehyde-cross-linked chitosan microspheres were able to be internalized by AMs, and the uptake was 3.6 times greater than blank powder. Chitosan is a biodegradable and bioadhesive polymer having positive charge constituted by monomers of N-acetyl-Dglucosamine and β-linked D-glucosamine. That accounts for electrostatic interaction with negative cell membrane. The macrophages could recognize and capture chitosan particles due to the presence of N-acetylglucosamine residues on chitosan.

In another research, it has been reported that coating of particles with chitosan offers stealth characteristic and consequently the phagocytosis by macrophages was reduced (Samento et al. 2011).

For effective delivery into alveolar macrophages, the surficial properties and size of the particles plays an imperative role. Globally, researchers are attempting enduringly to cultivate novel therapeutic systems for AMs targeting. Mannose receptors are mainly considered for targeting to AMs.

Macrophage Targeting in Lung Cancer

Tumor-associated macrophages (TAMs) are among the most profuse immune cells in the tumor microenvironment (TME) (Casetta et al. 2018). They were thought to have anti-tumor activity at first because of their capacity to destroy tumour cells in in vitro tests. In fact, in early cancer stage, the immune system is supposed to boost activation of T cells and macrophages seek to remove cancer cells. But once the tumor advances, the TME is affected by tumor cells to support for their growth. However, the antitumor macrophages may be present even in this condition and these may be "educated" to augment tumor progression and metastasis. Macrophages are important to mediate an immune response against tumor cells, by displaying tumor-associated antigens, cytokines release, and stimulation of antitumor lymphocytes, but these host-defense mechanisms can be circumvented by cancer cells. The influence of macrophages on tumor-induced immune suppression is still controversial. Tumor-associated macrophages (TAM) are also presented in lung cancer, having both pro-tumor and anti-tumor effects. The macrophage functions are determined by microenvironment location, tumor stage, and cancer type, and in most tumors, a high prevalence of TAMs is related to a poor prognosis. Macrophages exist as pro- and anti-inflammatory phenotypes. In the initial phase of cancer, the macrophage exist as pro-inflammatory phenotype but after progression of cancer, the phenotype changes to anti-inflammatory nature (Lewis et al. 2006).

Inventive research has been conducted on inflammatory diseases of the respiratory system, for example, tuberculosis and chronic obstructive pulmonary disease (COPD). In a study conducted in mouse having COPD initiated by cigarette smoke, the role of macrophages in the pathogenesis of COPD was established (Beckett et al. 2013). The intranasal administration of clodronate liposomes resulted in reduced smoke-induced epithelial thickening and emphysema growth. In a similar study made by Fritz et al. (2014), it was observed that administration of clodronate-encapsulated liposomes resulted in depletion of AMs and pulmonary macrophage in mice exposed to carcinogens. Alveolar macrophage populations declined to less than 50% of control levels after 4–6 weeks of liposomal clodronate treatment. Tumor load lessened by 50% as compared to mice treated with plain vehicle, and the spread of tumorous cell was also weakened. Tumor-associated macrophages (TAM) indicated markers of both M1 and M2 in vehicle and clodronate liposome-treated mice. Mice short of CCR2 (the receptor for macrophage chemotactic factor

CCL2) had equivalent counts of alveolar macrophages. There was no obvious variation in progression proportions of tumor when matched to similarly treated wild-type mice. This finding advocates that while CCL2 may employ macrophages to lung tumor microenvironments, redundant passageways can compensate when CCL2/CCR2 signaling is not active. Enfeeblement of pulmonary macrophages in spite of inhibition of their employment can possibly be a beneficial tactic for mitigating lung cancer progression.

7 Final Remark

An understanding of role of respiratory macrophages – alveolar macrophages, dendritic cells, interstitial macrophages (IMs), and monocytes; ontogeny and their functional activities has achieved wide recognition in improving the therapeutic efficacy of drugs in various respiratory diseases – from inflammation to cancer. A number of studies have been conducted on targeting the macrophages for therapeutic purpose. Nanotechnology-based approaches have been extensively appraised for targeting macrophages in therapy of cancer and other diseases. It also apparent that macrophage-targeted therapies are inherently characterized by certain limitations. The effectiveness is related to the pathogenesis progress and stages of disease. Therefore, studies on the dynamics of macrophage activation and movement in cancer and other inflammatory disease are needed to be honed for efficient targeting. Another challenge is phenotypic reversal and re-population of macrophages in re-education of macrophages. This may occur even after cessation of the therapy. Furthermore, the complete challenge that nanotechnology-based systems have to face is signified by their poor selectivity.

In macrophage-determined therapies, the creation of nano-carrier system adorned with macrophage-specific ligands and containing combination of suitable drug molecules is a strategic step for both macrophages abolishment and re-programming.

A future insight is demanded with respect to toxicity and stability of nanoparticles and feasibility of their scaling up is desirable.

References

Abboud RT, Vimalanathan S, et al. Pathogenesis of COPD. Part I. the role of protease-antiprotease imbalance in emphysema. Int J Tuberc Lung Dis. 2008;4:361–7.

Aleksandra Z, Filipa C, Ana MO, Andreia N, Bárbara P, Nagasamy VD, Alessandra D, Massimo L, Piotr E, Amélia MS, Antonello S, Eliana B, et al. Souto polymeric nanoparticles: production, characterization, toxicology and ecotoxicology. Molecules. 2020;25:3731.; Available at. https://doi.org/10.3390/molecules25163731.

Anatomy and Normal Microbiota of the Respiratory System. https://courses.lumenlearning.com/microbiology/chapter/anatomy-and-normal-microbiota-of-the-respiratory-tract/#:~:text=Of%20the%20organisms%20identified%20in,or%20if%20they%20are%20transients

Asthma – key facts. https://www.who.int/news-room/fact-sheets/detail/asthma

Barnes P, Burney P, Silverman E, et al. Chronic obstructive pulmonary disease. Nat Rev Dis Primers. 2015;1:15076. https://doi.org/10.1038/nrdp.

Beckett EL, Stevens RL, Jarnicki AG, Kim RY, Hanish I, Hansbro NG, Deane A, Keely S, Horvat JC, Yang M, et al. A new short-term mouse model of chronic obstructive pulmonary disease identifies a role for mast cell tryptase in pathogenesis. J Allergy Clin Immunol. 2013;131:752–62. Available at. https://doi.org/10.1016/j.jaci.2012.11.053.

Blasi F, et al. Atypical pathogens and respiratory tract infections. Eur Respir J. 2004;24:171–82.

Byrne AJ, Mathie SA, Gregory LG, Lloyd CM, et al. Pulmonary macrophages: key players in the innate defence of the airways. Thorax. 2015;70:1189–96.

Cassetta L, Pollard JW, et al. Targeting macrophages: therapeutic approaches in cancer. Nat Rev Drug Discov. 2018;17:904.

Dickson RP, Erb-Downward JR, Martinez FJ, Huffnagle GB, et al. The microbiome Rev Physiol. 2016; 78:481–504. Available at. https://doi.org/10.1146/annurev-physiol-021115-105238.

Fredric G, Markus GM, Michael HS, Miriam M, Klaus L, et al. Development of monocytes, macrophages and dendritic cells. Science. 2010;327(5966):656–61.

Fricker M, Gibson PG, et al. Macrophage dysfunction in the pathogenesis and treatment of asthma. Eur Respir J. 2017;50:1700196. Available at. https://doi.org/10.1183/13993003.00196-2017.

Fritz JM, Tennis MA, Orlicky DJ, Lin H, Ju C, Redente EF, Choo KS, Staab TA, Bouchard RJ, Merrick DT, Malkinson AM, Dwyer-Nield LD, et al. Depletion of tumor-associated macrophages slows the growth of chemically induced mouse lung adenocarcinomas. Front Immunol. 2014;25(5):587. Available at. https://doi.org/10.3389/fimmu.2014.00587.

Gordon SB, Read RC. Macrophage defences against respiratory tract infections. Br Med Bull. 2002;61:45–61.

Guochang H, John WC. Alveolar macrophages in lung inflammation and resolution. Front Immunol.

2019.10.2275. Available at. https://doi.org/10.3389/fimmu.2019.02275.

Guorong H, Mengfei G, Juanjuan X, Feng W, Jinshuo F, Huang Q, Guanghai Y, Zhilei L, Xuan W, Yang J, et al. Nanoparticles targeting macrophages as potential clinical therapeutic agents against cancer and inflammation. Front Immunol. 2019;10:1998. Available at. https://doi.org/10.3389/fimmu.2019.01998.

Jain A, Agarwal A, Majumder S, et al. Mannosylated solid lipid nanoparticles as vectors for site-specific delivery of an anti-cancer drug. J Control Release. 2010;148:359–67.

Joey S, Fabrice B, Thomas M, et al. Lung interstitial macrophages: past, present, and future. J Immunol Res. 2018:51–60. Available at. https://doi.org/10.1155/2018/5160794.

Kraal G, Rooijen NV, Reaching the macrophages: routes of delivery in the macrophage as therapeutic target. In: Siamon Gordon, editor. Volume 158 of handbook of experimental pharmacology. Springer; 2012. p. 58–58.

Lei Z, Yi W, Guorao W, Weining X, Weikuan G, Cong-Yi W, et al. Macrophages: friend or foe in idiopathic pulmonary fibrosis. Respir Res. 2018;19:170. Available at. https://doi.org/10.1186/s12931-018-0864-2.

Lewis CE, Polland JW. District role of macrophages in different tumor microenvironments. Cancer Res. 2006;66:605–12.

Lewis CE, Pollard JW, et al. Distinct role of macrophages in different tumor microenvironments. Cancer Res. 2006;66:605–12. Available at. https://doi.org/10.1158/0008-5472.CAN-05-4005.

Man W, Steenhuijsen PW, Bogaert D, et al. The microbiota of the respiratory tract: gatekeeper to respiratory health. Nat Rev Microbiol. 2017:259–70. Available at. https://doi.org/10.1038/nrmicro.

Park J-H, Jin H-E, Kim D-D, Chung S-J, Shim W-S, Shim C-K, et al. Chitosan microspheres as an alveolar macrophage delivery system of ofloxacin via pulmonary inhalation. Int J Pharm. 2013;441:562–9.

Pouniotis DS, Plebanski M, Apostolopoulos V, McDonald CF, et al. Alveolar macrophage function is altered in patients with lung cancer. Clin Exp Immunol. 2006;143:363–72.

Pulmonary fibrosis overview. https://www.pulmonaryfibrosis.org/life-with-pf/about-pf

Puttur F, Gregory LG, Lloyd CM, et al. Airway macrophages as the guardians of tissue repair in the lung. Immunol Cell Biol. 2019;97(3):246–57. Available at. https://doi.org/10.1111/imcb.12235.

Robbins SL, Cortran RS, editors. Pathological basis of disease. Holt: Saunders Company; 1979. p. 814–85.

Rosen H, et al. The macrophage as a validated pharmaceutical target in the macrophage as therapeutic target. In: Gordon S, editor. Volume 158 of handbook of experimental pharmacology. Springer; 2012.

Sahu PK, Mishra DK, Jain N, Rajoriya V, Jain AK. Mannosylated solid lipid nanoparticles for lung-targeted delivery of paclitaxel. Drug Dev Ind Pharm. 2015;41(4):640–9. Available at. https://doi.org/10.3109/03639045.2014.891130.

Sarmento B, Mazzaglia D, Bonferoni MC, Neto AP, Monteiro M, Seabra V, et al. Effect of chitosan coating in overcoming the phagocytosis of insulin loaded solid lipid nanoparticles by mononuclear phagocyte system. Carbohyd Polym. 2011;84:919–25.

Shao J, Ma JKH, et al. Characterization of a mannosylphospholipid liposome system for drug targeting to alveolar macrophages. Drug Deliv. 2008;4(1):43–8. Available at. https://doi.org/10.3109/10717549709033187.

Simons T, Shepherd DM, Moser M, et al. Dendritic cells. In: Comprehensive toxicology, vol. 5. 2nd ed; 2010. p. 155–70.

Smola M, Vandamme T, Sokolowski A, et al. Nanocarriers as pulmonary drug delivery systems to treat and to diagnose respiratory and non-respiratory diseases. Int J Nanomedicine. 2008;3:1–19.

Song E-H, Manganiello MJ, Chow Y-H, et al. In vivo targeting of alveolar macrophages via RAFT-based glycopolymers. Biomaterials. 2012;33:6889–97.

Stephen C, Ankit B, et al. Role of monocytes and macrophages in regulating immune response following lung transplantation. Curr Opin Organ Transplant. 2016;21:239–45.

The world health report 2000 – Health Systems: Improving Performance. https://www.who.int/whr/2000/en/

Tracy H, Bell TJ, Toshifumi F. In: Biswas SK, Mantovani A, editors. Airway macrophages: sentinels of health, macrophages: biology and role in the pathology of diseases. New York/Heidelberg/Dordrecht/London: Springer; 2014. p. 171–82.

Veaceslav B, Valeriu C, Natalia S, Serghei P, Marina Z, Eugenia S, Vsevolod P, Fliur M, et al. Molecular concepts of macrophage targeting. Chem J Mold. 2013;8(2):21–31.

Vyas SP, Katare YK, Mishra V, Sihorkar V, et al. Ligand directed macrophage targeting of amphotericin B loaded liposomes. Int J Pharm. 2000;210:1–14.

Vyas SP, Kannan ME, Jain S, Mishra V, Singh P, et al. Design of liposomal aerosols for improved delivery of rifampicin to alveolar macrophages. Int J Pharm. 2004;269:37–49.

Vyas SP, Quraishi S, Gupta S, Jaganathan KS, et al. Aerosolized liposome-based delivery of amphotericin B to alveolar macrophages. Int J Pharm. 2005;296:12–25.

WHO fact sheets. Chronic obstructive pulmonary disease (COPD). https://www.who.int/news-room/fact-sheets/detail/chronic-obstructive-pulmonary-disease-(copd)

Yongtao D, Abhishek D, Chetan P, Mehul K, Swarnali N, Prolay S, Nadavala Siva K, Rohit LV, et al. A brief review on solid lipid nanoparticles: part and parcel of contemporary drug delivery systems. RSC Adv. 2020;10:26777.

Yu W, Liu C, Liu Y, Zhang N, Xu W, et al. Mannan-modified solid lipid nanoparticles for targeted gene delivery to alveolar macrophages. Pharm Res. 2010;27:1584–96.

Macrophage Targeting for Therapy of Intraocular Diseases

Nagendra Bhuwane, Ishwari Choudhary, Shweta Ramkar, Narayan Hemnani, Abhishek K. Sah, and Preeti K. Suresh

Abstract

The eye is a unique and intricate organ with a persisting challenge to uphold optical clarity and sustain satisfactory neural retina function. The immune defenses in the eye occur at varying microenvironments including the corneal and conjunctival epithelia, uveal pigmented connective tissue, and even the highly protected neural retina. Ocular macrophages have been implicated in homeostasis and number of pathological conditions. In this chapter, the present understanding of ocular immunology with the distribution, phenotype, and the physiological role of the various ocular immune cells is discussed with their implication on ocular pathologies. Various novel strategies for potential macrophage targeting of drug in intraocular diseases are also discussed.

Keywords

Eye · Intraocular drug delivery · Macrophage · Ocular immunology · Ophthalmic

N. Bhuwane · I. Choudhary · S. Ramkar · N. Hemnani · P. K. Suresh (✉)
University Institute of Pharmacy, Pt. Ravishankar Shukla University, Raipur, Chhattisgarh, India

A. K. Sah
Department of Pharmacy, Shri Govindram Seksariya Institute of Technology & Science (SGSITS), Indore, Madhya Pradesh, India

1 Introduction

1.1 Anatomical and Physiological Features of the Eye

Anatomically and physiologically eye is an extremely complex and specialized organ. It is an isolated and exceedingly protected organ, primarily ordained for photoreception. The eye is anatomically divided into two segments, namely anterior and posterior (Mitra et al. 2014; Vadlapudi et al. 2012). The anterior segment includes cornea, conjunctiva, iris, ciliary body, aqueous humor and lens, and the posterior segment includes sclera, choroid, retina, and vitreous body (Agrahari et al. 2016). The front portion of the eye is bound by a clear cornea and a small part of the sclera. The cornea and sclera are connected with each other through the limbus. The cornea lacks blood vessels and sources its nourishment and oxygen from the aqueous humor and tear film, while the corneal periphery receives its sustenance through the limbal capillaries (Dawson et al. 2011). The human cornea is approx. 12 mm in diameter with a thickness of 520 μm and has six layers, namely the epithelium, Bowman's membrane, stroma, Dua's layer, Descemet's membrane, and endothelium (Sridhar 2018). The sclera is a sturdy outer coat mainly comprising connective tissue. The sclera has a protective function and also, by resisting intraocular pressure, maintains the shape of the eye-

S. Gupta, Y. V. Pathak (eds.), *Macrophage Targeted Delivery Systems*,
https://doi.org/10.1007/978-3-030-84164-5_19

ball. The choroid is basically a vascular layer, and in combination with a separate retinal blood supply delivers blood to provide a sustenance to the retinal cells. The retina is separated from the choroid by Bruch's membrane, and it is basically the sensory inner coat in the posterior segment (Goharian and Sehi 2016; Ljubimova 2009).

1.2 The Lachrymal System

The lachrymal system consists, basically, of lachrymal glands, upper and lower eyelids, conjunctival sac, lachrymal puncta, and ducts. The tears, after being secreted by the lachrymal glands, are distributed over the eyes by blinking of the eyelids and subsequently collected in the lower conjunctival sac. The fluid is drained into the lachrymal sac through the puncta and the lachrymal duct. Blinking promotes drainage, while the capillary effect in the lachrymal duct assists this movement into the lachrymal sac. From the lachrymal sac, the fluid passes through the nasolachrymal duct and empties into the highly vascularized inferior nasal meatus. The fluid is finally carried toward the nasopharynx and ultimately ends up into the GIT (Tong et al. 2012; Price and Richard 2009).

1.3 Barriers Restricting Intraocular Drug Transport

1.3.1 Tear

Tear film serves as one of the primary precorneal barriers and it reduces the effective concentration of the administered drugs as a fallout of incessant dilution by the tear turnover (approx. 1 mL/min), faster clearance, and drug binding to the tear proteins. Another limitation is posed by the instilled dosing volume, which is usually 20–50 mL, whereas the volume of the cul-de-sac permits retention of only 7–10 mL. The excess volume may either be spilled out of the eye onto the cheek or exit through the nasolacrimal pathway, whereby it is lost without absorption (Bachu et al. 2018; Gaudana et al. 2010) (Fig. 1).

1.3.2 Cornea

The cornea, with its triple layers of epithelium, stroma, and endothelium, presents a mechanical barrier that limits passage of exogenous constituents into the eye. Each individual layer of cornea has a distinct polarity and also a rate-limiting structure that regulates drug permeation. The corneal epithelium is lipophilic, and cellular tight junctions are formed to limit paracellular drug transport from the tear film. The stroma has an

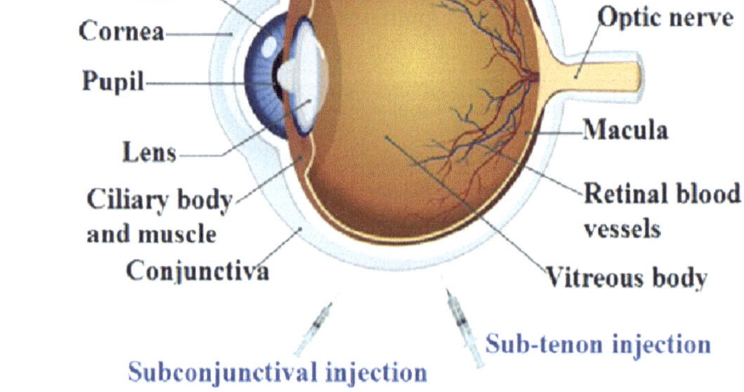

Fig. 1 Anatomical features of human eye and common routes of ocular drug administration

extracellular matrix with collagen fibrils in a lamellar alignment, is extremely hydrated, and presents a barrier for the permeation of lipophilic actives. Corneal endothelium, the innermost monolayer of hexagonal-shaped cells, is a barrier between the stroma and aqueous humor. The endothelial junctions are characteristically leaky and permit the movement of macromolecules amid the aqueous humor and stroma (Solanki et al. 2016; Kaluzhny et al. 2018).

1.3.3 Conjunctiva

The conjunctiva is a thin and transparent membrane of the eyelids and globe, and participates in the development and maintenance of the tear film. With a profuse supply of capillaries and lymphatics in the conjunctiva or episclera, drugs administered in the conjunctival or episcleral area may be emptied via blood and lymph (Singh et al. 2003). The blood vessels of conjunctiva are not engaged in a tight junction barrier indicating that drug moieties can gain access into the blood circulation by pinocytosis and/or convective transport via paracellular pores existing in the vascular endothelial layer (Raviola 1983). The conjunctival lymphatics function as an efflux system for the effective removal from the conjunctival space. Approximately 10% of sodium fluorescein, a hydrophilic and low molecular weight model compound, administered in the subconjunctival space, was eliminated within an hour in rat eyes lymphatically (Lee et al. 2008). Thus, drugs transported by lymphatics in conjunction with the elimination by blood circulation can attribute to systemic exposure, as the interstitial fluid moves back to the systemic circulation subsequent to filtration by lymph nodes.

1.3.4 Sclera

The sclera is primarily comprised of collagen fibers and proteoglycans entrenched within an extracellular matrix (Oyster 1999). Scleral permeability is governed by the molecular radius and scleral permeability shows an exponential decline with molecular radius (Ambati et al. 2006). Also, the posterior sclera has a relatively loose weave of collagen fibers unlike the anterior sclera and the human sclera is comparatively thick near the limbus (0.53 ± 0.14 mm), thinner at the equator (0.39 ± 0.17 mm), and much thicker near the optic nerve (0.9–1.0 mm) (Curtin 1969). Thus, the ideal site for transscleral drug delivery is adjacent to the equator at 12–17 mm posterior to the corneoscleral limbus (Myles et al. 2005). Hydrophobicity of drugs has an impact on scleral permeability with an increment in lipophilicity showing lower permeability and, on the other hand, hydrophilic drugs diffuse across the aqueous medium of proteoglycans in the fiber matrix pores more easily as compared to the lipophilic drugs (Maurice and Polgar 1977; Cruysberg et al. 2002). Also, the permeability of the drug molecule across the sclera is also influenced by its charge. Positively charged compounds bind to the negatively charged proteoglycan matrix, and thereby may be poorly permeable (Gaudana et al. 2010).

1.3.5 Choroid/Bruch's Membrane

Choroid is among the highly vascularized tissues for blood supply to the retina. In the choroid, blood flow/unit tissue weight is ten times greater than in the brain. An added feature is that the capillary endothelial cells of choroid are fenestrated and, in humans, are relatively large in diameter (20–40 μm). An optical coherence tomography (OCT) has been used to noninvasively measure the thickness of retina and choroid (Huang et al. 2018; Hee et al. 1998). The OCT studies indicated that choroidal thickness grow thinner with age (Margolis and Spaide 2009; Ikuno et al. 2010). Histological studies have demonstrated that choroidal thickness at birth is around 200 μm and reduces to approx. 80 μm by the age 90 (Nickla and Wallman 2010).

1.3.6 Retina

The drugs in the vitreous are eliminated by two main routes (Maurice and Polgar 1977). All the drugs can be eliminated through the anterior route, indicating that the drugs can transverse the vitreous to reach the posterior chamber and, subsequently, be eliminated through aqueous turnover and uveal blood flow. Elimination through the posterior route occurs by permeation across the retina. The internal limiting membrane (ILM) restricts drug penetration from the vitreous to the retina. It is reported that the ILM that separates the retina and the vitreous comprises of ten distinct extracellular matrix proteins (Candiello et al. 2007). A study with primates has indicated that molecules over 100 kDa fail to cross the retinal layers and enter the subretinal space, but immunohistochemical analysis indicates that a humanized, antivascular endothelial growth factor (VEGF) monoclonal antibody (Bevacizumab, Avastin®, Genentech Inc.), composed of 214 amino acids with a molecular weight of 149 kDa, injected into the vitreous cavity, can penetrate through the sensory retina and gain entry into retinal pigment epitheliums (RPE), subretinal and choroidal space, in monkey and rabbit (Heiduschka et al. 2007; Dib et al. 2008). Also, in rabbits, nanodimensional particles (<200 nm) could enter the sensory retina and into RPE following intravitreal injection (Sakurai et al. 2001). In the intact retina, the drug(s) present in the subretinal fluid may have two fates, namely absorbed by the sensory retinal blood vessels or transported across the RPE, from where it can gain entry into the choroidal vessels or pass through the sclera. The drug transport through the RPE can occur both by transcellular and paracellular routes. The outer transport of molecules from the subretinal spaces is driven by hydrostatic and osmotic forces, and small molecules might gain passage via paracellular inter-RPE cellular gaps and by active transport via the transcellular route (Pederson 1989).

1.3.7 Blood-Retinal Barrier

Blood-retinal barrier (BRB) places a control on the drug transport from blood into the retina. BRB has tight junctions of retinal capillary endothelial cells and RPE, referred to as iBRB for the inner and oBRB for the outer BRB, respectively (Cunha-Vaz 1979). The function of iBRB is supported by Müller cells and astrocytes. The capillary endothelial cells of retina lack fenestrations and do not have vesicles. These endothelial vesicles participate in endocytosis or transcytosis that may be receptor mediated or fluid phase requiring adenosine triphosphate (Schnitzer et al. 1995). Müller cells and retinal capillary vessels have a close spatial relationship to sustain the iBRB for the uptake of nutrients and in the disposal of metabolites under normal conditions (Distler and Dreher 1996). Müller cells support neuronal activity and sustain the proper functioning of the iBRB under normal conditions (Reichenbach et al. 2007). They have a crucial role in the control and homeostasis of K+ and other ions signaling molecules, and in the control of extracellular pH (Bringmann et al. 2000). Müller cells dysfunction may disturb the iBRB in several pathological conditions, like diabetes (Tretiach et al. 2005). They augment the secretion of VEGF under conditions of hypoxia and inflammation (Eichler et al. 2000). In vitro studies have indicated that VEGF-induced occluding phosphorylation and ubiquitination cause trafficking of tight junction leading to enhanced retinal vascular permeability (Murakami et al. 2009). The astrocytes originate from the optic nerve and migrate to the nerve fiber layer during development (Watanabe and Raff 1988). They are closely associated with the retinal capillary vessels and help to maintain the capillary integrity (Zhang and Stone 1997). Astrocytes also strengthen the barrier characteristics of the retinal vascular endothelium by raising the expression of the tight junction protein ZO-1 and modifying endothelial morphology (Gardner et al. 1997).

2 Ocular Drug Delivery

The first option for ocular drug administration in the management of diseases affecting the anterior segment of the eye is topical instillation. This route is extremely preferred owing to its low cost, effortless administration, noninvasiveness, and patient-friendly nature (Patel et al. 2013). Conventional ophthalmic formulations like eye drops, suspensions, gels, and ointments are topically administered. Topical eye drops are most popular among these and have a share of more than 90% of ocular formulations currently in the global market (Abdelkader and Alany 2012). However, drug delivery via topical routes involves several complex biological processes, which result in poor ocular bioavailability. Indeed, peculiar anatomical and physiological features of the eye are instrumental in low drug permeation of topical formulations to the internal tissues of the eye. Upon administration, precorneal factors, including solution drainage, blinking, lacrimation, tear turnover rate, and protein binding, along with static ocular barriers presented by the layers of the cornea, conjunctiva, and sclera reduce the efficiency of topically applied drug absorption through intraocular tissues. The small dimensions and customizable surface properties have assisted nanodrug carriers to emerge as a promising platform for topical drug delivery. Nanocarriers have the potential to improve corneal permeability and uptake of drugs across physiological membranes. For example, nanoemulsions deliver poorly soluble drugs more efficiently than conventional emulsions because of remarkably higher corneal and conjunctival penetration. Additionally, the nanocarriers have greater interactions with the corneal epithelial membranes by the virtue of their larger surface area, which can prolong the retention of the drug administered topically (Mudgil et al. 2012; Wadhwa et al. 2009; Zhang et al. 2012) (Fig. 2).

3 Immunoregulatory Role of Ocular Macrophages

Macrophages are known to be instrumental in causing tissue damage in several other experimental autoimmune diseases, including collagen-induced arthritis, nephritis, thyroiditis, and encephalomyelitis (Godiska et al. 1995). In uveoretinitis, macrophages are involved in the phagocytosis of rod outer segments. Macrophages are also considered to be potent antigen-presenting cells (APCs) during the induction of inflammatory autoimmune diseases (Navegantes et al. 2017). The recruitment of leukocytes is a key feature of ocular inflammation in experimental autoimmune uveoretinitis (EAU). Increased chemokine production is correlated with leukocyte recruitment in the eye. In a recent study of patients, interleukin-8 (IL-8), inflammatory protein-10 (IP-10), monocyte chemoattractant protein-1 (MCP-1), regulated upon activation, normal T cell expressed and presumably secreted (RANTES), and macrophage inflammatory protein-1 alpha (MIP-1 α) displayed an increment in the aqueous humor during the active stages of anterior uveitis (Strack et al. 2002). Chemokines are also associated with the neuroinflammatory processes in experimental allergic encephalomyelitis (EAE), a mouse model for multiple sclerosis (MS). In this model, increases in MCP-1 and MIP-1 mRNA were linked to severity of the disease and occurred along with or before disease onset (Banisor et al. 2005). Ocular-infiltrating macrophages are critical in the tissue injury in cases of EAU. A number of chemokines are formed in the inflamed eye (Sonoda et al. 2003). According to Mérida et al. (2015), in uveitis, resi-

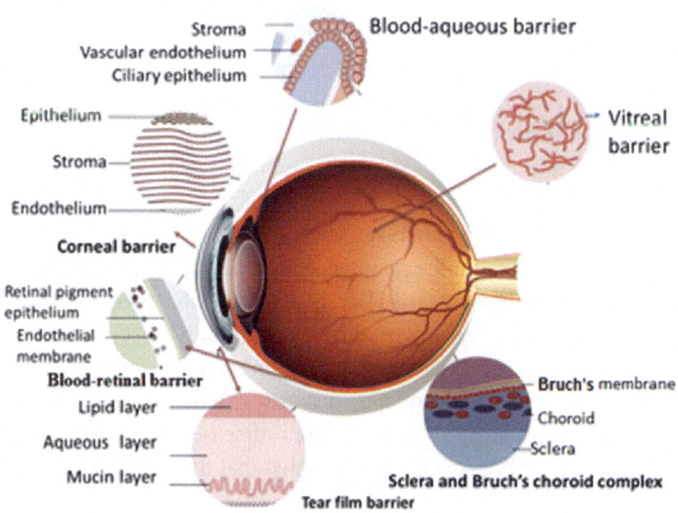

Fig. 2 A schematic representation of critical ocular barriers that impose challenges to ocular drug delivery

dent and infiltrated macrophages are crucial as effectors of innate immunity and inductors of acquired immunity. In uveitis, they are prime effectors of tissue damage and are also implicated as potent antigen-presenting cells. With the experimental animal models of uveitis, it has been possible to understand the important role of macrophages in eye inflammation processes, including macrophage polarization in EAU and the major role of Toll-like receptor 4 in endotoxin-induced uveitis (Mérida et al. 2015). Certain cellular components of the eye, like neural retina, are unable to regenerate and replicate after destructive inflammation. The distinct status of the eye, referred to as the ocular immune privilege, provides it with immune protection against intraocular inflammation in order to minimize the risk to vision integrity. The eye and immune system use strategies to maintain the ocular immune privilege by regulating the innate and adaptive immune response, which includes immunological ignorance, peripheral tolerance to eye-derived antigens, and intraocular immunosuppressive microenvironment (Keino et al. 2018). Progressive vascularization and fibroproliferation may involve the cornea and progress to blindness. Studies performed on conjunctival biopsies have reported an increase in CD4+ T cells and DCs in the epithelium and IL-2R-bearing T cells (CD4+ and CD8+), macrophages, and B cells in the stroma Th2 cells seem to amplify the fibrotic process through production of cytokines, such as IL-13, that stimulate resident fibroblasts to produce profibrotic cytokines, such as TGF- and platelet-derived growth factor, basic fibroblast growth factor, and connective tissue growth factor, and all of these are augmented in the conjunctiva (Stern et al. 2010). Dendritic cells (DC) and macrophages are components of the immune cell populations in the uveal tract, whose density, distribution, turnover, and function may be critical in the maintenance of ocular immunological homeostasis. Little is known of these cells in the mouse eye, despite this being the predominant experimental model in many studies of ocular immune responses and immunoinflammatory-mediated eye diseases (Kezic et al. 2008). A study has revealed that the mouse uveal tract, like the rat, contains rich networks of DC and resident tissue macrophages (McMenamin 1999). The closed systems of resident tissue macrophages in the mouse uveal tract thoroughly resemble similar networks in nonocular tissues. The phenotype of uveal tract DC suggests they are in the "immature" phase of their life cycle, similar to Langerhans cells of the skin, thus implying their part in situ inside the eye is antigen capture and not antigen presentation. Macrophages are the prime infiltrate in the corneas of mice infected ocularly with herpes simplex virus 1 (HSV-1). However, not much is known about the relative roles of M1 (classically activated or polarized) and M2 (alternatively activated or polarized) macrophages in ocular HSV-1 infection. The impact of directed M1 or M2 activation of RAW264.7 macrophages and peritoneal macrophages (PM) on subsequent HSV-1 infection was assessed. In both the RAW264.7 macrophage and PM in vitro models, HSV-1 replication in M1 macrophages was markedly lower than in M2 macrophages and unstimulated controls. The M1 macrophages expressed notably greater levels of 28 of the 32 tested cytokines and chemokines than M2 macrophages, with HSV-1 infection significantly increasing the levels of proinflammatory cytokines and chemokines in the M1 versus the M2 macrophages (Lee and Ghiasi 2017).

4 Macrophages at Diseased Intraocular Sites

4.1 Corneal Macrophages

Macrophages, the innate immune cells, are widely distributed in peripheral tissues and are involved in the development of numerous tissues and organs. For instance, microglia can facilitate innervation by regulating the survival and the programmed cell death of neurons and can secrete IGF-1 to promote the formation of the cortical region and a deficiency in these cells results in a decrease in the complexity and innervation of the central nervous system (Wu et al. 2020).

4.1.1 Distribution and Phenotype

The cornea acts as a defense to the foreign world; a dense, and regular connective tissue having precisely arranged collagen layers, squeezed between an anterior stratified, squamous, non-keratinized epithelium and a posterior endothelial monolayer, a principal site for peripheral infections. But transparency and vision may be compromised due to inflammation and scarring as a fallout of infections. Hence, the cornea has to strike a precarious balance between combating infections and safeguarding vision (Hu et al. 2015). Clear vision depends on an accurate refractive device (Liu et al. 2017). This corneal avascularity is indispensable for transparency and is dynamically sustained by numerous anti-angiogenic mechanisms (Ambati et al. 2006; Albuquerque et al. 2009; Cursiefen 2007; Cursiefen et al. 2006). The adult cornea comprises of discrete macrophage populations F4/80 and CD11b antigens have been earlier utilized to detect macrophages and monocytes in some tissues. Keratocytes, the mesenchymal-derived fibroblasts of the corneal stroma, are considered to contain only a few scattered lymphocytes in the stroma (Naumann 2012). It is recognized that the healthy cornea houses homeostatic populations of macrophages that exist throughout the anterior and posterior regions of the corneal stroma, sandwiched between the collagenous layers and closely associated with keratocytes (Sosnová et al. 2005).

In human corneas, the main difference between species is the possible higher representation of stromal dendritic cells as compared to the macrophages. The normal corneal stroma seems to only host macrophages and dendritic cells, members of the mononuclear phagocyte system. A lack in the complementary array of varied leukocyte subsets that normally lie beneath the epithelial barrier sites makes it imperative to comprehend the task of resident macrophages as the major immunological guardians of the corneal stroma (Wu et al. 2020; Hu et al. 2015; Liu et al. 2017; Albuquerque et al. 2009; Cursiefen 2007; Cursiefen et al. 2006; Naumann 2012; Sosnová et al. 2005; Kiesewetter et al. 2019).

4.1.2 Role in Homeostasis

In healthy human, a modest proportion of CD45+ resident cells express the hematopoietic stem cell marker CD34 (Bruno et al. 2009). The majority of studies of human and mouse corneal macrophages approve that most CD45+ cells in healthy corneas phenotypically are similar to macrophages (Brissette-Storkus et al. 2002; Knickelbein et al. 2014). These macrophages may be having a multipotent capacity to differentiate into keratocytes and contribute to collagen synthesis following injury, as established by in vitro studies with primary cultures of isolated human corneal stromal cells (Chinnery et al. 2017).

4.1.3 Role in Pathology

The cornea may be physiologically challenged by infection, injury, ocular surface stress, or during transplant rejection. Several studies using macrophage Fas-induced apoptosis (MaFIA) mice, in which macrophages expressing the colony-stimulating factor 1 receptor can be conditionally induced to undergo apoptosis, have observed the role of resident macrophages in producing innate inflammatory responses to pathogenic threats including the highly contagious ones.

Macrophages function as crucial mediators of injury-associated corneal hemangiogenesis (HA) and lymphangiogenesis (LA), but still molecular regulators of the hem- and lymphangiogenic potential of corneal wound macrophages lack much understanding. Two different mouse models of acute (perforating corneal incision injury) and chronic (corneal suture placement model) corneal injury were used to identify distinct functions of early- versus late-phase corneal wound macrophages in corneal HA and LA. It was established that while early-phase wound macrophages are essential for initiation and progression of injury-mediated corneal HA and LA, late-phase wound macrophages control maintenance of established corneal lymphatic vessels but not blood vessels. Furthermore, it was observed that the hem- and lymphangiogenic capacity of corneal wound macrophages is controlled by the nature of the corneal damage. Perforating corneal incision injury primarily activate wound macrophages with lymphangiogenic potential, while corneal suture placement triggered wound macrophages with both hem- and lymphangiogenic potential (Kiesewetter et al. 2019).

Also, the CCR2 was detected in the cornea at E12.5, while the CCR2þ population did not appear in the cornea until E17.5. These populations show diverse phenotypes, gene expression profiles, and maintenance mechanisms, and are also involved in corneal wound healing in various ways. CCR2þ macrophages display a proinflammatory capacity during the initial stage of corneal wound healing, and on the other hand, CCR2 macrophages wield antiinflammatory effects during the later stage. In the absence of either of these macrophages, a delay in the healing of corneal wound is observed. CD64þ macrophages in the cornea have been identified, and these cells were categorized into CCR2 and CCR2þ populations. Studies indicate that tissue-resident macrophages can originate from c-MybEMPs, situated in the yolk sac; c-Mybþ EMPs, positioned in the fetal liver; and HSCs, present in the fetal liver or bone marrow. Both c-Mybþ EMPs and HSCs differentiate into monocytes before becoming macrophages, while c-MybEMPs do not go through the monocyte stage and directly differentiate into macrophages (Wu et al. 2020; Liu et al. 2017).

4.2 Macrophages in the Uveal Tract

The uveal tract contains rich network of both resident cells. The latter appear to be strategically located to act as sentinels for capturing and sampling blood-borne and intraocular antigens.

4.2.1 Iris and Ciliary Body Macrophages

Iris, a circular ring of contractile connective tissue, dilates and constricts to regulate the passage of light across its central aperture. It has a highly vascularized stroma of connective tissue rich in melanocytes, and on its posterior side comprises of a modified epithelium which forms the dilator pupillae muscle and a pigmented epithelium. The major resident leukocytes in the iris are macrophages and dendritic cells (Yoshida et al. 2000) and only few resident mast cells are reported in human (May 1999). On the posterior of iris is the ciliary body, which produces aqueous humor that fills the anterior chamber of the eye. The inner layer of the ciliary epithelium has tight junctions, which is a key component of the blood ocular barrier, restraining the movement of large macromolecules and circulating immune cells that can have a bearing on the transparency of the ocular media. The iris and ciliary body macrophages primarily function to limit the exposure of the anterior chamber of the eye to exogenous antigens, and thereby unwarranted immune responses.

4.2.2 Role in Homeostasis and Pathology

In the iris as well as ciliary body, phagocytosis of melanin granules by resident macrophages is crucial to ensure that excess pigment granules are retained within the confines of tissue and not released into the anterior chamber, where they could possibly upset the clear visual axis and/or block the drainage of aqueous fluid from the anterior chamber. Some of the resident iris pigment-laden cells are related to their immense ability to ingest and store large amounts of pigment released from melanocytes and atrophic iris pigment epithelial cells (Wobmann and Fine 1972). Pigment dispersion syndrome is a condition, where excessive shedding of pigment granules from the epithelium occurs, that can lead to blinding ocular hypertension and glaucoma. (Schraermeyer et al. 2009). There are some other potentially blinding conditions that involve large debris-laden macrophages like phacolytic glaucoma (macrophages replete with lens debris), melanomalytic glaucoma (melanin-laden macrophages in uveal melanoma), and hemolytic and ghost cell glaucoma (erythrocytes or ghost erythrocyte-laden macrophages following hemorrhage in the eye) (Ueno et al. 1989; McMenamin and Lee 1986).

4.3 Choroidal Macrophages

4.3.1 Distribution and Phenotype

In the healthy choroid, macrophages are the resident cells. They basically function to remove debris, dead, and dying cells in the usual course of cell turnover. The choroid is the loose, vascular connective tissue rich in melanocytes, which lies behind the retina, and is separated by a monolayer of cuboidal retinal pigment epithelium (RPE), whose barrier properties are essential for normal functioning of the adjacent neural retina's (Chinnery et al. 2017) extensive network of ED1+ ED2+ ED3+ (CD68+ CD163+ and CD169+) macrophages in the perivascular space (Krause et al. 1996). A small proportion of these tissue-resident macrophages expressed MHC class II+ (OX6+), but a majority of the MHC class II+ cells were considered at the time to be putative dendritic cells, as ED2 or ED3 markers were not expressed (Butler and McMenamin 1996).

Normal-aged choroid has several tissue macrophages, but out of these only few (approximately 35% in submacula) are apparently activated (i.e., express HLA-DR). Most of the macrophages have ramified morphology and designated as antiinflammatory M2 macrophages (McWhorter et al. 2013). This hypothesis was supported by quantitative real-time polymerase chain reaction for representative M1 (CXCL11) and M2 (CCL22) transcripts (Cao et al. 2011), which demonstrated high M2 chemokine transcripts and a low M1-to-M2 chemokine transcript ratio in aging non-AMD eyes. (Condren et al. 2013; McLeod et al. 2016).

4.3.2 Role in Pathology

In early AMD, macrophages are restricted to pathologic areas of Bruch's membrane and here they introduce processes into Bruch's membrane deposits, probably to scavenge debris. These deposits have several molecules that can activate macrophages including C3a and C5a (Crabb et al. 2002; Nozaki et al. 2006). In normal human eyes, resident choroidal macrophages do not express inducible nitric oxide synthase (iNOS), but in the presence of soft drusen or continuous thick basal laminar deposits, macrophage recruitment to Bruch's membrane and iNOS expression (a characteristic of M1 macrophages) can take place (Cherepanoff et al. 2010). These choroids were incubated with antiionized calcium-binding adapter molecule 1 (anti-IBA1) to label macrophages, antihuman leukocyte antigen-antigen D-related (anti-HLADR) as a macrophage activation marker, and *Ulex europaeus* agglutinin lectin to label blood vessels. Whole mounts were imaged using confocal microscopy. IBA1- and HLA-DR-positive (activated) cells were counted in submacula, paramacula, and nonmacula, and cell volume and sphericity were observed. HLA-DRþ submacular macrophages were significantly increased in all stages of AMD, and they were significantly more round and smaller in size in the submacular AMD choroid, suggesting their activation (McLeod et al. 2016).

4.4 Retinal Microglia

4.4.1 Distribution and Phenotype

The retina is a complex tissue with multiple cell layers that are highly ordered. Its sophisticated structure makes it especially sensitive to external or internal perturbations that exceed the homeostatic range. This necessitates the continuous surveillance of the retina for the detection of harmful stimuli. This task is mainly performed by microglia cells, the resident tissue macrophages which confer neuroprotection against transient pathophysiological insults. However, under sustained pathological stimuli, microglial inflammatory responses become dysregulated, often worsening disease pathology (Rashid et al. 2019). Patients with uveitis often suffer serious visual loss after persistent inflammation due to immune-mediated damage in the targeted tissues. Retina has microglia and these are the resident immune cells, and have been implicated to be the key population that initiates retinal inflammation. Microglia are essential for instigating retinal autoimmune response, as microglial ablation completely blocks disease. Autoimmune uveitis is a serious sight-threatening condition defined by an autoreactive immune response to uveal tissues and the retina. As a consequence, microglia mediate autoreactive immune cell entry into the retina, and by depleting microglia, circulating immune cells cannot gain entry into the retina (Okunuki et al. 2019).

4.4.2 Targeting Microglia for the Treatment of Retinal Degenerative Diseases

Microglia participate in both physiological and pathophysiological functions in the retina. Therefore, despite the fact that neuroinflammatory responses from overly reactive microglia are instrumental in the onset and progression of retinal degenerative disorders, complete blocking of retinal microglial functions would result in undesirable effects. Hence, valid immunotherapeutic approaches for the treatment of retinal degeneration should be those that inhibit dysregulated microglial-mediated proinflammatory responses and/or simultaneously enhance their beneficial neuroprotective functions (Rathnasamy et al. 2019).

4.4.3 Functions of Retinal Microglia

Microglia-derived nerve growth factor is reported to be essential for developmental neuronal cell death (Rathnasamy et al. 2019). The finding that the processes of microglial cells are extremely motile and that they have the propensity to withdraw and extend led to the identification of a series of probable contribution of microglia in neural plasticity, neurogenesis, synaptic pruning maintenance of synaptic structure and function, and modulation of inflammatory reactions (Lee et al. 2008; Huang et al. 2012; Schafer et al. 2012; Wang et al. 2016; Wang et al. 2014). In addition, as microglia are often found in close juxtaposition to the developing vasculature, they may participate in the organization of retinal angiogenesis, and suppressing microglial reactivity has been suggested to be a promising strategy for curing retinal diseases (Checchin et al. 2006; Fischer et al. 2015).

4.4.4 Phagocytosis

The phagocytic tasks of microglia are further supported by their ability to ingest an intravenously administered tracer, rhodamine isothiocyanate, which leaks into the immature retina (Zeng et al. 2000). De Hoz et al. showed that microglia were activated in the adult retina to remove damaged neurons following interventions such as axotomy. Microglial cells possessing rod-shaped morphology were found to be closely associated with degenerating retinal ganglion cells (RGC) in a mouse model of laser-induced ocular hypertension or were engaged in phagocytosing such cells in a rat model of optic nerve transection (Yuan et al. 2015). It is evident from these studies that cellular debris in the retina is predominantly cleared by microglia.

Studies have indicated the molecular mechanisms of microglial phagocytosis. In the CNS, microbial pathogens are recognized through Toll-like receptors (TLRs), dectin-1, Fc receptors (FcRs), complement receptors (CRs), and scavenger receptors (SRs) (Maneu et al. 2011). Among these receptors, TLRs and dectin-1 are expressed by the resident microglia in the retina and promote the clearance of infectious microbes. For instance, neutralization of dectin-1 with its antibody resulted in a reduced clearance of *Candida albicans* by retinal microglia. Also, pretreatment of TLR2 receptors retinal microglia with their agonists increased the phagocytic activity of microglia against *Staphylococcus aureus (*Rathnasamy et al. 2019*).*

4.5 Other Ocular Macrophages

In the adult eye, the lens is devoid of macrophages (or indeed other cell types besides lens epithelial cells); macrophages are crucial for elimination of the tunica vasculosa lentis and pupillary membrane, the temporary vascular networks nourishing the developing lens till this task is performed by the aqueous humor (Chinnery et al. 2017; McMenamin et al. 2002). Many of the macrophages may be retained in the eye as vitreal macrophages or halocytes that express characteristic macrophage/myeloid cell markers like F4/80, Iba-1, CD169, and CD11b in mice, and ED2 in rats (Qiao et al. 2005). Hyalocytes are basically positioned amid the inner limiting membrane of retina and vitreous membrane that functions as the encapsulating coating of condensed vitreal collagen. The exact role of hyalocytes in the adult eye remains to be established, but it is reported that their morphology, activation status, and increased number may serve as early signs of pathological changes in models of diabetes and systemic exposure to infective agents (Vagaja et al. 2012). The sclera, the outer protective layer of the eye, also has resident macrophages (Xu et al. 2007). In the normal human sclera, the episclera has a high number of CD68+ CD11b+ MHC class II+ LYVE-1+ CCR7+ cells, and these are supposed to be M1 macrophages (Liu et al. 2017; Schlereth et al. 2016; McKay et al. 2019; Xiao et al. 2011). Whereas, simplex virus type-1 (HSV-1) is a leading cause of neurotrophic keratitis, characterized by decreased or absent corneal sensation due to damage to the sensory corneal innervation. Chucair-Elliott et al. (2017) studied MaFIA transgenic C57BL/6 in mice systemically administered AP20187 dimerizer or vehicle (VEH), and their corneas, lymph nodes, and blood were assessed for CD45þCD11bþGFPþ cell depletion by flow cytometry (FC). C57BL6 mice were used to compare to the MAFIA mouse model. It was observed that MaFIA mice treated with AP20187 had efficient depletion of CD45þCD11bþGFPþ cells in the tissues analyzed. The reduction in CD45þCD11bþGFPþ cells recruited to the infected corneas of AP20187-treated mice correlated with preservation of corneal nerve structure and function, decreased protein concentration of inflammatory cytokines, and decreased STAT3 activation, despite no changes in viral content in the cornea compared to VEH-treated animals. It was concluded that following HSV-1 infection, filtrated macrophages are the initial effectors in the nerve regression (Chucair-Elliott et al. 2017).

5 Nanocarriers for Macrophage Targeting in Ocular Diseases

Nanomedicine and nanodelivery systems are a relatively new but rapidly developing science with materials of nanodimensions serving as diagnostic tools and/or to carry therapeutics to specific target sites at a controlled rate. Nanotechnology offers multiple benefits in treating chronic diseases by facilitating site-specific and target-oriented delivery of precise medicines. In the last few years, a number of outstanding applications of the nanomedicine (chemotherapeutic agents, biological agents, immunotherapeutic agents, etc.) in the treatment of various diseases have been reported (Patra et al. 2018). However, certain challenges remain to be addressed and an advanced technology needs to be developed for successful delivery of drugs to its ocular target sites (Casettari and Illum 2014; Obeid et al. 2017). In the following section, the various drug delivery systems successfully employed in the recent times have been described.

5.1 Polymeric Nanoparticles

Nanoparticulates have emerged as a valuable technology to overcome the poor stability of the drug/bioactive in the biological milieu and deliver the drug across biological barriers. In spite of the opportunities offered by the nanoparticles, loading of high amounts of DNA is a challenge for successful gene delivery applications (Pannier and Segura 2013). Polymeric nanoparticles have demonstrated their potential as a viable therapeutic strategy for controlling intraocular inflammation. Autoimmune uveitis, an occoecatio disease, targets the posterior segment of the eye. Nanoparticle-based carriers have been reported for the treatments of immunological autoimmune uveitis with reduced side effects (Bittencourt et al. 2012). Nanocarriers facilitated indomethacin to gain entry into the inner eye structures via transmucosal route (Diebold and Calonge 2010). Dexamethasone (DEX)-loaded poly(lactic acid-co-glycolic acid) NPs following intravitreal injection were able to sustain drug concentrations over an extended period of time with the drug in all the layers of the eye ball, indicating their applications for the treatments of posterior segment diseases (Zhang et al. 2009). Stealth nanosteroids with betamethasone phosphate encapsulated within biocompatible and biodegradable blended nanoparticles of poly(lactic acid) (PLA) homopolymers and PEG-block-PLA copolymers were found to decrease the experimental scores of rats with EAU and reduced the retinal inflammatory cytokines of EAU by extended circulation in blood (Sakai et al. 2011). Dexamethasone-loaded nanoparticles prepared with sialyl-Lewis X conjugated liposome were able to target selectively the autoimmune uveoretinitis (Hashida et al. 2008). Poly(lactic acid) nanoparticles encapsulating betamethasone phosphate on intravenous administration to EAU rats remained for 7 days with decreased infiltration of activated T cells and macrophages along with hypertrophy of Muller cells (Sakai et al. 2006). Tamoxifen-loaded nanoparticles coated with polyethylene glycol (PEG) (NP-PEG-TAM) were injected in the rat's vitreous cavity with retinal soluble antigen (S-Ag)-induced EAU. It was observed that expressions of MHC class II (+) inflammatory cells, TNF-alpha, IL-1-beta, and RANTES mRNA diminished following this treatment (de Kozak et al. 2004).

Injection of biodegradable polymeric nanoparticles could reduce choroidal neovascularization originated from age-related macular degeneration (AMD) without any toxic effects (Kim and Csaky 2010). Intravitreal injection of basic fibroblast growth factor-impregnated NPs (bFGF-NPs) averted degeneration of photoreceptor by arresting apoptosis in rat retina (Sakai et al. 2007). Subretinal delivery of polyethylene glycol-substituted lysine peptide (CK30PEG)-compacted DNA nanoparticles led to effective gene expression in retinal cells. Inflammatory cytokines, monocyte chemotactic protein-1, chemokine KC, macrophage marker F4/80, or myeloid marker myeloperoxidase presented at control levels following subretinal delivery of polyethylene glycol-substituted lysine peptide-compacted DNA nanoparticles (Ding et al. 2009).

5.2 Polymeric Micelles

Polymeric micelles are defined as core-shell nanoparticles that self-assemble from amphiphilic copolymers. These carriers enable the encapsulation of hydrophobic drugs in their core. Polymeric micelles can be fabricated via direct dissolution, solvent evaporation, and cosolvent evaporation methods. Micellar nanoparticles have found promising applications in both the anterior and posterior ocular diseases owing to mucoadhesive profile, high water solubility, small size, ability to form clear solutions, enhanced stability, high loading capacity, biocompatibility, facile surface modification, simple and cost-effective production, and potential for achieving controlled-release profile of the drug. Various types of polymeric materials are used for fabrication of micellar system and formulation for ophthalmic DDSs (e.g., pluronic-chitosan, poly (hydroxyethyl) aspartamide, methoxypoly (ethylene glycol)-poly (ε-caprolactone), pluronic F127, and ginsenosides Rb1 among others). The hydrophilic nature of stroma makes the use of hydrophobic drugs difficult, while the hydrophilic shell of micelles promotes the stability and the half-life of the drug which efficiently prolong the circulation time and bioavailability of drugs (Attama et al. 2016; Aliabadi et al. 2007; Zhao et al. 2016). Curcumin-loaded nanomicelles have been prepared using polyvinyl caprolactam-polyvinyl acetate-polyethylene glycol to treat ocular inflammation (Gorantla et al. 2020). The polymeric micelles present a viable strategy for improving the corneal and conjunctival penetration of therapeutic drugs, for sustaining drug levels, and reducing systemic side effects. Micelles self-assembled from mPEG-PLA display higher drug penetration, and nonirritation to eye with only slight cytotoxicity to cells (Yu et al. 2018).

5.3 Dendrimers

Dendrimers are nanostructured polymers with a characteristic tree-like assembly, and have emerged as a viable drug and gene delivery vehicle with a varied array of potential applications. Dendrimers have a high density of surface functional groups that are able to moderate the local environment, permitting inimitable tissue interactions, and these can be valuable as emerging novel therapies (Menjoge et al. 2010). Microglial activation and the accompanying neuroinflammation are crucial in the pathogenesis of several retinal diseases. Intravitreal and systemically administered dendrimers can target activated microglia and display qualitatively analogous retinal biodistribution. Studies have provided proof-of-concept insights for developing dendrimer drug formulations as treatment options for retinal diseases associated with microglia or macrophage activation such as age-related macular degeneration, diabetic retinopathy, and retinal degenerations (Kambhampati et al. 2015).

PAMAM dendrimers provide a unique opportunity to selectively target reactive immune cell types, affording new opportunities to assess immune cell dynamics and function. Dendrimer conjugation of drugs has facilitated to gain insights specifically on the role of reactive microglia and macrophages on the regenerative process. In addition, conjugating drugs to dendrimers are reported to be comparatively less toxic to cells and organisms than comparable free drug treatments, presumably due to rapid kidney clearance in the absence of inflammation (Sharma et al. 2018). Dex conjugated to G4 PAMAM dendrimers was reported to have substantially reduced toxicity, superaccelerated neuronal regeneration kinetics, and targeted both dying rod cells and reactive microglia/macrophages on pericardial injection (Soiberman et al. 2017). The temporal dynamics of the latter result suggested that dendrimers target dying neurons first and, secondarily, accumulate in phagocytic microglia/macrophages. Each of these findings is novel and builds upon an expanding body of work demonstrating profound therapeutic potential of dendrimer nanoparticles (Iezzi et al. 2012; Sharma et al. 2017).

The PAMAM dendrimers can target one main cell form in retinal neuroinflammation-activated microglia. Following dendrimer administration by intravenous or intravitreal route, retention by activated microglia occurred. Also, while the microglia retained dendrimer, other cell types failed to take up the dendrimer. The dendrimers remained in microglia for ~21 days. Ischemia-reperfusion (I/R) injury has been employed as a prototype for some features of chronic glaucoma, diabetic retinopathy, and branch vein occlusion (BVO). I/R injury is implicated in occlusion of retinal and choroidal blood vessels, leading to a decrease in blood flow and tissue hypoxia (Kambhampati 2014).

5.4 Liposomes

In the last decade, liposomal formulations have been extensively studied as a carrier for the ophthalmic drug delivery applications. Liposomes are biodegradable and biocompatible carriers, primarily composed of phosphatidylcholine and other constituents like cholesterol and lipid-conjugated hydrophilic polymers. Liposomes for topical delivery are primed to enhance corneal adhesion and permeation by incorporating various bioadhesive and penetration enhancing polymers. In the case of posterior segment disorders, improvement in intravitreal half-life and targeted retinal drug delivery can be achieved by liposomes (Mishra et al. 2011). Rhodamine-conjugated liposomes loaded with vasoactive intestinal peptide (VIP) were intravenously administered to healthy rats to evaluate its potential to treat ocular inflammation. VIP is an immunomodulatory neuropeptide involved in the regulation of ocular immune response by modulating the activities of macrophages, T lymphocytes, and dendritic cells (Torchilin 2005; Taylor et al. 1994). Rhodamine-conjugated liposomes (Rh-Lip) alone and loaded with VIP (VIP-Rh-Lip) were examined in male Lewis rats. Following single intravitreal injection, liposomes were internalized by retinal Muller glial cells, resident macrophages, and rare infiltrating activated macrophages (Lajavardi et al. 2007). A large fraction of the liposomes was able to translocate to cervical lymph nodes via conjunctival lymphatics. VIP-Rh-Lip that were internalized via macrophages led to a slower release and long-term expression within the ocular tissues and cervical lymph nodes. Intravenous delivery of VIP by liposomes displayed effectiveness in the treatment of uveitis and other immune-mediated eye diseases by modulating the immune microenvironment of the ocular region (Camelo et al. 2007). Clodronate liposomes (CL2MDP-lip) were reported to inhibit infiltration of macrophages in the conjunctiva in blepharoconjunctivitis developed in Brown Norway rats. CL2MDP-lip could reduce the conjunctival count of ED2-positive macrophages, whereas ED1-positive macrophages infiltration was controlled only on injection just preceding the OVA challenge (Fukushima et al. 2005). The half-lives of liposomes in vitreous bodies may be prolonged with low toxicity (Bochot et al. 2002). However, their access to the deeper tissues is limited owing to their low bioadhesiveness.

A diverse range of biomacromolecules (such as polymers and ligand) have been used with liposomes with an objective to enhance transshipment efficiency (Ravar et al. 2016). PAMAM G3.0, the third-generation formulation of polyamidoamine dendrimer, with typical dendritic branches coupled with high-density functional amino groups, is an outstanding drug delivery platform (Wang et al. 2011; Choi et al. 2010; Biswas et al. 2013). It also presents a spherical architecture with massive internal hydrophobic cavities, permitting encapsulation of drug(s) like micelles, which could also alter the liposomal structure to enable targeting (Gopidas et al. 1991; Jansen and Meijer 1994; Jansen et al. 1995).

5.5 Nanostructured Lipid Carriers

Nanostructured lipid carriers (NLCs) are the next-generation solid lipid nanoparticles (SLNs) to dominate the probable difficulties of SLNs (Naseri et al. 2015). NLCs were developed considering their biocompatible and nontoxic nature due to lipids (Haider et al. 2020). Meanwhile, NLC could also promote the precorneal retention time and corneal absorption via the ocular tissues of the ocular drugs due to their nanoscale dimensions providing them mucoadhesive properties (Souto et al. 2010). The main disadvantage of traditional colloidal dispersion systems is their low viscosity, which can be swiftly eliminated by innate defense mechanisms of the ocular globe (Hao et al. 2014). NLC has been explored as a viable drug carrier for the treatment of wide range of ocular disorders, from the common ocular inflammation or infection to diseases confined to posterior segment (Luo et al. 2011; Balguri et al. 2016). Use of cationic surfactants or polymers allowed the carriers to bind with the negatively charged mucosal surface by electrostatic forces, thereby increasing the retention of carriers in the eyes to permit prolonged action of drugs (Andrade et al. 2016). The mucoadhesive property of NLC increases the interaction with the cornea with an extended retention time, improved permeability to the posterior segment, enhanced bioavailability, and reduced side effects. NLCs were also developed with a perspective to meet industrial needs like scale-up, qualification and validation, simple technology, and low cost among others (Shidhaye et al. 2008).

Palmitoylethanolamide (PEA), an endogenous congener of the endocannabinoid anandamide (AEA), displays potent antiinflammatory activity, and has the potential to be used in varied pathological states and biological systems including retina. PEA can have positive outcomes in several retinal diseases including diabetic retinopathy and glaucoma. PEA is reported to decrease the degree of retinal inflammation while protecting the blood-retinal barrier in diabetic rats (Petrosino and Di Marzo 2017). Also, systemic PEA treatment was found to improve visual field in glaucoma patients and reduce the intraocular pressure (Costagliola et al. 2014). Several studies have demonstrated superior ocular bioavailability of therapeutic agents from these colloidal nanoparticulate systems, possibly because of improved retention and phagocytosis by epithelial cells (Balguri et al. 2016). NLCs have been studied for treatment of ocular inflammation, infections, glaucoma, and also for various pathologies affecting the posterior eye segment (Araújo et al. 2011). The studies involving macrophage delivery of buparvaquone demonstrated that, when Tween 80 and Kolliphor® P188 are used together, more than 3% surfactant is required to reach low particle size having a solid-to-liquid lipid ratio of 2.75 and above (Subramaniam et al. 2020). NLCs have been coated with mannose since there is an overexpression of mannose receptors on the surfaces of macrophages at the inflammation sites (Xiao et al. 2011). Macrophage and lymphocytes have a significant role in the host acute or chronic inflammation through the release of a variety of proinflammatory cytokines (e.g., IL-6, TNF-α, and NO) in response to an activating stimulus (e.g., LPS) (Patel et al. 2013). As reported previously, the infiltration of macrophages and neutrophils had been mainly implicated in the initiation of inflammatory symptoms in uveitis and the breakdown of the blood-aqueous barrier. The mouse fibroblast cell line (L-929 cells), mouse macrophage cell line (RAW264.7 macrophages), and human corneal epithelial cell line (HCEC cells) were used to evaluate the in vitro cytotoxicity of the formed TA-SA/PECE nanoparticles (Huang et al. 2018). Topical administration of NLCs containing α-terpineol (αT) resulted in significant reduction in bacterial count in corneal tissue by 4 $\log_{10}$ on 5th postinfection day. The protective efficacy of αT-NLCs demonstrated improvement in corneal histopathology, and decreased the levels of various inflammatory markers including myeloperoxidase (MPO) and reactive nitrogen intermediates (RNI). Further, αT-NLCs treatment showed immunomodulatory effects by manipulating the production of inflammatory cytokines, tumor necrotic factor (TNF-α), macrophage inhibitory protein-2 (MIP-2), and interleukin-2 (IL-2) in infected eyes. In addition, ex vivo studies exhibited enhanced susceptibility of *P. aeruginosa* toward serum and macrophages in the presence of αT-NLCs. A potent antibiofilm effect was also observed by αT-NLCs against *P. aeruginosa,* which was confirmed by fluorescent microscopic analysis and was proposed for the treatment of biofilm-associated keratitis caused by *P. aeruginosa* (Bose et al. 2020).

5.6 Niosome

Niosomes are spherical and consist of microscopic lamellar (unilamellar or multilamellar) structures. The bilayer is formed by nonionic surfactants, with or without cholesterol and a charge inducer. Different types of surfactants at variable combinations and molar ratios are used to form niosomes (Yeo et al. 2018). Niosomes can be categorized into three groups based on their vesicle size, namely small unilamellar vesicles (0.025–0.05 mm), multilamellar vesicles (>0.05 mm), and large unilamellar vesicles (>0.10 mm) (Kazi et al. 2010). Pardakhty et al. (2011) reported that gentamicin sulfate, a water-soluble antibiotic, showed an extensive alteration in the release rate during its experimental studies. Moreover, in contrast to the regular drug sample solution, niosomal formulation of drug exhibits sluggish release (Pardakhty et al. 2011). Timolol maleate formulated as niosomes and coated with chitosan exhibited more effect on intraocular tension with fewer side effects as compared to the marketed products. The release profile of niosome-based complexes can be modulated by the cholesterol percentage of the formulation (Abdelbary and El-Gendy 2008). Farmoudeh et al. have reported methylene blue-loaded niosomes formulations with high encapsulation efficiency and adequate stability. The biochemical and macroscopic studies indicated a higher recovery rate in surgical wounds with the niosomal-treated group (Farmoudeh et al. 2020). Akbari et al. (2013) reported antibacterial activity against intracellular *Staphylococcus aureus* infection of murine macrophage-like, J774, cells. It was reported that the size and composition of niosomes can influence their in vitro biological properties. Vesicles in the 300–600 nm size range were phagocytosed to a greater degree by macrophages in comparison to other size vesicles. The minimum inhibitory concentrations (MICs) of CPFX-loaded niosomes were two- to eightfold lower than MICs of free CPFX. In addition, niosome encapsulation of CPFX provided high intracellular antimicrobial activities while free CPFX is ineffective for eradicating intracellular forms of *S. aureus* (Akbari et al. 2013). Studies indicate that natamycin niosomes are promising ocular nanocarriers with ketorolac tromethamine for treatment of candida keratitis (El-Nabarawi et al. 2019).

6 Conclusion

In the past decades, the studies in the field of ocular immunology have generated considerable scientific evidence through preclinical studies, human ocular tissue analyses, and clinical trials. These findings categorically indicate the specific molecular targets linked to macrophages, which could be crucial in translating the various potential strategies in clinical setup. But there are certain issues that must be addressed before these translations become a reality. These include experimental studies on separating retinal immune cell population, linking retinal microglia and macrophage phenotypes and functions with

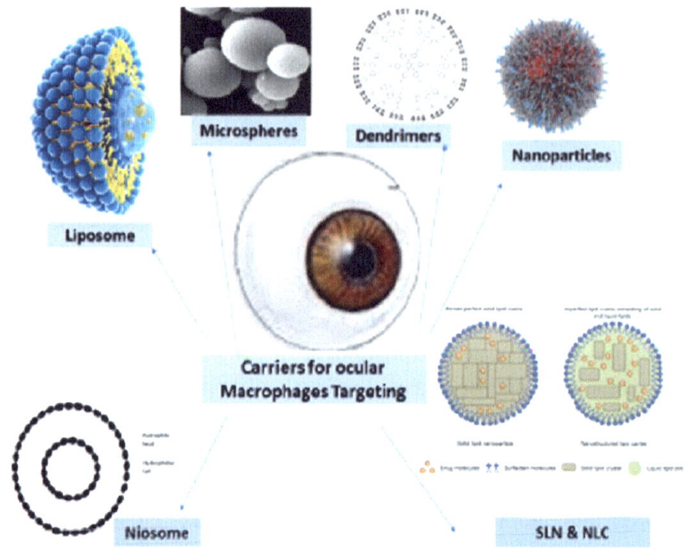

Fig. 3 Novel drug delivery systems for ocular macrophages targeting

disease outcome, and identifying optimal targets in the complement cascade. There is also a need to focus on clinical studies with immunomodulatory compounds. Consideration also needs to be placed on the choroid as it may be a protective immunological shield for the retina (Fig. 3).

Conflict of Interest The authors confirm that the content of this chapter has no conflict of interest.

Acknowledgments The authors are thankful to the Director, University Institute of Pharmacy, Pandit Ravishankar Shukla University, Raipur. The authors extend special thanks to the Librarian, Pt. Sundarlal Sharma Library of the University for making e-resources available through UGC-INFLIBNET. The authors are thankful to the Chhattisgarh Council of Science and Technology for MRP (1863/CCOST/MRP/2014).

References

Abdelbary G, El-Gendy N. Niosome-encapsulated gentamicin for ophthalmic controlled delivery. AAPS PharmSciTech. 2008;9(3):740–7.

Abdelkader H, Alany RG. Controlled and continuous release ocular drug delivery systems: pros and cons. Curr Drug Deliv. 2012;9(4):421–30.

Agrahari V, Mandal A, Agrahari V, Trinh HM, Joseph M, Ray A, Hadji H, Mitra R, Pal D, Mitra AK. A comprehensive insight on ocular pharmacokinetics. Drug Deliv Transl Res. 2016;6(6):735–54.

Akbari V, Abedi D, Pardakhty A, Sadeghi-Aliabadi H. Ciprofloxacin nano-niosomes for targeting intracellular infections: an in vitro evaluation. J Nanopart Res. 2013;15(4):1556.

Albuquerque RJ, Hayashi T, Cho WG, Kleinman ME, Dridi S, Takeda A, Baffi JZ, Yamada K, Kaneko H, Green MG, Chappell J. Alternatively spliced vascular endothelial growth factor receptor-2 is an essential endogenous inhibitor of lymphatic vessel growth. Nat Med. 2009;15(9):1023–30.

Aliabadi HM, Elhasi S, Mahmud A, Gulamhusein R, Mahdipoor P, Lavasanifar A. Encapsulation of hydrophobic drugs in polymeric micelles through co-solvent evaporation: the effect of solvent composition on micellar properties and drug loading. Int J Pharm. 2007;329(1-2):158–65.

Ambati BK, Nozaki M, Singh N, Takeda A, Jani PD, Suthar T, Albuquerque RJ, Richter E, Sakurai E, Newcomb MT, Kleinman ME. Corneal avascularity is due to soluble VEGF receptor-1. Nature. 2006;443(7114):993–7.

Andrade LM, Rocha KA, De Sá FA, Marreto RN, Lima EM, Gratieri T, Taveira SF. Voriconazole-loaded nanostructured lipid carriers for ocular drug delivery. Cornea. 2016;35(6):866–71.

Araújo J, Nikolic S, Egea MA, Souto EB, Garcia ML. Nanostructured lipid carriers for triamcinolone acetonide delivery to the posterior segment of the eye. Colloids Surf B: Biointerfaces. 2011;88(1):150–7.

Attama AA, Lovelyna C, Onuigbob EB. Nanotechnology for ocular and otic drug delivery and targeting. In: Nanotechnology and drug delivery, Volume two: nanoengineering strategies and nanomedicines against severe diseases. CRC Press; 2016. p. 165–90.

Bachu RD, Chowdhury P, Al-Saedi ZH, Karla PK, Boddu SH. Ocular drug delivery barriers—role of nanocar-

riers in the treatment of anterior segment ocular diseases. Pharmaceutics. 2018;10(1):28.
Balguri SP, Adelli GR, Majumdar S. Topical ophthalmic lipid nanoparticle formulations (SLN, NLC) of indomethacin for delivery to the posterior segment ocular tissues. Eur J Pharm Biopharm. 2016;109:224–35.
Banisor I, Leist TP, Kalman B, et al. Involvement of β-chemokines in the development of inflammatory demyelination. J Neuroinflammation. 2005;2(1):1–4.
Biswas S, Deshpande PP, Navarro G, Dodwadkar NS, Torchilin VP. Lipid modified triblock PAMAM-based nanocarriers for siRNA drug co-delivery. Biomaterials. 2013;34(4):1289–301.
Bittencourt MG, Sepah YJ, Do DV, Agbedia O, Akhtar A, Liu H, Akhlaq A, Annam R, Ibrahim M, Nguyen QD. New treatment options for noninfectious uveitis. Dev Ophthalmol. 2012;51:134–61.
Bochot A, Fattal E, Boutet V, Deverre JR, Jeanny JC, Chacun H, Couvreur P. Intravitreal delivery of oligonucleotides by sterically stabilized liposomes. Invest Ophthalmol Vis Sci. 2002;43(1):253–9.
Bose SK, Sharma K, Chhibber S, Harjai K. Therapeutic potential of nanolipoidal α-Terpineol in combating keratitis induced by pseudomonas aeruginosa in the Murine Model. Int J Pharm. 2020;120175
Bringmann A, Skatchkov SN, Pannicke T, Biedermann B, Wolburg H, Orkand RK, Reichenbach A. Müller glial cells in anuran retina. Microsc Res Tech. 2000;50(5):384–93.
Brissette-Storkus CS, Reynolds SM, Lepisto AJ, Hendricks RL. Identification of a novel macrophage population in the normal mouse corneal stroma. Invest Ophthalmol Vis Sci. 2002;43(7):2264–71.
Bruno S, Bussolati B, Grange C, Collino F, di Cantogno LV, Herrera MB, Biancone L, Tetta C, Segoloni G, Camussi G. Isolation and characterization of resident mesenchymal stem cells in human glomeruli. Stem Cells Dev. 2009;18(6):867–80.
Butler TL, McMenamin PG. Resident and infiltrating immune cells in the uveal tract in the early and late stages of experimental autoimmune uveoretinitis. Invest Ophthalmol Vis Sci. 1996;37(11):2195–210.
Camelo S, Lajavardi L, Bochot A, Goldenberg B, Naud MC, Fattal E, Behar-Cohen F, de Kozak Y. Ocular and systemic bio-distribution of rhodamine-conjugated liposomes loaded with VIP injected into the vitreous of Lewis rats. Mol Vis. 2007;13:2263.
Candiello J, Balasubramani M, Schreiber EM, Cole GJ, Mayer U, Halfter W, Lin H. Biomechanical properties of native basement membranes. FEBS J. 2007;274(11):2897–908.
Cao X, Shen D, Patel MM, Tuo J, Johnson TM, Olsen TW, Chan CC. Macrophage polarization in the maculae of age-related macular degeneration: a pilot study. Pathol Int. 2011;61(9):528–35.
Casettari L, Illum L. Chitosan in nasal delivery systems for therapeutic drugs. J Control Release. 2014;190:189–200.
Checchin D, Sennlaub F, Levavasseur E, Leduc M, Chemtob S. Potential role of microglia in retinal blood vessel formation. Invest Ophthalmol Vis Sci. 2006;47(8):3595–602.
Cherepanoff S, McMenamin P, Gillies MC, Kettle E, Sarks SH. Bruch's membrane and choroidal macrophages in early and advanced age-related macular degeneration. Br J Ophthalmol. 2010;94(7):918–25.
Chinnery HR, McMenamin PG, Dando SJ. Macrophage physiology in the eye. Pflügers Arch. 2017;469(3-4):501–15.
Choi SK, Thomas T, Li MH, Kotlyar A, Desai A, Baker JR Jr. Light-controlled release of caged doxorubicin from folate receptor-targeting PAMAM dendrimer nanoconjugate. Chem Commun. 2010;46(15):2632–4.
Chucair-Elliott AJ, Gurung HR, Carr MM, Carr DJ. Colony stimulating factor-1 receptor expressing cells infiltrating the cornea control corneal nerve degeneration in response to HSV-1 infection. Invest Ophthalmol Vis Sci. 2017;58(11):4670–82.
Condren AB, Kumar A, Mettu P, Liang KJ, Zhao L, Tsai JY, Fariss RN, Wong WT, et al. Perivascular mural cells of the mouse choroid demonstrate morphological diversity that is correlated to vasoregulatory function. PLoS One. 2013;8(1):e53386.
Costagliola C, Romano MR, Dell'Omo R, Russo A, Mastropasqua R, Semeraro F. Effect of palmitoylethanolamide on visual field damage progression in normal tension glaucoma patients: results of an open-label six-month follow-up. J Med Food. 2014;17(9):949–54.
Crabb JW, Miyagi M, Gu X, Shadrach K, West KA, Sakaguchi H, Kamei M, Hasan A, Yan L, Rayborn ME, Salomon RG. Drusen proteome analysis: an approach to the etiology of age-related macular degeneration. Proc Natl Acad Sci. 2002;99(23):14682–7.
Cruysberg LP, Nuijts RM, Geroski DH, Koole LH, Hendrikse F, Edelhauser HF. In vitro human scleral permeability of fluorescein, dexamethasone-fluorescein, methotrexate-fluorescein and rhodamine 6G and the use of a coated coil as a new drug delivery system. J Ocul Pharmacol Ther. 2002;18(6):559–69.
Cunha-Vaz J. The blood-ocular barriers. Surv Ophthalmol. 1979;23(5):279–96.
Cursiefen C. Immune privilege and angiogenic privilege of the cornea. Chem Immunol Allergy. 2007;92:50–7.
Cursiefen C, Chen L, Saint-Geniez M, Hamrah P, Jin Y, Rashid S, Pytowski B, Persaud K, Wu Y, Streilein JW, Dana R. Nonvascular VEGF receptor 3 expression by corneal epithelium maintains avascularity and vision. Proc Natl Acad Sci. 2006;103(30):11405–10.
Curtin BJ. Physiopathologic aspects of scleral stress-strain. Trans Am Ophthalmol Soc. 1969;67:417.
Dawson DG, Ubels JL, Edelhauser HF. Cornea and sclera. Adler's physiology of the eye. 2011;11,71–130.
de Kozak Y, Andrieux K, Villarroya H, Klein C, Thillaye-Goldenberg B, Naud MC, Garcia E, Couvreur P. Intraocular injection of tamoxifen-loaded nanoparticles: a new treatment of experimental autoimmune uveoretinitis. Eur J Immunol. 2004;34(12):3702–12.

Dib E, Maia M, Longo-Maugeri IM, Martins MC, Mussalem JS, Squaiella CC, Penha FM, Magalhães O, Rodrigues EB, Farah ME. Subretinal bevacizumab detection after intravitreous injection in rabbits. Invest Ophthalmol Vis Sci. 2008;49(3):1097–100.

Diebold Y, Calonge M. Applications of nanoparticles in ophthalmology. Prog Retin Eye Res. 2010;29(6):596–609.

Ding XQ, Quiambao AB, Fitzgerald JB, Cooper MJ, Conley SM, Naash MI. Ocular delivery of compacted DNA-nanoparticles does not elicit toxicity in the mouse retina. PLoS One. 2009;4(10):e7410.

Distler C, Dreher Z. Glia cells of the monkey retina—II. Müller cells. Vision Res. 1996;36(16):2381–94.

Eichler W, Kuhrt H, Hoffmann S, Wiedemann P, Reichenbach A. VEGF release by retinal glia depends on both oxygen and glucose supply. Neuroreport. 2000;11(16):3533–7.

El-Nabarawi MA, Abd El Rehem RT, Teaima M, Abary M, El-Mofty HM, Khafagy MM, Lotfy NM, Salah M. Natamycin niosomes as a promising ocular nanosized delivery system with ketorolac tromethamine for dual effects for treatment of candida rabbit keratitis; in vitro/in vivo and histopathological studies. Drug Dev Ind Pharm. 2019;45(6):922–36.

Farmoudeh A, Akbari J, Saeedi M, Ghasemi M, Asemi N, Nokhodchi A. Methylene blue-loaded niosome: preparation, physicochemical characterization, and in vivo wound healing assessment. Drug Deliv Transl Res. 2020;10(5):1428–41.

Fischer AJ, Zelinka C, Milani-Nejad N. Reactive retinal microglia, neuronal survival, and the formation of retinal folds and detachments. Glia. 2015;63(2):313–27.

Fukushima A, Ozaki A, Ishida W, Van Rooijen N, Fukata K, Ueno H. Suppression of macrophage infiltration into the conjunctiva by clodronate liposomes in experimental immune-mediated blepharoconjunctivitis. Cell Biol Int. 2005;29(4):277–86.

Gardner TW, Lieth E, Khin SA, Barber A, Bonsall DJ, Lesher T, Rice K, Brennan WA. Astrocytes increase barrier properties and ZO-1 expression in retinal vascular endothelial cells. Invest Ophthalmol Vis Sci. 1997;38(11):2423–7.

Gaudana R, Ananthula HK, Parenky A, Mitra AK. Ocular drug delivery. AAPS J. 2010;12(3):348–60.

Godiska R, Chantry D, Dietsch GN, Gray PW. Chemokine expression in murine experimental allergic encephalomyelitis. J Neuroimmunol. 1995;58(2):167–76.

Goharian I, Sehi M. Is there any role for the choroid in glaucoma? J Glaucoma. 2016;25(5):452–8.

Gopidas KR, Leheny AR, Caminati G, Turro NJ, Tomalia DA. Photophysical investigation of similarities between starburst dendrimers and anionic micelles. J Am Chem Soc. 1991;113(19):7335–42.

Gorantla S, Rapalli VK, Waghule T, Singh PP, Dubey SK, Saha RN, Singhvi G. Nanocarriers for ocular drug delivery: current status and translational opportunity. RSC Adv. 2020;10(46):27835–55.

Haider M, Abdin SM, Kamal L, Orive G. Nanostructured lipid carriers for delivery of chemotherapeutics: a review. Pharmaceutics. 2020;12(3):288.

Hao J, Wang X, Bi Y, Teng Y, Wang J, Li F, Li Q, Zhang J, Guo F, Liu J. Fabrication of a composite system combining solid lipid nanoparticles and thermosensitive hydrogel for challenging ophthalmic drug delivery. Colloids Surf B: Biointerfaces. 2014;114:111–20.

Hashida N, Ohguro N, Yamazaki N, Arakawa Y, Oiki E, Mashimo H, Kurokawa N, Tano Y. High-efficacy site-directed drug delivery system using sialyl-Lewis X conjugated liposome. Exp Eye Res. 2008;86(1):138–49.

Hee MR, Puliafito CA, Duker JS, Reichel E, Coker JG, Wilkins JR, Schuman JS, Swanson EA, Fujimoto JG. Topography of diabetic macular edema with optical coherence tomography. Ophthalmology. 1998;105(2):360–70.

Heiduschka P, Fietz H, Hofmeister S, Schultheiss S, Mack AF, Peters S, Ziemssen F, Niggemann B, Julien S, Bartz-Schmidt KU, Schraermeyer U. Penetration of bevacizumab through the retina after intravitreal injection in the monkey. Invest Ophthalmol Vis Sci. 2007;48(6):2814–23.

Hu K, Harris DL, Yamaguchi T, von Andrian UH, Hamrah P. A dual role for corneal dendritic cells in herpes simplex keratitis: local suppression of corneal damage and promotion of systemic viral dissemination. PLoS One. 2015;10(9):e0137123.

Huang T, Cui J, Li L, Hitchcock PF, Li Y. The role of microglia in the neurogenesis of zebrafish retina. Biochem Biophys Res Commun. 2012;421(2):214–20.

Huang J, Yu X, Zhou Y, Zhang R, Song Q, Wang Q, Li X. Directing the nanoparticle formation by the combination with small molecular assembly and polymeric assembly for topical suppression of ocular inflammation. Int J Pharm. 2018;551(1-2):223–31.

Iezzi R, Guru BR, Glybina IV, Mishra MK, Kennedy A, Kannan RM. Dendrimer-based targeted intravitreal therapy for sustained attenuation of neuroinflammation in retinal degeneration. Biomaterials. 2012;33(3):979–88.

Ikuno Y, Kawaguchi K, Nouchi T, Yasuno Y. Choroidal thickness in healthy Japanese subjects. Invest Ophthalmol Vis Sci. 2010;51(4):2173–6.

Jansen JF, Meijer EW. Encapsulation of guest molecules into a dendritic box. Science. 1994;266(5188):1226–9.

Jansen JF, Meijer EW, de Brabander-van den Berg EM. The dendritic box: shape-selective liberation of encapsulated guests. J Am Chem Soc. 1995;117(15):4417–8.

Kaluzhny Y, Kinuthia MW, Truong T, Lapointe AM, Hayden P, Klausner M. New human organotypic corneal tissue model for ophthalmic drug delivery studies. Investig Ophthalmol Vis Sci. 2018;59(7):2880–98.

Kambhampati SP. Dendrimer based nanotherapeutics for ocular drug delivery. 2014.

Kambhampati SP, Clunies-Ross AJ, Bhutto I, Mishra MK, Edwards M, McLeod DS, Kannan RM, Lutty G, et al. Systemic and intravitreal delivery of dendrimers to activated microglia/macrophage in ischemia/reperfusion mouse retina. Invest Ophthalmol Vis Sci. 2015;56(8):4413–24.

Kazi KM, Mandal AS, Biswas N, Guha A, Chatterjee S, Behera M, Kuotsu K. Niosome: a future of targeted drug delivery systems. J Adv Pharm Technol Res 2010;1(4):374.

Keino H, Horie S, Sugita S. Immune privilege and eye-derived T-regulatory cells. J Immunol Res. 2018;2018

Kezic J, Xu H, Chinnery HR, Murphy CC, McMenamin PG. Retinal microglia and uveal tract dendritic cells and macrophages are not CX3CR1 dependent in their recruitment and distribution in the young mouse eye. Invest Ophthalmol Vis Sci. 2008;49(4):1599–608.

Kiesewetter A, Cursiefen C, Eming SA, Hos D. Phase-specific functions of macrophages determine injury-mediated corneal hem-and lymphangiogenesis. Sci Rep. 2019;9(1):1–6.

Kim H, Csaky KG. Nanoparticle–integrin antagonist C16Y peptide treatment of choroidal neovascularization in rats. J Control Release. 2010;142(2):286–93.

Knickelbein JE, Buela KA, Hendricks RL. Antigen-presenting cells are stratified within normal human corneas and are rapidly mobilized during ex vivo viral infection. Invest Ophthalmol Vis Sci. 2014;55(2):1118–23.

Krause L, Coupland SE, Hoffmann F. The behaviour of ED1-and ED2-positive cells in the rat iris and choroid following penetrating keratoplasty and cyclosporin A therapy. Graefes Arch Clin Exp Ophthalmol. 1996;234(1):S149–58.

Lajavardi L, Bochot A, Camelo S, Goldenberg B, Naud MC, Behar-Cohen F, Fattal E, de Kozak Y. Downregulation of endotoxin-induced uveitis by intravitreal injection of vasoactive intestinal peptide encapsulated in liposomes. Invest Ophthalmol Vis Sci. 2007;48(7):3230–8.

Lee DH, Ghiasi H. Roles of M1 and M2 macrophages in herpes simplex virus 1 infectivity. J Virol. 2017;91(15):e00578.

Lee JE, Liang KJ, Fariss RN, Wong WT. Ex vivo dynamic imaging of retinal microglia using time-lapse confocal microscopy. Invest Ophthalmol Vis Sci. 2008;49(9):4169–76.

Liu J, Xue Y, Dong D, Xiao C, Lin C, Wang H, Song F, Fu T, Wang Z, Chen J, Pan H. CCR2− and CCR2+ corneal macrophages exhibit distinct characteristics and balance inflammatory responses after epithelial abrasion. Mucosal Immunol. 2017;10(5):1145–59.

Ljubimova D. Biomechanics of the human eye and intraocular pressure measurements (Doctoral dissertation, KTH). 2009

Luo Q, Zhao J, Zhang X, Pan W. Nanostructured lipid carrier (NLC) coated with Chitosan Oligosaccharides and its potential use in ocular drug delivery system. Int J Pharm. 2011;403(1-2):185–91.

Maneu V, Yáñez A, Murciano C, Molina A, Gil ML, Gozalbo D. Dectin-1 mediates in vitro phagocytosis of Candida albicans yeast cells by retinal microglia. FEMS Immunol Med Microbiol. 2011;63(1):148–50.

Margolis R, Spaide RF. A pilot study of enhanced depth imaging optical coherence tomography of the choroid in normal eyes. Am J Ophthalmol. 2009;147(5):811–5.

Maurice DM, Polgar J. Diffusion across the sclera. Exp Eye Res. 1977;25(6):577–82.

May CA. Mast cell heterogeneity in the human uvea. Histochem Cell Biol. 1999;112(5):381–6.

McKay TB, Seyed-Razavi Y, Ghezzi CE, Dieckmann G, Nieland TJ, Cairns DM, Pollard RE, Hamrah P, Kaplan DL. Corneal pain and experimental model development. Prog Retin Eye Res. 2019;71:88–113.

McLeod DS, Bhutto I, Edwards MM, Silver RE, Seddon JM, Lutty GA. Distribution and quantification of choroidal macrophages in human eyes with age-related macular degeneration. Invest Ophthalmol Vis Sci. 2016;57(14):5843–55.

McMenamin PG. Dendritic cells and macrophages in the uveal tract of the normal mouse eye. Br J Ophthalmol. 1999;83(5):598–604.

McMenamin PG, Lee WR. Ultrastructural pathology of melanomalytic glaucoma. Br J Ophthalmol. 1986;70(12):895–906.

McMenamin PG, Djano J, Wealthall R, Griffin BJ. Characterization of the macrophages associated with the tunica vasculosa lentis of the rat eye. Invest Ophthalmol Vis Sci. 2002;43(7):2076–82.

McWhorter FY, Wang T, Nguyen P, Chung T, Liu WF. Modulation of macrophage phenotype by cell shape. Proc Natl Acad Sci. 2013;110(43):17253–8.

Menjoge AR, Kannan RM, Tomalia DA. Dendrimer-based drug and imaging conjugates: design considerations for nanomedical applications. Drug Discov Today. 2010;15(5-6):171–85.

Mérida S, Palacios E, Navea A, Bosch-Morell F. Macrophages and uveitis in experimental animal models. Mediat Inflamm. 2015;2015

Mishra GP, Bagui M, Tamboli V, Mitra AK. Recent applications of liposomes in ophthalmic drug delivery. J Drug Deliv. 2011;2011

Mitra AK, Kwatra D, Vadlapudi AD. Drug delivery. Sudbury: Jones & Bartlett Publishers; 2014.

Mudgil M, Gupta N, Nagpal M, Pawar P. Nanotechnology: a new approach for ocular drug delivery system. Int J Pharm Pharm Sci. 2012;4(2):105–12.

Murakami T, Felinski EA, Antonetti DA. Occludin phosphorylation and ubiquitination regulate tight junction trafficking and vascular endothelial growth factor-induced permeability. J Biol Chem. 2009;284(31):21036–46.

Myles ME, Neumann DM, Hill JM. Recent progress in ocular drug delivery for posterior segment disease: emphasis on transscleral iontophoresis. Adv Drug Deliv Rev. 2005;57(14):2063–79.

Naseri N, Valizadeh H, Zakeri-Milani P. Solid lipid nanoparticles and nanostructured lipid carriers: structure, preparation and application. Adv Pharm Bull. 2015 Sep;5(3):305.

Naumann GO. Pathology of the eye. New York: Springer; 2012.

Navegantes KC, de Souza Gomes R, Pereira PA, Czaikoski PG, Azevedo CH, Monteiro MC. Immune modulation of some autoimmune diseases: the critical role of macrophages and neutrophils in the innate and adaptive immunity. J Transl Med. 2017;15(1):1–21.

Nickla DL, Wallman J. The multifunctional choroid. Prog Retin Eye Res. 2010;29(2):144–68.

Nozaki M, Raisler BJ, Sakurai E, Sarma JV, Barnum SR, Lambris JD, Chen Y, Zhang K, Ambati BK, Baffi JZ, Ambati J. Drusen. complement components C3a and C5a promote choroidal neovascularization. Proc Natl Acad Sci. 2006;103(7):2328–33.

Obeid MA, Al Qaraghuli MM, Alsaadi M, Alzahrani AR, Niwasabutra K, Ferro VA. Delivering natural products and biotherapeutics to improve drug efficacy. Ther Deliv. 2017;8(11):947–56.

Okunuki Y, Mukai R, Nakao T, Tabor SJ, Butovsky O, Dana R, Ksander BR, Connor KM. Retinal microglia initiate neuroinflammation in ocular autoimmunity. Proc Natl Acad Sci. 2019;116(20):9989–98.

Oyster CW. The human eye. Sunderland: Sinauer; 1999.

Pannier AK, Segura T. Surface-and hydrogel-mediated delivery of nucleic acid nanoparticles. In: Nanotechnology for nucleic acid delivery. Totowa: Humana Press; 2013. p. 149–69.

Pardakhty A, Moazeni E, Varshosaz J, Hajhashemi VA, Najafabadi AR. Pharmacokinetic study of niosome-loaded insulin in diabetic rats. DARU J Pharm Sci. 2011;19(6):404.

Patel A, Cholkar K, Agrahari V, Mitra AK. Ocular drug delivery systems: An overview. World J Pharmacol. 2013;2(2):47.

Patra JK, Das G, Fraceto LF, Campos EV, del Pilar Rodriguez-Torres M, Acosta-Torres LS, Diaz-Torres LA, Grillo R, Swamy MK, Sharma S, Habtemariam S. Nano based drug delivery systems: recent developments and future prospects. J Nanobiotechnol. 2018;16(1):1–33.

Pederson JE. Fluid physiology of the subretinal space. Retina. 1989:89–102.

Petrosino S, Di Marzo V. The pharmacology of palmitoylethanolamide and first data on the therapeutic efficacy of some of its new formulations. Br J Pharmacol. 2017;174(11):1349–65.

Peyman GA. Methods to regulate polarization of excitable cells. United States patent US 8,409,263. 2013.

Price KM, Richard MJ. The tearing patient: Diagnosis and management. Ed. Scott IU & Fekrat S. American Academy of Ophthalmology. www.aao.org. 2009.

Qiao H, Hisatomi T, Sonoda KH, Kura S, Sassa Y, Kinoshita S, Nakamura T, Sakamoto T, Ishibashi T. The characterisation of hyalocytes: the origin, phenotype, and turnover. Br J Ophthalmol. 2005;89(4):513–7.

Rashid K, Akhtar-Schaefer I, Langmann T. Microglia in retinal degeneration. Front Immunol. 2019;10: 1975.

Rathnasamy G, Foulds WS, Ling EA, Kaur C, et al. Retinal microglia–A key player in healthy and diseased retina. Prog Neurobiol. 2019;173:18–40.

Ravar F, Saadat E, Gholami M, Dehghankelishadi P, Mahdavi M, Azami S, Dorkoosh FA. Hyaluronic acid-coated liposomes for targeted delivery of paclitaxel, in-vitro characterization and in-vivo evaluation. J Control Release. 2016;229:10–22.

Raviola G. Conjunctival and episcleral blood vessels are permeable to blood-borne horseradish peroxidase. Invest Ophthalmol Vis Sci. 1983;24(6):725–36.

Reichenbach A, Wurm A, Pannicke T, Iandiev I, Wiedemann P, Bringmann A. Müller cells as players in retinal degeneration and edema. Graefes Arch Clin Exp Ophthalmol. 2007;245(5):627–36.

Sakai T, Kohno H, Ishihara T, Higaki M, Saito S, Matsushima M, Mizushima Y, Kitahara K. Treatment of experimental autoimmune uveoretinitis with poly (lactic acid) nanoparticles encapsulating betamethasone phosphate. Exp Eye Res. 2006;82(4):657–63.

Sakai T, Kuno N, Takamatsu F, Kimura E, Kohno H, Okano K, Kitahara K. Prolonged protective effect of basic fibroblast growth factor–impregnated nanoparticles in Royal College of Surgeons rats. Invest Ophthalmol Vis Sci. 2007;48(7):3381–7.

Sakai T, Ishihara T, Higaki M, Akiyama G, Tsuneoka H. Therapeutic effect of stealth-type polymeric nanoparticles with encapsulated betamethasone phosphate on experimental autoimmune uveoretinitis. Invest Ophthalmol Vis Sci. 2011;52(3):1516–21.

Sakurai E, Ozeki H, Kunou N, Ogura Y. Effect of particle size of polymeric nanospheres on intravitreal kinetics. Ophthalmic Res. 2001;33(1):31–6.

Schafer DP, Lehrman EK, Kautzman AG, Koyama R, Mardinly AR, Yamasaki R, Ransohoff RM, Greenberg ME, Barres BA, Stevens B. Microglia sculpt postnatal neural circuits in an activity and complement-dependent manner. Neuron. 2012;74(4):691–705.

Schlereth SL, Kremers S, Schrödl F, Cursiefen C, Heindl LM. Characterization of antigen-presenting macrophages and dendritic cells in the healthy human sclera. Invest Ophthalmol Vis Sci. 2016;57(11):4878–85.

Schnitzer JE, Liu J, Oh P. Endothelial caveolae have the molecular transport machinery for vesicle budding, docking, and fusion including VAMP, NSF, SNAP, annexins, and GTPases. J Biol Chem. 1995;270(24):14399–404.

Schraermeyer M, Schnichels S, Julien S, Heiduschka P, Bartz-Schmidt KU, Schraermeyer U. Ultrastructural analysis of the pigment dispersion syndrome in DBA/2J mice. Graefes Arch Clin Exp Ophthalmol. 2009;247(11):1493–504.

Sharma R, Kim SY, Sharma A, Zhang Z, Kambhampati SP, Kannan S, Kannan RM. Activated microglia targeting dendrimer–minocycline conjugate as therapeutics for neuroinflammation. Bioconjug Chem. 2017;28(11):2874–86.

Sharma A, Liaw K, Sharma R, Zhang Z, Kannan S, Kannan RM. Targeting mitochondrial dysfunction and oxidative stress in activated microglia using dendrimer-based therapeutics. Theranostics. 2018;8(20):5529.

Shidhaye SS, Vaidya R, Sutar S, Patwardhan A, Kadam VJ. Solid lipid nanoparticles and nanostructured lipid carriers-innovative generations of solid lipid carriers. Curr Drug Deliv. 2008;5(4):324–31.

Singh D, Singh RS, Singh K, Singh SK, Singh IR, Singh R, Fugo RJ. The conjunctival lymphatic system. Ann Ophthalmol. 2003;35(2):99–104.

Soiberman U, Kambhampati SP, Wu T, Mishra MK, Oh Y, Sharma R, Wang J, Al Towerki AE, Yiu S, Stark WJ, Kannan RM. Subconjunctival injectable dendrimer-dexamethasone gel for the treatment of corneal inflammation. Biomaterials. 2017;125:38–53.

Solanki A, Desai S, Grover A, Hirani A, Pathak Y, Sutariya V. Ocular drug delivery: impact of in vitro cell culture models. In: Nano-biomaterials for ophthalmic drug delivery. Cham: Springer; 2016. p. 483–94.

Sonoda KH, Sasa Y, Qiao H, Tsutsumi C, Hisatomi T, Komiyama S, Kubota T, Sakamoto T, Kawano YI, Ishibashi T. Immunoregulatory role of ocular macrophages: the macrophages produce RANTES to suppress experimental autoimmune uveitis. J Immunol. 2003;171(5):2652–9.

Sosnová M, Bradl M, Forrester JV. CD34+ corneal stromal cells are bone marrow–derived and express hemopoietic stem cell markers. Stem Cells. 2005;23(4):507–15.

Souto EB, Doktorovova S, Gonzalez-Mira E, Egea MA, Garcia ML. Feasibility of lipid nanoparticles for ocular delivery of anti-inflammatory drugs. Curr Eye Res. 2010;35(7):537–52.

Sridhar MS. Anatomy of cornea and ocular surface. Indian J Ophthalmol. 2018;66(2):190.

Stern ME, Schaumburg CS, Dana R, Calonge M, Niederkorn JY, Pflugfelder SC. Autoimmunity at the ocular surface: pathogenesis and regulation. Mucosal Immunol. 2010;3(5):425–42.

Strack A, Asensio VC, Campbell IL, Schlüter D, Deckert M. Chemokines are differentially expressed by astrocytes, microglia and inflammatory leukocytes in Toxoplasma encephalitis and critically regulated by interferon-γ. Acta Neuropathol. 2002;103(5):458–68.

Subramaniam B, Siddik ZH, Nagoor NH. Optimization of nanostructured lipid carriers: understanding the types, designs, and parameters in the process of formulations. J Nanopart Res. 2020;22:1–29.

Taylor AW, Streilein JW, Cousins SW. Immunoreactive vasoactive intestinal peptide contributes to the immunosuppressive activity of normal aqueous humor. J Immunol. 1994;153(3):1080–6.

Tong L, Lan W, Petznick A. Definition of the ocular surface. In: Ocular surface: anatomy and physiology, disorders and therapeutic care. 2012. 1.

Torchilin VP. Recent advances with liposomes as pharmaceutical carriers. Nat Rev Drug Discov. 2005;4(2):145–60.

Tretiach M, Madigan MC, Wen L, Gillies MC. Effect of Müller cell co-culture on in vitro permeability of bovine retinal vascular endothelium in normoxic and hypoxic conditions. Neurosci Lett. 2005;378(3):160–5.

Ueno H, Tamai A, Iyota K, Moriki T. Electron microscopic observation of the cells floating in the anterior chamber in a case of phacolytic glaucoma. Jpn J Ophthalmol. 1989;33(1):103–13.

Vadlapudi AD, Patel A, Cholkar K, Mitra AK. Recent patents on emerging therapeutics for the treatment of glaucoma, age related macular degeneration and uveitis. Recent Pat Biomed Eng. 2012;5(1):83.

Vagaja NN, Chinnery HR, Binz N, Kezic JM, Rakoczy EP, McMenamin PG. Changes in murine hyalocytes are valuable early indicators of ocular disease. Invest Ophthalmol Vis Sci. 2012;53(3):1445–51.

Wadhwa S, Paliwal R, Paliwal SR, Vyas SP. Nanocarriers in ocular drug delivery: an update review. Curr Pharm Des. 2009;15(23):2724–50.

Wang WY, Yao C, Shao YF, Mu HJ, Sun KX. Determination of puerarin in rabbit aqueous humor by liquid chromatography tandem mass spectrometry using microdialysis sampling after topical administration of Puerarin PAMAM dendrimer complex. J Pharm Biomed Anal. 2011;56(4):825–9.

Wang M, Wang X, Zhao L, Ma W, Rodriguez IR, Fariss RN, Wong WT. Macroglia-microglia interactions via TSPO signaling regulates microglial activation in the mouse retina. J Neurosci. 2014;34(10):3793–806.

Wang X, Zhao L, Zhang J, Fariss RN, Ma W, Kretschmer F, Wang M, Hua Qian H, Badea TC, Diamond JS, Gan WB. Requirement for microglia for the maintenance of synaptic function and integrity in the mature retina. J Neurosci. 2016;36(9):2827–42.

Watanabe T, Raff MC. Retinal astrocytes are immigrants from the optic nerve. Nature. 1988;332(6167):834–7.

Wobmann PR, Fine BS. The clump cells of Koganei: a light and electron microscopic study. Am J Ophthalmol. 1972;73(1):90–101.

Wu M, Downie LE, Grover LM, Moakes RJ, Rauz S, Logan A, Jiao H, Hill LJ, Chinnery HR. The neuro-regenerative effects of topical decorin on the injured mouse cornea. J Neuroinflammation. 2020;17:1–4.

Xiao X, Wu ZC, Chou KC. A multi-label classifier for predicting the subcellular localization of gram-negative bacterial proteins with both single and multiple sites. PLoS One. 2011;6(6):e20592.

Xu H, Chen M, Reid DM, Forrester JV. LYVE-1–positive macrophages are present in normal murine eyes. Invest Ophthalmol Vis Sci. 2007;48(5):2162–71.

Yeo PL, Lim CL, Chye SM, Ling AP, Koh RY. Niosomes: a review of their structure, properties, methods of preparation, and medical applications. Asian Biomed. 2018;11(4):301–14.

Yoshida M, Takeuchi M, Streilein W. Participation of pigment epithelium of iris and ciliary body in ocular immune privilege. 1. Inhibition of T-cell activation in vitro by direct cell-to-cell contact. Invest Ophthalmol Vis Sci. 2000;41(3):811–21.

Yu Y, Chen D, Li Y, Yang W, Tu J, Shen Y. Improving the topical ocular pharmacokinetics of lyophilized cyclosporine A-loaded micelles: formulation, in vitro and in vivo studies. Drug Deliv. 2018;25(1):888–99.

Yuan TF, Liang YX, Peng B, Lin B, So KF. Local proliferation is the main source of rod microglia after optic nerve transection. Sci Rep. 2015;5(1):1–8.

Zeng XX, Ng YK, Ling EA. Labelling of retinal microglial cells following an intravenous injection of a fluorescent dye into rats of different ages. J Anat. 2000;196(2):173–9.

Zhang Y, Stone J. Role of astrocytes in the control of developing retinal vessels. Invest Ophthalmol Vis Sci. 1997;38(9):1653–66.

Zhang L, Li Y, Zhang C, Wang Y, Song C. Pharmacokinetics and tolerance study of intravitreal injection of dexamethasone-loaded nanoparticles in rabbits. Int J Nanomed. 2009;4:175.

Zhang J, Zhou T, Wang L, He J, Xia H. Nanoemulsion as a vehicle to enhance the ocular absorption after topically applied Cyclosporine A in the rabbit eye. Invest Ophthalmol Vis Sci. 2012;53(14):488.

Zhao Y, He G, Guo W, Bao L, Yi M, Gong Y, Zhang S. Self-assembled micelles prepared from amphiphilic copolymers bearing cell outer membrane phosphorylcholine zwitterions for a potential anti-phagocytic clearance carrier. Polym Chem. 2016;7(36):5698–708.

Macrophage Targeting for Therapy of HIV

Sakshi Nainwani, Anushka Tyagi, Yashwant V. Pathak, and Swati Gupta

Abstract

HIV is a global threat, not only from today but for ages as nearly 25 million people have died due to this curse of nature and approximately 40 million population get infected globally.

Macrophages are the chosen ones here and are favourite site for HIV to live in; they attack the macrophages and the monocytes, and hence depletion of T lymphocytes begins where the cells provide immunity to the body. The worst part is that these macrophages are present in all parts and tissues; hence, this viral infection spreads rapidly and opens its branches in all parts of the body. Cells die and make an individual weak; story does not end here. Few cells remain alive and create more fuss by being the latent reservoir for the culprit, that is, HIV. Latent viral reservoirs are the cells which survive even after the HIV infection; most of them are killed or infected in the process, so basically the main problem in the complete eradication of HIV is these latent viral reservoirs, although the process through which these reservoirs survive is still not understood. So how to treat this deadly disease?

Here, in this chapter, we will study the physiology, how HIV enters the body, how it becomes the resident and open its branches, and how it can be treated using various conventional and current approaches. Conventional therapies include various ART and enzyme inhibitor therapies like RTIs, NNRTIs, NRTIs, protease inhibitors, fusion inhibitors, and various other blocking agent therapies, and various novel approaches include therapies like carbohydrate-binding agents, PI3K/Akt blocking agents, siRNA, immune-based therapies, flushing therapies and more.

Various studies of nanocarriers targeting approaches such as anti-HIV have also been included like liposomes, nanoparticles, dendrimers, bioconjugates, solid lipid nanoparticles and ethosomes.

Keywords

HIV-1 · Macrophages · Nef · Tat · Vpr · Antiretroviral therapy · Latency · Nanocarriers

S. Nainwani · A. Tyagi · S. Gupta (✉)
Department of Pharmaceutics, Amity Institute of Pharmacy, Amity University Uttar Pradesh, Noida, India

Y. V. Pathak
University of South Florida, Taneja College of Pharmacy, Tampa, FL, USA

Adjunct Professor University of Airlangga, Surabaya, Indonesia

S. Gupta, Y. V. Pathak (eds.), *Macrophage Targeted Delivery Systems*,
https://doi.org/10.1007/978-3-030-84164-5_20

1 Introduction

Human Immunodeficiency Virus (HIV) belongs to a group of viruses known as retrovirus, which are having a proteinaceous envelope enclosing its RNA (Ribonucleic acid) genome. If you have HIV, you have an infection that damages your immune system over the due course of time and causes AIDS. AIDS stands for Acquired Immunodeficiency Syndrome ('Syndrome' means a group of symptoms). It is the final stage of HIV when the immune system becomes too weak to fend off an ordinary infection. When invaders such as bacteria or viruses enter the body, they can cause illness. As the body's immune system is weak, other diseases can also occur easily; those diseases are known as opportunistic diseases. If a person has HIV and two or more opportunistic diseases, then it is said to be AIDS.

There are certain ways through which transmission of HIV infection occurs: (a) sexual contact with infected person, (b) by transfusion of contaminated blood and blood products, (c) by sharing infected needles as within the case of intravenous drug abusers, and (d) from infected mother to her child through placenta. So, people that are at high risk of getting this infection include individuals who have multiple sexual partners, drug addicts who take drugs intravenously, individuals that require blood transfusions regularly and youngsters born to an HIV-infected mother (Kumar and Herbein 2014).

There are two types of HIV that have been encountered so far, that is, HIV-1 and HIV-2.

HIV-1 was also known as the anorexigenic organism before long as the primary official recognition of HIV patients within the USA. HIV-2 was reported first in Africa in 1985 and is marked as totally different from HIV-1. It closely bears resemblance to simian virus, which infects macaques in captivity. Simian viruses that naturally infect African primates are suspected to succeed in humans via multiple cross-species transmissions leading to the unfold of viruses, that is, HIV-1 and HIV-2.

HIV infects various types of cells, though macrophages and T lymphocytes are its main target. Macrophages are functionally phagocytotic type of immune cells which have mobility and can travel to the location of infection. Their main function is to perform clearance of pathogens and cellular debris. Another function of macrophages is that they mimic antigen-presenting cells and mediators of both innate and acquired immunity. Macrophages act as a potent reservoir of HIV. Macrophages deal with pathogen antigen peptide and processed it to the cluster of differentiation 4 (CD4+) and T cells through major histocompatibility complex (MHC II) pathway. By this interchange of information between macrophages and T cells, transmission of HIV occurs to T cells from macrophages. Not only it spreads but it also releases cytotoxic factors that encourage the apoptosis of standard cells like CD4+ (cluster of differentiation 4) and CD8+ (cluster of differentiation 8) T cells. HIV virus ultimately causes the destruction of these cells (CD4+ and CD8+ T cells). Since the residence time of HIV is long in macrophages, they pretend to be the greatest source of HIV production in infected person for a sufficiently long duration. Macrophages are present in every organ, so they can infect the body thoroughly, including the brain and lungs, and can persist by avoiding immune system detection. In the central nervous system (CNS), various types of macrophages are present, like meningeal macrophages, parenchymal microglia, choroid plexus macrophages and perivascular macrophages, all of which express viral co-receptors and are susceptible to HIV infection.

Between all the cells stated above, perivascular macrophages are the prominent target of HIV in the CNS, and the viral DNA (Deoxyribonucleic Acid) can be eradicated by these cells throughout the infection which shows that these perivascular macrophages act as a reservoir for HIV. If we talk about lungs, then Alveolar macrophages are the primary viral targets in lungs (Jambo et al. 2014). Unlike the CNS and lungs, GIT macrophages downregulate CD4 (a glycoprotein for T cell development) and CCR5 (CC Chemokine Receptor 5, a GPCR) in response to HIV. Though this remains an open question, downregulation of CD4 proteins and CCR5 receptors may result in reduced viral infection in these GIT macrophages. HIV can also infect marginal zone macrophages in the spleen but conflicting data exist on whether these cells maintain a reservoir of virus (Zaritsky et al. 2013).

Currently combinational antiretroviral therapy, commonly known as "ART," has been prominently used in treating HIV-1 infection since it suppresses the infection to a significant level (Pomerantz and Horn 2003). ART is the most prominent and promising treatment for HIV as of now and can help in increasing life expectancy, as HIV is a syndrome which cannot be treated completely. Another approved treatment for HIV is reverse transcriptase inhibitors (RTIs). As of now, 25 compounds have been approved for treating HIV in infected patients. Out of those, nearly 50% are RTIs (Abbas and Herbein 2012a). There are two types of RTIs whichinclude NRTIs (nucleoside reverse transcriptase inhibitors) and NNRTIs (non-nucleoside reverse transcriptase inhibitors). NRTIs target reverse transcriptase enzyme due to which cDNA is formed from HIV genomic RNA where this enzyme coverts viral genomic RNA into cDNA, an important step in the life cycle of HIV (Fig. 1). NRTIs include emtricitabine, tenofovir, abacavir, lamivudine, stavudine, zalcitabine, didanosine and zidovudine (Perno et al. 2006).

Protease inhibitor (PI) is one more treatment method for HIV. Unlike reverse transcriptase inhibitors (RIs), PIs function at one stage of HIV, that is, the post-integration stage (Greenhead et al. 2000) (Fig. 1). PI basically binds on the HIV protease enzyme's active site and makes it non-functional. These PIs are effective in both types of infected macrophages and CD4+ either acute or chronic. So, in this chapter, we are going to see how HIV interacts with macrophages and replicates inside it and how therapy of HIV is done by targeting macrophages and will also study some novel therapies for the same. Figure 1 depicts the key events which occur in life cycle of HIV-1 when targeted by antiretroviral drug. These drugs target four crucial steps in viral life cycle that are entry of viral genome in the host cell, reverse transcription, proviral DNA integration into host chromatin and viral encoded protease processing polyprotein. On the basis of targeting steps, the antiretroviral drugs are named fusion inhibitors: (a) reverse transcription, (b) inhibition of integrase, (c) protease inhibitors and (d) Targeting occurs step by step, which result in resistant mutant's emergence. Antiretroviral therapy (ART) suppresses the growth of HIV-1 up to a prominent extent. A point to note is that the viral assembly in macrophage takes over on both plasma membrane and virus-containing compartments. Here, in the figure, only key protein involved is shown.

Abbreviations: RT—reverse transcription, MA—matrix protein, IN—integrase, VPr—viral

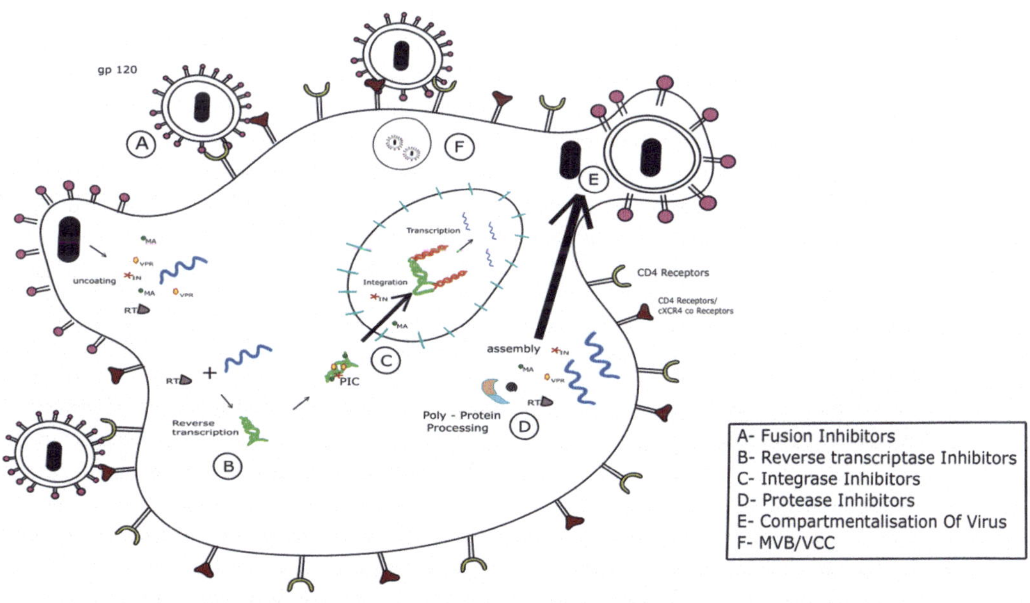

Fig. 1 Key events occurring in life cycle of HIV targeted by antiretroviral drug

protein R, P—virus-encoded protease, PIC—pre-integration complex, MVB—multivesicular bodies, LE—late endosomes and VCC—virus-containing compartment.

1.1 Entry of HIV-1 into Macrophages

It begins with the encounter of HIV-1 into host cells, which involves virus substance (virus surface compound gp120 protein) and its involvement with CD4 receptor that is present in each T cell additionally as in macrophages (Herbein and Varin 2010b). In the next step, fusion of viral envelope occurs with the membrane of host cell that is ruled by the interaction of the co-receptors (CCR5 or CXCR4), and it was believed that macrophages have several receptors like CCR5 receptor and CXCR4 receptor of the T cells which resulted in T cell tropic and phagocyte tropic HIV-1 terminology (Iordanskiy et al. 2013). Further in vivo studies discovered that each co-receptor is present on macrophages likewise as in T cells (Iordanskiy et al. 2013; Carter and Ehrlich 2008). Notably, the naturally transmitted HIV-1 viruses consume CCR5 for infection, despite the fact that their primary targets are not macrophages but T cells. Resident macrophages of brain (CNS, microglia) are infected via co-receptor CCR5. Common understanding is that these R5 and X4 viruses will replicate in each macrophage alongside T cells. However, their replication potency is different in different cells that rely on the cellular environment. Connection of infective agents from macrophages and T cells is identical, but they have different types of host protein integrated into their viral particle.

1.2 Reverse Transcription and Various Factors Restricting in Host

HIV-1 enters the host via CXCR4 or CCR5 co-receptors. In both cases, ribonucleoprotein complex which is the infective agent enters inside the cytoplasm (Carter and Ehrlich 2008) where virus-encoded reverse transcriptase exploits viral genomic RNA and generates single-stranded deoxyribonucleic acid which is followed by the formation of double-stranded DNA (Mougel et al. 2009) (Fig. 1). However, the speed of reverse transcription is somewhat slower in macrophages than what is determined in T cells. Macrophages being terminally differentiated cells (non-dividing) have restricted dNTP pools for proviral deoxyribonucleic acid production (Schmidtmayerova et al. 1998). Many reports claim that addition of deoxynucleosides to the primary human macrophage culture will improve the rate of reverse transcription HIV-1 proving that dNTP pool is a very important rate-limiting factor in macrophages (Furge and Guengerich 1997; Diamond et al. 2004).

Alongside, macrophages consist of bound repressing factors that intervene with virus life cycle and are known as host restriction factors. These host restriction factors embody tetherin, APOBEC3G and recently known sterile alpha motif (SAM) domain and HD domain-containing supermolecule 1 (SAMHD1) (Sheehy et al. 2002). APOBEC3G triggers G-to-A hypermutation in aborning deoxyribonucleic acid. CD317/BST-2, also referred to as tetherin, hinders the discharge of viral components from infected cells. HIV-1 employs several methods to beat these restrictive factors. HIV-1 has supermolecules like Vif and Vpu which restrain the APOBEC3G and tetherin (Sheehy et al. 2002; Koppensteiner et al. 2012), even there are reports that describe tetherin antagonism through HIV-1 Nef supermolecule.

SAMHD1 could be a phagocyte-specific host restriction factor that has triphosphohydrolase activity ensuing dNTPs hydrolysis in nucleosides and triphosphates. Therefore, SAMHD1 lessens pool of dNTPs in macrophages to a precise level leading to the inefficient transformation of HIV-1 genome from RNA into proviral deoxyribonucleic acid (reverse transcription) (Lahouassa et al. 2012). However, Vpx macromolecule of HIV-2 induces proteasome-dependent degradation of SAMHD1 through a ligase, that is, CRL4DCAF1 E3. Recently, McKnight and colleagues, looking for various host restriction factors, screened and identified 114 genes from several human genes with important impact on

HIV-1 replication. Moreover, these studies discovered that inhibition of PAF1 family will result in elevated HIV-1 replication. Particularly, PAF1 isn't restricted to macrophages solely; they're conjointly expressed in several monocytes and T cells, suggesting list of restriction factors against HIV-1 (Liu et al. 2011). Recently, Allouch and co-workers revealed through their studies that cyclin-dependent enzyme inhibitor p21 inhibits HIV-1 replication in MDMs, that is, monocyte-derived macrophages by intervening with reverse transcription of the viral genome by a mechanism independent of SAMHD1. In addition, they depict that p21 mitigates the synthesis of dNTP through the downregulation process of the expression of RNR2 which is mandatory for the biosynthesis of dNTPs (Allouch et al. 2013).

1.3 Nuclear Transport

HIV dsDNA (newly formed) is sent to the nucleus as pre-integration complicated (PIC) (as shown in Fig. 1). Unlike T cells, PIC gets transported to the nucleus in macrophages and performs cell division. PIC includes viral proteins that undergo reverse transcriptase (MA, p17) matrix, Vpr, integrase (IN), and capsid macromolecule (CA) additionally to freshly synthesized dsDNA. Nevertheless, CA detaches from PIC more before the nuclear entry. PIC is directed to importin alpha/beta by Vpr, MA, IN. However, precise performance of those PIC proteins in nuclear transport continues to be a matter of discussion (Carter and Ehrlich 2008). Contrasting MA and IN, nuclear localization signal is lacked by Vpr (Gallay et al. 1996; Gallay et al. 1997). Additionally, interaction between importin α and Vpr is essential not only for the nuclear transport of PIC, however, conjointly for the HIV-1 replication in macrophages (Nitahara-Kasahara et al. 2007). Moreover, in primary macrophages, macromolecule emerin (an integral nuclear inner membrane protein) of host cell plays an important role in viral DNA integration into the chromatin (Jacque and Stevenson 2006; Kobiler et al. 2012). Primary macrophages lack emerin which have bad rate of proviral DNA integration of HIV into the host chromatin granule, but lack of emerin does not inhibit PIC entry into the nucleus. Additionally, emerin's binding partner, the LEM (LAP2 (lamina-associated peptide 2)/emerin/ MAN1), is important for the interaction of viral complementary DNA with emerin and potential of emerin to support HIV-1 infection in macrophages (Jacque and Stevenson 2006). However, a study suggests that HIV-1 has efficiency to infect dividing cells despite the absence of emerin, hence proposing that the role of emerin in HIV-1 infection constricts to only macrophages. Besides many different host factors concerned, life cycle of HIV in macrophages is recently reviewed (Kobiler et al. 2012).

1.4 Transcription of HIV-1

HIV-1 transcription is ruled by binding to viral proteins and other host factors through the LTR (long terminal repeat), of the virus, whose function is to promote infectious viral agent (Kilareski et al. 2009). Factors included in host factors are nuclear factor kappa B (NF-κB) family, Sp family, activator protein 1 (AP-1), CCAAT enhancer-binding protein (C/EBP) and nuclear issue of activated T cells. LTR has particular binding sites present in which host factors bind. On the other hand, LTR also binds to viral proteins, that is, Tat and Vpr to control HIV-1 transcription (Kilareski et al. 2009; Herbein et al. 2010a). Host factors could also be cell-specific, for example, C/EBP proteins bind to specific binding sites and are important for replication of HIV-1 in macrophages although not in T cells (Henderson and Calame 1997). However, primary macrophages which get infected with HIV-1 infection can have mutation in binding sites of C/EBP which doesn't support replication of HIV-1. On the other hand, Jurkat and H9 cells primarily a type of CD4+ T cells support the replication of HIV-1C/EBP mutants (Henderson and Calame 1997).

1.5 Assembly of HIV-1 in Macrophages

Assembly of HIV-1 takes place in plasma membrane in the case of CD4+ T cells, while the other assembly site is not yet identified. Earlier finding suggests the HIV-1 virion's presence in multivesicular bodies (MVBs) or in late endosomes (LEs). Through immune-electron microscopy studies, latter finding supports the studies which disclose the presence of several MVB-specific markers, namely, tetraspanins, CD9, CD53, CD81 and MHC II (Tan and Sattentau 2013). Additionally, HIV-1 progeny free from infected macrophages conjointly possess these markers, further strengthening the fact that macrophages are free from LEs or MVBs (Tan and Sattentau 2013; Kramer et al. 2005). However, many studies disclosed that structures harbouring HIV-1 in infected macrophages have some distinct characters that don't seem to be LEs or MVBs characteristics. These distinctive characteristics include tubular connection to the extracellular area and neutral hydrogen ion concentration (Kramer et al. 2005). The term 'virus-containing compartments' (VCCs) has been appointed to the structures that act as the assembly of virus in macrophages (Tan and Sattentau 2013). Surprisingly, these VCCs have connection in uninfected macrophages, but they become much prominent upon HIV-1 infection (Welsch et al. 2011). Value mentioning, VCCs, have restricted access to the adaptive and innate immune effector molecules (Tan and Sattentau 2013). In distinction, many studies are in the favour of HIV-1 progeny budding in macrophages from plasma membrane. Taken together, all these different findings indicate that there is a fair risk that HIV-1 might bind to plasma membrane more prominently than VCCs. VCCs might behave like a safe house for HIV-1 in macrophages resulting in reservoirs of HIV-1. However, more prominent experiments and studies are required to support these hypothetical statements.

2 HIV Protein Interaction with Cell Signalling in Macrophages

Among HIV-1 proteins, the viral proteins Vpr, Tat and Nef interfere with signalling pathways in macrophages, which have been discussed in detail. Signalling pathways are a series of chemical reactions in which cell receives signal from its surrounding environment when a molecule, like growth factor or hormone, binds to a protein (here viral protein) or in the cell.

2.1 Tat (Trans-Activator of Transcription)

Trans-activator of transcription (Tat) protein is an 86–101aa virus-encoded pleiotropic protein that directly or indirectly moderates many pathways of HIV life cycle like replication, transcription and progeny release by controlling both cellular and viral gene expressions (Buonaguro et al. 1992). Additionally, Tat has been found in sera of HIV-infected patients similarly in cell culture settings, which indicates that its role is of a modulator for cellular performance in infected cells and additionally to target various uninfected cells (Ensoli et al. 1993). Moreover, microglia, monocytes and macrophages are the cells which are activated by Tat protein (Welsch et al. 2011). Additionally, Tat is understood to activate the expression of HIV co-receptors (CXCR4, CCR5 and CCR3) in macrophages in an exceedingly dose-dependent manner that could absolutely influence HIV-1 infection (Huang et al. 1998). Moreover, Tat acts as a potent chemoattractant for monocytes, macrophages and dendritic cells. Tat induces the formation and release TNF-α, namely, tumour necrosis factor from macrophages (Chen et al. 1997). Moving ahead, Tat mediates tumour necrosis factor induction which is NF-kappa B (NF-κB) dependent and mediates through activation of communication cascades as well as phospholipase C, macromolecule enzyme A and macromolecule amino acid kinase. Additionally, Tat amplifies the endogenous levels of Ca2+ in macrophages which can later induce the assembly of chemokines and pro-inflammatory cytokines (Mayne et al. 2000). Latter events are also being to blame for HIV-1 induced neuropathogenesis and inflammation.

2.2 Viral Protein R (Vpr)

Virion-associated protein is essential for infective agent replication in T cells but is not needed for viral replication in macrophages (Subbramanian et al. 1998). Vpr is generally localized in cytoplasm additionally in genome of the infected cells (Jacquot et al. 2007). Vpr is a multiskilled protein that controls replication of virus and other cellular events like NF-κB-mediated transcription, necrobiosis and protein production. Recombinant Vpr (rVpr) effect has been incontestable in macrophages. Although higher concentration of rVpr resulted in vital cytotoxicity in macrophages, low concentration of rVpr has been shown to extend biological activity of many transcription factors including together with c-Jun and AP-1 in promonocytic cells and first macrophages (Varin et al. 2005). Additionally, rVpr initiates replication of HIV-1 in extremely infected primary macrophages. Moreover, infection of macrophages by infective agents that are Vpr-deficient mutants results in lower production of p24 which might be rectified by the addition of rVpr (Eckstein et al. 2001). Moreover, Vpr stimulates the cyclin-dependent enzyme's expression of matter 1A (CDKN1A/p21) in macrophages, whereas mutants of Vpr manifest reduced viral replication and show lack of upregulation of p21 (Vazquez et al. 2005). Taken along, the above-discussed data refer that Vpr enhances the replication of virus in extremely and acutely infected macrophages.

2.3 Nef

Nef is expressed throughout the HIV-1 early life cycle. Nef is a 27 kDa myristoylated protein needed for effective viral replication in cells that are infected (Das and Jameel 2005). Additionally, Nef amplifies the survival rate of infected cells, which helps in the infectious viral population's expansion. Moreover, Nef hampers the mechanism of infected patients through several different mechanisms as well as by downregulating the expression of MHC I&II, CD28 and CD4 (Lama et al. 1999) and by activating PI3K. Nef also downregulates the CD4 receptor's expression in macrophages that serves two functions. Initially, CD4 downregulation occurs in infected cells which promote the discharge of microorganism relation by avoiding sequestration of microorganism enveloped by CD4 (Lama 2003). In the next step, it helps to avoid infection that otherwise may cause premature death (Das and Jameel 2005; Lama 2003). In monocyte-derived macrophages (MDMs), exogenously added recombinant Nef (rNef) regulates the expression of many genes in a very short time span of 2 hours. These findings indicate a sturdy transcriptional programming ruled by Nef macromolecule resulting in the assembly and soluble factors secretion, which successively activates STAT1&3 in primary monocytes/macrophages (Mangino et al. 2011). Similarly, addition of rNef to the culture of MDMs resulted in the fast induction of transcription factors AP-1, NF-κB and c-Jun N-terminal kinase and increased HIV-1 transcription. Moreover, in vitro treatment with rNef of macrophages has been found to trigger IKK/NF-κB, MAPK and IRF-3 signal cascades. Moreover, Nef induces sturdy phosphorylation of MAPKs, as well as ERK1/2, JNK and p38 (Mangino et al. 2007). The role of Nef has been recently described in HIV-HCV coinfected macrophages.

3 Macrophage Contribution in HIV-1 Pathogenesis

HIV-1 pathogenesis can be distinguished by progressive cell depletion associated with adaptive immunity along with CD4+ T and CD8+ T cells. Not alone HIV-infected CD4+ T cells are lysed but CD4+ T cells that are uninfected more significantly undergo apoptosis (Finkel et al. 1995) (Fig. 2). Nef plays dual role in HIV-1 pathologic process. On one hand, Nef protects HIV-infected cells from necrobiosis to favour economical infective agent production. On the other hand, Nef induces caspase-mediated cell death in observing CD4+ T cells. Moreover, it has been observed that macrophages that are Nef-expressing unleash paracrine factors involving soluble CD23I and CAM that elevates lymphocytes liberally for HIV-1 infection (Swingler et al. 2003) (Fig. 2). Furthermore, Nef induces Fas ligand (CD95L) expression on the surface of T cells which are infected. Moreover, interaction among CD95L and on its receptor cells in local surroundings triggers programmed cell death in uninfected cells (Oyaizu et al. 1997) (Fig. 2). Particularly, Nef's function is to protect infected cells from programmed cell death via CD95-CD95L cis interaction by apoptosis signal-regulating enzyme 1 (ASK1) inhibition, caspase 8 and proteolytic enzyme 3 activation (Huang et al. 1998) (Fig. 2). It is worth mentioning that ASK1 could be a common partner of Fas ligand and TNF-α-mediated death communication cascades (Geleziunas et al. 2001). In addition, uninfected macrophages are shown to consult resistance against programmed cell death in potentially infected CD4+ T cells. Although expression of Nef is through these infected CD4+ T cells which are critical for anti-apoptotic behaviour, macrophages' presence may enhance the amount of non-apoptotic cells via intercellular contacts mediated by tumour necrosis factor stimulation (Mahlknecht et al. 2000). This might be one of the processes of promotion of HIV-1 reservoir in T cells by macrophages. Another HIV protein, Tat, has been encountered to initiate the expression of TNF-related apoptosis-induced ligand (TRAIL) in U937, primary macrophages and monocytes that result in the programmed cell death of uninfected cells (Fig. 2). This finding provides understanding of another process of eviction of uninfected cells. Recombinant compound protein gp120 (rgp120) (from X4 strain) has been reported to induce cell death of cytotoxic T cells (CTLs, CD8+ T cells). Moreover, apoptosis is mediated by interaction of TNFR-2 receptors on the CD8+ T cells and TNF-α present on the surface of macrophages (Fig. 2). Additionally, expressions of TNF-α and TNFR-2 are regulated by rgp120 treatment upon HIV infection. Moreover, stimulation of TNFR-2 in T cells

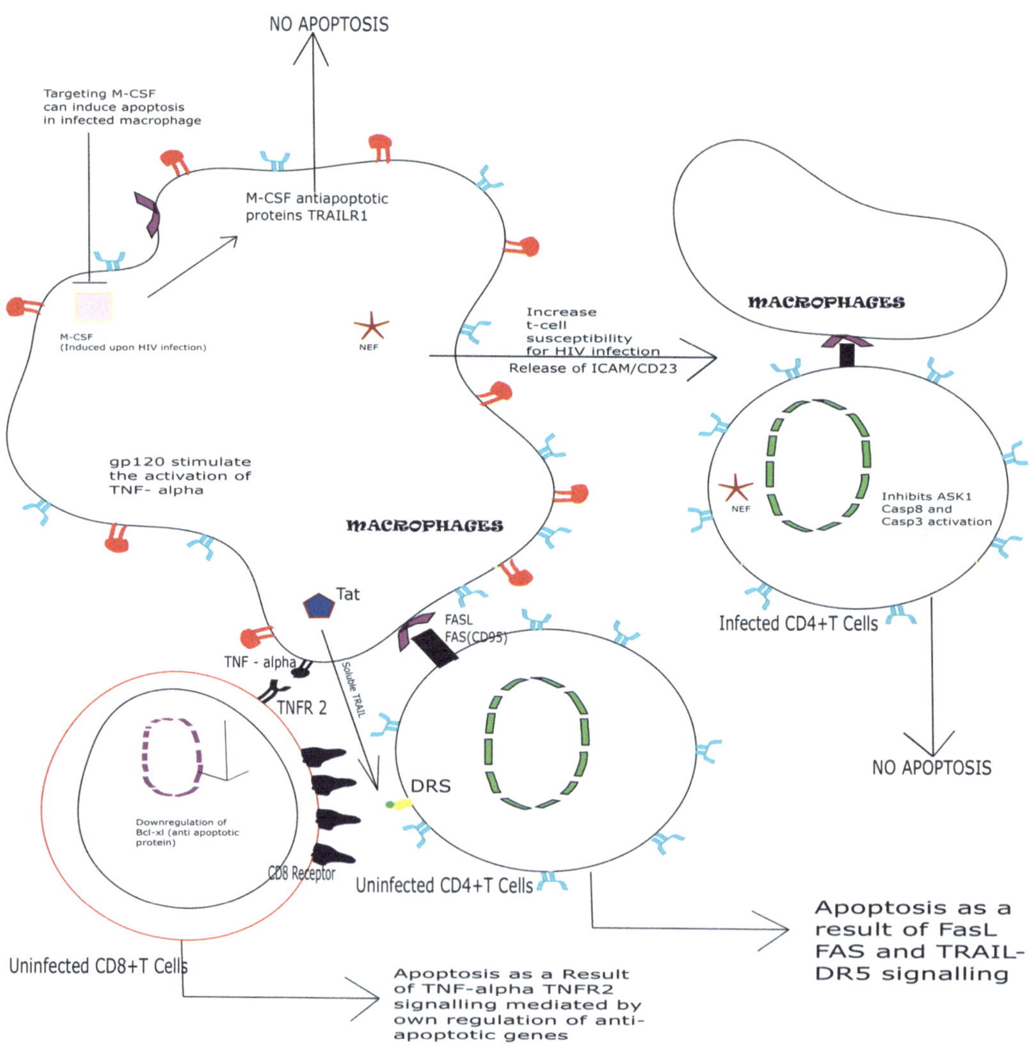

Fig. 2 Depiction of connection among macrophages and T lymphocytes in HIV-1 disease

resulted in the downregulation of anti apoptotic protein Bcl-XL which can make a case for CD8+ lymph cell elimination (Lin et al. 1997). Hence, we can conclude that macrophages have a critical role within the proliferation of HIV-1 infection and in depletion of CD4+& 8+ T cells.

Macrophage that harbours HIV-1 plays a crucial role in pathogenesis of HIV. Soluble factors CD23 and ICAM were stimulated by Nef, and these factors make uninfected CD4+ T cells more prone to HIV infection, hence favouring the growth of viral reservoir. (a) Additionally, Fas ligand is induced by Nef on infected HIV cells. Interaction of CD95L and its receptor (Fas) present on uninfected CD4+ T cells results in apoptosis. (b) However, in infected CD4+ T cells, Nef prevents the expression of proteins involved in apoptosis including ASK1, caspase 8 and caspase 3 and (c) protects infected CD4+ T cells from cell death and further expands the viral reservoir. HIV regulatory protein Tat stimulates the production and release of TRAIL from the infected macrophages. TRAIL binds with its receptor (DR5) present in uninfected CD4+ T cells and induces apoptosis (d). Furthermore, gp120 interaction with CXCR4 receptor increases the expression of TNF-α on macrophages which interacts with TNFR2 present on CD8+ T cells. This interaction results in the downregulation of the anti-apoptotic protein Bcl-XL and ultimately leads to apoptosis (e). Moreover, HIV infection in macrophages is known to induce macrophage colony-stimulating factor (M-CSF) which inhibits the expression of TRAILR1 on macrophages and upregulates the expression of anti-apoptotic proteins (f), favouring the resistance to apoptosis of infected macrophages. Therefore, targeting M-CSF has been suggested to increase apoptosis in infected macrophages. The role of macrophages in HIV-1 disease can be well understood as shown in Fig.2.

3.1 CTLs (Cytotoxic T cells) and Macrophages

Cytotoxic T cells (CTLs) of HIV-1 play the critical role in managing HIV-1 infection during early stage of infection. CTLs act on the data provided by antigen or CD4+ T cells (Pozzi et al. 2005). Although HIV-1 patients of effective CTLs response are additionally hampered. Studies confirm that Nef downregulates the HLA expression of category I molecule which is in infected CD4+ T cells leading to their escape from HIV-1-specific CTLs (Collins et al. 1998). Curiously, Fujiwara and Takiguchi stated in their in vitro study that HIV-1-specific CTLs are efficient in effectively repressing R5 virus replication in infected macrophages (Fujiwara and Takiguchi 2007). Furthermore, their data unconcealed that macrophages infected via HIV-1 induce a lot of spread of HIV-1 CTLs than CD4+ T cells. Combined data recommend effective response involvement of the macrophages throughout early part of HIV-1 infection (Eckstein et al. 2001). However, in in vivo studies, the role of macrophages that are HIV-1 infected is largely affected by their activation states. Notably, macrophages are planned to be in three types of activation states that are selected as M1 (pro-inflammatory in nature), M2 (anti-inflammatory in nature) and deactivated macrophages. Of note, M1 macrophages manufacture cytokines IL-23, IL-12, IL1-β, TNF-α and support Th1 response. On the other hand, in activation state of M2, macrophages support Th2 responses and secrete IL-10 (Cassol et al. 2009). In keeping with planned model, throughout the initial stage of HIV-1 infection, activation of M1 is the main that supports strong HIV-1 transcription and shaping of reservoirs that are viral. As the infection develops, M1 state shuts down and in M2 state activation occurs which is predominant and goes along with the deactivation of macrophages ensuing finally in failure in representing antigen to the CTLs.

4 Agents Inducing Apoptosis in HIV-Infected Macrophages

Initiation of apoptosis in persistently affected T cells has been recommended as a potential cure for HIV infection (Badley et al. 2000). In T cells, many novel targets are suggested, alteration of which may induce programmed cell death in infected T cells (Schnepple et al. 2011). Vigorous efforts are also needed to look for alike targets in macrophages that are infected. HIV-1 infection in macrophages has been rumoured to induce the assembly of macrophage colony-stimulating issue (M-CSF). Moreover, M-CSF completely regulates the articulation of anti-apoptotic proteins (Bfl-1 and Mcl-1) and prohibits the utterance of death receptor TRAIL-R1 (Fig. 2). In addition, targeting of M-CSF has been conjointly outlined to reinforce the programmed cell death in macrophages (Swingler et al. 2007). In one more new report, apoptotic result of microorganism protein Vpr has been inspected in MDMs and THP1 macrophages. Their finding is unconcealed that Vpr is not ready to induce cell death in MDMs and THP1. Contrary to undifferentiated cells, Vpr does not negatively regulate the expression of Bcl2 and inhibitors of apoptosis (IAPs) relations in macrophages (Busca et al. 2012). Moreover, negative regulation of IAP1 and IAP2 builds the macrophages vulnerable for Vpr-meditated altering. Sterilization of IAP activity has been advised as a possible way to induce cell death in infected macrophages (Busca et al. 2012).

5 Conventional HIV-1 Therapies in Macrophages

Antiretroviral therapy (ART) is widely utilized in suppressing HIV-1 infection to a major level. ART has created a stimulating contribution in rising and enhancing the sera of infected patients. HIV-1 growth mechanics is somewhat different in macrophages and T cells, which suggests varied impact of antiretroviral medication against HIV-1 in the target cells. Now here we will in short display the possible contribution of ART in macrophages which are infected by HIV.

5.1 RTIs (Reverse Transcriptase Inhibitors)

More than 25 compounds are authorized for treating HIV in infected patients (Abbas and Herbein 2012b). Out of them, nearly 50% are reverse transcriptase inhibitors (RTIs) (Abbas and Herbein 2012b). RTIs are of two varieties: nucleoside reverse transcriptase inhibitors (NRTIs) and non-nucleoside reverse transcriptase polymerase inhibitors (NNRTIs) (Gavegnano and Schinazi 2009). The conventional therapies used for treating HIV have been illustrated in Fig.3.

5.2 Nucleoside Reverse Transcriptase Inhibitors (NRTIs)

NRTIs prey on reverse transcriptase enzyme that is accountable for conversion of HIV genomic ribonucleic acid into complementary DNA, an important step within the life span of HIV (Fig. 1). NRTIs comprise emtricitabine, tenofovir, abacavir, lamivudine, stavudine, zalcitabine, didanosine and zidovudine (Abbas and Herbein 2012b). NRTIs compete and imitate with natural nucleotides pool for fusion in developing chain of nascent HIV DNA. Particularly, NRTIs need intracellular phosphorylation for transversion into functional inhibitors of HIV. As most of NRTIs lack 3′ OH moiety, therefore, their assimilation into nascent HIV deoxyribonucleic acid ends up in termination of deoxyribonucleic acid chain formation. Effectuality of those NRTIs is majorly based on the degree of dNTP pools (Diamond et al. 2004; Gavegnano and Schinazi 2009). As mentioned earlier, macrophages being terminally differentiated non-dividing cells which have restricted pools of dNTPs contrasted to actively dividing cells (Gavegnano and Schinazi 2009). Thus, in theory during this situation, NRTIs face less contesting with natural dNTPs in macrophages. This may be the one among the explanations for higher effectiveness of NRTIs in macrophages as compared to CD4+ T cells (Aquaro et al. 1998; Aquaro et al. 2006). If truth be told, NRTIs have shown promising results in reducing the neuropathological consequences of HIV encephalitis within the central nervous system and onset of HIV-associated dementia (HAD) (Aquaro et al. 2006). Notably, in CNS, macrophages represent the foremost HIV-infected population (Busca et al. 2012). Additionally, NRTI treatments in macrophages lead to very lesser emergency cases of resistant HIV mutants as compared to lymphocytes (Aquaro et al. 2005).

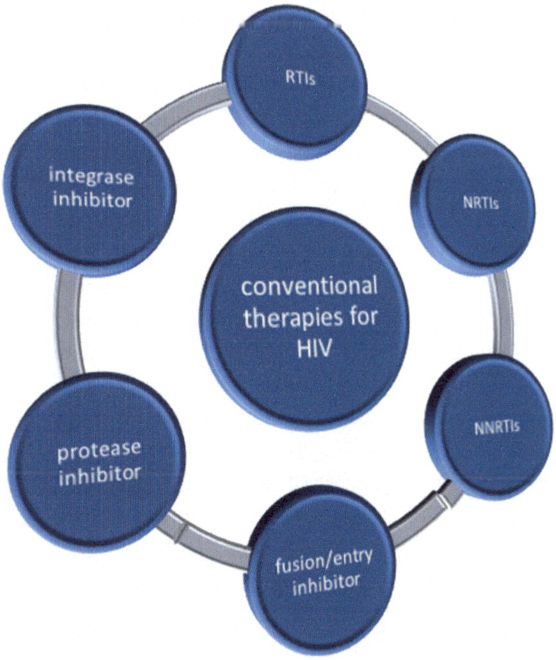

Fig. 3 Depiction of conventional therapies used in treating HIV

Strikingly, NRTIs' effectuality is remarkably altered in acutely and chronically infected macrophages. Exact procedure answerable for such observation is rarely understood. Seeing that dreadfully infected cells possess integrated HIV deoxyribonucleic acid into host chromatin, HIV ribonucleic acid made through integrated DNA utilizing transcription by host RNA polymerase is thus not receptive to NRTIs. Besides this, there should be many alternative mechanisms answerable for the distinction within the effectuality of NRTIs between chronically and acutely infected macrophages (Gavegnano and Schinazi 2009; Aquaro et al. 2006). Notably, NRTIs are related to many undesirable effects as well as their intrusion into cell cycle and mitochondrial surroundings and additionally induce apoptosis (Chariot et al. 1993; Viora et al. 1997).

5.3 Non-nucleoside Reverse Transcriptase Inhibitors (NNRTIs)

Licensed NNRTIs comprise rilpivirine, etravirine, delavirdine, efavirenz and nevirapine. In contrast to NRTIs, NNRTIs do not need phosphorylation and neither engage with natural dNTPs pools for their activity. NNRTIs perform by binding to the hydrophobic pocket close to the reverse transcriptase active site leading to the inhibition of polymerization reaction (Gavegnano and Schinazi 2009). Since NNRTIs' effectivity doesn't rely upon the cellular dNTP's pools, their effect on acutely infected macrophages and CD4+ T cells isn't significantly different. Macrophage colony-stimulating issue that absolutely regulates the dNTPs pool don't have any impact on the NNRTIs effectuality against HIV. Particularly, NNRTIs have somewhat less adverse effects as compared to NRTIs. Moreover, a study has reported the side effects of non-nucleoside reverse transcriptase inhibitor in Jurkat T cells and peripheral blood mononuclear cells (PBMCs). They ascertained the selection of caspase and mitochondrial-dependent programmed cell death by NNRTIs (Pilon et al. 2002). Similar to NRTIs, NNRTIs show anti-HIV activities exceptionally contradicted between acutely infected and chronically infected macrophages. To be a lot of precise, EC50 of NNRTIs as opposed to acutely infected macrophages varies from 10 to 50 nm. On the opposite hand, their outcome is very less against inveterately infected macrophages (Gavegnano and Schinazi 2009; Aquaro et al. 2006). Reasons for these observations are somewhat completely understood.

5.4 Inhibitors for Integrase

Chronic HIV infection is usually characterized by incorporation of proviral DNA into the host chromatin (Fig. 1). This method is known as strand transfer which is ruled by HIV-encoded enzyme known as integrase and is indispensable for the development of latency (Siliciano and Greene 2011) (Fig. 1). Until now, three integrase inhibitors (raltegravir, elvitegravir and dolutegravir) are approved for clinical use. Effectiveness of certain inhibitors like integrase inhibitors has been studied in MDMs and lymphocytes and showed similar results. Notably, single mutation in integrase confers resistance against the integrase inhibitor raltegravir. However, other integrase inhibitors are still effective therein scenario (Canducci et al. 2013). Concurrent selectivity of several parts of HIV is important to avoid emergence of resistant mutants.

5.5 Inhibitors for Protease Inhibitors (PIs)

To date, 10 protease inhibitors (PIs) are authorized for HIV-1 infection treatment. In contrast to reverse transcriptase inhibitors, Protease inhibitors act at one stage of HIV 1 lifecycle, that is, post-integration stage. HIV protease provides support in the construction of an infectious viral progeny. Protease inhibitor binds at one site which is of HIV proteases and makes them nonfunctional. As compared to polymerase inhibitors, protease inhibitors provide prominent effect in each acutely, moreover as chronically infected macrophage and also CD4+ T cells. Nevertheless, concentration needed for prominent HIV inhibition is a lot of just in case of inveterately infected macrophages as compared to CD4+ T cells (Perno et al. 1994; Perno et al. 1998). In clinical state of affairs, bioavailability of protease inhibitors in macrophage tissue-specific and in plasma is significantly altered. Therefore, HIV in macrophages might break loose from PIs. Furthermore, till now there is no as such result of protease inhibitor on integrated DNA of HIV has been reported, thus lapse of PIs treatment can quickly end in the production and unleash of infectious HIV virions.

5.6 Entry/Fusion Inhibitors

So far, there are two approved entry inhibitors, that is, enfuvirtide and maraviroc, against HIV (Oyaizu et al. 1997). Enfuvirtide also referred to as Fuzeon, T-20, could be derived from gp41, which is an HIV envelope protein, that inhibits hairpin formation essential for the fusion of viral envelope with host membrane (Gavegnano and Schinazi 2009; Perno et al. 1998). Enfuvirtide inhibits the entry of HIV into completely different target cells as well as macrophages, and immature dendritic cells. However, brief studies of these inhibitors in primary macrophages are more needed. On the other hand, maraviroc could be a tiny molecule which can bind reversibly with CCR5 receptor and prevents the interactions with host–virus (Gavegnano and Schinazi 2009) (Fig. 1). Notably, maraviroc is to date the sole CCR5 antagonist authorized for the treatment in patients having HIV (Dorr et al. 2005). Because of prominent side effects and clinical effectivity lack, other CCR5 inhibitors as well as aplaviroc and vicriviroc are not given thought of clinical development. Resistance to maraviroc has been rumoured, and accountable mechanisms are studied (Anastassopoulou et al. 2009). New CCR5 antagonists are in several stages of development, and CCR5 antagonists' cocktail has been with different ART which might improve the impact against HIV infection.

6 Novel Therapies Against HIV-1 in Macrophages

Various new approaches are present to completely eradicate HIV-1 from infected patients. Here, we will look over the novel approaches and novel molecular therapeutic tools which are pop up against HIV-1 in macrophages. The novel therapies used for treating HIV are shown in Fig.4.

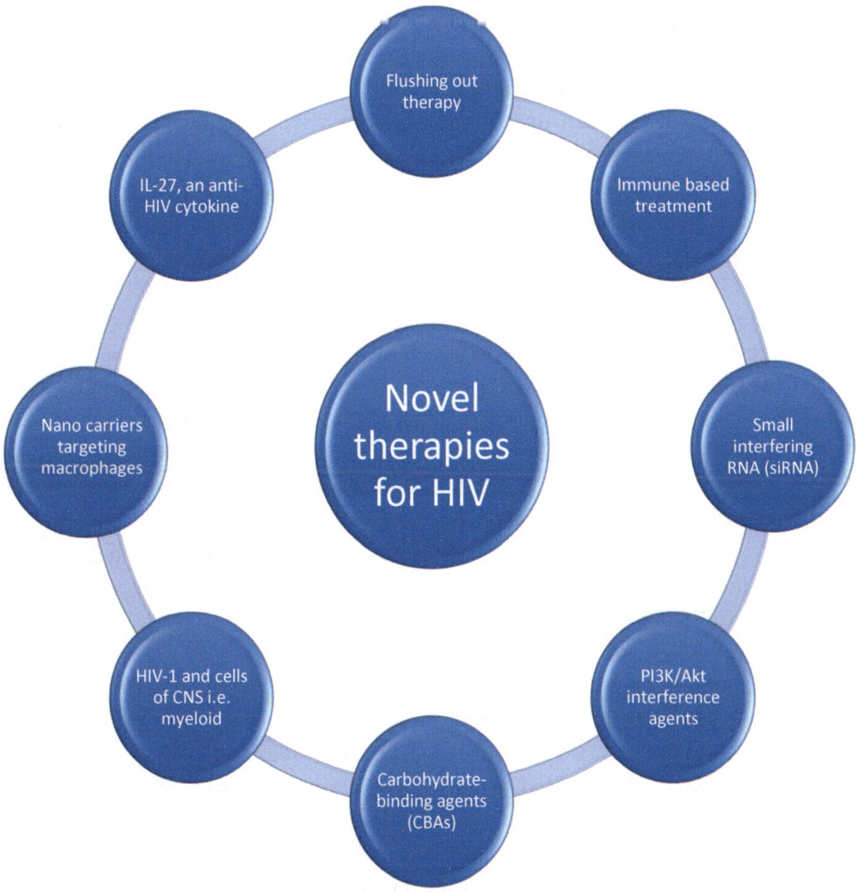

Fig. 4 Depiction of novel therapies used in treating HIV

6.1 Carbohydrate-Binding Agents (CBAs)

CBAs are represented as anti-HIV molecules that specifically target glycans of HIV-1 gp120 (Balzarini 2007). As a result of glycosylation of gp120, macrophages and dendritic cells lose their ability to acknowledge and present processed antigen to the CD4+ T cells to vital level, resulting in inefficient transfer of infection to the CD4+ T cells. Balzarini and colleagues disclosed that even short exposure of HIV-1 to CBA hampers the power of immature dendritic cells containing glycan-targeting C-type DC-SIGN lectin receptor to bind HIV-1 and stop syncytia formation once co-inoculated with T cells (Balzarini et al. 2007). Recently, Balzarini research lab has shown that griffithsin, an anti-HIV CBA, inhibits the interaction between DC-SIGN and HIV gp120 protein and expeditiously hampers the transfer of HIV-1 to CD4+ T cells (Hoorelbeke et al. 2013). Impact of CBAs in chronic HIV-1 infection has been poorly outlined.

6.2 PI3K/Akt Interference Agents

The PI3K/Akt communication cascades are well known as a positive target for anti-cancer strategies (El-Deiry 2001). Many research groups depict that PI3K/Akt inhibitors in cancer therapy are well tolerated and have minimum toxicologic profile in animal models and humans (Morgensztern and McLeod 2005). In the past few years, inhibitors of PI3K/Akt sign are utilized as anti-HIV-1 strategy. PI3K/Akt inhibitors are shown to effectively inhibit HIV-1 replication in acutely infected primary macrophages. PI3K/Akt inhibitors utilized by Chugh et al. were optimally effective at 200 nm, which is much on top of physiologically relevant concentrations (Chugh et al. 2008). Despite this, their results offer a valuable insight into a communication event specifically active in HIV-1-infected cells. In addition, the blockade of the PI3K/Akt pathway may favour programmed cell death and therefore the clearance of infected cells. The impact of PI3K/Akt inhibitors on inveterately infected macrophages has to be more investigated.

6.3 Small Interfering RNA (siRNA)

siRNAs are strong molecules which may potentially degrade any viral RNA species (Leonard and Schaffer 2006). siRNAs or shRNAs have been found to be effective in inhibiting HIV-1 replication in many types of cells together with primary macrophages (Song et al. 2003). Data of siRNAs against HIV have been compiled within the information types known as HIVsiDB. HIVsiDB has info of over 750 anti-HIV siRNAs (Tyagi et al. 2011). In in vivo toxicity, there is a lack of efficient delivery tools, and generation of viral escape mutants is the core hurdle within the development of siRNA as an efficient therapeutic tool against HIV.

6.4 Immune-Based Treatment

HIV-1 infection ultimately ends up in the depletion of CD4+ T and CD8+ T cells. Efforts are created within the direction of boosting immunity against HIV-1. For example, in numerous studies, applying IL-2, IL7, IL-12 and growth hormone results in increase in CD4+ T counts in HIV-1-infected individuals (Chun et al. 1999; Jacobson et al. 2000). Apparently, IL-2 besides ART significantly reduces HIV-1 replication in infected patients as compared to ART solely treated patients. However, upon treatment, virus stops to retrieve, indicating the shortcoming of IL-2 to boost immunity for the longer amount of time (Chun et al. 1999). Additionally, the role of IL-15 has been urged in rising functionality of anti-HIV CTLs and natural killer (NK) cells in vitro (Seder et al. 1995). Moreover, IL-15 enhances simian immunological disorder virus (SIV)-specific CD8+ T cells and NK cells, and decreases the amount of SIV-infected cells in lymph nodes in infected rhesus monkey macaque (Mueller et al. 2008). Astonishingly, viral load was found to be enhanced over twofold upon IL-15 treatment (Mueller et al. 2008). Notably, IL-21 treatment in SIV-infected monkeys resulted in elevated granzyme B and perforins in NK cells and CD8+ T cells (Pallikkuth et al. 2011). Benefits of such transient immunity evoke by interleukins and impact of continuous use of such immune-based therapeutics on the health of HIV-1-infected people have to be addressed carefully.

6.5 IL-27, an Anti-HIV Cytokine

IL-27 is a protein that belongs to the IL-12 cytokine family and plays a necessary role in innate and adaptive immunity. IL-27 is made up of epithelial cells, dendritic cells and macrophages (Hunter 2005). Many scientists have reported the anti-HIV properties of IL-27 in MDMs, CD4+ T cells, immature and mature nerve fibre cells (Chen et al. 2013). Mechanistic details of anti-HIV cytokine IL-27 have been recently unconcealed. IL-27 downregulates the expression of spectrin β non-erythrocyte 1 (SPTBN1), one of the host factors needed for HIV-1 infection in macrophages (Dai et al. 2013). Moreover, IL-27 downregulates the expression of SPTBN1 via TAK-1-mediated MAPK signal cascade (Dai et al. 2013). Significantly, their results indicate that SPTBN1 could be a crucial host part which may specifically target and inhibit HIV-1 replication in one of the principal HIV-1 reservoirs, the macrophages.

6.6 Nanocarriers Targeting HIV

For effective therapy, there must be certain complimenting delivery tools that can surpass the need for therapy. For HIV treatment, nanotechnology is one of the novel and potent therapies, and it serves the purpose of delivering the agents that are therapeutic to specific sites and to specific cells and also to anatomical location which might not be located or treated by conventional-type methods (Park 2007). Nanotechnology works wonderfully with lower side effects as it is supposed that it gets through the infected cell and accumulates there only leaving behind the unaffected cells, hence reducing side effects (Langer 2001). Wan along with his co-workers suggested and developed a nano-based carrier system for the delivery of drug in macrophage, utilizing formyl-methionine-leucine-phenylalanine (FMLP), which is a derivative of peptide PEG (Poly ethylene glycol) (Wan et al. 2007). This PEG derivative is used because these FMLP receptors are present on macrophages and on other phagocytotic cells as well and bind to the specific receptors on the phagocytotic cells along with macrophages that too with high affinity (Wan et al. 2007). In vivo studies have been conducted in which bio-distribution of nanocarrier FMLP is done which revealed that the accumulation of FMLP is greater on the macrophages of spleen, liver and kidney in comparison with PEG. Results are promising as it depicts targeted delivery on macrophages of HIV-1. Numerous nanocarriers have been developed to date for treating HIV as drug delivery applications.

Entry of HIV takes place via mucosal barrier, and then it gets spread locally cell to cell and then reaching to the tissues that too specifically in mucosal lamella propria and in lymph nodes, and then it can float as a free virus in blood circulation or can spread inside CD4+ cells that are infected. Nanocarrier drug faces many challenges in between the pathways before reaching to the target site while free virus does not have to face that much challenges apart from its shorter half-life. Generally, nanocarrier drugs are given either intravenously or orally to enhance their presence in the bloodstream. The challenges faced by nanocarriers during circulation are to maintain the extent of drug; basically, its bioavailability, biostability and other challenges include the avoidance of renal clearance and through RES (Reticuloendothelial system) notably from spleen and the liver. To overcome these challenges, various strategies have been developed and reviewed (Ehrlich et al. 1960) in which PEGylation use and effect of size have been included. Several anti-HIV drugs have potency to bind to human serum albumin which is a plasma component, and it can also bind within the tissue compartments, which greatly affects the transport and excretion in individual organs, basically affecting the whole pharmacokinetics. Nanocarrier system of anti-HIV drug needs to be formulated in such a manner that it minimizes or eliminates the non-specific bindings (Opanasopit et al. 2002). The various types of nanocarrier and surface modifiers used for drug delivery have been described below.

6.6.1 Liposomes

These act as a vehicle in which the drug is loaded. Generally, they have small size of 80–100 nm. Surface of liposomes can be modified according to the need or to improve the properties. Composition of lipid and its charge can influence drug properties which is encapsulated in the liposome. 2′,3′-didexycytidine (DDC) has been encapsulated in various formulations of liposomes, and their intracellular stability is evaluated on the basis of RAW 264.7 macrophage drug concentration. When liposome drug concentration was increased, then an enhanced uptake was observed in liposomal encapsulated DDC via macrophage. In addition, as we know that cholesterol is present in lipid composition, it is noted to be a prominent factor to claim higher intracellular DDC uptake. Liposomes are the nanocarriers having fatty acyl phospholipid long saturated fatty chain that contains 50% of cholesterol which depicts the higher stability in all compartments of intra-macrophage and at all DDC concentrations. Then again, although the spillage of DDC through liposomes sterically balanced out with polyethylene glycol chain was like that of other liposomes that are cholesterol-containing, the counter HIV drug was promptly delivered from cells for all liposomal concentrations that are DDC tested. These outcomes recommend that some lipid segments like cholesterol can adjust the liposomal stability of medications like DDC in response to the condition, the properties of the medication utilized and the idea of cooperation among liposomes and cells (Makabi-Panzu et al. 1998). In another examination, the impact of lipid utilized, concentration of cholesterol and the surface charge present on the bilayer of liposome, encapsulated with stavudine (d4T) marked stavudine, on the cell uptake M/M were researched. d4T uptake was discovered to be greatest (around 950 ng/million cells) in liposomes that are containing dipalmitoylphosphatidylcholine. The presence of sphingomyelin (which increments bilayer inflexibility) upgraded the uptake of d4T contrasted with positive charge and diminished cell uptake of d4T and the presence of negative charge on the bilayer. There is no obvious distinction in uptake when varying cholesterol measures was added in liposomal preparation. This examination shows that the affectability of macrophages to various charges and lipid types can be used to one or the other reduction or increment in cellular uptake as wanted (Katragadda et al. 2000). Subsequently, lipid formulation has been appeared to influence the steadiness of DDC-encapsulated liposomes (Makabi-Panzu et al. 1998), hence composition of liposome and charge to influence d4T-encapsulated uptake of liposomes cells which are macrophage-like (Katragadda et al. 2000). These outcomes affirm that generally accepted cargo drug is not able to change the destiny of in vivo nanocarriers.

CD4-derived peptide and soluble CD4 are used for surface modification of liposomes to enhance their cellular delivery in infected HIV-1 cells. One of the studies suggests that N-butyldeoxynojirimycin (NB-DNJ) is an HIV gp120 folding inhibitor, which is encapsulated as soluble CD4 liposome that is a surface modifier which targets PBMCs. Soluble CD4 along with peptide, CD4 derived, has been utilized to liposomes (surface-modify) to improve delivery of cell to infected HIV-1 cells. In one investigation, N-butyldeoxynojirimycin (NB-DNJ), an HIV gp 120 inhibitor, is incorporated within soluble CD4 liposome for targeting PBMCs. The improvement depends on focusing on viral gp120/41 complex communicated on HIV-1-contaminated cells. Low pH affectability has likewise been designed into the liposomes to encourage endosomal escape. A fivefold expansion in uptake into HIV-contaminated cells contrasted, and uninfected cells were noticed (Pollock et al. 2008). In another examination, a synthetic peptide which is from the complementary deciding area 2 (CDR-2) like area of CD4 could tie explicitly to HIV-contaminated cells and intercede the limiting of peptide-coupled liposomes from these cells. A CDR-3 peptide-like CD4 domain inhibits syncytia formation which is HIV-induced, however neglected to target liposomes to infected cells. This clear disparity might be because of the prerequisite conformational change in the CD4 receptor for CDR-3 locale to cooperate with the HIV envelope protein. These outcomes exhibit the achievability of utilizing manufactured peptides to target liposomes containing antiviral medications to HIV-contaminated cells. Low-density lipoprotein (LDL) has been utilized both as carrier for drug and targeting moiety to target macrophages. In ordinary individuals, LDL is about 25.5 nm in size (Kondo et al. 2001). In these contemplates, LDL is formed with AZT (Azidothymidine), and the loaded drug LDL might be viewed as a quasi-drug-modified liposome. LDL is perceived by scavenger receptors explicitly communicated on macrophages and taken into cells by endocytosis. A 10-fold expansion in cell take-up of AZT conveyed by LDL-altered liposome when contrasted with free AZT was noticed. The surface modification in various drugs has been summarized in Table 1.

6.6.2 Nanoparticles

Surface modification of nanoparticle (NP) can influence their properties identified with treatment of HIV-1 contamination. In an examination, poly(ethylene oxide) changed poly(epsilon caprolactone) (PEO-PCL) NP framework as an intracellular conveyance vehicle for saquinavir (SQV) was created. Cell uptake and dissemination of the NP were inspected in THP-1 human cell line M/M. The PEO-PCL NPs had a circular shape and smooth surface and showed a moderately uniform size circulation with a mean molecule width of ~200 nm and a zeta potential of ~ −25 mV. The surface charge and size are compli-

Table 1 Summary of effect of drugs on surface modification on cell/tissue/organ

Anti-HIV drug	Surface modification	Cell/tissue/organs	References
ddc	Induction of charge in lipid composition of liposomes	RAW 264.7 macrophage-like cells	Makabi-Panzu et al. (1998)
d4t	Presence of charge on the liposome bilayer Stearylamine or dicetyl phosphate used for induction of positive or negative charge on the bilayer	U937 M/M cells	Katragadda et al. (2000)
NB-DNJ	Soluble recombinant CD4 molecule and IgG antibody conjugated separately to liposomes	PBMCs	Pollock et al. (2008)
AZT prodrug	Acetylation of LDL	J7774.A and U937 macrophage-like cells	Hu et al. (2000)
FLT and AZT	Covalent coupling of AZT and FLT (3'-fluro-3'-deoxythymidine) to LDL	HIV-infected human macrophages and Molt 4/8 cells (CD4+ T cell line)	Mankertz et al. (1996)
ddCTP	Antibody attached to liposome surface	Human M/M cells	Betageri et al. (1993a)
ddITP	Antibody attached to liposome surface	Human M/M cells	Betageri et al. (1993b)
AZT	Induction of cations by stearylamine on liposome-like emulsomes	Liver cells	Vyas et al. (2006)
ddCTP	Induction of charge on liposome surface	Cells of Mononuclear phagocyte system (MPS)	Oussoren et al. (1999)

cated for most detailed NPs. A fast cell uptake of the (rhodamine-123) fluorescent probe incorporated in PEO-PCL NPs was seen in THP-1 cells. SQV intracellular fixations were fundamentally higher than free SQV (Shah and Amiji 2006). The impact of polymer kind of AZT-stacked polylactic corrosive (PLA) and PLA-PEG mix NPs on phagocytosis in rodent polymorphonuclear leucocyte was examined. NP size (around 300 nm) and zeta potential (~ −7 to −20 mV) were not significant. The PLA nanoparticles were more proficiently phagocytosed than PLA/PEG mixes principally depending on the PEG and its proportion in the mix. The mix with the most noteworthy PEG extent did not forestall phagocytosis, showing that the steric impact of PEG was fixation of subordinate (Mainardes et al. 2009). The impact of polymer engineering of PEGylated NPs on NP properties was assessed in RAW 264.7 macrophage-like cells. NP size (around 185 nm) and zeta potential (~−3 to −23 mV) did not change altogether between various designs. The united copolymer NPs decreased macrophage take-up when contrasted with multiblock copolymer, in spite of the fact that instruments extraordinary than phagocytosis were included. Consequently, polymer design is a significant element for deciding the surface properties and henceforth, protein restricting and cell connections of NPs. The impact of NP shape/size on cell take-up was explored. A round- and two tube-shaped NPs, whose lengths were particularly fluctuated, were built by the specific cross-connecting of amphiphilic block copolymer micelles. At the point when the NPs were changed with Tat cell entering peptide (CPP, otherwise called protein transduction space), the more modest, round NPs had a higher pace of cell incorporation into Chinese hamster ovary cells than did the bigger, round and hollow NPs. NPs were delivered after disguise, and the pace of cell exit was reliant on both the NP shape and the measure of surface-bound CPP.

As macrophages effectively phagocytose unfamiliar particles, this has become a significant component of targeting macrophage techniques. For instance, human M/M take-up of poloxamer surface-modified (Pluronic F68 and F108) AZT-stacked NP was examined. The impacts of size, arrangement, fixation and surface of the NP just as macrophage enactment state on phagocytosis were tried. NP made of polyhexylcyanoacrylate (PHCA) or HSA having a measurement of diameter of around 200 nm was discovered generally helpful for explicit conveyance to macrophages. Macrophages separated from HIV-contaminated patients additionally show great joining of NP. Extra focuses on moieties have been added to NP to upgrade drug conveyance to macrophages. Two examinations utilized mannan and mannose as the target modifiers (Kaur et al. 2008; Jain et al. 2008). Mannan-coated NPs for macrophage focusing on didanosine (ddI) were researched. Consequences of the ex vivo cell take-up investigation demonstrated fivefold higher take-up from the mannan-coated NP detailing of ddI by the macrophages in correlation with free medication. After-effects of the quantitative bio-distribution study showed 1.7, 12.6 and 12.4 occasions higher limitation of ddI in the spleen, LNs and mind, individually, after organization of mannan-coated NPs contrasted with that after infusion of ddI buffer solution. Along these lines, mannan-coated NPs appeared to effectively target ddI to macrophages by misusing mannose receptor-intervened endocytosis (Kaur et al. 2008). Mannosylated gelatin NPs (MN-G-NPs) for the specific conveyance of ddI to macrophage tissues were researched. Cell take-up by MN-G-NPs was 2.7 occasions higher when contrasted with the non-targeted G-NPs. Intravenous organization of free-drug arrangement brought about a high serum drug concentration though serum concentration was a lot of lower for G-NPs proposing local release and targeted delivery of medication. NPs coupled with mannose altogether improved the liver, lung and lymph nodes take-up of medication, which is reflected in the recuperation of a higher level of the portion from these organs following organization of MN-G-NPs in correlation with non-coupled G-NPs or free medication.

6.6.3 Dendrimers

Dendrimers belong to monodisperse polymers and can be differentiated through their repetitive branching structure coming out through a central core. Dendrimers are very much useful in the anti-HIV drug delivery. Its attractive features are its smaller size (less than 100 nm) and its narrow molecular weight distribution and also the ease of incorporation and targeting. Fifth-generation surface-modified poly(propyleneimine) (PPI) nanocarrier-based dendrimers were built for focusing on lamivudine (3TC) and efavirenz (EFV) to MT-2 cells and M/M cells, individually. Mannose was utilized as the surface-focusing modifier. The aftereffects of a ligand agglutination examined demonstrated that even after formation of PPI, mannose shows restricting explicitness towards Con A (an all-around researched lectin, known to tie explicitly with saccharides, for example, mannose). Huge expansion in uptake of 3TC by MT-2 cells was seen when mannosylated PPI (MPPI) was utilized, which was 21 and 8.3 times higher than that of free medication ($p < 0.001$) and PPI ($p < 0.001$) at 48 h individually. Altogether higher anti-HIV action of PPI and MPPI was noticed, and this was because of the upgraded cell uptake of 3TC in plan when contrasted with that of free drug (Dutta and Jain 1770). The cell take-up of EFV was discovered to be time dependent and on concentration. Huge expansion in cell take-up of EFV by M/M cells was seen in the event of mannose-formed dendrimer, which are multiple times higher (specifically 12 times) than free available drug and 5.5 times more than t-Boc-glycine formed PPI dendrimer. On the other hand, mannose-formed dendrimers were taken up by various lectin receptors present in the cells, and answer for the enhanced uptake is the phagocytosis of dendrimer formed from t-Boc-glycine. Curiously, the PPI dendrimer cytotoxicity was discovered insignificant when these were surface-modified dendrimers with t-Boc-glycine or mannose. Conclusion recommends that the stated nanocarriers that are dendrimer-based hold potential to build the viability and lessen the harmful effects of antiretroviral treatment (Dutta et al. 2007).

6.6.4 Bioconjugates

Polymeric bioconjugates have also been utilized for the HIV treatment and its drug delivery. Various surface targeting modifiers that are specific for macrophages consist of chemoattractant peptide, N-formyl-methionine-leucine-phenylalanine (FMLP) and mannose. Basically, biodegradation of target cell has been focused in this type of nanocarrier system. Tat peptide cell gets penetrated inside the cell and acts as surface modifier imparting endosomal escape (Gunaseelan et al. 2004).

6.6.5 Solid Lipid Nanoparticles (SLNs)

These are the nanocarriers having a solid lipid core having a size range of micrometres to nanometres. These carriers provide drug release in a controlled manner and are physically stable and are capable of protecting the drug incorporated from degradation. These are solid at room temperature as well as body temperature. These are preferred over other lipid particulate systems for providing modulated drug release. SLNs can be customized according to the need along with various polymers for combating the drawbacks associated like drug loading capacity can be increased, enhanced drug solubility and drug release profile can be predictable. In a study, atazanavir-loaded SLNs were prepared of a size range 167 nm, and the work was to provide enhanced brain delivery. Viability of cell and its uptake productivity in monolayer of endothelial cell were found to be higher than that of atazanavir which was distinguished utilizing [3H]-atazanavir (Joshy et al. 2017).

6.6.6 Ethosomes

Ethosomes are the type of nanocarriers that have lipid vesicles which consist of phospholipids and alcohol in comparatively higher concentration than that of water. These are potent for encapsulating various drugs either lipophilic or hydrophilic; even amphiphilic drug can be incorporated in ethosomes. If we talk about the size range, then it varies from nanometres to microns. In a study, lamivudine ethosomes were formulated and characterized in in vitro, in vivo and ex vivo studies. Cytotoxic studies and cellular uptake of these nanocarriers were done by using MT-2. Result showed that transdermal flux is 25 times higher across rat skin when compared to lamivudine solution. If we talk about ethosomal lamivudine, then it showed that cellular uptake is three times higher than that of lamivudine solution. Similar to above delivery systems, this study also concludes that these nanocarriers provide more cellular uptake than other systems, although the study shortfall is on the evaluation part of ethosomes for potential targeting to the lymphatic system and macrophages. Due to higher ethanol content present in the inner core of these vesicles, they have higher potency of ARV (Anti retro viral) drug entrapment and can perfume improvised drug delivery. This is the novel approach, yet very effective, and can be used as a viable carrier system for macrophage targeting ARV therapeutics (Chu and Murad 2017). The various types of nanosystems along with their properties and therapeutic outcome have been depicted in Tables 2 and 3.

Table 2 Summary of types of nanosystem along with their properties and therapeutic outcome in HIV

Type of nanosystem	Drug carrier	Therapeutic outcome/ result	Disease model used/cell line used	Size of drug release and drug entrapment (physicochemical properties)	References
Dendrimers	PPI terminated with tuftsin or mannose	Toxicity was reduced after functionalization of PPI dendrimers with targeting ligands	MT2 cell lines	The entrapment efficiency was observed to be 43.27 ± 0.13%	Dutta and Jain (2007)
Polymeric nanoparticles	PSC-RANTES was entrapped in the PLGA (Poly lactic-coglycolic Acid) nanoparticles	The vesicles containing PLGA NPs were loaded with the inhibitor peptide	HEC-1A cell line	Its zeta potential was observed to be −0.5 ± 0.3 The entrapment efficiency was found out to be 69 %	Sánchez-López et al. (2020)

(continued)

Table 2 (continued)

Type of nanosystem	Drug carrier	Therapeutic outcome/result	Disease model used/cell line used	Size of drug release and drug entrapment (physicochemical properties)	References
Quantum rods	Incorporation of saquinavir, within Tf-conjugated quantum rods (QRs)	Significant decrease in p24 production and LTR/RU5 gene expression in the HIV-1-infected PBMCs	Peripheral blood mononuclear cells (PBMCs)	They are having hydrodynamic diameter in the range of 130–140 nm	Mahajan et al. (2010)
Magnetic nanoparticles	ARV (nelfinavir) + sigmareceptor antagonist	Inhibition of HIV-1 infection	(RAG-hu) humanized mouse model	These are spherical particles with diameter of 100 nm	Zhou et al. (2011)
Liposomes	Indinavir added to ethosomes	More release is found in the ethosomal transdermal system	Human model	The unilamellar spheres are having diameter of 189 nm. It has 89% of entrapment efficiency	Dubey et al. (2010a)
Silver nanoparticles	Of MAb HIV-1 gp120	AgNPs at 1 mg/ml concentration have been shown to exert ~40% inhibition of cell-free HIV-1_{IIIB} virus infection	U373-MAGI-$CXCR4_{CEM}$ cells	The silver nanoparticles were 30–50 nm in size	Lara et al. (2011)
Gold nanoparticles	AuNPs coated with P-mercaptobenzoic acid were synthesized	AuNPs conjugation occurred with four molecules of antiviral drug raltegravir, and it inhibited HIV-1 replication	Primary PBMCs exhausted from CD8+cells	Average diameter of 1.8 ± 0.32 nm has been observed Cellular uptake reached up to 54% in 3 h	Garrido et al. (2015)
Solid lipid nanoparticles	ACV-loaded SLNs (ACV-SLNs) were prepared by double-emulsion process	ACV-SLN accumulated 15.17 times more than ACV cream	Sprague-Dawley rat skin was used here	The SLN was having diameter of 262 nm Its polydispersity index was found out to be 0.280 ± 0.01 Its entrapment efficiency was found out to be 40.08 ± 4.39%	Jain et al. (2011)
Ethosomes	Ethosomes prepared by cast film method consist of soya PC, ethanol and indinavir	Improved delivery of indinavir via transdermal route was observed	Human cadaver skin	The vesicles were having diameter of 147 nm Its entrapment efficiency was found out to be 96.71 ± 1.4%	Dubey et al. (2010b)
Nanoemulsion	SQV nanoemulsions contain 1 ml of flaxseed oil	The relative bioavailability was observed to be 108.1% with respect to aqueous suspension	Male Balb/c mice	The nanoemulsions were having a size of 218.0 ± 13.9 The zeta potential was −43.28 ± 3.79	Chakravarty and Vora (2020)

Table 3 Recent studies on various smart nanoparticles developed against HIV

Type of smart nanoparticle	Characteristic feature	Recent studies	References
Semen triggered nanoparticles	It is a form of pH-triggered drug delivery It can also be used for prevention of HIV infection	Zhang et al. formulated (TDF) loaded nanoparticles (NPs) prepared with a blend of PLGA and methacrylic acid copolymer pH-responsive release of anti-HIV microbicides occurred in the presence of human semen fluid simulant	Zhang et al. (2011)
Enzyme-triggered nanoparticles	Drug release occurs after coming in contact with enzymes The polymers get degraded by hyaluronidase and PSA which act like triggers	Asai et al prepared a polymeric gene regulator grafted with a cationic peptide containing the HIV-Tat peptide The protease cleaved the peptide causing the release of transgene and imparting the action in HIV-infected cells	Asai et al. (2010)
Temperature-triggered nanoparticle	Here, phase transition occurs due to the change in solvent property Sustained release of drug occurs Thermosensitive polymers were used	Date et al. synthesized a thermosensitive vaginal gel containing raltegravir and efavirenz loaded nanoparticles They were effective in long-term vaginal pre-exposure prophylaxis of heterosexual HIV-1	Date et al. (2012)
Magnetic-field triggered nanoparticles	Field-controlled drug delivery is possible by bypassing BBB (Blood Brain Barrier) Can kill HIV-infected T cells at elevated temperature They contain paramagnetic or superparamagnetic materials which are embedded into polymers	Saiyed et al. evaluated the specific drug targeting to the brain using zidovudine triphosphate bound with magnetic nanoparticles (MP-AZTTP) and encapsulated within liposomes A threefold increase in permeation was observed across BBB	Saiyed et al. (2010)
Virus mimetic nanoparticle	Non-replicating Non-infective Genetic material is absent Appears to body as a pathogen	Xu et al. demonstrated that fullerenol can assemble itself into virus-like particle The cells migrate towards the lymph node where they generate immune response by activating T cells	Xu et al. (2013)

7 HIV-1 and Cells of CNS, i.e., Myeloid

ART is the technique which has already reduced the mortality rate due to HIV-1. Despite all the number of patients, undergoing ART develops CNS disorders associated with HIV (Clifford 2008). ART can reduce viral in CNS of macaques having SIV, that is, simian immunodeficiency virus. Although they noticed the presence of such SIV DNA in the cerebrospinal fluid, HIV-1 has major reservoirs in CNS specifically the myeloid cells which consist of several cells including microglia, perivascular cells and meningeal macrophages. Therefore, the role between HIV-1 and these cells is of utmost importance. In recent in vivo and in vitro studies, Tat protein of HIV is shown in disrupting synaptical architecture (Kim et al. 2008). In another recent study, the role of secretion has been shown, namely, cathepsin B which secretes from HIV-1 infected macrophages. Low level of cathepsin B is noticed in the post-mortem cells of brain having HIV-1 along with HAD, but no such indication is noticed in healthy normal individual or in individual having normal cognition HIV. This study suggests that there is definitely the involvement of cathepsin B in HAD. Summing up together, all the above finding provides us enough data to form novel drugs or therapy for targeting and managing HAD.

8 Flushing Out Therapy

Although ART has wonderfully reduced the viral level in HIV-1 patients, interruption in the therapy leads to increase in rapid viremia. As we all know, HIV leads to rapid depletion of T cells, CD4+ and CD8+ T cells. Along with this, certain viral cell gets integrated with host chromatin and tries to form copies of virus in activation state but is silent in resting state (Chun et al. 1995).

This forms a pool of infection which is the main hurdle in getting rid of HIV-1 infection from the patient. Other than CD4+ T cells, various other cells like macrophages, monocytes, hematopoietic stem cells and dendritic cells can also be infected latently via HIV [187–189]. There are various experimental proofs for the same (McElrath et al. 1991).

Macrophages are unique viral reservoirs due to their resistance to cytopathic effect in HIV and due to their prolonged life span, though HIV latency association with macrophages is not that clear till now. Notably, patients of HIV-1 on ART treatment have very few macrophage-infected lymph nodes, though they can undergo reactivation in some cases of opportunistic infection (Caselli et al. 2005). Surprisingly, FDA has approved an antifungal drug, that is, amphotericin B which has been noticed to reactivate a model cell line of HIV, that is, THP89GFP in macrophages and not in T cells. Although when THP89GFP is induced via amphotericin B, they get co-cultured along with J89GFP which are infected T cells. Recently, in vitro studies have shown the role of polybacteria in activation of cells of monocyte lineage in HIV (Gonzalez et al. 2010). These studies depict that macrophages can be the latent site of HIV-1 infection.

Unlike the prior target cells, that is, CD4+ and T cells, latency in pre-integration macrophages takes part in the formation of viral reservoir formation on the next level [195], while post-integration mechanism is poorly understood. However, the presence of transcriptional repressors of host anti-HIV micro-RNA along with lack of Tat can play a vibrant role in macrophages for building post-integration latency, for example, C/EBPb which is a host factor well known for repressing HIV-1 transcription in phagocytic cells that contributes to latency of HIV. In addition, microglial cells in humans, that is, CTIP2, which is a strongly expressive transcriptional repressor of brain, are well known for inhibiting the replication of HIV-mediated chromatin altering complex which includes methylase SUV39H1, HDAC1 and HDAC2, while CT1P2's role has already been discussed in latency of post-integration in microglia cell (Marban et al. 2005).

As we know that ART is unable to completely remove the HIV from host cell, so recent efforts are made for reactivation of HIV with latent reservoir and also efforts are induced for the complete removal of HIV by ART. As stated by a hypothesis, cells having reactivated latency should perform apoptosis, may be by cytotoxic T cell recognition or by viral cytopathic effect (Siliciano and Greene 2011). Alongside, the new infection caused by viral progeny which is released from infected lysed cells will get stopped by ART.

Various new perspectives have been developed, which suggests histone deacetylase inhibitors (HDACi) are used for reactivating HIV, and various other components are used for reactivation, such as valproic acid (VPA), sodium butyrate, trichostatin and suberoylanilide hydroxamic acid; several methylation inhibitors are also included, such as chaetocin, BIX-01294, NFKB, 5-aza-2′ deoxycytidine(Aza-CdR), bryostatin and C modulators, and also there are various immune modulators, such as IL-7 and IL-5 (Abbas and Herbein 2012b; Kumar et al. 2013). The abovementioned compounds have shown remarkable impact in reactivation of HIV and the latency in CD4+ T cells; these cells are at variant stages of development. For inference, the first successful case of clinical trial was noted with VPA and HDACi, although the above fact is not confirmed in other variant trials (Siliciano et al. 2007).

Concerning effectiveness of these newly developed compounds for reactivating latency is not available in many reports. Although several studies have confirmed that HDACi and ACH2 along with U1 cell lines are equally efficacious in two of those cell lines, various tests on ITF2357 (givinostat), ACH2 and on U1 cell line were done, which discloses that latency of VPA is somewhat lesser potent than ITF2357. Particularly, givinostat was found to be safer in normal healthy people that in phase 1 of clinical trials. Combining all data together from in vitro studies provides information that the compounds which were potent to activate latency in CD4+ cells are also efficacious in reactivating macrophages and monocytes lineage. Nevertheless, in clinical trials, it depicts that viral load is mainly present in T lymphocytes. Above all, if we talk about isolation, then isolation of T cells is a shorter process than isolation of monocytes, which is followed by producing macrophages which are derived from monocytes (Shikuma et al. 2012).

Furthermore, we talk about macrophages which are now permanent residents in brain and in other anatomical structures where penetration of drug is rather difficult and also the determination of drug effectiveness is also a tough grind (Shikuma et al. 2012). In addition, we talk about the persistence of various arrays of metabolic enzymes and efflux pumps in BBB which altogether makes the phenomenon of checking drug permeation a tough process. Macrophages resident at brain play a crucial role in HAD, which is a severe illness of HIV-1 infection. Treatment of HIV has to be complete in which cells of macrophage as well as monocytes lineage should be taken into consideration rather than only considering T lymphocyte cells. Any viral reservoir should not be left behind; otherwise, no feasible outcome will be there.

9 Conclusion

After understanding the chapter, it is quite clear that macrophages are the chosen site for HIV-1 and it is the best suited latent viral reservoir which is potentially chronic. These cells also contribute as an inflammatory mediator in CNS diseases which are HIV-mediated. However, combinatorial therapies like ART are proven to be potent in supressing viremia in some infected patients. While complete removal of HIV from latent reservoirs is quite impossible, several novel therapies are being tested by the researchers for the same, and there are few novel approaches that are working on the clearance of these latent reservoirs, although the implementation of these novel approaches in vivo is still a tough grind.

Here, we have discussed the physiology, and how HIV enters the body and, in the macrophages, attacks CD4+ and CD8+ T cells and forms a reservoir, and then we have discussed various conventional therapies which include various ART and enzyme inhibitor therapies like RTIs, NNRTIs, NRTIs, protease inhibitors and fusion inhibitors, and various other blocking agent therapies, but as we concluded that these conventional therapies were not enough for the eradication of HIV, there is the need for the novel approaches and our main focus was on how macrophages are the site of HIV and how we can work on it, so that we can completely get rid of HIV.

Various novel approaches include therapies like carbohydrate-binding agents which focus on the glycans in HIV gp120; the second therapy which was discussed is PI3K/Akt blocking agents, where signalling cascades were widely identified as favourite target. Other strategies include small interfering RNA (siRNA), and these are potent molecules capable of degrading viral genome; similarly, shRNA works the same. Various other novel approaches include nanocarriers, which are found to be the most potent in targeting delivery and have shown prominent results in overcoming many physiological and anatomical barriers and have delivered the drug on the required site, hence improving HIV therapy. Then comes immune-based therapies; it promotes boosting the immune system through various techniques by which body can combat the loss caused due to HIV, and lastly flushing therapy is included which is quite recent and made to treat HIV. Although measures are being taken, still there is no such therapy which can completely eradicate the viral disease, that is, HIV, and scientists are working on the same and we can conclude by saying that, if precise tools along with combinatorial therapies and equipment's are used, or we have to mainly focus on the target of latent source, then only in the future we can be able to create such therapies which can completely eradicate HIV-1. Till now, we have to be satisfied with the prolonging life therapies.

References

Abbas W, Herbein G. Molecular understanding of HIV-1 latency. Adv Virol. 2012a;2012:574967.

Abbas W, Herbein G. Molecular understanding of HIV-1 latency. Adv Virol. 2012b;2012:574967.

Allouch A, David A, Amie SM, Lahouassa H, Chartier L, Margottin-Goguet F, Barré-Sinoussi F, Kim B, Sáez-Cirión A, Pancino G. p21-mediated RNR2 repression restricts HIV-1 replication in macrophages by inhibiting dNTP biosynthesis pathway. Proc Natl Acad Sci USA. 2013;110:E3997–4006.

Anastassopoulou CG, Ketas TJ, Klasse PJ, Moore JP. Resistance to CCR5 inhibitors caused by sequence changes in the fusion peptide of HIV-1 gp41. Proc Natl Acad Sci USA. 2009;106:5318–23.

Aquaro S, Calio R, Balestra E, Bagnarelli P, Cenci A, Bertoli A, Tavazzi B, Di Pierro D, Francesconi M, Abdelahad D, Perno CF. Clinical implications of HIV dynamics and drug resistance in macrophages. J Biol Regul Homeost Agents. 1998;12:23–7.

Aquaro S, Svicher V, Ceccherini-Silberstein F, Cenci A, Marcuccilli F, Giannella S, Marcon L, Caliò R, Balzarini J, Perno CF. Limited development and progression of resistance of HIV-1 to the nucleoside analogue reverse transcriptase inhibitor lamivudine in human primary macrophages. J Antimicrob Chemother. 2005;55:872–8.

Aquaro S, Svicher V, Schols D, Pollicita M, Antinori A, Balzarini J, Perno CF. Mechanisms underlying activity of antiretroviral drugs in HIV-1-infected macrophages: new therapeutic strategies. J Leukoc Biol. 2006;80:1103–10.

Asai D, Kuramoto M, Shoji Y, et al. Specific transgene expression in HIV-infected cells using protease-cleavable transcription regulator. J Control Release. 2010;141(1):52–61. https://doi.org/10.1016/j.jconrel.2009.08.025.

Badley AD, Pilon AA, Landay A, Lynch DH. Mechanisms of HIV-associated lymphocyte apoptosis. Blood. 2000;96:2951–64.

Balzarini J. Carbohydrate-binding agents: a potential future cornerstone for the chemotherapy of enveloped viruses? Antivir Chem Chemother. 2007;18:1–11.

Balzarini J, Van HY, Vermeire K, Vanham G, Schols D. Carbohydrate-binding agents efficiently prevent dendritic cell-specific intercellular adhesion molecule-3-grabbing nonintegrin (DC-SIGN)-directed HIV-1 transmission to T lymphocytes. Mol Pharmacol. 2007;71:3–11.

Betageri GV, Black CD, Szebeni J, Wahl LM, Weinstein JN. Fc-receptor-mediated targeting of antibody-bearing liposomes containing dideoxycytidine triphosphate to human monocyte/ macrophages. J Pharm Pharmacol. 1993a;45:48–53. (PubMed: 8094446)

Betageri GV, Jenkins SA, Ravis WR. Drug delivery using antibody-liposome conjugates. Drug Dev Ind Pharm. 1993b;19:2109–16.

Buonaguro L, Barillari G, Chang HK, Bohan CA, Kao V, Morgan R, Gallo RC, Ensoli B. Effects of the human immunodeficiency virus type 1 Tat protein on the expression of inflammatory cytokines. J Virol. 1992;66:7159–67.

Busca A, Saxena M, Kumar A. Critical role for anti-apoptotic Bcl-xL and Mcl-1 in human macrophage survival and cellular IAP1/2 (cIAP1/2) in resistance to HIV-Vpr-induced apoptosis. J Biol Chem. 2012;287:15118–33.

Canducci F, Ceresola ER, Saita D, Castagna A, Gianotti N, Underwood M, Burioni R, Lazzarin A, Clementi M. In vitro phenotypes to elvitegravir and dolutegravir in primary macrophages and lymphocytes of clonal recombinant viral variants selected in patients failing raltegravir. J Antimicrob Chemother. 2013;68:2525–32.

Carter CA, Ehrlich LS. Cell biology of HIV-1 infection of macrophages. Annu Rev Microbiol. 2008;62:425–43.

Caselli E, Galvan M, Cassai E, Caruso A, Sighinolfi L, Di LD. Human herpesvirus 8 enhances human immunodeficiency virus replication in acutely infected cells and induces reactivation in latently infected cells. Blood. 2005;106:2790–7.

Cassol E, Cassetta L, Rizzi C, Alfano M, Poli G. M1 and M2a polarization of human monocyte-derived macrophages inhibits HIV-1 replication by distinct mechanisms. J Immunol. 2009;182:6237–46.

Chakravarty M, Vora A. Nanotechnology-based antiviral therapeutics [published online ahead of print, 2020 Aug 3]. Drug Deliv Transl Res. 2020:1–40. https://doi.org/10.1007/s13346-020-00818-0.

Chariot P, Monnet I, Gherardi R. Cytochrome c oxidase reaction improves histopathological assessment of zidovudine myopathy. Ann Neurol. 1993;34:561–5.

Chen P, Mayne M, Power C, Nath A. The Tat protein of HIV-1 induces tumor necrosis factor-alpha production. Implications for HIV-1-associated neurological diseases. J Biol Chem. 1997;272:22385–8.

Chen Q, Swaminathan S, Yang D, Dai L, Sui H, Yang J, Hornung RL, Wang Y, da Huang W, Hu X, Lempicki RA, Imamichi T. Interleukin-27 is a potent inhibitor of cis HIV-1 replication in monocyte-derived dendritic cells via a type I interferon-independent pathway. PLoS One. 2013;8:e59194.

Chu GJ, Murad A. A case of ethanol-induced systemic allergic dermatitis. Contact Dermatitis. 2017;76(3):182–4.

Chugh P, Bradel-Tretheway B, Monteiro-Filho CM, Planelles V, Maggirwar SB, Dewhurst S, Kim B. Akt inhibitors as an HIV-1 infected macrophagespecific anti-viral therapy. Retrovirology. 2008;5:11.

Chun TW, Finzi D, Margolick J, Chadwick K, Schwartz D, Siliciano RF. In vivo fate of HIV-1-infected T cells: quantitative analysis of the transition to stable latency. Nat Med. 1995;1:1284–90.

Chun TW, Davey RT Jr, Engel D, Lane HC, Fauci AS. Re-emergence of HIV after stopping therapy. Nature. 1999;401:874–5.

Clifford DB. HIV-associated neurocognitive disease continues in the antiretroviral era. Top HIV Med. 2008;16:94–8.

Collins KL, Chen BK, Kalams SA, Walker BD, Baltimore D. HIV-1 Nef protein protects infected primary cells against killing by cytotoxic T lymphocytes. Nature. 1998;391:397–401.

Dai L, Lidie KB, Chen Q, Adelsberger JW, Zheng X, Huang D, Yang J, Lempicki RA, Rehman T, Dewar RL, Wang Y, Hornung RL, Canizales KA, Lockett SJ, Lane HC, Imamichi T. IL-27 inhibits HIV-1 infection in human macrophages by down-regulating host factor SPTBN1 during monocyte to macrophage differentiation. J Exp Med. 2013;210:517–34.

Das SR, Jameel S. Biology of the HIV Nef protein. Indian J Med Res. 2005;121:315–32.

Date A, Shibata A, Goede M, et al. Development and evaluation of a thermosensitive vaginal gel containing raltegravir+efavirenz loaded nanoparticles for HIV prophylaxis. Antiviral Res. 2012;96(3):430–6. https://doi.org/10.1016/j.antiviral.2012.09.015.

Diamond TL, Roshal M, Jamburuthugoda VK, Reynolds HM, Merriam AR, Lee KY, Balakrishnan M, Bambara RA, Planelles V, Dewhurst S, Kim B. Macrophage tropism of HIV-1 depends on efficient cellular dNTP utilization by reverse transcriptase. J Biol Chem. 2004;279:51545–53.

Dorr P, Westby M, Dobbs S, Griffin P, Irvine B, Macartney M, Mori J, Rickett G, Smith-Burchnell C, Napier C, Webster R, Armour D, Price D, Stammen B, Wood A, Perros M. Maraviroc (UK-427,857), a potent, orally bioavailable, and selective small-molecule inhibitor of chemokine receptor CCR5 with broad-spectrum anti-human immunodeficiency virus type 1 activity. Antimicrob Agents Chemother. 2005;49:4721–32.

Dubey V, Mishra D, Nahar M, Jain V, Jain NK. Enhanced transdermal delivery of an anti-HIV agent via ethanolic liposomes. Nanomedicine. 2010a;6(4):590–6. https://doi.org/10.1016/j.nano.2010.01.002.

Dubey V, Mishra D, Nahar M, Jain V, Jain N. Enhanced transdermal delivery of an anti-HIV agent via ethanolic liposomes. Nanomedicine. 2010b;6(4):590–6. https://doi.org/10.1016/j.nano.2010.01.002.

Dutta T, Jain NK. Targeting potential and anti-HIV activity of lamivudine loaded mannosylated poly (propyleneimine) dendrimer. Biochim Biophys Acta. 1770;2007:681–6. (PubMed: 17276009)

Dutta T, Jain NK. Targeting potential and anti-HIV activity of lamivudine loaded mannosylated poly (propyleneimine) dendrimer. Biochim Biophys Acta. 2007;1770(4):681–6. https://doi.org/10.1016/j.bbagen.2006.12.007.

Dutta T, Agashe HB, Garg M, Balakrishnan P, Kabra M, Jain NK. Poly (propyleneimine) dendrimer based nanocontainers for targeting of efavirenz to human monocytes/macrophages in vitro. J Drug Target. 2007;15:89–98. (PubMed: 17365278)

Eckstein DA, Sherman MP, Penn ML, Chin PS, De Noronha CM, Greene WC, Goldsmith MA. HIV-1 Vpr enhances viral burden by facilitating infection of tissue macrophages but not nondividing CD4+ T cells. J Exp Med. 2001;194:1407–19.

Ehrlich PR, Himmelweit F, Marquardt M, Dale HH. The collected papers of Paul Ehrlich. London: Pergamon Press; 1960.

El-Deiry WS. Akt takes centre stage in cell-cycle deregulation. Nat Cell Biol. 2001;3:E71–3.

Ensoli B, Buonaguro L, Barillari G, Fiorelli V, Gendelman R, Morgan RA, Wingfield P, Gallo RC. Release, uptake, and effects of extracellular human immunodeficiency virus type 1 Tat protein on cell growth and viral transactivation. J Virol. 1993;67:277–87.

Finkel TH, Tudor-Williams G, Banda NK, Cotton MF, Curiel T, Monks C, Baba TW, Ruprecht RM, Kupfer A. Apoptosis occurs predominantly in bystander cells and not in productively infected cells of HIV- and SIV-infected lymph nodes. Nat Med. 1995;1:129–34.

Fujiwara M, Takiguchi M. HIV-1-specific CTLs effectively suppress replication of HIV-1 in HIV-1-infected macrophages. Blood. 2007;109:4832–8.

Furge LL, Guengerich FP. Analysis of nucleotide insertion and extension at 8-oxo-7,8-dihydroguanine by replicative T7 polymerase exo- and human immunodeficiency virus-1 reverse transcriptase using steady-state and pre-steady-state kinetics. Biochemistry. 1997;36:6475–87.

Gallay P, Stitt V, Mundy C, Oettinger M, Trono D. Role of the karyopherin pathway in human immunodeficiency virus type 1 nuclear import. J Virol. 1996;70:1027–32.

Gallay P, Hope T, Chin D, Trono D. HIV-1 infection of nondividing cells through the recognition of integrase by the importin/karyopherin pathway. Proc Natl Acad Sci USA. 1997;94:9825–30.

Garrido C, Simpson C, Dahl N, et al. Gold nanoparticles to improve HIV drug delivery. Future Med Chem. 2015;7(9):1097–107. https://doi.org/10.4155/fmc.15.57.

Gavegnano C, Schinazi RF. Antiretroviral therapy in macrophages: implication for HIV eradication. Antivir Chem Chemother. 2009;20:63–78.

Geleziunas R, Xu W, Takeda K, Ichijo H, Greene WC. HIV-1 Nef inhibits ASK1-dependent death signalling providing a potential mechanism for protecting the infected host cell. Nature. 2001;410:834–8.

Gonzalez OA, Li M, Ebersole JL, Huang CB. HIV-1 reactivation induced by the periodontal pathogens Fusobacterium nucleatum and Porphyromonas gingivalis involves Toll-like receptor 2 [corrected] and 9 activation in monocytes/macrophages. Clin Vaccine Immunol. 2010;17:1417–27.

Greenhead P, Hayes P, Watts PS, Laing KG, Griffin GE, Shattock RJ. Parameters of human immunodeficiency virus infection of human cervical tissue and inhibition by vaginal virucides. J Virol. 2000;74:5577–86. https://doi.org/10.1128/JVI.74.12.5577-5586.2000.

Gunaseelan S, Debrah O, Wan L, Leibowitz MJ, Rabson AB, Stein S, Sinko PJ. Synthesis of poly (ethylene glycol)-based saquinavir prodrug conjugate and assessment of release and anti-HIV-1 bioactivity using a novel protease inhibition assay. Bioconj Chem. 2004;15:1322–33.

Henderson AJ, Calame KL. CCAAT/enhancer binding protein (C/EBP) sites are required for HIV-1 replication in primary macrophages but not CD4(+) T cells. Proc Natl Acad Sci USA. 1997;94:8714–9.

Herbein G, Varin A. The macrophage in HIV-1 infection: from activation to deactivation? Retrovirology. 2010a;7:33. https://doi.org/10.1186/1742-4690-7-33.

Herbein G, Varin A. The macrophage in HIV-1 infection: from activation to deactivation? Retrovirology. 2010b;7:33.

Herbein G, Gras G, Khan KA, Abbas W. Macrophage signaling in HIV-1 infection. Retrovirology. 2010;7:34.

Hoorelbeke B, Xue J, LiWang PJ, Balzarini J. Role of the carbohydratebinding sites of griffithsin in the prevention of DC-SIGN-mediated capture and transmission of HIV-1. PLoS One. 2013;8:e64132.

Hu J, Liu H, Wang L. Enhanced delivery of AZT to macrophages via acetylated LDL. J Control Release. 2000;69:327–35. (PubMed: 11102673)

Huang L, Bosch I, Hofmann W, Sodroski J, Pardee AB. Tat protein induces human immunodeficiency virus type 1 (HIV-1) coreceptors and promotes infection with both macrophage-tropic and T-lymphotropic HIV-1 strains. J Virol. 1998;72:8952–60.

Hunter CA. New IL-12-family members: IL-23 and IL-27, cytokines with divergent functions. Nat Rev Immunol. 2005;5:521–31.

Iordanskiy S, Santos S, Bukrinsky M. Nature, nurture and HIV: the effect of producer cell on viral physiology. Virology. 2013;443:208–13.

Jacobson MA, Hardy D, Connick E, Watson J, DeBruin M. Phase 1 trial of a single dose of recombinant human interleukin-12 in human immunodeficiency virus-infected patients with 100–500 CD4 cells/ microL. J Infect Dis. 2000;182:1070–6.

Jacque JM, Stevenson M. The inner-nuclear-envelope protein emerin regulates HIV-1 infectivity. Nature. 2006;441:641–5.

Jacquot G, Le RE, David A, Mazzolini J, Bouchet J, Bouaziz S, Niedergang F, Pancino G, Benichou S. Localization of HIV-1 Vpr to the nuclear envelope: impact on Vpr functions and virus replication in macrophages. Retrovirology. 2007;4:84.

Jain SK, Gupta Y, Jain A, Saxena AR, Khare P, Jain A. Mannosylated gelatin nanoparticles bearing an anti-HIV drug didanosine for site-specific delivery. Nanomed. 2008;4:41–8.

Jain S, Mistry MA, Swarnakar NK. Enhanced dermal delivery of acyclovir using solid lipid nanoparticles. Drug Deliv Transl Res. 2011;1(5):395–406. https://doi.org/10.1007/s13346-011-0036-0.

Jambo KC, Banda DH, Kankwatira AM, Sukumar N, Allain TJ, Heyderman RS, et al. Small alveolar macrophages are infected preferentially by HIV and exhibit impaired phagocytic function. Mucosal Immunol. 2014;7:1116–26. (PubMed: 24472847)

Joshy KS, et al. Gelatin modified lipid nanoparticles for anti- viral drug delivery. Chem Phys Lipids. 2017;207(Pt A):24–37.

Katragadda A, Bridgman R, Betageri G. Effect of liposome composition and cholesterol on the cellular uptake of stavudine by human monocyte/macrophages. Cell Mol Biol Lett. 2000;5:483–94.

Kaur A, Jain S, Tiwary AK. Mannan-coated gelatin nanoparticles for sustained and targeted delivery of didanosine: in vitro and in vivo evaluation. Acta Pharm. 2008;58:61–74. (PubMed: 18337208)

Kilareski EM, Shah S, Nonnemacher MR, Wigdahl B. Regulation of HIV-1 transcription in cells of the monocyte-macrophage lineage. Retrovirology. 2009;6:118.

Kim HJ, Martemyanov KA, Thayer SA. Human immunodeficiency virus protein Tat induces synapse loss via a reversible process that is distinct from cell death. J Neurosci. 2008;28:12604–13.

Kobiler O, Drayman N, Butin-Israeli V, Oppenheim A. Virus strategies for passing the nuclear envelope barrier. Nucleus. 2012;3:526–39.

Kondo A, Muranaka Y, Ohta I, Notsu K, Manabe M, Kotani K, Saito K, Maekawa M, Kanno T. Relationship between triglyceride concentrations and LDL size evaluated by malondialdehyde modified LDL. Clin Chem. 2001;47:893–900. (PubMed: 11325894)

Koppensteiner H, Brack-Werner R, Schindler M. Macrophages and their relevance in human immunodeficiency virus type I infection. Retrovirology. 2012;9:82.

Kramer B, Pelchen-Matthews A, Deneka M, Garcia E, Piguet V, Marsh M. HIV interaction with endosomes in macrophages and dendritic cells. Blood Cells Mol Dis. 2005;35:136–42.

Kumar A, Herbein G. The macrophage: a therapeutic target in HIV-1 infection. Mol Cell Ther. 2014;2:10. https://doi.org/10.1186/2052-8426-2-10.

Kumar A, Abbas W, Herbein G. TNF and TNF receptor superfamily members in HIV infection: new cellular targets for therapy? Mediators Inflamm. 2013;2013:484378.

Lahouassa H, Daddacha W, Hofmann H, Ayinde D, Logue EC, Dragin L, Bloch N, Maudet C, Bertrand M, Gramberg T, Pancino G, Priet S, Canard B, Laguette N, Benkirane M, Transy C, Landau NR, Kim B, Margottin-Goguet F. SAMHD1 restricts the replication of human immunodeficiency virus type 1 by depleting the intracellular pool of deoxynucleoside triphosphates. Nat Immunol. 2012;13:223–8.

Lama J. The physiological relevance of CD4 receptor down-modulation during HIV infection. Curr HIV Res. 2003;1:167–84.

Lama I, Mangasarian A, Trono D. Cell-surface expression of CD4 reduces HIV-1 infectivity by blocking Env incorporation in a Nef- and Vpu-inhibitable manner. Curr Biol. 1999;9:622–31.

Langer R. Drug delivery. Drugs on target. Science. 2001;293:58–9.

Lara HH, Ixtepan-Turrent L, Garza Treviño EN, Singh DK. Use of silver nanoparticles increased inhibition of cell-associated HIV-1 infection by neutralizing antibodies developed against HIV-1 envelope proteins. J Nanobiotechnol. 2011;9:38. https://doi.org/10.1186/1477-3155-9-38.

Leonard JN, Schaffer DV. Antiviral RNAi therapy: emerging approaches for hitting a moving target. Gene Ther. 2006;13:532–40.

Lin RH, Hwang YW, Yang BC, Lin CS. TNF receptor-2-triggered apoptosis is associated with the down-regulation of Bcl-xL on activated T cells and can be prevented by CD28 costimulation. J Immunol. 1997;158:598–603.

Liu L, Oliveira NM, Cheney KM, Pade C, Dreja H, Bergin AM, Borgdorff V, Beach DH, Bishop CL, Dittmar MT, McKnight A. A whole genome screen for HIV restriction factors. Retrovirology. 2011;8:94.

Mahajan SD, Roy I, Xu G, et al. Enhancing the delivery of anti-retroviral drug "Saquinavir" across the blood brain barrier using nanoparticles. Curr HIV Res. 2010;8(5):396–404. https://doi.org/10.2174/157016210791330356.

Mahlknecht U, Deng C, Lu MC, Greenough TC, Sullivan JL, O'Brien WA, Herbein G. Resistance to apoptosis in HIV-infected CD4+ T lymphocytes is mediated by macrophages: role for Nef and immune activation in viral persistence. J Immunol. 2000;165:6437–46.

Mainardes RM, Gremiao MPD, Brunetti IL, da Fonseca LM, Khalil NM. Zidovudine-loaded PLA and PLA-PEG blend nanoparticles: influence of polymer type on phagocytic uptake by polymorphonuclear cells. J Pharm Sci. 2009;98:257–67. (PubMed: 18425813)

Makabi-Panzu B, Gourde P, Desormeaux A, Bergeron MG. Intracellular and serum stability of liposomal 2′, 3′-dideoxycytidine. Effect of lipid composition. Cell Mol Biol. 1998;44:277–84. (PubMed: 9593578)

Mangino G, Percario ZA, Fiorucci G, Vaccari G, Manrique S, Romeo G, Federico M, Geyer M, Affabris E. In vitro treatment of human monocytes/macrophages with myristoylated recombinant Nef of human immunodeficiency virus type 1 leads to the activation of mitogenactivated protein kinases, IkappaB kinases, and interferon regulatory factor 3 and to the release of beta interferon. J Virol. 2007;81:2777–91.

Mangino G, Percario ZA, Fiorucci G, Vaccari G, Acconcia F, Chiarabelli C, Leone S, Noto A, Horenkamp FA, Manrique S, Romeo G, Polticelli F, Geyer M, Affabris E. HIV-1 Nef induces proinflammatory state in macrophages through its acidic cluster domain: involvement of TNF alpha receptor associated factor 2. PLoS One. 2011;6:e22982.

Mankertz J, Matthes E, Rokos K, von Baeyer H, Pauli G, Riedel E. Selective endocytosis of fluorothymidine and azidothymidine coupled to LDL into HIV infected mononuclear cells. BBAMol Basis of Dis. 1996;1317:233–7.

Marban C, Redel L, Suzanne S, Van Lint C, Lecestre D, Chasserot-Golaz S, Leid M, Aunis D, Schaeffer E, Rohr O. COUP-TF interacting protein 2 represses the initial phase of HIV-1 gene transcription in human microglial cells. Nucleic Acids Res. 2005;33:2318–31.

Mayne M, Holden CP, Nath A, Geiger JD. Release of calcium from inositol 1,4,5-trisphosphate receptor-regulated stores by HIV-1 Tat regulates TNF-alpha production in human macrophages. J Immunol. 2000;164:6538–42.

McElrath MJ, Steinman RM, Cohn ZA. Latent HIV-1 infection in enriched populations of blood monocytes and T cells from seropositive patients. J Clin Invest. 1991;87:27–30.

Morgensztern D, McLeod HL. PI3K/Akt/mTOR pathway as a target for cancer therapy. Anticancer Drugs. 2005;16:797–803.

Mougel M, Houzet L, Darlix JL. When is it time for reverse transcription to start and go? Retrovirology. 2009;6:24.

Mueller YM, Do DH, Altork SR, Artlett CM, Gracely EJ, Katsetos CD, Legido A, Villinger F, Altman JD, Brown CR, Lewis MG, Katsikis PD. IL-15 treatment during acute simian immunodeficiency virus (SIV) infection increases viral set point and accelerates disease progression despite the induction of stronger SIV-specific CD8+ T cell responses. J Immunol. 2008;180:350–60.

Nitahara-Kasahara Y, Kamata M, Yamamoto T, Zhang X, Miyamoto Y, Muneta K, Iijima S, Yoneda Y, Tsunetsugu-Yokota Y, Aida Y. Novel nuclear import of Vpr promoted by importin alpha is crucial for human immunodeficiency virus type 1 replication in macrophages. J Virol. 2007;81:5284–93.

Opanasopit P, Nishikawa M, Hashida M. Factors affecting drug and gene delivery: effects of interaction with blood components. Crit Rev Ther Drug Carrier Syst. 2002;19:191–233. (PubMed: 12627613)

Oussoren C, Magnani M, Fraternale A, Casabianca A, Chiarantini L, Ingebrigsten R, Underberg WJM, Storm G. Liposomes as carriers of the antiretroviral agent dideoxycytidine-5 -triphosphate. Int J Pharm. 1999;180:261–70. (PubMed: 10370196)

Oyaizu N, Adachi Y, Hashimoto F, McCloskey TW, Hosaka N, Kayagaki N, Yagita H, Pahwa S. Monocytes express Fas ligand upon CD4 cross-linking and induce CD4+ T cells apoptosis: a possible mechanism of bystander cell death in HIV infection. J Immunol. 1997;158:2456–63.

Pallikkuth S, Rogers K, Villinger F, Dosterii M, Vaccari M, Franchini G, Pahwa R, Pahwa S. Interleukin-21 administration to rhesus macaques chronically infected with simian immunodeficiency virus increases cytotoxic effector molecules in T cells and NK cells and enhances B cell function without increasing immune activation or viral replication. Vaccine. 2011;29:9229–38.

Park K. Nanotechnology: what it can do for drug delivery. J Control Release. 2007;120:1–3.

Perno CF, Aquaro S, Rosenwirth B, Balestra E, Peichl P, Billich A, Villani N, Caliò R. In vitro activity of inhibitors of late stages of the replication of HIV in chronically infected macrophages. J Leukoc Biol. 1994;56:381–6.

Perno CF, Newcomb FM, Davis DA, Aquaro S, Humphrey RW, Calio R, Yarchoan R. Relative potency of protease inhibitors in monocytes/ macrophages acutely and chronically infected with human immunodeficiency virus. J Infect Dis. 1998;178:413–22.

Perno CF, Svicher V, Schols D, Pollicita M, Balzarini J, Aquaro S. Therapeutic strategies towards HIV-1 infection in macrophages. Antiviral Res. 2006;71:293–300.

Pilon AA, Lum JJ, Sanchez-Dardon J, Phenix BN, Douglas R, Badley AD. Induction of apoptosis by a nonnucleoside human immunodeficiency virus type 1 reverse transcriptase inhibitor. Antimicrob Agents Chemother. 2002;46:2687–91.

Pollock S, Dwek RA, Burton DR, Zitzmann N. N-Butyldeoxynojirimycin is a broadly effective antiHIV therapy significantly enhanced by targeted liposome delivery. AIDS. 2008;22:1961–9. (PubMed: 18753929)

Pomerantz RJ, Horn DL. Twenty years of therapy for HIV-1 infection. Nat Med. 2003;9:867–73.

Pozzi LA, Maciaszek JW, Rock KL. Both dendritic cells and macrophages can stimulate naive CD8 T cells in vivo to proliferate, develop effector function, and differentiate into memory cells. J Immunol. 2005;175:2071–81.

Saiyed ZM, Gandhi NH, Nair MP. Magnetic nanoformulation of azidothymidine 5′-triphosphate for targeted delivery across the blood-brain barrier. Int J Nanomed. 2010;5:157–66. https://doi.org/10.2147/ijn.s8905.

Sánchez-López E, Paús A, Pérez-Pomeda I, Calpena A, Haro I, Gómara MJ. Lipid vesicles loaded with an HIV-1 fusion inhibitor peptide as a potential microbicide. Pharmaceutics. 2020;12(6):502. https://doi.org/10.3390/pharmaceutics12060502.

Schmidtmayerova H, Alfano M, Nuovo G, Bukrinsky M. Human immunodeficiency virus type 1 T-lymphotropic strains enter macrophages via a CD4- and CXCR4-mediated pathway: replication is restricted at a postentry level. J Virol. 1998;72:4633–42.

Schnepple DJ, Shepard B, Bren GD, Cummins NW, Natesampillai S, Trushin S, Algeciras-Schimnich A, Meng XW, Sainski AM, Rizza SA, Kaufmann SH, Badley AD. Isolation of a TRAIL antagonist from the serum of HIV-infected patients. J Biol Chem. 2011;286:35742–54.

Seder RA, Grabstein KH, Berzofsky JA, McDyer JF. Cytokine interactions in human immunodeficiency virus-infected individuals: roles of interleukin (IL)-2, IL-12, and IL-15. J Exp Med. 1995;182:1067–77.

Shah LK, Amiji MM. Intracellular delivery of saquinavir in biodegradable polymeric nanoparticles for HIV/AIDS. Pharm Res. 2006;23:2638–45. (PubMed: 16969696)

Sheehy AM, Gaddis NC, Choi JD, Malim MH. Isolation of a human gene that inhibits HIV-1 infection and is suppressed by the viral Vif protein. Nature. 2002;418:646–50.

Shikuma CM, Nakamoto B, Shiramizu B, Liang CY, DeGruttola V, Bennett K, Paul R, Kallianpur K, Chow D, Gavegnano C, Hurwitz SJ, Schinazi RF, Valcour VG. Antiretroviral monocyte efficacy score linked to cognitive impairment in HIV. Antivir Ther. 2012;17:1233–42.

Siliciano RF, Greene WC. HIV latency. Cold Spring Harb Perspect Med. 2011;1:a007096.

Siliciano JD, Lai J, Callender M, Pitt E, Zhang H, Margolick JB, Gallant JE, Cofrancesco J Jr, Moore RD, Gange SJ, Siliciano RF. Stability of the latent reservoir for HIV-1 in patients receiving valproic acid. J Infect Dis. 2007;195:833–6.

Song E, Lee SK, Dykxhoorn DM, Novina C, Zhang D, Crawford K, Cerny J, Sharp PA, Lieberman J, Manjunath N, Shankar P. Sustained small interfering RNA-mediated human immunodeficiency virus type 1 inhibition in primary macrophages. J Virol. 2003;77:7174–81.

Subbramanian RA, Kessous-Elbaz A, Lodge R, Forget J, Yao XJ, Bergeron D, Cohen EA. Human immunodeficiency virus type 1 Vpr is a positive regulator of viral transcription and infectivity in primary human macrophages. J Exp Med. 1998;187:1103–11.

Swingler S, Brichacek B, Jacque JM, Ulich C, Zhou J, Stevenson M. HIV-1 Nef intersects the macrophage CD40L signalling pathway to promote resting-cell infection. Nature. 2003;424:213–9.

Swingler S, Mann AM, Zhou J, Swingler C, Stevenson M. Apoptotic killing of HIV-1-infected macrophages is subverted by the viral envelope glycoprotein. PLoS Pathog. 2007;3:1281–90.

Tan J, Sattentau QJ. The HIV-1-containing macrophage compartment: a perfect cellular niche? Trends Microbiol. 2013;21:405–12.

Tyagi A, Ahmed F, Thakur N, Sharma A, Raghava GP, Kumar M. HIVsirDB: a database of HIV inhibiting siRNAs. PLoS One. 2011;6:e25917.

Varin A, Decrion AZ, Sabbah E, Quivy V, Sire J, Van Lint C, Roques BP, Aggarwal BB, Herbein G. Synthetic Vpr protein activates activator protein-1, c-Jun N-terminal kinase, and NF-kappaB and stimulates HIV-1 transcription in promonocytic cells and primary macrophages. J Biol Chem. 2005;280:42557–67.

Vazquez N, Greenwell-Wild T, Marinos NJ, Swaim WD, Nares S, Ott DE, Schubert U, Henklein P, Orenstein JM, Sporn MB, Wahl SM. Human immunodeficiency virus type 1-induced macrophage gene expression includes the p21 gene, a target for viral regulation. J Virol. 2005;79:4479–91.

Viora M, Di GG, Rivabene R, Malorni W, Fattorossi A. Interference with cell cycle progression and induction of apoptosis by dideoxynucleoside analogs. Int J Immunopharmacol. 1997;19:311–21.

Vyas SP, Subhedar R, Jain S. Development and characterization of emulsomes for sustained and targeted delivery of an antiviral agent to liver. Int J Clin Pharmacol Biopharm. 2006;58:321–6.

Wan L, Pooyan S, Hu P, Leibowitz MJ, Stein S, Sinko PJ. Peritoneal macrophage uptake, pharmacokinetics and biodistribution of macrophage-targeted PEG-fMLF (N-formyl-methionyl-leucyl-phenylalanine) nanocarriers for improving HIV drug delivery. Pharm Res. 2007;24:2110–9.

Welsch S, Groot F, Krausslich HG, Keppler OT, Sattentau QJ. Architecture and regulation of the HIV-1 assembly and holding compartment in macrophages. J Virol. 2011;85:7922–7.

Xu L, Liu Y, Chen Z, et al. Morphologically virus-like fullerenol nanoparticles act as the dual-functional Nanoadjuvant for HIV-1 vaccine. Adv Mater. 2013;25(41):5928–36. https://doi.org/10.1002/adma.201300583.

Zaritsky LA, Dery A, Leong WY, Gama L, Clements JE. Tissue-specific interferon alpha subtype response to SIV infection in brain, spleen, and lung. J Interferon Cytokine Res. 2013;33:24–33. (PubMed: 23050948)

Zhang T, Sturgis T, Youan B. pH-responsive nanoparticles releasing tenofovir intended for the prevention of HIV transmission. Eur J Pharm Biopharm. 2011;79(3):526–36. https://doi.org/10.1016/j.ejpb.2011.06.007.

Zhou J, Neff CP, Liu X, et al. Systemic administration of combinatorial dsiRNAs via nanoparticles efficiently suppresses HIV-1 infection in humanized mice. Mol Ther. 2011;19(12):2228–38. https://doi.org/10.1038/mt.2011.207.

Part V

Specific Therapeutics and Clinical Trials on Macrophage Targeted Delivery

Delivery of siRNA to Macrophages: Challenges and Opportunities

Divya Kaushal, Swati Gupta, and Yashwant V. Pathak

Abstract

The discovery of the RNA interference (RNAi) pathway in 1998 and small interfering RNA (siRNA) in 2001 have paved the wave for gene targeting and targeted drug delivery. The use of siRNA has been used to silence genes and target specific genes. Utilizing nanotechnology has allowed for detecting various cells, and therapies using siRNA have been used to target genes. The recent efforts to target macrophages have been successful due to their secretion of pro-inflammatory mediators such as cytokines that can activate other cells. In addition, targeting macrophages will reduce the pro-inflammatory response and phagocytosis triggered by antigens. The targeting of macrophages using siRNA has shown to be successful in gene silencing of macrophages and ocular use. Using targeted drug delivery via nanotechnology, siRNA, to target these antigens, will allow for a slow and controlled release of the drug treatment. Understanding the siRNA delivery mechanism and macrophages' biology will help identify the challenges and opportunities for improving the targeted therapies. The current siRNA delivery has consisted of targeting cancer cell lines through macrophages for gene transfer and ocular diseases.

Keywords

Nanodelivery · Drug delivery · Nanocarriers · Nanoparticles · Nanomaterials · Macrophages · Targeted delivery systems · siRNA

1 Introduction

Nanotechnology, the technology of nanoscience, refers to the development and research of science at an atomic and molecular level (Abiodun Solanke et al. 2014). With the new advancements of nanotechnology as a mode of targeted delivery, challenges and opportunities have arisen. With any new technology, an examination must

D. Kaushal
The University of South Florida, Judy Genshaft Honors College, Tampa, FL, USA

S. Gupta
Department of Pharmaceutics, Amity Institute of Pharmacy, Amity University Uttar Pradesh, Noida, India

Y. V. Pathak (✉)
University of South Florida, Taneja College of Pharmacy, Tampa, FL, USA

Adjunct Professor University of Airlangga, Surabaya, Indonesia
e-mail: ypathak1@health.usf.edu

S. Gupta, Y. V. Pathak (eds.), *Macrophage Targeted Delivery Systems*,
https://doi.org/10.1007/978-3-030-84164-5_21

thoroughly be completed to understand the effects. Recognizing the impact of manipulating the atoms and molecules of the atoms for targeted drug delivery has proven successful in treating diseases and emerged to broaden the term nanomedicine. Nanotechnology has been applied to many medical fields, such as ophthalmology, dentistry, pharmacological research, tissue engineering, cancer treatments, magnetic contrast imaging, contrast enhancement, supplementing of the immune system, etc. (Abiodun Solanke et al. 2014; Song et al. 2019). In nanotechnology, at the atomic level, there can be manipulation of atoms, chemical bonds, and molecules, allowing for controlling targeted drug delivery. (Table 1)

The understanding and progression of nanotechnology have paved the way for an advanced course of targeted therapies. Just as one uses one treatment to treat disease and a different therapy to treat another disease, similarly, there are multiple forms of nanocarriers for targeted drug therapy. Nanocarriers are colloidal drug carrier systems, less than 500 nm. They have been used for their high surface-area-to-volume ratio to alter basic drug properties and the bioactivity of drugs. Modifying nanocarriers' physicochemical properties can consist of the composition, sizes, shapes, and surface properties (Din et al. 2017). These modifications can aid in drug delivery to efficiently treat disease and reduce adverse side effects (Din et al. 2017). Nanocarriers can consist of micelles, cubosomes, and exosomes, liposomes, lipid nanoparticles, nanoemulsions, polymer-based self-assemblies, and macrocycle self-assemblies (Attia et al. 2019). Ongoing studies are testing the effectiveness of certain nanocarriers when combined with nanotechnologies, such as siRNA, to maximize the effect of the treatment (Zhao and Feng 2015). As this is still a new technique, further studies must be completed to understand the complexity. However, there have been successful studies in which various forms of nanotechnology have been used for targeted drug therapy. Nanotechnology has wide and industrial applications in drug delivery.

Utilizing nanotechnology has allowed for the detection of cancer cells, kidney damage, and drug delivery to certain areas in a higher concentration with minimal side effects as compared to normal drug delivery (Kang et al. 2018). One form of nanotechnology that has piqued the interest is small interfering ribonucleic acid (siRNA). The RNA interference (RNAi) mechanism was discovered in 1998 by Andrew Fire and Craig Mello in double-stranded RNA (dsRNA) in mammalian cells (Fire et al. 1998). Understanding the RNAi pathway allows for a more in-depth understanding of the effects of siRNA and the interference therapy. The use of siRNA has been shown to achieve a more effective therapeutic approach for treating diseases. (Kang et al. 2018). siRNA has shown to be effective in delivering to a targeted, site-specific tissue and cells to assure therapeutic therapy (Gupta et al. 2019). siRNA can be used to create a targeted therapy that can aid in angiogenic ocular diseases such as age-related macular degeneration (AMD) and diabetic macular edema.

siRNA has been used in vivo and in vitro specifically in gene silencing in primary human monocytes, dendritic cells, and macrophages (Troegeler et al. 2014). siRNA delivery in macrophages has been important to study due to their inflammatory response within the innate immune system. By examining and targeting macrophages, there is a possibility to silence the macrophage from secreting pro-inflammatory and antimicrobial mediators. In return, this can go a long way in the targeted treatment. The use of siRNA has shown to be successful in gene silencing and with the use of nanocarriers beneficial in targeted drug delivery. The various siRNA delivery systems have been looked at to identify the benefits and challenges. In comparison to other forms of therapies and therapeutic applications, siRNA when combined with nanocarriers has shown to be beneficial in targeting macrophages. Current investigation on targeted delivery systems, such as siRNA, to macrophages found in the human body and the eye, has shown to be successful.

2 Background

Tremendous advancements have been in the field of nanotechnology. It has become an attractive tool from materials science to biomedicine due to its characteristics. The main research area of nanotechnology has been in the diagnosis, treatment,

Table 1 Forms of nanotechnology delivery systems

Delivery system	Use	References
Nano-wafer	Antibacterial burn wound dressing	Mohebali et al. (2020)
Hydrogel	Contact lens material to treat fungal keratitis, a base for cell scaffolds in tissue engineering, in situ drug delivery systems	Huang et al. (2016; Vijayavenkataraman et al. (2020) and Wei et al. (2020)
Polymer hydrogel	Used a biosensor with a high permeability to biomolecules, biocompatibility, and electron transfer properties	Schwall and Banerjee (2009)
Nanosuspension	Used for drugs that are insoluble in water and organic solvents. Improves the overall solubility and bioavailability	Jacob et al. (2020)
Liposome	Used to decrease the poor pharmacokinetics, limited bioavailability, and high toxicity. More efficient at treating acute and chronic diseases than drug-alone formulations	Kelly et al. (2011)
Polymeric micelles	Low biocompatibility, toxicity, adaptable morphology, nanosize, and high stability. Has been used in cancer therapies	Jhaveri and Torchilin (2014)
Dendrimers	Ability to polydisperse products at different molecular weights. It contains polyvalency, self-assembling, electrostatic interactions, stability, low cytotoxicity, and solubility. A combination of artificial macromolecules containing a larger number of functional groups	Madaan et al. (2014)
Chitosan	Can bind to DNA and halt RNS synthesis. Contains a positive surface charge and mucoadhesive properties. Naturally occurring polysaccharide which is biodegradable and its biodistribution is crucial for in vitro administrations. High antimicrobial activity and when combined with metal nanoparticles	Rasul et al. (2020)
Alginate	Known as an anionic polysaccharide. Versatile physicochemical properties. Chemical modifications for site-specific targeting enabling biocompatibility and biodegradation	Abasalizadeh et al. (2020)
Xanthan gum	Versatile biopolymer, temperature, and ionic strength can be changed to work with drugs and proteins' carriers	Cortes et al. (2020)
Cellulose	The primary natural resource contains a semicrystalline structure. Biodegradable, high mechanical strength/weight performance, and low-cost polymer	Sheikhi et al. (2019)
Inorganic nanoparticles (silver, gold, iron oxide, silica)	Metal nanoparticles can enhance drug delivery and gene therapy. Used for magnetic resonance imaging as contrast enhancers. An extensive array of shapes and sizes for plasmonic properties	Anselmo and Mitragotri (2015)
Protein and polysaccharides nanoparticles	The macromolecules have been shown to give potential therapeutic biodegradability, biocompatibility, and high structural flexibility. High surface-to-volume ratio and the surface functionality	Hu et al. (2019)

The use of each nanotechnology delivery system is based on its use on size, safety, antimicrobial properties, and compatibility for the function of a particular use.

and prevention of diseases (Mir et al. 2017). The use of nanotechnology in medicine has influenced the field of nanomedicine. Nanomedicine has been influenced by drug delivery by using nanoparticle-based drug delivery platforms addressing the pharmacokinetic disadvantages associated with conventional drug delivery (Mir et al. 2017). Many applications of nanotechnology have been successful in vivo and in vitro. Nanoparticles and siRNA, branches of nanotechnology, have been used to target macrophages as clinical therapeutic agents against inflammation (Troegeler et al. 2014). The development of nanotechnology, the manufacture of nanoparticles (NPs), and the modification of siRNA have increased the understanding of macrophages' role as a drug target (Hu et al. 2019). Compared to the conventional treatments, siRNA is highly potent, contains fewer adverse side effects, and is designed to treat any gene of interest and act on "non-druggable" target sites (Lam et al. 2015; Daka and Peer 2012).

The tumor necrosis factor-α (TNF-α) has been associated with numerous neurological and age-related diseases. TNF-α is a cytokine associated with the cell signaling of the immune system (Idriss and Naismith 2000). This inflammatory cytokine is produced by activated macrophages, monocytes, and microglia during acute and persistent inflammation (Idriss and Naismith 2000). TNF-α plays a critical role in the varied range of signaling events within cells, leading to the necrosis or apoptosis of the cell (Idriss and Naismith 2000). Understanding the body's inflammatory response and the effects of the activated macrophages are crucial in examining new ways to inhibit inflammation. Using nanotechnology and siRNA to target macrophages, there can be a suppression of the TNF-α to reduce the lipopolysaccharide (LPS)-induced neuroinflammation in vitro and achieve efficient gene silencing (Kim et al. 2010).

2.1 Macrophages

Macrophages are large, versatile white blood cells of the immune system that accumulate in response to an inflammatory response or accumulation of dead/damaged cells. The word macrophage origin is combined using the Greek terms "makros," meaning large, and "phagein," meaning to eat. Macrophages are mononuclear cells that can undergo phagocytosis of microbes. Macrophages' physiological role is heterogeneous, meaning that they can be triggered by different cellular origins, tissue environments, and different adaptations of organisms (Gordon and Martinez-Pomares 2017). The function of macrophages has been used for in vivo practices and has been studied vigorously on their motility and ability to recognize and respond to host and foreign particles (Gordon and Martinez-Pomares 2017). Additionally, they can sense and respond to microbial constituents within the cytoplasm, such as DNA and RNA, to activate inflammasomes and the release of Interleukin (IL) proteins (Guo et al. 2015). They contain a wide range of systems that the tissue environment can influence, cellular origin, or the organism's physiological requirements that can cause the following: pattern recognition receptors, receptors for apoptotic cells, cytokine receptors, receptors for neurotransmitters, inhibitory receptors, and integrins (Gordon and Martinez-Pomares 2017). (Fig. 1)

For the past decades, there has been an emphasis on investigating macrophages due to their connection with diseases. Macrophages are crucial because of their role in chronic diseases and inflammation. The inflammatory response of macrophages consists of antigen presentation, phagocytosis, and immunomodulation through cytokines and growth factors (Fujiwara and Kobayashi 2005). In the early stage of inflammation, neutrophils will be attracted to a site. At this site, they will perform their function and die off. Then the neutrophils are phagocytized by macrophages. During this process, the macrophage will ingest the pathogen, which will become entrapped in a phagosome.

The importance of studying macrophages is to understand their role in the inflammatory process of cells further. Macrophages secrete pro-inflammatory and antimicrobial mediators meaning that they play a crucial part in immunity and immune responses. Their primary function is to be effective phagocytes and protect the body from foreign particles, dead/dying cells, and bacteria. They have three prominent roles in the inflammation process, which are initiation, maintenance, and resolution (Fujiwara and Kobayashi 2005). These roles play an essential factor in the activation and deactivation of the inflammatory process. They can activate signals including cytokines, bacterial lipopolysaccharide, extracellular matrix proteins, and other chemical mediators (Fujiwara and Kobayashi 2005). Macrophages participate in an autoregulatory cycle of the inflammatory process due to their constantly changing role.

Macrophages are crucial therapeutic targets because of their acquisition of different phenotypes and their response to constant environmental stimuli. These therapies' design has been

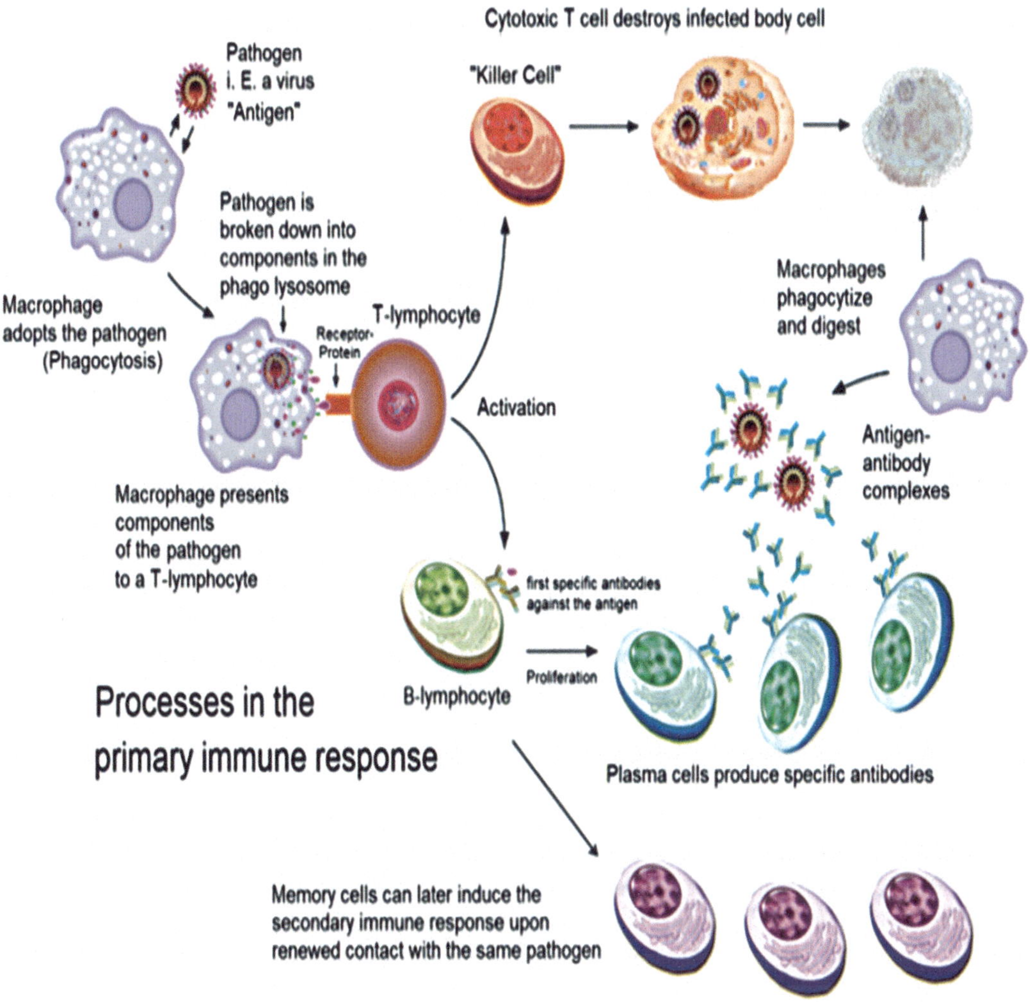

Fig. 1 The primary immune response when a pathogen or antigen is present; macrophages are then activated. When activated, they undergo phagocytosis to break down the pathogen using T lymphocytes. Macrophages interacted with the T cells to aid as an inflammatory messenger. The T lymphocytes' role is to kill the infected host cell directly and activate other immune cells, such as cytokines (cytotoxic T cells) or B lymphocytes (helps to secrete antibodies) to overall help regulating the immune response. [Credit of image: Title: "Simplified overview of the processes involved in the primary immune response"; Creator: "me=Sciencia58 an the makers of the single images Domdomegg"; Source: "Wikimedia Commons"; License: "CC BY-SA 4.0"]

challenging because they lack clear markers to identify pathological macrophages to beneficial counterparts (Ahsan 2002). Strategies and therapies have been pursued to manipulate macrophages through their re-education, reprogramming, and depletion, a direct killing of macrophages (Piaggio et al. 2016). The use of nanotechnology in these scenarios has been implicated and led to successful models. By implementing nanotechnology into the modification of macrophages, a direct therapy can be used to target them. Successful studies have been completed showing that with the use of nanoparticles and nanotechnology, such as small interfering RNA (siRNA), on macrophages, such as certain signaling pathways were restored; there was an induction of cytotoxic and immunostimulatory tumor-associated macrophages (TAMs); antitumor activities were stimulated and reduced gene expression of the macrophage migration inhibitory factor in primary macrophages (Ortega et al. 2015; Zanganeh et al. 2016; Huang et al. 2012; Zhang et al. 2015).

2.2 Small Interfering Ribonucleic Acid (siRNA)

Small interfering ribonucleic acid (siRNA) is a double-stranded RNA molecule and contains small 20–25 nucleotide non-coding RNAs that regulate genes. The strand comprises 5′ phosphate groups and 3′ hydroxyl groups. siRNA is produced from a viral double-stranded RNA (dsRNA) and is cleaved by RNAase III family enzyme Dicer, producing 19–27 base pairs (Zhao and Feng 2015). These base pairs allow for long molecules containing a complementary middle region with a 2-nt overhanging on both of the 3′ hydroxyl groups (Levanova and Poranen 2018). The short-stranded RNAs cause specific post-transcriptional gene silencing. This process begins with the RNA-induced silencing complex (RISC), allowing gene expression to occur by cleaving the mRNA to code targeted genes. The method includes the siRNA being associated with the RISC complex and being unwounded to form a single strand. This will allow the siRNA to be scanned and analyzed through RISC to find the complementary mRNA sequence.

The siRNA and mRNA's pairing leads to the enzymatic cleavage and degradation of the mRNA molecule. This is one specific way to engineer a pathway or get rid of any particular gene. The biological functions of siRNA are still being researched due to the wide range of therapeutic uses varying from gene silencing, defense of viral infections, gene expression, and autoimmune disorders like retinal degeneration (Tatiparti et al. 2017; Zhang et al. 2018). siRNA is a factor of the RNAi pathway, a cellular mechanism that inhibits gene expression in targeting and neutralize mRNA molecules. Previously researched, the RNAi pathway was found to be in the suppression of desired genes (Fire et al. 1998). The functions can help aid in many genome functions due to the interfering of either strand individually. These functions can include RNA stability, chromosome segregation, transcription, transcription, translation, and chromatin structure (Tatiparti et al. 2017).

siRNAs have been delivered by two major delivery systems: viral vectors and nonviral biopolymers. These delivery systems are dependent on the target site and the intended use of the siRNA (Tatiparti et al. 2017). The polymer-mediated system has previously been used for plasmid DNA. It has been used for siRNA as it can help protect contents from degrading, participate in targeted delivery, and control the released agent's release (Ray 2019). This polymer is advantageous because of its flexible chemistries and characteristics, allowing it to be modified to change the physicochemical properties (Dana et al. 2017a). It has been reported that the chitosan polymer is a successful vector because of its biocompatible materials and versatile chemical, physical, and biological properties. This polymer is a natural polysaccharide that has been researched and tested with other synthetic materials to lower the toxicity, a beneficial factor for in vivo and in vitro results (Serrano-Sevilla et al. 2019).

The peptide-based delivery system has been used to deliver siRNA due to the efficient cross over the cell membrane permeation. There has been evidence that the cell-penetrating peptides (CPPs) have high potential in the oligonucleotide delivery of nanoparticles (Konate et al. 2019). CPPs are short peptides that help to promote the covalent and non-covalent interactions within the cell through endocytosis. One of the main advantages is its ability to translocate in the plasma membrane and aid in the movement of molecular interactions to organelles or cytoplasm (Kurrikoff et al. 2011). This polymer is more suitable because of the design versatility and the sequence and function diversity (Konate et al. 2019). The lipid-based delivery system has been developed specifically for in vivo applications. Like the other delivery systems, lipid-based systems are beneficial in targeted delivery, specifically in oral delivery systems (Beloqui et al. 2017a). These delivery systems have been used because of their hydrophilic and lipophilic properties, aiding in using those types of drugs, proteins, and nucleic acids to produce a target delivery (Dolatabadi et al. 2015).

Nonviral delivery systems are chemical and physical systems. These systems can be used together or separately to deliver siRNA. Nonviral systems reduce the immune response and have no size limit for the transgenic DNA (Nayerossadat et al. 2012). Nonviral delivery systems affect the use of in vivo and in vitro delivery based on cell membrane penetration. The most common nonviral deliveries are divided among the chemical and physical systems. The chemical systems comprise cationic lipids, cationic polymers, and lipid polymers (Nayerossadat et al. 2012). The physical systems contain naked DNA, DNA bombardment, electroporation, hydrodynamic, ultrasound, and magnetoreception (Nayerossadat et al. 2012). The lipid-substituted polyethylene imine chemical system of nonviral delivery of siRNA has been targeted to acute myeloid leukemia cells (Landry et al. 2012).

2.3 siRNA-Targeted Therapies

Since the discovery of RNA interference (RNAi) back in 1998 by Craig C. Mello and Andrew Fire, there is now a strong understanding of RNA molecules' biological process inhibiting gene expression (Fire et al. 1998). By understanding this pathway, it has paved the way to develop targeted therapies. After discovering RNAi, Elbashir et al. developed a way to target and silence specific sequences instead of the whole gene by using small interfering RNA (siRNA) (Elbashir 2001). Both of these discoveries were crucial in developing therapeutic applications to perform sequence-specific silencing within a gene. The use of siRNA has been demonstrated within the field of nanomedicine from identifying viral infections, cancer and cancer treatment, diseases (ophthalmological, neurological), biotechnology, and the understanding of the two branches of the immune system.

The first use of targeted therapies using siRNA in 2001 had drawn attention for a new class of drug treatments to be produced, using siRNA for gene silencing and targeting many chronic diseases, such as age-related macular degeneration, diabetic retinopathy, glaucoma, and cancer-related genes. Since the discovery of siRNA, there has been more of an effort to utilize it for treatment. Currently (January 2021), there have been clinical trials using siRNA delivery systems to target solid tumors and the Anti-Vascular Endothelial Growth Factor (Anti-VEGF), which is associated with choroidal neovascularization in ocular diseases (Davis et al. 2010; Feng et al. 2015a).

The use of siRNA for targeted therapies has become popular in the past decades. siRNA has been applied in target drug therapy to be used as a new form of treatment for diseases. By targeting and silencing specific genes, we can understand further the pathways and how they play a role in a particular disease. Using siRNA or any form of targeted drug delivery to macrophages is an alluring proposition to improve drug delivery's overall therapeutic efficacy (Jain et al. 2013). By enabling controlled and targeted drug delivery for in vivo and in vitro, there is a new option of incorporating nanomedicine and cellular manipulation (Visser et al. 2019). Visser et al. discussed "cargo loading into macrophages" (Visser et al. 2019). Due to monocytes circulating within the innate immune system and recognizing the foreign materials and microbes within the cell, the macrophages engulf material, and it results in pseudopod formation and internalization of materials (Visser et al. 2019). Previous studies have shown macrophages' engulfment process because they can readily take purpose-designed materials or various particles through the endocytic pathways (Muthana et al. 2011; Fan et al. 2018).

As a form of target drug delivery, siRNA can be used for gene silencing but can target cells such as macrophages to decrease inflammation and prevent the progression of certain diseases (Kim et al. 2010). As discussed earlier, the tumor necrosis factor-a (TNF-α) can be suppressed by using siRNA to reduce the lipopolysaccharide (LPS)-induced neuroinflammation as well as the neuronal apoptosis in vivo (Kim et al. 2010) using siRNA in conjunction with drugs to target particular cells, such as macrophages, or specific glycoproteins, such as CD98, or antigen-presenting cells, such as dendritic cells (Xiao et al. 2016; Yang et al. 2006).

As siRNA is still being researched, the extent of its use targeted therapies and delivering efficiently and safely. To maximize the treatment and reduce adverse effects, there has been a great investigation into chemical modifications and the development of delivery systems compared to standard drug delivery (Hu et al. 2020). In past decades, siRNAs have been developed for therapeutic applications (Beloqui et al. 2017b). They can be easily modified by either creating a pathway through RNAi or inhibiting a specific gene expression. Since discovering the siRNA mechanism, it has been used as a delivery pathway to target a specific gene expression, especially in drug delivery for ophthalmic applications (Thakur et al. 2012). The benefits of siRNA allow for a more specific target delivery preventing DNA modifications in the genome.

The delivery to specific tissues, such as in the eye, is a challenge but can be extremely beneficial by target delivery. The use of siRNA can silence gene expression and target specific disease-related genes, such as age-related macular degeneration (AMD) (Raja et al. 2019a). Within specific tissues and cells, it can provide a high local concentration without using more drugs. In ocular tissues, it has been found that a siRNA-based anti-angiogenic agent drug, bevasiranib, has been used in treating wet AMD (Garba and Mousa 2010a). The use of bevasiranib helps to eliminate repeated intravitreal injections that could cause complications, such as retinal tears, intravitreal bleeding, or lens injury. The use of siRNA to target a gene will allow for fewer complications, bursting of blood vessels, and repeated injections (Garba and Mousa 2010a). By examining different delivery materials, such as the vectors, it can aid in targeting target drug therapies using siRNA.

3 Targeted Delivery System—siRNA

In the past couple of decades, there has been an increase in discovering and formulating targeted delivery systems to treat diseases. From the discovery of RNAi to the discovery of siRNA, nanotechnology has merged into the world of medicine for new and improved targeted delivery systems. As discussed previously, the use of targeted delivery systems such as siRNA can be beneficial tools to study and understand gene function both in vivo and in vitro. In doing so, researchers can create new therapeutics against cells, such as macrophages, and incurable diseases (Dana et al. 2017b).

As we progress in the understanding of diseases, we must complement that with newer treatments. Using siRNA is emerging to be the new form of treating diseases compared to standard treatments and therapies. By allowing for gene silencing using siRNA, synthetic siRNA delivery can create a naturally occurring RNAi pathway, making it predictable and consistent for therapeutics (de Fougerolles et al. 2007). Studies have already shown that synthetic siRNA to target macrophages helps reduce inflammation by silencing tumor necrosis factor-alpha (TNF-α), a central component of diseases (Kim et al. 2010). Like any treatment or drug, there will be benefits and side effects, calling for a more comprehensive understanding of the siRNA and its use.

3.1 Importance of Targeted Delivery Systems

As the field of targeted delivery is progressing, the forms of delivery systems are increasing as well. Targeted delivery systems can help to decrease or minimize the duration of diseases. With the use of these delivery systems, a higher concentration of medication or therapy can be delivered. By having a higher concentration, the delivery system is much more beneficial and successful.

Targeted delivery systems have various delivery vehicles, such as liposomes, lipoprotein-based drug carriers, nanoparticle drug carriers, micelles, hydrogel, and dendrimers (Singh et al. 2019). These different vehicles can be implemented with nanoparticles to create nanoparticle-mediated drug delivery. With the use of nanotechnology and nanoparticles, the unfavorable effects of conventional drug delivery are avoided (Muller and Keck 2004). The nanoparticles are employed with the drugs and have a specific target site to avoid off-targeted therapies. One of the significant nanoparticles targeted delivery systems discussed throughout this chapter and used in clinical settings is the siRNAs. The use of siRNAs with a delivery vehicle can silence and target specific gene sequences (Kowal et al. 2019). The ability to target and deliver a more concentrated version of a drug can help change drugs' administration. Treatment for diseases that cannot be managed with conventional drug therapy can be done using targeted delivery systems using nanotechnology.

Targeted delivery systems can come in a variety of shapes and sizes. The need for multiple vehicles is to enhance the efficacy and safety of the area that is being targeted. All of the vehicles are unique in their way, but they can ensure high efficacy, long duration of drug bioavailability, and minimal adverse effects when administered (Singh et al. 2019; Moghimi et al. 2005). Also, these vehicles can enhance the efficacy or reduce specific agents/drugs' toxicity (Tiwari et al. 2012). Many of these vehicles have been used in preclinical and clinical studies for the treatment of various diseases. The variety of different systems and vehicles are crucial to using targeted delivery systems because of their use.

The importance of using nanocarriers and nanoparticles targeted delivery systems is due to the advantages of treating a disease with a prolonged availability to the final site (Vega-Vásquez et al. 2020). The targeted systems are created to release treatment in a controlled manner with higher potency and efficacy than conventional treatments (Drug Delivery Systems 2016). By encapsulating molecules within a shell-like protective barrier, the active compound's bioavailability is increased, and the nonspecific distribution is reduced. In return, it is reducing the need for frequent treatments over time (Felice et al. 2014; Choudhury et al. 2017). The main advantage of using targeted delivery systems is administering the drug at a controlled release in a time-dependent manner by active or passive targeting. These two types of targeting can allow the targeted tissue's pathophysiological features to accumulate on it, or it refers to the assembly of surface-active ligands that recognize and interact with the target cell's receptor sites (Vega-Vásquez et al. 2020).

3.2 The Barriers of siRNA

The use of siRNA is still a new field that needs to be examined more to understand the effects of macrophage and disease targeted drug therapy. There are still many biological barriers that need to be examined to understand the direct delivery to the intended locations and carry out gene silencing. Direct accessibility to localized targets has its difficulties that can be either extracellular or intercellular.

When using siRNA therapeutics, the delivery mechanism is dependent on the targeted organs and the administration routes. One of the main factors that can limit siRNA use is a "loss-of-function approach," which can limit or prevent the intended use. The administration barrier is crucial, as seen in the previous target sites, such as cancer sites. Many target sites have two forms of administration routes, oral and subcutaneous injections. As seen previously, oral routes are not as effective because of their inability to maintain intestinal stability and permeability across the intestinal epithelium, defeating the purpose of using them for targeted treatments (Haussecker 2014). Subcutaneous injections are more effective in maintaining the intestinal epithelium and having access to the circulatory system (Haussecker 2014). Even in this administration route, there are challenges in avoiding phagocytosis, which depends on the lipophilicity and gene vector size. Another route that has been common for administration is intravenous or infusion injections. Both have been used in the clinical setting for drug administration rapidly (Malamed 2010).

Determining the administration route is crucial to establish how the siRNA delivery can cross the vascular barrier to reach the target site/systems. The high specificity of the siRNA action can cause high concentrations of toxicity to the non-targeted sites (Chernikov et al. 2019). The mistarget of siRNA can activate the innate immune system due to the siRNA sequence. The immune response is a primary barrier of siRNA because all delivery systems should be non-immunogenic and not elicit undesirable side effects. By using a delivery system, they should refrain from off-targeting genes of normal cells.

This can assure that the delivery is not intentionally silencing the wrong gene sequence (Jackson 2006). During formulation, they should not be identified as foreign particles within the immune system. This would prevent the innate immune system from attacking and destroying the particle before the target was reached (Judge et al. 2005). To prevent this from occurring, the proper formulation must occur to prompt an immune response. When interacting with the membrane surface, there can be a production of interferons and inflammatory cytokines to activate the immune response (Mansoori et al. 2015).

Managing the immune response is one of the biggest hurdles when formulating siRNA and its intended delivery location. The benefits of siRNA have been withheld due to their induction of inflammatory cytokines, interferons induction, tumor necrosis factor (TNF)-α, and interleukin (IL)-6 (Mansoori et al. 2015). With a possibility of off-target delivery, either be sequence-dependent or endosomal toll-like receptor (TLR), pathways are dependent, causing IL-6 and TNF-α (Jackson and Linsley 2010; Qiu et al. 2006). With the introduction of siRNA, the immature dendritic cells can trigger IL-12 production and stimulate CD83 (activation marker) expression. The pattern sequence-dependent recognition sites of siRNA are TLR7 and TLR8 (de Marcken et al. 2019). These are part of the antiviral defense response and are secondary responses driven by cytokines and type 1 interferon (de Marcken et al. 2019). Understanding this immune response can be one of the biggest obstacles to identifying the pathways and effects of siRNA within the innate immune system.

4 Therapeutic Applications

The use of therapeutic applications has become relevant as there has been an emergence of chronic diseases in the past decades. According to the World Health Organization (WHO), chronic diseases have been prevalent (cardiovascular diseases, cancer, chronic obstructive pulmonary disease, and type 2 diseases and conditions worldwide), causing a demand on healthcare systems to treat these diseases (World Health Organization 2010). It was predicted that in 2020 there would be a rise in the death of chronic diseases from 60% to 73% and an increase in the global burden of disease from 43% to 60% (World Health Organization 2010). With this emergence of chronic diseases, targeted efforts have been placed to create cost-effective ways to prevent and treat them. Due to the increase in chronic diseases, researchers have moved to use nanotechnology to use therapeutic applications as a supplemental or primary treatment form.

With the use of nanotechnology, the use of therapeutic applications has improved in the past decades. Using nanotechnology to curate treatment for diseases has shown that there is a new way that diseases can be treated. Currently, nanomedicine is progressing by offering new advantages for the facilitation of drugs and targeted therapies with fewer adverse effects (Mankamna Kumari et al. 2020). The importance of researching and understanding therapeutic applications will allow there to advance how diseases are treated. Therapeutic applications can help treat and manage diseases by curating a new treatment form compared to the conventional treatment. The advances in researching and understanding therapeutic applications will allow there to be an improved way to treat diseases, that is, creating a new form of drug delivery by using nanotechnology.

4.1 Current Therapeutic Applications

Therapeutic applications have become relevant to prevent and treat diseases. With the progression in the field of nanomedicine, it is drastically increasing. As technology is improving, so is the way that diseases are being treated. Treating disease with therapeutic applications can allow for diseases to be better treated by implementing a controlled long-term drug release mechanism using nanomaterials and nanotechnology (Deng et al. 2020). Using therapeutic applications is essential because it will provide drastically better clinical benefits to patients suffering from acute and chronic diseases. Using the therapeutic application will allow the treatment to be better delivered to the target site (Tilley et al. 2011).

The use of therapeutic applications is extremely diverse as they can be adapted and used for any setting to allow for targeted drug delivery to silence a gene or to a subset of cells. Researching nanoparticles' effects allow for a better understanding of improving the solubility across the membranes, improving the therapeutic index, and overall reducing the immunogenicity (Davis et al. 2008; Zhang et al. 2007). Additionally, the mechanisms of the molecular interactions of nanoparticles within the cells and tissues are essential to understand how the development process can be more efficient for the encapsulation of drugs and an activated, controlled release (Petros and DeSimone 2010). (Fig. 2)

Tilley et al. reviewed the different treatments for asthmatic smokers and patients with chronic obstructive pulmonary disease and cystic fibrosis because of the chronic inflammation caused by these diseases (2011). They investigated the pharmaceutical industry exploring new treatments to address Adenosine Deaminase (ADA)-related severe combined immunodeficiency (ADA-SCID). The current treatment was to receive an allogeneic bone marrow transplant, but if there is no histocompatible sibling, the patients would have to undergo irradiated purified erythrocytes giving rise to various complications. Such treatment's future directions include a genetic replacement of the ADA function using retroviral vectors that target lymphocytes and hematopoietic progenitors. This groundbreaking treatment is non-invasive and restores immune function, extends the survival period, and restores ADA activity.

Riglar and Silver reviewed the effects of engineering bacteria as a therapeutic application to target diseases such as diabetes mellitus, inflammatory bowel disease, HIV infection, and cancer (Riglar and Silver 2018). The advantages of using bacteria therapeutics were to deliver drugs using bacteria that would otherwise be degraded within the human body barriers, such as the bloodstream or upper gastrointestinal tract. Additionally, bacteria's use is beneficial because it can reach sites that oral or parenteral drug delivery would not reach. Natural and synthetic pathways of bacteria can aid in the bacteria sensing and signaling. They examined the difference between the one- and two-component bacterial sensing systems. The one component involved using transcription factors that bind to the regulatory DNA sequence and the ligand at interest, which can either activate or derepress the gene expression. The two-component involved the histidine kinase and a response regulator. After the ligand processing, the gene expression can get activated in a phosphorylated state. This engineered bacteria therapeutics is promising for long-term applications because they can have an integrated complex synthetic system to help link the expression of the therapeutic to the specific environment.

4.2 The Use of Targeted Delivery Systems to Macrophages

Using nanotechnology as a form of targeted delivery systems has been used to treat chronic diseases from ocular to cardiovascular to certain types of cancers. The benefit of using siRNA is its ability to target and silence certain gene expressions. Using siRNA as a nanocarrier as a targeted delivery system is helpful as it can decrease the effects of the diseases. Using siRNA to target macrophages will decrease the overall inflammation process caused when antigens and foreign particles are present in the innate immune system and phagocytosis occurs, by directly targeting the macrophages using nanotechnology-based systems such as siRNA to silence the cells or deliver targeted therapies (Ponzoni et al. 2018).

Macrophages are of interest to researchers to use as potential therapeutic targets, but due to the design and the lack of specific markers, strategies have been created. Being able to target macrophages directly, there can either be the re-education or depletion of the macrophages (Ponzoni et al. 2018). The advantages of using targeted delivery systems on macrophages improve the therapeutic efficacy of an enclosed

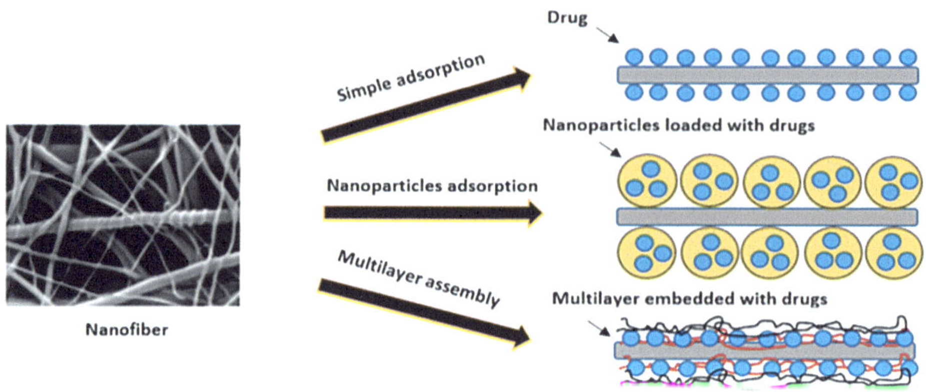

Fig. 2 The image depicts the difference between the drug delivery, the drug with nanoparticles, and the drugs embedded with multilayers of drugs. This is an example of the different therapeutic application examples that are present for drug delivery. [Credit of image: Title: "Diagram of drug delivery mechanism"; Creator: "Parksoh17"; Source: "Wikimedia Commons"; License: "CC BY-SA 4.0"]

drug, and using nanocarriers can better deliver over different membrane barriers (Jain et al. 2013).

Forbes and Peppas investigated the use of siRNA delivery systems to target murine macrophages (Forbes and Peppas 2014). They took polycationic nanocarriers bound to siRNA. In doing so, they investigated the cellular internalization of RAW264.7 murine macrophage cells. They found that the siRNA was able to bind to the nanocarriers and the knockdown experiments were successful using siRNA. It was shown that a higher concentration of the nanoparticles and siRNA was able to reduce the murine macrophages, but future studies must be completed to understand the effects of the siRNA.

Kaps et al. reviewed the effect of using in vivo siRNA to immunosuppressive liver macrophages through cationic nanohydrogel particles (Kaps et al. 2020). They investigated that macrophages can become immunosuppressive, and the M2-type macrophages can overexpress the mannose receptor CD206. The M2-type macrophages play a crucial role in chronic diseases such as autoimmune liver diseases or any persistent inflammation diseases. The M2-type macrophages can be re-educated using siRNA because specific phenotypes of macrophages before phagocytosis maintain their plasticity. The cellular uptakes of α-mannose nanohydrogels in M2-type macrophages were higher, and the nonfunctionalized counterparts were increased. The siRNA proved to be effective in preventing adverse effects and enhancing the uptake of the M2-polarized macrophages because of the high expression of CD206 in vivo.

Yong et al. investigated the effect of targeting human CD64 using a nonviral siRNA delivery system to examine blood monocyte gene modulation (Yong et al. 2017). siRNA was used as a targeted therapeutic for gene silencing and the ability to target genes before translation. Mononuclear phagocytes are a central component of the immune system as it helps with homeostasis and immunological effector functions. These phagocytes are also a hindrance to drug delivery because they can limit the drug efficacy. Using a nonviral siRNA delivery system, the human blood monocytes that were accumulating macrophages were used as delivery targets. The experiment was done to demonstrate the specific uptake of blood monocytes and gene silencing. Targeting CD46 for targeted siRNA delivery was successful because it was beneficial for human blood monocyte gene targeting and delivery.

As can be seen, the targeting of macrophages has considered being successful in targeting the cell without minimal side effects when using siRNA with a nanocarrier.

5 Current siRNA Targeted Deliveries for Therapeutic Systems

As discussed previously, after the discovery of the RNAi mechanism in 1998, siRNAs and their roles were discovered in the early 2000s. Understanding the role that siRNA plays in post-transcriptional gene silencing allows for research performed based on that a specific gene can be silenced using synthetic siRNA (Elbashir 2001). The research that was previously performed has shaped the way for targeted drug therapies using nanoparticles. The progression of using siRNA therapies with nanoparticles has been successful in enhancing drug delivery.

The current use of siRNA in targeted deliveries for therapeutic systems has shown to be promising in terms of the treatments that siRNA has been used to treat diseases, such as macular degeneration or cancer treatments. In comparison to standard drug delivery, these siRNA therapies have shown to be successful in targeting and silencing specific genes and cells. Examining the effects of siRNA can change therapeutic systems because siRNA can be used more regularly and possibly become the main form of treatment in diseases. From the early 2000s to now (2021), siRNA delivery has made a prominent place in nanomedicine to treat and reduce the effects of diseases. Continuing to study the use of siRNA therapies, in vivo and in vitro will allow them to understand the positive and adverse effects siRNA can cause. Doing so will allow there to be a better understanding of how the siRNA therapies can be used and the limits.

5.1 siRNA Delivery for Ocular Use

The use of siRNA for ocular use has been used quite frequently. It was first demonstrated in animal models that present with ocular neovascularization and scarring (Reich et al. 2003).

With the headways of siRNA and its applications for explicit problems, siRNA's utilization has provoked specialists to take a gander at the ramifications of it inside AMD. Recently examined, VEGF has been related to the choroidal neovascularization (CNV) related to AMD, which can cause a deficiency of vision in the old populace (Feng et al. 2015b). It was recognized in mice that intravitreal Ets1 siRNA could play a supportive angiogenic job in the favorable to provocative capacity. Ets1, also called record factor E26 change explicit 1, has been related to directing quality articulation through various development factors, attachment atoms, and chemokines (76). In this particular examination, the intravitreal Ets1 siRNA was confined to the mouse's CNV laser district of the macrophage and microglia locale. The Ets1 siRNA was infused in the CNV locale, improving the spillage and limiting retinal pigment epithelium (RPE) cells' brokenness and initiation of macrophages/microglia. Initially, a CNV mouse model was created utilizing laser photocoagulation. The Ets1 was infused intravitreally and was followed over a 7-day time frame. After the infusions and laser injury, it was identified that the Ets1 protein level in the RPE-choroid–retina had radically diminished when contrasted with vehicle and scramble siRNA gatherings. When the fundus fluorescein angiography and the indocyanine green angiography were performed, there was a diminishing in the spillage in the CNV region. The investigation has featured the adequacy of utilizing Ets1 siRNA as a focus on medication for CNV by diminishing the choroidal level mount.

Diabetic retinopathy (DR) is possibly the most widely recognized confusion because of diabetes and the primary source of visual deficiency worldwide. Albeit the two sorts of diabetes do not have an immediate fix, intravitreal infusions can be given to patients to diminish the defective veins. A new report has focused on the human antigen R (HuR) that can be a useful remedial methodology for diabetic retinopathy (Green et al. 2019). The HuR is a human protein encoded in the ELAL1 quality, containing 3′-untranslated districts of mRNA restricting areas, including in aggravation and cell development. This considers the quality articulation to be controlled and changed into the agreeing soundness and interpretation vital (Kataki et al. 2015). It was demonstrated that explicit nanosystems are stacked with siRNA to quiet the HuR articulation. This articulation contained potent lipid nanoparticles (SLN) and liposomes (SUV). The siRNA had the option to send to the rodent's retina utilizing lipid-based nanocarriers and lipoplexes to quiet the outflow of the retinal HuR and the VEGF. The examination exhibited that when the HuR is focused on, the overexpression of the VEGF is diminished. At that point, while directing intraocular siRNA, the HuR articulation is hushed, considering a revelation of another pathway for treating DR. In this specific investigation, the utilization of siRNA nanocarriers was utilized to improve the retinal entrance as opposed to utilizing exposed siRNA, as it did not secure the retinal layers. It was recently shown that the HuR protein could tie to the VEGF-encoding mRNA to improve protein articulation. It was causing an expansion in the VEGF of the retina in diabetic rodents. When the HuR siRNA intraocular was controlled, there was a diminishing in the retinal HuR in the diabetic rodents. This was because of the post-transcriptional level for the HuR-intervened guideline managing the VEGF articulation. It was found that the liposome restricting to the lipoplexes (SLN-based lipoplex L4 with liposomes with a similar N/P proportion of L1 and L3) brought about a positive charge taking into account a uniform surface inclusion of the lipid lattice. This demonstrated that the excess charge influenced the lipoplex L4 to fittingly connect with the focused cells to prompt a legitimate conveyance to oversee retinal sicknesses.

The delivery to specific tissues, such as in the eye, is a challenge but can be extremely beneficial by using target delivery. The use of siRNA can silence gene expression and target specific disease-related genes, such as age-related macular degeneration (AMD) (Raja et al. 2019b). Within specific tissues and cells, it can provide a high local concentration without using more drugs. In ocular tissues, it has been found that a siRNA-based anti-angiogenic agent drug, bevasiranib, has been used in treating wet AMD (Garba and Mousa 2010b). The use of bevasiranib helps to eliminate repeated intravitreal injections that could cause complications, such as retinal tears, intravitreal bleeding, or lens injury. The use of siRNA to target a gene will allow for fewer complications, bursting of blood vessels, and repeated injections (Garba and Mousa 2010b).

6 Conclusion

As discussed, using nanotechnology to target macrophages will allow there to prevent/reduce the inflammation process associated with acute and chronic diseases. The next step would be to further study to understand the therapeutic and side effects of targeted nanodelivery of siRNA to macrophages and the eye. Studies have shown to be promising to deliver anti-VEGF using siRNA to reduce the neovascularization and edema of the eye and target the macrophages to reduce the inflammation process caused by antigens in the innate immune system.

When examining macrophages, the process of preventing phagocytosis that activates B and T cells will reduce the inflammation process. Through the examples, it can be seen that the use of siRNA and nanocarriers have been and can be used to target macrophages of chronic diseases as a form of immunotherapy. The benefit of using siRNA specifically in macrophages targeted delivery systems is to undergo sequence-specific inhibitions of genes, gene expression, or translation. The use of siRNA with nanocarriers to target macrophages has been shown to decrease the pro-inflammatory response of macrophages and reduce the overall course of inflammation in chronic diseases.

For ocular use, VEGF is a crucial mediator in the angiogenesis of neovascular age-related macular degeneration and diabetic macular edema and has been targeted by anti-VEGF drugs to decrease the formation of the blood vessels within the eye. Targeting macrophages and signaling will allow for more insight into drug therapy, such as combining and delivering siRNA with loaded drugs. This would allow for a higher concentration delivery to the macrophages to increase the effectiveness and decrease the need for additional treatment, and reduced adverse side effects, as previous studies have shown. If these siRNA nanocarriers could be used to deliver anti-VEGF to silence the retinal VEGF, then it would be a novel therapy used for patients with wet age-related macular degeneration and diabetic macular edema. The nanocarriers' use would allow for a higher concentration of the anti-VEGF to be delivered to the retina. This would reduce the resistance of anti-VEGF drugs to reduce the long-term treatment of neovascular AMD.

To deliver siRNA will require research on the utilization of nanocarriers and the efficacy of the targeted size. Ideally, siRNA's effect with the loaded drug at the target site will be long-lasting at a controlled dose. Delayed-release forms would be a potential area for future research as siRNA is versatile in functions and targets specific sites of action. Also, more scalable and clinically applicable methods of siRNA delivery are becoming areas of extensive research. The future trends of using siRNA to target macrophages will be to examine the limits and understanding the long effects of using siRNA with nanocarriers to prevent and manage chronic diseases.

References

Abasalizadeh F, Moghaddam SV, Alizadeh E, et al. Alginate-based hydrogels as drug delivery vehicles in cancer treatment and their applications in wound dressing and 3D bioprinting. J Biol Eng. 2020;14:8. https://doi.org/10.1186/s13036-020-0227-7.

Abiodun Solanke IMF, Ajayi DM, Arigbede AO. Nanotechnology and its application in dentistry. Ann Med Health Sci Res. 2014;4(9):171. https://doi.org/10.4103/2141-9248.141951.

Ahsan F. Targeting to macrophages: role of physicochemical properties of particulate carriers—liposomes and microspheres—on the phagocytosis by macrophages. J Control Release. 2002;79(1–3):29–40. https://doi.org/10.1016/s0168-3659(01)00549-1.

Anselmo AC, Mitragotri S. A review of clinical translation of inorganic nanoparticles. AAPS J. 2015;17:1041–54. https://doi.org/10.1208/s12248-015-9780-2.

Attia MF, Anton N, Wallyn J, Omran Z, Vandamme TF. An overview of active and passive targeting strategies to improve the nanocarriers efficiency to tumour sites. J Pharm Pharmacol. 2019;71(8):1185–98. https://doi.org/10.1111/jphp.13098.

Beloqui A, Pozo-Rodríguez A del, Isla A, Rodríguez-Gascón A, Solinís MÁ. Nanostructured lipid carriers as oral delivery systems for poorly soluble drugs. (2017a, June 19). https://doi.org/10.1016/j.jddst.2017.06.013.

Beloqui A, del Pozo-Rodríguez A, Isla A, Rodríguez-Gascón A, Solinís MÁ. Nanostructured lipid carriers as oral delivery systems for poorly soluble drugs. J Drug Deliv Sci & Technol. 2017b;42:144–54. https://doi.org/10.1016/j.jddst.2017.06.013.

Chernikov IV, Vlassov VV, Chernolovskaya EL. Current development of siRNA bioconjugates: from research to the clinic. Front Pharmacol. 2019;10:444. https://doi.org/10.3389/fphar.2019.00444.

Choudhury SR, Ordaz J, Lo C-L, Damayanti NP, Zhou F, Irudayaraj J. From the cover: zinc oxide nanoparticles-induced reactive oxygen species promotes multimodal cyto- and epigenetic toxicity. Toxicol Sci. 2017;156(1):261–74. https://doi.org/10.1093/toxsci/kfw252.

Cortes H, Caballero-Florán IH, Mendoza-Muñoz N, Escutia-Guadarrama L, Figueroa-González G, Reyes-Hernández OD, González-Del Carmen M, Varela-Cardoso M, González-Torres M, Florán B, Del Prado-Audelo ML, Leyva-Gómez G. Xanthan gum in drug release. Cell Mol Biol (Noisy-le-Grand). 2020;66(4):199–207.

Daka A, Peer D. RNAi-based nanomedicines for targeted personalized therapy. Adv Drug Deliv Rev. 2012;64(13):1508–21. https://doi.org/10.1016/j.addr.2012.08.014.

Dana H, Chalbatani GM, Mahmoodzadeh H, Karimloo R, Rezaiean O, Moradzadeh A, Mehmandoost N, Moazzen F, Mazraeh A, Marmari V, Ebrahimi M, Rashno MM, Abadi SJ, Gharagouzlo E. Molecular mechanisms and biological functions of siRNA. Int J Biomed Sci: IJBS. 2017a;13(2):48–57.

Dana H, Chalbatani GM, Mahmoodzadeh H, Karimloo R, Rezaiean O, Moradzadeh A, Mehmandoost N, Moazzen F, Mazraeh A, Marmari V, Ebrahimi M, Rashno MM, Abadi SJ, Gharagouzlo E. Molecular mechanisms and biological functions of siRNA. Int J Biomed Sci: IJBS. 2017b;13(2):48–57.

Davis ME, Chen Z, Shin DM. Nanoparticle therapeutics: an emerging treatment modality for cancer. Nat Rev Drug Discov. 2008;7(9):771–82. https://doi.org/10.1038/nrd2614.

Davis ME, Zuckerman JE, Choi CHJ, Seligson D, Tolcher A, Alabi CA, Yen Y, Heidel JD, Ribas A. Evidence of RNAi in humans from systemically administered siRNA via targeted nanoparticles. Nature. 2010;464(7291):1067–70. https://doi.org/10.1038/nature08956.

de Fougerolles A, Vornlocher H-P, Maraganore J, Lieberman J. Interfering with disease: a progress report on siRNA-based therapeutics. Nat Rev Drug Discov. 2007;6(6):443–53. https://doi.org/10.1038/nrd2310.

de Marcken M, Dhaliwal K, Danielsen AC, Gautron AS, Dominguez-Villar M. TLR7 and TLR8 activate distinct pathways in monocytes during RNA virus infection. Sci Signal. 2019;12(605):eaaw1347. https://doi.org/10.1126/scisignal.aaw1347.

Deng Y, Zhang X, Shen H, He Q, Wu Z, Liao W, Yuan M. Application of the Nano-drug delivery system in treatment of cardiovascular diseases. Front Bioeng Biotechnol. 2020;7:489. https://doi.org/10.3389/fbioe.2019.00489.

Din F, Aman W, Ullah I, Qureshi OS, Mustapha O, Shafique S, Zeb A. Effective use of nanocarriers as drug delivery systems for the treatment of selected tumors. Int J Nanomedicine. 2017;12:7291–309. https://doi.org/10.2147/ijn.s146315.

Dolatabadi JEN, Valizadeh H, Hamishehkar H. Solid lipid nanoparticles as efficient drug and gene delivery Systems: recent breakthroughs. Adv Pharmaceutic Bullet. 2015;5(2):151–9. https://doi.org/10.15171/apb.2015.022.

Drug Delivery Systems. (2016). National Institute of Biomedical Imaging and Bioengineering. https://www.nibib.nih.gov/science-education/science-topics/drug-delivery-systems-getting-drugs-their-targets-controlled-manner

Elbashir SM. Functional anatomy of siRNAs for mediating efficient RNAi in Drosophila melanogaster embryo lysate. EMBO J. 2001;20(23):6877–88. https://doi.org/10.1093/emboj/20.23.6877.

Fan C-H, Lee Y-H, Ho Y-J, Wang C-H, Kang S-T, Yeh C-K. Macrophages as drug delivery carriers for acoustic phase-change droplets. Ultrasound Med Biol. 2018;44(7):1468–81. https://doi.org/10.1016/j.ultrasmedbio.2018.03.009.

Felice B, Prabhakaran MP, Rodríguez AP, Ramakrishna S. Drug delivery vehicles on a nano-engineering perspective. Mater Sci Eng C. 2014;41:178–95. https://doi.org/10.1016/j.msec.2014.04.049.

Feng W, Chumley P, Prieto MC, Miyada K, Seth DM, Fatima H, Hua P, Rezonzew G, Sanders PW, Jaimes EA. Transcription factor avian erythroblastosis virus E26 Oncogen Homolog-1 is a novel mediator of renal injury in salt-sensitive hypertension. Hypertension. 2015a;65(4):813–20. https://doi.org/10.1161/hypertensionaha.114.04533.

Feng W, Chumley P, Prieto MC, Miyada K, Seth DM, Fatima H, et al. Transcription factor avian erythroblastosis virus E26 Oncogen Homolog-1 is a novel mediator of renal injury in salt-sensitive hypertension. Hypertension. 2015b;65(4):813–20. https://doi.org/10.1161/hypertensionaha.114.04533.

Fire A, Xu SQ, Montgomery MK, Kostas SA, Driver SE, Mello CC. Potent and specific genetic interference by double-stranded RNA in Caenorhabditis elegans. Nature. 1998;391(6669):806–11. https://doi.org/10.1038/35888.

Forbes DC, Peppas NA. Polymeric Nanocarriers for siRNA delivery to murine macrophages. Macromol Biosci. 2014;14(8):1096–105. https://doi.org/10.1002/mabi.201400027.

Fujiwara N, Kobayashi K. Macrophages in inflammation. Curr Drug Target Inflam & Allergy. 2005;4(3):281–6. https://doi.org/10.2174/1568010054022024.

Garba AO, Mousa SA. Bevasiranib for the treatment of wet, age-related macular degeneration. Ophthalmology and eye diseases, 2, OED.S4878; 2010a. https://doi.org/10.4137/oed.s4878.

Garba AO, Mousa SA. Bevasiranib for the treatment of wet, age-related macular degeneration. (2010b, December 19). https://doi.org/10.4137/OED.S4878.

Gordon S, Martinez-Pomares L. Physiological roles of macrophages. Pflugers Arch - Eur J Physiol. 2017;469(3-4):365–74. https://doi.org/10.1007/s00424-017-1945-7.

Green LC, Anthony SR, Slone S, Lanzillotta L, Nieman ML, Wu X, Robbins N, Jones SM, Roy S, Owens AP 3rd, Aube J, Xu L, Lorenz JN, Blaxall BC, Rubinstein J, Benoit JB, Tranter M. Human antigen R as a therapeutic target in pathological cardiac hypertrophy. JCI Insight. 2019;4(4):e121541. https://doi.org/10.1172/jci.insight.121541.

Guo H, Callaway JB, Ting JP. Inflammasomes: mechanism of action, role in disease, and therapeutics. Nat Med. 2015;21(7):677–87. https://doi.org/10.1038/nm.3893.

Gupta N, Rai DB, Jangid AK, Pooja D, Kulhari H. Nanomaterials-based siRNA delivery: routes of administration, hurdles and role of Nanocarriers. Nanotechnol Modern Animal Biotechnol. 2019:67–114. https://doi.org/10.1007/978-981-13-6004-6_3.

Haussecker D. Current issues of RNAi therapeutics delivery and development. J Control Release. 2014;195:49–54. https://doi.org/10.1016/j.jconrel.2014.07.056.

Hu G, Guo M, Xu J, Wu F, Fan J, Huang Q, Yang G, Lv Z, Wang X, Jin Y. Nanoparticles targeting macrophages as potential clinical therapeutic agents against cancer and inflammation. Front Immunol. 2019;10:1998. https://doi.org/10.3389/fimmu.2019.01998.

Hu B, Zhong L, Weng Y, et al. Therapeutic siRNA: state of the art. Sig Transduct Target Ther. 2020;5:101. https://doi.org/10.1038/s41392-020-0207-x.

Huang Z, Zhang Z, Jiang Y, Zhang D, Chen J, Dong L, Zhang J. Targeted delivery of oligonucleotides into tumor-associated macrophages for cancer immunotherapy. J Control Release. 2012;158(2):286–92. https://doi.org/10.1016/j.jconrel.2011.11.013.

Huang JF, Zhong J, Chen GP, Lin ZT, Deng Y, Liu YL, Cao PY, Wang B, Wei Y, Wu T, Yuan J, Jiang GB. A hydrogel-based hybrid Theranostic contact lens for fungal keratitis. ACS Nano. 2016;10(7):6464–73. https://doi.org/10.1021/acsnano.6b00601.

Idriss HT, Naismith JH. TNF alpha and the TNF receptor superfamily: structure-function relationship(s). Microsc Res Tech. 2000;50(3):184–95. https://doi.org/10.1002/1097-0029(20000801)50:3<184::AID-JEMT2>3.0.CO;2-H.

Jackson AL. Widespread siRNA "off-target" transcript silencing mediated by seed region sequence complementarity. RNA. 2006;12(7):1179–87. https://doi.org/10.1261/rna.25706.

Jackson AL, Linsley PS. Recognizing and avoiding siRNA off-target effects for target identification and therapeutic application. Nat Rev Drug Discov. 2010;9(1):57–67. https://doi.org/10.1038/nrd3010.

Jacob S, Nair AB, Shah J. Emerging role of nanosuspensions in drug delivery systems. Biomaterial Res. 2020;24:3. https://doi.org/10.1186/s40824-020-0184-8.

Jain NK, Mishra V, Mehra NK. Targeted drug delivery to macrophages. Expert Opin Drug Deliv. 2013;10(3):353–67. https://doi.org/10.1517/17425247.2013.751370.

Jhaveri AM, Torchilin VP. Multifunctional polymeric micelles for delivery of drugs and siRNA. Front Pharmacol. 2014;5:77. https://doi.org/10.3389/fphar.2014.00077.

Judge AD, Sood V, Shaw JR, Fang D, McClintock K, MacLachlan I. Sequence-dependent stimulation of the mammalian innate immune response by synthetic siRNA. Nat Biotechnol. 2005;23(4):457–62. https://doi.org/10.1038/nbt1081.

Kang YJ, Cutler EG, Cho H. Therapeutic nanoplatforms and delivery strategies for neurological disorders. Nano Convergence. 2018;5:35. https://doi.org/10.1186/s40580-018-0168-8.

Kaps L, Leber N, Klefenz A, Choteschovsky N, Zentel R, Nuhn L, Schuppan D. In vivo siRNA delivery to immunosuppressive liver macrophages by α-Mannosyl-functionalized cationic Nanohydrogel particles. Cell. 2020;9(8):1905. https://doi.org/10.3390/cells9081905.

Kataki MS, Kakoti BB, Jameson M, Solanki A, Hirani A, Pathak Y. Nano platforms for delivery of siRNA to the eye. Curr Pharm Des. 2015;21(31):4587–93. https://doi.org/10.2174/138161282131151013191051.

Kelly C, Jefferies C, Cryan SA. Targeted liposomal drug delivery to monocytes and macrophages. J Drug Deliv. 2011;2011:1–11. https://doi.org/10.1155/2011/727241.

Kim S-S, Ye C, Kumar P, Chiu I, Subramanya S, Wu H, Shankar P, Manjunath N. Targeted delivery of siRNA to macrophages for anti-inflammatory treatment. Mol Ther. 2010;18(5):993–1001. https://doi.org/10.1038/mt.2010.27.

Konate K, Dussot M, Aldrian G, Vaissière A, Viguier V, Neira IF, et al. Peptide-based nanoparticles to rapidly and efficiently wrap n roll siRNA into cells. Bioconjug Chem. 2019:592–603. https://doi.org/10.1021/acs.bioconjchem.8b00776.s001.

Kowal J, Kornete M, Joyce JA. Re-education of macrophages as a therapeutic strategy in cancer. Immunotherapy. 2019;11(8):677–89. https://doi.org/10.2217/imt-2018-0156.

Kurrikoff K, Suhorut J, Langel Ü. Cell-penetrating peptides in cancer targeting. Drug Deliv Oncol. 2011:1187–217. https://doi.org/10.1002/9783527634057.ch37.

Lam JKW, Chow MYT, Zhang Y, Leung SWS. siRNA versus miRNA as therapeutics for gene silencing. Molecul Therapy Nucleic Acid. 2015;4:e252. https://doi.org/10.1038/mtna.2015.23.

Landry B, Aliabadi HM, Samuel A, Gül-Uludağ H, Jiang X, Kutsch O, Uludağ H. Effective non-viral delivery of siRNA to acute myeloid Leukemia cells with lipid-substituted Polyethylenimines. PLoS One. 2012;7(8):e44197. https://doi.org/10.1371/journal.pone.0044197.

Levanova A, Poranen MM. RNA interference as a prospective tool for the control of human viral infections; 2018, August 22. https://doi.org/10.3389/fmicb.2018.02151.

Madaan K, Kumar S, Poonia N, Lather V, Pandita D. Dendrimers in drug delivery and targeting: drug-dendrimer interactions and toxicity issues. J Pharm Bioallied Sci. 2014;6(3):139–50. https://doi.org/10.4103/0975-7406.130965.

Malamed SF. Intravenous sedation. Sedation. 2010: 269–70. https://doi.org/10.1016/b978-0-323-05680-9.00023-0.

Mankamna Kumari R, Goswami R, Nimesh S. Application of nanotechnology in diagnosis and therapeutics. Nanotechnol Energy & Environ Eng. 2020:413–40. https://doi.org/10.1007/978-3-030-33774-2_18.

Mansoori B, Mohammadi A, Shir Jang S, Baradaran B. Mechanisms of immune system activation in mammalians by small interfering RNA (siRNA). Artific Cell Nanomed & Biotechnol. 2015;44(7):1589–96. https://doi.org/10.3109/21691401.2015.1102738.

Mir M, Ishtiaq S, Rabia S, et al. Nanotechnology: from in vivo imaging system to controlled drug delivery. Nanoscale Res Lett. 2017;12:500. https://doi.org/10.1186/s11671-017-2249-8.

Moghimi SM, Hunter AC, Murray JC. Nanomedicine: current status and future prospects. FASEB J. 2005; 19(3):311–30. https://doi.org/10.1096/fj.04-2747rev.

Mohebali A, Abdouss M, Afshar Taromi F. Fabrication of biocompatible antibacterial nanowafers based on HNT/PVA nanocomposites loaded with minocycline for burn wound dressing. Mater Sci Eng C. 2020;110:110685. https://doi.org/10.1016/j.msec.2020.110685.

Muller RH, Keck CM. Challenges and solutions for the delivery of biotech drugs – a review of drug nanocrystal technology and lipid nanoparticles. J Biotechnol. 2004;113(1–3):151–70. https://doi.org/10.1016/j.jbiotec.2004.06.007.

Muthana M, Giannoudis A, Scott SD, Fang H-Y, Coffelt SB, Morrow FJ, Murdoch C, Burton J, Cross N, Burke B, Mistry R, Hamdy F, Brown NJ, Georgopoulos L, Hoskin P, Essand M, Lewis CE, Maitland NJ. Use of macrophages to target therapeutic adenovirus to human prostate Tumors. Cancer Res. 2011;71(5):1805–15. https://doi.org/10.1158/0008-5472.can-10-2349.

Nayerossadat N, Maedeh T, Ali PA. Viral and nonviral delivery systems for gene delivery. Adv Biomed Res. 2012;1:27. https://doi.org/10.4103/2277-9175.98152.

Ortega RA, Barham WJ, Kumar B, Tikhomirov O, McFadden ID, Yull FE, Giorgio TD. Biocompatible mannosylated endosomal-escape nanoparticles enhance selective delivery of short nucleotide sequences to tumor associated macrophages. Nanoscale. 2015;7(2):500–10. https://doi.org/10.1039/c4nr03962a.

Petros RA, DeSimone JM. Strategies in the design of nanoparticles for therapeutic applications. Nat

Rev Drug Discov. 2010;9(8):615–27. https://doi.org/10.1038/nrd2591.

Piaggio F, Kondylis V, Pastorino F, Di Paolo D, Perri P, Cossu I, Schorn F, Marinaccio C, Murgia D, Daga A, Raggi F, Loi M, Emionite L, Ognio E, Pasparakis M, Ribatti D, Ponzoni M, Brignole C. A novel liposomal Clodronate depletes tumor-associated macrophages in primary and metastatic melanoma: anti-angiogenic and anti-tumor effects. J Control Release. 2016;223:165–77. https://doi.org/10.1016/j.jconrel.2015.12.037.

Ponzoni M, Pastorino F, Di Paolo D, Perri P, Brignole C. Targeting macrophages as a potential therapeutic intervention: impact on inflammatory diseases and cancer. Int J Mol Sci. 2018;19(7):1953. https://doi.org/10.3390/ijms19071953.

Qiu Z, Kwon A-H, Tsuji K, Kamiyama Y, Okumura T, Hirao Y. FIBRONECTIN PREVENTS D-galactosamine/lipopolysaccharide-induced lethal hepatic failure in mice. Shock. 2006;25(1):80–7. https://doi.org/10.1097/01.shk.0000185797.04589.5c.

Raja MAG, Katas H, Amjad MW. Design, mechanism, delivery and therapeutics of canonical and dicer-substrate siRNA. Asian J Pharmaceut Sci. 2019a;14(5):497–510. https://doi.org/10.1016/j.ajps.2018.12.005.

Raja MAG, Katas H, Amjad MW. Design, mechanism, delivery and therapeutics of canonical and Dicer-substrate siRNA. (2019b, February 13). https://doi.org/10.1016/j.ajps.2018.12.005.

Rasul RM, Tamilarasi Muniandy M, Zakaria Z, Shah K, Chee CF, Dabbagh A, Rahman NA, Wong TW. A review on chitosan and its development as pulmonary particulate anti-infective and anti-cancer drug carriers. Carbohydr Polym. 2020;250:116800. https://doi.org/10.1016/j.carbpol.2020.116800.

Ray L. Polymeric nanoparticle-based drug/gene delivery for lung cancer. Nanotechnology-based targeted drug delivery Systems for Lung Cancer, 77–93; 2019. https://doi.org/10.1016/b978-0-12-815720-6.00004-6.

Reich SJ, Fosnot J, Kuroki A, Tang W, Yang X, Maguire AM, Bennett J, Tolentino MJ. Small interfering RNA (siRNA) targeting VEGF effectively inhibits ocular neovascularization in a mouse model. Mol Vis. 2003;9:210–6.

Riglar D, Silver P. Engineering bacteria for diagnostic and therapeutic applications. Nat Rev Microbiol. 2018;16:214–25. https://doi.org/10.1038/nrmicro.2017.172.

Schwall C, Banerjee I. Micro- and nanoscale hydrogel systems for drug delivery and tissue engineering. Materials. 2009;2(2):577–612. https://doi.org/10.3390/ma2020577.

Serrano-Sevilla I, Artiga Á, Mitchell SG, De Matteis L, de la Fuente JM. Natural polysaccharides for siRNA delivery: nanocarriers based on chitosan, hyaluronic acid, and their derivatives. Molecules. 2019;24(14):2570. Published 2019 Jul 15. https://doi.org/10.3390/molecules24142570.

Sheikhi A, Hayashi J, Eichenbaum J, Gutin M, Kuntjoro N, Khorsandi D, Khademhosseini A. Recent advances in nanoengineering cellulose for cargo delivery. J Control Release. 2019;294:53–76. https://doi.org/10.1016/j.jconrel.2018.11.024.

Singh AP, Biswas A, Shukla A, et al. Targeted therapy in chronic diseases using nanomaterial-based drug delivery vehicles. Sig Transduct Target Ther. 2019;4:33. https://doi.org/10.1038/s41392-019-0068-3.

Song W, Anselmo AC, Huang L. Nanotechnology intervention of the microbiome for cancer therapy. Nat Nanotechnol. 2019;14(12):1093–103. https://doi.org/10.1038/s41565-019-0589-5.

Tatiparti K, Sau S, Kashaw SK, Iyer AK. siRNA delivery strategies: a comprehensive review of recent developments. 2017, April 5; https://doi.org/10.3390/nano7040077.

Thakur A, Fitzpatrick S, Zaman A, et al. Strategies for ocular siRNA delivery: potential and limitations of non-viral nanocarriers. J Biol Eng. 2012;6:7. https://doi.org/10.1186/1754-1611-6-7.

Tilley S, Volmer J, Picher M. Therapeutic applications. Subcell Biochem. 2011;55:235–76. https://doi.org/10.1007/978-94-007-1217-1_9.

Tiwari G, Tiwari R, Sriwastawa B, Bhati L, Pandey S, Pandey P, Bannerjee SK. Drug delivery systems: an updated review. Int J Pharmaceutic Investigat. 2012;2(1):2–11. https://doi.org/10.4103/2230-973X.96920.

Troegeler A, Lastrucci C, Duval C, Tanne A, Cougoule C, Maridonneau-Parini I, Neyrolles O, Lugo-Villarino G. An efficient siRNA-mediated gene silencing in primary human monocytes, dendritic cells and macrophages. Immunol Cell Biol. 2014;92(8):699–708. https://doi.org/10.1038/icb.2014.39.

Vega-Vásquez P, Mosier NS, Irudayaraj J. Nanoscale drug delivery Systems: from medicine to agriculture. Front Bioeng Biotechnol. 2020;8:79. https://doi.org/10.3389/fbioe.2020.00079.

Vijayavenkataraman S, Garcia-Sabate A, Teo JC. Specialized multimaterial print heads for 3D hydrogel printing: tissue-engineering applications. IEEE Nanotechnol Mag. 2020;14(3):42–52. https://doi.org/10.1109/mnano.2020.2966065.

Visser JG, Van Staden ADP, Smith C. Harnessing macrophages for controlled-release drug delivery: lessons from microbes. Front Pharmacol. 2019;10:22. https://doi.org/10.3389/fphar.2019.00022.

Wei W, Li H, Yin C, Tang F. Research progress in the application of in situ hydrogel system in tumor treatment. Drug Deliv. 2020;27(1):460–8. https://doi.org/10.1080/10717544.2020.1739171.

World Health Organization. (2010, December 6). WHO | Integrated chronic disease prevention and control. https://www.who.int/chp/about/integrated_cd/en/

Xiao B, Ma P, Viennois E, Merlin D. Urocanic acid-modified chitosan nanoparticles can confer anti-inflammatory effect by delivering CD98 siRNA to macrophages. Colloids Surf B Biointerfaces. 2016;143:186–93. https://doi.org/10.1016/j.colsurfb.2016.03.035.

Yang R, Yang X, Zhang Z, et al. Single-walled carbon nanotubes-mediated in vivo and in vitro delivery of siRNA into antigen-presenting cells. Gene Ther. 2006;13:1714–23. https://doi.org/10.1038/sj.gt.3302808.

Yong SB, Kim H, Kim J, et al. Human CD64-targeted non-viral siRNA delivery system for blood monocyte gene modulation. Sci Rep. 2017;7:42171. https://doi.org/10.1038/srep42171.

Zanganeh S, Hutter G, Spitler R, Lenkov O, Mahmoudi M, Shaw A, Pajarinen JS, Nejadnik H, Goodman S, Moseley M, Coussens LM, Daldrup-Link HE. Iron oxide nanoparticles inhibit tumour growth by inducing pro-inflammatory macrophage polarization in tumour tissues. Nat Nanotechnol. 2016;11(11):986–94. https://doi.org/10.1038/nnano.2016.168.

Zhang L, Gu FX, Chan JM, Wang AZ, Langer RS, Farokhzad OC. Nanoparticles in medicine: therapeutic applications and developments. Clinic Pharmacol & Therapeutic. 2007;83(5):761–9. https://doi.org/10.1038/sj.clpt.6100400.

Zhang M, Gao Y, Caja K, Zhao B, Kim JA. Non-viral nanoparticle delivers small interfering RNA to macrophages in vitro and in vivo. PLoS One. 2015;10(3):e0118472. https://doi.org/10.1371/journal.pone.0118472.

Zhang R, Jing Y, Zhang H, et al. Comprehensive evolutionary analysis of the major RNA-induced silencing complex members. Sci Rep. 2018;8:14189. https://doi.org/10.1038/s41598-018-32635-4.

Zhao J, Feng S-S. Nanocarriers for delivery of siRNA and co-delivery of siRNA and other therapeutic agents. Nanomedicine. 2015;10(14):2199–228. https://doi.org/10.2217/nnm.15.61.

Role of Macrophages and Immunotherapy in Wound Healing

Ashley Oake, Swati Gupta, and Yashwant V. Pathak

Abstract

The body undergoes four phases of healing, which are hemostasis, inflammation, proliferation, and reprogramming/remodeling, when exposed to a wound. Hemostasis is the process of blood clotting to prevent the blood flow at a wound. The inflammatory phase is then initiated through the recruitment of macrophages with cytokines and other immune factors. They are responsible for the clearing of cells, releasing cytokines, cellular proliferation, tissue remodeling, and extracellular matrix growth. Macrophages may be polarized toward either M1, proinflammatory cells, or M2, antiinflammatory cells. The proliferation phase includes production of the extracellular matrix while the wound contracts with the help of micro fibroblasts. The skin is then remodeled with the cross-linking of collagen and removal of excess extracellular matrix. A wound may be considered chronic when there is an inability to heal or it is severely delayed. Treatment for these wounds focuses on the removal of the M1 macrophages and the addition of M2 macrophages. There are many factors affecting wound healing that can be divided into local factors (oxygen, infections, and foreign bodies) or systematic factors (age, stress, weight, additional health concerns, and alcohol consumption).

Keywords

Macrophages · Wound healing · Phagocytosis · Immune system · Tissue formation

1 Introduction

The body takes a number of steps to heal after a wound or other trauma that prevent the loss of excess blood and defend against bacteria that lead to infections (Gale 2011). The four phases of healing are hemostasis, inflammation, proliferation, and reprogramming/remodeling (Gale 2011). Hemostasis is the process of continuing the flow of blood through the body while preventing it through vasoconstriction and clotting at an open wound (Gale 2011). Initially, the body undergoes primary hemostasis triggered by components of the subendothelial extracellular matrix

A. Oake
Morsani College of Medicine, Tampa, FL, USA

S. Gupta
Department of Pharmaceutics, Amity Institute of Pharmacy, Amity University Uttar Pradesh, Noida, India

Y. V. Pathak (✉)
University of South Florida, Taneja College of Pharmacy, Tampa, FL, USA

Adjunct Professor University of Airlangga, Surabaya, Indonesia
e-mail: ypathak1@health.usf.edu

S. Gupta, Y. V. Pathak (eds.), *Macrophage Targeted Delivery Systems*,
https://doi.org/10.1007/978-3-030-84164-5_22

(ECM) where platelets are recruited to form a blood clot (Gale 2011). Platelets are small cellular fragments that bind together during an injury to essentially plug a damaged blood vessel and have an average lifespan of about 10 days (Varga-Szabo et al. 2008). During normal blood flow, an anticoagulant surface prevents them from congregating and sticking together (Varga-Szabo et al. 2008). The receptor glycoprotein (GP) Ib-V-IX complex binds with the von Willebrand factor found in the ECM while collagen binds to the receptor GPVI (Varga-Szabo et al. 2008). These reactions bind with such a low affinity that they are unable to initiate and control adhesion by themselves, but cause platelets to stay close to the surface of the vessels and lead to the release of secondary messengers (Varga-Szabo et al. 2008).

The activation of platelets is initiated by the binding of high-affinity β1 integrins and the major platelet integrin (αIIbβ3) to collagen, fibronectin, laminin, fibrinogen, and vWF (Gale 2011). The von Willebrand factor contains multiple binding sites for collagen and platelet receptors that aid in the adhesion of platelets (Varga-Szabo et al. 2008). Glycoprotein (GP) Ib-V-IX complex and collagen VI bind to vWF at the A1 domain to initiate the recruitment of platelets to create a plug at an open wound (Varga-Szabo et al. 2008). The binding of the A1 domain with collagen immobilizes it and creates a higher affinity for the GPIb-V-IX complex comprised of four genes from the leucine-rich repeat protein superfamily (Varga-Szabo et al. 2008). G protein-mediated platelet activation is initiated by the release of adenosine diphosphate, which binds to P2Y1 and P2Y12, and thromboxane A2, which binds to the thromboxane receptor (Varga-Szabo et al. 2008).

Once platelets are recruited and adhere to each other, fibrinogen is cleaved to fibrin by thrombin generating a mesh covering the open wound (Gale 2011). Tissue factor (TF) binds with serine protease factor VIIa to initiate coagulation by activating factors X and IX (Gale 2011). Prothrombin is then activated by factor X to produce thrombin through the extrinsic pathway (Gale 2011). The intrinsic pathway, or positive feedback loop, may be initiated by the cleaving of PAR1 and PAR4 by thrombin, the terminal protease (Gale 2011). Factor XI then leads to the activation of factor IX and cofactors VIII and V (Gale 2011). The extrinsic pathway is initiated by cofactor TF and a small amount of thrombin is generated, the intrinsic pathway is then initiated to amplify the process and coagulation complexes are formed (Gale 2011). Fibrinolysis, activated protein C, and serpins are anticoagulant systems that occur to prevent the over clotting of blood (Gale 2011).

The inflammatory phase is then initiated through the recruitment of inflammatory cells and macrophages by chemotaxis (Guo and DiPietro 2010). Macrophages are responsible for the clearing of apoptotic cells, releasing cytokines to promote inflammation, promoting cellular proliferation through growth factors, tissue remodeling, and extracellular matrix growth (Guo and DiPietro 2010). When macrophages respond to cytokines and other immune factors, they may be polarized toward either M1 or M2 activation (Wu et al. 2016). IFN-γ is associated with the M1 classical activation of macrophages leading to an inflammatory cellular response (Wu et al. 2016). These macrophages are polarized by nuclear factor kappa B, release high amounts of ROS, and possess antimicrobial activity (Wu et al. 2016). IL-4 and IL-13 lead to the M2 alternative activation of macrophages and are associated with wound healing (Ferrante and Leibovich 2012). They inhibit T-cell function and induce T-cell apoptosis to inhibit immune responses (Ferrante and Leibovich 2012). M1 macrophages are present during the inflammation phase of wound healing, but as the body transitions to the proliferation and reprogramming phase macrophages are mainly polarized toward the M2 activation state (Ferrante and Leibovich 2012).

As cells are removed, tissue regeneration is initiated by fibroblasts and recruitment of other leukocytes as they transition to the next stage of wound healing – proliferation (Guo and DiPietro 2010). Reactive oxygen species (ROS) formed through electron transfer reactions are released and lead to cell signaling pathways and migration of macrophages to initiate phagocytic activity

(Wu et al. 2016). The proliferation phase includes growth of blood vessels, granulation tissue formation, and production of collagen, glycosaminoglycans, and proteoglycans in order to form the extracellular matrix (Guo and DiPietro 2010). This causes the wound to contract by micro fibroblasts pulling the edges together as new epithelial cells form over the wound with the help of keratinocytes (Guo and DiPietro 2010). Skin may become lighter, pink, or uneven during the healing at this phase (Guo and DiPietro 2010). Growth factors, cytokines, and lipid mediators are released to reorganize the cytoskeleton and increase the permeability of the membrane (Guo and DiPietro 2010). The growth factors then bind to receptors to activate endothelial cells and secrete proteolytic enzymes (Reinke and Sorg 2012). The basal lamina is removed allowing formation of new epithelial cells and differentiation of blood vessels leading to the development of granulation tissue completing the phase (Reinke and Sorg 2012).

The final stage of wound healing is the reprogramming/remodeling phase, which can take up to a year to be completed (Reinke and Sorg 2012). Collagen that was produced during earlier phases is now reorganized from type III to type I and cross-linked while extra cells used to form the new tissue are removed by apoptosis (Reinke and Sorg 2012). This will reduce scarring of the skin, create stronger epithelial tissue, and slow down metabolic activity (Reinke and Sorg 2012). Unorganized collagen in wounds that are unable to heal correctly can lead to a bulge or buildup of collagen called a keloid (Reinke and Sorg 2012). After the final stages of wound healing, hair follicles and sweat glands will never grow back in the damaged area (Reinke and Sorg 2012).

2 Function of the Skin

The skin is the largest organ in the body and is made up of three distinct layers: the epidermis, dermis, and the hypodermis (Yousef et al. 2020). It is the first line of defense against the external environment while also regulating the temperature of the body through vasoactive dermal vessels and acting as a barrier from water with the help of a cellular and lipid envelop (Yousef et al. 2020). The skin exhibits endocrine functions by the production of vitamin D through keratinocytes and exocrine functions with the use of sweat glands (Yousef et al. 2020). Specific areas of the body, including the hands and soles of the feet are significantly thicker and contain an extra layer of the epidermis called the statue lucid (Yousef et al. 2020). The epidermis, derived from the ectoderm, is further divided into five layers, which are stratum basale, stratum spinosum, stratum granulosum, stratum lucidum, and stratum corneum (Yousef et al. 2020). (Table 1)

Each layer of the epidermis contains specialized cells that contribute to the overall maintenance of the body (Yousef et al. 2020). The stratum basale contains keratinocytes that regulate calcium to assist with the production of vitamin D, produce keratin, and secrete lipids to create a water barrier (Yousef et al. 2020). Kertinocytes also have the ability to produce lamellar bodies transferred to the stratum granulosum that contain glycosphingolipids, phospholipids, and ceramides (Yousef et al. 2020). Melanocytes are also found in the stratum basale and contribute to the pigment of the skin and protection from UV rays with the production of melanin (Yousef et al. 2020). Merkel cells are bound to keratinocytes and contain mechanoreceptors for light sensory function in the hands, feet, mouth, and genital area (Yousef et al. 2020). Following the stratum basale is the stratum spinosum which contains dendritic Langerhans cells that maintain homeostasis and defend against foreign invaders (Clayton et al. 2017). They are part of the mononuclear phagocytic system and are involved with antigen presentation with MHC I and II molecules (Clayton et al. 2017). The immune response provided by Langerhans cells can be regulated by the cytokines produced by keratinocytes leading to the activation of T-cell components (Clayton et al. 2017).

Between the epidermis and the hypodermis lies the dermis, derived from the mesodermal tissue, which is composed of collagen and elastic tissue and divided into the papillary layer and the reticular layer (Brown and Krishnamurthy 2020).

Table 1 Layers of the epidermis

	Order from the bottom	Specialized cell/function	Cellular shape
Stratum basale	1 (deepest)	Merkel cells melanocytes, and keratinocytes	Cuboidal to columnar
Stratum spinosum	2	Dendritic cells	8–10 cell layers, irregular and polyhedral cells
Stratum granulosum	3	Keratohyalin and lamellar granules, and glycolipids	3–5 cell layers, diamond-shaped cells
Stratum lucidum	4	Eleidin	2–3 cell layers, thin clear layer
Stratum corneum	5 (uppermost)	Keratinocytes and defensins	20–30 cell layers, anucleate squamous cells

Source: Yousef et al. (2020)

The papillary layer is thinner and more vascular than the reticular layer, which is located below and contains dense connective tissue (Brown and Krishnamurthy 2020). The reticular layer contains bundles of type I and III collagen fibers allowing for elasticity and tensile strength (Brown and Krishnamurthy 2020). Langerhans cells are the primary phagocytic cell in the epidermis, but the dermis contains dermal macrophages to maintain homeostasis (Brown and Krishnamurthy 2020). There are also two types of elastic fibers in the dermis for structural support; elaunin (horizontally arranged) and oxytalan (vertically arranged) (Brown and Krishnamurthy 2020). The dermis also contains many support cells including fibroblasts, Schwann cells, histocytes, adipocytes, mast cells, and stem cells, as well as the vasculature system, nerve endings, hair follicles, and glands (Brown and Krishnamurthy 2020).

The deepest layer of the skin is called the hypodermis, or subcutaneous fascia (Driskell et al. 2014). Also derived from the mesodermal tissue, it contains adipose tissue, hair follicles, sensory neurons, and blood vessels and is the thickest layer of the skin (Driskell et al. 2014). Adipose tissue helps with temperature regulation and can act as an insulator in the body (Driskell et al. 2014). It releases a hormone called leptin that alerts the body that it should stop eating (Driskell et al. 2014). Thickness of the hypodermis in areas around the body is dependent on gender, with males having a thicker hypodermis in their upper body and women having a thicker hypodermis in their lower body (Driskell et al. 2014). Subcutaneous injections absorb the drug at a slower rate than intravenous so medications for allergic reactions, vaccinations, and insulin may chose this pathway (Driskell et al. 2014).

The three layers of the skin contain nerves that contribute to the autonomic and somatic nervous system (Roosterman et al. 2006). Specialized receptors are present in the dermis that assist the body in sensing pain, touch, and temperature through the somatic sensory system (Roosterman et al. 2006). Merkel disks are present to sense discriminative touch throughout the body while Ruffini endings sense warmth, skin stretch, and deformations within joints (Roosterman et al. 2006). Meissner's corpuscles and Pacinian corpuscles are very similar as they both sense touch and vibration, but Pacinian corpuscles sense deep pressure and high-frequency vibrations and Meissner's corpuscles sense light touch and low-frequency vibrations (Roosterman et al. 2006). Nociceptors in the skin are responsible for sensing pain (Roosterman et al. 2006). Sweating, pilomotor stimulation, and the vascular system all fall under the autonomic sensory system as they are unconsciously controlled (Roosterman et al. 2006). The pilomotor stimulation is controlled by arrector pili muscles that are connected to hair follicles (Roosterman et al. 2006). When stimulated, they pull on the hair follicle and cause the erection of hair (Roosterman et al. 2006).

3 Wound Macrophages

Wound healing undergoes four distinct phases to prevent bleeding out and repair the tissue; hemostasis, inflammation, proliferation, and reprogramming/remodeling (Minutti et al. 2017). Macrophages are recruited during the inflammation stage and present during proliferation to remove apoptotic cell debris allowing the proliferation of new cells to restructure the wound (Minutti et al. 2017). At the initial instance of a wound, calcium and hydrogen peroxide are produced to signal immune cells to be recruited to a location (Minutti et al. 2017). Inflammation is initiated by the high-mobility group Box 1 protein (HMGB-1) that is released by hydrogen peroxide and is considered a danger-associated molecular pattern along with ATP (Minutti et al. 2017). The G protein-coupled receptor P2Y and ligand-gated ion channel P2X both recognize ATP to initiate an inflammatory response (Minutti et al. 2017). During the initiate stages, macrophages also are activated by the release of cytokines IL-1α and IL-33 (Minutti et al. 2017). Hemostasis, the first stage in the wound healing process, involves forming a blood clot with the help of platelets to stop the flow of blood at that location (Minutti et al. 2017). As a blood clot begins to form, macrophages will recruit additional leukocytes and monocytes to transition into the inflammatory stage of wound healing (Minutti et al. 2017). Those monocytes will differentiate into macrophages to add to the population of resident macrophages (Minutti et al. 2017).

Monocytes recruited express two types of chemokine receptor, CCR-2 and CX3CR-1, with CCR-2 being in the first wave of recruitment and CX3CR-1 being in the second wave (Minutti et al. 2017). Specific research has been done to examine the role of the CCR-2 monocyte phenotype during wound healing and what specific structures they affect (Willenborg et al. 2012). The study included the use of monocytes in mice lacking the myeloid cell-restricted CCR-2 signaling and CCR-2/GFP reporter mice that expressed CCR-2 (Willenborg et al. 2012). Results showed that CCR-2-deficient mice showed significantly less macrophages in the initial stages of wound healing and there was reduced blood vessel formation (Willenborg et al. 2012). This suggests that CCR-2 monocytes may aid in angiogenesis and the defense against foreign pathogens (Willenborg et al. 2012). The second wave of recruited monocytes in the later phases of wound healing includes the CX3CR-1 monocytes (Minutti et al. 2017). A deficiency in expressing CX3CR-1 in these monocytes can lead to issues with wound healing and scar formation, but may have no effect on the formation of blood vessels (Willenborg et al. 2012).

Neutrophils and mast cells are also recruited to the wound during the initial phase, with neutrophils being the most abundant and the first initial cells to appear (Koh and DiPietro 2011). As macrophages release reactive oxygen species and proteases during the initial stages of inflammation, there is destruction of normal tissues and components of the extracellular matrix that can lead to scarring and loss of function (Koh and DiPietro 2011; Ruytinx et al. 2018). M2 macrophages are not only responsible for the regulation of fibrosis and the proliferation/remodeling stage of wound healing but also secrete chemokines along with M1 macrophages (Ruytinx et al. 2018). Chemokines are proteins that play a role in cell migration and can be divided into four groups that bind to receptors; C, CC, CXC, and CX3C (Ruytinx et al. 2018). The presence of CCL2 on monocytes binds to CCR2 creating a signaling cascade for them to migrate, while CCR4 plays a role in the attraction of natural killer cells along with other immunity cells (Ruytinx et al. 2018).

4 Role of Macrophages in Inflammation and Tissue Formation

Once the blood clot is formed, leukocytes are recruited, and the new extracellular matrix is formed the proliferation stage is initiated (Minutti et al. 2017). The recruitment on cells is suppressed and macrophages start the production and release of growth factors (Minutti et al.

2017). Platelet-derived growth factors are produced during wound healing to initiate proliferation and stimulate fibroblasts (Pierce et al. 1991). They can speed up wound closure and increase the overall strength to decrease the chance of breakage (Pierce et al. 1991). TGF-β1 is involved in the initiation of inflammation and stimulation of angiogenesis (Penn et al. 2012). Two other TGF-β isoforms are combined with TGF-β1 to create a signaling pathway to regulate the production and activity of fibroblasts (Penn et al. 2012). Insulin-like growth factor-1 enhances protein production and proliferation of cells during tissue repair (Talebpour Amiri et al. 2014). Macrophages are the main source of vascular endothelial growth factor (VEGF), which is vital for repair (Bao et al. 2009). During the process of angiogenesis, capillaries are formed by branching off existing blood vessels to supply oxygen and nutrients to the new tissue (Bao et al. 2009). VEGF has the ability to promote vasodilation, increase vascular permeability, and mediate platelet adhesion (Bao et al. 2009). When platelet adhesion is initiated, VEGF also will lead to the production of thrombin and fibrin (Bao et al. 2009).

These growth factors assists with the proliferation of cells and are regulated by the extracellular matrix which contains heparan sulfate (Minutti et al. 2017). Heparan sulfate is a linear polysaccharide chain with a glucopyranose residue and is considered a glycosaminoglycan that has the ability to bind with many growth factors and cytokines (Olczyk et al. 2015). It can also bind to other macromolecules to assists the binding of other cells to the extracellular matrix (Olczyk et al. 2015). During the initial stages of wound healing it can act as a procoagulant and assists in the recruitment of inflammatory cells (Olczyk et al. 2015). They also can regulate the cytokine release from macrophages by acting as a sensor of injury leading to a proinflammatory response (Olczyk et al. 2015). As the body transitions to the reprogramming/remodeling stage, it binds to heparin binding growth factor to initiate proliferation of the cells, differentiation, and angiogenesis to repair the tissue (Olczyk et al. 2015).

Macrophages exhibit a specialized phagocytic function that allows them to neutralize and digest apoptotic neutrophils, clearing the way for new tissue formation (Minutti et al. 2017). This leads to the release of growth factors and TGF-beta as an antiinflammatory and cell proliferation mediator (Minutti et al. 2017). Reduction in inflammation can also be initiated by the binding of interleukin-1 receptor antagonist (IL-1Ra) to inhibit the proinflammatory effects of IL-1α and IL-1β that are released at the initial stages of wound healing (Minutti et al. 2017). The release of IL-1Ra inhibits the expression of chemokines and other factors responsible for the recruitment of neutrophils (Minutti et al. 2017). This activates the protein RELMα in order to initiate the production of lysyl hydroxylase 2 (LH2), which organizes collagen in cross-links for stability (Minutti et al. 2017). Amphiregulin is a growth factor that not only suppresses inflammation but also aids in the differentiation of cells and activation of T-cell function (Minutti et al. 2017). Arginase is responsible for the production of ornithine that is the precursor for many pathways leading to the generation of collagen and polyamides while also activating T cells (Minutti et al. 2017). Macrophages are responsible for organization of the new tissue and collagen, while fibroblasts are responsible for the production of cells (Minutti et al. 2017).

M(IL-4) macrophages are alternatively activated and found primarily in the later stages of wound healing during formation of new tissue (Caley et al. 2015). They are involved in regulating the major functions during this stage including proliferation, cross-linking, establishing an antiinflammatory environment, and collagen production (Caley et al. 2015). Mesenchymal stem cells have the ability to polarize M(IL-4) macrophages from the proinflammatory state to an antiinflammatory state, which creates a more suited environment for cell proliferation and tissue repair (Caley et al. 2015). This feedback loop assists in the regulation of the tissue regeneration in wound healing (Caley et al. 2015).

Myofibroblasts are formed to initiate the closure of the wound and form the extracellular matrix (Wynn and Vannella 2016). In the later stages of tissue repair, antiinflammatory macrophages become the dominant immune cells that release growth factors and cytokines to decrease inflammation throughout the body, allowing for proliferation (Wynn and Vannella 2016). Studies have shown that embryonic-derived macrophages are considered the resident tissue macrophages that play a larger role in the proliferation of cells and angiogenesis (Wynn and Vannella 2016). When bone marrow macrophages are recruited to the location of a wound, they lead to a breakdown of the tissue while the resident macrophages lead to tissue repair (Wynn and Vannella 2016). Recruited monocytes may also have the ability to act as reparative or inflammatory by specific mediators, in contrast to a second wave of recruited monocytes differentiating for a specific response (Wynn and Vannella 2016). During CNS remyelination and demyelination, microglia can begin as the M1 phenotype to promote inflammation and then convert into the M2 phenotype with other inflammatory macrophages to start the reparative process (Wynn and Vannella 2016). This supports the idea that both resident and recruited macrophages play multiple roles in internal and external repair throughout multiple areas in the body (Wynn and Vannella 2016).

Transitioning from M1 to M2 macrophage activation is the signaling factor in the switch from the inflammation stage to the proliferation stage (Kotwal and Chien 2017). Antiinflammatory cytokines are released including IL-4 and IL-13 along with glucocorticoids and prostaglandins to polarize macrophages toward the alternate activation (Kotwal and Chien 2017). The rate of the reaction may be increased by the introduction of ATP nanoliposomes and growth factors (Kotwal and Chien 2017). Treatment with ATP nanoliposomes leads to a chemokine increase and significant decrease in IL-10, which triggers the shift from M1 to M2 macrophages (Kotwal and Chien 2017). Posttranslational modifications of histones regulate the polarization between M1 and M2 macrophages as they are controlled by the proteins affecting the overall inflammatory state (Kotwal and Chien 2017). Reversal of the M2 macrophage to the M1 phase can be controlled by mitochondria (Kloc et al. 2019). Studies have shown that M1 macrophages have a more circular shape than the longer M2 macrophages due to the location of the mitochondria (Kloc et al. 2019). Disruption of the macrophage activity leads to the increase in the reactive oxygen species and signaling pathways as the mitochondria is relocated to an area of the cell where more energy is needed (Kloc et al. 2019). This action may reverse the polarization of M2 macrophages to overproduce oxygen species and increase inflammation resulting in the M1 polarization (Kloc et al. 2019).

5 Role of Macrophages in Tissue Reorganization

As the tissue regeneration concludes, the body is responsible for removing the excess components of the extracellular matrix and reorganizing the tissue (Xue and Jackson 2015a). This process can take weeks or months to complete and the tissue may never regain the strength of the previous tissue (Xue and Jackson 2015a). Fibroblasts are the specific cells that produce the structural protein, collagen, which play a major role in tissue repair (Xue and Jackson 2015a). Having 28 different subsets, they are responsible for maintaining skin structure, cell adhesion, and tissue scaffolding as well as many tissue repair functions (Xue and Jackson 2015a). They are produced as precursors that undergo a chain assembly and cross-link with the assistance of posttranslational factors to form a collagen fibril (Xue and Jackson 2015a). These collagen fibrils are then connected to the cell with the help of a protein called fibronectin (Xue and Jackson 2015a). Fibronectin also works with fibrin during blood clots to help seal the open wound and stop the bleeding (Xue and Jackson 2015a). Macrophages produce a profibrotic cytokine TGF-β1 that initiates fibrogenesis and may result from a heavy growth factor environment (Xue and Jackson 2015b). Mesenchymal repair happens when the wound is deeper than the epithelial tissue and can lead to scarring/stiffening of the organ tissue as the specific cells are replaced with connective tissue (Xue and Jackson 2015b). This can lead to suppression in organ function and prolonged inflammation (Xue and Jackson 2015b).

Macrophages can reduce the presence of scarring in these tissue by releasing and recruiting other factors, like prostaglandin E2, to remove excess extracellular matrix (Xue and Jackson 2015b). While TGF-β1 promotes fibrogenesis, TGF-β3 promotes fibrolysis and leads to healing with less scarring by suppressing the production of myofibroblasts (Xue and Jackson 2015b). If TGF-β1 is not available, IL-13 can initiate fibrosis as well by decreasing the activation of metalloproteinases (Xue and Jackson 2015b). If M2 macrophages are not suppressed, the overproduction of growth factors and activation of fibroblasts will lead to heavy scarring and insufficient repair (Xue and Jackson 2015b). When there is an unregulated amount of tissue formed during the scarring process, a keloid can form which is not harmful to overall health but can protrude out of the skin (Russell et al. 2010). There is currently no cure or treatment for this condition, but they can be surgically removed if the size increases (Russell et al. 2010). Individuals under 30 years old and from Asian or Latino descent may be at a higher risk to develop them (Russell et al. 2010).

Matrix metalloproteinases (MMPs) are enzymes produced by macrophages and other inflammatory cells to assist in the degradation of excess extracellular matrix proteins (Caley et al. 2015). They can be activated by a variety of growth factors, cytokines, and interferons and are primarily expressed as pro-MMP, the latent form (Caley et al. 2015). When activated by serine proteases, the zinc ion and prodomain PRCGVPD bond is broken and regulated by inhibitors and binding proteins (Caley et al. 2015). MMPs are responsible for regulating cell signaling, modifying receptors and proteins, and releasing cytokines (Caley et al. 2015). They are divided into eight different groups depending on their location and activity throughout the body (Caley et al. 2015). Gelatinases, comprised of MMP-2 and MMP-9, are involved in cell migration and keratinocytes (Caley et al. 2015). MMP-2 is linked to laminin-332 that when cleaved initiated cell migration and can act as a ligand while remodeling (Caley et al. 2015). MMP-9 is more involved in cell signaling and healing through wound closure in keratinocytes (Caley et al. 2015). They regulate proangiogenic cytokines and peptides, like VEGF and endostatin, further increasing the rate of angiogenesis in the wound healing process (Caley et al. 2015).

As macrophages are responsible for removing excess cells during the wound healing process, they also help with the reorganization and growth of hair follicles as well (Krzyszczyk et al. 2018). The removal of collagen clears up space and allows for the activation of hair follicle stem cells regulating hair growth (Krzyszczyk et al. 2018). Apoptosis signal-regulating kinase 1 is responsible for the recruitment of macrophages to the hair follicle, and without this enzyme there is a decrease in overall hair growth (Krzyszczyk et al. 2018). After an injury, the resulting tissue does not contain hair follicles and will be unable to grow hair like the palms of hands and the soles of feet (Krzyszczyk et al. 2018).

6 Chronic Wounds and Tumors

As the rate of obesity and diabetes rise in the United States, the incidence of chronic wounds has increased almost 10% in the previous few years (Krzyszczyk et al. 2018). Specific conditions including ulcers fall under the category of chronic wounds and tend to stay stagnant in the inflammation phase of wound healing for an extended period of time (Krzyszczyk et al. 2018). Macrophages around this area are primarily halted in the M1 polarization to induce the inflammation (Krzyszczyk et al. 2018). Because the wound is unable to enter the proliferation and remodeling stages, the wound is unable to fully heal which can lead to life-threatening conditions (Krzyszczyk et al. 2018). One of the major factors in the inability of a wound to heal is the lack of blood supply, called ischemia, to an area along with infection of damage to the nerves (Kotwal and Chien 2017). This lack of cellular energy and oxygen supply prevents the proliferation of cells, production of collagen, and cellular migration (Kotwal and Chien 2017).

When macrophages are held in an inflammatory state, there is no proper regulation leading these macrophages to become less effective (Kotwal and Chien 2017). Apoptotic cells will not be cleared as efficiently and it will cause a buildup of these nonfunctional immune cells increasing the inflammation at the site of the wound (Kotwal and Chien 2017). This is very common in diabetic patients as they may suffer from hyperglycemia, which is a condition where there are very high levels of glucose in the blood (Kotwal and Chien 2017). Studies have shown that there is a significantly higher amount of M1 macrophages surrounding a diabetic ulcers than the average nonchronic wound (Kotwal and Chien 2017). Other levels of immune cells may be affected including neutrophils, T cells, and B cells (Kotwal and Chien 2017). MMPs may be overproduced at this stage as well leading to a decrease in growth factors and proteins preventing the proliferation of cells (Kotwal and Chien 2017). Commonly in leg ulcers due to disfunction in venous valves, the level of intracellular storage of iron is significantly higher due to the fact that M1 macrophages store the iron while M2 macrophages release it (Kotwal and Chien 2017).

Studies regarding methods of treating chronic wounds are centered on either removal and reduction in M1 macrophages or the introduction of M2 macrophages to the wound (Kotwal and Chien 2017). Goren and colleges conducted a study on the effects of anti-TNF-α and anti-F4/80 antibodies on wounds with delayed healing with the use of diabetic mice (Goren et al. 2007). The antibodies were injected at day 7 of the wound healing process to mimic the timeline of an acute wound around late-stage inflammation (Goren et al. 2007). This resulted in a decrease in macrophages and an increase in wound healing compared to mice without the antibodies (Goren et al. 2007). Compared to the control group without antibodies, the mice also showed a decrease in TNF-α, IL-1β, and CCL2(MCP-1) protein (Goren et al. 2007).

A study done by Zuloff-Shani and colleges researched the effects of injecting ulcers with hypoosmotic shock-treated macrophages (Zuloff-Shani et al. 2010). This method increased the amount of wound healing genes significantly with an increase in IL-1 protein synthesis up to 123- and 175-fold (Zuloff-Shani et al. 2010). Overall healing increased by 69% compared to the control group without the injection which only increased 13.3% (Zuloff-Shani et al. 2010). Not only did the treatment show improved wound healing, the wounds treated with the hypoosmotic shock-treated macrophages healed at a much faster rate (Zuloff-Shani et al. 2010).

Other therapies that have been studied include the injection of mesenchymal stromal cells, growth factors, and oxygen therapy (Kotwal and Chien 2017). Mesenchymal stromal cells can be injected into the bloodstream and when they come in contact with macrophages it causes them to exhibit M2 activity (Kotwal and Chien 2017). They show increase in recruitment of macrophages and release prostaglandin E-2 to increase the amount of IL-10 (Kotwal and Chien 2017). Growth factors including PDGF-BB and GM-CSF can be used to elicit a M2-like response from macrophages (Kotwal and Chien 2017). PDGF-BB recruits macrophages/monocytes not at the surface to the location of the wound and increases collagen synthesis (Kotwal and Chien 2017). GM-CSF usually are found to create a proinflammatory environment, but in chronic wounds they are shown to increase the amount of VEGF leading to vascularization and signaling to other cells (Kotwal and Chien 2017). Wound healing may also be encouraged with constant exposure to oxygen with the help of the hyperbaric chambers (Kotwal and Chien 2017). This method does decrease the amount of macrophages, but only with a short-term exposure (Kotwal and Chien 2017).

A tumor is the result of an overgrown amount of cells that form a mass which can be cancerous or not (Hua and Bergers 2019). Although these cells do not heal, they are not considered chronic wounds (Hua and Bergers 2019). Unlike chronic wounds, tumors skip the inflammation stage of wound healing and go right to the proliferation stage by producing immunosuppressive cells (Hua and Bergers 2019). Tumors undergo the remodeling stages like chronic wounds by the development of blood vessels and connective tissue, but the production is significantly faster and more effective than the breakdown (Hua and Bergers 2019). Treatment methods have been directed toward antiangiogenic therapy in attempts to limit the ongoing growth, but the tumor resorts to other methods of growth in a short period of time (Hua and Bergers 2019).

7 Factors Affecting Wound Healing

Many factors can affect the overall process of healing whether it be local or systematic factors (Guo and DiPietro 2010). Local factors include oxygen, infections, and foreign bodies and are factors that directly impact the wound (Guo and DiPietro 2010). Oxygen helps prevent infections and promotes collagen synthesis (Guo and DiPietro 2010). If there is a lack of oxygen in the body, it can create a hypoxic wound leading to impaired healing (Guo and DiPietro 2010). Superoxide and hydrogen peroxide initiate cytokines and angiogenesis, but an excess cause tissue damage (Guo and DiPietro 2010). Infection by foreign microorganisms can hinder the wound healing process and lead to more invasive issue (Guo and DiPietro 2010). The inflammation process is important for the removal of foreign pathogens that have entered the wound causing an increase in cytokines (Guo and DiPietro 2010). If this inflammation time is ongoing for an extended period of time, proteases can degrade the extracellular matrix and growth factors (Guo and DiPietro 2010). An excess of bacteria can develop into a biofilm and create an immunity to antibiotics (Guo and DiPietro 2010).

Factors including age, stress, weight, and overall health fall under the category of systematic factors (Guo and DiPietro 2010). As individuals age, the risk of delays with wound healing increases due to growth factor decrease and issues with collagen development (Guo and DiPietro 2010). Exercise may aid in decreasing this risk by creating antiinflammatory responses (Guo and DiPietro 2010). Sex hormones can also have an effect as males tend to have delayed reactions to wound healing due to a lack of estrogen which can improve wound healing with age (Guo and DiPietro 2010). When individuals are more stressed, the sympathetic nervous system is not properly regulated causing many diseases throughout the body including delayed wound healing and cardiovascular issues (Guo and DiPietro 2010). In higher stress situations, the amount of hormones released increases throughout the body, including glucocorticoids, which can cause a decrease in the amount of proinflammatory cytokines leading to a delayed response in the initial phases of wound healing (Guo and DiPietro 2010).

Individuals with health concerns including diabetes can have additional delays or complications with wound healing (Villines 2019). When they are unable to control their blood sugar, the amount of glucose in the blood may decrease the effectivity of white blood cells (Villines 2019). This can also lead to a slower circulation, taking more time for a wound to heal and nutrients to be delivered to the damaged tissue (Villines 2019). An increase in glucose for an extended period of time may cause nerve damage, reduced angiogenesis, and impaired collagen production (Villines 2019). A wound that doesn't heal properly can lead to an infection that could spread throughout the body causing osteomyelitis, gangrene, or sepsis (Villines 2019). Proper glucose control, hand and feet washing, and monitoring wounds can help prevent or lessen these complications (Villines 2019).

Obesity affects over a third of the American population and may lead to serious complications with wound healing (Pence and Woods 2014). As there is an increase in adipose tissue, the circulatory system has to increase its workload to maintain homeostasis, leading to poor vascularity (Pence and Woods 2014). Vascularity complications including decreased oxygenation, impaired angiogenesis, delayed inflammation, and tissue death may arise, preventing wounds to heal in the body (Pence and Woods 2014). Adipose tissue produces a cytokine called adiponectin that will stimulate angiogenesis and proliferation, but when the amount of adipose tissue increases there is a significant decrease in the production of adiponectin (Pence and Woods 2014). Reepithelialization will not be properly completed and production of blood vessels will be impaired (Pence and Woods 2014). Capillaries under this stress cannot properly transport oxygen throughout the body and, with low oxygen levels, leukocytes are unable to kill bacteria (Pence and Woods 2014). Fibroblasts will underproduce collagen and tensile strength will not be sufficient enough to efficiently heal the wound (Pence and Woods 2014). Nutrition overall does affect wound healing as a deficiency in any vitamins, nutrients, and fatty acids will significantly impair the body's ability to heal (Guo and DiPietro 2010).

Factors like smoking and alcohol consumption may impair the wound healing process by increasing susceptibility to infection and decreasing the immune system of the individual (Guo and DiPietro 2010). Studies show that short-term exposure to alcohol will decrease neutrophil and proinflammatory cytokine recruitment during wound healing (Guo and DiPietro 2010). Alcohol consumption will reduce the process of angiogenesis up to 61%, leading to complications in the later stages of wound healing (Guo and DiPietro 2010). This will decrease the protease balance and production of collagen as well (Guo and DiPietro 2010). Smoking can cause infections and raise the risk of the wound leaking, affecting the oral area the most (Guo and DiPietro 2010). Nicotine is one of the major substances in cigarettes, with there being over 4000, and can negatively affect the body by decreasing blood flow, increasing blood viscosity, increase vasoconstriction, and decrease oxygen supply (Guo and DiPietro 2010). Other components include carbon monoxide and hydrogen cyanide that impair the binding and metabolism of the oxygen in the bloodstream (Guo and DiPietro 2010). (Table 2)

Table 2 Factors affecting wound healing

Local factors	Systematic factors
Oxygen	Age
Infections	Gender and hormones
Foreign bodies	Stress
	Weight and nutrition
	Alcohol and drug use
	Additional diseases

Source: Guo and DiPietro (2010)

8 Conclusion

The body undergoes four phases of healing, which are hemostasis, inflammation, proliferation, and reprogramming/remodeling, when exposed to a wound. Hemostasis is the process of preventing the blood flow at a wound by the help of platelets forming blood clots. The inflammatory phase is then initiated through the recruitment of macrophages with cytokines and other immune factors. They are responsible for the clearing of cells, releasing cytokines, cellular proliferation, tissue remodeling, and extracellular matrix growth. Macrophages may be polarized toward either M1, proinflammatory cells, or M2, antiinflammatory cells. M1 macrophages are present during the initial stages of wound healing, while M2 macrophages are present at later stages for tissue regeneration and remodeling. ROS are released to signal macrophages to initiate phagocytic activity and fibroblasts are activated for cell proliferation and collagen synthesis. The proliferation phase includes growth of blood vessels, granulation tissue formation, and production of the extracellular matrix, while the wound contracts with the help of micro fibroblasts. The skin is then remodeled with the cross-linking of collagen and removal of excess extracellular matrix.

A wound may be considered chronic when there is an inability to heal or it is severely delayed. Ulcers are one of the most common form of chronic wounds, where the wound is halted in the inflammatory stages and cells are unable to proliferate. Macrophages are primarily M1 polarized and may lead to destruction of the tissue or infection. Treatment for these wounds focuses on the removal of the M1 macrophages and the addition of or differentiation of M2 macrophages with mesenchymal stromal cells, growth factors, and oxygen therapy.

There are many factors affecting wound healing that can be divided into local or systematic. Local factors are outside influences that impact the wound and may act quickly, while systematic factors are related to overall lifestyle. Oxygen, infections, and foreign bodies are all local factors that may decrease the wound healing ability if not properly regulated. Systematic factors are age, stress, weight, alcohol consumption, and overall health. Wound healing decreases with age due to growth factor and collagen decrease, although men are more prone to complications with healing. As stress increases, hormones may be released which delay the initial inflammatory stages. Exercise level and weight can significantly impact an individuals' ability to heal as exercise can create an antiinflammatory response. As there is an increase in adipose tissue, there is a decrease in vascularity leading wounds to heal slower in the body. Other health concerns including diabetes may decrease the effectivity of white blood cells and overall healing process.

9 Future Trends

Recent studies have focused on further research regarding the presence of M2 macrophages in early stages of wound healing and the ability of M1 macrophages to polarize to M2 macrophages at different stages (Ferrante and Leibovich 2012). This understanding could aid in the development of drugs targeting infections and disease (Ferrante and Leibovich 2012). There is evidence as well that the profibrotic and pro-wound healing phenotypes may not just fit into a subset of proinflammatory or antiinflammatory (Caley et al. 2015). Research is being done comparing the surface markers and activity in order to create a more concise description in order to study their manipulation potential in future clinical trials (Caley et al. 2015). Ongoing research is focused on the treatment of chronic wounds focusing on the removal of M1 macrophages, addition of M2 macrophages, and injections with growth factors as well (Kotwal and Chien 2017).

References

Gale AJ. Continuing education course #2: current understanding of hemostasis. Toxicol Pathol. 2011;39(1):273–80.

Varga-Szabo D, Pleines I, Nieswandt B. Cell adhesion mechanisms in platelets. Arterioscler Thromb Vasc Biol. 2008;28:403–12.

Guo S, DiPietro LA. Factors affecting wound healing. J Dent Res. 2010;89(3):219–29.

Wu YS, Nan FH, Chen S, Chen SN. Polarization of macrophage and immunotherapy in the wound healing. Wound Healing – New insights into Ancient Challenges 2016; 8: 167–192.

Ferrante CJ, Leibovich SJ. Regulation of macrophage polarization and wound healing. Adv Wound Care (New Rochelle). 2012;1(1):10–6.

Reinke JM, Sorg H. Wound repair and regeneration. Eur Surg Res. 2012;49:35–43.

Yousef H, Alhajj M, Sharma S. Anatomy, skin (integument), epidermis. StatPearls. 2020.

Clayton K, Vallejo AF, Davies J, Sirvent S, Polak ME. Langerhans cells—programmed by the epidermis. Front Immunol. 2017;8:1676.

Brown TM, Krishnamurthy K. Histology, Dermis. StatPearls. 2020.

Driskell RR, Jahoda CA, Chuong CM, Watt FM, Horsley V. Defining dermal adipose tissue. Exp Dermatol. 2014;23(9):629–31.

Roosterman D, Goerge T, Schneider SW, Bunnett NW, Steinhoff M. Neuronal control of skin function: the skin as a Neuroimmunoendocrine organ. Physiol Rev. 2006;86(4):1309–79.

Minutti CM, Knipper JA, Allen JE, Zaiss DMW. Tissue-specific contribution of macrophages to wound healing. Semin Cell Dev Biol. 2017;61:3–11.

Willenborg S, Lucas T, van Loo G, Knipper JA, et al. CCR2 recruits an inflammatory macrophage subpopulation critical for angiogenesis in tissue repair. Blood. 2012;120(3):613–25.

Koh TJ, DiPietro LA. Inflammation and wound healing: the role of the macrophage. Expert Review. 2011;13(23):1–9.

Ruytinx P, Proost P, Van Damme J, Struyf S. Chemokine-induced macrophage polarization in inflammatory conditions. Front Immunol. 2018;9:1930.

Pierce GF, Mustoe TA, Altrock BW, Deuel TF, Thomason A. Role of platelet-derived growth factor in wound healing. J Cell Biochem. 1991;45(4):319–26.

Penn JW, Grobbelaar AO, Rolfe KJ. The role of the TGF-β family in wound healing, burns and scarring: a review. Int J Burns & Trauma. 2012;2(1):18–28.

Talebpour Amiri F, Fadaei Fathabadi F, Mahmoudi Rad M, et al. The effects of insulin-like growth factor-1 gene therapy and cell transplantation on rat acute wound model. Iran Red Crescent Med J. 2014;16(10):e16323.

Bao P, Kodra A, Tomic-Canic M, et al. The role of vascular endothelial growth factor in wound healing. J Surg Res. 2009;153(2):347–58.

Olczyk P, Mencner L, Komosinska-Vassev K. Diverse roles of Heparan Sulfate and heparin in wound repair. Biomed Res Int. 2015;2015(549417):7.

Caley MP, Martins VL, O'Toole EA. Metalloproteinases and wound healing. Adv Wound Care (New Rochelle). 2015;4(4):225–34.

Wynn TA, Vannella KM. Macrophages in tissue repair, regeneration, and fibrosis. Immunity. 2016;44(3):450–62.

Kotwal GJ, Chien S. Macrophage differentiation in Normal and accelerated wound healing. Results Probl Cell Differ. 2017;62:353–64.

Kloc M, Ghobrial RM, Wosik J, Lewicka A, et al. Macrophage functions in wound healing. J Tissue Eng Regen Med. 2019;13(1):99–109.

Xue M, Jackson CJ. Extracellular matrix reorganization during wound healing and its impact on abnormal scarring. Adv Wound Care (New Rochelle). 2015a;4(3):119–36.

Xue M, Jackson CJ. Extracellular matrix reorganization during wound healing and its impact on abnormal scarring. Adv Wound Care (New Rochelle). 2015b;4(3):119–36.

Russell SB, Russell JD, Trupin KM, Gayden AE, et al. Epigenetically altered wound healing in keloid fibroblasts. J Investig Dermatol. 2010;130(10):2489–96.

Krzyszczyk P, Schloss R, Palmer A, Berthiaume F. The role of macrophages in acute and chronic wound healing and interventions to promote pro-wound healing phenotypes. Front Physiol. 2018;9:419.

Goren I, Müller E, Schiefelbein D, Christen U, et al. Systemic anti-TNFalpha treatment restores diabetes-impaired skin repair in ob/ob mice by inactivation of macrophages. J Investig Dermatol. 2007;127(9):2259–67.

Zuloff-Shani A, Adunsky A, Even-Zahav A, Semo H, et al. Hard to heal pressure ulcers (stage III-IV): efficacy of injected activated macrophage suspension (AMS) as compared with standard of care (SOC) treatment controlled trial. Arch Gerontol Geriatr. 2010;51(3):268–72.

Hua Y, Bergers G. Tumors vs. chronic wounds: an immune cell's perspective. Front Immunol. 2019;10:2178.

Villines Z. How does diabetes affect wound healing? MedicalNewsToday. 2019.

Pence BD, Woods JA. Exercise, obesity, and cutaneous wound healing: evidence from rodent and human studies. Adv Wound Care (New Rochelle). 2014;3(1):71–9.

Effects of Mycotoxins on Macrophages and Their Possible Clinical Implications

David Cleary

Abstract

Fungi and their mycotoxins are ubiquitous, contaminating the air we breathe and the food we eat on a daily basis. Mycotoxins are considered a natural and unavoidable contaminant of food crops. Anthropogenic climate change is anticipated to increase mycotoxin contamination. Although mycotoxins are too small to be seen by the naked eye, uncontrolled mycotoxins can cause gangrene, miscarriage, cancer, and death. Mycotoxins in low amounts stimulate an immune response from macrophages. Mycotoxins in high amounts shut down the immune response preventing an adequate defence. Fumigaclavine C reduces toll-like receptor 4 expression and tumor necrosis factor alpha. Aflatoxins are carcinogenic and have been linked to tumor formation in the colon, kidneys, liver, and lungs. Aflatoxin B_1 is the strongest carcinogen of all the aflatoxins. Aflatoxin B_1 is removed by macrophage autophagy and by extracellular traps. Ochratoxin A has been found to be nephrotoxic, teratogenic, carcinogenic, and immunotoxic. Ochratoxin A acts as a competitive inhibitor for phenylalanine-tRNA synthetase which halts protein synthesis. Ochratoxin A's effects can be blocked by phenylalanine. Trichothecene mycotoxins are lethal enough to be utilized as biological chemical warfare agents, and they have been. Mycotoxins act synergistically with other mycotoxins, bacterial endotoxin, and virus causing a greater adverse effect than the sum of their individual effects. Synergy helps to explain the observed effects of damp building-related illness. Controlled amounts of specific mycotoxins can be clinically useful. Ergometrine has been used clinically to stimulate uterine contractions in order to induce childbirth. After child delivery, ergometrine can be used to help control bleeding via its vasoconstrictive action. Ergotamine has been used clinically to reduce pain associated with migraine headaches. The immunosuppressant activity of mycophenolic acid arises from its ability to reversibly inhibit the enzyme inosine 5′-monophosphate dehydrogenase thereby shutting down the biochemical *de novo* pathway which synthesizes guanine nucleotides. The immunosuppressive effects of mycophenolic acid have found utility in treating autoimmune diseases of rheu-

D. Cleary (✉)
Sullivan University, College of Pharmacy and Health Sciences, Louisville, Kentucky, USA
e-mail: dcleary@sullivan.edu

S. Gupta, Y. V. Pathak (eds.), *Macrophage Targeted Delivery Systems*,
https://doi.org/10.1007/978-3-030-84164-5_23

matoid arthritis and lupus nephritis. Mycophenolic acid helps prevent organ transplant rejection for kidney, heart, and liver recipients.

Keywords

Mycotoxin · Macrophage · Ergot · Aflatoxin · Trichothecene · Mycophenolic acid · Ochratoxin · Damp building

1 Introduction

The term mycotoxin refers generally to any toxic substance produced by a fungus. This is a very large and ever-expanding group of substances. Two terms often mistakenly interchanged are mold and mycotoxin. Although related, a clear understanding of the difference between mold and mycotoxin is required to appreciate the difficulty in preventing and reversing toxic effects. Mold is a subset of fungi and is characterized by hyphal growth. Mold is a living organism and thus can be killed. Mycotoxins are chemicals biosynthesized by fungi. They are not living. Once mycotoxins have been biosynthesized and have contaminated items such as food, buildings, air, clothing, etc., the items cannot be made nontoxic simply by killing the producing fungi.

Molds reproduce by producing spores. Spore size varies depending upon the type of mold, but their diameter is generally in the range from 2 microns to 100 microns. To put this into perspective, an average human hair cross section is around 50 microns. The average human eye cannot see anything smaller than 40 microns. Mycotoxins produced by molds are in the orders of magnitude smaller than mold spores. Mold spores are easily spread by the wind. Outdoor mold spores blow into crop fields contaminating farmland and food production. Air used for ventilation can bring mold spores into grain silos and buildings. Residential crawlspaces and attics are often ventilated using outdoor air. Trying to control mold spore movement is very much like trying to control the direction the wind blows. Of course, air movement is not the only method of travel utilized by mold spores. Mold spores can ride into buildings, on clothing, be tracked into buildings on the shoes of occupants, be carried by insects, carried by rodents, and more.

A mold spore requires more than just access in order to grow. Being a living organism, it requires organic nutrients for food, oxygen, and water. Molds are saprophytic; they feed off dead or decaying matter. In nature, molds help reclaim and degrade biological matter as part of the ecosystem. They are plentiful in the soil and compete with bacteria and other organisms for food. Molds do not discriminate between the wood of a dead tree on the forest floor and the wood of a house above the soil of a crawlspace. In their competition for food (primary metabolites), mold produces toxic chemicals designed to eliminate competitors vying for the same food. These toxic metabolites (secondary metabolites) are released into the environment to kill other organisms that would consume the mold's food.

Anthropogenic climate change is increasing both heat and humidity around the globe. Some examples of extreme heat include: 2003 European heat wave, 2010 Russian heat wave, 2012 and 2013 Australian heat wave, 2013 US Southwest heat wave, 2015 and 2016 Middle East heat wave, and 2017 European heat wave (Coffel et al. 2018). But climate change is more than just heat. Climate change brings extremes of weather conditions, including extreme cold, extreme rain, and extreme hurricanes. Extreme weather allows for increased incursions into inhabited building structures. This increases the environment for fungal growth. A building design may be alright for current weather conditions, but increased humidity, heat, wind, and flooding can increase water levels inside the building and support mold growth. Water to support mold growth can come from many sources. Water from flooding, roof leaks, plumbing leaks, leaking windows, and leaking doors are more obvious sources. Less obvious sources of water for mold can come in the form of air humidity condensing on cooler surfaces, such as windows, tile walls, air conditioner coils, and basement walls (United States Environmental Protection Agency Mold Course Chapter 1: Introduction to Molds n.d.). A damp building creates an environment that supports mold, bacteria, rodents, and insects. The United States Environmental Protection Agency recommends keeping indoor humidity below 60% to limit mold growth and preferably keeping the humidity between 30 and 50% (United States Environmental Protection Agency A Brief Guide to Mold, Moisture and Your Home n.d.). Heat generated from anthropogenic climate change increases the air's capacity to retain water vapor. When outdoor air enters a building and cools, its moisture carrying capacity is decreased. When the saturation limit is reached, water vapor will condense allowing liquid water to form which in turn can support mold growth.

Another mechanism by which heat generated from anthropogenic climate change increases mold growth is through farm crops. Barley, coffee, corn, oats, peanuts, rye, wheat, and other food staples are commonly grown in open fields exposed to weather elements as well as mold contamination. Under normal growing conditions, crop plants must defend themselves against mold, bacteria, and insects. Infections and infestations occur even under normal conditions; however, occurrences increase under heat stress. Extreme heat from climate change stresses plants making them more susceptible to mold infection. The plants' immune systems are weakened by heat stress (Desaint et al. 2021). Climate change is anticipated to increase mold and mycotoxin contamination of food crops. Out of all the harvested crops in the world, approximately 25% are contaminated with mycotoxins under current growing conditions (Alshannaq and Yu 2017). Increased mycotoxin contamination would be passed along the food chain to livestock in animal feed and to humans in grocery foods.

Mycotoxins pose a relevant health threat to humans. Over 300 mycotoxins have been identified (Alshannaq and Yu 2017). The chemical structures of mycotoxins are very diverse and so are the health hazards. Mycotoxins have been proven to have characteristics of being carcinogenic, embryotoxic, emetic, genotoxic, hemorrhagic, hepatotoxic, immunosuppressive, immunotoxic, mutagenic, neurotoxic, teratogenic, and ulcerative (Alshannaq and Yu 2017). Possessing these characteristics, mycotoxins have been linked to many human diseases, including cancer, Alzheimer's, nephropathy, lupus, rheumatoid arthritis, ergotism, aspergillosis, and stachybotryotoxicosis. The three most common entry routes for mycotoxins into humans are by inhalation, ingestion, and absorption through the skin. Farm workers handling grain crops are at risk for breathing air-borne mycotoxins as well as absorbing mycotoxins through their skin. A second group of people at risk for inhaling mycotoxins are people living in damp buildings. Aging inner city buildings are often damp and contaminated with mycotoxins (Barrett 2000). Inhalation of mycotoxins can collect the mycotoxins on the mucosal lining of the pharynx and end up being swallowed; thus, inhalation can become ingestion of mycotoxins. Ingestion of mycotoxins can occur by consuming contaminated crops directly but can also occur by consuming animals that have fed on contaminated feed. Domestic animals, such as chickens, cows, and pigs can concentrate mycotoxins from feed into their meat and milk. These products are then passed on to humans who ingest the animals.

Below is a table showing a few of the most common mycotoxins and the fungal genus that can produce each of them:

Genus	Toxin
Aspergillus	Aflatoxin B_1 Aflatoxin M_1 Ergot Gliotoxin Ochratoxin A Ochratoxin C
Chaetomium	Chaetoglobosin A
Claviceps	Ergot
Fusarium	Deoxynivalenol Enniatin B Fumonisin B_1 HT-2 T-2 Type A trichothecenes Type B trichothecenes Zearalenone
Penicillium	Mycophenolic Acid Ochratoxin A Patulin Sterigmatocystin
Stachybotrys	Roridin E trichothecene Satratoxin trichothecene Verrucarin A trichothecene

2 Ergot Toxin

Ergot refers to a collection of mycotoxins which, include bioderivatives of lysergic acid, clavine, and ergopeptine. Ergot mycotoxins include: agroclavine, ergotamine, ergometrine (ergonovine), ergocristine, ergosine, ergocryptine, ergocornine, ergovaline, ergonine, fumigaclavine C, lysergic acid amide (ergine), and others. Lysergic acid has sedative properties as well as effects on the autonomic nervous system, such as hypersalivation, emesis, dizziness, and diarrhea. Ergot toxins can be produced by the fungal species *Claviceps purpurea, Claviceps fusiformis, Aspergillus fumigatus, Neotyphodium coenophialum,* and more. Ergot outbreaks have been reported throughout history. An Assyrian tablet from 600 BC references a noxious pustule in the ear of grain (Haarmann et al. 2009). In 857 AD, an ergot outbreak in the Rhine Valley occurred. People affected felt a burning sensation in the extremities (arms, hands, legs, and feet), which was said to be a punishment from god and was called "Holy Fire" (Holstege and Traven 2014). Symptoms can range from tingling or burning sensations to gangrene of phalanges, limbs, and death. During the Middle Ages, the Holy Fire was sometimes relieved by a pilgrimage to the shrine of St. Anthony. Holy Fire became known as St. Anthony's Fire (Aronson 2014). The monks of St. Anthony's order treated the pilgrims with some success. The true cause of St. Anthony's Fire was not known at that time; by chance, the area where St. Anthony's shrine was located did not happen to be infected with ergot producing fungi. While pilgrims lived in the area, they consumed the local non-toxic foods. The immune system could sometimes clear enough toxins for relief while the continuation of ingested toxin was ceased.

In 1670, a French physician proposed that the ergot-contaminated grain was responsible for St. Anthony's Fire (Holstege and Traven 2014). After the discovery of ergot as the true cause of St. Anthony's Fire, the disease became known as ergotism and also became less common. Farmers are now able to be more watchful for sclerotia on grain crops. Ergot toxins are able to survive grinding into flour and baking into goods such as bread. Sclerotia are clusters of dark cylindrical fungal growths called sclerotium. Each sclerotium measures approximately 0.2 inches in diameter and 0.6 inches long (Holstege and Traven 2014). The sclerotia was said to resemble the shape of a rooster spur. The name ergot is of French origin and means cock's spur (Online Etymology Dictionary n.d.).

Although less often, outbreaks of ergotism still occur throughout the world. In Southern Russia during 1926 and 1927, a severe outbreak of ergotism occurred. Over 10,000 cases of convulsive ergotism were reported during this outbreak. The source of ergot toxin was traced to contaminated rye used to make bread (Belser-Ehrlich et al. 2013). Manchester, England, also had an ergotism outbreak in 1927. Over 200 cases of gangrenous ergotism were reported. The source of the toxin was traced to rye grown in South Yorkshire. An ergotism outbreak in Pont St. Esprit, France, affected 150 people in 1951. In southern Mumbai, India, an ergot outbreak affected over 23 people from 1956 to 1958. A second outbreak in western India occurred during 1976 affecting 78 people. Both of the Indian outbreaks were traced to ergot contaminated bajra (pearl millet) grain. Ethiopia had two large outbreaks during 1977–1978 and 2001–2002 located in the Wollo Administrative Region and in the Arsi Zone, respectively. These two outbreaks affected over 140 and 20 people, respectively. The source of ergot toxin was traced to wild oats in both of the Ethiopian outbreaks. Evidence from the Ethiopian outbreaks indicates that ergot toxin can accumulate over time in the human body. A low level of toxin-contaminated grain can slowly build to pathological levels by repeated consumption (Belser-Ehrlich et al. 2013).

Ergot poisoning in humans manifests itself in two main types: gangrenous ergotism and convulsive ergotism. Some ergot toxins cause extreme vasoconstriction in the extremities. Cutting off the blood flow to the hands and feet is responsible for paresthesia and the extreme burning sensation of ergotism. Continued vasoconstriction causes gangrene. If enough toxin is present, vasoconstriction and gangrene can progress throughout the entire limb. The gangrenous limbs can spontaneously fall off without any blood loss (Bardal et al. 2011). The gangrene can cause death. The same vasoconstriction which causes gangrene also causes blood flow through the placenta of pregnant women to be restricted. In addition to reduced blood flow, ergot toxin also causes the muscles of the uterus to contract. Restricted blood flow along with muscle contraction induces miscarriage. Convulsive ergotism is caused by ergot toxins that act as serotonergic agonists. Overstimulation of serotonin receptors can cause muscle spasms, tremors, convulsive seizures, delusions, hallucinations, fever, and coma. Ergotamine is an example of an ergot alkaloid that acts as a serotonin receptor agonist. The quantity of each ergot alkaloid is extremely variable changing depending upon parameters, such as the genus and species of the fungi, the type of host, the growing temperature, the growing moisture, and more. The toxic effects manifested are therefore also extremely variable. Fungal growth producing ergot toxin favors severely cold winters and cool, wet growing seasons.

Several ergot toxins have found clinical applications. Ergometrine has been used clinically to stimulate uterine contractions in order to induce childbirth. The mechanism of action for ergometrine involves alpha-adrenoreceptors, serotonin receptors, and dopamine receptors. After child delivery, ergometrine can be used to help control bleeding via its vasoconstrictive action. Ergotamine has been used clinically to reduce pain associated with migraine headaches. The mechanism of action is unclear but is believed to be due to the vasoconstrictive properties of the ergotamine. The ergot alkaloid lysergic acid amide is known to induce hallucinations. This ergot toxin led to the fully synthetic derivative lysergic acid diethylamide (LSD) which became a widely used recreational psychedelic.

Ergot mycotoxins have been found to suppress the immune response of human and murine macrophages (Du et al. 2011). Du et al. investigated the effects of the ergot mycotoxin fumigaclavine C on human macrophages and on RAW 264.7 murine-derived macrophages. Bacterial endotoxins in the form of lipopolysaccharides would normally elicit an inflammatory response from macrophages. Expression and activation of toll-like receptor 4 (TLR4) are normally increased which increases transcription of nuclear factor kappa-light-chain-enhancer of activated B cells (NFκB) in response to inflammatory lipopolysaccharides. Transcription of the pro-inflammatory cytokine tumor necrosis factor α (TNF-α) is also normally upregulated in response to lipopolysaccharides. Du et al. used quantitative real-time polymerase chain reaction to determine the mRNA expression of toll-like receptor 4. Western blot analysis was used to determine the actual protein level of toll-like receptor 4. When macrophages were primed with lipopolysaccharides, toll-like receptor 4 increased as expected; however, when fumigaclavine C was also present, toll-like receptor 4 overexpression was decreased. Enzyme-linked immunosorbent assays (ELISA) were used to follow the production of tumor necrosis factor α. In similar fashion, fumigaclavine C reduced the production of tumor necrosis factor α from lipopolysaccharide-primed macrophages. Even though the inflammatory agent lipopolysaccharide was present, the macrophages did not mount the appropriate immune response when the mycotoxin fumigaclavine C was present.

Fumigaclavine C is an ergot alkaloid produced by Claviceps species but is also produced by *Aspergillus fumigatus* and causes nausea, vomiting, delirium, and coma (Mitchell et al. 1997). The mold *Aspergillus fumigatus* is commonly found in the air and is capable of infecting human lungs forming aspergillosis. Aspergillosis can become systemic in immunocompromised individuals (Paulussen et al. 2017). *Aspergillus fumigatus* is capable of biosynthesizing many hyphal mycotoxins, such as gliotoxin, helvolic acid, fumagillin, fumigaclavine C, and aurasperone C. *Aspergillus fumigatus* is the most common cause of invasive aspergillosis, a frequently fatal lung disease. *Aspergillus fumigatus* gains an advantage over other spores infecting human lungs by a non-hyphal toxin which is diffusible from the spore surface (Mitchell et al. 1997). This diffusible, less than 10 kilodalton toxin can inhibit macrophages from moving toward the spore, chemotaxis, as well as inhibit macrophages from spreading (Robertson et al. 1987a). This inhibitory action is in contrast to spores from the nonpathogenic fungus *Penicillium ochrochloron* which did not inhibit macrophage spreading and chemotaxis. This diffusible *A. fumignatus* toxin was further shown to inhibit the formation of superoxide anions and hydrogen peroxide reactive oxidative species in phagocytotic cells, thus preventing the host's immune system from defending against the fungal infection (Robertson et al. 1987b). *A. fumignatus* was found to bind with the surface of phagocytic cells but was not phagocytosed (Robertson et al. 1987c). *A. fumignatus* is capable of biosynthesizing gliotoxin which can inhibit phagocytosis, but *A. fumignatus* must grow for approximately 3 days in order to reach the maturity level necessary for gliotoxin production. The mechanism protecting *A. fumignatus* from phagocytosis prior to gliotoxin production was identified as a diffusible toxin from the spore surface. It is this diffusible spore surface toxin that gives *A. fumignatus* a pathogenic advantage over other fungal species.

3 Aflatoxin

Aflatoxins are a group of mycotoxins biosynthesized by some members of the *Aspergillus* genus (Barrett 2000; An et al. 2017). Aflatoxins are found worldwide. They are a common food crop contaminant and are found very often in corn and peanuts. Aflatoxins are very toxic to the liver and kidneys; in fact aflatoxin B_1 is the strongest natural carcinogen being extremely hepatocarcinogenic (Squire 1981). Aflatoxins have been linked to tumor formation in the colon, kidney, liver, and lungs. Aflatoxins are such a significant human health hazard that not only has the United States Food and Drug Administration set maximum allowable aflatoxin limits for food commodities, but world commerce also regulates their maximum allowable limit.

Aflatoxin B_1 is the most potent carcinogen of the aflatoxins. This mycotoxin does not exhibit carcinogenic properties until it is activated by cytochrome p450s. Once activated, aflatoxin B_1 forms adducts with deoxyribonucleic acid, allowing mutations to occur (Barrett 2000). These mutations explain the extreme carcinogenicity of aflatoxin B_1. The effects of aflatoxin B_1 are compounded by the presence of other mycotoxins or virus. For example, aflatoxin B_1 along with the carcinogenic hepatitis B virus exhibits a synergistic carcinogenic effect (Kew 2003).

Yanan An et al. studied the effects of aflatoxin B_1 on macrophages using the BALB/c mouse-derived cell line RAW264.7 and the human-derived cell line THP-1 (An et al. 2017). Light-chain-3′ (LC3) is a protein that associates with autophagosome membranes. Red fluorescent protein (RFP) and green fluorescent protein (GFP) are markers that can be used to visualize their location within a cell. GFP is sensitive to pH and will be quenched in the acidic environment of a lysosome. A plasmid containing the tandem autophagosome reporter RFP-GFP-LC3 was transfected into RAW264.7 cells. Aflatoxin B_1 treated cells showed an increase in autophagosome formations and quenching of GFP, indicating lysosomal fusion.

Autophagy of ubiquitinated proteins was mediated by the adapter protein p62/sequestosome 1 (SQSTM1). Decreasing SQSTM1 can be followed as a reporter for the completion of autophagy. Western blotting was used to monitor SQSTM1 levels. In both RAW264.7 cells and THP-1 cells, SQSTM1 was seen to decrease in a dose- and time-dependent response in aflatoxin B_1 treated cells. The results indicate that aflatoxin B_1 is removed by macrophage cells via autophagy and lysosomal digestion.

In addition to phagocytosis, macrophages can utilize extracellular traps to remove offending agents. Extracellular traps are composed of extracellular deoxyribonucleic acids (DNA), elastase, histone, and myeloperoxidase. When RAW264.7 cells or THP-1 cells were treated with aflatoxin B_1, a network of extracellular structures was formed. Imaging using Hoechst 33342, SYTOX Orange, anti-elastase antibody, anti-histone H3 antibody, and anti-myeloperoxidase antibody indicated that all the components of extracellular traps were aggregated upon treatment with aflatoxin B_1. Extracellular traps can be activated through a pathway using nicotinamide adenine dinucleotide phosphate (NADPH) oxidase 2 or by other means. The addition of the NADPH oxidase 2 inhibitor diphenylene iodonium (DPI) abolished the formation of aflatoxin B_1-induced extracellular traps. This indicates that the reactive oxygen species generated by NADPH oxidase 2 are required for the aflatoxin B_1-induced extracellular traps (An et al. 2017).

Aflatoxin B_1 has been tested on the skin of white mice and found to initiate papilloma growth. The effects of aflatoxin are magnified when combined with the trichothecene mycotoxin T-2. The combination of aflatoxin B_1 and T-2 produced a synergistic lethal effect on mice in acute toxicity tests (Lindenfelser et al. 1974).

4 Trichothecene Toxin

During 1943 and 1944 in the Orenburg district of the Soviet Union, thousands of people died from alimentary toxic aleukia (ATA). Out of the entire Orenburg population, 30% were affected by ATA. A total of 10% of the Orenburg population died from ATA. Symptoms of ATA include vomiting, serous hemorrhagic inflammation, necrosis and ulceration in the digestive tract, diarrhea, leukopenia, myelosuppression, and immunosuppression. Soviet involvement during World War II depleted manpower needed to harvest food crops. Without enough workers for harvest, many crops were overwintered in fields. Unfortunately, World War II also caused a food shortage. People were forced to survive on whatever food they could obtain. Overwintered grains were linked to the ATA occurrences; samples of millet, wheat, rye, oats, barley, and buckwheat were collected from the fields in Orenburg as well as from the homes of the deceased. The Orenburg grains were found to contain the fungi *Fusarium poae* and *Fusarium sporotrichioides*. These *Fusarium* species were tested via thin-layer chromatography, gas-liquid chromatography, mass spectrometry, and rabbit skin test. The analyses showed that the *Fusarium* species were producing the toxic trichothecene T-2 which produced edema, hemorrhage, necrosis of the epidermis, necrosis of the dermis, and necrosis of the hair follicles on the skin of rabbits (Yagen and Joffe 1976). The T-2 toxin is chemically 4β, 15diacetoxy-8α-(3-methylbutyryloxy)-12,13-epoxytrichothec-9-en-3α-ol.

The Soviet Union learned how deadly the T-2 toxin can be on a population from the Orenburg experience. Unfortunately, the Soviet Union put this knowledge to malicious use. Evidence indicates that the Soviet forces allied with Vietnamese forces may have sprayed T-2 toxin on villages containing resistance fighters in Kampuchea near the Thailand boarder using the T-2 toxin as a biological chemical warfare agent. There were similar other attacks in Laos reported during mid 1970s. The Soviet Union was involved in a technical advisory and supply role during the Kampuchea and Laos attacks. Additional reports of similar attacks in Afghanistan occurred when the Soviet Union was more directly involved. The chemical attacks continued for years and became referred to as "yellow rain." A Defence Intelligence Agency report which was declassified and released in 2007 revealed the chemical analysis of agents used in one Kampuchea attack. Analysis of the chemical warfare agent sprayed was found to contain the trichothecenes T-2, nivalenol, and deoxynivalenol (Special Report No. 104 1982). The production of these trichothecenes by *Fusarium* requires low temperatures which are not found in Southeast Asia; this indicates a non-natural occurrence of these trichothecenes. The trichothecene concentrations collected at the site of attack were chemically quantified and were determined to be many times more concentrated than the levels found in nature. In 1982, then Secretary of the United States George P. Schultz presented special report No. 104 to the members of the United States Congress and to the members of the United Nations providing scientific evidence collected at the sites of attacks. The report details sites of attack, methods of attack, number of people killed, number of people ill, and mycotoxins found in victims' blood and urine (Special Report No. 104 1982). Special report No. 104 also reveals the toxin analysis from two gas masks taken off from two dead

Soviet soldiers near Kabul, Afghanistan. The trichothecene toxins, T-2, diacetoxyscirpenol (DAS), and verricarol along with the mycotoxin zearalenone were identified on the Soviet gas masks, which indicated that the soldiers were wearing the masks when biological chemical warfare was used. Mycotoxins are lethal enough to be utilized as biological chemical warfare agents.

Many human illnesses which include respiratory, immunologic, and neurologic components have been linked under the umbrella term damp building-related illness. Damp indoor environments amplify fungal growth. One genus of particular interest in damp building-related illness is *Stachybotrys* which has been commonly termed as toxic black mold. *Stachybotrys* preferentially grows on wet cellulose building materials, such as drywall, ceiling tiles, and floor joists. *Stachybotrys* is known to biosynthesize trichothecene toxins. Air samples from *Stachybotrys* contaminated buildings have detected trichothecene which could be inhaled into the respiratory tracts of inhabitants (Pestka et al. 2008). Trichothecenes can bind to ribosomes, stopping translation, and thus stopping protein biosynthesis (Barrett 2000). Trichothecenes can also activate three mitogen-activated protein kinase (MAPK) pathways: Jun N-terminal kinase (JNK), extracellular signal-regulated kinase (ERK), and p38. Via MAPKs, trichothecenes can activate pro-inflammatory cytokines and apoptotic cytokines. Activation of multiple pathways yields a hormetic response depending upon factors, such as trichothecene dose, frequency, time since exposure, and other variables. A low level of trichothecene can elicit an immune response, whereas a higher level can trigger an immunosuppressive response (Haschek and Beasley 2009). The activation of inflammatory pathways can lead to stimulating immune cells. Continued trichothecene exposure allows apoptotic pathways to dominate, destroying immune cells and suppressing an immune response. Trichothecenes have been studied and found to be cytotoxic in neuroepithelial cells, olfactory neurons, lymphocytes in lymphoid tissues, leukocytes of bone marrow, macrophages, and more.

The observed effects of damp building-related illness may be explained in part by synergistic effects from both mycotoxins and endotoxins. Wang and Yadav investigated the effects of *Stachybotrys* extracts on the murine alveolar macrophage cell line MH-S (Wang and Yadav 2006; Wang and Yadav 2007). The extract-treated macrophages showed DNA strand breaks within 15 minutes as evidenced by the comet gel electrophoresis assay. Activation of macrophage apoptosis was measured by morphologic changes, DNA ladder formation, activation of caspase 3, and activation of caspase 7. Apoptosis could be detected within 3 hours of treatment with *Stachybotrys* extract. The *Stachybotrys* extract caused lactate dehydrogenase release and inhibited cell proliferation without producing detectable levels of nitric oxide or the pro-inflammatory cytokines interleukin (IL)-1β, IL-6, or TNF-α. Intracellular glutathione showed a significant decrease 9 hours after extract treatment. This was interpreted as oxidative stress lagging apoptosis under the assay conditions. Lipopolysaccharides are bacterial endotoxins commonly released from the outer membrane of gram-negative bacteria. Lipopolysaccharides were tested in the assay at levels which could induce inflammation but not induce apoptosis. The combination of both mycotoxin *Stachybotrys* extract and endotoxin lipopolysaccharides increased macrophage cytotoxicity. Toxin-treated alveolar macrophages showed activation of p53, JNK, and p38 pathways. Also working with a mouse model, Islam and Pestka investigated the effects of lipopolysaccharides and the trichothecene mycotoxin deoxynivalenol (Islam and Pestka 2006). Mice were intraperitoneally injected with lipopolysaccharide followed by oral deoxynivalenol, 8 hours later. Deoxynivalenol elicited a much greater cytokine response after lipopolysaccharide priming compared to vehicle priming. Lipopolysaccharide priming sensitized the mice to deoxynivalenol which both decreased the time to onset and increased the magnitude and durations of IL-1α, IL- 1β, IL-6, TNF-α, and splenic mRNA responses. Lipopolysaccharide primed mice exhibited much more apoptosis of the thymus compared to mice that received only deoxynivalenol or only lipopolysaccharide. The synergy between mycotoxin and endotoxin seems to hold true for murine cells and for whole animals.

Päivi Kankkunen et al. have investigated the effects of trichothecene mycotoxins and bacterial endotoxins on human macrophages (Kankkunen et al. 2009). Bacterial endotoxins can activate the innate immune system. The sequence of events involved can include endotoxins, such as lipopolysaccharides from the outer membrane of gram-negative bacteria activating pattern recognition receptors of the innate immune system and nucleotide oligomerization domain (NOD)-like receptors (NLRs). NOD1 and NOD2 receptors activate the nuclear factor kappa-light-chain-enhancer of activated B cells (NF-κB) pathway and the mitogen-activated protein kinase (MAPK) pathway. These pathways activate transcription of genes, including IL-1β, IL-6, and TNF-α. Päivi Kankkunen et al. prepared suspensions containing a strain of *Stachybotrys chartarum* which produces the trichothecene satratoxin. The satratoxin samples were gamma-irradiated before being combined with cultures of live human macrophages. IL-1β is very pyrogenic and is not expressed in macrophages under normal condition. IL-18 is constitutively expressed in macrophages but requires activation by an inflammasome. IL-1β and IL-18 messenger RNA were quantified using RT-PCR for satratoxin-stimulated macrophages and for lipopolysaccharide-stimulated macrophages. IL-1β mRNA expression was greatly increased by lipopolysaccharide stimulation but only modestly increased by satratoxin stimulation. The constitutive mRNA level of IL-18 was not affected much by either lipopolysaccharide alone or satratoxin alone stimulation. When human macrophages were co-stimulated by lipopolysaccharide and satratoxin combined, the effect was dramatically different. Co-stimulation greatly increased mRNA levels of both IL-1β and IL-18. The increase was much greater than simply adding the two effects together. Co-stimulation by lipopolysaccharide and satratoxin shows synergistic cooperativity. An enzyme-linked immunosorbent assay was used to measure mature processed IL-1β and IL-18 secreted into the cell culture supernatant. The results for mature secreted IL-1β and IL-18 were similar to the results for intracellular mRNA. Macrophage stimulation by lipopolysaccharide alone or by satratoxin alone produced little secreted IL-1β or IL-18; however, co-stimulation showed a large release of both mature IL-1β and mature IL-18. The endogenous pyrogen IL-1β is biosynthesized as the inactive 31 kDa precursor protein p31. Caspase-1 can cleave IL-1β into its active form. IL-18 is biosynthesized as the inactive 24 kDa precursor protein p24. The enzyme caspase-1 can cleave IL-18 into its active form of p18. The enzyme caspase-3 can cleave IL-18 into p15 and p16. Caspase-1 and caspase-3 are themselves inactive zymogens which require cleavage to become activated. Päivi Kankkunen et al. used Western blot analysis and specific antibodies to analyze the presence of IL-1β, IL-18, caspase-1, and caspase-3 cleavage products after stimulation by lipopolysaccharide alone, by satratoxin alone, or by both. Lipopolysaccharides alone were able to stimulate the expression of inactive IL-1β precursor but were not able to induce activation of IL-1β on their own. In fact, lipopolysaccharide stimulation alone cleaved caspase-1 into p22 which is an inactive product with no enzymatic activity; therefore, caspase-1did not activate the IL-1β precursor which was induced. A second event such as exposure to mycotoxin satratoxin, roridin A, verrucarin A, or T-2 was necessary to convert caspase-1 to its p20 active subunit. Mycotoxins by themselves were able to activate caspase-1 and caspase-3 which generated an inflammatory and apoptotic response. Lipopolysaccharides and mycotoxins together were found to synergistically activate macrophage inflammasomes yielding a much greater response than the simple sum of their individual responses (Kankkunen et al. 2009).

5 Mycophenolic Acid Toxin

Records date back as far as 1735 in Spain describing a disease affecting many thousands of people. The disease was characterized by dermatitis, diarrhea, dementia, and death. The disease which would become known as pellagra afflicted mostly lower class peasants throughout the world and by 1893 was associated with the consumption of corn. The Italian physician and scientist Bartolomeo Gosio collected fungi from spoiled corn samples in search of the cause for pellagra. Gosio cultured the various fungi and in 1893 discovered a new species which he named *Penicillium glaucum*. The species was later renamed *Penicillium brevicompactum*. From *P. brevicompactum*, Gosio isolated a secondary metabolite that possessed antibiotic properties. After 3 years of work, Gosio was able to isolate the pure antibiotic and demonstrate its ability to inhibit anthrax bacterial growth. The metabolite was mycophenolic acid. Gosio had made an important discovery even though he had not found the cause of the disease pellagra. Pellagra continued to claim countless lives. In the Unites States during 1912, the state of South Carolina reported 30,000 cases of pellagra with a 40% fatality rate. The true cause of pellagra was not determined until 1937; a deficiency of vitamin B3 (niacin) was the cause (Morabia 2008).

Mycophenolic acid has been found to have several useful properties in addition to being an antibiotic. Mycophenolic acid was approved as a prescription drug under the brand name myfortic. In order to improve bioavailability, mycophenolic acid was condensed with 2-(morpholin-4-yl) ethanol to form mycophenolate mofetil (National Center for Biotechnology Information 2021). Mycophenolate mofetil is sold under the prescription brand name cellcept. Mycophenolate mofetil is a prodrug which contains a carboxylic ester. The carboxylic ester is cleaved in the liver, releasing the active drug mycophenolic acid. Mycophenolic acid has been found to possess antiviral properties preventing viral RNA synthesis and accumulation (Borroto-Esoda et al. 2004). Mycophenolic acid has also been found as a useful antifungal, antitumor, and antipsoriasis agent (Regueira et al. 2011). Mycophenolic acid is used clinically as an immunosuppressant.

The immunosuppressant activity of mycophenolic acid arises from its ability to reversibly inhibit the enzyme inosine 5′-monophosphate dehydrogenase (IMPDH) (National Center for Biotechnology Information 2021; Jonsson and Carlsten 2002). Inosine 5′-monophosphate dehydrogenase is necessary for functioning of the biochemical *de novo* pathway which synthesizes guanine nucleotides. Cells need guanine nucleotides to form strands of ribonucleic acids (RNA) as well as strands of deoxyribonucleic acids (DNA). Lymphocyte cells rely heavily on the *de novo* pathway for RNA and DNA synthesis; therefore, lymphocytes are very sensitive to mycophenolic acid. Other cells utilize both the *de novo* pathway and the salvage pathway to provide guanine nucleotides. By utilizing the alternate salvage pathway to provide guanine nucleotides, other cells are not affected as greatly by the level of mycophenolic acid. Immunosuppression is of great value in the field of organ transplantation. After a kidney, heart, or liver organ transplant, the recipient's immune system usually recognizes the new organ as a foreign tissue. The recipient's immune system attacks and attempts to remove the foreign invader; this is known as organ rejection. Mycophenolic acid suppresses lymphocytes (natural-killer cells, T-lymphocytes, and B-lymphocytes) thereby reducing organ transplant rejection (Behrend 2001). Mycophenolic acid immunosuppression has also been found clinically useful for treating autoimmune diseases, such as rheumatoid arthritis and lupus nephritis (Goldblum 1993; Chan et al. 2000).

The effect of mycophenolic acid on monocytes has also been explored. It was found to reduce guanosine triphosphate levels and to induce monocytic differentiation (Sokoloski et al. 1986). The effects of mycophenolic acid on macrophages were investigated by Jonsson and Carlsten using a murine model (Jonsson and Carlsten 2002). Murine macrophages were grown in microtiter plates. Proliferation was measured using incorporation of the radiolabel [^{3}H]thymidine followed by β-counter analysis. Macrophage proliferation was reduced by the addition of mycophenolic acid. Proliferation was not affected when both mycophenolic acid and guanosine were added to the culture indicating that the reduced proliferation was due to a lack of guanosine production. As a control, etoposide was added to a macrophage culture and reduced proliferation was observed. Etoposide is a DNA topoisomerase II inhibitor which induces apoptosis. When both etoposide and guanosine were added to a culture, the culture showed reduced proliferation indicating that reduced proliferation was not due to a lack of guanosine production. The proliferation assays were each repeated with the addition of lipopolysaccharide plus recombinant interferon gamma (IFN-γ) stimulating the macrophages. Stimulation did not change the outcomes of the proliferation assays. Macrophage intracellular and extracellular nitrite levels were measured. Nitrite can serve as a physiological reservoir, able to be reduced to nitric oxide (Shiva 2013). Nitric oxide has been found to be increased in patients with the autoimmune disease systemic lupus erythematosus (Belmont et al. 1997). Macrophages are potential contributors to the nitric oxide involved in lupus. Extracellular nitric oxide levels were measured in the supernatant of cultured macrophages. Stimulated macrophages measured extracellular nitrite release which was inhibited by the addition of mycophenolic acid. The addition of guanosine abrogated the effects of mycophenolic acid. Macrophage cells were lysed to determine the intracellular nitrite content. There was no measurable intracellular nitrite under any of the test conditions. The results indicate that stimulated macrophages do release nitrite contributing to the overall nitric oxide level. Mycophenolic acid inhibits the release of nitrite which helps to explain the beneficial effects observed when mycophenolic acid is administered to patients with lupus. Macrophages normally biosynthesize the pro-inflammatory cytokines, tumor necrosis factor alpha, and interleukin-1-beta in response to lipopolysaccharide stimulation (Hogquist et al. 1991). Both of these cytokines were measured using an enzyme-linked immunosorbent assay conducted on both the cell supernatant and the cell lysate. Stimulated macrophage cells treated with mycophenolic acid measured a decreased total tumor necrosis factor alpha and a decreased total interleukin-1-beta compared to stimulated macrophages not treated with mycophenolic acid. The addition of guanosine abrogated the effects of mycophenolic acid.

Mycophenolic acid has been proven to be a very potent immunosuppressant acting on macrophages, natural-killer cells, T-lymphocytes and B-lymphocytes; however, a doctor's prescription is not the only path for mycophenolic acid entry into a person. Mycophenolic acid produced by *Penicillium*-contaminated foods such as corn can be ingested and suppress an unsuspecting persons immunity. *Penicillium* also is commonly found growing in damp buildings. The occupants of damp buildings can inhale and ingest mycophenolic acid unaware of their immune systems being suppressed.

6 Ochratoxin

Ochratoxin A is a mycotoxin produced by *Aspergillus ochraceus*, *Penicillium viridicatum*, and other molds. Ochratoxin A is commonly found on food crops, such as cereal grains, fruit, and livestock feed (e.g., corn, barley, and grapes). Ingestion is often the route of entry for ochratoxin A into animals (e.g., pigs, chickens, and rats) and humans. Ochratoxin A has been found to be nephrotoxic, teratogenic, carcinogenic, and immunotoxic (Boorman et al. 1984). Ochratoxin A can be ingested by humans directly with food grains or indirectly via animals ingesting contaminated feed, then humans consuming the contaminated animals. Pigs feeding on contaminated grains accumulate ochratoxin A in their kidneys. Ochratoxin A contaminated grains have been found throughout the world, including Bulgaria, Canada, Denmark, Egypt, France, Germany, Korea, Poland, Sweden, the UK, and Yugoslavia (Bui-Klimke and Wu 2015; Madsen et al. 1982). Denmark condemns porcine carcasses having kidneys exceeding 25 μg/kg ochratoxin A content. Once ingested, ochratoxin A is absorbed via the stomach and small intestine and is bound to albumin in the blood. Ochratoxin A generally concentrates in the order of kidneys, muscle, liver, and fat tissues.

The chemical structure of ochratoxin A contains a phenylalanine moiety. During normal biochemistry, the ligase phenylalanine-tRNA synthetase connects phenylalanine to its corresponding transfer RNA. The charged tRNA is then utilized for protein synthesis. Ochratoxin A acts as a competitive inhibitor for phenylalanine-tRNA synthetase which halts protein synthesis. Inhibiting protein synthesis is the primary effect of ochratoxin. In rat hepatoma tissue culture cells, protein synthesis was monitored by incorporation of [^{3}H]leucine (Creppy et al. 1979). Increasing ochratoxin concentration first showed cytostatic effects and then cytotoxic effects at higher concentrations. Protein synthesis was observed to halt first, followed by cessation of RNA synthesis as measured by [^{3}H]uridine incorporation. The effects of ochratoxin were abrogated by simultaneous addition of phenylalanine along with ochratoxin.

Using BALB/c mice, Haubeck et al. investigated ochratoxin's immunosuppressive abilities (Haubeck et al. 1981). Sheep red blood cells via intraperitoneal injection were used as antigens to elicit an immune response. An ochratoxin A injection of 0.005 μg/kg body weight was found to suppress the immune response by 50%. For comparison, 16% of slaughtered pigs in a sample from Sweden were found to contain 2 to 187 μg/kg blood weight, which is orders of magnitude larger than required to suppress the immune system. This ochratoxin A would be passed on to humans consuming the pigs. Phenylalanine injected concurrently with ochratoxin A reduced the observed immunosuppression. Phenylalanine, at approximately twice the ochratoxin weight, was able to completely reverse immunosuppression. Creppy et al. investigated the lethal dose of ochratoxin A on mice (Creppy et al. 1980). A dose of 0.8 mg ochratoxin A injected intraperitoneally caused death within 24 h. The lethal effect of ochratoxin A was completely prevented by injection of 1.0 mg phenylalanine along with the 0.8 mg ochratoxin A injection. When ochratoxin A was injected, but the phenylalanine injection was delayed 30 min, a much greater amount of phenylalanine was necessary for the mice to survive. A total of 25 milligrams of phenylalanine was necessary to bring the survival rate up to 92% for the delayed injection.

The BALB/c mouse-derived macrophage cell line J774A.1 has been used to study the toxic effects of ochratoxin (Ferrante et al. 2008). Ochratoxin alone applied from 30 nM to 100 μM showed a hormetic response, first increasing inflammatory response, then decreasing cell survival. Lipopolysaccharide alone applied at 100 ng/ml did not change cell viability. Co-stimulation by lipopolysaccharide and ochratoxin showed a reduced inflammatory response while cell death increased. Ochratoxin A toxicity was demonstrated again using the human-derived cell line THP-1 (Müller et al. 2003). THP-1 is a human monocyte/macrophage cell line. When crude mycotoxin extract containing ochratoxin A and other fungal contaminants was applied to THP-1 cells, the cells showed reduced viability evidenced by suppression of metabolic activity, cell proliferation, cell membrane integrity, cell differentiation, and phagocytic behavior.

7 Summary

Mycotoxins are a very diverse group of chemicals. They have been in existence longer than human records. Their effects are various and their potential for use is hardly tapped. They have been shown to have negative effects causing great suffering and death to countless humans and animals. Humans have sometimes been collateral victims inadvertently consuming mycotoxin-laden crops. At other times, humans have been intended targets on the receiving end of chemical warfare agents. Mycotoxins have also been harnessed for positive purposes in clinical applications. Their utility in treating organ transplantation, childbirth, and rheumatoid arthritis has benefitted society. There is a chemical war for survival being waged by fungi, and most people are unaware that they are in the middle of it.

References

Alshannaq A, Yu JH. Occurrence, toxicity, and analysis of major mycotoxins in food. Int J Environ Res Pub Health. 2017;14(6):632.

An Y, Shi X, Tang X, et al. Aflatoxin B1 induces reactive oxygen species-mediated autophagy and extracellu-

lar trap formation in macrophages. Front Cell Infect Microbiol. 2017;7:53.
Aronson JK. Plant poisons and traditional medicines. In: Farrar J, Hotez PJ, Junghanss T, Kang G, Lalloo D, White NJ, editors. Manson's tropical infectious diseases. 23rd ed. W.B. Saunders; 2014. p. 1128–50.
Bardal SK, Waechter JE, Martin DS. Neurology and the neuromuscular system. In: Bardal SK, Martin DS, Waechter JE, editors. Applied pharmacology. St. Louis: Elsevier Saunders; 2011. p. 325–65.
Barrett JR. Mycotoxins: of molds and maladies. Environ Health Perspect. 2000;108(1):A20–3.
Behrend M. Mycophenolate mofetil: suggested guidelines for use in kidney transplantation. BioDrugs. 2001;15(1):37–53.
Belmont HM, Levartovsky D, Goel A, et al. Increased nitric oxide production accompanied by the upregulation of inducible nitric oxide synthase in vascular endothelium from patients with systemic lupus erythematosus. Arthritis Rheum. 1997;40(10):1810–6.
Belser-Ehrlich S, Harper A, Hussey J, Hallock R. Human and cattle ergotism since 1900: symptoms, outbreaks, and regulations. Toxicol Ind Health. 2013;29(4):307–16.
Boorman GA, Hong HL, Dieter MP, et al. Myelotoxicity and macrophage alteration in mice exposed to ochratoxin A. Toxicol Appl Pharmacol. 1984;72:304–12.
Borroto-Esoda K, Myrick F, Feng J, Jeffrey J, Furman P. In vitro combination of amdoxovir and the inosine monophosphate dehydrogenase inhibitors mycophenolic acid and ribavirin demonstrates potent activity against wild-type and drug-resistant variants of human immunodeficiency virus type 1. Antimicrob Agents Chemother. 2004;48(11):4387–94.
Bui-Klimke TR, Wu F. Ochratoxin A and human health risk: a review of the evidence. Crit Rev Food Sci Nutr. 2015;55(13):1860–9.
Chan TM, Li FK, Tang CSO, et al. Efficacy of mycophenolate Mofetil in patients with diffuse proliferative lupus nephritis. N Engl J Med. 2000;343:1156–62.
Coffel ED, Horton RM, Sherbinin A. Temperature and humidity based projections of a rapid rise in global heat stress exposure during the 21st century. Environ Res Lett. 2018;13:014001.
Creppy EE, Lugnier AAJ, Beck G, Röschenthaler R, Dirheimer G. Action of Ochratoxin A on cultured hepatoma cells reversion of inhibition by phenylalanine. FEBS Lett. 1979;104:287–90.
Creppy EE, Schlegel M, Röschenthaler R, Dirheimer G. Phenylalanine prevents acute poisoning by Ochratoxin A in mice. Toxicol Lett. 1980;6:77–80.
Desaint H, Aoun N, Deslandes L, Vailleau F, Roux F, Berthome R. Fight hard or die trying: when plants face pathogens under heat stress. New Phytol. 2021;229(2):712–34.
Du RH, Li EG, Cao Y, Song YC, Tan RX. Fumigaclavine C inhibits tumor necrosis factor α production via suppression of toll-like receptor 4 and nuclear factor κB activation in macrophages. Life Sci. 2011;89(7-8):235–40.
Ferrante MC, MattaceRaso G, Bilancione M, Esposito E, Iacono A, Meli R. Differential modification of inflammatory enzymes in J774A.1 macrophages by Ochratoxin A alone or in combination with lipopolysaccharide. Toxicol Lett. 2008;181(1):40–6.
Goldblum R. Therapy of rheumatoid arthritis with mycophenolate mofetil. Clin Exp Rheumatol. 1993;11(Suppl 8):S117–9.
Haarmann T, Rolke Y, Giesbert S, Tudzynski P. Ergot: from witchcraft to biotechnology. Mol Plant Pathol. 2009;10(4):563–77.
Haschek WM, Beasley VR. Trichothecene Mycotoxins. In: Gupta RC, editor. Handbook of toxicology of chemical warfare agents. Academic Press; 2009. p. 353–69.
Haubeck HD, Lorkowski G, Kölsch E, Röschenthaler R. Immunosuppression by Ochratoxin A and its prevention by phenylalanine. Appl Environ Microbiol. 1981;41(4):1040–2.
Hogquist KA, Unanue ER, Chaplin DD. Release of IL-1 from mononuclear phagocytes. J Immunol. 1991;147(7):2181–6.
Holstege CP, Traven SA. Ergot. In: Wexler P, editor. Encyclopedia of toxicology. 3rd ed. Academic Press; 2014. p. 444–7.
Islam Z, Pestka JJ. LPS priming potentiates and prolongs proinflammatory cytokine response to the trichothecene deoxynivalenol in the mouse. Toxicol Appl Pharmacol. 2006;211(1):53–63.
Jonsson CA, Carlsten H. Mycophenolic acid inhibits inosine 5′-monophosphate dehydrogenase and suppresses production of pro-inflammatory cytokines, nitric oxide, and LDH in macrophages. Cell Immunol. 2002;216(1-2):93–101.
Kankkunen P, Rintahaka J, Aalto A, et al. Trichothecene Mycotoxins activate inflammatory response in human macrophages. J Immunol. 2009;182(10):6418–25.
Kew MC. Synergistic interaction between aflatoxin B1 and hepatitis B virus in hepatocarcinogenesis. Liver Int. 2003;23(6):405–9.
Lindenfelser LA, Lillehoj EB, Burmeister HR. Aflatoxin and Trichothecene toxins: skin tumor induction and synergistic acute toxicity in white mice. J Natl Cancer Inst. 1974;52(1):113–6.
Madsen A, Mortensen HP, Hald B. Feeding experiments with ochratoxin A contaminated barley for bacon pigs I. influence on pig performance and residues. Acta Agric Scand. 1982;32:225–39.
Mitchell CG, Slight J, Donaldson K. Diffusible component from the spore surface of the fungus Aspergillus fumigatus which inhibits the macrophage oxidative burst is distinct from gliotoxin and other hyphal toxins. Thorax. 1997;52:796–801.
Morabia A. Joseph Goldberger's research on the prevention of pellagra. J R Soc Med. 2008;101(11):566–8.
Müller G, Rosner H, Rohrmann B, et al. Effects of the mycotoxin Ochratoxin A and some of its metabolites on the human cell line THP-1. Toxicology. 2003;184(1):69–82.

National Center for Biotechnology Information. PubChem compound summary for CID 5281078, Mycophenolate mofetil. 2021. https://pubchem.ncbi.nlm.nih.gov/compound/Mycophenolate-mofetil

Online Etymology Dictionary.. https://www.etymonline.com/word/ergot

Paulussen C, Hallsworth JE, Alvarez-Perez S, et al. Ecology of aspergillosis: insights into the pathogenic potency of Aspergillus fumigatus and some other Aspergillus species. Microbial Biotechnol. 2017;10(2):296–322.

Pestka JJ, Yike I, Dearborn DG, Ward MD, Harkema JR. Stachybotrys chartarum, trichothecene mycotoxins, and damp building-related illness: new insights into a public health enigma. Toxicol Sci. 2008;104(1):4–26.

Regueira TB, Kildegaard KR, Hansen BG, Mortensen UH, Hertweck C, Nielsen J. Molecular basis for Mycophenolic acid biosynthesis in *Penicillium brevicompactum*. Appl Environ Microbiol. 2011;77(9):3035–43.

Robertson MD, Seaton A, Raeburn A, Milne LJR. Inhibition of phagocyte migration and spreading by spore diffusates of Aspergillus fumigatus. J Med Vet Mycol. 1987a;25(6):389–96.

Robertson MD, Seaton A, Milne LJ, Raeburn JA. Suppression of host defences by Aspergillus fumigatus. Thorax. 1987b;42(1):19–25.

Robertson MD, Seaton A, Milne LJR, Raeburn A. Resistance of spores of Aspergillus fumigatus to ingestion by phagocytic cells. Thorax. 1987c;42:466–72.

Shiva S. Nitrite: A physiological store of nitric oxide and modulator of mitochondrial function. Redox Biol. 2013;1(1):40–4.

Sokoloski JA, Blair OC, Salterelli AC. Alterations in glycoprotein synthesis and guanosine triphosphate levels associated with the differentiation of HL-60 leukemia cells produced by inhibitors of inosine 5′-phosphate dehydrogenase. Cancer Res. 1986;46:2314–9.

Special Report No. 104 United States Department of State Chemical Warfare in Southeast Asia and Afghanistan: An Update Report from Secretary of State George P. Shultz November 1982. http://insidethecoldwar.org/sites/default/files/documents/Chemical%20Warfare%20in%20Southeast%20Asia%20and%20Afghanistan%20an%20Update%20November%201982.pdf

Squire RA. Ranking animal carcinogens: a proposed regulatory approach. Science. 1981;214:877–80.

United States Environmental Protection Agency A Brief Guide to Mold, Moisture and Your Home. https://www.epa.gov/mold/brief-guide-mold-moisture-and-your-home

United States Environmental Protection Agency Mold Course Chapter 1: Introduction to Molds. https://www.epa.gov/mold/mold-course-chapter-1

Wang H, Yadav JS. DNA damage, redox changes, and associated stress-inducible signaling events underlying the apoptosis and cytotoxicity in murine alveolar macrophage cell line MH-S by methanol-extracted Stachybotrys chartarum toxins. Toxicol Appl Pharmacol. 2006;214(3):297–308.

Wang H, Yadav JS. Global gene expression changes underlying *Stachybotrys chartarum* toxin-induced apoptosis in murine alveolar macrophages: Evidence of multiple signal transduction pathways. Apoptosis. 2007;12:535–48.

Yagen B, Joffe AZ. Screening of toxic isolates of *Fusarium poae* and *Fusarium sporotrichioides* involved in causing alimentary toxic aleukia. Appl Environ Microbiol. 1976;32(3):423–7.

Recent Trends in Clinical Studies on Macrophage-Targeted Delivery

Ashim Malhotra

Abstract

A discussion of macrophage diversity has taken a central stage over the past decade with preclinical and clinical investigations, suggesting the role of macrophages in the regulation of inflammation in almost every organ system in the body. Even for organs such as the heart, where macrophages were not traditionally thought to be present, recent literature demonstrates macrophage-mediated inflammation during disease and also macrophage-mediated general maintenance of organs including muscle repair and tissue building. The subdivision of macrophages into the M1 (proinflammation) and M2 (anti-inflammation) phenotypes has allowed a detailed characterization of their roles in a variety of human diseases, including cardiovascular diseases such as myocarditis, myocardial infarction, and autoimmune diseases such as rheumatoid arthritis and tumors. In fact, the discovery of tumor-associated macrophages (TAMs) has ignited a debate regarding the regulation of the M1/M2 macrophage polarization axis, a concept that has been further tested in other diseases. This chapter presents select examples of preclinical testing of interventional strategies and drug delivery systems for regulating the M1/M2 polarization axis in a variety of diseases. Specific examples of clinical trials are cited, along with future directions for areas where only preclinical data are currently available. Macrophage targeting through specialized drug delivery systems has transitioned from the bench to clinical trials in humans and offers a new and exciting toolkit for combating human disease.

Keywords

Macrophage · Drug delivery · Targeting · Clinical trials · M1 · M2

1 Introduction: The Central Role of Macrophages in Human Biology

First discovered in 1884 by the Russian zoologist, Metchnikoff, macrophages comprise an important immune defense mechanism in humans. Derived in humans by differentiation of the circulating monocytes, the monocyte-macrophage system was earlier known as the mononuclear phagocyte system. Macrophages are large professional phagocytic cells that engulf and destroy pathogens, such as bacteria, remove cellular debris, such as cellular components from necrotic, apoptotic, or T-cell-targeted cell death,

A. Malhotra (✉)
California Northstate University College of Pharmacy, Elk Grove, CA, USA
e-mail: ashim.malhotra@cnsu.edu

S. Gupta, Y. V. Pathak (eds.), *Macrophage Targeted Delivery Systems*,
https://doi.org/10.1007/978-3-030-84164-5_24

and play a quintessential role in mounting nonspecific immune responses as a part of the panoply of antigen-presenting cells (APCs) through a well-characterized molecular pathway known as the exogenous pathway of antigen presentation. As can be seen, macrophages play a very important role in a variety of molecular, cellular, and systemic processes, making it an urgent need to understand and characterize their detailed functions.

In humans, macrophages are resident within a variety of tissues where they play an essential role in tissue maintenance and homeostasis. They are derived from monocytes, which are agranulocytic leukocytes, mainly responsible for viral and bacterial immune responses. For example, following a triggering chemotaxis event such as the bacterial invasion of an organ, monocytes extravasate from veins in the organ by a process known as diapedesis and subsequently differentiate into macrophages. In doing so, they become enlarged and tissue-resident with a half-life of many weeks to months to much longer timespans in some organs. In fact, as Table 1 shows, there are many well-characterized tissue-resident macrophages in various organs of the body.

However, in addition to Table 1, macrophages may be recruited to special cites during the etiological onset or progression of a disease, and may as a consequence become resident in organs or other anatomical locations following recruitment. This is an important observation because often following recruitment to such sites, macrophages determine the fate of disease progression.

For instance, many investigators have documented the orchestrating role of macrophages in (1) the progression of a variety of solid tumors and in cancers; (2) an abundance of autoimmune disorders including rheumatoid arthritis (RA), inflammatory bowel disease (IBD), multiple sclerosis (MS), systemic lupus erythematosus (SLE), and others; (3) type 1 diabetes; and (4) cardiovascular diseases, such as myocarditis, stroke, pulmonary artery hypertension, and others. The sheer preponderance of macrophage involvement in human disease makes them central candidates for drug therapy. Over the years, targeted drug delivery to macrophages to alter their function has emerged as an important strategy to modulate and, in many cases, ameliorate disease progression as one of the success stories of translational molecular research. Many instances exist where targeted drug delivery for macrophage modulation has successfully transitioned from the bench to clinical trials in humans. Before we discuss some of these clinical trials and their outcomes, we need to examine some aspects of the biological response of macrophages, including receptor expression and distribution. This is important because in many instances, specialized delivery systems have been invented and deployed to target macrophages localized to specific microenvironments. An exposition of clinical trials therefore must include commentary on the drug delivery system employed.

Table 1 Examples of tissue-resident macrophages by body organs

Body organ or location	Name of tissue-resident macrophage
Liver	Kupffer cells
Brain and spinal cord	Microglial cells
Lungs	Dust cells
Adipose tissue	Macrophages
Placenta in the fetus	Hofbauer cells

2 Select Aspects of Macrophage Biology

2.1 The Role of the Macrophage in the Phagocytic Response

As professional phagocytes, macrophages are essential for nonspecific immune responses involving the endocytosis and consequent removal of extracellular tissue pathogens, such as the prokaryotic bacteria and other eukaryotic unicellular and multicellular pathogens. They are also important for tissue maintenance by the removal of cell debris.

Macrophages express a variety of cell surface receptors that play an important role in initiating contact as a first step in the attachment and subsequent binding of the object to be engulfed. For example, in the case of the removal of tissue cells dying through apoptosis as a part of the regular cell turnover process, macrophages have a complicated receptor system for sensing the externalized phosphatidylserine on the surface of dying cells. This is achieved by first combining phosphatidylserine with a variety of proteins which is followed by binding to the $\alpha_v\beta_3$ integrin on the surface of the macrophage. Billions of cells in the human body are removed by this process daily as a part of turnover. On the other hand, macrophages express a variety of pattern recognition receptors (PRRs) that recognize pathogen-associated molecular patterns (PAMPs) which are clusters of molecular signals on the surfaces of invading pathogens. PRRs constitute an important signal mechanism for the attachment and initiation of phagocytosis for a variety of extracellular pathogens. Specifically, macrophages express (1) the mannose receptor Dectin-1 that recognizes PAMPs on fungal cell surfaces, (2) the scavenger receptor A (SR-A) and the macrophage receptor with collagenous structure (MARCO) system for the recognition and attachment of gram-negative and gram-positive bacteria, (3) Toll-like receptors (TLRs) with further subcategories of TLRs for sensing bacterial PAMPs such as flagellin protein (TLR 2, 4, and 5) and eventual viral processing in endosomes (TLR 3, 7, and 9), and a variety of other receptors (Hirayama et al. 2017).

Following attachment, endocytosis begins and is mediated through clathrin and caveolin-coated pits forming invaginations of the surface membrane of macrophages which envelops the attached ligand, eventually forming a vesicle called a phagosome. In subsequent steps, the phagosome merges with a lysosome forming a phagolysosome where oxygen-free radicals such as peroxide destroy the antigen. In a tangent of immune system response, certain other APCs can subsequently display the digested peptide fragments of the antigen on their surfaces and initiate antigen recognition responses through the recruitment of macrophages.

However, for professional phagocytic cells like the macrophages, the molecular details governing the exact steps constituting stepwise engulfment are important to demarcate. Following chemotaxis, the macrophage cell must adhere to the cell to be engulfed, extend pseudopodia, and subsequently internalize the target cell, followed by entrapment within the phagosome. A thorough elucidation of the molecular process of engulfment will help identify and ultimately target specific cell-surface proteins and signal transduction pathways in the macrophage that facilitate this process. For instance, binding to the target cells generally occurs after the target is opsonized by antibodies which aid in identifying the cell to be engulfed, sparing surrounding cells. Following identification, the macrophage cell surface alters the expression of the Fc receptor (FcγRs), which in turn orchestrates a series of molecular processes that result in the formation of a "phagocyte synapse," which is a precursor for the engulfment. A variety of proteins, including nonmuscle myosin proteins, are engaged in creating this local contractile response at the phagocyte synapse (Allen and Aderem 1996).

However, this process may also be halted if the target cell presents a "do not eat me" molecular signal to the macrophage. The CD47 cell surface protein, discovered in murine knockout loss-of-function models, may be expressed by a variety of human cells. When expressed on the cell surface, CD47 inhibits the ability of macrophages to engulf the target cell. This occurs because the CD47 protein binds to the signal regulatory protein (SIRPα), an inhibitory molecule located on the cell surface of macrophages (Fujioka et al. 1996). The CD47-SIRPα interaction is conserved within a species, and this is an important escape mechanism by which phagocyte-mediated engulfment may be evaded. This evasion mechanism plays an essential role by which certain tumors and cancers evade the nonspecific immune response mounted by macrophages by displaying a "do not eat me" signal. Thus, delineation of the molecular pathways such as the CD47-SIRPα inhibitory signal interaction allows the development of targeted drug delivery systems to abrogate this pathway. As discussed later, many clinical trials are underway to test blocking the CD47-SIRPα interaction in tumors and cancers.

The above general discussion illustrates a number of features that are critical to understanding macrophage-targeted drug delivery and related clinical trials. It is important to recognize that there is a vast panoply of multiple receptors and complicated sensing systems on macrophage surfaces. Targeted therapeutic strategies may therefore include a variety of approaches, including (1) antibody or peptide-guided delivery of nanoparticle systems specific to any of the receptors discussed thus far, (2) modulation strategies that alter (either increase or decrease) surface expression of some of the abovementioned macrophage receptors, and (3) the construction of generic or organ-specific receptor modulation delivery systems.

2.2 Diversity and Adaptation of the Macrophage Response

Another classic feature of the human macrophage system is its diversity. While macrophages were initially thought to only regulate proinflammatory responses such as those arising from infections or in the cases of tumors and cancers, not only by responding to inflammatory cues but also by proactively secreting inflammatory signals, a body of evidence has successfully documented that a subset of macrophages evince an anti-inflammatory phenotype. Macrophages are now classified into two categories: M1 that have the class proinflammatory phenotype and secrete IFN-1β, IL-12, IL-23, and TNF-α (Arora et al. 2018) and M2 that inhibit inflammation by secreting anti-inflammatory cytokines such as IL-10 and reduce the secretion of proinflammatory IL-12 (Arora et al. 2018). Adding to their diversity, M2 macrophages may be further classified based on their functions and roles in tissues as M2a, M2b, and M2c macrophages. M2b macrophages regulate Th2 responses, while M2c macrophages orchestrate tissue rebuilding, such as muscle-repair types of responses (Fig. 1).

2.3 Clinical Implications of Macrophage Diversity

To contextualize the above discussion, it is important to think in terms of an M1/M2 axis of differentiation and response where macrophage-dependent responses are concerned. This approach clarifies the roles of macrophages in a variety of human diseases and provides a toolbox to design, implement, and assess clinical strategies that have been tested in clinical trials. An immediate example of the approach of delineating the M1/M2 axis is in solid tumors where macrophages play an important role as tumor-associated

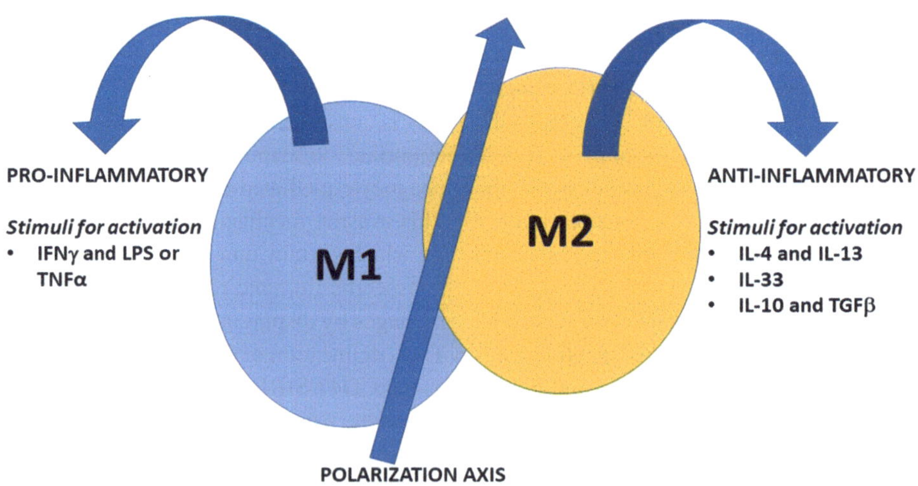

Fig. 1 This schematic summarizes the M1/M2 polarization axis and the resulting balance between proinflammatory and anti-inflammatory status

macrophages (TAMs). In general, TAMs drive an anti-inflammatory phenotype in solid tumors which prevents the recruitment of monocytes in response to the growing tumor and stabilizes the tumor by preventing its destruction. One strategy tested in a variety of clinical trials and discussed in detail later in this chapter is an intervention to skew to the M1/M2 axis more toward the M1 phenotype to drive a proinflammatory response which has been shown to abrogate tumor growth and viability (Cassetta and Pollard 2018).

Similar approaches have been documented to work for other diseases such as the autoimmune disorder RA. During its progression, RA results in the recruitment and retention of macrophages to the synovial tissues of joints, where they drive inflammation especially by secreting IL-6 and TNF-α (Udalova et al. 2016), a hallmark associated with this disease. This indicates the skewing of the M1/M2 axis heavily toward the M1 phenotype, driving RA progression, making the strategy of intervening to reverse polarization to induce an M2-predominant phenotype a viable strategy.

This basic premise can be extended to seemingly disparate organ-related diseases as well. For example, it has been demonstrated that suppressing an active M1 phenotype or augmenting polarization of the M2 phenotype causes advantageous therapeutic outcomes for multiple sclerosis (MS) (Ma et al. 2019). An increase in M1 macrophages is known to cause exacerbation of MS as a result of the recruitment of B and T cells which result in demyelination of neurons (Bramow et al. 2010; Nylander and Hafler 2012). Thus, here too, repolarization of the M1/M2 axis toward an overall M2 phenotype is considered advantageous, and clinical trials for such interventions through the targeted delivery of modulator molecules have documented the usefulness of this approach.

From the standpoint of cardiovascular diseases (CVDs), macrophages are known to play a role in almost all forms including atherosclerosis, pulmonary artery hypertension, myocarditis, and myocardial fibrosis. Plausibly, this is a consequence of the universality of inflammation in CVD. For example, during atherosclerosis, monocyte deposition and subsequent oxidation of cholesterol (HLD) result in the formation of foam cells which accrue in the walls of the arteries. Later, during the episodic narrowing of the arterial lumen through repeated fibrosis and capping, macrophages play an important role in plaque rupture of late plaques. Overall, the macrophage-dependent secretion of proinflammatory cytokines such as IL-1β, IL-18, and macrophage inflammatory protein-1α (MIP-1α) plays a not as yet well-characterized role in atherosclerosis (Moore and Tabas 2011).

The deleterious effect of macrophage-mediated alterations in the inflammatory milieu of the lungs has been documented in the literature from a plurality of causation. Investigations report the damaging effect of the accumulation of mannose-receptor-positive (MR-positive) macrophages in the lungs, leading to the development of pulmonary artery hypertension (PAH) (Park et al. 2019). While others have shown that the hypoxia-induced secretion of IL-6 in the lungs causes the M2 polarization of the M1/M2 axis (Hashimoto-Kataoka et al. 2015). Thus, macrophage-driven inflammatory responses seem to influence the overall clinical outcomes in PAH patients. Various interventional strategies for suppression of the inflammatory phenotype responses of macrophages in PAH are underway. Some examples include the (1) blockage of IL-6-mediated downstream effects through the clinical use of IL-6 antibodies, (2) reduction in MR-positive macrophages, (3) reduction in CD68+ cells, where CD68 is a surface marker of macrophages, and the (4) blockage of macrophage-derived cytokines such as macrophage-derived leukotriene B4 (LTB4).

3 Clinical Trials of Drug Delivery Systems Targeting Macrophages in Tumors and Cancer

The main premise in the design of drug delivery systems to target macrophages is the strategy of altering the inflammatory milieu depending on the disease and the organ. This is done to achieve the lost homeostasis between anti-inflammatory and proinflammatory signals through a readjustment of the M1/M2 axis polarization. While a number of macrophage-targeting strategies and drug delivery systems have been employed to achieve macrophage-specific delivery of drugs, generally lipid nanodelivery vehicles are the most popular due to the manufacturing flexibility of being able to add macrophage receptors, peptides, or antibodies that would permit their uptake. In general, either the extracellular matrix in the organs where the macrophages are located may be targeted to alter the chemotaxis and recruitment cues for macrophage engagement, or macrophages themselves may be targeted to influence their secretion of inflammatory cytokines.

We will discuss some clinical trials in the context of cancers, autoimmune diseases such as rheumatoid arthritis, and diseases of the lung and a few other organs. In some cases, such as for cancers, macrophages play a direct role since some tumors contain tumor-associated macrophages (TAMs) which are known to be opalized to the M2 phenotype. The anti-inflammatory M2 phenotype assists tumor growth and viability by exerting an anti-inflammatory effect on the surrounding tissues. Thus, clinical trials are aimed at drug delivery systems that alter the tumor microenvironment by ameliorating the M2 TAM phenotype, as discussed below. In the other instances, clinical trials using drug delivery systems have tested the effect of suppressing the inflammatory phenotype to prevent exacerbation of the disease.

3.1 Clinical Trials Targeting Macrophages in Cancer

Clinical trials targeting macrophages in cancer are usually aimed at TAMs and are designed to achieve any of the following three objectives: (1) reduce the recruitment of macrophages to tumors, (2) block the anti-inflammatory M2 phenotype, and (3) promote the proinflammatory M1 phenotype.

3.1.1 The CD47-SIRPα Pathway

As detailed earlier, the CD47-SIRPα interaction between a target cell and macrophage cell surface provides an inhibitory "do not eat me" signal to the macrophage. Tumors are known to exploit this system to evade macrophage-mediated nonspecific immune responses. Overexpression of CD47 on the surface of tumor cells will prevent their engulfment and removal by macrophages. Thus, therapeutic strategies to downregulate or block the CD47 protein may be advantageous. One way to block the CD47 molecule on tumor cells would be to design and deliver a CD47 antibody which by binding to CD47 on target cells would allow their engulfment and removal by macrophages. Two such antibodies have been recently tested in human clinical trials and are discussed in detail below. These are Hu5F9-G4 (5F9) and CC-90002.

Hu5F9-G4 (5F9) is a humanized antibody that targets the CD47 molecule and enables macrophage-mediated phagocytosis. Gholamin et al. tested the efficacy of the Hu5F9-G4 (5F9) antibody in vitro and in vivo using patient-derived orthotopic xenograft tumor models. The included five aggressive and etiologically distinct brain tumors: group 3 medulloblastoma (primary and metastatic), atypical teratoid rhabdoid tumor, primitive neuroectodermal tumor, pediatric glioblastoma, and diffuse intrinsic pontine glioma. The antibody was also administered through direct intraventricular injection and proved therapeutically effective in the treatment groups. The authors noted that in their study, the antibody spared normal neural tissues and only affected the tumor cells overexpressing CD47 and thus was safe for use (Gholamin et al. 2017).

Subsequently, in 2019, Sikic et al. of Gilead Sciences reported a first-in-human, first-in-class, open-label, interventional human clinical trial of the Hu5F9-G4 (5F9) antibody in patients with advanced cancers. The authors' main intent was to characterize the pharmacokinetics and pharmacodynamics of the antibody. The clinical trial included 62 patients with 11 in the group to determine a priming dose, 14 patients in the group to determine the weekly maintenance dose, 22 patients in the cohort to determine a loading dose in week 2, and finally, 15 patients included in the tumor-biopsy group. Observed toxicities included "transient anemia (57% of patients), hemagglutination on peripheral blood smear (36%), fatigue (64%), headaches (50%), fever

(45%), chills (45%), hyperbilirubinemia (34%), lymphopenia (34%), infusion-related reactions (34%), and arthralgias (18%)" (Sikic et al. 2019).

CC-90002 is a humanized anti-CD47 monoclonal antibody that skews the TAM phenotype toward a high M2/M1 ratio. Zeidan et al. conducted a phase 1 multicenter clinical trial of CC-90002 in patients with relapsed and/or refractory (R/R) acute myeloid leukemia (AML) and high-risk myelodysplastic syndromes (MDSs) (Zeidan et al. 2019). "CC-90002 was administered intravenously once a week for 4 weeks of each 42-day cycle during cycles 1–4, then once every 4 weeks during a maintenance phase of 28-day cycles. Patients were enrolled in cohorts of escalating dose levels using a modified 3 + 3 design. The primary objectives were to determine preliminary safety and tolerability, nontolerated dose, maximum tolerated dose, and/or recommended phase 2 dose. Secondary objectives were to measure preliminary efficacy, pharmacokinetics, and the presence and frequency of antidrug antibodies." However, the authors reported that the antibody showed "a lack of objective response" in these patients.

3.1.2 The CD40 Pathway

The CD40 protein is located on the surfaces of antigen-presenting cells and the CD40L ligand is located on CD4+ T cells, with engagement resulting in activation of the immune response. The anti-human CD40 antibody, the agonist CP-870,893 was demonstrated in 2007 in a first clinical trial in 29 patients with advanced solid tumors, was found to be effective against cancers and was biologically active with select toxicities (Vonderheide et al. 2007).

However, it was later noted that the CD40-CD40L interaction resulted in a skewing of the M1/M2 axis toward the M1 proinflammatory phenotype which, as documented above, is preferable in tumors with TAMs (Khalil and Vonderheide 2007). The CP-870,893 antibody has been tried in clinical trials both alone and in combination with other drugs such as FOLFIRNOX.

3.1.3 Drug Delivery Strategies

Interestingly, a variety of drug delivery systems have been employed in clinical trials to alter the M1/M2 macrophage axis for TAMs. For instance, Weigel et al. investigated the use of intravenous administration of the molecule 852a, also known as S-32865, *N*-[4-(4-amino-2-ethyl-1*H*-imidazo[4,5-*c*]quinolin-1-yl)butyl]methanesulfonamide). S-32865 is a novel immune response modifier related to the imidazoquinoline molecule imiquimod, and acts as a TLR7 agonist. The authors conducted a first-in-human hematologic malignancy phase II clinical trial of a subcutaneously (SC) delivered TLR7 agonist (852A) using a prolonged dosing schedule and reported evidence of sustained tolerability and clinical activity in hematologic malignancies (Weigel et al. 2012).

Another strategy to deliver drug cargo to TAMs is to exploit the fact that TAMs express high levels of mannose receptor 1 (CD206). Thus, mannosylated nanoparticles may be used to deliver nucleic acids, siRNA, micro-RNA, and other agents directly into the TAMs. These in turn can directly dictate the outcome by changing the inflammatory phenotype and the secretion of inflammatory and inhibitory cytokines. For example, injectable mannose-modified PLGA-based nanoparticles can be used to deliver in vitro-transcribed mRNA encoding transcription factors that favor the M1 phenotype in TAMs.

Zhang et al. showed that by taking advantage of electrostatic interactions between the cationic poly(β-amino ester) (PbAE) polymers and anionic mRNA, they created stable nanoparticles enclosing the mRNA of Interferon Regulatory Factor 5 (IRF5) and IKKβ (a kinase that phosphorylates and activates IRF5). These were targeted to the TAMs by encapsulating them in nanoparticles coated with CD206 (Zhang et al. 2017).

4 The Role of Macrophages in Driving Cardiovascular Diseases: Myocardial Infarction, Myocarditis, Arrhythmia, and Atherosclerosis

Macrophages have been discovered to orchestrate a variety of deleterious events in the heart following adverse events such as the development of myocardial infarction. Suppression of inflammation has been shown to have promising effects in the early stages of the development of most cardiovascular diseases, in some of which macrophages play a clear early role in pathogenesis and disease progression. Below, we first summarize the role of inflammation and macrophages in heart disease, primarily from experimental animal model data, and some limited human clinical trials. Following that, we discuss the strategies for drug delivery to macrophages in heart disease and cite an example of a clinical trial.

Macrophages play a well-established role in the development and pathological progression of atherosclerosis, which is a vascular disease resulting in a narrowing of the blood vessel wall due to years of damage to the wall. Atherosclerosis initiates due to triggered damage to the endothelium of arteries, resulting in the recruitment of monocytes and macrophages to the injured area, which store and oxidize cholesterol, eventually dying as foam cells which are deposited in the vessel walls. Macrophages secrete a variety of proinflammatory signals, such as interleukin-1β (IL-1β), which hypothetically produce an overall inflammatory milieu that supports the progression of atherosclerosis.

Similarly, macrophages are known to be recruited to the heart following episodes of myocardial infarction (MI). The massive infiltration of macrophages following MI has been linked to the creation of a harmful inflammatory microenvironment in the heart which is thought to facilitate inflammation-related fibrosis and damage. However, interestingly, macrophages have also been reported to be resident in the healthy heart. The role, and indeed the origin, of macrophages in the healthy heart is not known, though experimental observations have been noted using murine models that these macrophages are located in direct contact with myocytes and endothelial cells (Pinto et al. 2012).

Importantly, in the heart, as in other organs, and in solid tumors, the M1/M2 macrophage axis seems to play an important role. At least, it may be able to distinguish the functions of macrophages in the healthy versus the diseased heart. For example, in the healthy heart, as can be expected, macrophages express 22 genes associated with the anti-inflammatory M2 phenotype and do not express the cell surface marker called Ly6C (Pinto et al. 2012; Frantz and Nahrendorf 2014). Ly6C is expressed on the surface of highly inflammatory monocytes as they are recruited from the bone marrow in response to tissue injury and damage. By contrast, within a short time following an adverse cardiovascular event such as MI, the numbers of Ly6C^{HIGH} monocytes in the mouse and CD14^{+}CD16^{-} in humans and subsequently macrophages recruited to the infarcted areas of the heart increases (Jung et al. 2013), suggesting a vital role played by macrophages in exacerbating the inflammatory conditions leading to permanent or sustained damaged to heart tissues following MI.

Common to both atherosclerosis and MI disease process above may be the recruitment of CD14^{+}CD16^{-} monocytes or directly macrophages. Similarly, myocarditis, which is a general inflammation of the heart, and may be either acute or chronic, is another significant heart disease where macrophages and monocytes are newly thought to play an important role in disease progression. The role of macrophages in myocarditis is now established enough to be considered a clinical diagnostic for the disease. Specifically, a cytology hallmark for myocarditis in humans is a biopsy of the diseased heart showing positive staining for CD68^{+} histiocytes, which are classic macrophages (Sagar et al. 2012). It is documented that during myocarditis, the numbers of CD11b + monocytes/macrophages tend to increase compared with a lack of similar increases in other leukocytes, suggesting a preferred route for monocyte/macrophage recruitment following injury or damage (Afanasyeva et al. 2004). This observation would be consistent with other cardiovascular diseases such as atherosclerosis and myocardial infarction discussed above.

A landmark study by Leuschner et al. published in 2015 documented the role played by the chemokine (C-C motif) receptor 2 (CCR2) in the recruitment and subsequent establishment of macrophages to the heart during experimental myocarditis in the murine disease model. CCR-2 is known to also facilitate the exodus of monocytes from the bone marrow (Leuschner et al. 2015). CCR-2 depletion was linked with amelioration of myocarditis prior to Leuschner's work (Chen et al. 2019). Two things are interesting and ought to be noted about their work: (1) they used a unique drug delivery system composed of lipidoid nanoparticles enclosing a cargo of siRNA for CCR2 (siCCR2) and prepared from distearoylphosphatidylcholine (DSPC), cholesterol, and polyethylene glycol/dimyristoyl-rac-*glycero* (PEG-DMG). (2) Furthermore, the authors also evaluated $CCR2^+$ levels in human patients with myocarditis and reported that $CCR2^+$ cells accumulated over time in the hearts of human patients with myocarditis based on immunohistochemical staining of biopsy samples obtained from the patients. Thus, their work, which constitutes a seminal study in the field of the exploration of macrophage biology in the context of human heart diseases, succeeded in providing both a human correlation and a unique drug delivery vehicle for manipulating heart macrophages.

Thus, in conclusion for this section, both the recruitment and polarization of the M1/M2 macrophage axis seem to be viable strategies for targeting macrophages in cardiovascular compartments, though much remains to be discovered about the exact function and mechanisms of macrophages in the human heart. Albeit a promising line of inquiry with more recent discoveries, macrophage biology, and clinical targeting remain a nascent field, perhaps explaining the paucity of clinical trials.

4.1 Clinical Trials for Canakinumab: Targeting the Cardiovascular Inflammatory Milieu

A landmark clinical trial history exploring the role of inflammation in cardiovascular disease, particularly as a consequence of macrophage cytokine production of IL-1β, is the story of the clinical development and use of the anti-human IL-1β antibody called canakinumab (Alten et al. 2008). Canakinumab, a human IgG1/κ antibody against IL-1β was first produced by Novartis. Canakinumab was shown to bind human IL-1β with an equilibrium binding constant of 40 pM (Alten et al. 2008). It was also shown to inhibit the in vitro biological activity of IL-1β with an IC_{50} of about 43 pM (Rondeau et al. 2015).

As discussed above, IL-1β is an important diver of atherosclerosis as it regulates late plaque stability. The first evidence of the role of IL-1β in influencing macrophage-dependent late plaque biology came from the work by Gomez et al. (2018). These authors showed that the long-term use of the anti-IL-1β antibody in the preclinical murine model of atherosclerosis resulted in the repolarization of the M1/M2 macrophage axis toward the anti-inflammatory M2 phenotype, which caused relative ablation of the advanced atherosclerotic plaques (Gomez et al. 2018). This important preclinical work laid the foundation for clinical exploration of the use of anti-human IL-1β antibody (canakinumab) in the famous clinical study called the Canakinumab Anti-inflammatory Thrombosis Outcomes Study (CANTOS) (Gram 2020).

The CANTOS trial was a very important experiment in the form of a human clinical trial that asked a vital biological question: what is the effect of overall inflammation in mediating heart disease? The design of the clinical trial included a clever readout of overall systemic inflammation – the C-reactive protein. Thus, the authors measured the effect of administering canakinumab and abrogating the IL-1β response on a direct readout of the blood levels of circulating C-reactive protein. In the CANTOS trial, three different doses of canakinumab, specifically, 50, 150, and 300 mg, were administered every 3 months versus placebo to 10,061 patients with a prior myocardial infarction and persistent inflammation evidenced by a C-reactive protein at a consistent concentration greater than 2 mg/L in serum. The authors measured the outcome in the context of cardiovascular disease with an analysis of the effect of reduction in circulating IL-1β on nonfatal stroke, nonfatal myocardial infarction, and cardiovascular death. Interestingly, in the group administered 150 mg dose of canakinumab, a 15% reduction in adverse cardiovascular events was noted, with a hazard ratio of 0.85, 95% CI

0.74–0.98; and an adjusted $p = 0.021$, indicating that the finding was statistically significant. This constituted the first direct clinical and in-human evidence of the linkage between macrophage-mediated inflammatory cytokines and the progression of cardiovascular disease.

5 Conclusion

The M1/M2 polarization of macrophage response has come to be recognized as a critical regulatory element in overall human health and disease. The variety of clinical trials included in this chapter serves to hone the important point of the clinical relevance of macrophage biology to a vast array of human diseases. While in some instances such as tumor and cancer progression, due to the direct involvement of inflammatory pathways, pharmaceutical companies and research groups have intensified the sear for modulation of the M1/M2 macrophage axis, and many different drug delivery systems are being tried out in clinical trials; in other areas such as myocarditis, atherosclerosis, and myocardial infarction, preclinical evidence regarding the role of macrophages has built a strong foundation for future clinical trials in humans.

References

Afanasyeva M, Georgakopoulos D, Belardi DF, et al. Quantitative analysis of myocardial inflammation by flow cytometry in murine autoimmune myocarditis: correlation with cardiac function. Am J Pathol. 2004;164:807–15.

Allen LA, Aderem A. Molecular definition of distinct cytoskeletal structures involved in complement- and Fc receptor-mediated phagocytosis in macrophages. J Exp Med. 1996;184:627–37.

Alten R, Gram H, Joosten LA, et al. The human anti-IL-1 beta monoclonal antibody ACZ885 is effective in joint inflammation models in mice and in a proof-of-concept study in patients with rheumatoid arthritis. Arthritis Res Ther. 2008;10:R67.

Arora S, Dev K, Agarwal B, Das P, Syed M. Macrophages: their role, activation and polarization in pulmonary diseases. Immunobiology. 2018;223:383–96.

Bramow S, Frischer JM, Lassmann H, et al. Demyelination versus remyelination in progressive multiple sclerosis. Brain. 2010;133:2983–98.

Cassetta L, Pollard JW. Targeting macrophages: therapeutic approaches in cancer. Nat Rev Drug Discov. 2018;17:1–18.

Chen B, Brickshawana A, Frangogiannis NG. The functional heterogeneity of resident cardiac macrophages in myocardial injury CCR2+ cells promote inflammation, whereas CCR2- cells protect. Circ Res. 2019;124(2):183–5.

Frantz S, Nahrendorf M. Cardiac macrophages and their role in ischaemic heart disease. Cardiovasc Res. 2014;102(2):240–8.

Fujioka Y, Matozaki T, Noguchi T, et al. A novel membrane glycoprotein, SHPS-1, that binds the SH2-domain-containing protein tyrosine phosphatase SHP-2 in response to mitogens and cell adhesion. Mol Cell Biol. 1996;16(12):6887–99.

Gholamin S, Mitra SS, Feroze AH, et al. Disrupting the CD47-SIRPα anti-phagocytic axis by a humanized anti-CD47 antibody is an efficacious treatment for malignant pediatric brain tumors. Sci Transl Med. 2017;9(381):eaaf 2968.

Gomez D, Baylis RA, Durgin BG, et al. Interleukin-1β has atheroprotective effects in advanced atherosclerotic lesions of mice. Nat Med. 2018;24:1418–29.

Gram H. The long and winding road in pharmaceutical development of canakinumab from rare genetic autoinflammatory syndromes to myocardial infarction and cancer. Pharmacol Res. 2020;154:104139.

Hashimoto-Kataoka T, Hosen N, Sonobe T, et al. Interleukin-6/interleukin-21 signaling axis is critical in the pathogenesis of pulmonary arterial hypertension. Proc Natl Acad Sci U S A. 2015;112(20):E2677–86.

Hirayama D, Iida T, Nakase H. The phagocytic function of macrophage-enforcing innate immunity and tissue homeostasis. Int J Mol Sci. 2017;19(1):92. https://doi.org/10.3390/ijms19010092.

Jung K, Kim P, Leuschner F, et al. Endoscopic time-lapse imaging of immune cells in infarcted mouse hearts. Circ Res. 2013;112:891–9.

Khalil M, Vonderheide RH. Anti-CD40 agonist antibodies: preclinical and clinical experience. Update Cancer Ther. 2007;2:61–5.

Leuschner F, Courties G, Dutta P, et al. Silencing of CCR2 in myocarditis. Eur Heart J. 2015;36(23): 1478–88.

Ma WT, Gao F, Gu K, Chen DK. The role of monocytes and macrophages in autoimmune diseases: a comprehensive review. Front Immunol. 2019;10:1140.

Moore KJ, Tabas I. Macrophages in the pathogenesis of atherosclerosis. Cell. 2011;145:341–55.

Nylander A, Hafler DA. Multiple sclerosis. J Clin Invest. 2012;122:1180.

Park JB, Suh M, Park JY, et al. Assessment of inflammation in pulmonary artery hypertension by 68Ga-mannosylated human serum albumin. Am Respir Crit Care Med. 2019; https://doi.org/10.1164/rccm.201903-0639OC.

Pinto AR, Paolicelli R, Salimova E, et al. An abundant tissue macrophage population in the adult murine heart with a distinct alternatively-activated macrophage profile. PLoS One. 2012;7:e36814.

Rondeau J-M, Ramage P, Zurini M, Gram H. The molecular mode of action and species specificity of canakinumab, a human monoclonal antibody neutralizing IL-1β. MAbs. 2015;7:1151–60.

Sagar S, Liu PP, Cooper LT. Myocarditis. Lancet. 2012;379:738–47.

Sikic BI, Lakhani N, Patnaik A, et al. First-in-human, first-in-class phase I trial of the anti-CD47 antibody Hu5F9-G4 in patients with advanced cancers. J Clin Oncol. 2019;37(12):946–53.

Udalova IA, Mantovani A, Feldmann M. Macrophage heterogeneity in the context of rheumatoid arthritis. Nat Rev Rheumatol. 2016;12:472–85.

Vonderheide RH, Flaherty KT, Khalil M, et al. Clinical activity and immune modulation in cancer patients treated with CP-870,893, a novel CD40 agonist monoclonal antibody. J Clin Oncol. 2007;25(7):876–83.

Weigel BJ, Cooley S, DeFor T, et al. Prolonged subcutaneous administration of 852A, a novel systemic toll-like receptor 7 agonist, to activate innate immune responses in patients with advanced hematologic malignancies. Am J Hematol. 2012;87:953–6.

Zeidan AM, DeAngelo DJ, Palmer JM, et al. A phase I study of CC-90002, a monoclonal antibody targeting CD47, in patients with relapsed and/or refractory (R/R) acute myeloid leukemia (AML) and high-risk myelodysplastic syndromes (MDS): final results. Blood. 2019;134(Supplement_1):1320.

Zhang H, Bi J, Yi H, et al. Silencing c-Rel in macrophages dampens Th1 and Th17 immune responses and alleviates experimental autoimmune encephalomyelitis in mice. Immunol Cell Biol. 2017;92:593–600.

Index

S. Gupta, Y. V. Pathak (eds.), *Macrophage Targeted Delivery Systems*,
https://doi.org/10.1007/978-3-030-84164-5

B

C

E

F

O

Q

R

The manufacturer's authorised representative in the EU is Springer Nature Customer Service Centre GmbH, Europaplatz 3, 69115 Heidelberg, Germany. If you have any concerns regarding our products, please contact ProductSafety@springernature.com

Printed and bound by CPI Group (UK) Ltd, Croydon, CR0 4YY
25/09/2026
02222830-0002